AGENTS NETWORKS EVOLUTION

A Quarter Century of Advances in Complex Systems

Recommended Titles in Related Topics

Modeling Complexity in Economic and Social Systems
edited by Frank Schweitzer
ISBN: 978-981-238-034-0
ISBN: 978-981-238-035-7 (pbk)

Hypernetworks in the Science of Complex Systems
by Jeffrey Johnson
ISBN: 978-1-86094-972-2

Complexity Science: An Introduction
edited by Mark A Peletier, Rutger A van Santen and Erik Steur
ISBN: 978-981-3239-59-3

AGENTS NETWORKS EVOLUTION

A Quarter Century of Advances in Complex Systems

Preface by Stefan Thurner and Luis M. A. Bettencourt

Frank Schweitzer

ETH Zürich, Switzerland

World Scientific

NEW JERSEY • LONDON • SINGAPORE • BEIJING • SHANGHAI • HONG KONG • TAIPEI • CHENNAI • TOKYO

Published by

World Scientific Publishing Co. Pte. Ltd.
5 Toh Tuck Link, Singapore 596224
USA office: 27 Warren Street, Suite 401-402, Hackensack, NJ 07601
UK office: 57 Shelton Street, Covent Garden, London WC2H 9HE

Library of Congress Control Number: 2022946214

British Library Cataloguing-in-Publication Data
A catalogue record for this book is available from the British Library.

AGENTS, NETWORKS, EVOLUTION
A Quarter Century of Advances in Complex Systems

ISBN 978-981-126-781-9 (hardcover)
ISBN 978-981-126-782-6 (ebook for institutions)
ISBN 978-981-126-783-3 (ebook for individuals)

For any available supplementary material, please visit
https://www.worldscientific.com/worldscibooks/10.1142/13184#t=suppl

Preface

A lot has happened over the last 25 years in the scientific quest for a systematic understanding of complex systems. Complexity science has gradually transitioned from a stage dominated by conceptualization, exploration in simple models, and theory at the turn of the millennium, to a truly empirical endeavor tackling science's deepest questions – a science as it ought to be: quantitative, predictive, and testable.

A quarter-century ago, the concepts of evolutionary dynamics, adaptation, emergence, co-evolution, networks, agent-based models, self-organization, scaling, hypergraphs, or artificial life – all seemed to have been at their infancy. Similarly, the statistics of strongly correlated systems, the physics of small stochastic systems, the mathematics of driven and non-ergodic systems, algorithmic vs. analytic dynamics were practically non-existent. Networks just started to enter the scene, expanding early applications in sociology and geography. Some highlights from that time were the deep and original understanding of some aspects of non-linearity (chaos theory), fractals, replicator dynamics, or the discovery of self-organization in simple sand-pile models. Most of the complexity science then relied heavily on conceptualization and theory applied to toy models and was far from being "useful" in real systems. It consisted of many loosely connected interesting bits and pieces and it was unclear if it would ever develop into a coherent and useful scientific framework applicable in biology, or society. Many scientists were critical of the achievements of complex systems: How should it ever be possible to treat heavily interconnected systems that have co-evolving boundaries, are non-linear, non-ergodic, non-separable, that are driven and out of equilibrium? In short, complexity scientists were interested in exotic problems that a mainstream scientist would never touch.

The crucial game changer over the past two decades was the explosion of data. Because of underlying technological trends, many different kinds of data quickly became large and pervasive for many different complex systems, such as social, living, ecological, medical, traffic, urban, economic, financial, and climate systems. In more and more cases, relevant "big" data about the structure and dynamics of any given system has becomes available at high resolution and with high accuracy.

The fast expansion of data forced all fields dealing with complex systems to operate differently, engaging anew with questions of heterogeneity, inequality, non-linear interactions and collective organization. The time was then ripe for the convergence of bigger and better data with the concepts developed all along by complexity science. Novel modeling techniques, network and hypergraph methods, new mathematical analysis tools and more appropriate statistics, finally became not only useful but necessary. This convergence also created new possibilities for putting complexity science to the test, and for its fast development in interaction of appropriate evidence closing the loop of the scientific method for the first time in many complex systems. In other words, we currently see many areas of complexity science turning into an empirical science that is effectively testable and that allows us to calibrate higher dimensional models with data from real and important systems. This empirical progress in turn tremendously accelerates the development towards a "useful" science in the sense that it helps us better understand the complexity around us and maybe – for the first time – manage it on scientific grounds. This practical potential and some remarkable successes became obvious during the recent COVID-19 pandemic, when many complexity scientists played an active role all over the world.

This success of complexity science over the past 25 years would not have been possible without pioneering institutions, like the Santa Fe Institute, or journals, such as the Advances of Complex Systems (ACS). These pioneering venues provided the fora for communication, coordination, selection of ideas and cooperation between scientists. ACS, as one of the traditionally relevant journals in the field, covered the history and the remarkable transformation of complexity science over time. This is also reflected in the present book.

Agents, Networks, Evolution – a Quarter Century of Advances of Complex Systems presents a comprehensive overview of some of the most relevant developments in complex systems science in the past two decades. In particular, four different modeling approaches are covered in detail: (1) Agent based models, as one of the central tools of linking microscopic update dynamics with systemic (system-wide) macroscopic effects, (2) networks as the central book keeping tool of complexity science that specifies the interactions between the constituent elements of the systems, (3) system dynamics, which covers biological phenomena, traffic, ecosystems, climate, and disease spreading, and, finally, (4) evolutionary dynamics, a chapter that spans a range of questions from replicator dynamics to the tremendous challenge of co-evolution, a topic that is still far from being fully understood.

All four chapters start with an overview of the topical areas by Frank Schweitzer, the Editor in Chief and the driving force behind ACS for many years. These insightful introductions make the book more accessible to non-experts, a fact that should help to spread recent developments in complexity science also to a younger audience and scientists outside the field. Experts will enjoy (re-)reading the contributions of a selected crowd of renown complexity scientists. Finally, the book takes an explicit focus on a number of concrete applications in different scientific areas, which makes it particularly attractive as it gives a glimpse of what the science of complex systems might bring to how we fundamentally understand and manage our societies and natural environments in the decades ahead.

Stefan Thurner, Complexity Science Hub Vienna & Santa Fe Institute
Luís M. A. Bettencourt, University of Chicago & Santa Fe Institute

Contents

Chapter 1

Introduction

A book about a journal

Advances in Complex Systems, the multidisciplinary journal to promote the sciences of complexity, has been successfully acting in the market for 25 years. It is a reason to celebrate. It is also a unique opportunity to reflect upon the development of complex systems research during the last quarter century. For this book, I selected 25 papers published in *Advances in Complex Systems* during the last 25 years, one paper for each year, to highlight the main research trends. *Pars pro toto*, these papers shall illustrate how our scientific focus has evolved during these years, how new topics and tools have entered complex systems research and how the sciences of complexity established themselves as one of the most dynamic research areas internationally.

This selection of papers is grouped into four parts: 1. Agent-Based Models, 2. Network Models, 3. System Dynamics Models, 4. Models of Evolution. Each part captures a broader research topic that I deem relevant and persisting over the last 25 years. It was difficult enough to pick 25 papers out of almost 1000 publications. Therefore, this selection is somewhat subjective; many other papers could have been chosen. The assignment of these papers to the four different parts concerning their models is not free of contradictions. For instance, some evolution papers build on system dynamics models, and some agent-based models also use network approaches. Instead of arguing about the dominant modeling perspective, we should realize that all four parts contribute to one overarching goal: model complex systems. The papers presented illustrate possible approaches.

A few selected papers cannot fully reflect the many facets of each modeling perspective. Therefore, for each part I provide a short introduction that summarizes the main problems of each approach. I also refer to other papers published in *Advances in Complex Systems* on the same subject but not included in this selection. It shall help to illustrate the bigger picture without claiming to be a comprehensive review. I want to emphasize that a selection concerning *topics*, rather than for *models*, could have been an alternative way of ordering. Here I can refer to the impressive list of *topical issues*, below.

At this point, it may be handy to shortly remind on the history of *Advances in Complex Systems*. The journal was founded in 1997 as a "quarterly journal that aims to provide a unique medium of communication for multidisciplinary approaches, either empirical or theoretical, to the study of complex systems." The first volume was published in 1998 by the publisher Hermes (Paris), and Eric Bonabeau was the first editor-in-chief. Papers with a biological focus dominate the early published issues. It is remarkable because, in recent years, *Advances in Complex Systems* published much less on biological topics. One of the reasons, in addition to more vigorous competition among journals, is the decision of authors to concentrate on journals with a more considerable disciplinary impact. At the same time, papers focusing on socio-economic topics were present in the journal from the beginning. Notably, already volume 3 is entirely devoted to problems of social and economic simulations. In fact, *Advances in Complex Systems* has been an advocate journal for *agent-based modeling* since its launch. The first publication in *Advances in Complex Systems* is about an agent-based model of economic agglomeration. A selection of 21 papers from *Advances in Complex Systems* about *Modeling Complexity in Economic and Social Systems* was published as a book by World Scientific in 2002.

From volume 4 on, *Advances in Complex Systems* is published with World Scientific, and Peter Stadler took over as editor-in-chief after serving as co-editor-in-chief with Eric Bonabeau for two years. In 2007, the increasing number of submissions each year and the wider topical variety required a change in the organizational structure and the editorial board. In this situation, I became the editor-in-chief to undertake this significant overhaul of the journal. I have the privilege to serve the scientific community by taking the lead for *Advances in Complex Systems* for more than 15 years. During this long time, I also had responsibilities as editor-in-chief for other journals and, in addition, founded a new one. This experience allows me to highlight some features that make *Advances in Complex Systems* a unique journal.

Small is beautiful. *Advances in Complex Systems* is a rather small journal by design. The number of published papers per year has been stable since the beginning. This development differs from other journals in the field now serving as container journals for thousands of papers with hundreds of editors. Just increasing publication numbers without a clear focus will likely increase the revenue. However, in my view, it will also dilute the scientific profile of the journal at a time when we need to sharpen it. I am thankful to the publisher World Scientific that they respected this attitude. As one of the many advantages, *Advances in Complex Systems* is driven by a recognized scientific community. Authors know their editors, and editors know their reviewers and do not need to rely on profile matching provided by anonymous recommender systems.

Mind the gap. Already in his editorial for the first issue, Eric Bonabeau pointed out a tension between disciplinary and multidisciplinary research: "A genuine field of Complex Systems, that is, a multidisciplinary field of research dedicated to the mutual enrichment and cross-fertilization of more traditional disciplines dealing with their own complex systems, is not only possible but desirable. Possible because many researchers in different fields face complex systems that share common properties at some level of description. Desirable because it is important to establish solid and durable bridges between disciplines and possibly a common framework or language that will enhance collaborations. Paradoxically, the daily practice of science is becoming more and more narrowly focused on tiny details and at the same time requires more and more multidisciplinary collaboration among scientists." This tension still persists today, and one might rightly argue that the gap between the need for a multidisciplinary perspective and the ability to develop it has increased. Thus, are 25 years not enough to recognize the value of multidisciplinary research?

It is one of the problems in science that multidisciplinary research is less rewarded than disciplinary research on the institutional and individual levels. The situation has slightly improved for biological and environmental sciences, but less for social and economic research, where publications in disciplinary journals still dominate scientific reputation. Multidisciplinary research requires additional skills, such as learning other scientific languages and communicating across scientific disciplines. *Advances in Complex Systems* offers young scientists the opportunity to practice these skills and get recognized for them with their publications. As specified in the submission requirements and double-checked by editors and reviewers, publications in *Advances in Complex Systems* should be accessible to a readership from a wide range of scientific disciplines. Thus, *Advances in Complex Systems* has the potential to enhance knowledge transfer across scientific boundaries and should leverage it.

No theory of everything. When deciding the suitability of submissions for *Advances in Complex Systems*, I was often confronted with the authors' stance that everything is complex. It confuses "complex" with "complicated" and modeling approaches with real systems. Complex systems are comprised of many strongly interacting elements, which does not allow decomposing into parts to understand them better. They can self-organize such that new systemic properties emerge. Complex systems research builds on several established theoretical concepts, mathematical techniques, modeling and simulation tools to study these systems. They have their roots, not in one, but in many disciplines, as the prime example of agent-based modeling illustrates. *Advances in Complex Systems* has always tried to facilitate this common understanding of complex systems, focusing on quantitative and analytical approaches and computer simulations. It is the common denominator of the publications in *Advances in Complex Systems*.

The selection of papers for this book shall demonstrate our philosophy. They highlight the advancements of research on complex systems concerning methods and scientific questions from different perspectives. Most impressive, their topical variety can inspire multidisciplinary research in different scientific fields. I hope that this book, with its overarching view on the complex systems research of the past 25 years, will further the development of common approaches to complex systems and impact scientific research beyond the disciplinary horizon.

Selected 25 publications ordered by year

Advances in Complex Systems (**1998**), vol. 01, pp. 115-127
Andrew J. Spencer, Iain D. Couzin and Nigel R. Franks: *The Dynamics of Specialization and Generalization within Biological Populations*

Advances in Complex Systems (**1999**), vol. 02, pp. 371–393
Hugues Juillé and Jordan B. Pollack: *Coevolutionary Learning and the Design of Complex Systems*

Advances in Complex Systems (**2000**), vol. 03, pp. 283–297
M. G. Zimmermann, V. M. Eguíluz, M. San Miguel and A. Spadaro: *Cooperation in an Adaptive Network*

Advances in Complex Systems (**2001**), vol. 04, pp. 79–96
W. Ebeling, Karmeshu and A. Scharnhorst: *Dynamics of Economic and Technological Search Processes in Complex Adaptive Landscapes*

Advances in Complex Systems (**2002**), vol. 05, pp. 73–89
A. J. Palmer, T. L. Schneider and L. A. Benjamin: *Inference Versus Imprint in Climate Modeling*

Advances in Complex Systems (**2003**), vol. 06, pp. 47–77
Bärbel M. R. Stadler and Peter F. Stadler: *Molecular Replicator Dynamics*

Advances in Complex Systems (**2004**), vol. 07, pp. 139–155
Alan Kirman: *Economics and Complexity*

Advances in Complex Systems (**2005**), vol. 08, pp. 87–116
Dirk Helbing, Martin Schönhof, Hans-Ulrich Stark and Janusz A. Hołyst: *How Individuals Learn to Take Turns: Emergence of Alternating Cooperation in a Congestion Game and the Prisoner's Dilemma*

Advances in Complex Systems (**2006**), vol. 09, pp. 315–335
Christoph Hauert: *Cooperation, Collectives Formation and Specialization*

Advances in Complex Systems (**2007**), vol. 10, pp. 5–28
Marc Barthélemy, Alain Barrat and Alessandro Vespignani: *The Role of Geography and Traffic in the Structure of Complex Networks*

Advances in Complex Systems (**2008**), vol. 11, pp. 927–950
F. Pozzi, T. Di Matteo and T. Aste: *Centrality and Peripherality in Filtered Graphs from Dynamical Financial Correlations*

Advances in Complex Systems (**2009**), vol. 12, pp. 99–112
Luiz H. Gomes, Virgilio A. F. Almeida, Jussara M. Almeida, Fernando D. O. Castro and Luís M. A. Bettencourt: *Quantifying Social and Opportunistic Behavior in Email Networks*

Advances in Complex Systems (**2010**), vol. 13, pp. 327–338
Robin C. Ball, Marina Diakonova and Robert S. Mackay: *Quantifying Emergence in Terms of Persistent Mutual Information*

Advances in Complex Systems (**2011**), vol. 14, pp. 317–339
David M. D. Smith, Jukka-Pekka Onnela, Chiu Fan Lee, Mark D. Fricker and Neil F. Johnson: *Network Automata: Coupling Structure and Function in Dynamic Networks*

Advances in Complex Systems (**2012**), vol. 15, pp. 1250043
Stephanie Keller-Schmidt and Konstantin Klemm: *A Model of Macroevolution as a Branching Process Based on Innovations*

Advances in Complex Systems (**2013**), vol. 16, pp. 1250089
Malte Harder and Daniel Polani: *Self-Organizing Particle Systems*

Advances in Complex Systems (**2014**), vol. 17, pp. 1450005
Jianjun Wu, Mingtao Xu and Ziyou Gao: *Modeling the Coevolution of Road Expansion and Urban Traffic Growth*

Advances in Complex Systems (**2015**), vol. 18, pp. 1550017
Hugo C. Mendes, Alberto Murta and R. Vilela Mendes: *Long Range Dependence and the Dynamics of Exploited Fish Populations*

Advances in Complex Systems (**2016**), vol. 19, pp. 1650005
Sandra D. Prado, Silvio R. Dahmen, Ana L.C. Bazzan, Padraig Mac Carron and Ralph Kenna: *Temporal Network Analysis of Literary Texts*

Advances in Complex Systems (**2017**), vol. 20, pp. 1750001
Floriana Gargiulo, Yerali Gandica and Timoteo Carletti: *Emergent Dense Suburbs in a Schelling Metapopulation Model: A Simulation Approach*

Advances in Complex Systems **(2018)**, vol. 21, pp. 1850014
Alessandro Pluchino, Alessio Emanuele Biondo and Andrea Rapisarda: *Talent Versus Luck: The Role of Randomness in Success and Failure*

Advances in Complex Systems **(2019)**, vol. 22, pp. 1850023
Inga Ivanova, Øivind Strand and Loet Leydesdorff: *An Eco-Systems Approach to Constructing Economic Complexity Measures: Endogenization of the Technological Dimension Using Lotka–Volterra Equations*

Advances in Complex Systems **(2020)**, vol. 23, pp. 2050022
Till D. Frank: *Simplicity from Complexity: On the Simple Amplitude Dynamics Underlying Covid-19 Outbreaks in China*

Advances in Complex Systems **(2021)**, vol. 24, pp. 2150006
Alejandro Dinkelberg, David Jp O'Sullivan, Michael Quayle and Pádraig Maccarron: *Detecting Opinion-Based Groups and Polarization in Survey-Based Attitude Networks and Estimating Question Relevance*

Advances in Complex Systems **(2022)**, vol. 25, pp. 2250003
Frank Schweitzer, Antonios Garas, Mario V. Tomasello, Giacomo Vaccario and Luca Verginer: *The Role of Network Embeddedness on the Selection of Collaboration Partners: An Agent-Based Model with Empirical Validation*

Topical issues and their guest editors ordered by year

2000 *Applications of Simulations to Social Sciences*
Gérard Ballot and Gérard Weisbuch

2001 *Complex Dynamics in Economics*
Frank Schweitzer and Dirk Helbing

Challenges in Granular Physics
Anita Mehta and Thomas C. Halsey

2002 *Applications of Celullar Automata in Complex Systems*
Frank Schweitzer

2003 *Emergence in Chemical Systems*
Jerzy Maselko

Agent-Based Approaches in Complex Systems
Akira Namatame and Yuji Aruka

2004 *Agent-based Computational Approach to Industrial and Labor Dynamics*
Bruno Contini, Roberto Leombruni and Matteo Richiardi

2006 *Collectives Formation and Specialization in Biosystems*
Peter T. Hraber

2007 *Complex Systems Methodologies in Econcomics and Geography*
Koen Frenken and Gerald Silverberg

Modeling of Complex Systems by Cellular Automata
Jiří Kroc

Physics and the City
Bruno Giorgini, Armando Bazzani and Sandro Rambaldi

2008 *Social Simulation*
Frédéric Amblard and Wander Jager

Social Systems and Complexity
Dirk Helbing, Jürgen Jost and David Lane

Econophysics
Fabio Clementi

2009 *Complex Networks*
Enrico Scalas and Frank Schweitzer

Artificial Life
Konstantin Klemm, Daniel Merkle and Eckehard Olbrich

2010 *Beyond Small-World and Scale-Free Networks*
Gaoxi Xiao and Janos Kertesz

European Conference on Complex Systems 2009
Markus Kirkilionis, Francois Kepes and Colm Connaughton

European Social Simulation Conference 2009
Bruce Edmonds

2011 *Agents and multi-agent systems*
Frank Schweitzer and Matthew E. Taylor

New Developments in Complex Systems Science
Jorge Louçã

2012 *Cultural Evolution*
Anne Kandler and James Steele

Language Dynamics
Andrea Baronchelli, Vittorio Loreto and Francesca Tria

Physics of Competition, Cooperation and Conflicts
Peter Richmond, Angel Sanchez, Andrea Scharnhorst, Janusz Holyst, Vittorio Rosato and Stefan Thurner

Modeling Socio-Technical Complexity
Stefan Thurner, Guillaume Deffuant and Timoteo Carletti

Managing Financial Instability in Capitalist Economies
Marco Raberto

2013 *Information and Self-Organization of Behavior*
Daniel Polani, Mikhail Prokopenko and Larry S. Yaeger

Agent-Based Modeling and Techno-Social Systems
Dirk Helbing and Thomas U. Grund

2014 *Structure and Dynamics of Collaborative Software Engineering*
Marcelo Cataldo, Ingo Scholtes and Giuseppe Valetto

2015 *Artificial Economics*
Philippe Mathieu and Frédéric Amblard

2017 *Artificial Economics*
Paola D'Orazio and Annalisa Fabretti

2018 *Quantifying Success*
Roberta Sinatra and Renaud Lambiotte

Opinion Dynamics and Collective Decisions
Jan Lorenz and Martin Neumann

2019 *Controlling Network Dynamics*
Aming Li and Yang-Yu Liu

2020 *Game Theory and Mechanism Design*
Frank Schweitzer

2021 *Success in Science*
Luca Verginer, Giacomo Vaccario and Alexander M. Petersen

2022 *Heterogeneity, Evolution and Networks in Economics*
Leonardo Bargigli and Giorgio Ricchiuti

Cultural Complexity
Ramona Roller, Maximilian Schich, Mikhail Tamm and Hyejin Youn

PART 1
Agent-Based Models

Chapter 2

Overview

Complex systems and agents. Complex systems are commonly defined as systems comprising many strongly interacting elements. These system elements are denoted as *agents*, a term with roots in computer science and economics. Today, agent-based modeling is used in various disciplines, notably in the social sciences. In biology and ecology, it is also known as individual-based modeling. Early concepts in distributed artificial intelligence used the term actor-oriented modeling.

The ultimate aim of complex systems research is to explain *systemic* properties on the "macro" level based on the properties of the agents and their interactions on the "micro" level. The systemic properties are often denoted as *emergent* properties because they result from the *collective* interactions of many agents. Consequently, agent-based models are *multi-agent* models.

Designing the model. To define a model, one needs to specify the system: (i) what is *inside* the system, i.e. what are system elements, and (ii) what is *outside* the system, i.e. what are boundary conditions or what belongs to the environment. Every system can be described by a variety of models, dependent on the research interest and the available information. A system dynamics model would be appropriate if a few representative agents should be modeled instead of many interacting agents. If the internal dynamics of agents do not play a role, but the interaction structure matters, one would consider a network model. Evolutionary models are needed if the focus is on adaptive processes and their feedback on agent properties. There is no sharp distinction between these models, but rather a shift of focus on specific model ingredients.

Agent-based modeling is tricky because of the many degrees of freedom in specifying the model. We need to decide about the internal complexity of agents and their interactions. This points to the central question of how much complexity is needed on the *micro level* to generate specific systemic dynamics on the *macro level.* To fully understand what drives the system dynamics, we should refrain from increasing the model complexity on the micro level. Getting a life-like model is the desired feature of computer games. Instead, agent-based models have to be designed such that they can be (i) studied by means of reproducible computer simulations, or (ii) treated analytically, to derive a system dynamics from interactions, or (iii) calibrated and validated against available data. Consequently, there are various ways to design the model.

Generic versus data driven-models. Generally, we can distinguish generic models from data-driven models. Generic models are proposed to investigate the impact of a particular interaction or feedback mechanism. They are not meant to be the most realistic ones. Instead, they follow the KISS principle stated by Robert Axelrod: "Keep It Simple, Stupid". They often allow for mathematical investigations to derive the macro dynamics or obtain critical parameter values. Agent-based models of cooperation [19, 49, 108, 140, 163] or opinion dynamics [92, 101, 141], discussed below, are good representatives for this approach.

On the other hand, data-driven models aim to reproduce an empirical observation. Hence, already from their outset, they provide interfaces for calibration. For example, agents follow a given rule with some probability, which then needs to be calibrated using aggregated data. It does not guarantee that a desired systemic behavior is obtained. For example, the rules could be wrong or incomplete. But in the positive case, this procedure would lend some evidence to the underlying interaction rules. Examples of this approach are agent-based models of pedestrian behavior [124], of road traffic [54] or of firm collaborations [166].

Limits of validation. The many degrees of freedom in designing agent-based models makes all models "unique" and is one of the reasons why results can hardly be compared across models.

Agent-based models aim to study the relation between modeling assumptions and their resulting systemic outcome for a *specific* model. These models can hardly be falsified or proven. Instead, we can state whether certain modeling assumptions under defined conditions would lead to an expected systemic outcome or not. Therefore, I have turned the statement "All models are wrong, but some are useful." by Georges Box into: "All models are right, but most are useless." Here, the term useless refers to the inability to understand a phenomenon deeper beyond the chosen modeling assumptions.

Before judging agent-based models, it is essential to understand this modeling philosophy and its limitations. These limitations, in turn, make agent-based models difficult to accept in some disciplines. This motivates attempts to link agent-based models also to qualitative observations [83].

Agent complexity. Agents are characterized by different *dynamic* variables that specify their internal complexity. They describe, for instance, internal degrees of freedom such as their opinion in a social or their strategy in an economic setting. Utilizing this internal complexity, rational agents can calculate their utilities and make decisions based on available information. Reactive agents, on the other hand, apply less sophisticated dynamics, e.g. they use social herding in opinion dynamics or avoid collisions in models of pedestrian behavior.

To make agent-based models more realistic, one could argue to increasing the agent's internal complexity. But this would not only hamper the possibility of understanding specific modeling assumptions' impact. It also violates the paradigm of complex systems modeling. Complexity is an emergent property, i.e. it is the *output* of an agent-based model, not the *input*. One of the fundamental insights of complex systems research is that it does not need complex agents to obtain complex behavior.

Despite this insight, the problem remains how much complexity of agents is needed for a given systemic behavior. It is reflected in general discussions about the design of agent-based models [36, 55, 150]. But there also concrete proposals, for instance for intelligent agents [11] where a mental layer is used to model cognition, or for mechanisms of learning [88, 100] to allow for adaptive dynamics. But most models try to reduce the agent complexity while focusing on the emerging systemic properties.

Agent interactions. The model complexity not only depends on the agents' internal complexity but also on their interactions' complexity. Generally, we distinguish between direct and indirect interactions. Direct interactions imply a specific counterparty, for instance, a neighbor. It can be the adjacent neighbor in cellular automata, or an agent to whom a link exists, in a network. Game theoretical models, for instance, describe the direct interaction of two players who follow a strategy without knowing the strategy of their counterparty [116]. The possibility of direct interactions is often constrained, for instance, in the bounded confidence model where agents need to have their opinions within a defined tolerance range.

Indirect interactions are based on agents' coupling, e.g. via a common resource. Agents know about the actions of others from the depletion of the resource. Many types of communication models apply indirect interactions via an information field. The information can be broadcast to reach everyone simultaneously, but also diffuse through the system, coupling all agents gradually. A more common assumption is the coupling via global or local densities. For instance, in opinion dynamics models, agents choose their opinion depending on a given opinion's local frequency.

Direct interactions can be also *directed* interactions, i.e. agent i influences agent j, but not vice versa. The classic example is the Boolean network, often represented as a cellular automaton. A Boolean function, i.e. a "rule", defines which agents' states $\{0, 1\}$ are considered as input, to map this to an output $\{0, 1\}$. It allows for varying the inputs chosen but also the output function across agents. What seems to be a very flexible setup for an agent-based model quickly turns into an unfeasible approach because of the combinatorial explosion resulting from a large number of possible rules and inputs. Therefore, it is a wise modeling decision to restrict the interactions in a way that allows studying the model systematically.

Agent heterogeneity. Because of external conditions or internal parameters, agents differ in the values of their characteristic variables, e.g. their wealth, preferences, efficiency, etc. This heterogeneity is vital in understanding the emergence of systemic features, such as failure cascades, polarization of opinions, or cooperation. Identical agents with the same number of interactions would respond to the same input similarly. Taking the example of a failure cascade, means they would all fail or stay alive. This is not very realistic, but, more importantly, we would not learn from such models why specific failure cascades stop, and others amplify.

A modeling approach that explicitly includes individual differences has an advantage over, e.g., system dynamics models. It can address the *path dependence* of dynamic phenomena, i.e., the temporal sequence of interactions and the specific local conditions at a particular time determine whether e.g. an epidemic disease spreads or whether a traffic jam emerges at a particular place. The downside of this approach is the limited generalization. To get the average picture, one would need to run many simulations to average over initial configurations and random interaction sequences But in many realistic cases, the average does not help explain a concrete observation, like a power blackout, a stock market crash, or a mass panic. Regarding heterogeneity in conditions, agent-based models can fully play out their strengths.

Cellular automata. Historically, cellular automata (CA) models played a vital role in understanding the relation between agent interactions and the resulting macroscopic pattern. They were first studied in the form of two-dimensional regular lattices, with the famous Ising model of ferromagnetism as a prime example [21, 23]. But already in the end of the 1940s CA models were used to study social phenomena. Schelling's model of urban segregation from the 1970s has inspired many researchers until today, as the reprinted paper [152] demonstrates.

Advances in Complex Systems has published about CA models from its very beginning. Two topical issues about CA models appeared in 2002 and 2007. Publications address the role of such models in simulating complex systems [28, 156], but also fundamental questions of a probabilistic description of CA [30, 47, 48]. Extensions of CA models include learning [40, 43, 53] or synchronization [151].

In early years, papers were often related to biological questions, like gene regulatory networks [67] or tumor growth [29] and micro cellular pattern formation [52, 139]. Spatial problems, such as urban dynamics [18, 152] or the spatial navigation of robots [127, 149] have been analyzed by means of CA models. Also socio-economic models of attitude formation [42], marketing [126] or contagious financial fragility [117] were built on CA models.

Another type of application focuses on the usability of CA in computing complex problems [57, 60, 61]. But many of the publications using CA were devoted to fundamental questions, such as transient dynamics [62], output aggregation [86] or stability [20].

Application: Economic systems. Many agent-based models published in *Advances in Complex Systems* deal with economic systems. Such models are often grouped under the headline of *Artificial Economics*, and two topical issues about this subject were published in 2015 and 2017. Our journal has certainly benefited from the fact that agent-based models were less accepted in economics journals for a long time.

Many models address market interactions. These can be labour markets [41, 46, 76], gamers markets [129], auction markets [31], product markets [73, 98, 119] or financial markets [24, 25, 27, 122, 158]. Other papers consider specifically agents' attitudes [14, 17], or the impact of resources and regulations [8, 128] in markets.

Emergent systemic properties such as financial fragility [34], economic instability [77] or resilience [138] are studied with agent-based models. Such frameworks also allow to address problems of policy design on the governmental level [97, 153, 163, 164]. Some papers use agent-based models to study the spread of innovations [81, 143] or the adoption of behavior, such as tax compliance [136]. Others try to explain aggregate economic phenomena such as the firm-size distribution [161], accumulation of human capital in firms [7], or the regional distribution of goods [147].

Advances in Complex Systems has published various topical issues on economic agent-based models (2001), industrial and labor dynamics (2004), econophysics (2008), on financial instabilities (2012) and economic heterogeneity (2022).

Application: Spatial systems. Agent-based models are best suited to simulate spatial problems because the spatial heterogeneity and the migration of agents can be explicitly captured. Already the very first paper published in *Advances in Complex Systems* explored the impact of migration on economic agglomeration [3]. Other papers focus on human dispersals at a global scale [130], residential segregation [79, 102, 152] or other types of urban dynamics [56, 59, 66, 68].

Human mobility is a particular application area of agent-based modeling. This can include road traffic [54, 84, 162] or pedestrian dynamics [63, 69, 109, 124]. But there are also applications of agent models to biology, e.g. to spatial population dynamics [32, 115, 123] or the spatial navigation of neurons [4].

More general approaches investigate the role of social representations of space [10] or of neighborhood effects [6, 70] or of spatial coordination [35].

Application: Social systems. A very broad range of models target social interactions and their resulting emergent phenomena. Communication can be seen as one of the fundamental mechanisms of agent interactions [33, 65, 94]. It plays an important role in the emergence of norms and conventions, in the social psychology of groups [58, 71, 103], in coordination and consensus formation [91], but also in other types of self-organizing social processes [5, 85].

A second fundamental mechanism is learning, which occurs both at the individual agent and the multi-agent organizational level [22, 87, 88, 100, 104]. Social learning mechanisms can be even applied to the coordination of mobile robots [1] or to adaptive routing and rewiring in communication networks [2, 111]. In combination with filtering algorithms, they can be used to build social recommender systems [112, 132].

Agent-based models have also been used to understand prehistoric human societies [114, 125, 134, 137], or research communities [39, 95]. Eventually, publications also discuss business applications of social agent-based simulations [9, 37] and agent-based models of techno-social systems [133].

Application: Opinion dynamics. Processes of opinion formation and decision-making are the main topic of social agent-based models. We distinguish between continuous opinion dynamics, with the bounded confidence model as the most prominent example and discrete opinion dynamics, with the voter model as a paragon.

Advances in Complex Systems has not only published the initial paper on the bounded confidence model [12], which is today the most cited paper from this journal, but also various follow-up studies [65, 75, 92, 93, 101, 120, 121]. On the other end of the spectrum are studies of the voter model and its variants [13, 80, 141, 146, 155]. There is also the attempt to mix both the continuous and the discrete opinion dynamics [96].

Closely related to opinion dynamics are models of social influence [118, 142] because this mechanism impacts decisions [51]. Opinion leader influence is only one example [64]. Additionally, multi-agent models are used to simulate voting systems [15, 106], the impact of social influence on market inequalities [72] or decision processes in particular social groups [82, 135, 148].

More refined models of decision-making try to incorporate the cognitive abilities of agents [110] and consider that agents' behavior and cognition are informed by qualitative data from case studies [74, 83]. Model calibration can also use methods of indirect inference [131]. A very different approach to decision making is provided by quantum decision theory [99], which allows a fresh look at prospect theory and behavioral economics [145]. Other discrete choice models use a generalized energy function to map individual and global influences [107].

Application: Games Different from opinion formation, interactions in games involve a strategic component. Agents have to choose a strategy under incomplete information because their payoff depends on the unknown strategic choice of others. Many game theoretic models are inspired by evolutionary game theory and therefore discussed in the Chapter "Models of Evolution". But very often the classic alternative between cooperation and defection is extended with additional mechanisms, such as social herding [140], openess [108], emotional influence [144], collective monitoring [45] or learning [19, 49].

The outcome of interactions in games crucially depends on the interaction topology [113] and on the payoff function, hence the problem of optimal rewards is addressed in empirical and simulation approaches [26, 105, 163]. These discussions also extend to other types of games, such as the Minority Game [16, 38, 50, 89, 90].

Comments on the reprinted publications

Economics and complexity [44]. Taking the perspective of complexity science, sees the economy as yet another complex adaptive system. How this insight translates into modeling approaches that are widely accepted in economics is a very different question. Economists have long dealt with the problem of aggregated variables. Do we need concepts of emergence and collective interactions to understand the distinction between micro and macroeconomics?

The paper discusses this question, taking the profound view of an economist. Selected examples demonstrate that individual behavior, decisions and learning do not allow for simple aggregation. The paper advocates an agent-based modeling approach rooted in a better understanding of economic agents and their mutual dependencies. It is worth noticing that the problem of how economic complexity should be modeled is also taken up in another paper of a leading economist published in *Advances in Complex Systems* [78].

Cooperation in an adaptive network [19]. This publication would also fit the selections about networks or evolutionary game theory. This underlines the philosophy of *Advances in Complex Systems* combining approaches from different scientific areas. This paper's main idea is to compare two scenarios, one with fixed interaction partners and one with changing partners, both defined by a network topology. For the interaction, a weak Prisoner's Dilemma game is chosen, i.e. agents either cooperate or defect. The term weak refers to the fact that defectors gain nothing from interacting with other defectors, while they would usually receive a small payoff.

Agents have two degrees of freedom: (i) they can change their strategy if their payoff from interacting with all partners is less than the payoff of these partners, and (ii) different from the evolutionary Prisoner's Dilemma game, if these agents are defectors, they can choose new partners to avoid interacting with defectors. In plain words, defectors always get a chance to find another cooperator to exploit in the adaptation scenario. Because of this rule, one would expect that the level of cooperation becomes lower in the scenario where agents can change their interaction partners. But counter-intuitively, in the long run, cooperation is enhanced provided the right set of payoff values. It is one of the examples where agent-based computer simulations are needed to understand the dynamics fully.

How individuals learn to take turns [49]. This paper is a lucid example of how to combine agent-based modeling with experiments. Similar to the previous paper, a game theoretic setup, called the Route Choice game, is chosen for the agent interactions. It bears similarities with public good games. Agents must choose between two routes, a fast one on a freeway and a slow one on a side road. Capacity constraints will lead to traffic congestion if everyone takes the freeway. Thus, the user equilibrium is obtained if a fraction of agents decides to use the side road. Indeed, if always the same agents take the side road, they have a disadvantage. So, the optimal solution is that agents alternate. But will they do that? Even more, will they learn to alternate in a synchronized manner? This would mean agents not only learn to cooperate, they also learn to reciprocate.

This innovative and exiting game is studied both experimentally and analytically. It contrasts the game theoretical predictions with the behavior of small groups excellently. We learn under which conditions the ideal solution can emerge and why this could take a long time. Punishment of unfair behavior can help. The paper tries to link the findings to real-world traffic scenarios, to illustrate the applicability of agent-based modeling.

Emergent dense suburbs in a Schelling metapopulation model [152]. The model of spatial segregation of agents of two different colors published by Schelling in 1971 has inspired many researchers to propose variations. This paper goes beyond the classical two-dimensional cellular automaton in introducing an additional spatial structure on the meso level. There is the local neighborhood represented by a cell, but there is also the larger neighborhood, i.e., an urban district composed of adjacent neighbors of a cell. Agents evaluate their utility by taking the composition of both neighborhoods into account.

The results of the extended model can be best understood when referring to the basic segregation model. Dependent on the global occupation density we observe perfect segregation of subpopulations with different colors in the case of low density when considering only local neighborhoods. The extended model allows the formation of segregated clusters on the global scale, whereas the local neighborhood still comprises a mixture of agents with different colors. What sets this paper apart from similar segregation simulations is the detailed investigation of the impact of tolerance thresholds on the various segregation patterns observed. Also, the role of different time scales in the segregation dynamics is discussed.

The role of randomness in success and failure [154]. *Advances in Complex Systems* was lucky to publish this paper about talent versus luck, also because it induced several follow-up publications with modifications and detailed investigations of the original model [159, 160, 165]. At the same time, these papers continued earlier discussions about the illusion of success [157]. In 2018 also a topical issue about "Quantifying Success" was published.

From a technical perspective, the paper is about the mismatch between the Gaussian distribution of an input variable, here denoted as "talent" and the power-law distribution of an output variable, here meant as "success". Compared to the narrow and scale-dependent normal distribution, the power law represents a very broad scale-invariant distribution. Thus, the question arises about the hidden mechanisms at work to turn originally similar agents into very different ones.

The answer of the authors is: pure randomness. Agent-based simulations support their findings. More importantly, these findings lead to an in-depth discussion how to counterbalance pure luck, e.g. by using efficient funding strategies. Thus, the paper addresses a relevant problem and bridges the gap between agent-based simulations and policy design. And it inspires other authors to provide their explanations.

Selection of collaboration partners [166]. Publications combining agent-based modeling with large-scale data analyses are still rare. The challenges are addressed in the overview above. Data-driven modeling requires an interface between agent interactions and available data, which is used to calibrate the model. But for the application discussed in this paper, the emergence of economic and scientific collaboration networks, data about interactions is scarce. We may know which partners were selected and when, but hardly why. Only the resulting collaboration networks and their evolution over time are observed.

To solve this problem, the authors define rules for agent interactions and then simulate their outcome to compare it with the empirical networks. The interaction parameters providing the best match are obtained from calibration. The agent-based model is then validated against its ability to reproduce macroscopic features of the collaboration networks not used during the calibration.

With the calibrated model, this paper addresses a specific question: to explain how agents improve their position in the collaboration network. Do they follow specific strategies in selecting their collaboration partners? The answer proposed from analyzing the agent-based model: It's more chance than choice, which reminds a bit of the "talent versus luck" discussion above. But this is one of the biggest advantages of agent-based modeling: to find out whether and to what extent a simple hypothesis about agent interactions can generate complex systemic features. This does not rule out more complex interaction rules but demonstrates that these are not needed to get the gist of an observed phenomenon.

Selected publications

[1] Chang, C. and Gaudiano, P., *Application of Biological Learning Theories to Mobile Robot Avoidance and Approach Behaviors*, Advances in Complex Systems **01** (**1998**) 79–114.

[2] Heusse, M., Snyers, D., Guérin, S., and Kuntz, P., *Adaptive Agent-Driven Routing and Load Balancing in Communication Networks*, Advances in Complex Systems **01** (**1998**) 237–254.

[3] Schweitzer, F., *Modelling Migration and Economic Agglomeration with Active Brownian Particles*, Advances in Complex Systems **01** (**1998**) 11–37.

[4] Segev, R. and Ben-Jacob, E., *From Neurons to Brain: Adaptive Self-Wiring of Neurons*, Advances in Complex Systems **01** (**1998**) 67–78.

[5] Ishii, H., Page, S. E., and Wang, N., *A Day At the Beach: Human Agents Self-Organizing on the Sand Pile*, Advances in Complex Systems **02** (**1999**) 37–63.

[6] Aschan-Leygonie, C., Mathian, H., Sanders, L., and Mäkilä, K., *A Spatial Microsimulation of Population Dynamics in Southern France: a Model Integrating Individual Decisions and Spatial Constraints*, Advances in Complex Systems **03** (**2000**) 109–125.

[7] Ballot, G. and Taymaz, E., *Competition, Training, Heterogeneity Persistence, and Aggregate Growth in A Multi-Agent Evolutionary Model*, Advances in Complex Systems **03** (**2000**) 335–351.

[8] Beckenbach, F., *Multi-Agent Modelling of Resource Systems and Markets: Theoretical Consideration and Simulation Results*, Advances in Complex Systems **03** (**2000**) 221–241.

[9] Bonabeau, E., *Business Applications of Social Agent-Based Simulation*, Advances in Complex Systems **03** (**2000**) 451–461.

[10] Bonnefoy, J.-L., Le Page, C., Rouchier, J., and Bousquet, F., *Modelling Spatial Practices and Social Representations of Space Using Multi-Agent Systems*, Advances in Complex Systems **03** (**2000**) 155–168.

[11] Conte, R., *The Necessity of Intelligent Agents in Social Simulation*, Advances in Complex Systems **03** (**2000**) 19–38.

[12] Deffuant, G., Neau, D., Amblard, F., and Weisbuch, G., *Mixing Beliefs Among Interacting Agents*, Advances in Complex Systems **03** (**2000**) 87–98.

[13] Galam, S., *Democratic Voting in Hierarchical Structures or How to Build a Dictatorship*, Advances in Complex Systems **03** (**2000**) 171–180.

[14] Iwamura, T. and Takefuji, Y., *An Artificial Market Based on Agents with Fluid Attitude Toward Risks and Returns*, Advances in Complex Systems **03** (**2000**) 385–397.

[15] Lepelley, D., Louichi, A., and Valognes, F., *Computer Simulations of Voting Systems*, Advances in Complex Systems **03** (**2000**) 181–194.

[16] Savit, R., Li, Y., and Vandeemen, A., *Variable Payoffs in the Minority Game*, Advances in Complex Systems **03** (**2000**) 271–281.

[17] Terna, P., *The "Mind Or No-Mind" Dilemma in Agents Behaving in a Market*, Advances in Complex Systems **03** (**2000**) 257–269.

[18] Vanbergue, D., Treuil, J.-P., and Drogoul, A., *Modelling Urban Phenomena with Cellular Automata*, Advances in Complex Systems **03** (**2000**) 127–140.

[19] Zimmermann, M. G., Eguíluz, V. M., San Miguel, M., and Spadaro, A., *Cooperation in an Adaptive Network*, Advances in Complex Systems **03** (**2000**) 283–297.

[20] Ahmed, E. and Hegazi, A. S., *On Persistence and Stability of Coupled Map Lattices*, Advances in Complex Systems **04** (**2001**) 191–205.

[21] Berg, J. and Mehta, A., *Spin-Models of Granular Compaction: From One-Dimensional Models To Random Graphs*, Advances in Complex Systems **04** (**2001**) 309–319.

[22] Crutchfield, J. P. and Feldman, D. P., *Synchronizing to the Environment: Information-Theoretic Constraints on Agent Learning*, Advances in Complex Systems **04** (**2001**) 251–264.

[23] Dean, D. S. and Lefevre, A., *The Steady State of the Tapped Ising Model*, Advances in Complex Systems **04** (**2001**) 333–343.

[24] Ilyinsky, A., *Spectral Regularization, Data Complexity and Agent Behavior*, Advances in Complex Systems **04** (**2001**) 57–70.

[25] Marsili, M. and Challet, D., *Trading Behavior and Excess Volatility in Toy Markets*, Advances in Complex Systems **04** (**2001**) 3–17.

[26] Wolpert, D. H. and Tumer, K., *Optimal Payoff Functions for Members of Collectives*, Advances in Complex Systems **04** (**2001**) 265–279.

[27] Zimmermann, G., Neuneier, R., and Grothmann, R., *Multi-Agent Market Modeling of Foreign Exchange Rates*, Advances in Complex Systems **04** (**2001**) 29–43.

[28] Chopard, B., Dupuis, A., Masselot, A., and Luthi, P., *Cellular Automata and Lattice Boltzmann Techniques: An Approach to Model and Simulate Complex Systems*, Advances in Complex Systems **05** (**2002**) 103–246.

[29] Moreira, J. and Deutsch, A., *Cellular Automaton Models of Tumor Development: A Critical Review*, Advances in Complex Systems **05** (**2002**) 247–267.

[30] Mühlenbein, H. and Höns, R., *Stochastic Analysis of Cellular Automata with Application to the Voter Model*, Advances in Complex Systems **05** (**2002**) 301–337.

[31] Chen, S.-H. and Tai, C.-C., *Trading Restrictions, Price Dynamics and Allocative Efficiency in Double Auction Markets: Analysis Based on Agent-Based Modeling and Simulations*, Advances in Complex Systems **06** (**2003**) 283–302.

[32] Chivers, W. J. and Herbert, R. D., *The Effects of Varying Parameter Values and Heterogeneity in an Individual-Based Model of Predator-Prey Interaction*, Advances in Complex Systems **06** (**2003**) 441–456.

[33] Darbyshire, P., *Effects of Communication on Group Learning Rates in a Multi-Agent Environment*, Advances in Complex Systems **06** (**2003**) 405–426.

[34] Gallegati, M., Giulioni, G., and Kichiji, N., *Complex Dynamics and Financial Fragility in an Agent-Based Model*, Advances in Complex Systems **06** (**2003**) 267–282.

[35] Kubo, M. and Sasakabe, Y., *Re-Formation of Mobile Agents Structure By the Synchronization of Coupled Chaotic Oscillators*, Advances in Complex Systems **06** (**2003**) 427–440.

[36] Leon Suematsu, Y. I., Takadama, K., Nawa, N. E., Shimohara, K., and Katai, O., *Analyzing the Agent-Based Model and Its Implications*, Advances in Complex Systems **06** (**2003**) 331–347.

[37] Mizuno, M. and Nishiyama, N., *Interacting TV Viewers: A Case of Empirical Agent-Based Modeling and Simulation for Business Application*, Advances in Complex Systems **06** (**2003**) 361–373.

[38] Nakayama, S., *A Café Choice Problem as an Extended Arthur's El Farol Problem*, Advances in Complex Systems **06** (**2003**) 393–404.

[39] Tanimoto, J. and Fujii, H., *A Study on Research Performance in Japanese Universities: Which is More Efficient — a Professor Who is Leading His Research Group or One Who is Working Alone? The Multi-Agent Simulation Knows*, Advances in Complex Systems **06** (**2003**) 375–391.

[40] Beigy, H. and Meybodi, M. R., *A Mathematical Framework for Cellular Learning Automata*, Advances in Complex Systems **07** (**2004**) 295–319.

[41] Gilbert, N., *Open Problems in Using Agent-Based Models in Industrial and Labor Dynamics*, Advances in Complex Systems **07** (**2004**) 285–288.

[42] He, M., Li, X., Yan, S., and Sun, S., *Evolution of Students' Studying Attitudes on the Cellular Automata Model*, Advances in Complex Systems **07** (**2004**) 321–327.

[43] He, M.-F., Deng, C.-R., Feng, L., and Tian, B.-W., *A Cellular Automata Model for a Learning Process*, Advances in Complex Systems **07** (**2004**) 433–439.

[44] Kirman, A., *Economics and Complexity*, Advances in Complex Systems **07** (**2004**) 139–155.

[45] Mendes, R. V., *Network Dependence of Strong Reciprocity*, Advances in Complex Systems **07** (**2004**) 357–368.

[46] Neugart, M., *Endogenous Matching Functions: An Agent-Based Computational Approach*, Advances in Complex Systems **07** (**2004**) 187–201.

[47] Baas, N. A. and Helvik, T., *Higher Order Cellular Automata*, Advances in Complex Systems **08** (**2005**) 169–192.

[48] Hansson, A. A., Mortveit, H. S., and Reidys, C. M., *On Asynchronous Cellular Automata*, Advances in Complex Systems **08** (**2005**) 521–538.

[49] Helbing, D., Schönhof, M., Stark, H.-U., and Hołyst, J. A., *How Individuals Learn to Take Turns: Emergence of Alternating Cooperation in a Congestion Game and the Prisoner's Dilemma*, Advances in Complex Systems **08** (**2005**) 87–116.

[50] Płatkowski, T. and Ramsza, M., *Multimarket Minority Game*, Advances in Complex Systems **08** (**2005**) 65–74.

[51] Guala, S. D., *Model of Influence: From Individual Decisions to Locked-In Markets*, Advances in Complex Systems **09** (**2006**) 59–67.

[52] Jiang, Y., Sozinova, O., and Alber, M., *On Modeling Complex Collective Behavior in Myxobacteria*, Advances in Complex Systems **09** (**2006**) 353–367.

[53] Beigy, H. and Meybodi, M. R., *Open Synchronous Cellular Learning Automata*, Advances in Complex Systems **10** (**2007**) 527–556.

[54] Beuck, U., Nagel, K., Rieser, M., Strippgen, D., and Balmer, M., *Preliminary Results of a Multiagent Traffic Simulation for Berlin*, Advances in Complex Systems **10** (**2007**) 289–307.

[55] Bosse, T., Jonker, C. M., and Treur, J., *Simulation and Analysis of Adaptive Agents: An Integrative Modeling Approach*, Advances in Complex Systems **10** (**2007**) 335–357.

[56] Cascetta, E., Pagliara, F., and Papola, A., *Governance of Urban Mobility: Complex Systems and Integrated Policies*, Advances in Complex Systems **10** (**2007**) 339–354.

[57] Darabos, C., Giacobini, M., and Tomassini, M., *Performance and Robustness of Cellular Automata Computation on Irregular Networks*, Advances in Complex Systems **10** (**2007**) 85–110.

[58] Fent, T., Groeber, P., and Schweitzer, F., *Coexistence of Social Norms Based on In- and Out-Group Interactions*, Advances in Complex Systems **10** (**2007**) 271–286.

[59] Fleury, D., *Urban Safety Management: How To Deal With Complexity*, Advances in Complex Systems **10** (**2007**) 327–338.

[60] González, L., *Algorithm Comparing Binary String Probabilities in Complex Stochastic Boolean Systems Using Intrinsic Order Graph*, Advances in Complex Systems **10** (**2007**) 111–143.

[61] Guisado, J. L., Jiménez-Morales, F., and Fernández De Vega, F., *Cellular Automata and Cluster Computing: An Application to the Simulation of Laser Dynamics*, Advances in Complex Systems **10** (**2007**) 167–190.

[62] Hiebeler, D. E., *Transient Dynamics and Quasistationary Equilibria of Continuous-Time Linear Stochastic Cellular Automata Voter Models with Multiscale Neighborhoods*, Advances in Complex Systems **10** (**2007**) 145–165.

[63] Johansson, A., Helbing, D., and Shukla, P. K., *Specification of the Social Force Pedestrian Model By Evolutionary Adjustment To Video Tracking Data*, Advances in Complex Systems **10** (**2007**) 271–288.

[64] Liu, F. C. S., *Constrained Opinion Leader Influence in an Electoral Campaign Season: Revisiting the Two-Step Flow Theory with Multi-Agent Simulation*, Advances in Complex Systems **10** (**2007**) 233–250.

[65] Lorenz, J. and Urbig, D., *About the Power to Enforce and Prevent Consensus By Manipulating Communication Rules*, Advances in Complex Systems **10** (**2007**) 251–269.

[66] Martinotti, G., *Gone With the Wind: Physical Spaces in the Third Generation Metropolis*, Advances in Complex Systems **10** (**2007**) 233–253.

[67] Pan, Z., Reggia, J., and Gao, D., *Properties of Self-Replicating Cellular Automata Systems Discovered Using Genetic Programming*, Advances in Complex Systems **10** (**2007**) 61–84.

[68] Semboloni, F., *From Spatially Explicit to Multiagents Simulation of Urban Dynamic*, Advances in Complex Systems **10** (**2007**) 355–362.

[69] Zanlungo, F., *A Collision-Avoiding Mechanism Based On a Theory of Mind*, Advances in Complex Systems **10** (**2007**) 363–371.

[70] Araújo, T. and Aubyn, M. S., *Education, Neighborhood Effects and Growth: An Agent-Based Model Approach*, Advances in Complex Systems **11** (**2008**) 99–117.

[71] Conte, R., Paolucci, M., and Sabater-Mir, J., *Reputation for Innovating Social Networks*, Advances in Complex Systems **11** (**2008**) 303–320.

[72] Delre, S. A., Broekhuizen, T. L. J., and Jager, W., *The Effect of Social Influence on Market Inequalities in the Motion Picture Industry*, Advances in Complex Systems **11** (**2008**) 273–287.

[73] García-Díaz, C., Van Witteloostuijn, A., and Péli, G., *Market Dimensionality and the Proliferation of Small-Scale Firms*, Advances in Complex Systems **11** (**2008**) 231–247.

[74] Geller, A. and Moss, S., *Growing Qawm: An Evidence-Driven Declarative Model of Afghan Power Structures*, Advances in Complex Systems **11** (**2008**) 321–335.

[75] Huet, S., Deffuant, G., and Jager, W., *A Rejection Mechanism in 2d Bounded Confidence Provides More Conformity*, Advances in Complex Systems **11** (**2008**) 529–549.

[76] Lewkovicz, Z. and Kant, J.-D., *A Multiagent Simulation of a Stylized French Labor Market: Emergences At the Micro Level*, Advances in Complex Systems **11** (**2008**) 217–230.

[77] Posada, M., Hernández, C., and López-Paredes, A., *Testing Marshallian and Walrasian Instability with an Agent-Based Model*, Advances in Complex Systems **11** (**2008**) 249–260.

[78] Rosser, J. B., *Econophysics and Economic Complexity*, Advances in Complex Systems **11** (**2008**) 745–760.

[79] Shin, J. K. and Fossett, M., *Residential Segregation By Hill-Climbing Agents on the Potential Landscape*, Advances in Complex Systems **11** (**2008**) 875–899.

[80] Stark, H.-U., Tessone, C. J., and Schweitzer, F., *Slower Is Faster: Fostering Consensus Formation By Heterogeneous Inertia*, Advances in Complex Systems **11** (**2008**) 551–563.

[81] Thiriot, S. and Kant, J.-D., *Using Associative Networks to Represent Adopters' Beliefs in a Multiagent Model Of Innovation Diffusion*, Advances in Complex Systems **11** (**2008**) 261–272.

[82] Uchida, M. and Shirayama, S., *Eigenmode of the Decision-By-Majority Process in Complex Networks*, Advances in Complex Systems **11** (**2008**) 565–579.

[83] Yang, L. and Gilbert, N., *Getting Away From Numbers: Using Qualitative Observation for Agent-Based Modeling*, Advances in Complex Systems **11** (**2008**) 175–185.

[84] Zanlungo, F., Arita, T., and Rambaldi, S., *Emergence of a Traffic Flow Convention in a Multiagent Model*, Advances in Complex Systems **11** (**2008**) 789–802.

[85] Araújo, T. and Mendes, R. V., *Innovation and Self-Organization in a Multi-Agent Model*, Advances in Complex Systems **12** (**2009**) 233–253.

[86] Jacobi, M. N. and Görnerup, O., *A Spectral Method for Aggregating Variables in Linear Dynamical Systems With Application to Cellular Automata Renormalization*, Advances in Complex Systems **12** (**2009**) 131–155.

[87] Tumer, K. and Agogino, A., *Multiagent Learning for Black Box System Reward Functions*, Advances in Complex Systems **12** (**2009**) 475–492.

[88] Tumer, K. and Khani, N., *Learning From Actions Not Taken in Multiagent Systems*, Advances in Complex Systems **12** (**2009**) 455–473.

[89] Wawrzyniak, K. and Wislicki, W., *Multi-Market Minority Game: Breaking the Symmetry of Choice*, Advances in Complex Systems **12** (**2009**) 423–437.

[90] Wawrzyniak, K. and Wislicki, W., *Phenomenology of Minority Games in Efficient Regime*, Advances in Complex Systems **12** (**2009**) 619–639.

[91] Agogino, A. and Tumer, K., *A Multiagent Coordination Approach to Robust Consensus Clustering*, Advances in Complex Systems **13** (**2010**) 165–197.

[92] Banisch, S., Araújo, T., and Louçã, J., *Opinion Dynamics and Communication Networks*, Advances in Complex Systems **13** (**2010**) 95–111.

[93] Huet, S. and Deffuant, G., *Openness Leads to Opinion Stability and Narrowness to Volatility*, Advances in Complex Systems **13** (**2010**) 405–423.

[94] Marchione, E., Salgado, M., and Gilbert, N., *"What Did You Say?" Emergent Communication in a Multi-Agent Spatial Configuration*, Advances in Complex Systems **13** (**2010**) 469–482.

[95] Martins, A. C. R., *Modeling Scientific Agents for a Better Science*, Advances in Complex Systems **13** (**2010**) 519–533.

[96] Martins, A. C. R. and Kuba, C. D., *The Importance of Disagreeing: Contrarians and Extremism in the Coda Model*, Advances in Complex Systems **13** (**2010**) 621–634.

[97] Mayor, E. and Sartor, G., *Why Are Lawyers Nice Or Nasty? Insights From Agent-Based Modeling*, Advances in Complex Systems **13** (**2010**) 535–558.

[98] Sengupta, A. and Glavin, S. E., *Volatility in the Consumer Packaged Goods Industry — a Simulation Based Study*, Advances in Complex Systems **13** (**2010**) 579–605.

[99] Yukalov, V. I. and Sornette, D., *Mathematical Structure of Quantum Decision Theory*, Advances in Complex Systems **13** (**2010**) 659–698.

[100] Banerjee, B. and Kraemer, L., *Action Discovery for Single and Multi-Agent Reinforcement Learning*, Advances in Complex Systems **14** (**2011**) 279–305.

[101] Carletti, T., Righi, S., and Fanelli, D., *Emerging Structures in Social Networks Guided By Opinions' Exchanges*, Advances in Complex Systems **14** (**2011**) 13–30.

[102] Carvalho, J., Lopes, R. L., and Tojo, J., *Modeling Settlement Patterns in Real Territories*, Advances in Complex Systems **14** (**2011**) 549–565.

[103] Caticha, N. and Vicente, R., *Agent-Based Social Psychology: From Neurocognitive Processes to Social Data*, Advances in Complex Systems **14** (**2011**) 711–731.

[104] Chang, M.-H., *Emergent Social Learning Networks in Organizations with Heterogeneous Agents*, Advances in Complex Systems **14** (**2011**) 169–199.

[105] Devlin, S., Kudenko, D., and Grześ, M., *An Empirical Study of Potential-Based Reward Shaping and Advice in Complex, Multi-Agent Systems*, Advances in Complex Systems **14** (**2011**) 251–278.

[106] Fosco, C., Laruelle, A., and Sánchez, A., *Turnout Intention and Random Social Networks*, Advances in Complex Systems **14** (**2011**) 31–53.

[107] Grauwin, S., Hunt, D., Bertin, E., and Jensen, P., *Effective Free Energy For Individual Dynamics*, Advances in Complex Systems **14** (**2011**) 529–536.

[108] Howley, E. and Duggan, J., *Investing in the Commons: A Study of Openness and the Emergence of Cooperation*, Advances in Complex Systems **14** (**2011**) 229–250.

[109] Kretz, T., Große, A., Hengst, S., Kautzsch, L., Pohlmann, A., and Vortisch, P., *Quickest Paths in Simulations of Pedestrians*, Advances in Complex Systems **14** (**2011**) 733–759.

[110] Quattrociocchi, W., Conte, R., and Lodi, E., *Opinions Manipulation: Media, Power and Gossip*, Advances in Complex Systems **14** (**2011**) 567–586.

[111] Taylor, M. E., Jain, M., Tandon, P., Yokoo, M., and Tambe, M., *Distributed On-Line Multi-Agent Optimization Under Uncertainty: Balancing Exploration and Exploitation*, Advances in Complex Systems **14** (**2011**) 471–528.

[112] Yamashita, A., Kawamura, H., and Suzuki, K., *Adaptive Fusion Method for User-Based and Item-Based Collaborative Filtering*, Advances in Complex Systems **14** (**2011**) 133–149.

[113] Antonioni, A. and Tomassini, M., *Cooperation on Social Networks and Its Robustness*, Advances in Complex Systems **15** (**2012**) 1250046.

[114] Barton, C. M. and Riel-Salvatore, J., *Agents of Change: Modeling Biocultural Evolution in Upper Pleistocene Western Eurasia*, Advances in Complex Systems **15** (**2012**) 1150003.

[115] Blythe, R. A., *Random Copying in Space*, Advances in Complex Systems **15** (**2012**) 1150012.

[116] Brede, M., *Preferential Opponent Selection in Public Goods Games*, Advances in Complex Systems **15** (**2012**) 1250074.

[117] Cirillo, P., Gallegati, M., and Hüsler, J., *A Pólya Lattice Model to Study Leverage Dynamics and Contagious Financial Fragility*, Advances in Complex Systems **15** (**2012**) 1250069.

[118] Edmonds, B., *Modeling Belief Change in a Population Using Explanatory Coherence*, Advances in Complex Systems **15** (**2012**) 1250085.

[119] Fan, H., *Distribution of Producer Size in Globalized Market*, Advances in Complex Systems **15** (**2012**) 1250076.

[120] Gargiulo, F. and Huet, S., *New Discussions Challenge the Organization of Societies*, Advances in Complex Systems **15** (**2012**) 1250033.

[121] Gómez-Serrano, J. and Le Boudec, J.-Y., *Comment on "Mixing Beliefs Among Interacting Agents"*, Advances in Complex Systems **15** (**2012**) 1250028.

[122] Kampouridis, M., Chen, S.-H., and Tsang, E., *Microstructure Dynamics and Agent-Based Financial Markets: Can Dinosaurs Return?*, Advances in Complex Systems **15** (**2012**) 1250060.

[123] Kandler, A. and Smaers, J. B., *An Agent-Based Approach to Modeling Mammalian Evolution: How Resource Distribution and Predation Affect Body Size*, Advances in Complex Systems **15** (**2012**) 1150014.

[124] Kemloh Wagoum, A. U., Seyfried, A., and Holl, S., *Modeling the Dynamic Route Choice of Pedestrians to Assess the Criticality Of Building Evacuation*, Advances in Complex Systems **15** (**2012**) 1250029.

[125] Kohler, T. A., Cockburn, D., Hooper, P. L., Bocinsky, R. K., and Kobti, Z., *The Coevolution of Group Size and Leadership: An Agent-Based Public Goods Model for Prehispanic Pueblo Societies*, Advances in Complex Systems **15** (**2012**) 1150007.

[126] Kowalska-Styczeń, A. and Sznajd-Weron, K., *Access to Information in Word of Mouth Marketing Within a Cellular Automata Model*, Advances in Complex Systems **15** (**2012**) 1250080.

[127] Rosenberg, A. L., *Cellular Antomata*, Advances in Complex Systems **15** (**2012**) 1250070.

[128] Teglio, A., Raberto, M., and Cincotti, S., *The Impact of Banks' Capital Adequacy Regulation on the Economic System: An Agent-Based Approach*, Advances in Complex Systems **15** (**2012**) 1250040.

[129] Adriaansen, T., Armbruster, D., Kempf, K., and Li, H., *An Agent Model for the High-End Gamers Market*, Advances in Complex Systems **16** (**2013**) 1350028.

[130] Callegari, S., Weissmann, J. D., Tkachenko, N., Petersen, W. P., Lake, G., De León, M. P., and Zollikofer, C. P. E., *An Agent-Based Model of Human Dispersals At a Global Scale*, Advances in Complex Systems **16** (**2013**) 1350023.

[131] Ciampaglia, G. L., *A Framework For the Calibration of Social Simulation Models*, Advances in Complex Systems **16** (**2013**) 1350030.

[132] Cimini, G., Zeng, A., Medo, M., and Chen, D., *The Role of Taste Affinity in Agent-Based Models for Social Recommendation*, Advances in Complex Systems **16** (**2013**) 1350009.

[133] Helbing, D. and Grund, T. U., *Agent-Based Modeling and Techno-Social Systems*, Advances in Complex Systems **16** (**2013**) 1303002.

[134] Kashif, A., Dugdale, J., and Ploix, S., *Simulating Occupants' Behavior for Energy Waste Reduction in Dwellings: A Multiagent Methodology*, Advances in Complex Systems **16** (**2013**) 1350022.

[135] Kitto, K. and Boschetti, F., *Attitudes, Ideologies and Self-Organization: Information Load Minimization In Multi-Agent Decision Making*, Advances in Complex Systems **16** (**2013**) 1350029.

[136] Llacer, T., Miguel, F. J., Noguera, J. A., and Tapia, E., *An Agent-Based Model of Tax Compliance: An Application to the Spanish Case*, Advances in Complex Systems **16** (**2013**) 1350007.

[137] Menezes, T. and Roth, C., *Automatic Discovery of Agent Based Models: An Application to Social Anthropology*, Advances in Complex Systems **16** (**2013**) 1350027.

[138] Namatame, A. and Tran, H. A. Q., *Enhancing the Resilience of Networked Agents Through Risk Sharing*, Advances in Complex Systems **16** (**2013**) 1350006.

[139] Nandi, S., Johnson, N. F., and Cohn, J. L., *Persistent Patterns in Microtubule Dipole Lattices*, Advances in Complex Systems **16** (**2013**) 1350033.

[140] Schweitzer, F., Mavrodiev, P., and Tessone, C. J., *How Can Social Herding Enhance Cooperation?*, Advances in Complex Systems **16** (**2013**) 1350017.

[141] Banisch, S., *From Microscopic Heterogeneity to Macroscopic Complexity in the Contrarian Voter Model*, Advances in Complex Systems **17** (**2014**) 1450025.

[142] Faletra, M., Palmer, N., and Marshall, J. S., *Effectiveness of Opinion Influence Approaches in Highly Clustered Online Social Networks*, Advances in Complex Systems **17** (**2014**) 1450008.

[143] Przybyła, P., Sznajd-Weron, K., and Weron, R., *Diffusion of Innovation Within an Agent-Based Model: Spinsons, Independence And Advertising*, Advances in Complex Systems **17** (**2014**) 1450004.

[144] Righi, S. and Takács, K., *Emotional Strategies As Catalysts for Cooperation in Signed Networks*, Advances in Complex Systems **17** (**2014**) 1450011.

[145] Yukalov, V. I. and Sornette, D., *Self-Organization in Complex Systems As Decision Making*, Advances in Complex Systems **17** (**2014**) 1450016.

[146] Banisch, S. and Lima, R., *Markov Chain Aggregation for Simple Agent-Based Models on Symmetric Networks: The Voter Model*, Advances in Complex Systems **18** (**2015**) 1550011.

[147] Lindgren, K., Jonson, E., and Lundberg, L., *Projection of a Heterogenous Agent-Based Production Economy Model to a Closed Dynamics of Aggregate Variables*, Advances in Complex Systems **18** (**2015**) 1550012.

[148] Troitzsch, K. G., *Extortion Racket Systems As Targets for Agent-Based Simulation Models. Comparing Competing Simulation Models and Emprical Data*, Advances in Complex Systems **18** (**2015**) 1550014.

[149] Calvo, C., Villacorta-Atienza, J. A., Mironov, V. I., Gallego, V., and Makarov, V. A., *Waves in Isotropic Totalistic Cellular Automata: Application to Real-Time Robot Navigation*, Advances in Complex Systems **19** (**2016**) 1650012.

[150] Lamarche-Perrin, R., Banisch, S., and Olbrich, E., *The Information Bottleneck Method for Optimal Prediction of Multilevel Agent-Based Systems*, Advances in Complex Systems **19** (**2016**) 1650002.

[151] De Abreu, J., García, P., and García, J., *A Deterministic Approach to the Synchronization of Nonlinear Cellular Automata*, Advances in Complex Systems **20** (**2017**) 1750006.

[152] Gargiulo, F., Gandica, Y., and Carletti, T., *Emergent Dense Suburbs in a Schelling Metapopulation Model: A Simulation Approach*, Advances in Complex Systems **20** (**2017**) 1750001.

[153] Schasfoort, J., Godin, A., Bezemer, D., Caiani, A., and Kinsella, S., *Monetary Policy Transmission in a Macroeconomic Agent-Based Model*, Advances in Complex Systems **20** (**2017**) 1850003.

[154] Pluchino, A., Biondo, A. E., and Rapisarda, A., *Talent Versus Luck: the Role of Randomness in Success and Failure*, Advances in Complex Systems **21** (**2018**) 1850014.

[155] Jacobs, F. and Galam, S., *Two-Opinions-Dynamics Generated By Inflexibles and Non-Contrarian and Contrarian Floaters*, Advances in Complex Systems **22** (**2019**) 1950008.

[156] Kroc, J., Jiménez-Morales, F., Guisado, J. L., Lemos, M. C., and Tkáč, J., *Building Efficient Computational Cellular Automata Models of Complex Systems: Background, Applications, Results, Software, and Pathologies*, Advances in Complex Systems **22** (**2019**) 1950013.

[157] Sornette, D., Wheatley, S., and Cauwels, P., *The Fair Reward Problem: the Illusion of Success and How to Solve It*, Advances in Complex Systems **22** (**2019**) 1950005.

[158] Beikirch, M., Cramer, S., Frank, M., Otte, P., Pabich, E., and Trimborn, T., *Robust Mathematical Formulation and Probabilistic Description of Agent-Based Computational Economic Market Models*, Advances in Complex Systems **23** (**2020**) 2050017.

[159] Challet, D., Pluchino, A., Biondo, A. E., and Rapisarda, A., *The Origins of Extreme Wealth Inequality in the Talent Versus Luck Model*, Advances in Complex Systems **23** (**2020**) 2050004.

[160] Elvidge, S., *The Luck in "Talent Versus Luck" Modeling*, Advances in Complex Systems **23** (**2020**) 2050007.

[161] González-López, R., Gómez, J. B., and Pacheco, A. F., *A Minimal Agent-Based Model for the Size-Frequency Distribution of Firms*, Advances in Complex Systems **23** (**2020**) 2050002.

[162] Redwan, C. S. and Bazzan, A. L., *How Hard Is for Agents to Learn the User Equilibrium? Characterizing Traffic Networks By Means of Entropy*, Advances in Complex Systems **23** (**2020**) 2050011.

[163] Schweitzer, F., Verginer, L., and Vaccario, G., *Should the Government Reward Cooperation? Insights From an Agent-Based Model Of Wealth Redistribution*, Advances in Complex Systems **23** (**2020**) 2050018.

[164] Silvestre, J., *Sovereign Default Contagion and Monetary Policy in an Agent-Based Model*, Advances in Complex Systems **23** (**2020**) 2050010.

[165] Simão, R., Rosendo, F., and Wardil, L., *The Talent Versus Luck Model As An Ensemble of One-Dimensional Random Walks*, Advances in Complex Systems **24** (**2021**).

[166] Schweitzer, F., Garas, A., Tomasello, M. V., Vaccario, G., and Verginer, L., *The Role of Network Embeddedness on the Selection of Collaboration Partners: An Agent-Based Model with Empirical Validation*, Advances in Complex Systems **25** (**2022**) 2250003.

Advances in Complex Systems, Vol. 7, No. 2 (2004) 139–155

ECONOMICS AND COMPLEXITY

ALAN KIRMAN
CGREQAM, EHESS, Universite d'Aix Marseille,
2 Rue de la Charite, 13002 Marseille, France
kirman@ehess.vcharite.univ-mrs.fr

Received 1 March 2004
Revised 20 April 2004

This paper presents a view of the economy as a complex system with heterogeneous interacting agents who collectively organize themselves to generate aggregate phenomena which cannot be regarded as the behavior of some average or representative individual. There is an essential difference between the aggregate and the individual and such phenomena as bubbles and crashes, herd behavior, the transmission of information and the organization of trade are better modeled in the sort of framework suggested here than in more standard economic models.

Keywords: Aggregation; interaction; self-organization; herd behavior; rationality.

1. Introduction

Complexity and complex systems are terms which are widely used but not always carefully defined. Few would argue that the economy is not a complex system, but without a clear specification of what we mean by this, we are not much further forward. What might be thought of as the characteristics of a complex system and how many of these characteristics are shared by economies? Three of the charateristics most frequently found in the literature on complex systems are the following: they are composed of interacting "agents," these agents may have simple behavioral rules, the interaction among the agents means that aggregate phenomena are intrinsically different from individual behavior. The idea that a large collection of interacting objects can produce behavior at the aggregate level which could not be thought of as corresponding to some blown up version of individual behavior is far from new. What is newer is the idea that such systems may tend to organize themselves and, perhaps more, that there may be common features of that behavior in many, apparently very different, types of system. Thus, features of the behavior of collections of neurons may share properties with air masses and with social systems. It is the "emergence" of organization and the associated aggregregate features that is emphasized by the founders of what has come to be known as the science of "complexity." Among the leading exponents of what might loosely be called complexity theory are Anderson, Gell-Mann, and Kauffman, all of whom are closely identified with the Santa Fe Institute. Several collections of papers applying these ideas to economics have appeared [1, 4]. Reference 18 presents a simple and clear account of the basic notions.

How are these developments related to the standard view of economics? In economics, many of the features of interaction mentioned are what have been referred to as "externalities," the fact that the behavior of some individuals has direct consequences for others. Such externalities are typically thought of as "market imperfections" whereas the main theme of this paper is that they should be considered as central and not marginal. Rather than regard externalities as an annoying distortion of the basic model the idea here is that our models should be built around the interactions and the externalities generated by those interactions. Hans Föellmer [10] in a path-breaking paper showed how important direct interactions were in determining whether or not an economy would have well determined aggregate behavior as the number of agents becomes large. His important contribution was to show that this was not necessarily the case if the interactions between agents were strong enough.

In the more conventional economic literature, the first substantial plea for taking externalities more seriously was that of Schelling [23]. He argued that in the presence of externalities there might be many equilibria of the economic system some of which might be highly unsatisfactory from a collective point of view. More recent work on complexity puts less emphasis on the notion of equilibrium in the classical sense and more weight on the idea of the economy as an evolving, open-ended system. Hayek's ideas of the emergence of order bear a resemblance to the sort of ideas considered here.

Another important related thread running through the economic literature is the "evolutionary approach." Here the interest is in the collective result of a situation in which myopic individuals with limited comprehension and rationality grope their way forward. This sort of idea discussed by Nelson and Winter [20] is clearly related to the view of the economy as having self organizing properties. Reference 9 presents a series of papers which might be thought of as in the Nelson Winter tradition.

One of the standard criticisms of this approach has been that the analysis was not rigorous, in the mathematical sense, but the strong recent interest in evolutionary game theory, (see Refs. 22 and 28) reveals that this sort of approach has now penetrated into areas which could hardly be accused of being less than analytically rigorous. We will not attempt to expound the basic ideas of complexity theory and self organization since these are dealt with at length elsewhere. Our aim is to show that these ideas can play an important role in developing economic theory. There are two essential things to examine: how the organization of the interaction between the individuals and the component parts of the system affects aggregate behavior and how that organization itself emerges. The purpose of this paper is to look at these two aspects of economic systems. The first question is of particular interest in the economic context since organization is rarely considered directly in economic models. The basic model underlying most modern economic analysis remains the General Equilibrium model. This model seems to have the great merit of considering all markets as interdependent and having a well-defined equilibrium notion. The essential feature of the model is, from the point of view of this project, that only one type of organization is considered and, even that consideration is implicit. Yet more and more dissatisfaction is expressed with this model. Our aim is to show that by building rather different models we can explain economic phenomena which seem to be inconsistent with the standard model.

It should be said at the outset that the current benchmark model is not as its originators intended it to be. From Adam Smith and his "invisible hand" to Walras, economists had in mind a complicated interactive system in which individuals acting in their own interest came to organize themselves. Contrary to the purified form of the model they did not have in mind any centralized price determining system. Prices themselves are part of the organization of the system and where those prices come from is a necessary part of the explanation of economic activity.

The standard view of the economy revolves around two themes, the rationality of the individuals and the specific view of the way in which agents interact. In that model, agents are "rational" optimizers and are isolated. We suggest that both of these features should be modified and that once we allow for direct interaction between agents we can assume less about their rationality and still observe interesting aggregate behavior which is no longer, however, the behavior of an average economic agent. As Forni and Lippi [11] have emphasized, aggregate behavior is not the behavior of a representative individual and trying to test models based on this idea leads to erroneous conclusion. Such an idea is so familiar to physicists and biologists that it seems banal. Nevertheless, it is still far from being generally accepted in economics. Why should we be dissatisfied with the General Equilibrium model as a benchmark? Simply because it has become clear that it fails to satisfy certain important criteria, even if one accepts the notion of equilibrium involved. As the results of Sonnenschein [27] and Debreu [8] show, there is no guarantee that the economy will converge to an equilibrium if it starts in an out-of-equilibrium situation, and furthermore, the equilibrium may not be unique. Put more briefly, the equilibrium is not well determined.

A radical reaction to this situation was that of Hildenbrand [14], one of the leading scholars of General Equilibrium Theory who said,

> When I read in the seventies publications of Sonnenschein, Mantel and Debreu on the structure of the excess demand function of an exchange economy, I was deeply consternated [*sic*]. Up to that time I had the naïve illusion that the microeconomic foundation of the general equilibrium model, which I admired so much, does not only allow us to prove that the model and the concept of equilibrium are logically consistent, but also allows us to show that the equilibrium is well determined. This illusion, or should I say rather this hope, was destroyed, once and for all, at least for the

traditional model of exchange economies. I was tempted to repress this insight and continue to find satisfaction in proving existence of equilibrium for more general models under still weaker assumptions. However, I did not succeed in repressing the newly gained insight because I believe that a theory of economic equilibrium is incomplete if the equilibrium is not well determined.

There are two ways out of this dilemma. Either one changes the foundations of the model, and this is what Hildenbrand has proposed, or one makes the heroic assumption that the average behavior of the market or economy can be assimilated to that of an individual. This avoids both of the problems mentioned and reduces the aggregate behavior to a problem that economists know how to solve. How appropriate is this, however, when we have no theoretical reason to believe that this can be done? Let us, at this point, start out with an example.

2. Aggregate and Individual Behavior: An Example, the Marseille Fish Market

What we wish to show is that, once one allows for direct interaction among agents, macro behavior cannot be thought of as reflecting the behavior of a "typical" or "average" individual. There is no simple direct correspondence between individual and aggregate regularity. As an illustration, consider the following simple empirical example, the behavior of agents on a market for a perishable product, fish. In Ref. 13 we showed that, although the transactions of individuals, on the wholesale market for fish, do not necessarily reveal any of the standard properties of a demand curve, nevertheless in aggregate there is a nice downward sloping relationship between prices and quantities transacted.

To understand this, assume for a moment that changes in the prices of fish do not result in a large amount of intertemporal substitution by consumers. This will lead fishmongers and other buyers on the wholesale market to behave in a relatively myopic way. This, in turn, justifies considering each day as a separate observation. This is particularly true since fish is evidently perishable. Indeed, this explains why, when considering particular markets, fish has been so widely used as an example, (for example by Marshall, Pareto, and Hicks) since with no stocks, successive markets can be thought of as independent. In our case, when fitting our price quantity relations we are implicitly treating price changes as resulting from random shocks to the supply of fish although the amount available is, at least in part, a result of strategic choice.

If we fit a demand system in the usual way, we are assuming that market behavior corresponds to that of an individual, or rather, that the aggregate relation has the same properties as the individual one. In our models we assume that individual relations have certain properties, yet, examination of individual data reveals none of the properties that can be derived for standard individual demand.

The main problem here, then becomes: can aggregation restore some of the properties that are lacking at the individual level? Individuals whose behavior does not satisfy the criteria that can be derived from standard assumptions may indeed, in the aggregate, satisfy such criteria. Aggregation may add structure. This is one side of the problem of aggregation. The other is that even if individuals did happen to satisfy certain properties, it is by no means necessary that these properties carry over to the aggregate level (see e.g. Refs. 8 and 27). Aggregation may destroy structure. The two taken together emphasize the basic theme of this paper: there is no direct connection between micro and macro behavior. This basic difficulty in the testing of aggregate models has been insisted upon in the past (see Refs. 15 and 19) when discussing representative individual macro models but as Lewbel observed, this has not, and is unlikely to, stop the profession from testing individually derived hypotheses at the aggregate level. Hence in our example, although some empirical properties of the aggregate relationships between prices charged and quantities purchased can be established, we would suggest that these should be viewed as independent of standard maximizing individual behavior.

If the market exhibits such features and one claims that they do not correspond to classical individual maximizing behavior, then one has to try to explain how the market organizes itself so that this comes about. On the Marseille market no prices are posted. Thus, information is highly dispersed among agents. Yet, a particular feature that one does observe on the Marseille fish market, is that over the day markets do more or less clear in the sense that the surplus left unsold never exceeds 4%. Furthermore, since sellers become aware, from the reactions of buyers to their offers, of the amount available on the market and *vice versa*, it would not be unreasonable to expect average prices to be lower on those days where the quantity is higher. However, the situation is not simple. For example, some buyers have to transact early, before such information becomes available, and others only make one transaction for a given fish on a given day. Thus, to deduce such a property formally from a complete model of the individuals' behavior and interactions on this market, would require very strong and unrealistic assumptions.

The important thing to re-emphasize here is that the "nice" monotonicity property of the aggregate price quantity curves does **not** reflect and is **not** derived from the corresponding characteristics of individual behavior. Nor indeed, given the previous discussion, should we expect it to be. But what is the lesson here? As we said previously, we are not looking at a standard demand curve but what we can say is that the market has organized itself in such a way that essentially all the fish is sold and, on those days where there is less supply, the prices are, on average, higher. These classic results arise from a complicated set of interactions in which agents know each other, price discriminate and attend the market with different frequencies. Thus, market organization simplifies the calculations that agents have to make. This was the point of Gode and Sunder's [12] famous "zero intelligence" agents who finally arrived at the competitive price on a double auction market, even though they bid at random. The way in which the market was organized led unthinking agents to the result that would have been achieved by highly rational forward looking economic agents. Furthermore, in our case, this occurs on a market which is not organized according to the standard model. Prices are not posted and individuals have to infer their information from their own observations, since they cannot observe the prices at which other transactions are occurring. The basic idea that we are trying to convey should, by now, be clear: aggregation across agents can add structure and in so doing, generate coherent aggregate behavior from the interaction of agents who interact directly with each other.

A second question is how prices are distributed. Most economists, faced with a situation in which some 500 buyers are faced with some 50 sellers under the same roof, would suggest that a unique price would be established for each good and that there would be little variation around that price. This is far from being the case: price distributions for each day show considerable variations and they do not have the sort of unimodal characteristic that standard analysis would lead us to expect. The question then arises whether, over time, some sort of structure emerges. Is there, for example, some sort of stability of the distributions if they are taken over a longer period of time? To answer this question, we tested the hypothesis that for each individual fish the monthly observed price distribution is stable over time. The statistical analysis could not reject it. Here then is a second message. Aggregation of a system with complicated patterns of interaction between agents may show little organization in the very short run but aggregation over time of this behavior may yield stability. Once again the behavior of the aggregate evolves in a rather stable way, something which is not true for individuals.

3. Trading Relationships

One of the reasons for the separation of aggregate and individual behavior in the previous discussion is that agents interact with only a subset of the others. Thus, information is being transmitted in an asymmetric way. Indeed, it is well known that in many markets individuals typically trade with very few partners. Thus, markets are characterized by small trading groups. How the choice of trading partners is made is one of the fundamental problems in economics. Either those with whom one trades are given exogenously or one assumes some anonymous mechanism through which trade takes place. One of the major challenges is to show how agents can learn to choose their trading partners in a simple market framework.

Some models of economic network formation attribute a great deal of rationality to the agents concerned and permit very limited conclusions. Others use simpler learning structures such as a reinforcement learning rule which can be justified theoretically and which produces stronger conclusions. Another type of learning model uses Holland's "classifier system" approach and makes very limited demands on the agents. Both of the latter class of models produce striking results as to the type of groups that emerge. We will give an example from the work based on the Marseille wholesale fish market by Weisbuch *et al.* [29], who used methods from statistical physics. We will show how the sorts of group that emerge correspond to those found in the data concerning the trading relationships on the Marseille wholesale fish market used as an example in the previous discussion.

One of the important features of the market is that individuals establish different sorts of relationships with each other. On the one hand, there are those buyers who regularly buy from the same seller and are extremely loyal, and on the other hand, there are people who shift between their sellers all of the time. This, of itself, seems to be a feature that one should try to explain. If one tries to go back to the framework that we outlined earlier with a full game theoretic model, this becomes extremely complicated because one now has to develop a dynamic game in which the experience of playing with each seller is taken into account or one has to think of a situation in which people have strategies which are so complicated that they can take into account of all the prices that they may face from different sellers. So the idea here is to develop a much simpler theoretical model in which people simply learn from their previous experience and, in consequence, they change their probability of visiting different sellers as a result of their experience. In Ref. 29 we consider a market in which buyers update their probability of visiting sellers on the basis of the profit that they obtained in the past from those sellers. If we denote by $J_{ij}(t)$ the cumulated profit, up to period t, that buyer i has obtained from trading with seller j then the probability p that i will visit j in that period is given by

$$P = \frac{\exp(\beta J_{ij})}{\sum_j \exp(\beta J_{ij})}, \tag{1}$$

where β is a reinforcement parameter which describes how sensitive the individual is to past profits. The non-linear updating rule 1 will be familiar, for example, from the model developed by Blume [6], and, as the logit decision, or the quantal response rule, and is widely used in statistical physics. Given this rule one can envisage the case of 3 sellers and 30 buyers, for example, as corresponding to the simplex in Fig. 1 below. Each buyer has certain probabilities of visiting each of the sellers and thus can be thought of as a point in the simplex. If he is equally likely to visit each of the three sellers then he can be represented as a point in the centre

Fig. 1. Sellers' simplex.

of the triangle. If, on the other hand, he visits one of the sellers with probability 1 then he can be shown as a point at one of the apexes of the triangle.

Thus, at any one point in time, the market is described by a cloud of points in the triangle and the question is how will this cloud evolve? If buyers all become loyal to particular sellers then the result will be be that all the points, corresponding to the buyers will move rapidly over time to the apexes of the triangle as in Fig. 1. This might be thought of as a situation in which the market is "ordered." On the other hand, if buyers learn to search randomly amongst the sellers, then the result will be a cluster of points at the centre of the triangle, which will persist over time. What we show in Ref. 29, is that which of these situations will develop depends crucially on the parameter β in (1), the discount rate of the buyers, and the profit per transaction. The stronger the reinforcement, the slower the individual forgets and the higher the profit, the more likely is it that order will emerge. In particular, the transition from disorder to order, as β changes, is very sharp. In our model this sort of "phase transition" is derived analytically using the "mean field" approach. The latter is open to the objection that random variables are replaced by their means and, in consequence, the process derived is only an approximation. The alternative is to consider the full stochastic process but this is often not tractable, and that is why we resorted to the simulations illustrated above. We wished to see whether the theoretical results from the approximation capture the features of the simulated stochastic process.[a]

A natural question to ask is what happens when there is a mixture of types in the market. The reason that the system consolidates itself on a network with highly loyal customers is that there is a co-evolution of the learning on both sides of the market. When buyers are loyal, sellers learn exactly how much fish to provide, and buyers are also sure to get what they want. Yet the presence of customers who are not loyal will interfere with this process since they may buy and prevent a loyal customer from getting what he wants. This could undermine the process by which loyalty develops. To see if this was the case, we simulated the process with two types of agents, those with a low critical value of β and those with a high one. The result is shown in Fig. 2. There is a clear separation between the two types. Those who have a low critical value of β become loyal and are represented by the dots at the corner of the simplex, and the others continue to shop around.

This is precisely what happens on the actual fish market. There is a clear separation between very loyal customers and those who search. Furthermore, loyal customers pay higher prices than casual shoppers but get what they want.

Even in a situation where the pay-offs from buying from the different sellers are the same and there seems to be no good *a priori* reason for loyalty to develop, the waste in a market with loyalty is much less than in one in which buyers search and

[a] A detailed discussion of this sort of problem is given by Aoki [3], who thinks of the buyers in the markets as being partitioned between the sellers and each buyer as having a probability of transiting from one seller to another. He looks at the limit distributions of such a process.

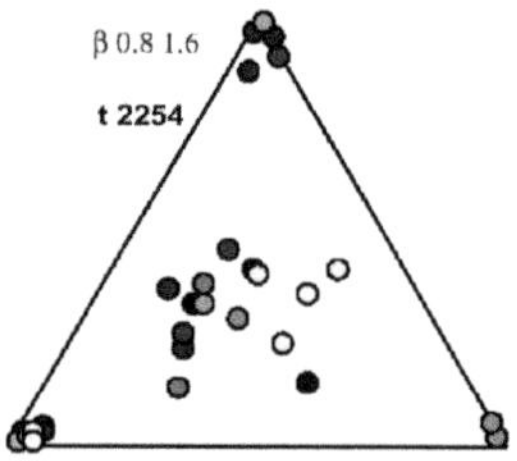

Fig. 2. Sellers' simplex with different types of buyers.

the market learns to organize itself appropriately. Thus, organization is an emergent property of the aggregate and not something directly intended by the agents.

4. Learning in Public Goods Experiments

We will now turn to another example where we see clearly the difference between individual and aggregate behavior.[b] The problem concerns that of people deciding how much of their money they should keep and how much they should contribute to a public good which everyone benefits from. In such games there are two reference points. There is a cooperative solution or collective optimum (CO), in which the average payoff of the agents is maximized. But there is also the non-cooperative solution in which everyone chooses his contribution as a best reply to the amounts chosen by the others. This involves individuals trying to "free-ride" and is the Nash Equilibrium (NE) with a much lower level of contribution than in the social optimum.

There is a wealth of information from public goods experiments showing that, although people start out by contributing generously, with repetition, people contribute less to public goods. A typical explanation is that people "learn to play Nash" or something approaching it. Saijo and Yamaguchi [21], for example, classified people, from their behavior, as Nash, and found that at the beginning of their experiments 50% of players were Nash and, at the end, 69% fell into this category. This makes it tempting to believe that the people who switched had "learned to play Nash." We will examine this idea by analyzing average and individual behavior in a series of public goods experiments.

In the basic game of private contribution to a public good, each subject i ($i = 1, \ldots, N$), has to split an initial endowment E into two parts: the first part $(E - C_i)$ represents his private share and the other part C_i represents his contribution to the public good. The payoff of each share depends on and varies with the

[b]This section is based on joint work with Walid Hichri, who ran the series of experiments described here.

experimental design, but in most experiments is taken to be linear [2]. If π_i is the total payoff of individual i, it can be represented by

$$\pi_i = E - C_i + \theta \sum_{j=1}^{N} C_j. \tag{2}$$

This linear case gives rise to a corner solution. In fact, assuming that it is common knowledge that players are rational payoff maximisers, such a function gives a Nash Equilibrium (NE) at zero and full contribution as a Collective Optimum (CO). Nevertheless, experimental studies show that there is generally overcontribution (30–70% of the initial endowments) in comparison to the NE.

The theoretical model and design used for the experiments, we discuss here, concerns a public goods game modified slightly so that the NE and CO solutions are interior. This allows us to see if individuals can contribute below the non-cooperative level or above the cooperative level. The individual payoff function is

$$\pi_i = E - C_i + \theta \left(\sum_{j=1}^{N} C_j \right)^{1/2}. \tag{3}$$

It is easy to show that for a group of N subjects, at the CO, the total contribution is given by

$$\bar{Y} = \sum_{i=1}^{N} y_i = N^2 \frac{\theta^2}{4} \tag{4}$$

and, at the NE, is equal to

$$Y^* = N y^* = \frac{\theta^2}{4}, \tag{5}$$

where y^* is the symmetric individual NE.

There were six groups of four players who played 25 rounds of the game and after each round they were told the total contribution to the public good of the group. The average contribution over all the groups is illustrated in Fig. 3.

As is easily seen from Fig. 3, the contributions start out just below the social optimum and decline towards the NE without actually attaining it. This is the sort of evidence that has been used to test the idea that there is learning of a certain type going on. This learning could be of a reinforcement type in which individuals simply update their choices on the basis of their experience with those choices in the past. Alternatively, it could be one in which agents learn to anticipate the contributions of the other agents. These two types of learning are incorporated in Camererer and Ho's Expected Weighted Attraction (EWA) learning model [7] and this cannot be rejected on the average data.

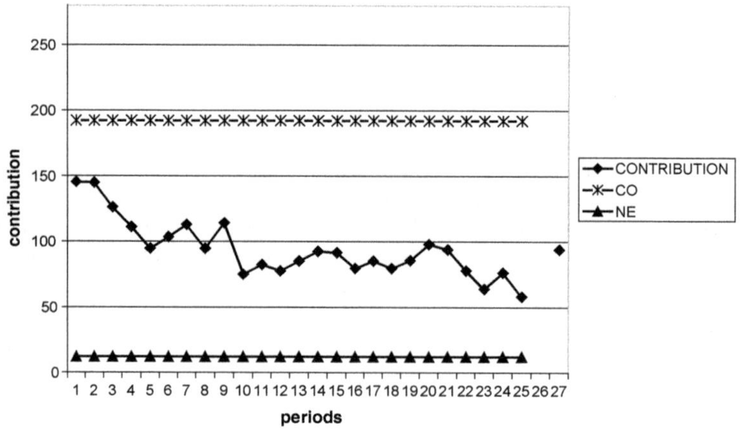

Fig. 3. Total contribution.

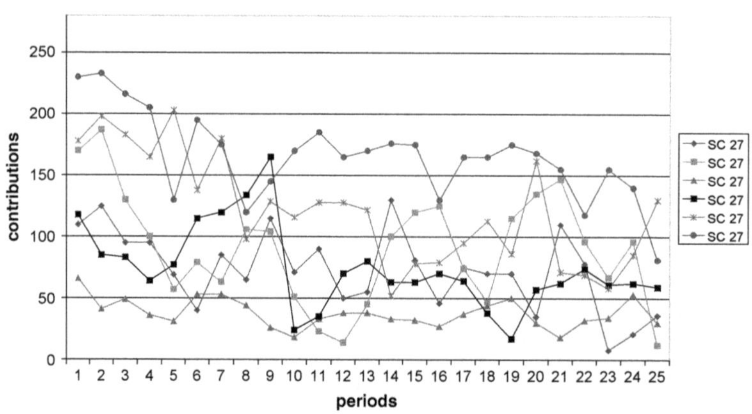

Fig. 4. Group contributions without communication.

Yet, the EWA model is a model of individual learning and it does not seem appropriate to apply it to average data. If we now look at the contributions of the groups in Fig. 4, we see that the picture is very different than in the aggregate.

Here, for certain groups, we rejected even the simple hypothesis that contributions declined during the experiments. Rather complicated behavior is going on in

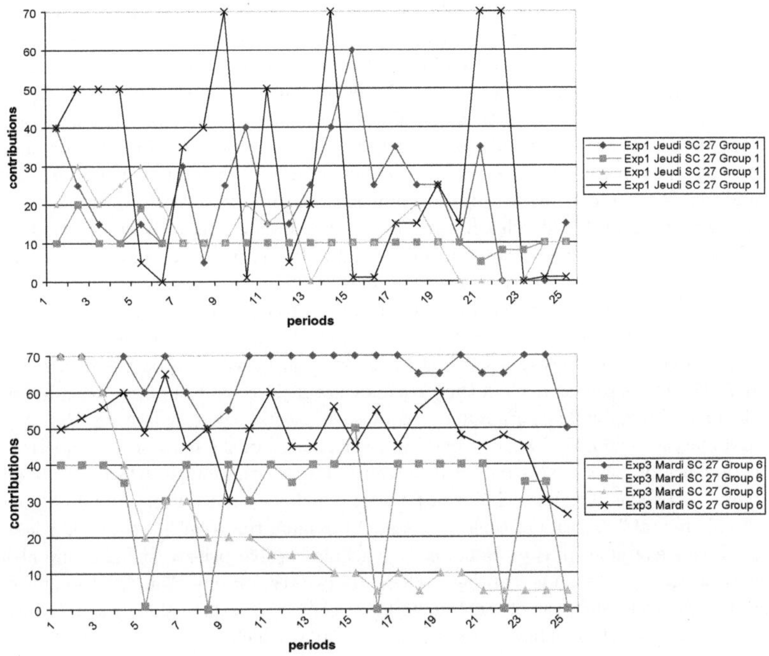

Fig. 5. Individual contributions for two groups.

the different groups and it is clear that agents in different groups behave differently and that the monotonic declining behavior of the average is an artifact of the aggregation process and does not reflect the actual behavior of groups. This is emphasized when we look at individuals within the groups (Fig. 5 gives the contributions of the individuals in two groups). We can clearly see the differences between the individuals and there is a clear indication that certain individuals try to signal in order to achieve a certain coordination. The overall message could not be simpler, the average behavior is the result of very different individual behavior and the apparent aggregate regularity does not reflect the variability of the individuals' behavior.

5. Information Cascades

Upto this point, the examples we have given reflect the idea that there is more structure at the aggregate level than at the individual level, but it can also lead to a loss of information and thus a worse collective outcome may occur than would have happened had the individuals not interacted directly. To see this, remember that an important consequence of interaction is the sort of "herd behavior" that may arise as agents are influenced by what other agents do, and, indeed, a number of phenomena corresponding to Keynes' "beauty queen" contest can arise.[c] To see how this can happen, recall that one of the most important features of markets is that the actions taken by individuals reveal something about the information they possess. This feature of markets is poorly incorporated in most economic models and yet is an important feature of many financial markets. Within the efficient-markets framework, this idea has no importance because the private information of individuals is immediately transmitted into the central price signal. However, there are many situations in which this does not happen. Indeed one of the main problems in analyzing financial markets has been to explain why the movement of stock prices is so much more volatile than that of the dividend process on which those prices are supposed, according to the standard theory, to be based (see e.g. Ref. 25). In seeking to explain this "excess volatility," economists have been led to the idea that individuals are influenced by each others' behavior (see e.g. Ref. 26), and may, for example, be led to modify their choices in the light of the choices or experience of others. This may lead to self fulfilling situations in which agents all "herd" on some particular choice or forecast which may not reflect any underlying "fundamental." A series of such events might explain the volatility of stock prices.

In an example, due to Banerjee [5], agents receive private signals, but also observe the choices made by others. There are two restaurants, A and B and one is, in fact "better" than the other. Individuals receive two sorts of signals as to which of the two is better. They receive a public signal which is not very reliable and which, say, has 55% probability of being right and a private, independently drawn, signal which has 95% probability of being correct. Suppose that restaurant A is actually better and that 95 out of the 100 potential clients of the two restaurants receive a signal to that effect and 5 get a signal indicating restaurant B as being superior. However, the public signal recommends B. Now, suppose that one of the 5, who received a signal indicating B, chooses first. The second client, observing the first, realizes that the latter must have received a B signal. He is aware that all private signals are equally reliable and that his own signal, if it indicated A, is cancelled out. He will therefore have to follow the public signal and enter restaurant B. Thus, whatever the private signal of the second agent, he will enter restaurant B. The third client is now in the same situation as the second and will enter restaurant B. Thus, all the clients will end up in B, and this is an inferior outcome. In this particular example there is a 5% probability of this happening, but Banerjee's result is to show that such a result will always occur, with positive probability.

A criticism that is frequently made of such models is that they depend, in an essential way, on the sequential nature of the decision making process. This, it is argued, is not a common feature of actual markets. Yet, in financial markets, for

[c]See, for example, Refs. 5, 16 and 24.

example, in addition to any information acquired from a private source, a trader observes what other participants are doing, or at least, proposing to do. Consider the market for foreign exchange, for example. Traders try to anticipate the direction of the move of market prices and they gain a great deal of information by listening to brokers, watching the bids and asks on the screens, and by telephoning other traders to ask for a quote. Each such piece of information modifies their individual information set. However, since there is no central equilibrium price, this information cannot be incorporated and become public through the price. It can only be inferred from the observable actions of the individuals. Thus, the action of one individual, based on some private piece of information may give rise to a whole sequence of actions by others and may, as a result, lead to significant moves in exchange rates.

Again, on the stock exchange we do not observe "equilibrium prices," we see the price of each successive transaction. Thus, each price change reflects an action by some individual. Hence any change of opinion by an agent in the light of observed price changes corresponds to observing the actions of others. When agents change their actions, in the light of the information they obtain from observing others, a so-called "information cascade" may arise just as in the restaurant example. In such a situation, individuals progressively attach more importance to the information they infer from the actions of others. They gradually abandon their private information. Thus, as the number of people involved grows, the cascade reinforces itself. Whilst quite fragile to start with, cascades later become almost immune to relevant, private, information. Hence, as more and more individuals act in this way, a trader would have to have almost unbounded confidence in his own information not to conform, particularly if such cascades lead to self-fulfilling outcomes. There is a significant loss of efficiency here. The private information, acquired by all the early agents, would be of use to their later counterparts, but, if they choose to follow what others do, this information is not made available. In this way, possibly relevant information about fundamentals, for example, could never be used and prices could get detached from these fundamentals. Thus, the conclusion to be drawn from this work is that the information obtained by observing the actions of others can outweigh the information obtained by the individuals themselves and lead to inefficient outcomes. Thus, interaction generates a result other than that which would have obtained had individuals acted on their own signals and not on the behavior of the others. In this case again the aggregate result does not directly reflect what happens at the individual level. Furthermore, the aggregate result can be inefficient from a social point of view.

6. Bubbles and Fluctuations in Financial Markets

Bubbles and crashes are phenomena typically associated with mass or herd behavior and not with the aggregation of isolated individual optimizers. This is fertile territory for the view of the economy that we have been advancing. A number of us have been building models based on rather simple characterizations of individual behavior but where the individuals change their expectations as a result of the influence of other agents. An important feature is that agents change their forecasts as a function of their experience with these rules. Furthermore, the success of rules depends on how many people follow it. Such models are capable of reproducing some of the features found in the empirical data from financial markets. Fat tails, long memory, and the periodic emergence of bubbles for example. There are a number of such models which reproduce these features when simulated (for references to this literature and for examples, see Ref. 17 for example). The emphasis in those models, as here, is on the importance of interaction and the mutual influencing of economic agents. There are many ways of modeling this and many different ways of characterizing the interaction between different agents. The basic idea of the models referred to is that agents modify their expectations over time and that there is a positive feedback as more agents converge on one forecasting rule. This generates self-reinforcing swings and if, for some of the time, the agents follow extrapolatory rules then many of the features that we observe in market data will emerge. One can think of many other ways in which interaction could take place and it should be clear that many variants can be analyzed by changing the rules available or by changing the horizon and utility functions of the actors in the markets. However, the rather simple interaction in the models referred to between agents who believe that prices follow some "fundamentals" and those who extrapolate, captures many of the observed features of financial time series. This is surely not the whole story but may prove to be a useful start.

Recall again that the basic goal from the economist's point of view, in this context, is to produce market models that are capable of reproducing the statistical properties of financial time series. More ambitiously, we would like to prove, formally, results which enable us to pin down the characteristics of the series. This is the goal pursued by Foellmer *et al.* (2004) for example. Where does the difficulty come from in the sort of models that we have mentioned? Note again, that an important feature of these models is the inclusion of chartists or trend chasers who may induce temporary bubbles and crashes. Whilst this introduces such characteristics as "excess volatility" and "long memory" in the price process (see Ref. 17), it presents a major difficulty from an analytical point of view. When chartists predominate, they may generate bubbles but there is no *a priori* bound on the levels that prices may attain during a bubble. Thus it is not obvious that the price process will not explode. The interest of the approach adopted by Foellmer *et al.*, is that one can develop a notion of equilibrium for the sort of time series of prices actually observed which is basically different from that usually envisaged in economics. Prices never settle to an equilibrium price, they change all the time, but the distribution of prices in the long run exists and is unique. Thus prices do have a well characterized structure in the long run and we do not have to content ourselves with such statements as "anything can happen." Even if we have some notion of what the fundamental value of an asset is we should not expect to see the distribution of prices concentrating itself more and more around this price. Prices will only stray far from underlying fundamentals for relatively short periods of time but they will never settle to those fundamental values. Being near to the latter is no guarantee of staying near and no exogenous shocks are required to move the prices. This seems to be a reasonable representation of economic reality and, what is more, seems to be consistent with the statistical properties of financial time series and survives the tests recently developed by econometricians.

7. Conclusions

In this paper we have suggested, with examples, that the economy should be conceived of as a complex system. Such systems are characterized by having interacting agents who may have rather limited powers of reasoning and it is the aggregation of these interacting agents that generates a number of macroeconomic phenomena which are difficult to explain with standard models. The key distinction that we have emphasized is the difference between micro and macroeconomic behavior. Complex systems such as the economy are characterized by the fact that aggregates cannot be treated as individuals. It is a fundamental mistake to reduce the behavior of the system to the behavior of an individual and efforts to do so prevent us from getting at the heart of many economic phenomena.

References

[1] Anderson, P. W., Arrow, K. J. and Pines, D. (eds.), *The Economy as an Evolving Complex System* (Addison-Wesley, Redwood City, CA, 1988).
[2] Andreoni, J., Warm-glow versus cold-prickle: The effects of positive and negative framing on cooperation in experiments, *Quarterly J. Econ.* **110**, 1–22 (1995).
[3] Aoki, M., *New Approaches to Macroeconomic Modeling: Evolutionary Stochastic Dynamics, Multiple Equilibria, and Externalities as Field Effects* (Cambridge University Press, Cambridge, 1996).
[4] Arthur, W. B., Asset pricing under endogenous expectations in an artificial stock market, in *The Economy as an Evolving Complex System II*, eds., W. B. Arthur, S. N. Durlauf and D. Lane, Vol. XXVII of *Santa Fe Institute Studies in the Sciences of Complexity Proc.* (Addison Wesley, Reading, MA, 1997).
[5] Banerjee, A., A simple model of herd behavior, *Quarterly J. Econ.* **107**, 797–817 (1992).
[6] Blume, L., The statistical mechanics of strategic interaction. *Games and Economic Behaviour* **5**, 387–424 (1993).
[7] Camererer, C. and Ho, T. H., Experienced weighted attraction in normal form games, *Econometrica* **67**, 827–873 (1999).
[8] Debreu, G., Excess demand functions, *J. Math. Econ.* **1**, 15–23 (1974).
[9] Dosi, G., *Innovation, Organization and Economic Dynamics: Selected Essays* (Edward Elgar, Cheltenham, 2000).
[10] Föllmer, H., Random economies with many interacting agents, *J. Math. Econ.* **1**(1), 51–62 (1974).
[11] Forni, M. and Lippi, M., *Aggregation and the Micro-foundations of Microeconomics* (Oxford University Press, Oxford, 1997).

[12] Gode, D. K. and Sunder, S., Allocative efficiency of markets with zero-intelligence traders: Markets as a partial substitute for individual rationality, *J. Political Econ.* **101**, 119–137 (1993).
[13] Härdle, W. and Kirman, A., Non-classical demand: A model-free examination of price quantity relations in the marseille fish market, *J. Econometrics* **67**, 227–257 (1995).
[14] Hildenbrand, W., *Market Demand: Theory and Empirical Evidence* (Princeton University Press, Princeton, NJ, 1994).
[15] Kirman, A. P., Whom or what does the representative individual represent, *J. Econ. Perspectives* **6**(2), 117–136 (1992).
[16] Kirman, A. P., Ants, rationality and recruitment, *Quarterly J. Econ.* **108**, 137–156 (1993).
[17] Kirman, A. P. and Teyssiere, G., Micro-economic models for long memory in the volatility of financial time series, in *The Theory of Markets*, eds., P. J. J. Herings, G. V. der Laan and A. J. J. Talman (North Holland, Amsterdam, 2002).
[18] Krugman, P., *The Self-Organizing Economy* (Blackwell, Cambridge, MA, 1996).
[19] Lewbel, A., Exact aggregation and a representative consumer, *Quarterly J. Econ.* **104**, 622–633 (1989).
[20] Nelson, R. and Winter, S., *An Evolutionary Theory of Economic Change* (Harvard University Press, Cambridge, MA, 1982).
[21] Saijo, T. and Yamaguchi, T., The spite dilemma in voluntary contribution mechanism experiments, paper presented at the New Orleans Public Choice Meetings, March 1992.
[22] Samuelson, L., *Evolutionary Games and Equilibrium Selection* (MIT Press, Cambridge, MA, 1997).
[23] Schelling, T., *Micromotives and Macrobehaviour* (W.W. Norton, New York, 1978).
[24] Sharfstein, D. S. and Stein, J. C., Herd behaviour and investment, *Am. Econ. Rev.* **80**, 465–479 (1990).
[25] Shiller, R. J., Do stock prices move too much to be justified by subsequent changes in dividends? *Am. Econ. Rev.* **71**(3), 421–436 (1981).
[26] Shiller, R. J., Comovements in stock prices and comovements in dividends, *NBER Working Papers*, 2846 (1989).
[27] Sonnenschein, H., Market excess demand functions, *Econometrica* **40**, 549–563 (1972).
[28] Weibull, J. W., *Evoutionary Game Theory* (MIT Press, Cambridge, MA & London, 1996).
[29] Weisbuch, G., Kirman, A. P. and Herreiner, D., Market organisation and trading relationships, *Econ. J.* **110**, 411–436 (2000).

Cooperation in an Adaptive Network

M.G. Zimmermann[1,2], V.M. Eguíluz[1,3], M. San Miguel[1], A. Spadaro[4]

Abstract: We study the dynamics of a set of agents distributed in the nodes of an adaptive network. Each agent plays with all its neighbors a weak prisoner's dilemma collecting a total payoff. We study the case where the network adapts locally depending on the total payoff of the agents. In the parameter regime considered, a steady state is always reached (strategies and network configuration remain stationary), where co-operation is highly enhanced. However, when the adaptability of the network and the incentive for defection are high enough, we show that a slight perturbation of the steady state induces large oscillations (with cascades) in behavior between the nearly all-defectors state and the all-cooperators outcome.

Keywords: Social organisation, Networks, Game theory, Agent based models, Weak Prisoner's Dilemma

[1]Instituto Mediterráneo de Estudios Avanzados IMEDEA(CSIC-UIB), E07071 Palma de Mallorca, Spain. Supplementary information at http:www.imedea.uib.es.

[2]Present address: Departamento de Física, Universidad de Buenos Aires, Pabellón I Ciudad Universitaria, 1428 Buenos Aires, Argentina.

[3]Corresponding author email: victor@delta.ens.fr. Present address: DELTA (Joint Research Unit CNRS-ENS-EHESS), 48 Bd Jourdan, 75014 Paris, France.

[4]DELTA (Joint research unit CNRS-ENS-EHESS) and Department of Economics and Business, Universitat de les Illes Balears, E07071 Palma de Mallorca, Spain.

Résumé : Nous étudions la dynamique d'un groupe d'agents distribués dans les noeuds d'un réseau. Chaque agent joue avec tous ses voisins un Faible Dilemme du Prisonnier en collectant un pay-off total. Nous étudions le cas d'un réseau qui s'adapte localement selon les pay-off des agents. Premièrement, un état stationnaire est toujours trouvé pour les paramètres étudiés, où la fraction d'agents qui coopère est stationnaire. Pour quelques paramètres même on a trouvé un état presque complètement coopérative. Cependant, quand l'adaptabilité du réseau et l'incitation á la défection sont assez grands, on montre que une petite perturbation de l'état stationnaire induit des grandes oscillations (avec cascades) dans le comportement entre 1 état quasi-tout-defection et quasi-tout-cooperation.

Mots-clés : Organisation sociale, Réseaux, Théorie des jeux, Models d'agents, Faible Dilemme du Prisonnier

1. Introduction

A subject which has intrigued many economists is how the organization of an economy arises and evolves. Agents interact in multiple ways, as for example information transmission in financial markets[5], or firms competition or collusion in an oligopoly[6]. These interactions can be described by a set economic agents which sit in the nodes of a network. Among other approaches (for a review see [KIR 99]), the mechanisms of interactions and the emergence of collaboration in a group of agents have been analyzed by the use of evolutionary game theory [WEI 96]. Using the Prisoner's Dilemma (PD) game, [AXE 81] and [AXE 84] showed how cooperation may be sustained by a population of agents meeting repeatedly and having certain degree of rationality. Strategies were allowed to mutate and reproduce in proportion to the difference between the agent's payoff and the population's average payoff. Cooperation was shown to be sustained by the use of the evolutionary stable strategy Tit-For-Tat. This approach assumes that the game is carried out by randomly matching a pair of agents from a fixed population. This assumption seems plausible for systems with a large number of agents where the probability of playing several times with the same agent is extremely low, and for systems where the agents cannot create links or preferences between them.

However, in many social and economic environments this assumption does not hold, and each agent interacts only with a small subset of the whole population[7]. One reason for this might be that agents have imperfect information on the whole population, except for a small subset which can be considered as "the neighbors".

[5]See for example [BAN 92], [KIR 93], [CON 99], [EGU 99].

[6]See for example [FER 98], [GOY 99].

[7]For a deep study of the dynamics of a PD game with different strategies, evolution of the strategies and networks see [COH 99].

In this paper we introduce a spatial Weak Prisoner's Dilemma (WPD)[8] model played on an endogenously adaptive network where cooperation is promoted and sustained by local interactions *and* the adaptation of the network. To simplify matters we do not allow the strategies to evolve and consider only zero-memory strategies, *i.e.* all-C [all-D] strategy means doing C [D] *all* the time irrespective of the outcome.

Each agent plays the same strategy either C (cooperate) or D (defect) with all its local neighbors, as in [NOW 92]. Each agent revises his strategy at each iteration of the game and imitates the strategy of the neighbor with highest aggregate payoff. Finally we consider the adaptation of the network by allowing the possibility of changing the neighborhood whenever an agent was unsatisfied and imitated its best local neighbor. Specifically, if this best local neighbor is a D-neighbor, the imitating agent cuts the link with the D-neighbor, with some probability, and establishes a new link selecting randomly a new agent to become its neighbor from the whole population of agents.

As possible scenarios where these cooperative networks might arise we can mention a network of individuals or firms that have some agreements among them but with some risk involved in the cooperation. The agreements correspond to the links in the network, and respecting their agreements results in playing Cooperation (C), while not respecting their agreement corresponds to Defection (D). An interesting application might be a network of firms which share their research and development outcome. Another application might be a network of scientist which agree to collaborate in different projects.

We have determined from numerical experiments the following main results. First, for the range of parameters studied, the network always reaches a steady state where the fraction of cooperating agents (C-agents) is high. In these states the network remains stationary. However, most of the agents are unsatisfied and thus are continuously imitating their best neighbor's strategy, which is the same strategy they are using. This does not change the local payoffs and thus it remains in a stationary state. We find also that although the dynamics always converges to a steady state, when the incentive to defect is sufficiently high, a perturbation may induce large oscillations in the fraction of C-agents together with a large reorganization of the network. These oscillations stretch between the *quasi*-all C-agents, to the *quasi*-all D-agents networks. We stress the fact that these are transient and they last for some time before a new steady state is reached. In most cases high cooperation is again reached, but there is a small probability to reach the *quasi*-all D-network. Thus we show that although cooperation is greatly enhanced by such a network update, the system may organize in a state where an exogenous or stochastic perturbation

[8] We will explain in the next section the difference between the classical Prisoner's Dilemma and the WPD we use in this paper.

may produce drastic changes on a finite time. The oscillations can be triggered by a change of strategy of a single agent with a large number of links. This identifies the importance of the highly-connected agents which play a leadership role in the collective dynamics of the system. Finally, it is interesting to study the characteristics of the network that emerges from the interaction between agents. Such structure is far from trivial in the sense that presents a given pattern.

The paper is organized as follows. The next section defines the adaptive model. Section 3 deals with the conducted numerical experiments. First the case of a fixed network is revisited, and then the full adaptive network is presented and discussed. Finally, in Section 4, we discuss our results and open problems.

2. The model

We consider an adaptive game where N agents play a WPD game on a network Γ. Each agent is located in a *node* of the network (and there is a single agent per node). Two agents are *neighbors* if they are directly connected by *one* link. We define the *neighborhood* of agent i as the subset of Γ which are neighbors of i, and we represent it as neigh(i); its cardinal is k_i. Each agent plays only with those other agents connected by one *link*[9]. If N_l is the total number of links and k_i is the number of links of node i, then the *average connectivity* of a network, k, is defined as the average number of links per node

$$k = \frac{\sum_{i=1}^{N} k_i}{N} = \frac{2N_l}{N} \,. \qquad [1]$$

In this paper we consider two different kind of initial networks on which the agents start the game: regular lattices in two-dimensions and random networks. A two-dimensional regular lattice correspond to a Manhattan-like grid, where the nodes are the intersections of streets and avenues. We will consider first-neighbor interaction that corresponds to the four possible moves a chess King can make (thus $k_i = 4$, $\forall i$ and $k = 4$), while the 2nd-neighbor interaction corresponds to the 4 extra neighbors which will complete a square around the King (thus $k_i = 8$, $\forall i$ and $k = 8$)[10]. Random networks of average connectivity k are formed by distributing $N_l = kN/2$ bidirectional links between pairs of nodes (i, j), with the constraint that $(i, j) = (j, i)$ (bidirectional links). The resulting distribution of the number of links in the network is Gaussian with the maximum located at the average connectivity k.

[9] We assume that the links are bidirectional. More general situations can be considered with uni-directional links, but we do not explore this further.

[10] Previous results of PD game in regular lattices are investigated in for example [NOW 92].

Let us denote by $s_i(t) = \{0, 1\}$ the strategy of agent i at time step t, where $s_i = 1$ corresponds to play cooperation (C), and $s_i = 0$ corresponds to defection (D). These will be referred to as C-agents or D-agents, respectively. The payoff matrix for a 2-agents game is:

	C	D
C	σ, σ	$0, b$
D	$b, 0$	δ, δ

where we take $b > \sigma > \delta > 0$, and $b/2 < \sigma$ for a Prisoner's Dilemma game. In this paper we study the case of a Weak Prisoner's Dilemma (WPD): a Prisoner's Dilemma is Weak when $\delta = 0$. In a standard Prisoner's Dilemma there exists a unique Nash equilibrium (D, D) while in WPD either a (C,D), (D,C) or (D,D) may be attained as a Nash equilibrium.

We use in this Paper the WPD for which the analysis is much simpler, taking into account that [NOW 92, LIN 94] showed that, at least for a fixed regular network, the results do not change qualitatively when using $1 \gg \delta > 0$.

We consider the situation in which agents always tries to maximize their utility, and therefore seeks the largest possible benefit from their local interactions in the network Γ. We assume each agent plays the same strategy with all its neighbors neigh(i) and only with them. The game is played *synchronously*, *i.e.* the players decide their strategy in advance and they all play at the same time. The strategy update of agent i is as follows:

1. Each agent i plays the WPD game with each neighbor using the *same* strategy s_i and collecting a total individual payoff Π_i.

2. Agent i revises its current strategy at each iteration of the game (i.e., at every time step), and updates it by *imitating* the strategy of its neighbor with a highest pay-off. Agent i is said to be *satisfied* if his pay-off is the maximum of its neighbors; otherwise it will be *unsatisfied* and it will revise its strategy.

3. The agents have also the possibility of an extra action which adapts its neighborhood. Namely we consider:

   Network Rule: *if agent i is unsatisfied and imitates from a D-agent j, then with probability p, i breaks the link with j and establishes a new link with another agent chosen randomly in the network Γ.*

This rule leads to a time evolution of the local connectivity of the network, leaving the global connectivity k, as defined in Eq. 1, constant. For each agent i which imitates a D-agent j and decides to break the link and choose a new agent β, neigh(i) changes by replacing $j \to \beta$ and $k_i(t+1) = k_i(t)$. However j will lose a link, $k_j(t+1) = k_j(t) - 1$, and the new agent β will increase its local connectivity $k_\beta(t+1) = k_\beta(t) + 1$. Thus, the network adaptation introduces a diversity in the agents local neighborhoods.

Agreements between satisfied agents do not change. This does not mean that new agreements with other agents are not possible (these agents may always receive new links), but those which exist remain untouched. The same is true for unsatisfied C-agents imitating another C-agent. However neighborhoods of D-agents that have the maximum pay-off in the neighborhood could change abruptly in just one iteration.

The network rule can be understood as a risk minimization. If agent i is a D-agent and is unsatisfied he minimizes the risk of cooperation by taking chances and selecting a new neighbor from the whole network. In the context of game theory, this can be seen as a retaliation, because is highly unlikely that he will play with those agents in the future.

The probability p represents a transaction cost of breaking *one* agreement *and* establishing a new agreement with a new partner. It can also be understood as a measure of the tolerance of being exploited. We would like to stress that the transaction cost has two components: first, the cost of breaking an agreement and second, the cost of finding a partner and that this new partner accepts the agreement[11]. The case $p = 0$ corresponds to an infinite transaction cost for breaking the link, while $p = 1$ corresponds to the limiting case of no transaction cost. It is clear that breaking at the same time more than one link, and finding their respective partners is more unlikely. The probability p is also a measure of the adaptability of the network to the results of the game at each iteration.

It will be useful to define the *looking function* of agent i which will be denoted by $l(i)$. This function points to i's neighbor with highest payoff, including himself; thus if i is not imitating any neighbor then $l(i) = i$. From this we can define an agent being a *local maximum* in payoff as the one which satisfies $i = l(i)$ and at least one of its neighbors is looking at it.

Suppose that at a given time step there is a D-agent which is a local maximum. This implies that at the next time step, the D-strategy will be replicated on all of its neighbors, *and* its links will be destroyed with probability p. Thus, a D-local maximum at one time step, is an unstable situation where the agent looses a fraction p of all its links on the next time step, and these links will be replaced by new neighbors.

With the network rule implemented in this Paper *the total number of links*

[11]One could separate these two costs, and would have a process of breaking links (with a given probability q) and another process of generation of links (with a probability r). This line of research is not explored in this paper.

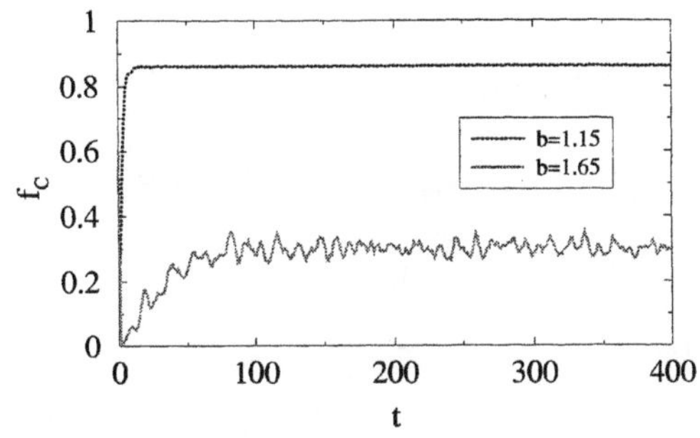

Figure 1: Time evolution of the fraction of C-agents f_C for a non-adaptive network $p = 0$, in a two-dimensional regular lattice with second-neighbors interaction ($k = 8$: (top trajectory) asymptotic periodic trajectory for $b = 1.15$ and (bottom trajectory) chaotic trajectory for $b = 1.65$. Payoff matrix: $\sigma = 1$, $\delta = 0$.

in the network Γ *is conserved.* We do not take into account more complicated network dynamics as spontaneous creation or destruction of links, that will break the conservation of total number of links.

3. Numerical studies

We have characterized numerically the model described above using as parameter the incentive to defect b. We have used as initial networks random networks with an average connectivity $k = 4$ and $k = 8$ and two-dimensional regular lattices with first- ($k = 4$) and second-neighbor ($k = 8$) interaction. We have also fixed the adaptability $p = 1$. The statistical measures that we have studied are:

- The fraction (normalized to the whole population N) of cooperating agents (C-agents), denoted by $f_C = (\sum_{i=1}^{N} s_i)/N$.
- The average payoff per agent $\Pi = (\sum_{i=1}^{N} \Pi_i)/N$ of the whole network.
- The probability of having a link between two C-agents, p_{CC}, between a C-agent and a D-agent, p_{CD}, and between two D-agents, p_{DD}. These probabilities satisfy:

$$1 = p_{CC} + 2p_{CD} + p_{DD} \qquad [2]$$

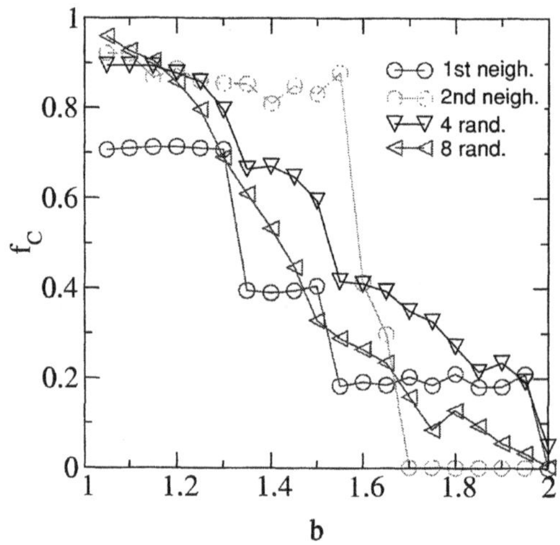

Figure 2: No network adaptation ($p = 0$). Average fraction of C-agents, f_C, vs. defecting incentive, b, in 1st and 2nd neighbors regular lattices, and in random networks with $k = \{4, 8\}$.

In the numerical simulations we maintain fixed the parameters $\delta = 0$ and $\sigma = 1$ of the payoff matrix and we vary b in the range $1 < b < 2$. Finally, in most simulations we take $N = 10000$ agents, and we start with an initial population of $0.6N$ C-agents randomly distributed in the network Γ.

The game in a fixed network ($p = 0$) and regular lattices, has been previously studied by [NOW 94, NOW 92, NOW 93]. Typically the behavior of the fraction of C-agents, f_C, can show different features. The simplest is an asymptotic stationary or periodic state, where f_C remains stationary or fluctuates periodically. A more complex behavior is the spatio-temporal chaotic regime, where f_C fluctuates in time around an average value while the spatial distribution of C- and D- agents present evolving patterns at each iteration. Finally, if the incentive to defect b, is high enough the asymptotic state of the system is all D-agents. Also the introduction of elements that disrupt the spatial correlations present when the game is played in a regular lattices (e.g., random lattices, noise and errors in the imitating process), was shown to destroy the periodic fluctuations and regular patterns observed, as expected.

One of the main result of these studies is that partial cooperation can be sustained by *local interactions*, together with a very simple choice of strategies which does not include memory. This can be illustrated by studying the average fraction of C-agents for an increasing value of the incentive to defect b (see

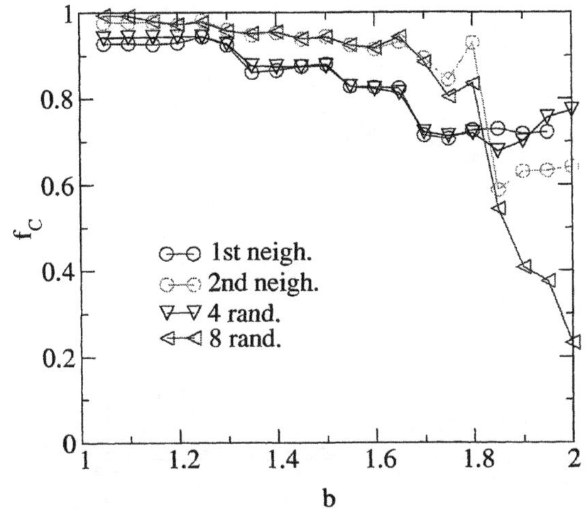

Figure 3: Network adaptation ($p = 1$). Fraction of cooperative agents vs. b in different initial networks.

Fig. 2). We have extended previous results by explicitly considering random networks, which is a more natural assumption in an economic context.

The numerical results also show that increasing the average size of the neighborhood, *i.e.* average number of links per agent, k, the average fraction of C-agents f_C decreases faster with b. Several extensions of this model have been studied in the literature: introducing asynchronous updates [HUB 93] or introducing errors in the imitation process [MUK 96], where the basic results shown above persist [NOW 94].

It is also worth noting that the fraction of C-agents depends strongly on the network. As can be seen from Fig. 2, both the regular lattice with first neighbor interaction and the random network with $k = 4$ have the same average connectivity, however f_C behaves differently with b. Same applies for the regular lattice with second neighbor interaction and random network with $k = 8$.

In Fig. 3, we plotted the averaged asymptotic value of f_C for the fully adaptive network $p = 1$ and for different initial networks. The average is over 10 different random initial conditions with a density of 0.6 C-agents. Unlike the non-adaptive case, the fraction f_C seems to be quite independent of the incentive to defect b, and the initial network chosen (regular or random). Only the average connectivity k seems to play a role for determining the asymptotic

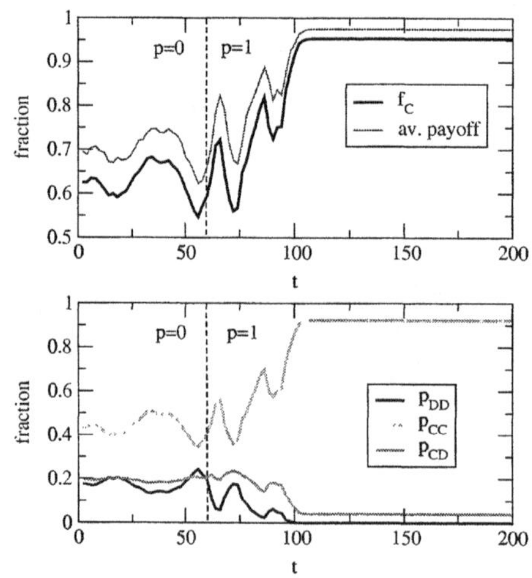

Figure 4: (Top) Time series of C-fraction f_C and average payoff in a random network: without network adaptation ($p = 0$) for $t < 60$, and thereafter in the full adaptive network ($p = 1$). (Bottom) Corresponding time series of the different link probabilities: two C-agents (p_{CC}), a C and a D-agent (p_{CD}), and two D-agents (p_{DD}) having a link. Parameter values: $b = 1.35$, $k = 8$.

fraction of cooperation. To illustrate how cooperation is enhanced against the non-adapting network, we show in Fig. 4 a simulation where the system first evolves with $p = 0$ until at $T = 60$ time iterations the adaptation is switched on ($p = 1$). It is clear from the figure that the fraction of C-agents is highly enhanced with respect to the static network.

One of the most interesting features encountered when the network is able to adapt with the game ($p = 1$), is that throughout the whole range of parameters studied ($1 < b < 2$), the dynamics settles onto a ***steady state*** after some transient time. Figure 4 also shows the time series of the different links probabilities (p_{CC}, p_{CD}, p_{DD}). The steady state corresponds to a stationary network structure and individual payoffs Π_i. Notice how the network adaptation clearly favors having links between C-agents. Most agents are unsatisfied[12] in this steady state and they continuously imitate the strategy of their neighbors with highest payoff (all of them C-agents). Also note that at the steady state, the probability of having a link between a C and a D-agent p_{CD} is finite, while for two D-agents $p_{DD} \to 0$. This shows that D-agents end up without

[12]In a weak sense, after all they are benefiting from cooperation.

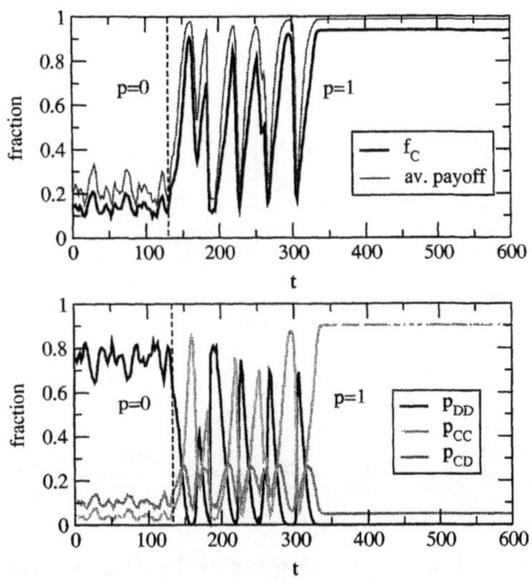

Figure 5: Time series for high b. During $0 < t < 120$, the network is left fixed ($p = 0$), while for $t > 120$, $p = 1$. Below the corresponding time series of the link probabilities p_{CC}, p_{CD}, p_{DD}. $b = 1.75, k = 8$.

links to other D-agents and exploiting other C-agents. Also note that when $p_{DD} = 0$, there is no D-agent which will change their neighborhood, so the system may reach a stationary state. In this case D-agents will be satisfied "passive" local maxima, in the sense that nobody is imitating their strategy[13].

A stationary situation with individual payoffs Π_i may arise with a number of agents forming, what we shall denote a ***chain***, such that: $s_{l^m(i)} = s_{l^{m-1}(i)} = \ldots = s_{l(i)} = s_i = 1$, where $l^m(i)$ is the only agent which is satisfied. All others in the chain are actually unsatisfied, but as they imitate the same strategy they were playing on the previous step, they never change their relative payoff, and the chain becomes a steady state.

An interesting result arises when one looks at the time series for large values of p ($p \sim 1$) and a high value of the incentive to defect b. Before reaching the asymptotic steady state, one observes that ***large oscillations*** occur for different choices of initial conditions. Figure 5 shows an example of the dynamics starting from the fixed network. From other initial conditions, similar oscillations are observed. The time evolution of the system before the network is allowed to

[13]A highly unlikely situation where two D-agents are neighbors in a steady state would occur if they have exactly the same payoff.

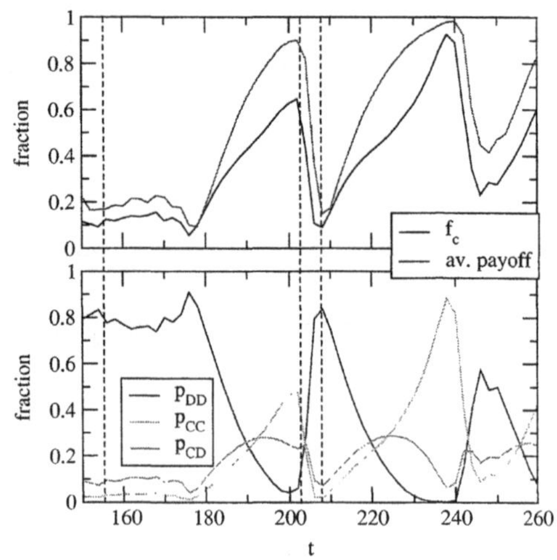

Figure 6: Blow-out of a time series showing avalanches. Parameter values as in Fig. 5.

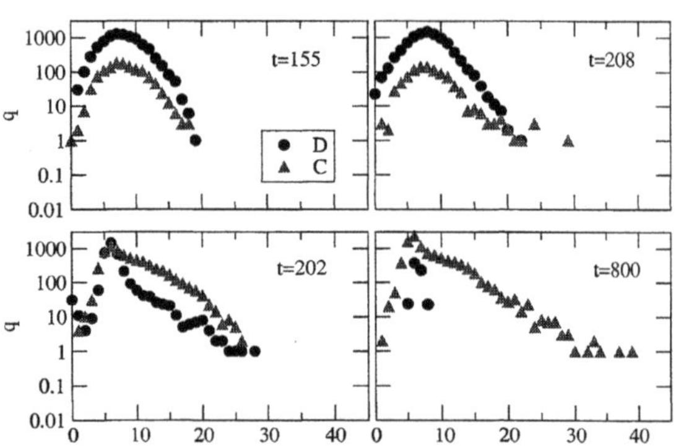

Figure 7: Histogram of number of links, q, for a C and D-agents, at different time steps (from time series in Fig. 6) during large oscillations. $t = 155$ corresponds to the distributions of the initial random network with $k = 8$. $t = 202$ corresponds to the maximum of p_{CC}, while at $t = 208$ to the maximum of p_{DD}. Finally at $t = 800$ the distributions of the asymptotic steady state are shown.

adapt is governed by the high reward of D-agents. However, when the network starts the adaptation ($p = 1$ for $t > 120$) the dynamics promotes the creation of more links between C-agents, while decreasing dramatically the links between D-agents, as can be seen at time step $t \sim 150$. However, first attempts to build a global cooperative behavior are unsuccessful because the system frustrates. The defecting behavior is so rewarding, that the cooperation has to be built in a specific network configuration in order to be robust against eventual changes of strategy. This correspond to the successive oscillations shown in Fig. 5. The systems reaches a similar fraction of C-agents as in the stationary solution, but several oscillations occur before the stationary regime settles.

In Fig. 6 we show a blow-out of two oscillations. Note the drastic change in the connectivity between C and D-agents. To quantify this phenomenon we determined the distribution of links for the population of each strategy at different time steps as shown in Fig. 7. The initial distribution ($t = 155$) shows a Gaussian distribution around the starting $k = 8$ value. Then at the maximum of p_{CC} ($t = 202$) it is observed that the tails of both distributions extend up to 28 links. Then very rapidly the network switches to the almost defective solution ($t = 208$). However there are a small number of C-agents with a large payoff, which permits the gradual build-up of cooperation. Finally for large times ($t = 800$), the D-agents distribution shrinks to a very narrow distribution, while the C-agents distribution displays a long exponential decay. The stationary network configuration is thus dominated by a few C-agents with a large number of links (the tails of the histogram of links for C-agents). These highly-connected agents dominate the collective behavior of the network.

To illustrate the relevance of the highly connected agents we have built a numerical experiment in which after the system reaches a steady state, the best connected agent, that is the one with a largest number of links switches strategy (from C to D). Figure 8 shows the resulting oscillations, before the system reaches again a (possibly different) steady state. This makes very clear the dominant role of these best connected agents.

4. Discussion

The main results of this work are the following. We have introduced a model of cooperation on an adaptive network, where the cooperation is highly enhanced. The network adaptation involves exclusively the D-agents, which in some sense are allowed to "search" for new neighbors, in the hope of finding C-agents to exploit. However our study reveals that this mechanism benefits in the long run cooperators. The asymptotic state reached by the system is a steady state in which the network structure and the average payoff Π remain stationary. However, most agents are unsatisfied, and continuously imitate the strategy of their neighbors with highest payoff (most of them C-agents). Also

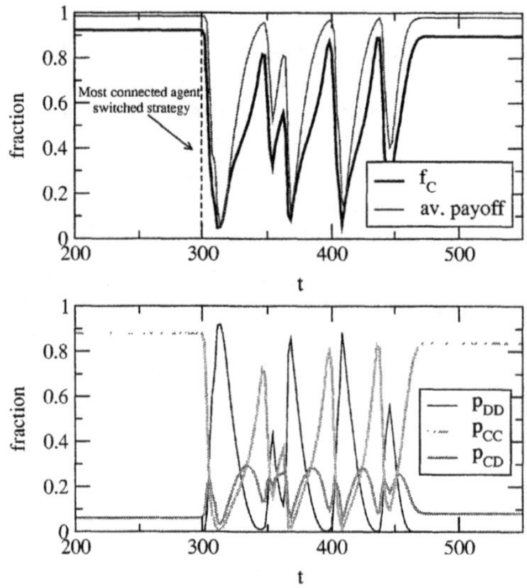

Figure 8: Time series of f_C, where at $t = 300$ the agent with most links changes strategy from C to D. Parameter values: $b = 1.75$ and $k = 8$.

the structure of the stationary network presents interesting characteristics. The distribution of links for C-agents has a long exponential tail, with a very few number of highly connected C-agents having up to 4 times the average number of the links in the whole network. These agents dominate the cooperative network structure.

We have also obtained that, for sufficiently high values of the incentive to defect b, the induced network structure may suffer large reorganizations. These manifest themselves as large oscillations in the fraction of C-agents, where the network visits for a short time the nearly full cooperative regime, followed by a short time of nearly full defecting regime. In the current dynamical model, these large oscillations are long lived transients, but the system reaches a final stationary state. The interesting aspect is that these large oscillations might be easily triggered by the spontaneous change of the strategy of a highly connected agent.

Acknowledgment. We acknowledge useful discussions with D. Cardona-Coll, P. Battigalli, A. Kirman and J. Weibull. M.G.Z., V.M.E. and M. S. M acknowledge financial support for DGYCIT (Spain) project PB94-1167.

References

[AXE 81] R. AXELROD AND W. D. HAMILTON. The evolution of cooperation. *Science*, 211:1390–1396, 1981.

[AXE 84] R. AXELROD. *The Evolution of Cooperation.* Basic Books, New York, 1984.

[BAN 92] A. BANNERJEE. A simple model of herd behaviour. *Quarterly Journal of Economics*, 108:797–817, 1992.

[COH 99] M. COHEN, R. RIOLO AND R. AXELROD. The emergence of social organization in the prisoner's dilemma: how context-preservation and other factors promote cooperation. Santa Fe Institute Working Paper 99-01-002, 1999.

[CON 99] R. CONT AND J. P. BOUCHAUD. Herd behavior and aggregate fluctuations in financial markets. *Macroeconomic Dynamics*, 1999. In press.

[EGU 99] V. M. EGUILUZ AND M. G. ZIMMERMANN. Dispersion of rumors and herd behavior. Los Alamos e-print archive (www.lanl.gov): cond-mat/9908069, 1999.

[FER 98] C. ALÓS FERRER, A. B. ANIA AND F. VEGA-REDONDO. An evolutionary model of market structure. Preprint from http://merlin.fae.ua.es/fvega/#rp, 1998.

[GOY 99] S. GOYAL AND S. JOSHI. Networks of collaboration in oligopoly. Mimeo, 1999.

[HUB 93] B. A. HUBERMAN AND N. S. GLANCE. Evolutionary games and computer simulations. *Proc. Natl. Acad. Sci. USA*, 90:7716–7718, 1993.

[KIR 93] A. KIRMAN. Ants, rationality and recruitment. *Quarterly Journal of Economics*, 108:137–156, 1993.

[KIR 99] A. KIRMAN. Aggregate activity and economic organisation. *Revue Economique des sciences sociales*, 113:189–230, 1999.

[LIN 94] K. LINDGREN AND M. G. NORDAHL. Evolutionary dynamics of spatial games. *Physica D*, 75:292–309, 1994.

[MUK 96] A. MUKHERJI, V. RAJAN AND J. R. SLAGLE. Robustness of cooperation. *Nature*, 379:125–126, 1996.

[NOW 92] M. A. NOWAK AND R. M. MAY. Evolutionary games and spatial chaos. *Nature*, 359:826–829, 1992.

[NOW 93] M. A. NOWAK AND R. M. MAY. The spatial dilemmas of evolution. *Int. Jour. of Bif. and Chaos*, 3(1):35–78, 1993.

[NOW 94] M. A. NOWAK, S. BONHOEFFER AND R. M. MAY. Spatial games and the maintenance of cooperation. *Proc. Natl. Acad. Sci. USA*, 91:4877–4881, 1994.

[WEI 96] J. WEIBULL. *Evolutionary Game Theory.* MIT University Press, 1996.

Advances in Complex Systems, Vol. 8, No. 1 (2005) 87–116

HOW INDIVIDUALS LEARN TO TAKE TURNS: EMERGENCE OF ALTERNATING COOPERATION IN A CONGESTION GAME AND THE PRISONER'S DILEMMA

DIRK HELBING, MARTIN SCHÖNHOF and HANS-ULRICH STARK

Institute for Transport & Economics, Dresden University of Technology, Andreas-Schubert-Str. 23, 01062 Dresden, Germany

JANUSZ A. HOŁYST

Faculty of Physics and Center of Excellence for Complex Systems Research, Warsaw University of Technology, Koszykowa 75, PL-00-662 Warsaw, Poland

Received 16 March 2005

In many social dilemmas, individuals tend to generate a situation with low payoffs instead of a system optimum ("tragedy of the commons"). Is the routing of traffic a similar problem? In order to address this question, we present experimental results on humans playing a route choice game in a computer laboratory, which allow one to study decision behavior in repeated games beyond the Prisoner's Dilemma. We will focus on whether individuals manage to find a cooperative and fair solution compatible with the system-optimal road usage. We find that individuals tend towards a user equilibrium with equal travel times in the beginning. However, after many iterations, they often establish a coherent oscillatory behavior, as taking turns performs better than applying pure or mixed strategies. The resulting behavior is fair and compatible with system-optimal road usage. In spite of the complex dynamics leading to coordinated oscillations, we have identified mathematical relationships quantifying the observed transition process. Our main experimental discoveries for 2- and 4-person games can be explained with a novel reinforcement learning model for an arbitrary number of persons, which is based on past experience and trial-and-error behavior. Gains in the average payoff seem to be an important driving force for the innovation of time-dependent response patterns, i.e. the evolution of more complex strategies. Our findings are relevant for decision support systems and routing in traffic or data networks.

Keywords: Game theory; reinforcement learning; multi-agent simulation.

1. Introduction

Congestion is a burden of today's traffic systems, affecting the economic prosperity of modern societies. Yet, the optimal distribution of vehicles over alternative routes is still a challenging problem and uses scarce resources (street capacity) in an inefficient way. Route choice is based on interactive, but decentralized individual decisions, which cannot be well described by classical utility-based decision models [27]. Similarto the minority game [16, 39, 43], it is reasonable for different people to react to the same situation or information in *different* ways. As a consequence, individuals tend to develop characteristic response patterns or roles [26]. Thanks to this differentiation process, individuals learn to coordinate better in the course of time. However, according to current knowledge, selfish routing does not establish the system optimum of minimum overall travel times. It rather tends to establish the Wardrop equilibrium, a special user or Nash equilibrium characterized by equal travel times on all alternative routes chosen from a certain origin to a given destination (while routes with longer travel times are not taken) [71].

Since Pigou [53], it has been suggested to resolve the problem of inefficient road usage by congestion charges, but are they needed? Is the missing establishment of a system optimum just a problem of varying traffic conditions and changing origin-destination pairs, which make route-choice decisions comparable to one-shot games? Or would individuals in an *iterated* setting of a day-to-day route choice game with identical conditions spontaneously establish cooperation in order to increase their returns, as the folk theorem suggests [6]?

How would such a cooperation look? Taking turns could be a suitable solution [62]. While simple symmetrical cooperation is typically found for the repeated Prisoner's Dilemma [2, 3, 44–46, 49, 52, 55, 59, 64, 67, 69], emergent alternating reciprocity has been recently discovered for the games Leader and Battle of the Sexes [11].[a] Note that such coherent oscillations are a time-dependent but deterministic form of individual decision behavior, which can establish a persistent phase-coordination, while mixed strategies, i.e. statistically varying decisions, can establish cooperation only by chance or on statistical average. This difference is particularly important when the number of interacting persons is small, as in the particular route choice game discussed below.

Note that oscillatory behavior has been found in iterated games before:

- In the rock-paper-scissors game [67], cycles are predicted by the game-dynamical equations due to unstable stationary solutions [28].
- Oscillations can also result from coordination problems [1, 29, 31, 33], at the cost of reduced system performance.
- Moreover, blinker strategies may survive in repeated games played by a mixture of finite automata [5] or result through evolutionary strategies [11, 15, 16, 38, 39, 42, 43, 74].

However, these oscillation-generating mechanisms are clearly distinguishable from the establishment of phase-coordinated alternating reciprocity we are interested in (coherent oscillatory cooperation to reach the system optimum).

Our paper is organized as follows: In Sec. 2, we will formally introduce the route choice game for N players, including issues like the Wardrop equilibrium [71] and

[a]See Fig. 2 for a specification of these games.

the Braess paradox [10]. Section 3 will focus on the special case of the 2-person route choice game, compare it with the minority game [1, 15, 16, 38, 39, 42, 43, 74], and discuss its place in the classification scheme of symmetrical 2×2 games. This section will also reveal some apparent shortcomings of the previous game-theoretical literature:

- While it is commonly stated that among the 12 ordinally distinct, symmetrical 2×2 games [11, 57] only 4 archetypical 2×2 games describe a strategical conflict (the Prisoner's Dilemma, the Battle of the Sexes, Chicken, and Leader) [11, 18, 56], we will show that, for specific payoffs, the route choice game (besides Deadlock) also represents an interesting strategical conflict, at least for iterated games.
- The conclusion that conservative driver behavior is best, i.e. it does not pay off to change routes [7, 65, 66], is restricted to the special case of route-choice games with a system-optimal user equilibrium.
- It is only half the truth that cooperation in the iterated Prisoner's Dilemma is characterized by symmetrical behavior [11]. Phase-coordinated asymmetric reciprocity is possible as well, as in some other symmetrical 2×2 games [11].

New perspectives arise by less restricted specifications of the payoff values.

In Sec. 4, we will discuss empirical results of laboratory experiments with humans [12, 18, 32]. According to these, reaching a phase-coordinated alternating state is only one problem. Exploratory behavior and suitable punishment strategies are important to establish asymmetric oscillatory reciprocity as well [11, 20]. Moreover, we will discuss several coefficients characterizing individual behavior and chances for the establishment of cooperation. In Sec. 5, we will present multi-agent computer simulations of our observations, based on a novel win-stay, lose-shift [50, 54] strategy, which is a special kind of reinforcement learning strategy [40]. This approach is based on individual historical experience [13] and, thereby, clearly differs from the selection of the best-performing strategy in a set of hypothetical strategies as assumed in studies based on evolutionary or genetical algorithms [5, 11, 15, 16, 39, 42, 43]. The final section will summarize our results and discuss their relevance for game theory and possible applications such as data routing algorithms [35, 72], advanced driver information systems [8, 14, 30, 37, 41, 63, 70, 73], or road pricing [53].

2. The Route Choice Game

In the following, we will investigate a scenario with two alternative routes between a certain origin and a given destination, say, between two places or towns A and B (see Fig. 1). We are interested in the case where both routes have different capacities, say a freeway and a subordinate or side road. While the freeway is faster when it is empty, it may be reasonable to use the side road when the freeway is congested.

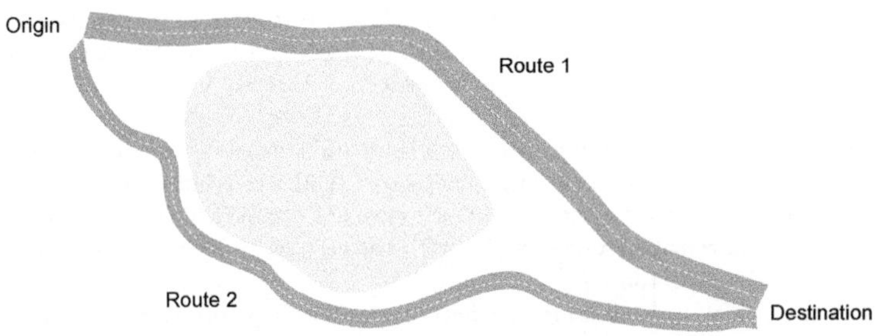

Fig. 1. Illustration of the investigated day-to-day route choice scenario. We study the dynamic decision behavior in a repeated route choice game, where a given destination can be reached from a given origin via two different routes, a freeway (route 1) and a side road (route 2).

The "success" of taking route i could be measured in terms of its inverse travel time $1/T_i(N_i) = V_i(N_i)/L_i$, where L_i is the length of route i and $V_i(N_i)$ the average velocity when N_i of the N drivers have selected route i. One may roughly approximate the average vehicle speed V_i on route i by the linear relationship [24]

$$V_i(N_i) = V_i^0\left(1 - \frac{N_i(t)}{N_i^{\max}}\right), \tag{1}$$

where V_i^0 denotes the maximum velocity (speed limit) and $N_i^{\max}$ the capacity, i.e. the maximum possible number of vehicles on route i. With $A_i = V_i^0/L_i$ and $B_i = V_i^0/(N_i^{\max} L_i)$, the inverse travel time then obeys the relationship

$$1/T(N_i) = A_i - B_i N_i, \tag{2}$$

which is linearly decreasing with the road occupancy N_i. Other monotonously falling relationships $V_i(N_i)$ would make the expression for the inverse travel times non-linear, but they would probably not lead to qualitatively different conclusions.

The user equilibrium of equal travel times is found for a fraction

$$\frac{N_1^{\mathrm{e}}}{N} = \frac{B_2}{B_1 + B_2} + \frac{1}{N}\frac{A_1 - A_2}{B_1 + B_2} \tag{3}$$

of persons choosing route 1. In contrast, the system optimum corresponds to the maximum of the overall inverse travel times $N_1/T_1(N_1) + N_2/T_2(N_2)$ and is found for the fraction

$$\frac{N_1^{\mathrm{o}}}{N} = \frac{B_2}{B_1 + B_2} + \frac{1}{2N}\frac{A_1 - A_2}{B_1 + B_2} \tag{4}$$

of 1-decisions. The difference between both fractions vanishes in the limit $N \to \infty$. Therefore, only experiments with a few players allow one to find out whether the test persons adapt to the user equilibrium or to the system optimum. We will see that both cases have completely different dynamical implications: While the most successful strategy to establish the user equilibrium is to stick to the same decision in subsequent iterations [27, 65, 66], the system optimum can only be reached by a time-dependent strategy (at least, if no participant is ready to pay for the profits of others).

Note that alternative routes can reach comparable travel times only when the total number N of vehicles is large enough to fulfil the relationships $P_1(N) < P_2(0) = A_2$ and $P_2(N) < P_1(0) = A_1$. Our route choice game will address this traffic regime and additionally assume $N \leq N_i^{\max}$. The case $N_i = N_i^{\max}$ corresponds to a complete gridlock on route i.

Finally, it may be interesting to connect the previous quantities with the vehicle densities ρ_i and the traffic flows Q_i: If route i consists of I_i lanes, the relation with the average vehicle density is $\rho_i(N_i) = N_i/(I_i L_i)$, and the relation with the traffic flow is $Q_i(N_i) = \rho_i V_i(N_i) = N_i/[I_i T_i(N_i)]$.

In the following, we will linearly transform the inverse travel time $1/T_i(N_i)$ in order to define the so-called payoff

$$P_i(N_i) = C_i - D_i N_i \tag{5}$$

for choosing route i. The payoff parameters C_i and D_i depend on the parameters A_i, B_i, and N, but will be taken as constant. We have scaled the parameters so that we have the payoff $P_i(N_i^{\text{e}}) = 0$ (zero payoff points) in the user equilibrium and the payoff $N_1 P_1(N_1^{\text{o}}) + N_2 P_2(N - N_1^{\text{o}}) = 100\,N$ (an average of 100 payoff points) in the system optimum. This serves to reach generalizable results and to provide a better orientation to the test persons.

Note that the investigation of social (multi-person) games with linearly falling payoffs is not new [33]. For example, Schelling [62] has discussed situations with "conditional externality," where the outcome of a decision depends on the independent decisions of potentially many others [62]. Pigou has addressed this problem, which has been recently focused on by Schreckenberg and Selten's project SURVIVE [7, 65, 66] and others [8, 41, 58].

The route choice game is a special congestion game [22, 47, 60]. More precisely speaking, it is a multi-stage symmetrical N-person single commodity congestion game [68]. Congestion games belong to the class of "potential games" [48], for which many theorems are available. For example, it is known that there always exists a Wardrop equilibrium [71] with essentially unique Nash flows [4]. This is characterized by the property that no individual driver can decrease his or her travel time by a different route choice. If there are several alternative routes from a given origin to a given destination, the travel times on all used alternative routes in the Wardrop equilibrium are the same, while roads with longer travel times are not used. However, the Wardrop equilibrium as the expected outcome of selfish routing does not generally reach the system optimum, i.e. minimize the total travel times. Nash flows are often inefficient, and selfish behavior implies the possibility of decreased network performance.[b] This is particularly pronounced for

[b]For more details, see the work by T. Roughgarden.

the Braess paradox [10, 61], according to which additional streets may sometimes increase the overall travel time and reduce the throughput of a road network. The reason for this is the possible existence of badly performing Nash equilibria, in which no single person can improve his or her payoff by changing the decision behavior.

In fact, recent laboratory experiments indicate that, in a "day-to-day route choice scenario" based on selfish routing, the distribution of individuals over the alternative routes is fluctuating around the Wardrop equilibrium [27, 63]. Additional conclusions from the laboratory experiments by Schreckenberg, Selten *et al.* are as follows [65, 66]:

- Most people, who change their decision frequently, respond to their experience on the previous day (i.e. in the last iteration).
- There are only a few different behavioral patterns: direct responders (44%), contrarian responders (14%), and conservative persons, who do not respond to the previous outcome.
- It does not pay off to react to travel time information in a sensitive way, as conservative test persons reach the smallest travel times (the largest payoffs) on average.
- People's reactions to short-term travel forecasts can invalidate these. Nevertheless, travel time information helps to match the Wardrop equilibrium, so that excess travel times due to coordination problems are reduced.

A closer experimental analysis based on longer time series (i.e. more iterations) for smaller groups of test persons reveals a more detailed picture [26]:

- Individuals do not only show an adaptive behavior to the travel times on the previous day, but also change their response pattern in time [26, 34].
- In the course of time, one finds a differentiation process which leads to the development of characteristic, individual response patterns, which tend to be almost deterministic (in contrast to mixed strategies).
- While some test persons respond to small differences in travel times, others only react to medium-sized deviations, still others people respond to large deviations, etc. In this way, overreactions of the group to deviations from the Wardrop equilibrium are considerably reduced.

Note that the differentiation of individual behaviors is a way to resolve the coordination problem to match the Wardrop equilibrium exactly, i.e. which participant should change his or her decision in the next iteration in order to compensate for a deviation from it. This implies that the fractions of specific behavioral response patterns should depend on the parameters of the payoff function. A certain fraction of "stayers," who do not respond to travel time information, can improve the coordination in the group, i.e. the overall performance. However, stayers can also prevent the establishment of a system optimum, if alternating reciprocity is needed, see Eq. (14).

3. Classification of Symmetrical 2 × 2 Games

In contrast to previous laboratory experiments, we have studied the route choice game not only with a very high number of repetitions, but also with a small number $N \in \{2, 4\}$ of test persons, in order to see whether the system optimum or the Wardrop equilibrium is established. Therefore, let us shortly discuss how the two-person game relates to previous game-theoretical studies.

Iterated symmetrical two-person games have been intensively studied [12, 18], including Stag Hunt, the Battle of the Sexes, or the Chicken Game (see Fig. 2). They can all be represented by a payoff matrix of the form $\mathbf{P} = (P_{ij})$, where P_{ij} is the success ("payoff") of person 1 in a one-shot game when choosing strategy $i \in \{1, 2\}$ and meeting strategy $j \in \{1, 2\}$. The respective payoffs of the second person are given by the symmetrical values P_{ji}. Figure 2 shows a systematics of the previously mentioned and other kinds of symmetrical two-person games [21]. The relations

$$P_{21} > P_{11} > P_{22} > P_{12}, \tag{6}$$

for example, define a Prisoner's Dilemma. In this paper, however, we will mainly focus on the two-person route choice game defined by the conditions

$$P_{12} > P_{11} > P_{21} > P_{22} \tag{7}$$

(see Fig. 3). Despite some common properties, this game differs from the minority game [16, 39, 43] or El Farol bar problem [1] with $P_{12}, P_{21} > P_{11}, P_{22}$, as a minority decision for alternative 2 is less profitable than a majority decision for

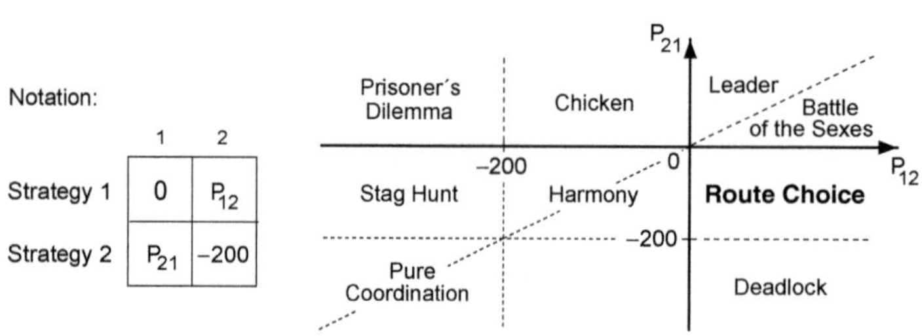

Fig. 2. Classification of symmetrical 2 × 2 games according to their payoffs P_{ij}. Two payoff values have been kept constant as payoffs may be linearly transformed and the two strategies of the one-shot game renumbered. Our choice of $P_{11} = 0$ and $P_{22} = -200$ was made to define a payoff of 0 points in the user equilibrium and an average payoff of 100 in the system optimum of our investigated route choice game with $P_{12} = 300$ and $P_{21} = -100$.

(a)

	1	2
Strategy 1	0	P_{12}
Strategy 2	P_{21}	-200

(b)

	Coop.	Def.
Cooperation	0	-300
Defection	100	-200

(c)

	Coop.	Def.
Cooperation	0	-300
Defection	500	-200

(d)

	1	2
Route 1	0	300
Route 2	-100	-200

Fig. 3. Payoff specifications of the symmetrical 2×2 games investigated in this paper. (a) General payoff matrix underlying the classification scheme of Fig. 2. (b) and (c) Two variants of the Prisoner's Dilemma. (d) Route choice game with a strategical conflict between the user equilibrium and the system optimum.

alternative 1. Although oscillatory behavior has been found in the minority game as well [9, 15, 16, 36, 43], an interesting feature of the route choice experiments discussed in the following is the regularity and phase-coordination (coherence) of the oscillations.

The two-person route choice game fits well into the classification scheme of symmetrical 2×2 games. In Rapoport and Guyer's taxonomy of 2×2 games [57], the two-person route choice game appears on page 211 as game number 7 together with four other games with strongly stable equilibria. Since then, the game has almost been forgotten and did not have a commonly known interpretation or name. Therefore, we suggest naming it the two-person "route choice game." Its place in the extended Eriksson–Lindgren scheme of symmetrical 2×2 games is graphically illustrated in Fig. 2.

According to the game-theoretical literature, there are 12 ordinally distinct, symmetric 2×2 games [57], but after excluding strategically trivial games in the sense of having equilibrium points that are uniquely Pareto-efficient, there remain four archetypical 2×2 games: the Prisoner's Dilemma, the Battle of the Sexes, Chicken (Hawk-Dove), and Leader [56]. However, this conclusion is only correct if the four payoff values P_{ij} are specified by the four values $\{1, 2, 3, 4\}$. Taking different values would lead to a different conclusion: If we name subscripts so that $P_{11} > P_{22}$, a strategical conflict between a user equilibrium and the system optimum results when

$$P_{12} + P_{21} > 2P_{11}. \tag{8}$$

Our conjecture is that players tend to develop alternating forms of reciprocity if this condition is fulfilled, while symmetric reciprocity is found otherwise. This has the following implications (see Fig. 2):

- If the 2×2 games Stag Hunt, Harmony, or Pure Coordination are repeated frequently enough, we always expect a symmetrical form of cooperation.
- For Leader and the Battle of the Sexes, we expect the establishment of asymmetric reciprocity, as has been found by Browning and Colman with a computer simulation based on a genetic algorithm incorporating mutation and crossing-over [11].

- For the games Route Choice, Deadlock, Chicken, and Prisoner's Dilemma both, symmetric (simultaneous) and asymmetric (alternating) forms of cooperation are possible, depending on whether condition (8) is fulfilled or not. Note that this condition cannot be met for some games, if one is restricted to ordinal payoff values $P_{ij} \in \{1, 2, 3, 4\}$ only. Therefore, this interesting problem has been largely neglected in the past (with a few exceptions, e.g. Ref. 51). In particular, convincing experimental evidence of alternating reciprocity is missing. The following sections of this paper will, therefore, not only propose a simulation model, but also focus on an experimental study of this problem, which promises interesting new results.

4. Experimental Results

Altogether we have carried out more than 80 route choice experiments with different experimental setups, all with different participants. In the 24 two-person (12 four-person) experiments evaluated here (see Figs. 4–15), test persons were instructed to choose between two possible routes between the same origin and destination. They

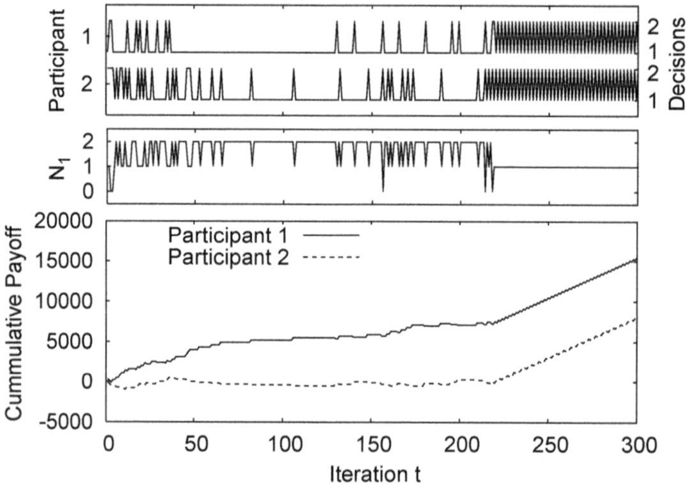

Fig. 4. Representative example for the emergence of coherent oscillations in a two-person route choice experiment with the parameters specified in Fig. 3(d). Top: Decisions of both participants over 300 iterations. Center: Number $N_1(t)$ of 1-decisions over time t. Note that $N_1 = 1$ corresponds to the system optimum, while $N_1 = 2$ corresponds to the user equilibrium of the one-shot game. Bottom: Cumulative payoff of both players in the course of time t (i.e. as a function of the number of iterations). Once the coherent oscillatory cooperation is established ($t > 220$), both individuals have high payoff gains on average.

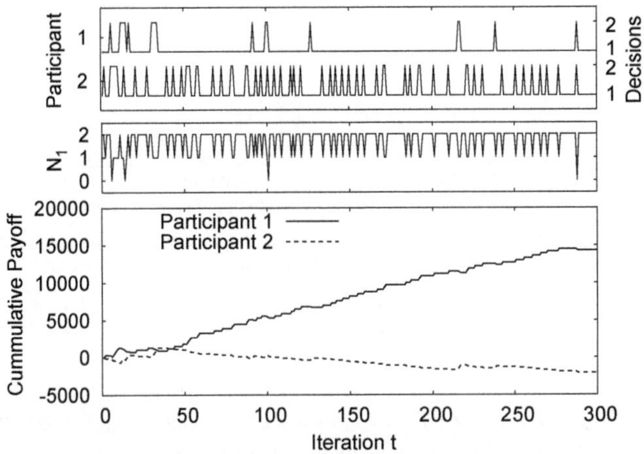

Fig. 5. Representative example for a two-person route choice experiment, in which no alternating cooperation was established. Due to the small changing frequency of participant 1, there were not enough cooperative episodes that could have initiated coherent oscillations. Top: Decisions of both participants over 300 iterations. Center: Number $N_1(t)$ of 1-decisions over time t. Bottom: The cumulative payoff of both players in the course of time t shows that the individual with the smaller changing frequency has higher profits.

knew that route 1 corresponds to a "freeway" (which may be fast or congested), while route 2 represents an alternative route (a "side road"). Test persons were also informed that, if two [three] participants chose route 1, everyone would receive 0 points, while if half of the participants chose route 1, they would receive the maximum average amount of 100 points, but 1-choosers would profit at the cost of 2-choosers. Finally, participants were told that everyone could reach an average of 100 points per round with variable, situation-dependent decisions, and that the (additional) individual payment after the experiment would depend on their cumulative payoff points reached in at least 300 rounds (100 points = 0.01 EUR).

Let us first focus on the two-person route-choice game with the payoffs $P_{11} = P_1(2) = 0, P_{12} = P_1(1) = 300, P_{21} = P_2(1) = -100$, and $P_{22} = P_2(2) = -200$ (see Fig. 3(d)), corresponding to $C_1 = 600, D_1 = 300, C_2 = 0$, and $D_2 = 100$. For this choice of parameters, the best individual payoff in each iteration is obtained by choosing route 1 (the "freeway") and have the co-player(s) choose route 2. Choosing route 1 is the dominant strategy of the one-shot game, and players are tempted to use it. This produces an initial tendency towards the "strongly stable" user equilibrium [57] with 0 points for everyone. However, this decision behavior is not Pareto efficient in the repeated game. Therefore, after many iterations, the players often learn to establish the Pareto optimum of the multi-stage supergame by selecting

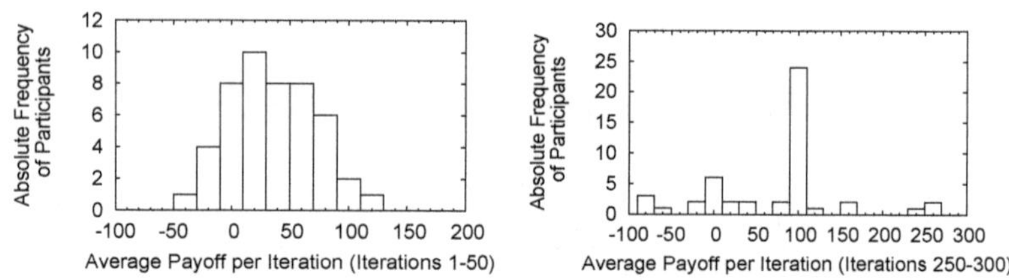

Fig. 6. Frequency distributions of the average payoffs of the 48 players participating in our 24 two-person route choice experiments. Left: Distribution during the first 50 iterations. Right: Distribution between iterations 250 and 300. The initial distribution with a maximum close to 0 points (left) indicates a tendency towards the user equilibrium corresponding to the dominant strategy of the one-shot game. However, after many iterations, many individuals learn to establish the system optimum with a payoff of 100 points (right).

route 1 in turns (see Fig. 4). As a consequence, the experimental payoff distribution shows a maximum close to 0 points in the beginning and a peak at 100 points after many iterations (see Fig. 6), which clearly confirms that the choice behavior of test persons tends to change over time. Nevertheless, in 7 out of 24 two-person experiments, persistent cooperation did not emerge during the experiment. Later on, we will identify reasons for this.

4.1. *Emergence of cooperation and punishment*

In order to reach the system optimum of $(-100+300)/2 = 100$ points per iteration, one individual has to leave the freeway for one iteration, which yields a reduced payoff of -100 in favor of a high payoff of $+300$ for the other individual. To be profitable also for the first individual, the other one should reciprocate this "offer" by switching to route 2, while the first individual returns to route 1. Establishing this oscillatory cooperative behavior yields 100 extra points on average. If the other individual is not cooperative, both will be back to the user equilibrium of 0 points only, and the uncooperative individual has temporarily profited from the offer by the other individual. This makes "offers" for cooperation and, therefore, the establishment of the system optimum unlikely.

Hence, the innovation of oscillatory behavior requires intentional or random changes ("trial-and-error behavior"). Moreover, the consideration of multi-period decisions is helpful. Instead of just two one-stage (i.e. one-period) alternative decisions 1 and 2, there are 2^n different n-stage (n-period) decisions. Such multi-stage strategies can be used to define higher-order games and particular kinds of supergame strategies. In the two-person second-order route choice game, for example, an encounter of the two-stage decision 12 with 21 establishes the system optimum and yields equal payoffs for everyone (see Fig. 8). Such an optimal and fair solution is not possible for one-stage decisions. Yet, the encounter of 12 with

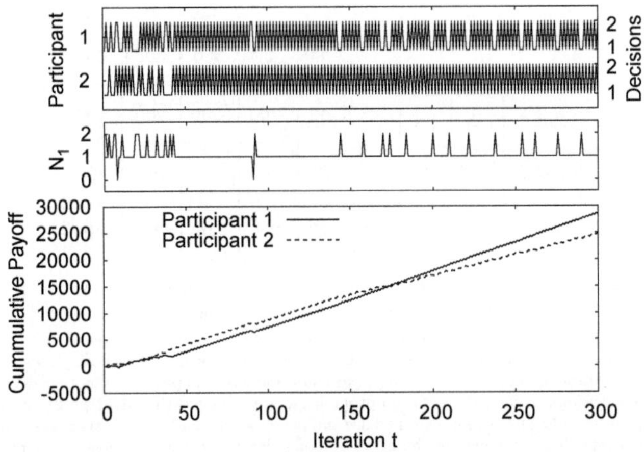

Fig. 7. Representative example for a two-person route choice experiment, in which participant 1 leaves the pattern of oscillatory cooperation temporarily in order to make additional profits. Note that participant 2 does not "punish" this selfish behavior, but continues to take routes in an alternating way. Top: Decisions of both participants over 300 iterations. Center: Number $N_1(t)$ of 1-decisions over time t. Bottom: Cumulative payoff of both players as a function of the number of iterations. The different slopes indicate an unfair outcome despite the high average payoffs of both players.

21 ("cooperative episode") is not a Nash equilibrium of the two-stage game, as an individual can increase his or her own payoff by selecting 11 (see Fig. 8). Probably for this reason, the first cooperative episodes in a repeated route choice game (i.e. encounters of 12-decisions with 21-decisions in two subsequent iterations) do often not persist (see Fig. 9). Another possible reason is that cooperative episodes may be overlooked. This problem, however, can be reduced by a feedback signal that indicates when the system optimum has been reached. For example, we have experimented with a green background color. In this setup, a cooperative episode could be recognized by a green background that appeared in two successive iterations together with two different payoff values.

The strategy of taking route 1 does not only dominate on the first day (in the first iteration). Even if a cooperative oscillatory behavior has been established, there is a temptation to leave this state, i.e. to choose route 1 several times, as this yields more than 100 points on average for the uncooperative individual at the cost of the participant continuing an alternating choice behavior (see Figs. 7 and 8). That is, the conditional changing probability $p_l(2|1, N_1 = 1; t)$ of individuals l from route 1 to route 2, when the system optimum in the previous iteration was established (i.e. $N_1 = 1$) tends to be small initially. However, oscillatory cooperation of period 2 needs $p_l(2|1, N_1 = 1; t) = 1$. The required transition in the decision behavior

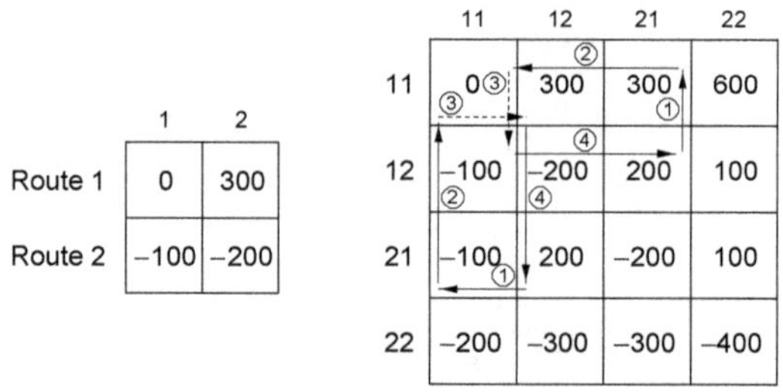

Fig. 8. Illustration of the concept of higher-order games defined by n-stage strategies. Left: Payoff matrix $\mathbf{P} = (P_{ij})$ of the one-shot 2×2 route choice game. Right: Payoff matrix $(P^{(2)}_{(i_1 i_2),(j_1 j_2)}) = (P_{i_1 j_1} + P_{i_2 j_2})$ of the second-order route choice game defined by two-stage decisions (right). The analysis of the one-shot game (left) predicts that the user equilibrium (with both persons choosing route 1) will establish and that no single player could increase the payoff by another decision. For two-period decisions (right), the system optimum (strategy 12 meeting strategy 21) corresponds to a fair solution, but one person can increase the payoff at the cost of the other (see arrow 1), if the game is repeated. A change of the other person's decision can reduce losses and punish this egoistic behavior (arrow 2), which is likely to establish the user equilibrium with payoff 0. In order to leave this state again in favor of the system optimum, one person will have to make an "offer" at the cost of a reduced payoff (arrow 3). This offer may be due to a random or intentional change of decision. If the other person reciprocates the offer (arrow 4), the system optimum is established again. The time-averaged payoff of this cycle lies below the system optimum.

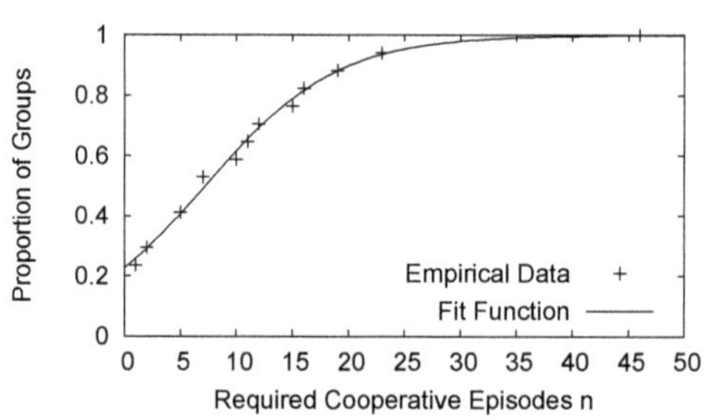

Fig. 9. Cumulative distribution of required cooperative episodes until persistent cooperation was established, given that cooperation occured during the duration of the game as in 17 out of 24 two-person experiments. The experimental data are well approximated by the logistic curve (9) with the fit parameters $c_2 = 3.4$ and $d_2 = 0.17$.

can actually be observed in our experimental data (see Fig. 10, left). With this transition, the average frequency of 1-decisions goes down to 1/2 (see Fig. 10, right). Note, however, that alternating reciprocity does not necessarily require oscillations of period 2. Longer periods are possible as well (see Fig. 11), but have occured only in a few cases (namely, 3 out of 24 cases).

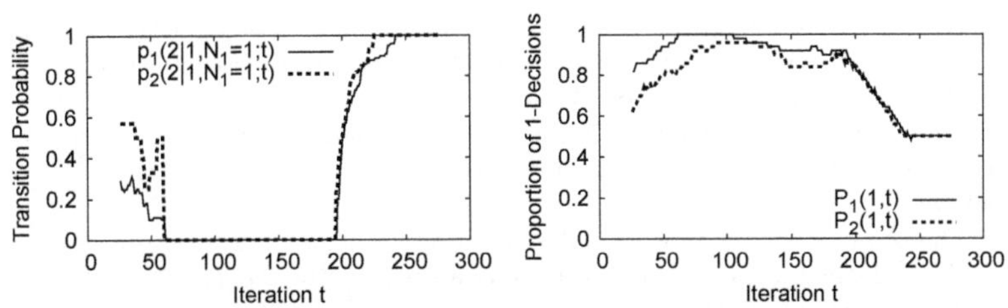

Fig. 10. Left: Conditional changing probability $p_l(2|1, N_1 = 1; t)$ of person l from route 1 (the "freeway") to route 2, when the other person has chosen route 2, averaged over a time window of 50 iterations. The transition from initially small values to 1 (for $t > 240$) is characteristic and illustrates the learning of cooperative behavior. In this particular group (cf. Fig. 4) the values started even at zero, after a transient time period of $t < 60$. Right: Proportion $P_l(1, t)$ of 1-decisions of both participants l in the two-person route choice experiment displayed in Fig. 4. While the initial proportion is often close to 1 (the user equilibrium), it reaches the value 1/2 when persistent oscillatory cooperation (the system optimum) is established.

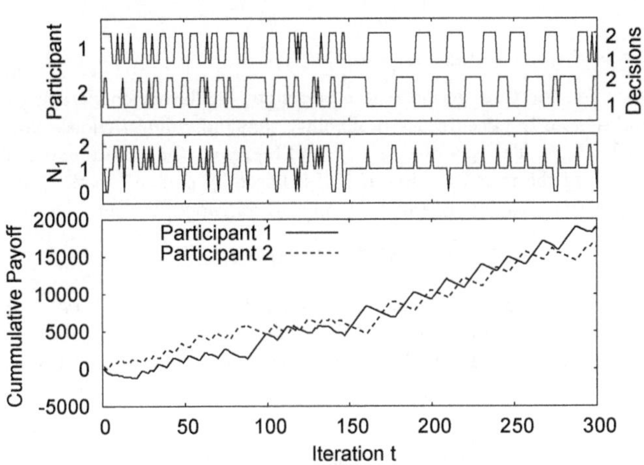

Fig. 11. Representative example for a two-person route choice experiment with phase-coordinated oscillations of long (and varying) time periods larger than 2. Top: Decisions of both participants over 300 iterations. Center: Number $N_1(t)$ of 1-decisions over time t. Bottom: Cumulative payoff of both players as a function of the number of iterations. The sawtooth-like increase in the cumulative payoff indicates gains by phase-coordinated alternations with long oscillation periods.

How does the transition to oscillatory cooperation come about? The establishment of alternating reciprocity can be supported by a suitable punishment strategy: If the other player should have selected route 2, but has chosen route 1 instead, he or she can be punished by changing to route 1 as well, since this causes an average payoff of less than 100 points for the other person (see Fig. 8). Repeated punishment of uncooperative behavior can, therefore, reinforce cooperative oscillatory behavior. However, the establishment of oscillations also requires costly "offers" by switching to route 2, which only pay back in the case of alternating reciprocity. It does not matter whether these "offers" are intentional or due to exploratory trial-and-error behavior.

Due to punishment strategies and similar reasons, persistent cooperation is often established after a number n of cooperative episodes. In the 17 of our 24 two-person experiments in which persistent cooperation was established, the cumulative distribution of required cooperative episodes could be mathematically described by the logistic curve

$$F_N(n) = 1/[1 + c_N \exp(-d_N n)] \tag{9}$$

(see Fig. 9). Note that, while we expect that this relationship is generally valid, the fit parameters c_N and d_N may depend on factors like the distribution of participant intelligence, as oscillatory behavior is apparently difficult to establish (see below).

4.2. *Preconditions for cooperation*

Let us focus on the time period before persistent oscillatory cooperation is established and denote the occurrence probability that individual l chooses alternative $i \in \{1, 2\}$ by $P_l(i)$. The quantity $p_l(j|i)$ shall represent the conditional probability of choosing j in the next iteration, if i was chosen by person l in the present one. Assuming stationarity for reasons of simplicity, we expect the relationship

$$p_l(2|1)P_l(1) = p_l(1|2)P_l(2), \tag{10}$$

i.e. the (unconditional) occurrence probability $P_l(1, 2) = p_l(2|1)P_l(1)$ of having alternative 1 in one iteration and 2 in the next agrees with the joint occurrence probability $P_l(2, 1) = p_l(1|2)P_l(2)$ of finding the opposite sequence 21 of decisions:

$$P_l(1, 2) = P_l(2, 1). \tag{11}$$

Moreover, if r_l denotes the average changing frequency of person l until persistent cooperation is established, we have the relation

$$r_l = P_l(1, 2) + P_l(2, 1). \tag{12}$$

Therefore, the probability that all N players simultaneously change their decision from one iteration to the next is $\prod_{l=1}^{N} r_l$. Note that there are 2^N such realizations of N decision changes 12 or 21, which have all the same occurrence probability because of Eq. (11). Among these, only the ones where $N/2$ players change from 1 to 2 and the other $N/2$ participants change from 2 to 1 establish cooperative episodes, given that the system optimum corresponds to an equal distribution over both alternatives. Considering that the number of different possibilities of selecting $N/2$ out of N persons is given by the binomial coefficient, the occurrence probability of cooperative events is

$$P_{\text{c}} = \frac{1}{2^N} \binom{N}{N/2} \prod_{l=1}^{N} r_l \tag{13}$$

(at least in the ensemble average). Since the expected time period T until the cooperative state incidentally occurs equals the inverse of P_{c}, we finally find the formula

$$T = \frac{1}{P_{\text{c}}} = 2^N \frac{(N/2)!^2}{N!} \prod_{l=1}^{N} \frac{1}{r_l}. \tag{14}$$

This formula is well confirmed by our two-person experiments (see Fig. 12). It gives the lower bound for the expected value of the minimum number of required iterations until persistent cooperation can spontaneously emerge (if already the first cooperative episode is continued forever).

Obviously, the occurrence of oscillatory cooperation is expected to take much longer for a large number N of participants. This tendency is confirmed by our four-person experiments compared to our two-person experiments. It is also in agreement with intuition, as coordination of more people is more difficult. (Note that mean first passage or transition times in statistical phyisics tend to grow exponentially in the number N of particles as well.)

Besides the number N of participants, another critical factor for the cooperation probability are the changing frequencies r_l; they are needed for the exploration of innovative strategies, coordination and cooperation. Although the instruction of test persons would have allowed them to conclude that taking turns would be a

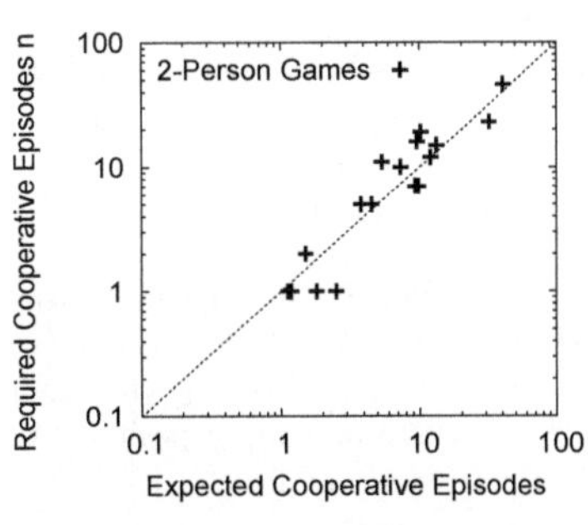

Fig. 12. Comparison of the required number of cooperative episodes y with the expected number x of cooperative episodes (approximated as occurrence time of persistent cooperation, divided by the expected time interval T until a cooperative episode occurs by chance). Note that the data points support the relationship $y = x$ and, thereby, formula (14).

good strategy, the changing frequencies r_l of some individuals was so small that cooperation within the duration of the respective experiment did not occur, in accordance with formula (14). The unwillingness of some individuals to vary their decisions is sometimes called "conservative" [7, 65, 66] or "inertial behavior" [9]. Note that, if a player never reciprocates "offers" by other players, this may discourage further "offers" and reduce the changing frequency of the other player(s) as well (see the decisions 50 through 150 of player 2 in Fig. 4).

Our experimental time series show that most individuals initially did not know a periodic decision behavior would allow them to establish the system optimum. This indicates that the required depth of strategic reasoning [19] and the related complexity of the game for an average person are already quite high, so that intelligence may matter. Compared to control experiments, the hint that the maximum average payoff of 100 points per round could be reached "by variable, situation-dependent decisions," increased the average changing frequency (by 75 percent) and with this the occurrence frequency of cooperative events. Thereby, it also increased the chance that persistent cooperation established during the duration of the experiment.

Note that successful cooperation requires not only coordination [9], but also innovation; in their first route choice game, most test persons discover the oscillatory cooperation strategy only by chance in accordance with formula (14). The changing frequency is, therefore, critical for the establishment of innovative strategies; it determines the exploratory trial-and-error behavior. In contrast, cooperation is easy when test persons *know* that the oscillatory strategy is successful; when two teams, who had successfully cooperated in two-person games, had afterwards to play a four-person game, cooperation was *always* and quickly established (see Fig. 13). In contrast, unexperienced co-players suppressed the establishment of oscillatory cooperation in four-person route choice games.

4.3. *Strategy coefficients*

In order to characterize the strategic behavior of individuals and predict their chances of cooperation, we have introduced some strategy coefficients. For this, let us introduce the following quantities, which are determined from the iterations before persistent cooperation is established:

- c_l^k = relative frequency of a *changed* subsequent decision of individual l if the payoff was negative ($k = -$), zero ($k = 0$), or positive ($k = +$).
- s_l^k = relative frequency of individual l to *stay* with the previous decision if the payoff was negative ($k = -$), zero ($k = 0$), or positive ($k = +$).

The Yule coefficient

$$Q_l = \frac{c_l^- s_l^+ - c_l^+ s_l^-}{c_l^- s_l^+ + c_l^+ s_l^-} \tag{15}$$

104 *D. Helbing et al.*

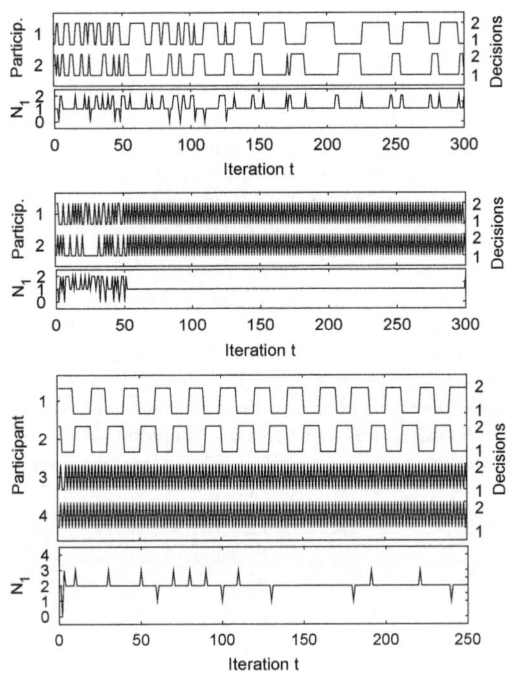

Fig. 13. Experimentally observed decision behavior when two groups involved in two-person route choice experiments afterwards played a four-person game with $C_1 = 900, D_1 = 300, C_2 = 100, D_2 = 100$. While oscillations of period 2 emerged in the second group (center), another alternating pattern corresponding to n-period decisions with $n > 2$ emerged in the first group (top). Bottom: After all persons had learnt oscillatory cooperative behavior, the four-person game just required coordination, but not the invention of a cooperative strategy. Therefore, persistent cooperation was quickly established (in contrast to four-person experiments with new participants). It is clearly visible that the test persons continued to apply similar decision strategies (bottom) as in the previous two-person experiments (top/center).

with $-1 \leq Q_l \leq 1$ was used by Schreckenberg, Selten *et al.* [65] to identify direct responders with $0.5 < Q_l \approx 1$ (who change their decision after a negative payoff and stay after a positive payoff), and contrarian responders with $-0.5 > Q_l \approx -1$ (who change their decision after a positive payoff and stay after a negative one). A random decision behavior would correspond to a value $Q_l \approx 0$. However, a problem arises if one of the variables c_l^-, s_l^+, c_l^+, or s_l^- assumes the value 0. Then, we have $Q_l \in \{-1, 1\}$, independently of the other three values. If two of the variables become zero, Q_l is sometimes even undefined. Moreover, if the values are small, the resulting conclusion is not reliable. Therefore, we prefer to use the percentage difference

$$S_l = \frac{c_l^-}{c_l^- + s_l^l} - \frac{c_l^+}{c_l^+ + s_l^+} \tag{16}$$

for the assessment of strategies. Again, we have $-1 \leq S_l \leq 1$. Direct responders correspond to $S_l > 0.25$ and contrarian responders to $S_l < -0.25$. For $-0.25 \leq S_l \leq 0.25$, the response to the previous payoff is rather random.

In addition, we have introduced the Z-coefficient

$$Z_l = \frac{c_l^0}{c_l^0 + s_l^0}, \tag{17}$$

for which we have $0 \leq Z_l \leq 1$. This coefficient describes the likely response of individual l to the user equilibrium. $Z_l = 0$ means that individual l does not change routes, if the user equilibrium was reached. $Z_l = 1$ implies that person l always changes, while $Z_l \approx 0.5$ indicates a random response.

Figure 14 shows the result of the two-person route choice experiments (cooperation or not) as a function of S_1 and S_2, and as a function of Z_1 and Z_2. Moreover, Figure 15 displays the result as a function of the average strategy coefficients

$$Z = \frac{1}{N}\sum_{l=1}^{N} Z_l \tag{18}$$

and

$$S = \frac{1}{N}\sum_{l=1}^{N} S_l. \tag{19}$$

Our experimental data indicate that the Z-coefficient is a good indicator for the establishment of cooperation, while the S-coefficient seems to be rather insignificant (which also applies to the Yule coefficient).

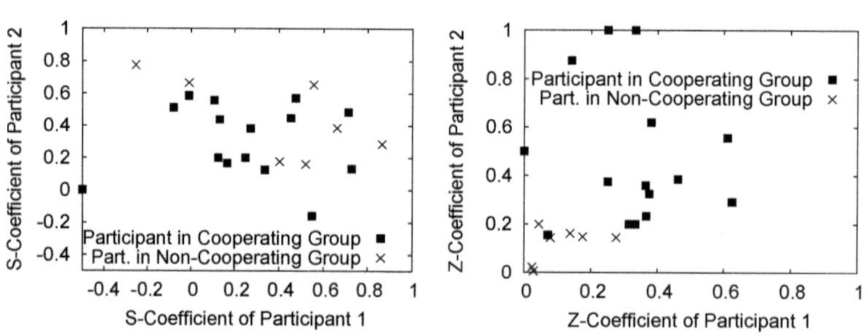

Fig. 14. Coefficients S_l and Z_l of both participants l in all 24 two-person route choice games. The values of the S-coefficients (i.e. the individual tendencies towards direct or contrarian responses) are not very significant for the establishment of persistent cooperation, while large enough values of the Z-coefficient stand for the emergence of oscillatory cooperation.

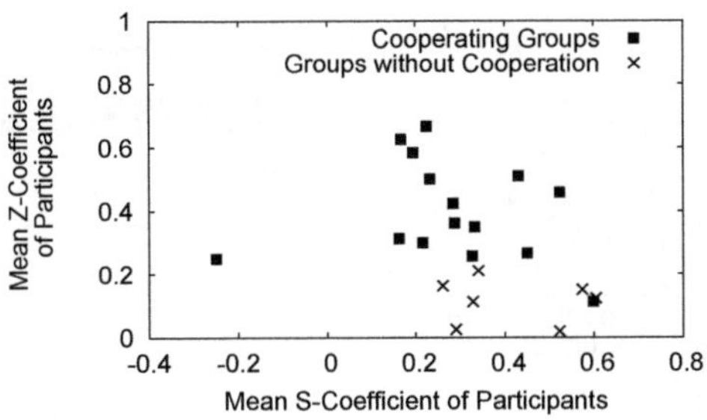

Fig. 15. S- and Z-coefficients averaged over both participants in all 24 two-person route choice games. The mainly small, but positive values of S indicate a slight tendency towards direct responses. However, the S-coefficient is barely significant for the emergence of persistent oscillations. A good indicator for their establishment is a sufficiently large Z-value.

5. Multi-Agent Simulation Model

In a first attempt, we have tried to reproduce the observed behavior in our two-person route choice experiments by game-dynamical equations [28]. We have applied these to the 2×2 route choice game and its corresponding two-, three- and four-stage higher-order games (see Sec. 4.1). Instead of describing patterns of alternating cooperation, however, the game dynamical equations predicted a preference for the dominant strategy of the one-shot game, i.e. a tendency towards choosing route 1.

The reason for this becomes understandable through Fig. 8. Selecting routes 2 and 1 in an alternating way is not a stable strategy, as the other player can get a higher payoff by choosing two times route 1 rather than responding with 1 and 2. Selecting route 1 all the time even guarantees that the own payoff is never below the one by the other player. However, when both players select route 1 and establish the related user equilibrium, no player can improve his or her payoff in the next iteration by changing the decision. Nevertheless, it is possible to improve the long-term outcome, if *both* players change their decisions, and if they do it in a coordinated way. Note, however, that a strict alternating behavior of period 2 is an optimal strategy only in infinitely repeated games, while it is unstable to perturbations in finite games.

It is known that cooperative behavior may be explained by a "shadow of the future" [2, 3], but it can also be established by a "shadow of the past" [40], i.e. experience-based learning. This will be the approach of the multi-agent simulation model proposed in this section. As indicated before, the emergence of phase-coordinated strategic alternation (rather than a statistically independent application of mixed strategies) requires an almost deterministic behavior (see Fig. 16). Nevertheless, some weak stochasticity is needed for the establishment of asymmetric cooperation, both for the exploration of innovative strategies and for phase coordination. Therefore, we propose the following reinforcement learning model, which could be called a generalized win-stay, lose-shift strategy [50, 54].

Emergence of Taking Turns in a Congestion Game 107

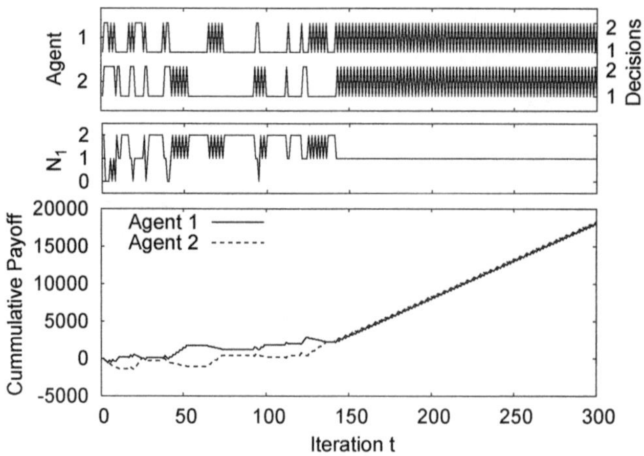

Fig. 16. Representative example for a two-person route choice simulation based on our proposed multi-agent reinforcement learning model with $P_{\text{av}}^{\text{max}} = 100$ and $P_{\text{av}}^{\text{min}} = -200$. The parameter ν_l^1 has been set to 0.25. The other model parameters are specified in the text. Top: Decisions of both agents over 300 iterations. Center: Number $N_1(t)$ of 1-decisions over time t. Bottom: Cumulative payoff of both agents as a function of the number of iterations. The emergence of oscillatory cooperation is comparable with the experimental data displayed in Fig. 4.

Let us presuppose that an individual approximately memorizes or has a good feeling of how well he or she has performed on average in the last n_l iterations and since he or she has last responded with decision j to the situation (i, N_1). In our success- and history-dependent model of individual decision behavior, $p_l(j|i, N_1; t)$ denotes agent l's conditional probability of taking decision j at time $t+1$, when i was selected at time t and $N_1(t)$ agents had chosen alternative 1. Assuming that p_l is either 0 or 1, $p_l(j|i, N_1; t)$ has the meaning of a deterministic response strategy: $p_l(j|i, N_1; t) = 1$ implies that individual l will respond at time $t+1$ with the decision j to the situation (i, N_1) at time t.

Our reinforcement learning strategy can be formulated as follows. The response strategy $p_l(j|i, N_1, t)$ is switched with probability $q_l > 0$, if the average individual payoff since the last comparable situation with $i(t') = i(t)$ and $N_1(t') = N_1(t)$ at time $t' < t$ is less than the average individual payoff $\bar{P}_l(t)$ during the last n_l iterations. In other words, if the time-dependent aspiration level $\bar{P}_l(t)$ [40, 54] is not reached by the agent's average payoff since his or her last comparable decision, the individual is assumed to substitute the response strategy $p_l(j|i, N_1; t)$ by

$$p_l(j|i, N_1; t+1) = 1 - p_l(j|i, N_1; t) \tag{20}$$

with probability q_l. The replacement of dissatisfactory strategies orients at historical long-term profits (namely, during the time period $[t', t]$). Thereby, it avoids short-sighted changes after temporary losses. Moreover, it does not assume a comparison of the performance of the actually applied strategy with hypothetical ones as in most evolutionary models. A readiness for altruistic decisions is also not required, while exploratory behavior ("trial and error") is necessary. In order to reflect this, the decision behavior is randomly switched from $p_l(j|i, N_1; t+1)$ to $1 - p_l(j|i, N_1; t+1)$ with probability

$$\nu_l(t) = \max\left(\nu_l^0, \nu_l^1 \frac{P_{\text{av}}^{\text{max}} - \bar{P}_l(t)}{P_{\text{av}}^{\text{max}} - P_{\text{av}}^{\text{min}}}\right) \ll 1. \tag{21}$$

Herein, $P_{\text{av}}^{\text{min}}$ and $P_{\text{av}}^{\text{max}}$ denote the minimum and maximum average payoff of all N agents (simulated players). The parameter ν_l^1 reflects the mutation frequency for $\bar{P}_l(t) = P_{\text{av}}^{\text{min}}$, while the mutation frequency is assumed to be $\nu_l^0 \leq \nu_l^1$ when the time-averaged payoff $\bar{P}_l$ reaches the system optimum $\bar{P}_{\text{av}}^{\text{max}}$.

In our simulations, no emergent cooperation is found for $\nu_l^0 = \nu_l^1 = 0$. $\nu_l^0 > 0$ or odd values of n_l may produce intermittent breakdowns of cooperation. A small, but finite value of ν_l^1 is important to find a transition to persistent cooperation. Therefore, we have used the parameter value $\nu_l^1 = 0.25$, while the simplest possible specification has been chosen for the other parameters, namely $\nu_l^0 = 0, q_l = 1$, and $n_l = 2$.

The initial conditions for the simulation of the route choice game were specified in accordance with the dominant strategy of the one-shot game, i.e. $P_l(1, 0) = 1$ (everyone tends to choose the freeway initially), $p_l(2|1, N_1; 0) = 0$ (it is not attractive to change from the freeway to the side road) and $p_l(1|2, N_1; 0) = 1$ (it is tempting to change from the side road to the freeway). Interestingly enough, agents learnt to acquire the response strategy $p_l(2|1, N_1 = 1; t) = 1$ in the course of time, which established oscillatory cooperation with higher profits (see Figs. 16 and 17).

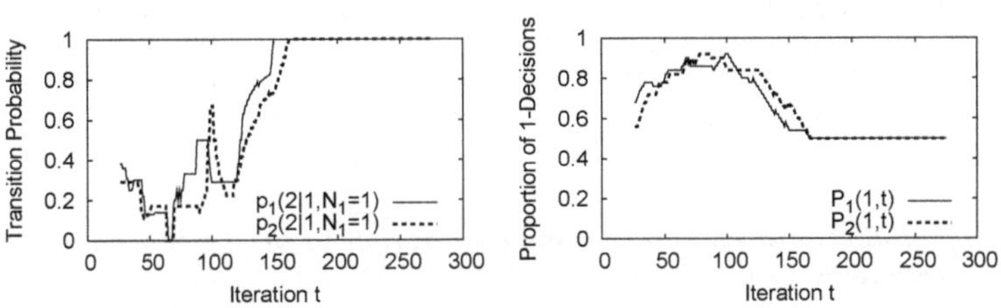

Fig. 17. Left: Conditional changing probability $p_l(2|1, N_1 = 1; t)$ of agent l from route 1 (the "freeway") to route 2, when the other agent has chosen route 2, averaged over a time window of 50 iterations. The transition from small values to 1 for the computer simulation displayed in Fig. 16 is characteristic and illustrates the learning of cooperative behavior. Right: Proportion $P_l(1, t)$ of 1-decisions of both participants l in the two-person route choice experiment displayed in Fig. 16. While the initial proportion is often close to 1 (the user equilibrium), it reaches the value 1/2 when persistent oscillatory cooperation (the system optimum) is established. The simulation results are compatible with the essential features of the experimental data (see, for example, Fig. 10).

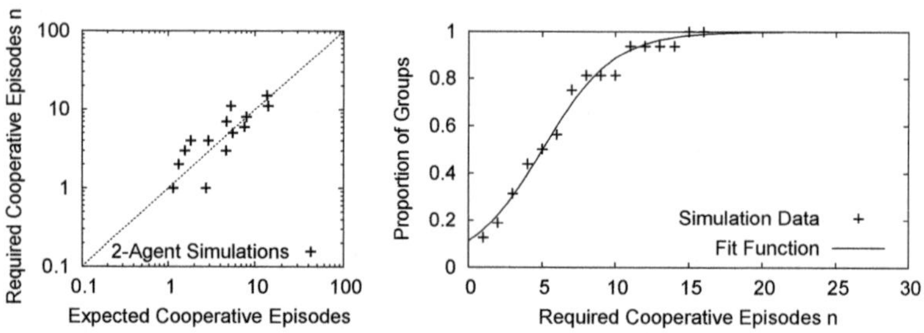

Fig. 18. Left: Comparison of the required number of cooperative episodes with the expected number of cooperative episodes in our multi-agent simulation of decisions in the route choice game. Note that the data points support formula (14). Right: Cumulative distribution of required cooperative episodes until persistent cooperation is established in our two-person route choice simulations, using the simplest specification of model parameters (not calibrated). The simulation data are well approximated by the logistic curve (9) with the fit parameters $c_2 = 7.9$ and $d_2 = 0.41$.

Note that the above described reinforcement learning model [40] responds only to the own previous experience [13]. Despite its simplicity (e.g. the neglect of more powerful, but probably less realistic k-move memories [11]), our "multi-agent" simulations reproduce the emergence of asymmetric reciprocity of two or more players, if an oscillatory strategy of period 2 can establish the system optimum. This raises the question why previous experiments of the N-person route choice game [27,63] have observed a clear tendency towards the Wardrop equilibrium [71] with $P_1(N_1) = P_2(N_2)$ rather than phase-coordinated oscillations? It turns out that the payoff values must be suitably chosen [see Eq. (8)] and that several hundred repetitions are needed. In fact, the expected time interval T until a cooperative episode among $N = N_1 + N_2$ participants occurs in our simulations by chance is well described by formula (14); see Fig. 18. The empirically observed transition in the decision behavior displayed in Fig. 10 is qualitatively reproduced by our computer simulations as well (see Fig. 17). The same applies to the frequency distribution of the average payoff values (compare Fig. 19 with Fig. 6) or to the number of expected and required cooperative episodes (compare Fig. 18 with Figs. 9 and 12).

5.1. *Simultaneous and alternating cooperation in the Prisoner's Dilemma*

Let us finally simulate the dynamic behavior in the two different variants of the Prisoner's Dilemma indicated in Figs. 3(b) and (c) with the above experience-based reinforcement learning model. Again, we will assume $P_{11} = 0$ and $P_{22} = -200$. According to Eq. (8), a simultaneous, symmetrical form of cooperation is expected for $P_{12} = -300$ and $P_{21} = 100$, while an alternating, asymmetric cooperation is expected for $P_{12} = -300$ and $P_{21} = 500$. Figure 20 shows simulation results for the two different cases of the Prisoner's Dilemma and confirms the two predicted forms

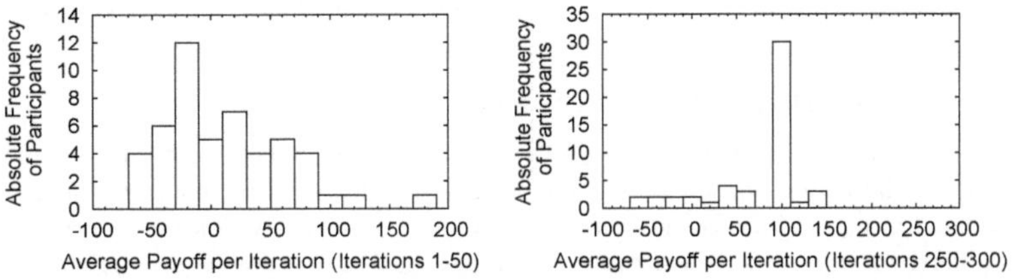

Fig. 19. Frequency distributions of the average payoffs in our computer simulations of the two-person route choice game. Left: Distribution during the first 50 iterations. Right: Distribution between iterations 250 and 300. Our simulation results are compatible with the experimental data displayed in Fig. 6.

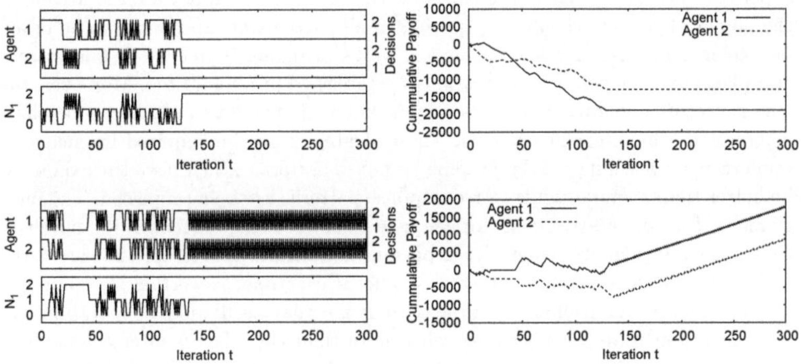

Fig. 20. Representative examples for computer simulations of the two different forms of the Prisoner's Dilemma specified in Figs. 3(b) and (c). The parameter ν_l^1 has been set to 0.25, while the other model parameters are specified in the text. Top: Emergence of simultaneous, symmetrical cooperation, where decision 2 corresponds to defection and decision 1 to cooperation. The system optimum corresponds to $P_{\text{av}}^{\max} = 0$ payoff points, and the minimum payoff to $P_{\text{av}}^{\min} = -200$. Bottom: Emergence of alternating, asymmetric cooperation with $P_{\text{av}}^{\max} = 100$ and $P_{\text{av}}^{\min} = -200$. Left: Time series of the agents' decisions and the number $N_1(t)$ of 1-decisions. Right: Cumulative payoffs as a function of time t.

of cooperation. Again, we varied only the parameter ν_l^1, while we chose the simplest possible specification of the other parameters $\nu_l^0 = 0, q_l = 1$, and $n_l = 2$. The initial conditions were specified in accordance with the expected non-cooperative outcome of the one-shot game, i.e. $P_l(1,0) = 0$ (everyone defects in the beginning), $p_l(2|2, N_1; 0) = 0$ (it is tempting to continue defecting), $p_l(1|1, N_1 = 1; 0) = 0$ (it is unfavorable to be the only cooperative player), and $p_l(1|1, N_1 = 2; 0) = 1$ (it is good to continue cooperating, if the other player cooperates). In the course of time, agents learn to acquire the response strategy $p_l(2|2, N_1 = 0; t) = 0$ when simultaneous cooperation evolves, but $p_l(2|2, N_1 = 1; t) = 0$ when alternating cooperation is established.

6. Summary, Discussion, and Outlook

In this paper, we have investigated the N-person day-to-day route-choice game. This special congestion game has not been thoroughly studied before in the case of small groups, where the system optimum can considerably differ from the user equilibrium. The two-person route choice game gives a meaning to a previously uncommon repeated symmetrical 2×2 game and shows a transition from the dominating strategy of the one-shot game to coherent oscillations, if $P_{12} + P_{21} > 2P_{11}$. However, a detailed analysis of laboratory experiments with humans reveals that the establishment of this phase-coordinated alternating reciprocity, which is expected to occur in other 2×2 games as well, is quite complex. It needs either strategic experience or the invention of a suitable strategy. Such an innovation is driven by the potential gains in the average payoffs of all participants and seems to be based on exploratory trial-and-error behavior. If the changing frequency of one or several players is too low, no cooperation is established in a long time. Moreover, the emergence of cooperation requires certain kinds of strategies, which can be characterized by the Z-coefficient (18). These strategies can be acquired by means of reinforcement learning, i.e. by keeping response patterns which have turned out to be better than average, while worse response patterns are being replaced. The punishment of uncooperative behavior can help to enforce cooperation. Note, however, that punishment in groups of $N > 2$ persons is difficult, as it is hard to target the uncooperative person, and punishment affects everyone. Nevertheless, computer simulations and additional experiments indicate that oscillatory cooperation can still emerge in route choice games with more than two players after a long time period (rarely within 300 iterations) (see Fig. 21).

Altogether, spontaneous cooperation takes a long time. It is, therefore, sensitive to changing conditions reflected by time-dependent payoff parameters. As a consequence, emergent cooperation is unlikely to appear in real traffic systems. This is the reason why the Wardrop equilibrium tends to occur. However, cooperation could be rapidly established by means of advanced traveller information systems (ATIS) [8, 14, 30, 37, 41, 63, 70, 73], which would avoid the slow learning process described by Eq. (14). Moreover, while we do not recommend conventional congestion charges, a charge for *unfair* usage patterns would support the compliance with individual route choice recommendations. It would supplement the inefficient individual punishment mechanism.

Different road pricing schemes have been proposed, each of which has its own advantages and disadvantages or side effects. Congestion charges, for example, could discourage the taking of congested routes, which is actually required to reach minimum *average* travel times. Conventional tolls and road pricing may reduce the trip frequency due to budget constraints, which potentially interferes with economic growth and fair chances for everyone's mobility.

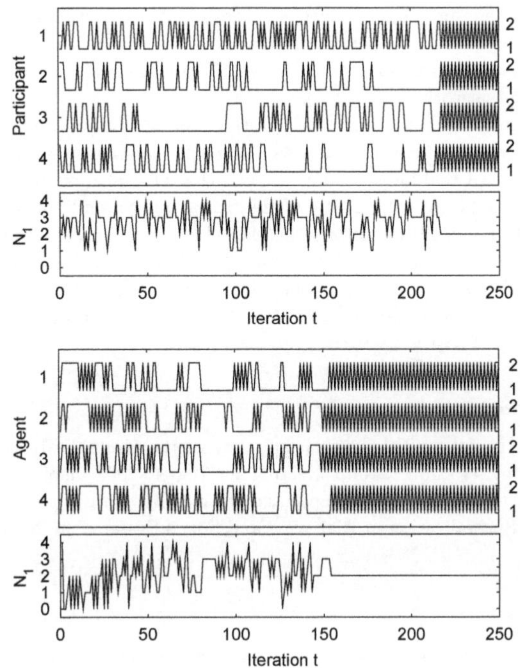

Fig. 21. Emergence of phase-coordinated oscillatory behavior in the four-person route choice game with the parameters specified in Fig. 13. Top: Experimental data of the decisions of four unexperienced participants over 250 iterations. Bottom: Computer simulation with the reinforcment learning model.

In order to activate capacity reserves, we therefore propose an automated route guidance system based on the following principles. After specification of their destination, drivers should get individual (and, on average, fair) route choice recommendations in agreement with the traffic situation and the route choice proportions required to reach the system optimum. If an individual selects a faster route instead of the recommended route that should be used, he/she will have to pay an amount proportional to the decrease in the overall inverse travel time compared to the system optimum. Moreover, drivers not in a hurry should be encouraged to take the slower route i by receiving the amount of money corresponding to the related increase in the overall inverse travel time. Altogether, such an ATIS could support the system optimum while allowing for some flexibility in route choice. Moreover, the fair usage pattern would be cost-neutral for everyone, i.e. traffic flows of potential economic relevance would not be suppressed by extra costs.

In systems with many similar routing decisions, a Pareto optimum characterized by asymmetric alternating cooperation may even emerge spontaneously. This could help to enhance the routing in data networks [72] and generally to resolve Braess-like paradoxes in networks [17].

Finally, it cannot be emphasized enough that taking turns is a promising strategy to distribute scarce resources in a fair and optimal way. It could be applied to a huge number of real-life situations due to the relevance for many strategical conflicts, including Leader, the Battle of the Sexes, and variants of Route Choice, Deadlock, Chicken, and the Prisoner's Dilemma. The same applies to their N-person generalizations, in particular social dilemmas [23, 25, 40]. It will also be interesting to find out whether and where metabolic pathways, biological supply networks, or information flows in neuronal and immune systems use alternating strategies to avoid the wasting of costly resources.

Acknowledgments

D. Helbing is grateful for the warm hospitality of the Santa Fe Institute, where the Social Scaling Working Group Meeting in August 2003 inspired many ideas in this paper. The results will be presented during the workshop on "Collectives Formation and Specialization in Biological and Social Systems" in Santa Fe (April 20–22, 2005).

References

[1] Arthur, W. B., Inductive reasoning and bounded rationality, *Am. Econ. Rev.* **84**, 406–411 (1994).
[2] Axelrod, R. and Dion, D., The further evolution of cooperation, *Science* **242**, 1385–1390 (1988).
[3] Axelrod, R. and Hamilton, W. D., The evolution of cooperation, *Science* **211**, 1390–1396 (1981).
[4] Beckmann, M., McGuire, C. B. and Winsten, C. B., *Studies in the Economics of Transportation* (Yale University Press, New Haven, 1956).
[5] Binmore, K. G., Evolutionary stability in repeated games played by finite automata, *J. Econ. Theory* **57**, 278–305 (1992).
[6] Binmore, K., *Fun and Games: A Text on Game Theory* (Heath, Lexington, MA, 1992), pp. 373–377.
[7] Bohnsack, U., Uni DuE: Studie SURVIVE gibt Einblicke in das Wesen des Autofahrers, Press release by *Informationsdienst Wissenschaft* (January 21, 2005).
[8] Bonsall, P. W. and Perry, T., Using an interactive route-choice simulator to investigate driver's compliance with route guidance information, *Transpn. Res. Rec.* **1306**, 59–68 (1991).
[9] Bottazzi, G. and Devetag, G., Coordination and self-organization in minority games: Experimental evidence, Working Paper 2002/09, Sant'Anna School of Advances Studies, May, 2002.
[10] Braess, D., Über ein Paradoxon der Verkehrsplanung [A paradox of traffic assignment problems], *Unternehmensforschung* **12**, 258–268 (1968). For about 100 related references see http://homepage.ruhr-uni-bochum.de/Dietrich.Braess/#paradox

[11] Browning, L. and Colman, A. M., Evolution of coordinated alternating reciprocity in repeated dyadic games, *J. Theor. Biol.* **229**, 549–557 (2004).

[12] Camerer, C. F., *Behavioral Game Theory: Experiments on Strategic Interaction* (Princeton University Press, Princeton, 2003).

[13] Camerer, C. F., Ho, T.-H. and Chong, J.-K., Sophisticated experience-weighted attraction learning and strategic teaching in repeated games, *J. Econ. Theory* **104**, 137–188 (2002).

[14] Cetin, N., Nagel, K., Raney, B. and Voellmy, A., Large scale multi-agent transportation simulations, *Computer Physics Communications* **147**(1–2), 559–564 (2002).

[15] Challet, D. and Marsili, M., Relevance of memory in minority games, *Phys. Rev.* **E62**, 1862–1868 (2000).

[16] Challet, D. and Zhang, Y.-C., Emergence of cooperation and organization in an evolutionary game, *Physica* **A246**, 407–418 (1997).

[17] Cohen, J. E. and Horowitz, P., Paradoxial behaviour of mechanical and electrical networks, *Nature* **352**, 699–701 (1991).

[18] Colman, A. M., *Game Theory and its Applications in the Social and Biological Sciences*, 2nd edn. (Butterworth-Heinemann, Oxford, 1995).

[19] Colman, A. M., Depth of strategic reasoning in games, *Trends Cogn. Sci.* **7**(1), 2–4 (2003).

[20] Crowley, P. H., Dangerous games and the emergence of social structure: evolving memory-based strategies for the generalized hawk-dove game, *Behav. Ecol.* **12**, 753–760 (2001).

[21] Eriksson, A. and Lindgren, K., Cooperation in an unpredictable environment, in *Proc. Artificial Life VIII* (eds. Standish, R. K., Bedau, M. A. and Abbass, H. A.) (MIT Press, Sidney, 2002), pp. 394–399; and poster available at http://frt.fy.chalmers.se/cs/people/eriksson.html

[22] Garcia, C. B. and Zangwill, W. I., *Pathways to Solutions, Fixed Points, and Equilibria* (Prentice Hall, New York, 1981).

[23] Glance, N. S. and Huberman, B. A., The outbreak of cooperation, *J. Math. Soc.* **17**(4), 281–302 (1993).

[24] Greenshield, B. D., A study of traffic capacity, in *Proc. Highway Research Board*, Vol. 14 (Highway Research Board, Washington, D. C., 1935), pp. 448–477.

[25] Hardin, G., The tragedy of the commons, *Science* **162**, 1243–1248 (1968).

[26] Helbing, D., Dynamic decision behavior and optimal guidance through information services: Models and experiments, in *Human Behaviour and Traffic Networks*, eds. Schreckenberg, M. and Selten, R. (Springer, Berlin, 2004), pp. 47–95.

[27] Helbing, D., Schönhof, M. and Kern, D., Volatile decision dynamics: Experiments, stochastic description, intermittency control, and traffic optimization, *New J. Phys.* **4**, 33.1–33.16 (2002).

[28] Hofbauer, J. and Sigmund, K., *The Theory of Evolution and Dynamical Systems* (Cambridge University Press, Cambridge, 1988).

[29] Hogg, T. and Huberman, B. A., Controlling chaos in distributed systems, *IEEE Trans. Syst. Man Cy.* **21**(6), 1325–1333 (1991).

[30] Hu, T.-Y. and Mahmassani, H. S., Day-to-day evolution of network flows under real-time information and reactive signal control, *Transport. Res.* **C5**(1), 51–69 (1997).

[31] Iwanaga, S. and Namatame, A., The complexity of collective decision, *Nonlinear Dynam. Psychol. Life Sci.* **6**(2), 137–158 (2002).

[32] Kagel, J. H. and Roth, A. E. (eds.), *The Handbook of Experimental Economics* (Princeton University, Princeton, NJ, 1995).

[33] Kephart, J. O., Hogg, T. and Huberman, B. A., Dynamics of computational ecosystems, *Phys. Rev.* **A40**(1), 404–421 (1989).
[34] Klügl, F. and Bazzan, A. L. C., Route decision behaviour in a commuting scenario: Simple heuristics adaptation and effect of traffic forecast, *JASSS* **7**(1), Jan. (2004).
[35] Korilis, Y. A., Lazar, A. A. and Orda, A., Avoiding the Braess-paradox in noncooperative networks, *J. Appl. Prob.* **36**, 211–222 (1999).
[36] Laureti, P., Ruch, P., Wakeling, J. and Zhang, Y.-C., The interactive minority game: A Web-based investigation of human market interactions, *Physica* **A331**, 651–659 (2004).
[37] Lee, K., Hui, P. M., Wang, B. H. and Johnson, N. F., Effects of announcing global information in a two-route traffic flow model, *J. Phys. Soc. Japan* **70**, 3507–3510 (2001).
[38] Lo, T. S., Chan, H, Y., Hui, P. M. and Johnson, N. F., Theory of networked minority games based on strategy pattern dynamics, *Phys. Rev.* **E70**, 056102 (2004).
[39] Lo, T. S., Hui, P. M. and Johnson, N. F., Theory of the evolutionary minority game, *Phys. Rev.* **E62**, 4393–4396 (2000).
[40] Macy, M. W. and Flache, A., Learning dynamics in social dilemmas, in *Proc. National Academy of Sciences USA*, Vol. 99, Suppl. 3 (2002), pp. 7229–7236.
[41] Mahmassani, H. S. and Jou, R. C., Transferring insights into commuter behavior dynamics from laboratory experiments to field surveys, *Transport. Res.* **A34**, 243–260 (2000).
[42] Mansilla, R., Algorithmic complexity in the minority game, *Phys. Rev.* **E62**, 4553–4557 (2000).
[43] Marsili, M., Mulet, R., Ricci-Tersenghi, F. and Zecchina, R., Learning to coordinate in a complex and nonstationary world, *Phys. Rev. Lett.* **87**, 208701 (2001).
[44] McNamara, J. M., Barta, Z. and Houston, A. I., Variation in behaviour promotes cooperation in the prisoner's dilemma game, *Nature* **428**, 745–748 (2004).
[45] Michor, F. and Nowak, M. A., The good, the bad and the lonely, *Nature* **419**, 677–679 (2002).
[46] Milinski, M., Semmann, D. and Krambeck, H.-J., Reputation helps solve the 'tragedy of the commons', *Nature* **415**, 424–426 (2002).
[47] Monderer, D. and Shapley, L. S., Fictitious play property for games with identical interests, *J. Econ. Theory* **1**, 258–265 (1996).
[48] Monderer, D. and Shapley, L. S., Potential games, *Games Econ. Behav.* **14**, 124–143 (1996).
[49] Nowak, M. A., Sasaki, A., Taylor, C. and Fudenberg, D., Emergence of cooperation and evolutionary stability in finite populations, *Nature* **428**, 646–650 (2004).
[50] Novak, M. and Sigmund, K., A strategy of win-stay, lose-shift that outperforms tit-for-tat in the Prisoner's Dilemma game, *Nature* **364**, 56–58 (1993).
[51] Nowak, M. A. and Sigmund, K., The alternating prisoner's dilemma, *J. Theor. Biol.* **168**, 219–226 (1994).
[52] Nowak, M. A. and Sigmund, K., Evolution of indirect reciprocity by image scoring, *Nature* **393**, 573–577 (1998).
[53] Pigou, A. C., *The Economics of Welfare* (Macmillan, London, 1920).
[54] Posch, M., Win-stay, Lose-shift strategies for repeated games — Memory length, aspiration levels and noise, *J. Theor. Biol.* **198**, 183–195 (1999).
[55] Queller, D. C., Kinship is relative, *Nature* **430**, 975–976 (2004).
[56] Rapoport, A., Exploiter, leader, hero, and martyr: The four archtypes of the 2×2 game, *Behav. Sci.* **12**, 81–84 (1967).

[57] Rapoport, A. and Guyer, M., A taxonomy of 2×2 games, *Gen. Systems* **11**, 203–214 (1966).
[58] Reddy, P. D. V. G. *et al.*, Design of an artificial simulator for analyzing route choice behavior in the presence of information system, *J. Math. Comp. Mod.* **22**, 119–147 (1995).
[59] Riolo, R. L., Cohen, M. D. and Axelrod, R., Evolution of cooperation without reciprocity, *Nature* **414**, 441–443 (2001).
[60] Rosenthal, R. W., A class of games possessing pure-strategy Nash equilibria, *Int. J. Game Theory* **2**, 65–67 (1973).
[61] Roughgarden, T. and Tardos, E., How bad is selfish routing?, *J. ACM* **49**(2), 236–259 (2002).
[62] Schelling, T. C., *Micromotives and Macrobehavior* (WW Norton and Co, New York, 1978), pp. 224–231, 237.
[63] Schreckenberg, M. and Selten, R. (eds.), *Human Behaviour and Traffic Networks* (Springer, Berlin, 2004).
[64] Schweitzer, F., Behera, L. and Mühlenbein, H., Evolution of cooperation in a spatial prisoner's dilemma, *Adv. Complex Syst.* **5**(2/3), 269–299 (2002).
[65] Selten, R. *et al.*, Experimental investigation of day-to-day route-choice behaviour and network simulations of autobahn traffic in North Rhine-Westphalia, in *Human Behaviour and Traffic Networks*, eds. Schreckenberg, M. and Selten, R. (Springer, Berlin, 2004), pp. 1–21.
[66] Selten, R., Schreckenberg, M., Pitz, T., Chmura, T. and Kube, S., Experiments and simulations on day-to-day route choice-behaviour, see http://papers.ssrn.com/sol3/papers.cfm?abstract_id=393841
[67] Semmann, D., Krambeck, H.-J. and Milinski, M., Volunteering leads to rock-paper-scissors dynamics in a public goods game, *Nature* **425**, 390–393 (2003).
[68] Spirakis, P., Algorithmic aspects of congestion games, Invited talk at the *11th Colloquium on Structural Information and Communication Complexity*, Smolenice Castle, Slovakia, June 21–23, 2004.
[69] Szabó, G. and Hauert, C., Phase transitions and volunteering in spatial public goods games, *Phys. Rev. Lett.* **89**, 118101 (2002).
[70] Wahle, J., Bazzan, A. L. C., Klügl, F. and Schreckenberg, M., Decision dynamics in a traffic scenario, *Physica* **A287**, 669–681 (2000).
[71] Wardrop, J. G., Some theoretical aspects of road traffic research, in *Proc. the Institution of Civil Engineers II*, Vol. 1, 1952, pp. 325–378.
[72] Wolpert, D. H. and Tumer, K., Collective intelligence, data routing and Braess' paradox, *J. Artif. Int. Res.* **16**, 359–387 (2002).
[73] Yamashita, T., Izumi, K. and Kurumatani, K., Effect of using route information sharing to reduce traffic congestion, *Lect. Notes Comput. Sci.* **3012**, 86–104 (2004).
[74] Yuan, B. and Chen, K., Evolutionary dynamics and the phase structure of the minority game, *Phys. Rev.* **E69**, 067106 (2004).

Advances in Complex Systems
Vol. 20, No. 1 (2017) 1750001 (17 pages)

DOI: 10.1142/S0219525917500011

EMERGENT DENSE SUBURBS IN A SCHELLING METAPOPULATION MODEL: A SIMULATION APPROACH

FLORIANA GARGIULO
GEMASS, CNRS — Université de Paris-Sorbonne,
20 Rue Berbier du Metz,
Paris 75013, France
floriana.gargiulo@cnrs.fr

YERALI GANDICA* and TIMOTEO CARLETTI†
NaXys, University of Namur,
8 rempart de la Vierge,
Namur 5000, Belgium
**ygandica@gmail.com*
†*timoteo.carletti@unamur.be*

Received 4 April 2016
Revised 24 March 2017
Accepted 11 April 2017
Published 29 May 2017

The Schelling model describes the formation of spatially segregated clusters starting from individual preferences based on tolerance. To adapt this framework to an urban scenario, characterized by several individuals sharing very close physical spaces, we propose a metapopulation version of the Schelling model defined on the top of a regular lattice whose cells can be interpreted as a bunch of buildings or a district containing several agents. We assume the model to contain two kinds of agents relocating themselves if their individual utility is smaller than a tolerance threshold. While the results for large values of the tolerances respect the common sense, namely coexistence is the rule, for small values of the latter we obtain two non-trivial results: first we observe complete segregation inside the cells, second the population redistributes highly heterogeneously among the available places, despite the initial uniform distribution. The system thus converges toward a complex heterogeneous configuration after a long quasi-stationary transient period, during which the population remains in a well mixed phase. We identify three possible global spatial regimes according to the tolerance value: microscopic clusters with local coexistence of both kinds of agents, macroscopic clusters with local coexistence (hereafter called soft segregation) and macroscopic clusters with local segregation but homogeneous densities (hereafter called hard segregation).

Keywords: Schelling model; metapopulation models; segregation; spatial heterogeneity.

1. Introduction

Modern societies often have to deal with social inequality and segregation dictated by race, religion, social status or incomes differences. The understanding of the rise of segregation patterns has thus attracted a lot of attention from economists, politicians and sociologists [1–4].

In two papers written in the late 60s, Thomas Schelling [5, 6] proposed a stylized model to describe the emergence of segregation in an initially well mixed society: T agents belonging to two different populations are randomly distributed on a square lattice with periodic boundary conditions and composed by D^2 cells, each one containing at most one agent. Agents calculate their utility as the fraction of own kind agents in their 4 neighboring cells. If the utility is lower than $1 - \varepsilon$, being $\varepsilon \in (0, 1)$ the tolerance threshold, the agent is unhappy and relocates to another empty cell in order to maximize her utility. The larger ε, the more tolerant is the agent by accepting a large fraction of agents different from her. This modeling framework displays a counterintuitive but widely observed outcome: a well integrated society can evolve into a rather segregated one even if, at individual level, nobody strictly prefers this final outcome (namely for quite low tolerance values $\varepsilon \leq \varepsilon_c = 1/3$). Even when individuals are quite tolerant to neighbors of their opposite kind, the basic principle of the model, allowing the agents to relocate themselves in order to optimize their local utility, will cause the emergence of spatial segregation patterns as a global aggregated phenomenon not directly designed from the individual choices.

Starting by the pioneering works of Schelling, this "social" model has also attracted the attention of physicists and mathematicians, interested in its simplicity and in the richness of the emergent behaviors recalling the models of Ising and Potts [8, 9, 11–16]. In this sense, the Schelling model is one of the cornerstones of the interdisciplinary field named socio-physics [12].

Several variations have been performed onto the original Schelling model (see [7] for a review), introducing, for instance, different ways for computing the agents' utility [16] or rules to choose the destination (swap agents positions, restrict moves to strictly increase the utility [8, 9], choices constrained by geographical distance or network proximity [2]). The Schelling model essentially contains two parameters: the tolerance threshold and the population density (T/D^2). Until now the physics community has addressed a major attention to the density parameter that has been shown to produce a phase transition in the model [8, 9, 14–17]: for low values of the density the system reaches a steady segregated state where all the agents are happy while for high values the system is frozen in a state where not all the agents are happy but no more moves are possible.

Metapopulation models ground their basis on the pioneering work by Robert May [18] in the framework of theoretical ecology [19]. More recently, they have proved their validity applied to human beings in the framework of global epidemic spreading, being able to better reproduce the dynamics and the outcome of the infections [20–22]. For this reason, we believe the Schelling model can be improved in realism, by adapting it to a society living in urban areas, by considering a metapopulation scheme, as the one proposed in [14, 15, 23]. In this case, the lattice cells have a large but finite carrying capacity representing thus an urban district whose spatial extension can be neglected, and then the population can be considered to be well mixed. In these papers, the utility and thus the willingness to move is conditioned by the fraction of agents of the opposite kind only inside a given cell (urban district).

The *bounded neighborhood model*, proposed by Schelling himself in [6], exploits similar ideas by adding the notion of distribution of tolerance in the population. This model, where the notion of space is abandoned, could be compared to ours once one focuses on a given cell and its neighborhood, and considers the remaining part of the system as a reservoir allowing to exchange agents, because of their willingness to leave. The choice of not considering spatial interactions simplifies the model enough, allowing to derive analytical results, however it completely loses the spatial context, upon which the original Schelling model is build.

To retrieve the spatial context, in this paper we consider an utility function based not only on the local neighborhood (single cell) but also on cells in the spatial proximity. We show that this new ingredient in the model definition leads to new dynamical properties in the system as the emergence of strong heterogeneity in cells' population and the complete segregation inside each cell. This formulation also allows to observe the emergence of spatially segregated clusters, that cannot emerge considering only the local (inside the cell) utility as in the previous papers. The combination of these peculiar emerging phenomena designs the formation of a complex urban shape. Extremely dense zones and (almost) empty areas coexist, as in real urban scenarios.

In Sec. 2, we describe the details of the simulation settings. While Sec. 3 is devoted to the presentation of the differences between the classical model and the metapopulation framework. In the following section, we show the different model outcomes: the cell-level segregation (4.1), the dynamical evolution (4.2), the formation of and the population heterogeneity (4.3) and finally the spatial segregation (4.4). In the last section, we summarize the results and we give some perspectives for a future work.

2. The Model

The basic feature of the metapopulation version of the Schelling model analyzed in this paper is the fact that the dynamics takes place on a two-dimensional (2D) lattice, composed by $N = D^2$ cells, with periodic boundary conditions where each cell can be occupied by at most L agents (carrying capacity) at any given time (for a pictorial representation, see Fig. 1).

For sake of simplicity, we assume a constant carrying capacity for each cell, but of course one could improve the model by considering a different value of L_i for each cell i. The cells can ideally represent the districts of a city, while links stand for the connections among these urban structures, for instance streets and squares. Two

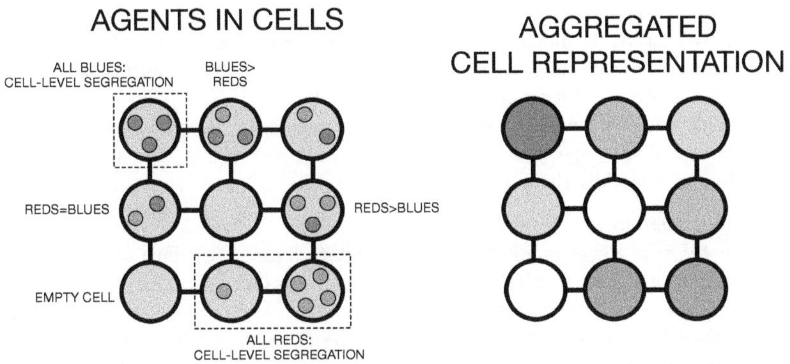

Fig. 1. (Color online) Visualizations of the metapopulation framework. On the left panel, agents' positions inside each cell are showed. On the right panel, the aggregated representation of all the possible states is presented: blue represents cells where only blue agents are present, red represents cells containing only red agents, cyan denotes cells where there is a majority of blue agents with respect to red one, similarly orange represents cells where more red than blue agents are present, gray represents cells with an equal number of red and blue agents, finally white represents empty cells.

kinds of agents, say Red (A) and Blue (B), can move across $N = D^2$ cells, choosing to move if not happy enough by the composition of their neighborhood, i.e., they have low utility. We assume that in the system there are ρLN vacancies, i.e., empty spaces ($\rho \in (0,1)$), and an equal number ($(1-\rho)NL/2$) of Red and Blue agents.

The metapopulation framework adds a relevant complexity to the traditional Schelling setup. The specific aspects of this approach will be discussed deeply in the following section, where we will describe the dynamics of the model.

The metapopulation approach has been firstly introduced in [23]. The main differences between this approach and our setup is the way in which we calculate the agents utility. In [23], the agent utility is computed using single cell information, namely the number of agents in a given site; for example the utility for A individuals at site i is given by $f_{0,i}^A = n_i^B/(n_i^A + n_i^B)$, where the notation f_0 refers to the fact that only the cell at distance 0 from i (namely i itself) has been considered. This particular choice, although very useful in order to have an analytical treatment, does not allow to obtain global spatial structures, because it completely lacks any notion of space. Notice that, with this choice for the utility calculation, the topology of the support system (1D-lattice, 2D-lattice, complex network, etc.) does not have any relevance on the model. This representation can be sufficiently close to realistic cases once the spatial density is low, namely the geographical distance among the cells is high, but this is not usually the case in urban scenarios. Moreover in [23], authors focused on the limit case where the number of cells N goes to infinity with fixed and small cell capacity L. On the contrary, using a numerical approach, we are able to study the model for large L values, to deep explore the role of the tolerance-based individual choices on the spontaneous growth and differentiation of urban districts. The basic aim of this study is indeed to analyze how a population of individuals, whose movements derive from the specific need of maximizing their utility, occupy the urban space: do we observe spatial segregation? How the space is used (are empty spaced leaved, do people have the tendency to concentrate in some location)?

The rest of the paper will be devoted to provide answers to these questions, to explain and motiviate the model rules.

Utility calculation

Differently from the setup of [23], we firstly reintroduce as Schelling originally did, the concept of spatial proximity [6]: agent utility takes into account the number of agents of the same own kind in a given cell and in the neighboring ones, namely cells at distance 1 from the cell where the agent lays (1-distance utility, f_1, for short). More precisely, in our case, the utility of an agent is computed as the fraction of agents of the own kind to her, living in her neighborhood with respect to the total number of agents living in the same neighborhood. The neighborhood of an agent is given by the topological neighborhood (on the lattice) of the cell where she lives, including the latter. Mathematically, assuming she is a A agent living in cell i, then her utility is given by:

$$f^A_{1,i} = \frac{\sum_{j\in\mathcal{V}(i)} n^A_j}{\sum_{j\in\mathcal{V}(i)}(n^B_j + n^A_j)} \equiv 1 - \frac{\sum_{j\in\mathcal{V}(i)} n^B_j}{\sum_{j\in\mathcal{V}(i)}(n^B_j + n^A_j)}, \tag{1}$$

where n^X_j, $X = A, B$, is the number of agent of X-kind in cell j, and we used the notation $j \in \mathcal{V}(i)$ to denote all cells j belonging to a neighborhood ($\mathcal{V}(i)$) of cell i, including the latter, that is the set of cells at distance smaller or equal to 1 from i.

Relocation choice

In the original Schelling model, an additional constraint is added, allowing agents to move only to cells where their utility is higher than in their actual cell. This constraint is fundamental in the classical Schelling model, since it can lead to stable regimes where not all the agents are happy. This gives rise to a phase transition in terms of the density of agents in the system: for low densities the system reaches a status where all the agents are happy, while for high densities a state where not all the agents are happy, although no more moves are allowed, is obtained [9].

Following [24], we relax the constraint based on the agents rationality assumption, of seeking for an utility increase and we hypothesize the behavioral assumption for the agents' relocation choice of reinforcement-learning (agents evaluate decisions by their consequences after the actions). Namely the agents, in this context will not calculate *a priori* the future utility at the destination ("forward-looking") but rather will learn *a posteriori* (after moving), from the new surrounding environment, their new utility ("backward-looking"). Because of the changing environment and the absence of cooperation among agents, the outcome of such decisions could result into a random-like move.

Operatively, we assume that an agent, once she decides to move based on her utility in the actual location, will move to any random, non-complete ($n_i^A + n_i^B < L$), position in the lattice.

Model setup

To summarize, the basic ingredients of the model are the metapopulation structure, the utility function based on spatial proximity and the reinforcement learning assumption for the relocation choice. At each time step and until the system reaches a steady state, we perform:

- An agent is randomly selected and her utility f_1 is computed, using informations from her actual location.
- The "happiness" or not of the agent (intended as a binary choice) at the actual location is decided. As in the original Schelling model, agents are "happy" if their utility is large enough with respect to a given *tolerance threshold*, $\varepsilon \in (0,1)$, hereby assumed to be the same for all agents of both kinds.
$$\text{Hence if } f_{1,i}^A > 1 - \varepsilon \Rightarrow \text{agent } A \text{ is happy at cell } i. \quad (2)$$
- Unhappy agents move by choosing uniformly at random another not completely occupied cell k, i.e., such that $n_k^A + n_k^B < L$.

Let us observe that using the definition of utility for a A agent given by Eq. (1), we can rewrite the previous statement as: agent A is happy at cell i if $\frac{\sum_{j \in \mathcal{V}(i)} n_j^B}{\sum_{j \in \mathcal{V}(i)} (n_j^B + n_j^A)} < \varepsilon$, namely if the relative amount of agents of a different kind of is small enough. This allows to consider ε as a tolerance threshold, if it is large then agents are tolerant because they can accept a large fraction of agents of the opposite kind.

One time step is defined as the random selection and the (eventual) relocation of an agent.

Notice that because all the positions are allowed once an unhappy agent moves, this guarantees that, in a finite time the system will reach a state where all the agents are happy (if $\rho < 1$). On the contrary, this could not happen if the agents are forced to configurations increasing their utility. In this case, in fact, the system can remain frozen in different equilibria where no more moves are possible but not all the agents are happy.

If not differently specified in the text, cells are assumed to be arranged in regular, finite 2D-lattices with periodic boundary conditions and each cell has 4 neighboring cells, as initially assumed by Schelling. The extension of such model to a complex network has been considered in [10].

3. The Metapopulation Framework: The $L = 1$ versus the $L > 1$ Model

In this section, we point out the structural differences between the standard Schelling framework, corresponding to the case $L = 1$, and the metapopulation structure $L > 1$.

In the standard Schelling model, agents are only relocated to positions with a larger utility than in the actual position and its outcome is that for high densities segregated structures do emerge, namely few macroscopic clusters (containing only agents of the same kind) of cells are formed. More precisely, we define a cluster ($\mathcal{C}$) to be set of non-empty neighboring cells, all containing an agent of the same kind; let us observe that $L = 1$ and thus a non-empty cell can only contain a A or a B agent.

Let us define $S_i = \#\mathcal{C}_i/T$ to be the size of the cluster C_i, namely the number of cells forming it, relative to the total number of agents in any cluster, namely the total population T. Then, we can measure the spatial segregation as the sum of the relative sizes of the two largest clusters ($\mathcal{C}_1$, $\mathcal{C}_2$), averaged over several replicas:

$$\sigma = \langle S_1 \rangle + \langle S_2 \rangle = \left\langle \frac{\#\mathcal{C}_1}{T} \right\rangle + \left\langle \frac{\#\mathcal{C}_2}{T} \right\rangle. \tag{3}$$

The rationale of this measure is that in a strongly spatially segregated population, agents will distribute themselves into very large clusters, namely groups of contiguous cells with agents of the same kind, and thus σ will be very close to one, value reached in the extreme case of only two clusters, one made by agents of kind A and the other of kind B. On the other hand, in a well mixed population, agents of one kind are on average surrounded by agents of different kinds and thus they tend to form small clusters, as a consequence σ will be quite small.

Let us recall that in the original Schelling model ($L = 1$ and agents moves finalized to increase the agent utility) the system passes from a strongly segregated behavior for small ε to a well mixed one for large values of the tolerance threshold. Let us observe that the same result is maintained if we introduce the reinforcement learning assumption for the relocation choice (random choice of the destination): As we can observe in Fig. 2, in this case the transition to a segregated state takes place at $\varepsilon_c = 0.6$. Therefore, the assumption of reinforced learning does not change the basic results of the Schelling model where the relocation choice is driven by the rational choice of increasing the utility. On the contrary, the metapopulation framework presents a higher complexity than the $L = 1$ case, whose outcome is the impossibility to have a direct comparison with this basic case.

In the metapopulation case, two different segregation levels should be considered. We call *cell-level segregation* once all the agents inside a cell are of the same kind (see Fig. 1). This kind of segregation will be analyzed in Sec. 4.1. In the $L = 1$ case, the cells can have only three states: occupied by an agent of kind A, occupied by an agent of kind B and to be empty. In the metapopulation framework, we can observe six different states (see Fig. 1, right panel): cell-level segregation of kind A (or B), empty cells, cells with a majority of kind A (or B) and cells with an equal number of A and B. We now define a spatial cluster as a group of non-empty contiguous cells each one containing a majority of agents of the same kind, namely more than half of the agents present in the cell are of the same kind. We can define *spatial segregation* as the situation where the spatial clusters become macroscopic. Spatial segregation is thus

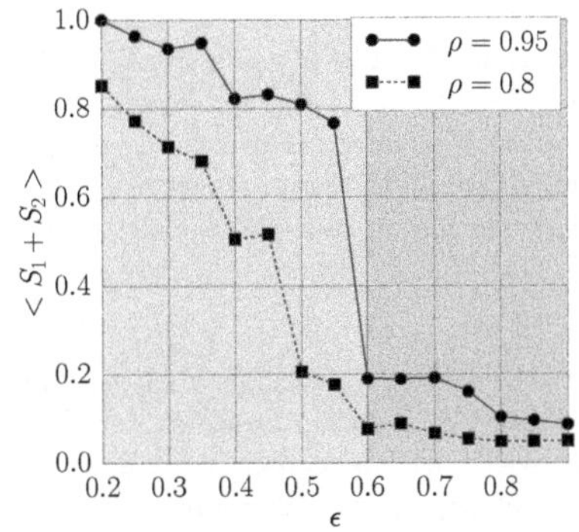

Fig. 2. Average relative size of the first cluster and of the second cluster as a function of the tolerance threshold ε for two values of the emptiness in the case $L = 1$ once the system settled into the equilibrium state. The result is the average of 50 simulations of a system evolving on a 2D lattice with periodic boundary conditions made by 50×50 cells.

the direct analogue of the segregation in the classical $L = 1$ Schelling model. This feature will be studied in Sec. 4.4.

Finally, we should notice that, differently from the $L = 1$ case, where the number of empty cells is fixed by the system emptiness, ρ, in the metapopulation framework the number of empty cells is determined by the dynamics of the system: according to the tolerance value, agents can prefer either to equally distribute across the available space (leaving almost none empty cells) or to occupy as much as possible some preferred cells leaving several empty locations. We will analyze this aspect in Sec. 4.3.

4. Results

In our simulation setting, the system is initialized with $T = 4000$ agents ($T_A = 2000$ Red and $T_B = 2000$ Blue agents) placed according to a random uniform distribution among the $N = 400$ cells of a (20×20) lattice with periodic boundary conditions. Once the carrying capacity is fixed to L, the emptiness of the system is therefore $\rho = 1 - T/(NL)$. With the chosen value of $T = 4000$, the constraint $\rho \in [0, 1]$ requires therefore to have $L > 10$. In most of the analyses, we will consider four values of $L = 100, 80, 50, 12$, corresponding respectively to $\rho = 0.9, 0.87, 0.8, 0.16$. We will study in particular the case $L = 100$ associated to a large enough ρ in order to allow agents to easily move in the lattice. At the initialization, we check that the initial conditions satisfy the local constraint $n_i^A + n_i^B \leq L$. Let us observe that this is always possible because of the chosen numbers of A and B agents and the emptiness, so we will draw configurations of agents as long as we find one satisfying the above constraint, then the dynamics of the system will respect such constraint for all time.

The following section containing our results is organized in four paragraphs. First, we will present the formation of cell-level segregation. Secondly, we will explore the dynamical evolution of the system. In the third part, we will explore the central feature of the metapopulation scenario, that is the formation of population heterogeneity. Finally, we will gather all the results together, in the classical framework of the Schelling model, showing the possible global spatial segregation scenarios.

4.1. *Cell-level segregation*

The aim of this section is to present the local properties of the system, namely at the level of single nodes. We start by showing the cell content heterogeneity (the relative imbalance of A and B agents within each cell) averaged over all the (non-empty) cells, hereby named the *cell segregation indicator*:

$$\langle\mu\rangle = \frac{1}{\sum_{i|(n_i^B+n_i^A)>0} 1} \sum_{i|(n_i^B+n_i^A)>0} \frac{|n_i^B - n_i^A|}{n_i^B + n_i^A}. \tag{4}$$

Small values of $\langle\mu\rangle$ mean that, on average, each cell is populated by the same number of agents of both kinds, while large values are associated to scenarios where most of the cells are filled with agents of only one kind (for the limit value $\langle\mu\rangle = 1$ all the cells contain a single kind of agent).

Notice that this indicator is specific to the metapopulation framework, indeed, independently from the system configuration parameters, for $L = 1$ this indicator would be identically $\langle\mu\rangle = 1$.

For $\varepsilon \leq 0.5$, we observe (see Fig. 3) that asymptotically $\langle\mu\rangle \to 1$, meaning that the system stabilizes into a frozen state where *cell-level segregation* is present in all

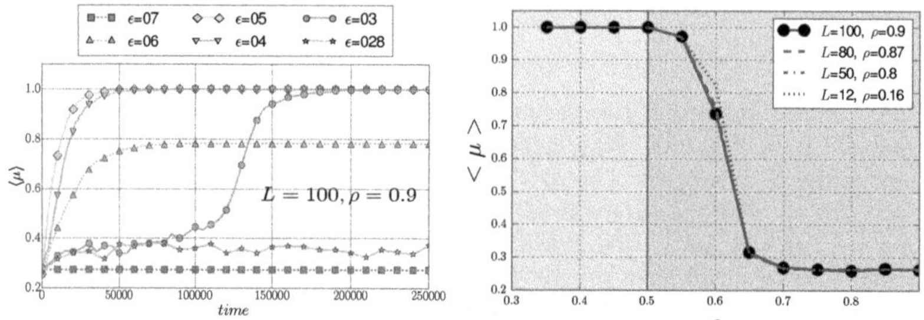

Fig. 3. (Color online) Local behavior. Left panel: cell-segregation indicator $\langle\mu\rangle = \frac{1}{N}\sum_i \frac{|n_A^i - n_B^i|}{n_A^i + n_B^i}$ as a function of time for a single generic simulations with $L = 100$. For $\varepsilon \leq 0.5$ the cell-segregation indicator goes to 1 indicating that cell-segregation is present inside each cell. Right panel: Asymptotic average cell-segregation indicator as a function of ε and for different values of L. Let us observe that $\langle\mu\rangle$ is almost independent from L being all the curves very close, moreover for $\varepsilon \leq 0.5$, the asymptotic average cell-segregation indicator is constantly equal to 1, independently from the value of L. The averages shown in the right panel have been obtained performing 50 replicas of the model.

the cells: each cell contains only one kind of agents. Notice moreover (right panel of Fig. 3) that the transition to a system where in all the cells cell-segregation is observed ($\langle \mu \rangle = 1$), always happens at $\varepsilon = 0.5$ independently from the maximal capacity of the cell (and consequently from the emptiness of the system).

Main findings:

- At $\varepsilon = 0.5$, we observe the transition to a cell-segregation scenario (single color in each cell, for all the cells).
- The transition is the same independently from the maximal cell capacity.

4.2. *Convergence time*

Another interesting feature of the system concerns the time needed to reach the steady state. As we can observe from the red circles and the orange stars curves in the left plot of Fig. 3, for low values of the tolerance ($\varepsilon \to 0$), the system exhibits a transient phase where it remains stuck for very long time into a *quasi-stationary non-segregated state* before suddenly jump to the final steady state with cell-segregation.

The convergence time (see Fig. 4) does not have a monotonic behavior as a function of ε. Passing from $\varepsilon \sim 1$ to $\varepsilon \sim 0.6$, we observe an exponential growth of the convergence time (note the logarithmic vertical scale). Then for $0.3 \leq \varepsilon < 0.6$, the convergence time exhibits an almost stable value — horizontal plateau — with a local minimum at $\varepsilon = 0.5$. Finally, for $\varepsilon \leq 0.3$, where the dynamics exhibits quasi-stationary states, the convergence times starts to increase faster than exponentially. Notice that the convergence time shows a large variance for small tolerances ε (see inset in Fig. 4). This is due to the fact that the quasi-stationary state is an unstable equilibrium of the system from which the system can get out through the formation of stochastic fluctuations that, as we will see in the following paragraph, produce the emergence of cells with an extremely large population.

Main findings:

- For $\varepsilon < 0.3$, the system dynamics passes through quasi-stationary states and the convergence time increases faster than exponentially.

4.3. *Population heterogeneity*

In the regime, where cell-segregation is experienced ($\varepsilon \leq 0.5$), a second fundamental self-organized phenomenon emerges, more and more pronounced for lower values of ε. At the initialization, the $T = 4000$ agents are randomly distributed on the $N = 400$ cells of the lattice, therefore each cell contains on average $\langle n_{\text{ini}} \rangle \sim T/N = 10$ agents. Exploiting the system dynamics, this initially homogeneously distributed population self-organizes into an heterogeneous state across the cells.

As we can observe in the right panel of Fig. 5, the maximal cell occupancy in the system ($n_{\max} = \max_i(n_i^A + n_i^B)$), increases for $\varepsilon \to 0$. Notice that the maximal cell capacity ($L = 100$) is never reached.

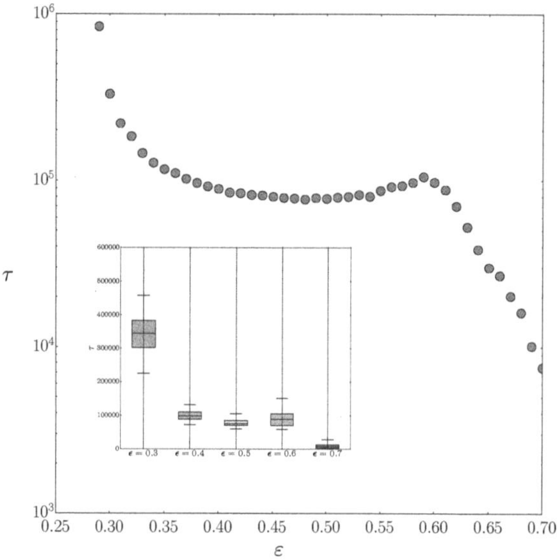

Fig. 4. (Color online) Convergence time as a function of ε. The blue line represents the average convergence time as a function of ε for 100 replicas of the system with same initial conditions and parameters values. The inset shows the box-plots for the convergence times for some representative values of $\varepsilon = 0.3, 0.4, 0.5, 0.6, 0.7$. The remaining model parameters have been set to $L = 100$, $T = 4000$ and $N = 400$.

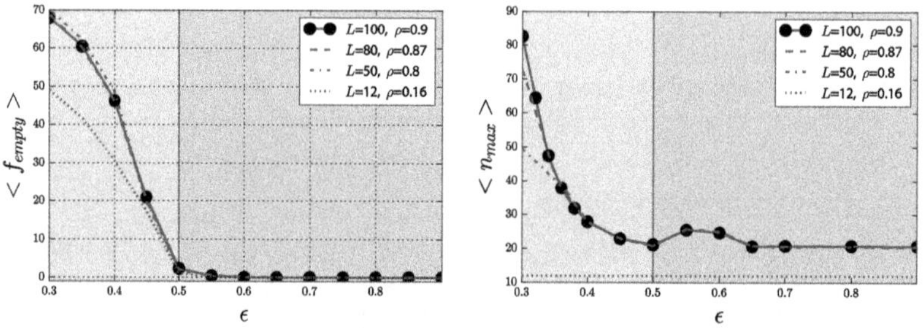

Fig. 5. Left panel: average fraction of empty cells f_{empty} as a function of ε for different values of L. The results slightly depend on L, only for small enough values the average fraction of empty cells shows lower values. Right panel: average maximal cell population $n_{\max} = \max_i (n_i^A + n_i^B)$ as a function of ε for different values of L. As $\varepsilon \to 0$ agents of the same kind tend to select and fill all the cells compatible with respect to the utility, creating thus an important fraction of empty cells. Also in this case, the results slightly depend on L. The averages shown have been obtained performing 100 replicas of the model.

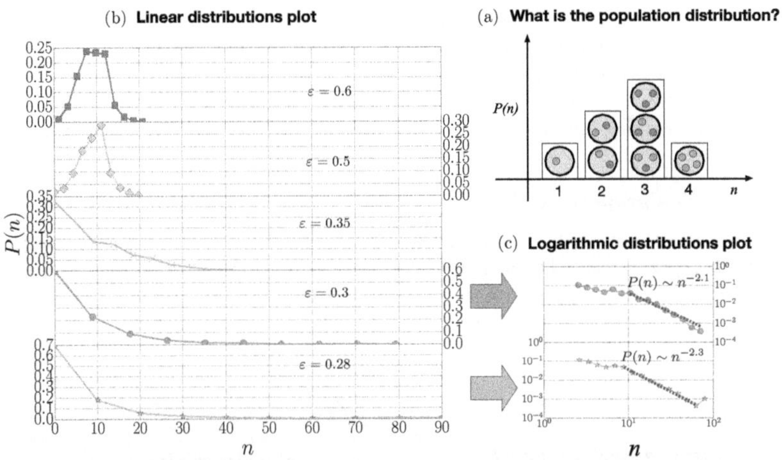

Fig. 6. Plot (a): Simplified visualization to explain the cells population distribution, with reference to Fig. 1. Plot (b): Distribution of cells population ($n = n_A + n_B$) for few values of $\varepsilon = 0.28, 0.3, 0.35, 0.5, 0.7$ using linear axes. Plot (c): Distribution of cells population ($n = n_A + n_B$) for $\varepsilon = 0.28, 0.3$ using log–log axes to emphasise the broad distribution. The plots B and C have been obtained performing 100 replicas of the model.

At the same time, for low tolerances we observe the formation of a relevant fraction of completely empty cells, i.e., cells for which $n_i^A + n_i^B = 0$ (left panel of Fig. 5).

These basic hints derived from Fig. 5 can be better understood studying the statistical distribution of the cells population $P(n)$ (see Fig. 6(a)). For this analysis, we focus on a system with an large emptiness, in order to leave the agents an high potential to move in the space.

In Fig. 6(b), we show the asymptotic cells population distribution for different values of the tolerance. For high values of the tolerance, the cells-occupancy do not change from the initialization with small fluctuation around the average value $\langle n \rangle = 10$. For low values of the tolerance, the variance of the distribution significantly increases. We can argue that the shape of the distribution is correlated with the time, the system spends in the quasi-stationary non-segregated state: the longer this time, the more the population distribution across cells moves from a Poissonian distribution (Fig. 6(b)) to a power-law shape as $\varepsilon \to 0$ (see Fig. 6(c)).

Main findings:

- For $\varepsilon \to 0$, stronger and stronger heterogeneous population distributions between cells emerges.
- For $\varepsilon \to 0$, a larger and larger number of empty cells is observed.

The outcomes of this model (the emergence of empty areas, of low populated zones and of extremely highly populated buildings, starting from an uniform distribution) can ideally mimic the formation of an urban area, characterized by heterogeneous densities (residential zones with few inhabitants, highly dense zones with high building and finally empty areas, like parks, working districts, etc.).

4.4. *Global properties*

To analyze the spatial structures, beyond the single cell description, we use the notion of *spatial cluster* based on cell's majority. As previously stated two non-empty neighboring cells belong to the same cluster ($\mathcal{C}$) if they both contain a majority of agents of the same kind (see Fig. 7(a)):

$$i, j \in \mathcal{C} \quad \text{if } n_i \neq 0, n_j \neq 0 \quad \text{and} \quad (n_i^A - n_i^B)(n_j^A - n_j^B) > 0. \tag{5}$$

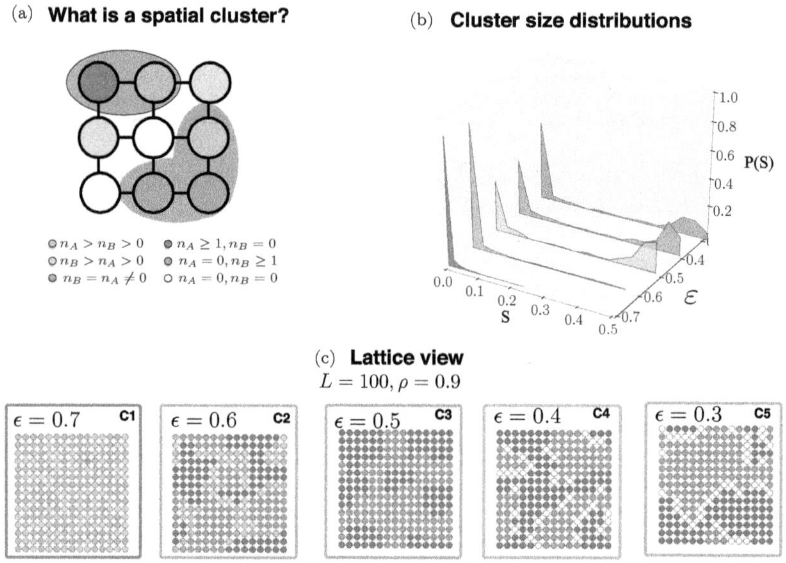

Fig. 7. (Color online) Global behavior. Panels (a): Simplified visualization to explain the cells states (white for empty cells, blue for only blue agents ($n_A \geq 1$ and $n_B = 0$), red for only Red agents ($n_A = 0$ and $n_B \geq 1$), gray for red and blue in equal number and positive, cyan for cells with blue majority and orange for red majority) and the clusters (brown areas), with reference to Fig. 1. Panel (b): Relative cluster size distributions $P(S)$, being S the relative size of the cluster, for several values of $\varepsilon = 0.3, 0.4, 0.5, 0.6, 0.7$. Results presented in panels are based on 100 replicas of the simulations. Panels (c(1-5)): single replica representation of cells occupancy at equilibrium (for $\varepsilon = 0.7, 0.6, 0.5, 0.4, 0.3$), each cell has been colored according to the definitions in panel (a).

Similarly to the $L = 1$ case we define $S_i = \#C_i/(\sum_{j=1}^{n_{\text{clust}}} \#C_j)$ to be relative size of the cluster C_i, being the size of the cluster the number of cells forming it, with respect the total number of occupied cells. Notice however that for $L > 1, \sum_{j=1}^{n_{\text{clust}}} \#C_j \neq T$. The relative largest cluster size is defined as S_1 and the second one as S_2.

On the other hand, an edge between two neighboring cells i and j is considered *interface*, Ξ, if:

$$(i,j) \in \Xi \quad \text{if } n_{i,j} \neq 0 \quad \text{and} \quad (n_i^A - n_i^B)(n_j^A - n_j^B) \leq 0. \tag{6}$$

that is the interface edge connects two cells each one populated with a different majority or cells with a definite majority and empty or parity cells.

In Fig. 7(c) (see also the movies at [25]), we show a prototypical evolution of the aggregated cell scenario, for the $L = 100$ case. As we can observe, for high values of the tolerance $\varepsilon = 0.7$ well defined patterns (large clusters) are not formed, instead there are cells containing a mixture of agents of both kinds, while they start to appear at $\varepsilon = 0.6$. At $\varepsilon = 0.5$, we have both spatial and cell segregation (only red and blue cells are visible and macroscopic cluster structures did emerge). As $\varepsilon \to 0$, the separation of the spatial clusters is more marked and several empty cells appear, that is the interface is growing to increase the spatial distance between the agents' communities.

In Fig. 7(b), we display the size distribution of the clusters, in the $L = 100$ case, for different values of the tolerance. This plot confirms the intuitive idea that macroscopic spatial structures start to emerge at $\varepsilon = 0.6$. To give a more quantitative argument, we use the fraction of the population that resides in the first two largest clusters (Eq. (3)), $S_1 + S_2$, as indicator of the emergence of such regime. We observe that, independently from the maximal cell capacity, for $\varepsilon \leq 0.6$ macroscopic percolating clusters are always formed (left panel of Fig. 8). Notice that this critical threshold is the same observed in the original Schelling model for $L = 1$ (Fig. 2). In this range of tolerance parameter, the average sizes of the first and second largest cluster added together cover more than 75% of the available cells, in this case we define "ghettos" such large percolating clusters. Notice that for $\varepsilon = 0.5$, the first and second clusters cover almost completely the lattice, containing more than 95% of cells.

Summarizing together the results for cell-level segregation and spatial *segregation*, we can define the following scenarios:

- for $\varepsilon > 0.6$ nor cell-level segregation nor spatial segregation is observed.
- for $0.5 < \varepsilon \leq 0.6$, the two largest clusters contain a majority of agents of the same kind, but different agents can coexist in the same cell. We call this scenario *soft segregation* because it allows a small mixing in the population.
- For $\varepsilon \leq 0.5$ each spatial cluster contains only one kind of agent, resulting in a *hard segregation* of the population. For $\varepsilon \to 0$ clusters of empty cells are created increasing the distance between the two monochromatic structures by interposing interfaces.

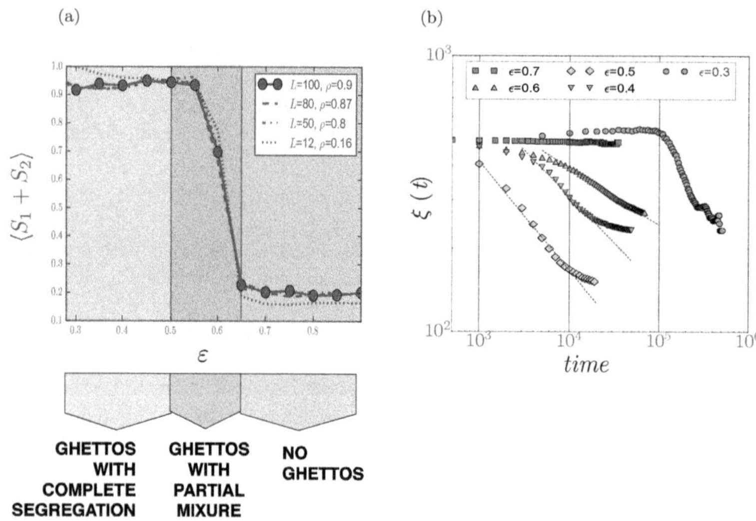

Fig. 8. Largest clusters and interfaces. Panel (a): First and second largest cluster as a function of ε for different values of L. Panel (b): Interface size as a function of time for several values of $\varepsilon = 0.3, 0.4, 0.5, 0.6, 0.7$. Results presented in both the panels are based on 100 replicas of the simulations.

Finally, to characterize the dynamics of the pattern growth in the different regimes, we also study the time evolution of the interfaces for the $L = 100$ case, $\xi(t) = \#\Xi(t)$, namely the number of cells forming the total interface. We can observe, (see Fig. 8, Panel (b)), another remarkable self-organized phenomenon: for $0.3 < \varepsilon \leq 0.6$ the size of the interface, monotonically decreases in time as $t^{-1/z}$ ($z = 4$ for $\varepsilon = 0.6$ and $z = 3$ for $\varepsilon \geq 0.5$). This is the typical signature of a coarsening phenomenon, that has already been observed in the classical Schelling model [9]. On the other side, for low values of the tolerance, where the system remains for long time in the quasi-stationary state (for instance $\varepsilon \leq 0.3$), the size of the interface has a slowly increasing phase (red curve Fig. 8(b)), corresponding to the formation of temporal segregated domains followed by an abrupt decrease when the system reaches the local segregation equilibrium. In this case, the mechanism of domain formation cannot be ascribed to the coarsening framework, signifying that the dynamics toward the stabilization of the system, in this case, is deeply different from the previous situations (described by a coarsening mechanism). For low values of the tolerance, the spatial segregation indeed is reached through a mechanism of aggregation of agents around few high dense cells (see Sec. 4.3). These few, highly populated, cells are indeed statistical instabilities that allow the system to exit the quasi-stationary state. Once the clusters are formed, the complete equilibrium is reached as the interface between Blue and Red zones becomes empty.

While the creation of heterogeneous distribution of agents among cells is due to the metapopulation mechanism, the formation of the monochromatic clusters was already present in the classical Schelling model ($L = 1$) and thus it is mainly due to the behavioral rules of the latter. We can therefore imagine the process at play in our model as the combined outcome of a local birth and death process (migration in-out) on the single cell and a mesoscale cell dynamics, namely several connected cells exhibiting a synchronised behavior as already observed in [9].

Summary of the global properties:

- For $0.5 < \varepsilon < 0.7$, we observe the *soft segregation scenario*: macroscopic segregated patterns are formed but each cell, on the other hand can still contain a mixture of blue and red agents.
- For $\varepsilon < 0.5$, we observe the *hard segregation scenario*: macroscopic segregated patterns are formed and each cell contains only one kind of agents.
- For $\varepsilon \to 0$ larger and larger empty interfaces between the clusters are observed.

5. Discussion

An extension of the classical Schelling model to a metapopulation framework has been analyzed in this paper. The main outcome of our model is the spontaneous emergence, for low values of the tolerance threshold, of heterogeneously populated cells without any exogenous preferential attachment mechanism. This behavior is a consequence of the permanence of the system for long time in a quasi-stationary non-segregated state, where in each cell the two populations are equally distributed. The system reaches the stabilization toward the locally segregated state through the creation (by random agents moves) of few highly populated cells.

At the same time, global spatial patterns emerge as in the classical Schelling model. For $\varepsilon < 0.7$, we observe the emergence of ghettos structures (namely the first and the second clusters cover more than the 75% of the cells). For $0.5 < \varepsilon < 0.7$, the global clusters are described as neighborhoods formed by a strong majority of individuals of the same color (soft segregation). For $\varepsilon \leq 0.5$, we have the formation of single-colored ghettos (hard segregation).

The mechanism of formation of the clusters is a typical coarsening phenomenon for high values of ε. On the contrary, for low tolerance cases, the clusters are formed around the towers that become stable points for a certain type of cells (once an agent, whose kind corresponds to the majority already inside the tower, enters she never gets out). Once a higher density zone starts to exist, this mechanism reinforces the (majority color) population growth in this cell and in the neighborhood. The global patterns start therefore to stabilize around the towers (see moves in the SI).

Acknowledgments

The work of F. G., Y. G. and T. C. presents research results of the Belgian Network DYSCO (Dynamical Systems, Control, and Optimization), funded by the Interuniversity

Attraction Poles Programme, initiated by the Belgian State, Science Policy Office. T. C. is grateful to Cyril Vargas, student at the ENSEEIHT — INP Toulouse (France), involved in a preliminary study of this subject during his Master Degree.

References

[1] Cutler, D. and Glaeser, E., *Quart. J. Econ.* **112** (1997) 827–872.
[2] Fagiolo, G., Valente, M. and Vriend, N. J., *Econ. Behav. Org.* **64** (2007) 316–336.
[3] Frey, W. and Farley, R., *Demography* **33** (1996) 35–50.
[4] Conradt, L., Krause, J., Couzin, I. D. and Roper, T. J., *Am. Nat.* **173** (2009) 305–312.
[5] Schelling, T. C., *Am. Econ. Rev.* **59** (1969) 488–493.
[6] Schelling, T. C., *J. Math. Soc.* **1** (1971) 143–186.
[7] Pancs, R. and Vriend, N., *J. Publ. Econ.* **91** (2007) 1–24.
[8] Vinković, D. and Kirman, A., *Proc. Natl. Acad. Sci. USA* **103**(51) (2006) 19261–19265.
[9] Dall'Asta, L., Castellano, C. and Marsili, M., *J. Stat. Mech.* **7**(L07002) (2008) 1–10.
[10] Gandica, Y., Gargiulo, F. and Carletti, T., *Chaos, Solitons Fractals* **90** (2016) 46–54.
[11] Schulze, C., *Int. J. Mod. Phys. C* **16** (2005) 351.
[12] Castellano, C., Fortunato, S. and Loreto, V., *Rev. Mod. Phys.* **81**(2) (2009) 591.
[13] Stauffer, D. and Solomon, S., *Eur. Phys. J. B* **57** (2007) 473–479.
[14] Grauwin, S., Bertin, E., Lemoy, R. and Jensen, P., *Proc. Natl. Acad. Sci. USA* **106**(49) (2009) 20622–20626.
[15] Rogers, T. and McKane, A., *Phys. Rev. E* **85**(1–5) (2012) 041136.
[16] Rogers, T. and McKane, A., *J. Stat. Mech.* **7**(P07006) (2011) 1–17.
[17] Gauvin, L., Vannimenus, J. and Nadal, J.-P., *Eur. Phys. J. B* **70**(2) (2009) 293–304.
[18] May, R., *Nature* **238** (1972) 413–414.
[19] Hanski, I., *Metapopulation Ecology* (Oxford University Press, 1999).
[20] Colizza, V., Pastor-Satorras, R. and Vespignani, A., *Nat. Phys.* **3**(4) (2007) 276–282.
[21] Colizza, V. and Vespignani, A., *J. Theor. Biol.* **251**(3) (2008) 450–467.
[22] Colizza, V., Gargiulo, F., Ramasco, J. J., Barrat, A. and Vespignani, A., in *BIOMAT 2008*, Chap. 5 (World Scientific, 2009), pp. 91–113.
[23] Durrett, R. and Zhang, Y., *Proc. Natl. Acad. Sci. USA* **111**(39) (2014) 14036–14041.
[24] Macy, M. W. and Flache, A., *Proc. Natl. Acad. Sci. USA* **99**(3) (2002) 7229–7236.
[25] https://directory.unamur.be/research/publications/ff110fe7-d4d5-42ba-83c9-b04cd13309b5/overview

Advances in Complex Systems
Vol. 21, Nos. 3 & 4 (2018) 1850014 (31 pages)

DOI: 10.1142/S0219525918500145

World Scientific
www.worldscientific.com

TALENT VERSUS LUCK: THE ROLE OF RANDOMNESS IN SUCCESS AND FAILURE

ALESSANDRO PLUCHINO
Department of Physics and Astronomy
University of Catania
Via S.Sofia 64, Catania 95123, Italy
INFN-CT, Via S. Sofia 64, Catania 95123, Italy
alessandro.pluchino@ct.infn.it

ALESSIO EMANUELE BIONDO
Department of Economics and Business
University of Catania
Corso Italia 55, Catania 95129, Italy
ae.biondo@unict.it

ANDREA RAPISARDA*
Department of Physics and Astronomy
University of Catania
Via S.Sofia 64, Catania 95123, Italy
INFN-CT, Via S. Sofia 64, Catania 95123, Italy
Complexity Science Hub Vienna, Austria
andrea.rapisarda@ct.infn.it

Received 28 May 2018
Revised 9 July 2018
Accepted 10 July 2018
Published 27 July 2018

The largely dominant meritocratic paradigm of highly competitive Western cultures is rooted on the belief that success is mainly due, if not exclusively, to personal qualities such as talent, intelligence, skills, smartness, efforts, willfulness, hard work or risk taking. Sometimes, we are willing to admit that a certain degree of luck could also play a role in achieving significant success. But, as a matter of fact, it is rather common to underestimate the importance of external forces in individual successful stories. It is very well known that intelligence (or, more in general, *talent* and personal qualities) exhibits a Gaussian distribution among the population, whereas the distribution of wealth — often considered as a proxy of success — follows typically a power law (Pareto law), with a large majority of poor people and a very small number of billionaires. Such a discrepancy between a Normal distribution of inputs, with a typical scale

*Corresponding author.

(the average talent or intelligence), and the scale-invariant distribution of outputs, suggests that some hidden ingredient is at work behind the scenes. In this paper, we suggest that such an ingredient is just randomness. In particular, our simple agent-based model shows that, if it is true that some degree of talent is necessary to be successful in life, almost never the most talented people reach the highest peaks of success, being overtaken by averagely talented but sensibly luckier individuals. As far as we know, this counterintuitive result — although implicitly suggested between the lines in a vast literature — is quantified here for the first time. It sheds new light on the effectiveness of assessing merit on the basis of the reached level of success and underlines the risks of distributing excessive honors or resources to people who, at the end of the day, could have been simply luckier than others. We also compare several policy hypotheses to show the most efficient strategies for public funding of research, aiming to improve meritocracy, diversity of ideas and innovation.

Keywords: Success; talent; luck; agent-based models; serendipity.

1. Introduction

The ubiquity of power-law distributions in many physical, biological or socio-economical complex systems can be seen as a sort of mathematical signature of their strongly correlated dynamic behavior and their scale-invariant topological structure [1–4]. In socio-economic context, after Pareto's work [5–9], it is well known that the wealth distribution follows a power-law, whose typical long-tailed shape reflects the deep existing gap between the rich and the poor in our society. A very recent report [10] shows that today this gap is far greater than it had been feared: eight men own the same wealth as the 3.6 billion people constituting the poorest half of humanity. In the last 20 years, several theoretical models have been developed to derive the wealth distribution in the context of statistical physics and probability theory, often adopting a multi-agent perspective with a simple underlying dynamics [11–16].

Moving along this line, if one considers the individual wealth as a proxy of success, one could argue that its deeply asymmetric and unequal distribution among people is either a consequence of their natural differences in talent, skill, competence, intelligence, ability or a measure of their willfulness, hard work or determination. Such an assumption is, indirectly, at the basis of the so-called *meritocratic paradigm*: it affects not only the way our society grants work opportunities, fame and honors, but also the strategies adopted by Governments in assigning resources and funds to those who are considered as the most deserving individuals.

However, the previous conclusion appears to be in strict contrast with the accepted evidence that human features and qualities cited above are normally distributed among the population, i.e., follow a symmetric Gaussian distribution around a given mean. For example, intelligence, as measured by IQ tests, follows this pattern: average IQ is 100, but nobody has an IQ of 1000 or 10,000. The same holds for efforts, as measured by hours worked: someone works more hours than the average and someone less, but nobody works a billion times more hours than anybody else.

On the other hand, there is nowadays an ever greater evidence about the fundamental role of chance, luck or, more in general, random factors, in determining successes or failures in our personal and professional lives. In particular, it has been shown that scientists have the same chance along their career of publishing their biggest hit [17]; that those with earlier surname initials are significantly more likely to receive tenure at top departments [18]; that the distributions of bibliometric indicators collected by a scholar might be the result of chance and noise related to multiplicative phenomena connected to a *publish or perish* inflationary mechanism [19]; that one's position in an alphabetically sorted list may be important in determining access to over-subscribed public services [20]; that middle name initials enhance evaluations of intellectual performance [21]; that people with easy-to-pronounce names are judged more positively than those with difficult-to-pronounce names [22]; that individuals with noble-sounding surnames are found to work more often as managers than as employees [23]; that females with masculine monikers are more successful in legal careers [24]; that roughly half of the variance in incomes across persons worldwide is explained only by their country of residence and by the income distribution within that country [25]; that the probability of becoming a CEO is strongly influenced by your name or by your month of birth [26–28]; that the innovative ideas are the results of a random walk in our brain network [29]; and that even the probability of developing a cancer, maybe cutting a brilliant career, is mainly due to simple bad luck [30, 31]. Recent studies on lifetime reproductive success further corroborate these statements showing that, if trait variation may influence the fate of populations, luck often governs the lives of individuals [32, 33].

In recent years many authors, among whom the statistician and risk analyst Taleb [34, 35], the investment strategist Mauboussin [36] and the economist Frank [37], have explored in several successful books the relationship between luck and skill in financial trading, business, sports, art, music, literature, science and in many other fields. They reach the conclusion that chance events play a much larger role in life than many people once imagined. Actually, they do not suggest that success is independent of talent and efforts, since in highly competitive arenas or "winner-takes-all" markets, like those where we live and work today, people performing well are almost always extremely talented and hard-working. Simply, they conclude that talent and efforts are not enough: you have to be also in the right place at the right time. In short: luck also matters, even if its role is almost always underestimated by successful people. This happens because randomness often plays out in subtle ways, therefore it is easy to construct narratives that portray success as having been inevitable. Taleb calls this tendency "narrative fallacy" [35], while the sociologist Lazarsfeld adopts the terminology "hindsight bias". In his recent book "Everything Is Obvious: Once You Know the Answer" [38], the sociologist and network science pioneer Watts suggests that both narrative fallacy and hindsight bias operate with particular force when people observe unusually successful outcomes and consider them as the necessary product of hard work and talent, while they mainly emerge from a complex and interwoven sequence of steps, each depending on precedent ones: if any of them had been different, an entire career or life trajectory would almost surely differ too. This argument is also based on the results of a seminal experimental study, performed some years before by Watts himself in collaboration with other authors [39], where the success of previously unknown songs in an artificial music market was shown not to be correlated with the quality of the song itself. And this clearly makes any kind of prediction very difficult, as also shown in another more recent study [40].

In this paper, by adopting an agent-based statistical approach, we try to realistically quantify the role of luck and talent in successful careers. In Sec. 2, building on a minimal number of assumptions, i.e., a Gaussian distribution of talent [41] and a multiplicative dynamics for both successes and failures [42], we present a simple model, that we call "Talent versus Luck" (TvL) model, which mimics the evolution of careers of a group of people over a working period of 40 years. The model shows that, actually, randomness plays a fundamental role in selecting the most successful individuals. It is true that, as one could expect, talented people are more likely to become rich, famous or important during their life with respect to poorly equipped ones. But — and this is a less intuitive rationale — ordinary people with an average level of talent are statistically destined to be successful (i.e. to be placed along the tail of some power law distribution of success) much more than the most talented ones, provided that they are more blessed by fortune along their life. This fact is commonly experienced, as pointed in [34, 35, 37], but, to our knowledge, it is modeled and quantified here for the first time.

The success of the averagely talented people strongly challenges the "meritocratic" paradigm and all those strategies and mechanisms, which give more rewards, opportunities, honors, fame and resources to people considered the best in their field [43, 44]. The point is that, in the vast majority of cases, all evaluations of someone's talent are carried out *a posteriori*, just by looking at his/her performances — or at reached results — in some specific area of our society like sport, business, finance, art, science, etc. This kind of misleading evaluation ends up switching cause and effect, rating as the most talented people those who are, simply, the luckiest ones [45, 46]. In line with this perspective, in previous works, it was advanced a warning against such a kind of "naive meritocracy" and it was shown the effectiveness of alternative strategies based on random choices in many different contexts, such as management, politics and finance [47–54]. In Sec. 3, we provide an application of our approach and sketch a comparison of possible public funds attribution schemes in the scientific research context. We study the effects of several distributive strategies, among which the "naively" meritocratic one, with the aim of exploring new ways to increase both the minimum level of success of the most talented people in a community and the resulting efficiency of the public expenditure. We also explore, in general, how opportunities offered by the environment, as the education and income levels (i.e., external factors depending on the country and the social context where individuals come from), do matter in increasing probability of success. Final conclusive remarks close the paper.

2. The Model

In what follows, we propose an agent-based model, called TvL model, which builds on a small set of very simple assumptions, aiming to describe the evolution of careers of a group of people influenced by lucky or unlucky random events.

We consider N individuals, with talent T_i (intelligence, skills, ability, etc.) normally distributed in the interval $[0, 1]$ around a given mean m_T with a standard deviation σ_T, randomly placed in fixed positions within a square world (see Fig. 1) with periodic boundary conditions (i.e., with a toroidal topology) and surrounded by a certain number N_E of "moving" events (indicated by dots), someone lucky, someone else unlucky (neutral events are not considered in the model, since they have not relevant effects on the individual life). In Fig. 1, we report these events as colored points: lucky ones, in green and with relative percentage p_L, and unlucky ones, in red and with percentage $(100 - p_L)$. The total number of event-points N_E are uniformly distributed, but of course, such a distribution would be perfectly uniform only for $N_E \to \infty$. In our simulations, typically will be $N_E \sim N/2$: thus, at the beginning of each simulation, there will be a greater random concentration of lucky or unlucky event-points in different areas of the world, while other areas will be more neutral. The further random movement of the points inside the square lattice, the world, does not change this fundamental features of the model, which exposes different

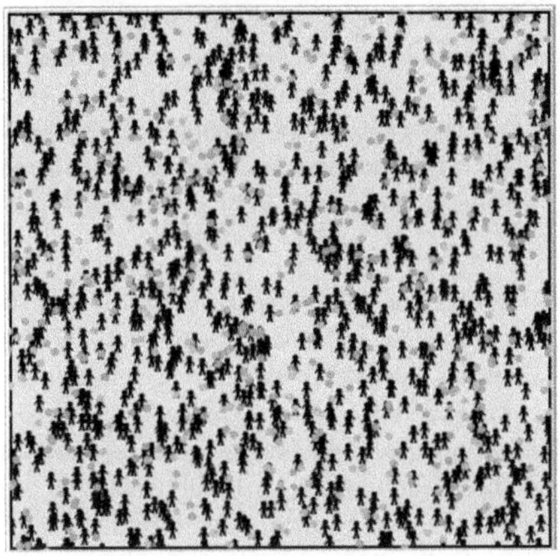

Fig. 1. (Color online) An example of initial setup for our simulations. All the simulations presented in this paper were realized within the NetLogo agent-based model environment [55]. $N = 1000$ individuals (agents), with different degrees of talent (intelligence, skills, etc.), are randomly located in their fixed positions within a square world of 201×201 patches with periodic boundary conditions. During each simulation, which covers several dozens of years, they are exposed to a certain number N_E of lucky (green circles) and unlucky (red circles) events, which move across the world following random trajectories (random walks). In this example, $N_E = 500$.

individuals to different amount of lucky or unlucky events during their life, regardless of their own talent.

For a single simulation run, a working life period P of 40 years (from the age of 20 to the age of 60) is considered, with a time step δ_t equal to six months. At the beginning of the simulation, all agents are endowed with the same amount $C_i = C(0)\ \forall i = 1, \ldots, N$ of capital, representing their starting level of success/wealth. This choice has the evident purpose of not offering any initial advantage to anyone. While the agents' talent is time-independent, agents' capital changes in time. During the time evolution of the model, i.e., during the considered agents' life period, all event-points move randomly around the world and, in doing so, they possibly intersect the position of some agent. More in detail, at each time, each event-point covers a distance of two patches in a random direction. We say that an intersection does occur for an individual when an event-point is present inside a circle of radius 1 patch centered on the agent (the event-point does not disappear after the intersection). Depending on such an occurrence, at a given time step t (i.e., every six months), there are three different possible actions for a given agent A_k:

(1) No event-point intercepts the position of agent A_k: this means that no relevant facts have happened during the last six months; agent A_k does not perform any action.
(2) A lucky event intercepts the position of agent A_k: this means that a lucky event has occurred during the last six months (note that, in line with [29], also the production of an innovative idea is here considered as a lucky event occurring in the agent's brain); as a consequence, agent A_k doubles her capital/success with a probability proportional to her talent T_k. It will be $C_k(t) = 2C_k(t-1)$ only if rand$[0,1] < T_k$, i.e., if the agent is smart enough to profit from his/her luck.
(3) An unlucky event intercepts the position of agent A_k: this means that an unlucky event has occurred during the last six month; as a consequence, agent A_k halves her capital/success, i.e., $C_k(t) = C_k(t-1)/2$.

The previous agents' rules (including the choice of dividing the initial capital by a factor 2, in case of unlucky events, and the choice of doubling it, in case of lucky ones, proportionally to the agent's talent) are intentionally simple and can be considered widely shareable, since they are based on the common sense evidence that success, in everyone life, has the property to both grow or decrease very rapidly. Furthermore, these rules give a significant advantage to highly talented people, since they can make much better use of the opportunities offered by luck (including the ability to exploit a good idea born in their brains). On the other hand, a car accident or a sudden disease, for example, is always an unlucky event where the talent plays no role. In this respect, we could more effectively generalize the definition of "talent" by identifying it with "any personal quality which enhances the chance to grab an opportunity". In other words, by the term "talent", we broadly mean intelligence, skill, smartness, stubbornness, determination, hard work, risk taking and so on.

What we will see in the following is that the advantage of having a great talent is a *necessary, but not a sufficient*, condition to reach a very high degree of success.

2.1. *Single run results*

In this section, we present the results of a typical single run simulation. Actually, such results are very robust so, as we will show later, they can be considered largely representative of the general framework emerging from our model.

Let us consider $N = 1000$ agents, with a starting equal amount of capital $C(0) = 10$ (in dimensionless units) and with a fixed talent $T_i \in [0, 1]$, which follows a normal distribution with mean $m_T = 0.6$ and standard deviation $\sigma_T = 0.1$ (see Fig. 2). As previously written, the simulation spans a realistic time period of $P = 40$ years, evolving through time steps of six months each, for a total of $I = 80$ iterations. In this simulation, we consider $N_E = 500$ event-points, with a percentage $p_L = 50\%$ of lucky events.

At the end of the simulation, as shown in panel (a) of Fig. 3, we find that the simple dynamical rules of the model are able to produce an unequal distribution of capital/success, with a large amount of poor (unsuccessful) agents and a small number of very rich (successful) ones. Plotting the same distribution in log–log scale in panel (b) of the same figure, a Pareto-like power-law distribution is observed, whose tail is well fitted by the function $y(C) \sim C^{-1.27}$. Therefore, despite the normal distribution of talent, the TvL model seems to be able to capture the first important feature observed in the comparison with real data: the deep existing gap between rich and poor and its scale-invariant nature. In particular, in our simulation, only

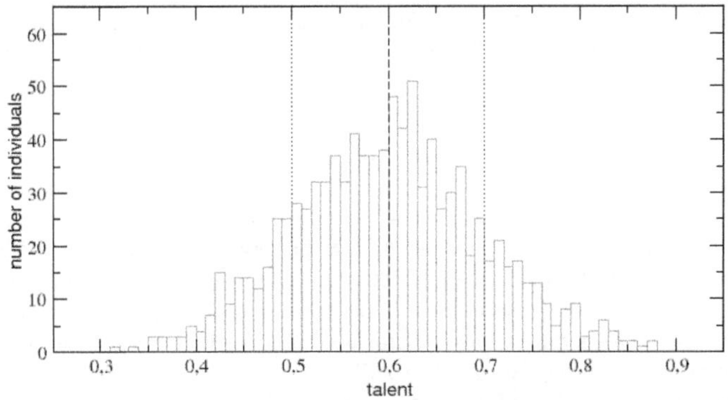

Fig. 2. Normal distribution of talent among the the population (with mean $m_T = 0.6$, indicated by a dashed vertical line, and standard deviation $\sigma_T = 0.1$ — the values $m_T \pm \sigma_T$ are indicated by two dotted vertical lines). This distribution is truncated in the interval $[0, 1]$ and does not change during the simulation.

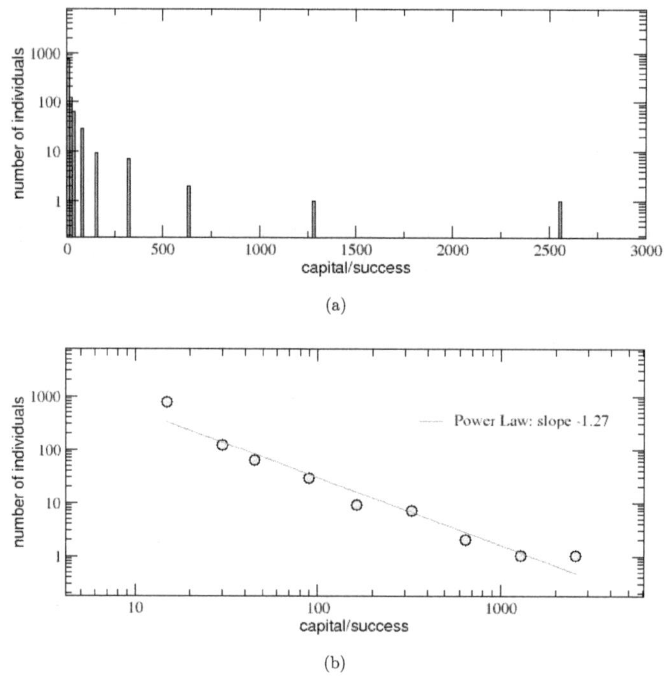

Fig. 3. Final distribution of capital/success among the population, both in (a) log-lin and in (b) log–log scale. Despite the normal distribution of talent, the tail of distribution of success - as visible in panel (b) — can be well fitted with a power-law curve with slope −1.27. We also verified that the capital/success distribution follows the Pareto's "80–20" rule, since 20% of the population owns 80% of the total capital, while the remaining 80% owns the 20% of the capital.

4 individuals have more than 500 units of capital and the 20 most successful individuals hold the 44% of the total amount of capital, while almost half of the population stay under 10 units. Globally, the Pareto's "80–20" rule is respected, since the 80% of the population owns only the 20% of the total capital, while the remaining 20% owns the 80% of the same capital. Although this disparity surely seems unfair, it would be to some extent acceptable if the most successful people were the most talented one, so deserving to have accumulated more capital/success with respect to the others. But are things really like that?

In panels (a) and (b) of Fig. 4, respectively, talent is plotted as function of the final capital/success and vice versa (note that, in panel (a), the capital/success takes only discontinuous values: this is due to the choice of having used an integer initial capital equal for all the agents). Looking at both panels, it is evident that, on one

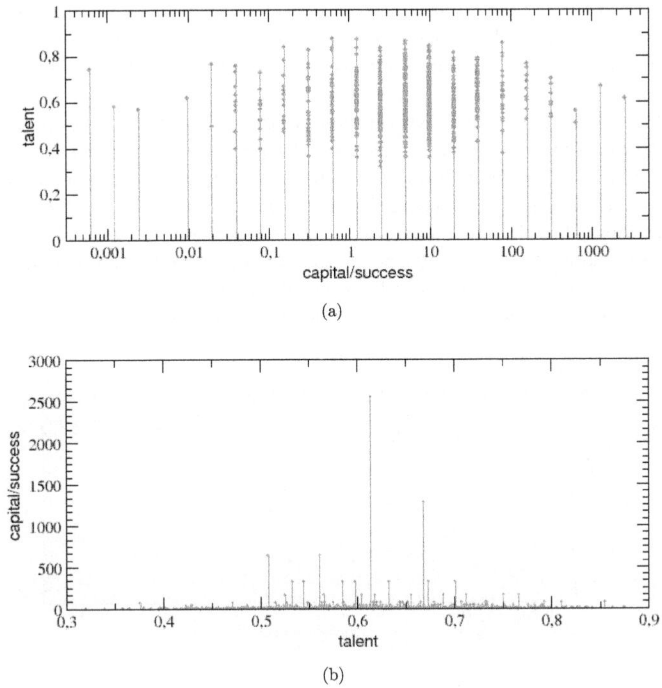

Fig. 4. In panel (a), talent is plotted as function of capital/success (in logarithmic scale for a better visualization): it is evident that the most successful individuals are not the most talented ones. In panel (b), vice versa, capital/success is plotted as function of talent: here, it can be further appreciated the fact that the most successful agent, with $C_{\text{max}} = 2560$, has a talent only slightly greater than the mean value $m_T = 0.6$, while the most talented one has a capital/success lower than $C = 1$ unit, much less of the initial capital $C(0)$. See text for further details.

hand, the most successful individuals are not the most talented ones and, on the other hand, the most talented individuals are not the most successful ones. In particular, the most successful individual, with $C_{\text{max}} = 2560$, has a talent $T^* = 0.61$, only slightly greater than the mean value $m_T = 0.6$, while the most talented one ($T_{\text{max}} = 0.89$) has a capital/success lower than 1 unit ($C = 0.625$).

As we will see more in detail in the next subsection, such a result is not a special case, but it is rather the rule for this kind of system: the maximum success never coincides with the maximum talent and vice versa. Moreover, such a misalignment between success and talent is disproportionate and highly nonlinear. In fact, the average capital of all people with talent $T > T^*$ is $C \sim 20$: in other words, the capital/success of the most successful individual, who is moderately gifted, is 128 times greater than the average capital/success of people who are more talented than him. We can conclude that, if there is not an exceptional talent behind the enormous success of some people, another factor is probably at work. Our simulation clearly shows that such a factor is just pure luck.

In Fig. 5, the number of lucky and unlucky events occurred to all people during their working lives is reported as a function of their final capital/success. Looking at panel (a), it is evident that the most successful individuals are also the luckiest ones (note that it in this panel are reported all the lucky events occurred to the agents and not just those that they took advantage of, proportionally to their talent). On the contrary, looking at panel (b), it results that the less successful individuals are also the unluckiest ones. In other words, although there is an absence of correlation between success and talent coming out of the simulations, there is also a very strong

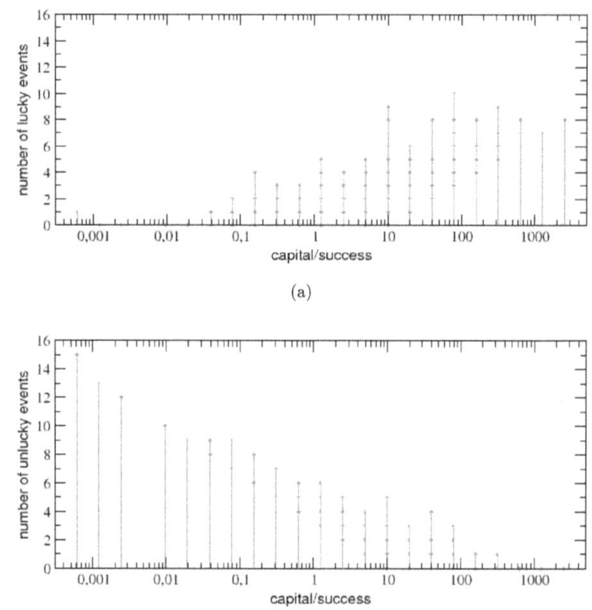

(b)

Fig. 5. Total number of (a) lucky events or (b) unlucky events as function of the capital/success of the agents. The plot shows the existence of a strong correlation between success and luck: the most successful individuals are also the luckiest ones, while the less successful are also the unluckiest ones. Again, having used an initial capital equal for all the agents, it follows that several events are grouped in discontinuous values of the capital/success. In panels (c) and (d), the frequency distributions of, respectively, the number of lucky and unlucky events are reported in log-linear scale. As visible, both the distributions can be well fitted by exponential distributions with similar negative exponents.

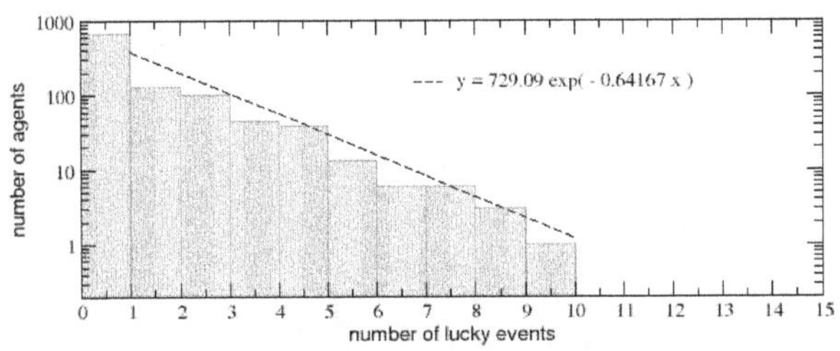

(c)

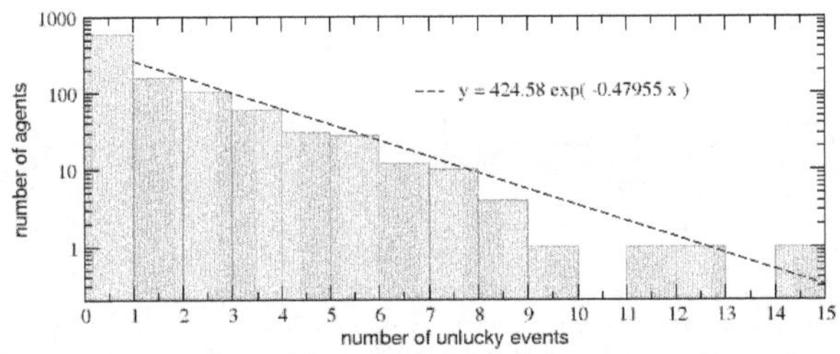

(d)

Fig. 5. (*Continued*)

correlation between success and luck. Analyzing the details of the frequency distributions of the number of lucky or unlucky events occurred to individuals, we found — as shown in panels (c) and (d) — that both of them are exponential, with exponents 0.64 and 0.48, and averages 1.35 and 1.66, respectively, and that the maximum numbers of lucky or unlucky events occurred were, respectively, 10 and 15. Moreover, about 16% of people had a "neutral" life, without lucky or unlucky events at all, while about 40% of individuals exclusively experienced only one type of events (lucky or unlucky).

It is also interesting to look at the time evolution of the success/capital of both the most successful individual and the less successful one, compared with the corresponding sequence of lucky or unlucky events occurred during the 40 years (80 time steps, one every 6 months) of their working life. This can be observed, respectively, in the left and the right part of Fig. 6. Different from the panel (a) of Fig. 5, in the bottom panels of this figure, only lucky events that agents have taken advantage of thanks to their talent, are shown.

In panel (a), concerning the moderately talented but most successful individual, it clearly appears that, after about a first half of his working life with a low occurrence of lucky events (bottom panel), and then with a low level of capital (top panel),

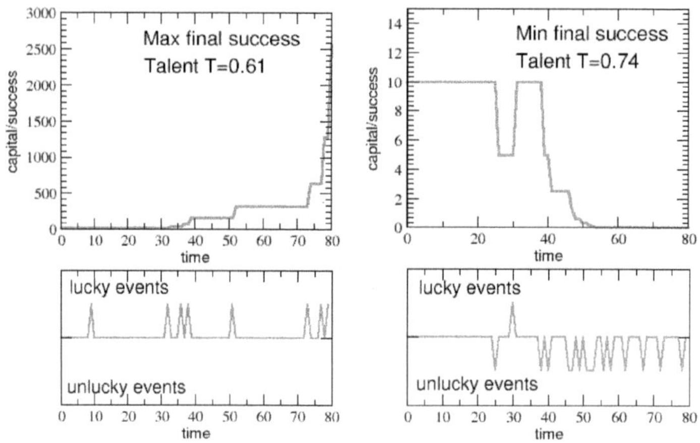

(a) (b)

Fig. 6. (a) Time evolution of success/capital for the most successful individual and (b) for the less successful one, compared with the corresponding sequences of lucky or unlucky events occurred during their working lives (80 semesters, i.e., 40 years). The time occurrence of these events is indicated, in the bottom panels, with upward or downwards spikes.

a sudden concentration of favorable events between 30 and 40 time steps (i.e., just before the age of 40 of the agent) produces a rapid increase in capital, which becomes exponential in the last 10 time steps (i.e., the last 5 years of the agent's career), going from $C = 320$ to $C_{\max} = 2560$.

On the other hand, looking at (top and bottom) panel (b), concerning the less successful individual, it is evident that a particularly unlucky second half of his working life, with a dozen of unfavorable events, progressively reduces the capital/success bringing it at its final value of $C = 0.00061$. It is interesting to note that this poor agent had, however, a talent $T = 0.74$ which was greater than that of the most successful agent. Clearly, good luck made the difference. And, if it is true that the most successful agent has had the merit of taking advantage of all the opportunities presented to him (in spite of his average talent), it is also true that if your life is as unlucky and poor of opportunities as that of the other agent, even a great talent becomes useless against the fury of misfortune.

All the results shown in this subsection for a single simulation run[a] are very robust and, as we will see in the next subsection, they persist, with small differences, if we repeat many times the simulations starting with the same talent distribution, but with a different random positions of the individuals.

[a] A demo version of the NetLogo code of the TvL model used for the single run simulations can be found on the Open ABM repository — https://www.comses.net/codebases/.

2.2. *Multiple runs results*

In this subsection, we present the global results of a simulation averaging over 100 runs, each starting with different random initial conditions. The values of the control parameters are the same of those used in the previous subsection: $N = 1000$ individuals, $m_T = 0.6$ and $\sigma_T = 0.1$ for the normal talent distribution, $I = 80$ iteration (each one representing $\delta_t = 6$ months of working life), $C(0) = 10$ units of initial capital, $N_E = 500$ event-points and a percentage $p_L = 50\%$ of lucky events.

In panel (a) of Fig. 7, the global distribution of the final capital/success for all the agents collected over the 100 runs is shown in log–log scale and it is well fitted by a power-law curve with slope −1.33. The scale-invariant behavior of capital and the consequent strong inequality among individuals, together with the Pareto's "80–20"

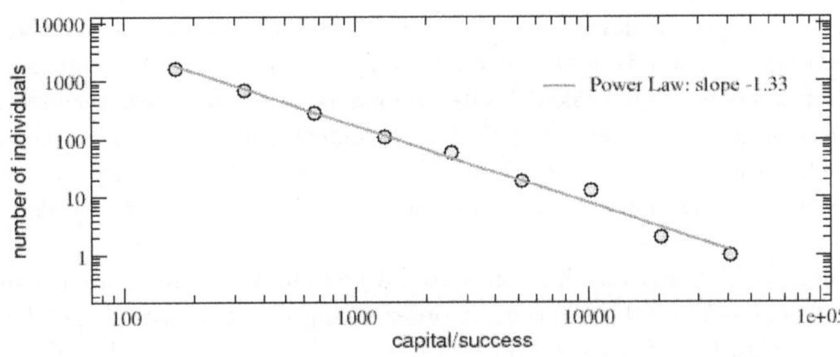

(a)

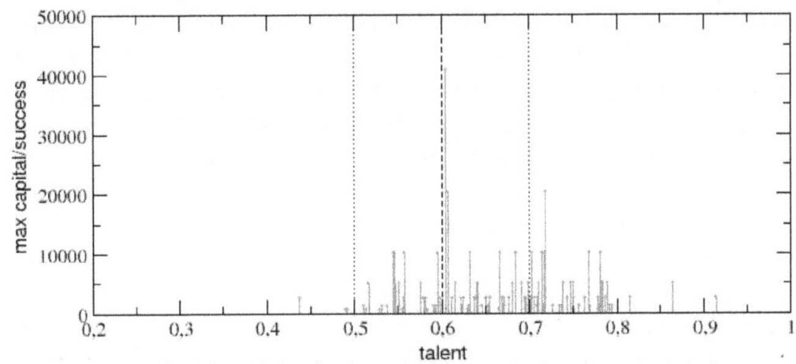

(b)

Fig. 7. (a) Distribution of the final capital/success calculated over 100 runs for a population with different random initial conditions. The distribution can be well fitted with a power-law curve with a slope −1.33. (b) The final capital $C_{\max}$ of the most successful individual in each of the 100 runs is reported as function of their talent. People with a medium–high talent result to be, on average, more successful than people with low or medium–low talent, but very often the most successful individual is a moderately gifted agent and only rarely the most talented one. The m_T values, together with the values $m_T \pm \sigma_T$, also reported as vertical dashed and dot lines, respectively.

rule observed in the single run simulation, are therefore conserved also in the case of multiple runs. Indeed, the gap between rich (successful) and poor (unsuccessful) agents has been increased, since the capital of the most successful people surpasses now the 40,000 units.

This last result can be better appreciated looking at panel (b), where the final capital C_{max} of the most successful individuals only, i.e., of the best performers for each one of the 100 runs, is reported as a function of their talent. The best score was realized by an agent with a talent $T_{best} = 0.6048$, practically coinciding with the mean of the talent distribution ($m_T = 0.6$), who reached a peak of capital $C_{best} = 40,960$. On the other hand, the most talented among the most successful individuals, with a talent $T_{max} = 0.91$, accumulated a capital $C_{max} = 2560$, equal to only 6% of C_{best}.

To address this point in more detail, in Fig. 8(a), we plot the talent distribution of the best performers calculated over 100 runs. The distribution seems to be shifted to the right of the talent axis, with a mean value $T_{av} = 0.66 > m_T$: this confirms, on one hand, that a medium–high talent is often necessary to reach a great success; but, on the other hand, it also indicates that it is almost never sufficient, since agents with the highest talent (e.g., with $T > m_T + 2\sigma_T$, i.e., with $T > 0.8$) result to be the best performers only in 3% of cases, and their capital/success never exceeds the 13% of C_{best}.

In Fig. 8(b), the same distribution (normalized to unitary area in order to obtain a PDF) is calculated over 10,000 runs, in order to appreciate its true shape: it appears to be well fitted by a Gaussian $G(T)$ with average $T_{av} = 0.667$ and standard deviation 0.09 (solid line). This definitely confirms that the talent distribution of the best performers is shifted to the right of the talent axis with respect to the original distribution of talent. More precisely, this means that the conditional probability $P(C_{max}|T) = G(T)dT$ to find among the best performers an individual with talent in the interval $[T, T + dT]$ increases with the talent T, reaches a maximum around a medium–high talent $T_{av} = 0.66$, then rapidly decreases for higher values of talent. In other words, the probability to find a moderately talented individual at the top of success is higher than that of finding there a very talented one. Note that, in a ideal world in which talent was the main cause of success, one expects $P(C_{max}|T)$ to be an increasing function of T. Therefore, we can conclude that the observed Gaussian shape of $P(C_{max}|T)$ is the proof that luck matters more than talent in reaching very high levels of success.

It is also interesting to compare the average capital/success $C_{mt} \sim 63$, over 100 runs, of the most talented people and the corresponding average capital/success $C_{at} \sim 33$ of people with talent very close to the mean m_T. We found in both cases quite small values (although greater than the initial capital $C(0) = 10$), but the fact that $C_{mt} > C_{at}$ indicates that, even if the probability to find a moderately talented individual at the top of success is higher than that of finding there a very talented one, the most talented individuals of each run have, on average, more success than moderately gifted people. On the other hand, looking at the average percentage, over the 100 runs, of

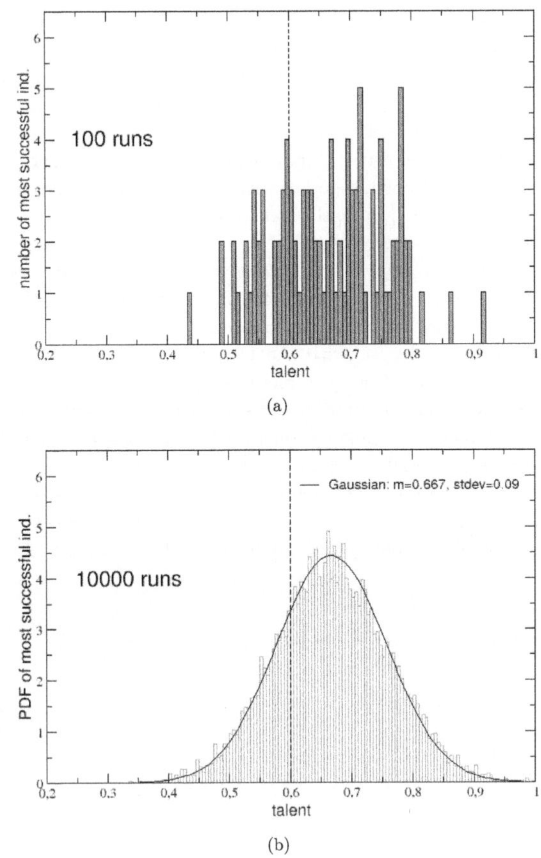

Fig. 8. (a) Talent distribution of the most successful individuals (best performers) in each of the 100 runs. (b) Probability distribution function of talent of the most successful individuals calculated over 10,000 runs: it is well fitted by a normal distribution with mean 0.667 and standard deviation 0.09 (solid line). The mean $m_T = 0.6$ of the original normal distribution of talent in the population is reported for comparison as a vertical dashed line in both panels.

individuals with talent $T > 0.7$ (i.e., greater than one standard deviation from the average) and with a final success/capital $C_{\text{end}} > 10$, calculated with respect to all the agents with talent $T > 0.7$ (who are, on average for each run, ~ 160), we found that this percentage is equal to 32%: this means that the aggregate performance of the most talented people in our population remains, on average, relatively small since only one-third of them reaches a final capital greater than the initial one.

In any case, it is a fact that the absolute best performer over the 100 simulation runs is an agent with talent $T_{\text{best}} = 0.6$, perfectly aligned with the average, but with a final success $C_{\text{best}} = 40,960$ which is 650 times greater than C_{mt} and more than 4000 times greater than the success $C_{\text{end}} < 10$ of 2/3 of the most talented people. This occurs just because, at the end of the story, she was just luckier than the others. Indeed, very lucky, as it is shown in Fig. 9, where the increase of her capital/success during her working life is shown, together with the impressive sequence of lucky (and only lucky) events of which, despite the lack of particular talent, she was able to take advantage of during her career.

Summarizing, what has been found up to now is that, in spite of its simplicity, the TvL model seems able to account for many of the features characterizing, as discussed in the introduction, the largely unequal distribution of richness and success in our society, in evident contrast with the Gaussian distribution of talent among human beings. At the same time, the model shows, in quantitative terms, that a great talent is not sufficient to guarantee a successful career and that, instead, less talented people are very often able to reach the top of success — another "stylized fact" frequently observed real life [34, 35, 37].

The key point, which intuitively explains how it may happen that moderately gifted individuals achieve (so often) far greater honors and success than much more talented ones, is the hidden and often underestimated role of luck, as resulting from

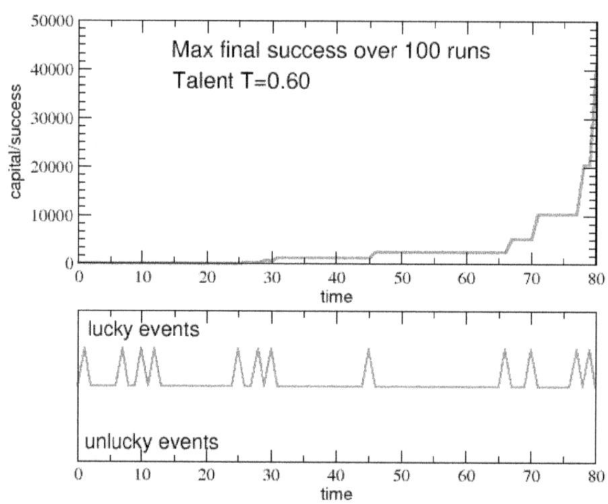

Fig. 9. Time evolution of success/capital for the most successful (but moderately gifted) individual over the 100 simulation runs, compared with the corresponding unusual sequence of lucky events occurred during her working life.

our simulations. But to understand the real meaning of our findings, it is important to distinguish the *macro* from the *micro* point of view.

In fact, from the *micro* point of view, following the dynamical rules of the TvL model, a talented individual has a greater *a priori* probability to reach a high level of success than a moderately gifted one, since she has a greater ability to grasp any opportunity will come. Of course, luck has to help her in yielding those opportunities. Therefore, from the point of view of a single individual, we should therefore conclude that, being impossible (by definition) to control the occurrence of lucky events, the best strategy to increase the probability of success (at any talent level) is to broaden the personal activity, the production of ideas, the communication with other people, seeking for diversity and mutual enrichment. In other words, to be an open-minded person, ready to be in contact with others, exposes to the highest probability of lucky events (to be exploited by means of the personal talent).

On the other hand, from the *macro* point of view of the entire society, the probability to find moderately gifted individuals at the top levels of success is greater than that of finding there very talented ones, because moderately gifted people are much more numerous and, with the help of luck, have — globally — a statistical advantage to reach a great success, in spite of their lower individual *a priori* probability.

In the next section, we will address such a *macro* point of view by exploring the possibilities offered by our model to investigate in detail new and more efficient strategies and policies to improve the average performance of the most talented people in a population, implementing more efficient ways of distributing prizes and resources. In fact, being the most talented individuals, the engine of progress and innovation in our society, we expect that any policy being able to improve their level of success will have a beneficial effect on the collectivity.

3. Effective Strategies to Counterbalance Luck

The results presented in the previous section are strongly consistent with largely documented empirical evidences, discussed in the introduction, which firmly question the naively meritocratic assumption claiming that the natural differences in talent, skill, competence, intelligence, hard work or determination are the only causes of success. As we have shown, luck also matters and it can play a very important role. The interpretative point is that, being individual qualities difficult to be measured (in many cases hardly defined in rigorous terms), the meritocratic strategies used to assign honors, funds or rewards are often based on individual performances, valued in terms of personal wealth or success. Eventually, such strategies exert a further reinforcing action and pump up the wealth/success of the luckiest individuals through a positive feedback mechanism, which resembles the famous "rich get richer" process (also known as "Matthew effect" [56–58]), with an unfair final result.

Let us consider, for instance, a publicly funded research granting council with a fixed amount of money at its disposal. In order to increase the average impact of research, is it more effective to give large grants to a few apparently excellent researchers, or small grants to many more apparently ordinary researchers? A recent study [43], based on the analysis of four indices of scientific impact involving publications, found that impact is positively, but only weakly, related to funding. In particular, impact per dollar was lower for large grant-holders and the impact of researchers who received increases in funding did not increase in a significant way. The authors of the study conclude that scientific impact (as reflected by publications) is only weakly limited by funding and suggest that funding strategies targeting diversification of ideas, rather than "excellence", are likely to be more productive. A more recent contribution [59] showed that, both in terms of the quantity of papers produced and of their scientific impact, the concentration of research funding generally produces diminishing marginal returns and also that the most funded researchers do not stand out in terms of output and scientific impact. Actually, such conclusions should not be a surprise in the light of the other recent finding [17] that impact, as measured by influential publications, is randomly distributed within a scientist's temporal sequence of publications. In other words, if luck matters, and if it matters more than we are willing to admit, it is not strange that meritocratic strategies reveal less effective than expected, in particular, if we try to evaluate merit *ex-post*. In the previous studies [47–54], there was already a warning against this sort of "naive meritocracy", showing the effectiveness of alternative strategies based on random choices in management, politics and finance. Consistently with such a perspective, the TvL model shows how the minimum level of success of the most talented people can be increased, in a world where luck is important and serendipity is often the cause of important discoveries.

3.1. *Serendipity, innovation and efficient funding strategies*

The term "serendipity" is commonly used in the literature to refer to the historical evidence that very often researchers make unexpected and beneficial discoveries by chance, while they are looking for something else [60, 61]. There is a long anecdotal list of discoveries made just by lucky opportunities: from penicillin by Alexander Fleming to radioactivity by Marie Curie, from cosmic microwave background radiation by radio astronomers Arno Penzias and Robert Woodrow Wilson to the graphene by Andre Geim and Kostya Novoselov. Just to give a very recent example, a network of fluid-filled channels in the human body, that may be a previously-unknown organ and that seems to help transport cancer cells around the body, was discovered by chance, from routine endoscopies [62]. Therefore, many people think that curiosity-driven research should always be funded because nobody can really know or predict where it can lead to [63].

Is it possible to quantify the role of serendipity? Which are the most efficient ways to stimulate serendipity? Serendipity can take on many forms, and it is difficult to constrain and quantify. That is why, so far, academic research has focused on serendipity in science mainly as a philosophical idea. But things are changing.

The European Research Council has recently given to the biochemist Yaqub a 1.7 million US dollars grant to quantify the role of serendipity in science [64]. Yaqub found that it is possible to classify serendipity into four basic types [65] and that there may be important factors affecting its occurrence. His conclusions seem to agree with ideas developed in earlier works [66–71] which argue that the commonly adopted — apparently meritocratic — strategies, which pursuit excellence and drive out variety, seem destined to be loosing and inefficient. The reason is that they cut out *a priori* researches that initially appear less promising but that, thanks also to serendipity, could be extremely innovative *a posteriori.*

From this perspective, we want to use the TvL model, which naturally incorporates luck (and therefore also serendipity) as a quantitative tool for policy, in order to explore, in this subsection, the effectiveness of different funding scenarios. In particular, in contexts where, as above discussed, averagely-talented-but-lucky people are often more successful than more-gifted-but-unlucky individuals, it is important to evaluate the efficiency of funding strategies in preserving a minimum level of success also for the most talented people, who are expected to produce the most progressive and innovative ideas.

Starting from the same parameters setup used in Sec. 2.2, i.e., $N = 1000$, $m_T = 0.6$, $\sigma_T = 0.1$, $I = 80$, $\delta_t = 6$, $C(0) = 10$, $N_E = 500$, $p_L = 50\%$ and 100 simulation runs, let us imagine that a given total funding capital F_T is periodically distributed among individuals following different criteria. For example, funds could be assigned as follows:

(1) in equal measure to all (*egalitarian criterion*), in order to foster research diversification;
(2) only to a given percentage of the most successful ("best") individuals (*elitarian criterion*), which has been previously referred to "naively" meritocratic, for it distributes funds to people according to their past performance;
(3) by distributing a "premium" to a given percentage of the most successful individuals and the remaining amount in smaller equal parts to all the others (*mixed criterion*);
(4) only to a given percentage individuals, randomly selected (*selective random criterion*);

We realistically assume that the total capital F_T will be distributed every 5 years, during the 40 years spanned by each simulation run, so that $F_T/8$ units of capital will be allocated from time to time. Thanks to the periodic injection of these funds, we intend to maintain a minimum level of resources for the most talented agents. Therefore, a good indicator, for the effectiveness of the adopted funding strategy, could be the percentage P_T, averaged over the 100 simulation runs, of individuals with talent $T > m_T + \sigma_T$ whose final success/capital is greater than the initial one, i.e. $C_{\text{end}} > C(0)$.

This percentage has already been calculated, in the multiple runs simulation presented in Sec. 2.2. There, we have shown that, in the absence of funding, the best

A. Pluchino, A. E. Biondo and A. Rapisarda

performance was scored by very lucky agents with a talent close to the mean, while the capital/success of the most talented people always remained very low. In particular, only a percentage $P_{T0} \sim 32\%$ of the total number of agents with $T > 0.7$ reached, at the end of the simulation, a capital/success greater than the initial one. Hence, in order to compare the efficiency of different funding strategies, the increment in the average percentage P_T of talented people which, during their career, increases their initial capital/success should be calculated with respect to P_{T0}. Let us define this increment as $P_T^* = P_T - P_{T0}$. The latter quantity is a very robust indicator: we have checked that repeating the set of 100 simulations, the variation in the value of P_T^* remains under 2%. Finally, if one considers the ratio between P_T^* and the total capital F_T distributed among all the agents during the 40 years, it is possible to obtain an *efficiency index* E, which quantifies the increment of sufficiently successful talented people per unit of invested capital, defined as $E = P_T^*/F_T$.

In the table shown in Fig. 10, we report the efficiency index (2nd column) obtained for several funding distribution strategies, each one with a different funding target (1st column), together with the corresponding values of P_T (3rd column) and P_T^* (4th column). The total capital F_T invested in each run is also reported in the last column. The efficiency index E has been normalized to its maximum value $E_{\max}$ and the various records (rows) have been ordered for decreasing values of

FUNDING-TARGET	E_{norm}	P_T	$P^*_T = P_T - P_{T0}$	F_T
ALL EQUAL 1u	1.00	69.48	37.43	8000
10% RANDOM 5u	0.85	49.83	17.78	4000
25% RANDOM 5u	0.79	68.00	35.95	10000
ALL EQUAL 2u	0.74	84.02	51.97	16000
50% RANDOM 5u	0.58	82.91	50.86	20000
25% BEST 5u, OTHERS 1u	0.55	70.83	38.78	16000
25% BEST 10u, OTHERS 1u	0.37	73.44	41.39	26000
ALL EQUAL 5u	0.37	94.40	62.35	40000
25% RANDOM 20u	0.31	84.74	52.69	40000
50% BEST 5u	0.25	54.26	22.21	20000
25% BEST 10u, OTHERS 5u	0.21	94.82	62.77	70000
25% BEST 5u	0.20	41.08	9.03	10000
25% BEST 10u	0.12	42.33	10.28	20000
10% BEST 5u	0.10	34.14	2.09	4000
25% BEST 15u	0.09	43.51	11.46	30000
25% BEST 20u	0.07	44.26	12.21	40000
10% BEST 10u	0.06	34.41	2.36	8000
10% BEST 20u	0.04	34.98	2.93	16000
NO FUNDING	0.00	32.05	0.00	0

Fig. 10. Funding strategies table. The outcomes of the normalized efficiency index $E_{\rm norm}$ are reported (2nd column) in the decreasing order, from top to bottom, for several funding distribution strategies with different targets (1st column). The corresponding values of both the percentage P_T of successful talented people and its net increase P_T^* with respect to the "no funding" case, averaged over the 100 simulation runs, are also reported in the third and fourth columns, respectively. Finally, the total capital F_T invested in each run is visible in the last column.

$E_{\text{norm}} = E/E_{\text{max}}$. For the no funding case, by definition, $E_{\text{norm}} = 0$. The same scores for E_{norm} are also reported in the form of a histogram in Fig. 11, as a function of the adopted funding strategies. Thanks to the statistical robustness of P_T, which shows fluctuations smaller than 2%, the results reported for the efficiency index E_{norm} are particularly stable.

Looking at the table and at the relative histogram of Fig. 11, it is evident that, if the goal is to reward the most talented persons (thus increasing their final level of success), it is much more convenient to distribute periodically (even small) equal amounts of capital to all individuals rather than to give a greater capital only to a small percentage of them, selected through their level of success — already reached — at the moment of the distribution.

On one hand, the histogram shows that the "egalitarian" criterion, which assigns 1 unit of capital every 5 years to all the individuals is the most efficient way to distribute funds, being $E_{\text{norm}} = 1$ (i.e., $E = E_{\text{max}}$): with a relatively small investment F_T of 8000 units, it is possible to double the percentage of successful talented people with respect to the "no funding" case, bringing it from $P_{T0} = 32.05\%$ to $P_T = 69.48\%$, with a net increase $P_T^* = 37.43\%$. Considering an increase of the total invested capital (for example, setting the egalitarian quotas to 2 or 5 units), this strategy also ensures a further increment in the final percentage of successful talented

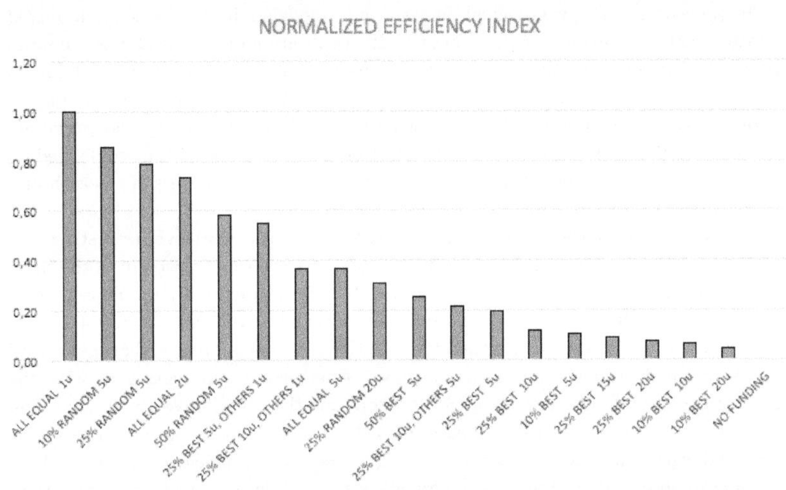

Fig. 11. Normalized efficiency index for several funding strategies. The values of the normalized efficiency index E_{norm} are reported as function of the different funding strategies. The figure shows that for increasing the success of a larger number of talented people with $C_{\text{end}} > C(0)$, it is much more efficient to give a small amount of funds to many individuals instead of giving funds in other more selective ways.

people P_T (from 69.48% to 84.02% and to 94.40%), even if the normalized efficiency progressively decreases from $E_{\rm norm} = 1$ to $E_{\rm norm} = 0.74$ and to $E_{\rm norm} = 0.37$.

On the other hand, the "elitarian" strategies which assign every 5 years more funds (5, 10, 15 or 20 units) only to the best 50%, 25% or even 10% of the already successful individuals are all at the bottom of the ranking, with $E_{\rm norm} < 0.25$: in all of these cases, the net increase P_T^* in the final number of successful talented people with respect to the "no funding" case remains very small (in almost all the cases smaller than 20%), often against a much larger invested capital if compared to that of the egalitarian strategy. These results do reinforce the thesis that this kind of approach is only apparently — i.e., naively — meritocratic.

It is worth noting that the adoption of a "mixed" criterion, i.e., assigning a "meritocratic" funding share to a certain percentage of the most successful individuals, for instance 25%, and distributing the remaining funds in equal measure to the rest of people, gives back better scores for the efficiency index values with respect to the "naively meritocratic" approach. However, the performance of this strategy is not able to overtake the "egalitarian" criterion. As it clearly appears — for example — by the comparison between the sixth and the fourth rows of the funding table, in spite of the same overall investment of 16,000 units, the value of P_T obtained with the mixed criterion stays well below the one obtained with the egalitarian approach (70.83% against 84.02%), as also confirmed by the values of the corresponding efficiency index $E_{\rm norm}$ (0.55 against 0.74).

If one considers psychological factors (not modeled in this study), a mixed strategy could be revalued with respect to the egalitarian one. Indeed, the premium reward — assigned to the more successful individuals — could induce all agents towards a greater commitment, while the equally distributed part would play a twofold role: at the individual level, it would act in fostering variety and providing unlucky talented people with new chances to express their potential, while feeding serendipity at the aggregate level, thus contributing to the progress of research and of the whole society.

Looking again at the funding strategy table, it is also worthwhile to stress the surprising high efficiency of the random strategies, which occupy two out of the three best scores in the general ranking. It results that, for example, a periodic reward of 5 units for only the 10% of randomly selected individuals, with a total investment of just 4000 units, gives a net increase $P_T^* = 17,78\%$, which is greater than almost all those obtained with the elitarian strategies. Furthermore, increasing to 25% the percentage of randomly funded people and doubling the overall investment (bringing it to 10,000 units), the net increase $P_T^* = 35.95\%$ becomes comparable to that obtained with the best egalitarian strategy, first in the efficiency ranking. It is striking to note that this latter score for P_T^* is approximately four times grater than the value ($P_T^* = 9.03\%$) obtained with the elitarian approach (see 12th row in the table), distributing exactly the same capital (10,000 units) to exactly the same number of individuals (25% of the total). The latter is a further confirmation that, in complex social and economical contexts where chance plays a relevant role, the efficiency of alternative strategies based on random choices can easily overtake that of standard strategies based on the "naively meritocratic" approach. Such a counterintuitive phenomenon, already observed in management, politics and finance [47–54], finds therefore new evidence also in the research funding context.

To further corroborate these findings, in Fig. 12, the results of another set of simulations are presented. At variance with the previous simulations, the total capital invested in each one of the 100 runs is now fixed to $F_T = 80,000$, so that $F_T/8 = 10,000$ units are distributed every 5 years among the agents following the main funding strategies already considered. Looking at the table, the egalitarian strategy results again the most efficient in rewarding the most talented people, with a percentage P_T close to 100%, immediately followed by the random strategy (with 50% of randomly funded individuals) and by the mixed one, with half of the capital distributed to the 25% of the most successful individuals and the other half in equal measure to the remaining people. On the contrary, all the elitarian strategies are placed again at the bottom of the ranking, thus further confirming the inefficiency of the "naively meritocratic" approach in rewarding real talent.

The results of the TvL model simulations presented in this subsection have focused on the importance of external factors (as, indeed, efficient funding policies) in increasing the opportunities of success for the most talented individuals, too often penalized by unlucky events. In the next subsection, we investigate to what extent new opportunities can be originated by the changes in the environment as for example the level of education or other stimuli received by the social context where people live or come from.

3.2. *The importance of the environment*

First, let us estimate the role of the average level of education among the population. Within the TvL model, the latter could be obtained by changing the parameters of the normal distribution of talent. Actually, assuming that talent and skills of individuals, if stimulated, could be more effective in exploiting new opportunities,

FUNDING-TARGET	E_{norm}	P_T	$P'_T = P_T - P_{T0}$	F_T
ALL EQUAL	1.00	98.14	67.68	80000
50% RANDOM	0.98	97.12	66.66	80000
HALF 25% BEST, HALF TO OTHERS	0.97	96.13	65.67	80000
25% RANDOM	0.85	87.67	57.21	80000
10% RANDOM	0.54	66.73	36.27	80000
50% BEST	0.45	61.19	30.73	80000
25% BEST	0.22	45.31	14.85	80000
10% BEST	0.06	34.83	4.37	80000
NO FUNDING	0.00	30.46	0.00	0

Fig. 12. Funding strategies table with fixed funds. The outcomes of the normalized efficiency index $E_{\rm norm}$ are reported again in the decreasing order, from top to bottom, for several funding distribution strategies with different targets (1st column). At variance with Fig. 10, now the total capital invested in each run was fixed to $F_T = 80,000$. The egalitarian strategy is, again, at the top of the ranking.

an increase in either the mean m_T or the standard deviation σ_T of the talent distribution could be interpreted as the effect of policies targeted, respectively, either at raising the average level of education or at reinforcing the training of the most gifted people.

In the two panels of Fig. 13, we report the final capital/success accumulated by the best performers in each of the 100 runs, as a function of their talent. The parameters setup is the same than in Sec. 2.2 ($N = 1000$, $I = 80$, $\delta_t = 6$, $C(0) = 10$, $N_E = 500$ and $p_L = 50\%$) but with different moments for the talent distributions. In particular, in panel (a), we left unchanged $m_T = 0.6$ but increased $\sigma_T = 0.2$, while in panel (b), we made the opposite, leaving $\sigma_T = 0.1$ but increasing $m_T = 0.7$. In both cases, a shift on the right of the maximum success peaks can be appreciated, but with different details.

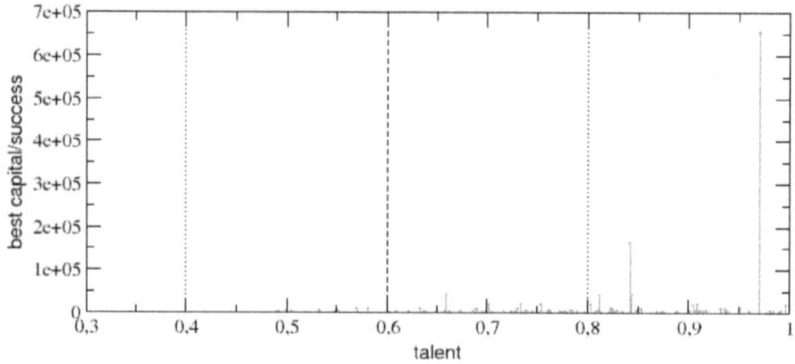

(a)

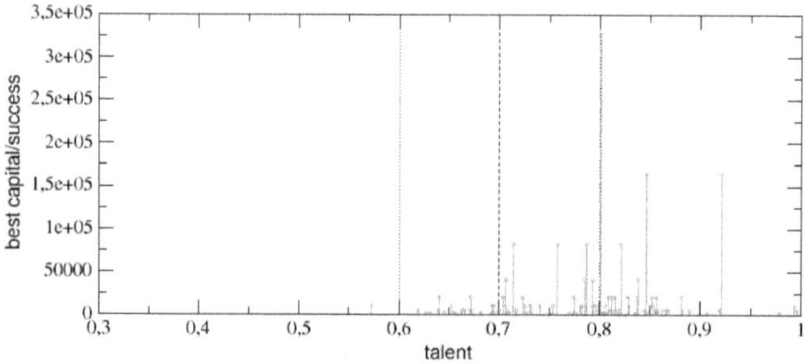

(b)

Fig. 13. The final capital of the most successful individuals in each of the 100 runs is reported as function of their talent for populations with different talent distribution parameters: (a) $m_T = 0.6$ and $\sigma_T = 0.2$ (which represents a training reinforcement for the most gifted people); (b) $m_T = 0.7$ and $\sigma_T = 0.1$ (which represents an increase in the average level of education). The corresponding m_T and $m_T \pm \sigma_T$ values are also indicated as, respectively, vertical dashed and dot lines.

Actually, it results that increasing σ_T without changing m_T, as shown in panel (a), enhances the chances for more talented people to get a very high success: the best performer is, now, a very talented agent with $T = 0.97$, who reaches an incredible level of capital/success $C_{\text{best}} = 655,360$. This, on one hand, could be considered positive but, on the other hand, it is an isolated case and it has, as a counterpart, an increase in the gap between unsuccessful and successful people.

Looking now at panel (b), it results that increasing m_T without changing σ_T produces a best performer, with $C_{\text{best}} = 327,680$ and a talent $T = 0.8$, followed by other two with $C = 163,840$ and, respectively, $T = 0.85$ and $T = 0.92$. This means that also in this case, the chances for more talented people to get a very high success are enhanced, while the gap between unsuccessful and successful people is lower than before.

Finally, in both considered examples, the average value of the capital/success for the most talented people over the 100 runs is increased with respect to the value $C_{\text{mt}} \sim 63$ found in Sec. 2.2. In particular, we found $C_{\text{mt}} \sim 319$ for panel (a) and $C_{\text{mt}} \sim 122$ for panel (b), but these values are quite sensitive to the specific set of simulation runs. A more reliable parameter in order to quantify the effectiveness of the social policies investigated here is, again, the indicator P_T introduced in the previous subsection, i.e., the average percentage of individuals with talent $T > m_T + \sigma_T$ and with final success/capital $C_{\text{end}} > 10$, over the total number of individuals with talent $T > m_T + \sigma_T$ (note that now, in both the cases considered, $m_T + \sigma_T = 0.8$). In particular, we found $P_T = 38\%$ for panel (a) and $P_T = 37.5\%$ for panel (b), with a slight net increment with respect to the reference value $P_{T0} = 32\%$ (obtained for a talent distribution with $m_T = 0.6$ and $\sigma_T = 0.1$).

Summarizing, our results indicate that strengthening the training of the most gifted people or increasing the average level of education produce, as one could expect, some beneficial effects on the social system, since both these policies raise the probability, for talented individuals, to grasp the opportunities that luck presents to them. On the other hand, the enhancement in the average percentage of highly talented people who are able to reach a good level of success seems to be not particularly remarkable in both the cases analyzed, therefore the result of the corresponding educational policies appears mainly restricted to the emergence of isolated extreme successful cases.

Of course, once a given level of education has been fixed, it is quite obvious that the abundance of opportunities offered by the social environment, i.e., by the country where someone accidentally is born or where someone choose to live, it is another key ingredient being able to influence the global performance of the system.

In Fig. 14, we show results analogous to those shown in the previous figure, but for another set of simulations, with 100 runs each, with the same parameters setup as in Sec. 2.2 ($N = 1000$, $m_T = 0.6$, $\sigma_T = 0.1$, $I = 80$, $C(0) = 10$, $N_E = 500$) and with different percentages p_L of lucky events (we remind that, in Sec. 2.2, this percentage was set to $p_L = 50\%$). In panels (a) we set $p_L = 80\%$, in order to simulate a very

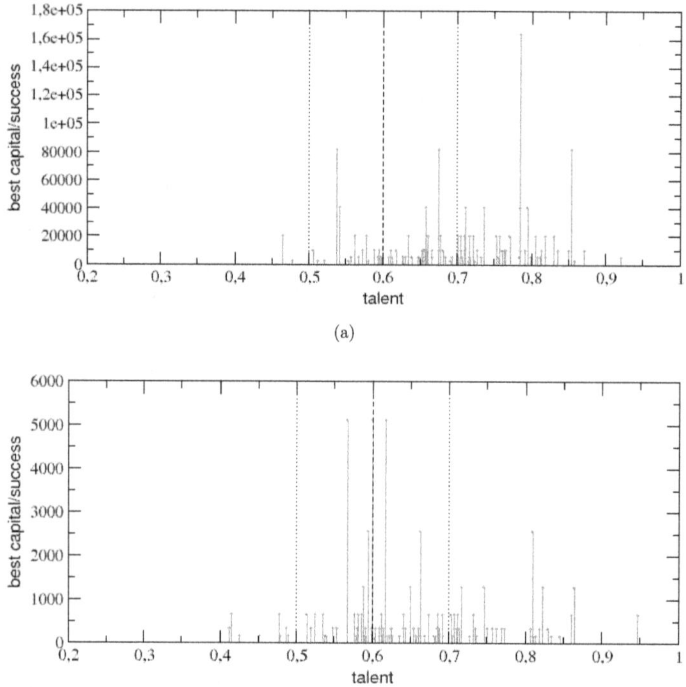

Fig. 14. The final capital of the most successful individuals in each of the 100 runs is reported as function of their talent for populations living in environments with a different percentage p_L of lucky events: (a) $p_L = 80\%$; (b) $p_L = 20\%$. The values of $m_T = 0.6$ and $m_T \pm \sigma_T$, with $\sigma_T = 0.1$ are also indicated as, respectively, vertical dashed and dot lines.

stimulating environment, rich of opportunities, like that of rich and industrialized countries such as the US [25]. On the other hand, in panel (b), the value $p_L = 20\%$ reproduces the case of a much less stimulating environment, with very few opportunities, like for instance that of Third World countries.

As visible in both panels, the final success/capital of the most successful individuals as function of their talent strongly depends on p_L.

When $p_L = 80\%$, as in panel (a), several agents with medium–high talent are able to reach higher levels of success compared to the case $p_L = 50\%$, with a peak of $C_{\text{best}} = 163,840$. On the other hand, the average value of the capital/success for the most talented individuals, $C_{\text{mt}} \sim 149$, is quite high and, what is more important, the same holds for the indicator $P_T = 62.18\%$ (about twice with respect to the reference value $P_{T0} = 32\%$), meaning that, as expected, talented people benefits of the higher percentage of lucky events.

Completely different outcomes are obtained with $p_L = 20\%$. Indeed, as visible in panel (b), the overall level of success is now very low, if compared to that found in the simulations of Sec. 2.2, with a peak value C_{best} of only 5120 units: it is a footprint of a reduction in the social inequalities, which is an expected consequence of the flattening of success opportunities. According with these results, also the P_T indicator reaches a minimal value, with an average percentage of only 8.75% of talented individuals being able to increase their initial level of success.

In conclusion, in this section, we have shown that a stimulating environment, rich of opportunities, associated to an appropriate strategy for the distribution of funds and resources, are important factors in exploiting the potential of the most talented people, giving them more chances of success with respect to the moderately gifted, but luckier, ones. At the macro-level, any policy being able to influence those factors and to sustain talented individuals, will have the result of ensuring collective progress and innovation.

4. Conclusive Remarks

In this paper, starting from few very simple and reasonable assumptions, we have presented an agent-based model which is able to quantify the role of talent and luck in the success of people's careers. The simulations show that although talent has a Gaussian distribution among agents, the resulting distribution of success/capital after a working life of 40 years follows a power law which respects the "80–20" Pareto law for the distribution of wealth found in the real world. An important result of the simulations is that the most successful agents are almost never the most talented ones, but those around the average of the Gaussian talent distribution — another stylized fact often reported in the literature. The model shows the importance, very frequently underestimated, of lucky events in determining the final level of individual success. Since rewards and resources are usually given to those that have already reached a high level of success, mistakenly considered as a measure of competence/talent, this result is even a more harmful disincentive, causing a lack of opportunities for the most talented ones. Our results highlight the risks of the paradigm that we call "naive meritocracy", which fails to give honors and rewards to the most competent people, because it underestimates the role of randomness among the determinants of success. In this respect, several different scenarios have been investigated in order to discuss more efficient strategies, which are able to counterbalance the unpredictable role of luck and give more opportunities and resources to the most talented ones — a purpose that should be the main aim of a truly meritocratic approach. Such strategies have also been shown to be the most beneficial for the entire society, since they tend to increase the diversity of ideas and perspectives in research, thus also fostering the innovation.

Acknowledgments

We would like to thank Robert H. Frank, Pawel Sobkowicz and Constantino Tsallis for their fruitful discussions and comments.

References

[1] Bak, P., Tang, C. and Wiesenfeld, K., Self-organized criticality, *Phys. Rev. A* **38** (1988) 364–374.
[2] Barabási, A.-L. and Albert, R., Emergence of scaling in random networks, *Science* **286**(5439) (1999) 509–512.
[3] Newman, M. E. J., Power laws, Pareto distributions and Zipf's law, *Contemp. Phys.* **46**(5) (2005) 323–351.
[4] Tsallis, C., *Introduction to Nonextensive Statistical Mechanics: Approaching a Complex World* (Springer, 2009).
[5] Pareto, V., *Cours d'Economique Politique*, Vol. 2 (Lausanne, F. Rouge Editeur, 1896).
[6] Steindl, J., *Random Processes and the Growth of Firms — A Study of the Pareto Law*, Charles Griffin and Company (London, 1965).
[7] Atkinson, A. B. and Harrison, A. J., *Distribution of Total Wealth in Britain* (Cambridge University Press, 1978).
[8] Persky, J., Retrospectives: Pareto's law, *J. Econ. Perspect.* **6** (1992) 181–192.
[9] Klass, O. S., Biham, O., Levy, M., Malcai, O. and Solomon, S., The Forbes 400 and the Pareto wealth distribution, *Econ. Lett.* **90** (2006) 290–295.
[10] Hardoon, D., An economy for the 99%, Oxfam GB, Oxfam House, John Smith Drive, Cowley, Oxford, OX4 2JY, UK (2017).
[11] Bouchaud, J.-P. and Mézard, M., Wealth condensation in a simple model of economy, *Physica A* **282** (2000) 536–554.
[12] Dragulescu, A. and Yakovenko, V. M., Statistical mechanics of money, *Eur. Phys. J. B* **17** (2000) 723–729.
[13] Chakraborti, A. and Chakrabarti, B. K., Statistical mechanics of money: How saving propensity affects its distribution, *Eur. Phys. J. B* **17** (2000) 167–170.
[14] Patriarca, M., Chakraborti, A. and Germano, G., Influence of saving propensity on the power law tail of wealth distribution, *Physica A* **369**(2) (2006) 723–736.
[15] Scalas, E., Random exchange models and the distribution of wealth, *Eur. Phys. J.* **225** (2016) 3293–3298.
[16] During, B., Georgiou, N. and Scalas, E., A stylised model for wealth distribution, in *Economic Foundations of Social Complexity Science*, Akura, Yuji and Kirman, Alan (eds.) (Springer, 2017), pp. 95–117.
[17] Sinatra, R., Wang, D., Deville, P., Song, C. and Barabási, A.-L., Quantifying the evolution of individual scientific impact, *Science* **354** (2016) 6312.
[18] Einav, L. and Yariv, L., What's in a surname? The effects of surname initials on academic success, *J. Econ. Perspect.* **20**(1) (2006) 175–188.
[19] Ruocco, G., Daraio, C., Folli, V. and Leonetti, M., Bibliometric indicators: The origin of their log-normal distribution and why they are not a reliable proxy for an individual scholar's talent, *Palgrave Commun.* **3** (2017) 17064, doi: 10.1057/palcomms.2017.64.
[20] Jurajda, S. and Munich, D., Admission to selective schools, alphabetically, *Econ. Educ. Rev.* **29**(6) (2010) 1100–1109.
[21] Van Tilburg, W. A. P. and Igou, E. R., The impact of middle names: Middle name initials enhance evaluations of intellectual performance, *Eur. J. Soc. Psychol.* **44**(4) (2014) 400–411.

[22] Laham, S. M., Koval, P. and Alter, A. L., The name-pronunciation effect: Why people like Mr. Smith more than Mr. Colquhoun, *J. Exp. Soc. Psychol.* **48** (2012) 752–756.

[23] Silberzahn, R. and Uhlmann, E. L., It pays to be Herr Kaiser: Germans with noble-sounding last names more often work as managers, *Psychol. Sci.* **24**(12) (2013) 2437–2444.

[24] Coffey, B. and McLaughlin, P., From lawyer to judge: Advancement, sex, and name-calling, *SSRN Electron. J.* (2009) doi:10.2139/ssrn.1348280, https://papers.ssrn.com/sol3/papers.cfm?abstract_id=1348280.

[25] Milanovic, B., Global inequality of opportunity: How much of our income is determined by where we live? *Rev. Econ. Stat.* **97**(2) (2015) 452–460.

[26] Du, Q., Gao, H. and Levi, M. D., The relative-age effect and career success: Evidence from corporate CEOs, *Econ. Lett.* **117**(3) (2012) 660–662.

[27] Deaner, R. O., Lowen, A. and Cobley, S., Born at the wrong time: Selection bias in the NHL draft, *PLoS One* **8**(2) (2013) e57753.

[28] Brooks, D., *The Social Animal: The Hidden Sources of Love, Character, and Achievement* (Random House, 2011), p. 424.

[29] Iacopini, I., Milojevic, S. and Latora, V., Network dynamics of innovation processes, *Phys. Rev. Lett.* **120** (2018) 048301.

[30] Tomasetti, C., Li, L. and Vogelstein, B., Stem cell divisions, somatic mutations, cancer etiology, and cancer prevention, *Science* **355** (2017) 1330–1334.

[31] Newgreen, D. F. *et al.*, Differential clonal expansion in an invading cell population: Clonal advantage or dumb luck? *Cells Tissues Organs* **203** (2017) 105–113.

[32] Snyder, R. E. and Ellner, S. P., We happy few: Using structured population models to identify the decisive events in the lives of exceptional individuals, *Am. Nat.* **188**(2) (2016) E28–E45.

[33] Snyder, R. E. and Ellner, S. P., Pluck or luck: Does trait variation or chance drive variation in lifetime reproductive success? *Am. Nat.* **191**(4) (2018) E90–E107.

[34] Taleb, N. N., *Fooled by Randomness: The Hidden Role of Chance in Life and in the Markets* (Random House, 2001).

[35] Taleb, N. N., *The Black Swan: The Impact of the Highly Improbable* (Random House, 2007).

[36] Mauboussin, M. J., *The Success Equation: Untangling Skill and Luck in Business, Sports, and Investing* (Harvard Business Review Press, 2012).

[37] Frank, R. H., *Success and Luck: Good Fortune and the Myth of Meritocracy* (Princeton University Press, 2016).

[38] Watts, D. J., *Everything Is Obvious: Once You Know the Answer* (Crown Business, 2011).

[39] Salganik, M. J., Dodds, P. S. and Watts, D. J., Experimental study of inequality and unpredictability in an artificial cultural market, *Science* **311** (2006) 854–856.

[40] Travis, M., Hofman, J. M., Sharma, A., Anderson, A. and Watts, D. J., Exploring limits to prediction in complex social systems, in *Proc. 25th ACM Int. World Wide Web Conf.* (2016), pp. 683–694. arXiv:1602.01013 [cs.SI], https://dl.acm.org/citation.cfm?doid=2872427.2883001.

[41] Stewart, J., The distribution of talent, Marilyn Zurmuehlin Working Papers in Art Education, Vol. 2 (1983), pp. 21–22.

[42] Sinha, S. and Pan, R. K., How a "Hit" is born: The emergence of popularity from the dynamics of collective choice, in *Econophysics and Sociophysics: Trends and Perspectives*, Chakrabarti, B. K., Chakraborti, A. and Chatterjee, A. (Wiley-VCH Verlag GmbH & Co. KGaA, 2006), doi: 10.1002/9783527610006.ch15.

[43] Fortin, J.-M. and Curr, D. J., Big science vs. little science: How scientific impact scales with funding, *PLoS One* **8**(6) (2013) e65263.
[44] Jacob, B. A. and Lefgren, L., The impact of research grant funding on scientific productivity, *J. Public Econ.* **95** (2011) 1168–1177.
[45] O'Boyle, J. E. and Aguinis, H., The best and the rest: Revisiting the norm of normality of individual performance, *Pers. Psychol.* **65** (2012) 79–119, doi: 10.1111/j.1744-6570.2011.01239.x.
[46] Denrell, J. and Liu, C., Top performers are not the most impressive when extreme performance indicates unreliability, *Proc. Nat. Acad. Sci.* **109**(24) (2012) 9331–9336.
[47] Pluchino, A., Rapisarda, A. and Garofalo, C., The Peter principle revisited: A computational study, *Physica A* **389**(3) (2010) 467–472.
[48] Pluchino, A., Garofalo, C., Rapisarda, A., Spagano, S. and Caserta, M., Accidental politicians: How randomly selected legislators can improve parliament efficiency, *Physica A* **390**(21) (2011) 3944–3954.
[49] Pluchino, A., Rapisarda, A. and Garofalo, C., Efficient promotion strategies in hierarchical organizations, *Physica A* **390**(20) (2011) 3496–3511.
[50] Biondo, A. E., Pluchino, A., Rapisarda, A. and Helbing, D., Reducing financial avalanches by random investments, *Phys. Rev. E* **88**(6) (2013) 062814.
[51] Biondo, A. E., Pluchino, A., Rapisarda, A. and Helbing, D., Are random trading strategies more successful than technical ones, *PLoS One* **8**(7) (2013) e68344.
[52] Biondo, A. E., Pluchino, A. and Rapisarda, A., The beneficial role of random strategies in social and financial systems, *J. Stat. Phys.* **151**(3–4) (2013) 607–622.
[53] Biondo, A. E., Pluchino, A. and Rapisarda, A., Micro and macro benefits of random investments in financial markets, *Contemp. Phys.* **55**(4) (2014) 318–334.
[54] Biondo, A. E., Pluchino, A. and Rapisarda, A., Modeling financial markets by self-organized criticality, *Phys. Rev. E* **92**(4) (2015) 042814.
[55] Wilensky, U., NetLogo. Center for Connected Learning and Computer-Based Modeling, Northwestern University, Evanston, IL (1999), http://ccl.northwestern.edu/netlogo/.
[56] Merton, R. K., The Matthew effect in science, *Science* **159** (1968) 56–63.
[57] Merton, R. K., The Matthew effect in science, II: Cumulative advantage and the symbolism of intellectual property, *Isis* **79** (1988) 606–623.
[58] Bol, T., de Vaan, M. and van de Rijt, A., The Matthew effect in science funding, *Proc. Natl. Acad. Sci.* **115**(19) (2018) 4887–4890, doi:10.1073/pnas.1719557115.
[59] Mongeon, P., *et al.*, Concentration of research funding leads to decreasing marginal returns, *Res. Eval.* **25** (2016) 396–404.
[60] Merton, R. K. and Barber, E., *The Travels and Adventures of Serendipity* (Princeton University Press, 2004).
[61] Murayama, K. *et al.*, Management of science, serendipity, and research performance, *Res. Policy* **44**(4) (2015) 862–873.
[62] Benias, P. C. *et al.*, Structure and distribution of an unrecognized interstitium in human tissues, *Sci. Rep.* **8** (2018) 4947.
[63] Flexner, A., *The Usefulness of Useless Knowledge* (Princeton University Press, 2017).
[64] Lucky science. Scientists often herald the role of serendipity in research. A project in Britain aims to test the popular idea with evidence, *Nature Editorial*, Vol. 554 (2018), https://www.nature.com/magazine-assets/d41586-018-01405-7/d41586-018-01405-7.pdf.
[65] Yaqub, O., Serendipity: Towards a taxonomy and a theory, *Res. Policy* **47** (2018) 169–179.
[66] Page, S. E., *The Diversity Bonus: How Great Teams Pay Off in the Knowledge Economy* (Princeton University Press, 2017).

[67] Cimini, G., Gabrielli, A. and Sylos Labini, F., The scientific competitiveness of nations, *PLoS One* **9**(12) (2014) e113470, https://doi.org/10.1371/journal.pone.0113470.
[68] Curry, S., Let's move beyond the rhetoric: It's time to change how we judge research, *Nature* **554** (2018) 147.
[69] Nicholson, J. M. and Ioannidis, J. P. A., Research grants: Conform and be funded, *Nature* **492** (2012) 34–36.
[70] Bollen, J., *et al.*, An efficient system to fund science: From proposal review to peer-to-peer distributions, *Scientometrics* **110** (2017) 521–528.
[71] Garner, H. R., McIver, L. J. and Waitzkin, M. B., Research funding: Same work, twice the money? *Nature* **493** (2013) 599–601.

Advances in Complex Systems
Vol. 24, No. 7 (2022) 2250003 (18 pages)

DOI: 10.1142/S0219525922500035

THE ROLE OF NETWORK EMBEDDEDNESS ON THE SELECTION OF COLLABORATION PARTNERS: AN AGENT-BASED MODEL WITH EMPIRICAL VALIDATION

FRANK SCHWEITZER*, ANTONIOS GARAS, MARIO V. TOMASELLO,
GIACOMO VACCARIO and LUCA VERGINER

*Chair of Systems Design, Department of Management
Technology and Economics, ETH Zürich
Weinbergstrasse 56/58, CH-8092 Zürich, Switzerland
fschweitzer@ethz.ch

Received 19 January 2022
Revised 23 March 2022
Accepted 7 April 2022
Published 25 May 2022

We use a data-driven agent-based model to study the core–periphery structure of two collaboration networks, R&D alliances between firms and co-authorship relations between scientists. To characterize the network embeddedness of agents, we introduce a coreness value obtained from a weighted k-core decomposition. We study the change of these coreness values when collaborations with newcomers or established agents are formed. Our agent-based model is able to reproduce the empirical coreness differences of collaboration partners and to explain why we observe a change in partner selection for agents with high network embeddedness.

Keywords: R&D network; agent-based modeling; scientific collaborations; R&D alliances; firm alliances.

1. Introduction

Collaboration is a pervasive phenomenon in the animated world. We find it at various organismic levels [3, 6], ranging from cancer cells to bacteria, from gregarious insects to bats and fish [4, 7, 11, 13, 14, 21]. We observe collaboration also in humans and even between human-generated higher-level structures, e.g., between economic firms [10, 25] or political parties [18]. In abstract terms, a collaborative effort most often leads to better results than the additive outcome of isolated efforts. In economics, phenomena such as the division of labor or the establishment of global supply chains are based on this rationale [23, 24]. In social systems, we find, for instance, that publications are written by a larger group of co-authors or strategic alliances between political actors are formed [19, 31, 32].

*Corresponding author.

Complexity science addresses the question of whether we can detect overarching principles to characterize collaborative systems and, subsequently, mechanisms to establish collaboration. The challenge comes with the abstraction: Instead of focusing on the specificity that distinguishes collaborating firms from, e.g., collaborating scientists, complexity science is primarily interested in the commonalities in the general principles that constitute collaborative systems. This aspiration entails two methodological problems: (i) We have to define quantitative measures that allow us to characterize and compare collaboration structures across systems. (ii) We have to identify dynamic principles that generate collaboration irrespective of the entities, or agents in general, that form the system.

As one successful approach to problem (i), the network representation has been established. Agents as the system elements are represented by nodes and interactions between agents as links. The network structure then allows to define topological measures to characterize the network position of agents and to relate it to their collaboration effort. From this perspective, successful collaboration networks should display similar features and collaborative agents could be identified by their network position.

Such an approach would merely state a relation between observed collaborations and certain network features. It does not explain why agents collaborate and how they establish their collaborations. To address problem (ii), agent-based modeling has been proposed. While this approach is deeply rooted in economics, as well as in computer science, there is no general way of developing agent-based models. We can distinguish at least two different directions [20]. The first one starts from the economic perspective [1, 12, 15–17]. Agents collaborate because they obtain a benefit. This requires defining utility functions, i.e., costs and benefits for agents. Further, one must define how agents evaluate current and expected outcomes, what information they take into account and how they make decisions. This is mostly done in a formal manner that allows us to analyze the mathematical properties of the model, albeit with restrictions for the chosen mathematical expressions. Such an approach can replicate certain topological features of observed collaboration networks, which lends some evidence to the underlying assumptions about utilities and decision rules. At the same time, all results crucially depend on these assumptions. Thus, instead of obtaining a general picture of how collaboration structures evolve, we mostly learn what distinguishes the utilities and decisions of firms from, e.g., scientists.

Therefore, in the second approach to agent-based modeling agents have a set of possible rules, which they follow with a certain probability [28]. This set of rules is neither complete, nor exclusive. It is rather motivated by empirical observations of possible actions that agents can choose in a given situation. Importantly, the probabilities to follow certain rules are obtained from data. We therefore call this a data-driven modeling approach. The rules are in some sense "universal", i.e., they apply to different collaborative systems, while the probabilities reflect the specifics of the system.

Similar to the first perspective, the validity of the modeling approach is measured against its ability to reproduce real-world collaboration structures. Succinctly, if problem (i) is solved by means of a network representation that gives us structural measures to compare different collaboration systems, then solving problem (ii) by means of data-driven agent-based modeling allows to compare mechanisms to establish collaborations across systems.

In our paper, we illustrate the power of our approach by modeling and analyzing two very different collaboration systems: R&D (research and development) collaborations between firms and co-authorship relations between scientists. To show that the same quantitative characterization and the same dynamic assumptions to form collaborations can be applied to systems from different domains, we use an agent-based model, i.e., firms and scientists are abstracted as agents in the following. The details of our agent-based model and its calibration for the two different systems are explained in Sec. 2.

In this paper, we focus on one specific question, namely how the network position of agents affects the selection of collaboration partners. Our study is motivated by the empirical observation that collaboration networks show a pronounced core–periphery structure, where a small, but highly integrated core of agents coexists with a large and sparse periphery. To quantify the network embeddedness of agents, in Sec. 3, we introduce the coreness value to measure the distance from the core. We then study how differences in coreness values evolve if agents start new collaborations.

From a dynamic perspective agents entering the network usually do not start from the core, but from the periphery. That means, during the evolution of the network some agents manage to better integrate themselves into the network, but others not. This brings up an important question: do agents follow specific strategies to improve their network embedding? If they do so, then in an agent-based model we should find that their actions could not be captured by the simple probabilistic rules we apply for the "normal" agents. If, on the other hand, they do *not* follow specific strategies to enter the core, then we could argue that their better network integration is the result of *chance* more than of strategic choice. Our detailed discussion in Secs. 3.4 and 4 shows that our agent-based model is able to reveal the feedback mechanisms that lead to the observed practice in partner selection.

The remainder of the paper is divided as follows. In Sec. 2, we present the agent-based model together with an overview of the collaboration data used to calibrate the model parameters. In Sec. 3, we investigate the empirical network embeddedness of firms and of scientists and compare it with the outcome of our agent-based model. This is followed by a discussion of the results in Secs. 3.4 and 4.

2. Modeling the Formation of Collaboration Networks

2.1. *Agent-based model of collaboration networks*

In the following, we utilize a recently proposed agent-based model which was already applied to collaborations between firms [26] and between scientists [29]. We consider a multi-agent system with N agents. They represent either firms in an R&D network or scientists in a co-authorship network, and links between agents represent collaborations. Collaborating agents form groups of various sizes m, i.e., R&D alliances or co-authorship teams, which appear as fully connected cliques in the collaboration network.

Our model uses two macroscopic features of empirical collaborations as input, the distribution of agents' activities, $P(a)$ (see Sec. 2.2), and the distribution of alliance sizes, $P(m)$. Activity defines the propensity of each agent to initiate a collaboration. From the distribution $P(a)$, we initially sample without replacement an *activity* value a^i for each agent. During the simulations at every time step agent i initiates a collaboration with probability $p^i \propto a^i dt$. Thus, at each time step the number of active agents is $N_A \propto \langle a \rangle N dt$, where $\langle a \rangle$ is the average agent activity. Upon activation, an agent becomes an *initiator*, i.e., selects the number of partners, m, with whom the collaboration is formed. This value of m is sampled without replacement from the empirical distribution of collaboration sizes, $P(m)$. By sampling without replacement, the number of created links is exactly equal to the number of links in the empirical network.

The second fundamental attribute of agents is their *label* l^i. The *label* attribute is used to model the participation of an agent in different groups with shared practices and/or behaviors. For the case of firms forming R&D alliances, labels translate to membership, "clubs" or "circles of influence". For co-authorship teams, labels indicate specific scientific specializations.

Labels do not change over time, but can propagate to other agents. We assume that collaborations allow the transfer of labels to those agents that are not labeled yet. Specifically, at the beginning of a simulation, all agents are *non-labeled*, i.e., they are *newcomers* with a blank membership attribute. Once they received a label, we denote them as established agents, or *incumbents*. A newcomer can obtain its label in two ways: (i) the agent either receives the label from another agent, if the latter initiates an collaboration (label propagation), or (ii) it takes an arbitrary and unique label when it becomes active for the first time (label generation).

This label propagation process is mapped to the formation of collaborations by means of five probabilities for link creation. If the initiator of a collaboration, chosen by its activity, is a newcomer (non-labeled), it links to a labeled agent with probability p_l^{NL}, or to another non-labeled agent with probability p_{nl}^{NL}. If the initiator is an established agent (labeled), it has three options to form a link. It can (i) link to an agent with the *same* label with probability p_s^L, (ii) link to an agent with a *different* label with probability p_d^L, or (iii) link to an agent without a label with probability p_n^L.

Because the number of collaboration partners, m, is already given from the sampling, the above five probabilities decide how many of the m partners come from each of the three *partner categories*: *same (s)/ different (d)/ no label* (nl). But the link probabilities alone are not sufficient to reproduce the features of empirical collaboration networks. We need an additional dynamic rule to actually *select* the partners *within* the three partner categories. Here, we use a linear preferential attachment rule, where the probability to attach to a node j linearly scales with its degree d^j, i.e., $\Pi(d^j) \propto d^j$. This preferential selection affects only incumbents that are already assigned to a partner category, as by definition newcomers are non-labeled and have no previous partners ($d^j = 0$). Therefore, if the initiator connects to a newcomer, the partner is selected among all non-labeled nodes with equal probability.

Because the preferential attachment rule applies to agents with very different activities, this results in a reinforcement dynamics which is important to understand the emergence of the core–periphery structure of the network. If agents have a high activity, they also have more collaborations over time, and therefore a higher (weighted) degree in the collaboration network. The linear preferential attachment rule implies that within the partner categories "*same/different* label" agents with a higher degree are also chosen with higher probability. This further increases their degree or at least the weight of the link in case of repeated collaborations, which eventually improves the network embeddedness of these agents.

When the partner selection process is complete, all m partners are mutually connected, forming a fully connected clique of size $m + 1$. This reflects the meaning of R&D consortia or of co-authorship teams. Once an agent has established a collaboration, it will remain in the system until the end of the simulation. It is grounded in the fact that the data set do not contain any information about the duration of the alliances or the exit of agents from the network. Therefore, we could not implement a reasonable exit dynamics into the model.

To summarize, our agent-based model is an *activity-driven model*, i.e., from the empirical distribution of activities agents get assigned a (fixed) activity a^i to form collaborations. Obviously, in a stochastic simulation agents with a higher activity are on average chosen *earlier* and *more often*. This generates a first mover advantage because such agents can increase their degree, i.e., the number of collaborations, early on. In the beginning, they also get a higher chance to propagate their *label* to other (unlabeled) agents.

We note that our agent-based model does not make strategic assumptions about collaborations. Instead, the decision of agents in establishing links with newcomers or incumbents are modeled only by means of the mentioned five probabilities, which need to be calibrated for firms and for scientists, separately. Therefore, this is an ideal null model to test whether the observed dynamics of the resulting collaboration network needs strategic agent considerations as an explanation.

2.2. *Calibration of the agent-based model*

In order to calibrate the mentioned probabilities for link formation, we need to use different data sets about collaborations of firms and of scientists. This calibration procedure was carried out and described in detail in previous publications, therefore, we only summarize it here.

Data sets. For firm collaborations we use Thomson Reuters' *SDC Platinum alliances database.* This contains all publicly announced R&D partnerships ("alliances") between 1984 and 2009 with a resolution of 1 year. In total, we have $E = 14,829$ alliance reports, referred to as collaboration events E, involving $N = 14,561$ firms.

For the co-authorship collaborations of scientists, we use the data set from the American Physical Society (APS) about papers published in any APS journal, namely Physical Review Letters, Reviews of Modern Physics, and all Physical Review journals.[a] This data set is quite large, it spans 110 years (1895–2009) and contains $N = 226,724$ unique authors and $E = 1,567.084$ publications, i.e., collaboration events. For the empirical study of network embeddedness we used the full data set. But for the calibration of the agent-based model we only selected papers published between 1984 and 2009 in specific research areas identified by their PACS code. We have restricted ourselves such that the time periods for both data sets are the same and the collaboration networks are of comparable size. In Table 1, we present the example of PACS 42 (Optics), for which we have in total $E = 20,105$ publications involving $N = 27,436$ scientists.

Distributions. From these data sets, we calculate the distribution of collaboration sizes, $P(m)$, as well as the activity distribution, $P(a)$, both for the firms and for the scientists. These distributions, which are used as an input for the agent-based simulations, are very broad [29]. In particular, the activity distributions span several orders of magnitude. Here, the empirical activity of a given agent i at time t is the number of collaboration events, $e_{i,t}^{\Delta t}$, involving agent i during a time window Δt (in years) ending at time t divided by the total number of collaboration events, $E_t^{\Delta t}$, involving any agent during the same period of time. We mention that the activity distributions are very stable regardless of the chosen Δt.

Network reconstruction. In a next step, we reconstruct the aggregated collaboration networks for firms and for scientists. We emphasize that these are undirected, but *weighted* networks, where the weight w^{ij} gives the number of collaborations between agents i and j over the whole time. Further, we note that the number of links, L, is not the same as the number of collaboration events, E. A publication co-authored by 4 scientists, for instance, would count as one collaboration event of size $m + 1 = 4$, but it generates 6 links between the 4 involved scientists.

Figure 1 shows these two networks as unweighted networks. There are two important observations: (i) Both collaboration networks have a largest connected component (LCC) and a large number of small disconnected clusters. (ii) The LCC itself shows a prominent *core–periphery* structure, which may be difficult to see because the LCC is quite dense. But for the R&D collaborations, for instance, we note that the inner core contains less than 250 out of 14,000 firms. When we analyze the evolution of network positions of agents in the following, these obviously refer to the core–periphery structure of the LCC only.

[a] http://www.aps.org/.

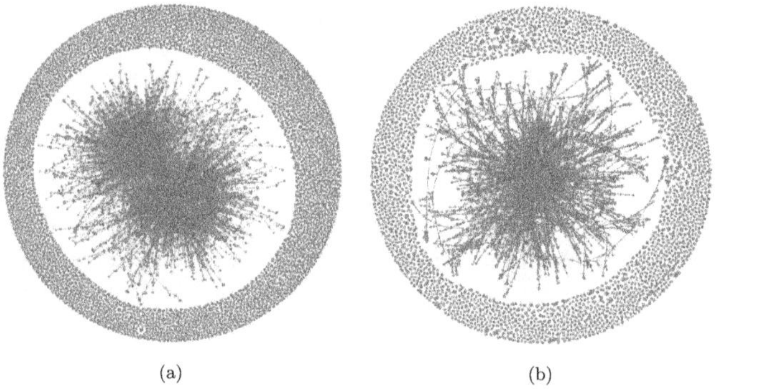

Fig. 1. Illustration of the collaboration networks of firms (a) and of scientists (b). Data: (a) complete R&D network with about 14,000 nodes and 21,000 links, (b) co-authorship network sampled from the full data set with about 11,000 nodes and 32,000 links.

Calculating average quantities. From the observed aggregated networks, we calculate three mean quantities: (i) Average degree $\langle d\rangle^{\text{obs}} = 2N/L$, where N is the total number of nodes and L is the total number of links. The normalization factor of 2 results from the fact that each link connects two nodes. (ii) Average path length $\langle l\rangle^{\text{obs}}$. A path is formally defined as a sequence of nodes, where any pair of consecutive nodes is connected by a link, i.e., loosely speaking the path length is the number of steps to reach a node over the network from a given starting point. (iii) Average clustering coefficient $\langle c\rangle^{\text{obs}}$. The local clustering coefficient of a node captures the fraction of its neighbors that are directly connected, i.e., loosely speaking it counts the fraction of triangles in a neighborhood. $\langle c\rangle^{\text{obs}}$ is the mean of all local clustering coefficients.

From the R&D network, we further obtain the degree distributions $P^{\text{obs}}(d)$, which give the number of collaboration links of firms, to later compare it with our simulation results in Sec. 3.3.

Calculation of link probabilities. To determine the quantities p_l^{NL}, p_{nl}^{NL}, p_d^L, p_s^L, p_n^L, we run agent-based simulations with all possible combinations of values. We take N and E as input, further we sample agent activities a^i and collaboration sizes m from the respective distributions. From each simulation, we construct the respective network on which we calculate the mean values $\langle d\rangle^{\text{sim}}$, $\langle l\rangle^{\text{sim}}$, $\langle c\rangle^{\text{sim}}$. We can then determine the error ϵ from the differences between the observed and the simulated mean values: $\epsilon_d = \left|\langle d\rangle^{\text{obs}} - \langle d\rangle^{\text{sim}}\right|$, $\epsilon_l = \left|\langle l\rangle^{\text{obs}} - \langle l\rangle^{\text{sim}}\right|$, and $\epsilon_c = \left|\langle c\rangle^{\text{obs}} - \langle c\rangle^{\text{sim}}\right|$. We require that these three errors have to be smaller than a threshold ϵ^0. For all probability combinations we perform 25 simulations and select the combination that gives us the highest fraction of networks that match the criterion $\epsilon < \epsilon^0$.

Table 1. Networks for R&D and for co-authorship collaborations (PACS 42): Number of nodes N, links L, collaboration events E. Optimal sets of link probabilities to simulate the collaboration networks are indicated by *. The probabilities of a labeled and an unlabeled agent both sum to 1. These two constraints reduce the number of free parameters from 5 to 3.

Collaboration	N	L	E	p_s^{*L}	p_d^{*L}	p_n^{*L}	$p_l^{*\mathrm{NL}}$	$p_{nl}^{*\mathrm{NL}}$
Firms	14,561	21,572	14,829	0.30	0.30	0.40	0.75	0.25
Scientists	27,436	94,961	20,105	0.60	0.05	0.35	0.35	0.65

Optimal simulation values. From our calibration procedure, we have determined for each data set those link probabilities (indicated by *) that would generate an *optimal* network in the sense that the expected error between observations and simulations is minimized. We emphasize that this only refers to the *averaged* quantities defined above. We have only used *global information* for the calibration, no specific knowledge about link preferences, etc.

Table 1 summarizes our values obtained, together with some characteristics of the data.

Interpretation of probabilities. We note that the values of the link probabilities allow for an interpretation [29]. For R&D collaborations between firms, for instance, we found that incumbents follow a balanced collaboration strategy. 30% of their collaborations are with agents in the same circle of influence ($p_s^{*L} = 0.3$), 30% with agents in a different circle of influence ($p_d^{*L} = 0.3$) and 40% with newcomers ($p_n^{*L} = 0.4$), represented by non-labeled agents. At the same time, newcomers show a strong tendency to connect to incumbents ($p_l^{*\mathrm{NL}} = 0.75$), as opposed to a low linking probability with other newcomers ($p_{nl}^{*\mathrm{NL}} = 0.25$).

Comparing these values with our findings for co-authorship networks, we note both similar and different tendencies. First, established agents prefer to form links with other established agents ($p_s^{*L} + p_d^{*L} \geq 0.6$). Second, when forming a link with an established agent, the initiator tends to select an agent with the same label ($p_s^{*L} > p_d^{*L}$). This tendency is much more pronounced in co-authorship networks, i.e., to choose a co-author from a different community is less likely than to choose a firm from a different circle of influence. Third, in co-authorship networks newcomers have a stronger tendency to link with other newcomers ($p_{nl}^{*\mathrm{NL}} > p_l^{*\mathrm{NL}}$), while for R&D collaboration the opposite is true: newcomers preferably link to incumbents ($p_l^{*\mathrm{NL}} > p_{nl}^{*\mathrm{NL}}$). This difference could be explained with different entry barriers for publications and patents. Scientists who are newcomers to the publication market can still find opportunities to write papers together, whereas firms that are newcomers to R&D activities may find it more difficult to generate patents.

Validation criteria. We can now use the optimal values for the link probabilities to simulate the evolution of synthetic networks over time. The results, averaged over many runs, are then compared to findings from the empirical network. A mere match between empirics and simulations would not be sufficient to conclude that the underlying rules of link formation are correct, because we cannot rule out that other sets of rules would lead to equally good results. However, a good agreement lends strong evidence to our agent-based approach, in particular if the *same* model is also able to replicate *different* empirical findings.

In fact, our agent-based model reproduces well the empirical *distributions* of path lengths, clustering coefficients, degrees, component sizes, etc., in two different domains, collaborations between firms and between scientists [29]. Note that the model calibration only uses information about the *mean values* of these distributions, i.e., no complete specification. Reproducing the empirical distributions is remarkable because it means that we are not simply (over)fitting free parameters to available observations. Second, this suggests that the proposed model rules capture the essence of the analyzed collaboration interactions.

3. Dynamics of Network Embeddedness

3.1. *Measuring network embeddedness*

In this paper, we focus on a quantitative measure for network embeddedness, which we take as a benchmark here. To characterize the topological embedding of nodes in a network, various centrality measures have been proposed that are also partially correlated [8]. Because our networks show a clear core–periphery structure, we have introduced a centrality measure for weighted networks, called *coreness* C_C^i [9]. Versions for unweighted networks have been used earlier in social network analysis [2, 22]. In a recent paper [30], we have shown that our coreness measure is well suited to quantify network embeddedness, in particular when compared to other centrality measures. It also strongly correlates with the success of agents, as quantified by non-topological measures such as the number of patents for firms or the number of citations for scientists. Hence, we have argued that our measure of embeddedness can be used to characterize the innovation potential of firms [30].

We do not replicate the argumentation here, but only explain how coreness values for agents are obtained from the reconstructed network. We use the so-called k-core decomposition (see Fig. 2), which *recursively* removes all nodes with a degree less than d from the network, similar to a *cascade*. It starts with $d = 1$, i.e., it removes all nodes that have only *one* neighbor in the networks. The removal may leave these neighboring nodes with one additional neighbor, hence in the second step of the cascade such nodes are also removed. Their removal again may leave other nodes with one remaining neighbor. Thus, in the third step they are also removed and so forth, unless the cascade *stops*. Then, *all* nodes that have been removed during this cascade are assigned a shell number k_s equal to d.

Nodes with a small k_s obviously are not well integrated in the network, whereas nodes with the largest $k_s = k_s^{\max}$ are in the *core* of the network. The coreness of each node is then defined as $C_C^i = k_s^{\max} - k_s^i$, i.e., it measures the distance from the core. *Low* coreness values characterize the *core*, whereas *high* coreness values characterize the *periphery* of the network.

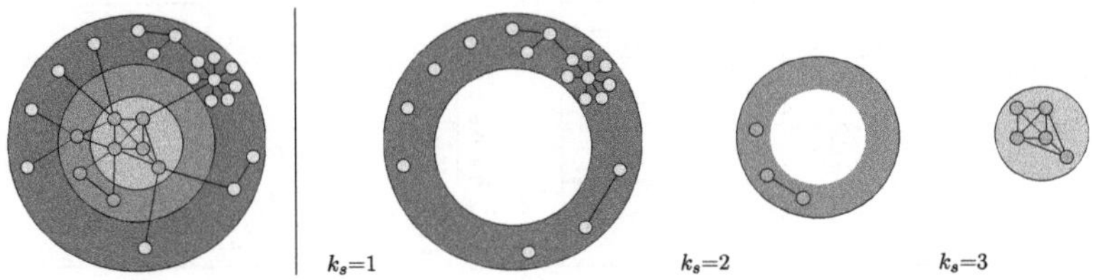

Fig. 2. Illustration of the weighted k-core decomposition.

Instead of the unweighted k-core decomposition, in this paper we apply the *weighted k-core decomposition* [9]. It uses the *weighted degree*, d', defined as follows:

$$d'^i = \left[(d^i)^\alpha \left(\sum_j^{d^i} w^{ij} \right)^\beta \right]^{\frac{1}{\alpha+\beta}}. \tag{1}$$

Here, d^i is the degree of node i and w^{ij} is the weight of the link between nodes i and j. The summation goes over all neighbors of i. The free parameters α and β can balance the influence of the weights w^{ij}. Following [30], we set $\alpha = 1$ and $\beta = 0.2$.

3.2. *Empirical dynamics of network embeddedness*

In the following, our focus is on the *evolution* of network embeddedness, as quantified by the coreness values $C^i_C(t)$ of individual agents i. These values change over time either because new collaborations are established that involve agent i or because the network as a whole grows. The latter means that new agents enter the network and form new links to incumbents, i.e., established agents, or to other newcomers. As the result of network growth, new k-shells appear, which instantaneously affect the coreness values of all agents, even if they not establish new collaborations.

To compare coreness values at different times, we introduce the *relative* coreness $c^i(t) = C^i_C(t)/C_m(t)$, i.e., the ratio between the current coreness $C^i_C(t)$ and the maximum coreness $C_m(t)$ at the same time. $c^i(t)$ can have values between 0, which indicates the very *core*, and 1, which indicates the outermost *periphery*.

Figure 3 shows two examples, one for a firm, the other one for a scientist, of how these relative coreness values change over time in the respective R&D or co-authorship network. We see that in both cases agents start with high relative coreness values because the collaboration network still has to be established. In the two examples, for both agents the relative coreness values then decline over time, indicating that they become part of the core as the network evolves.

We note that agents do not always keep their network position in the very core ($c^i = 0$), as can be seen for the firm GlaxoSmithKline. It has entered the R&D network in 1990 and reached the minimal distance to the core in 1994. Then, it slowly

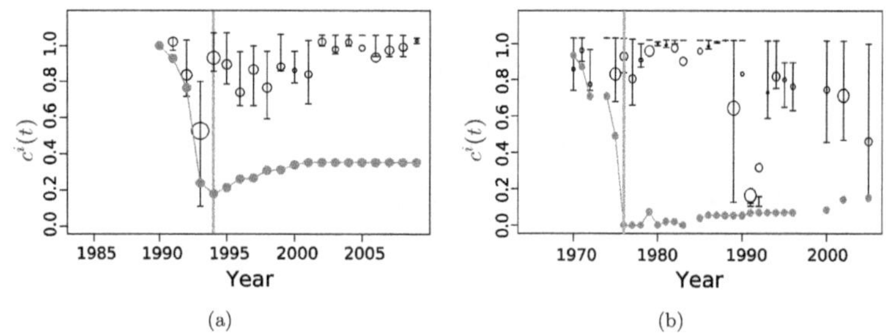

Fig. 3. (Color online) Evolution of relative coreness in blue for (a) Glaxo and (b) Feldmann. The year in which the minimum relative coreness value is marked with a red vertical line. The average relative coreness of the partners of the two selected agents is plotted using black circles. The size of the circle is proportional to the number of partners in that year.

moved away from the core, in particular because *other* firms managed to become better integrated into the core. Only in 2001, the firm reached a stable network position. Hence, for each agent we can identify a minimum relative coreness value $c^i(t^i_c)$ corresponding to the best overall position in the collaboration network and a time t^i_c at which the maximum embeddedness of agent i in the network is obtained. This is indicated by a red line in Fig. 3.

As an additional information, Fig. 3 also shows for the two selected agents their relative coreness together with the relative coreness of their collaboration partners. This allows an interesting observation. In the early period when the two agents are rather new to the network and are thus still part of the periphery, they have a strong tendency to collaborate with partners that have a *comparable* coreness value. This continues until the agents reach their state of minimal coreness, $c^i(t^i_c)$. For times $t > t^i_c$ we observe a change: once agents have reached the core, i.e., are well embedded in the network, they collaborate with partners of *high coreness*, i.e., newcomers or agents from the periphery.

We verify this finding by taking into account all collaborations of firms and of scientists over time. For each agent i, we calculate the relative coreness $c^i(t)$ for every year t and the time t^i_c of minimal coreness. We further calculate the number of collaborations with each of their partners j, i.e., $w^{ij}(t)$, and the total number of collaborations $A^i(t) = \sum_j w^{ij}(t)$ in the given year. Eventually, we calculate the relative coreness $c^j(t)$ of each of their partners j. Combining all these information, we obtain the weighted average of *coreness differences*:

$$\langle dc^i(t)\rangle = \frac{1}{A^i(t)} \sum_j w^{ij}(t)[c^i(t) - c^j(t)]. \tag{2}$$

After calculating $\langle dc^i(t)\rangle$ for all times t between t_{start} and t_{end}, i.e., between 1984 and 2009, we divide the values according to two time periods, `before` and `after` t^i_c and average for each of these periods separately:

$$\langle dc^i_{\text{before}}\rangle = \frac{1}{t^i_c - t_{\text{start}}} \sum_{t=t_{\text{start}}}^{t^i_c} \langle dc^i(t)\rangle; \quad \langle dc^i_{\text{after}}\rangle = \frac{1}{t_{\text{end}} - t^i_c} \sum_{t=t^i_c}^{t_{\text{end}}} \langle dc^i(t)\rangle. \quad (3)$$

For each agent i, these two values are related to the final coreness value of that agent, C_F at t_{end}, i.e., $\langle dc^i_{\text{before}}\rangle(C_F)$ and $\langle dc^i_{\text{after}}\rangle(C_F)$. Then, for each value of C_F, e.g., between 0 and 17 for the case of firms, we average the $\langle dc^i\rangle$ with the same C_F separately `before` and `after` t_c.

The results are shown in Fig. 4. We observe that, for the two periods `before` and `after` t_c, the averaged coreness differences *decrease* monotonously with final coreness C_F and even become negative. Positive values mean that the initiating agent has, on average, a higher coreness than its chosen partners. This applies for initiators with high coreness, i.e., newcomers or agents in the periphery that strive to get a better network position by choosing better integrated partners. Negative values mean that this relation switches: initiating agents have on average a lower coreness, i.e., they are better integrated than their partners. This applies for initiators with low coreness that made it to the core and confirms the previous discussion that agents closer to the core have more collaborations with newcomers or agents with high coreness.

Looking particularly at differences between the two time periods, we find that this shift from positive to negative coreness differences becomes much stronger in the period `after` t_c, i.e., for agents that have already reached their best network position. This means that established core agents choose even more partners with high coreness (newcomers, periphery) than agents in the period `before` t_c that are still striving for a better network position.

To test the robustness of this finding, we performed a random reshuffling of the collaboration links of firms while preserving the degree sequence of the empirical R&D network. The dashed curves in Fig. 4(a) show the averaged coreness differences

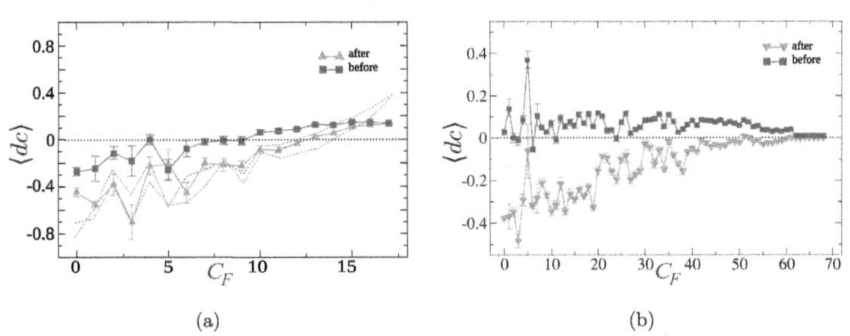

Fig. 4. (Color online) Average partner coreness deviation. Plot of the average normalized partner coreness deviation $\langle dc\rangle$ against C_F before and after t_c. Firms (a) and Scientists (b). The dashed blue lines in panel (a) are obtained by randomly reshuffling the collaboration links of firms while preserving the degree sequence of the empirical R&D network.

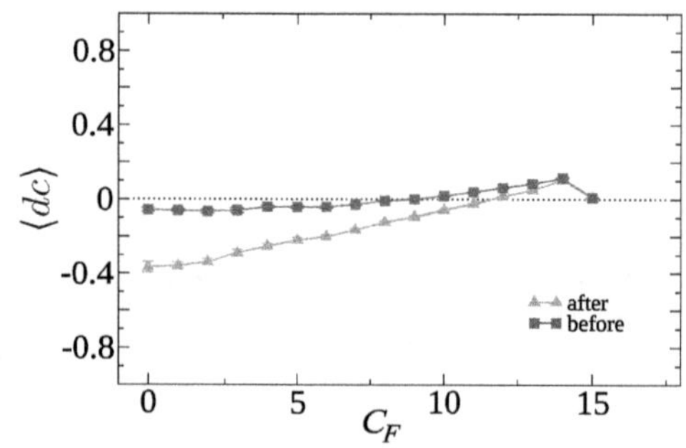

Fig. 5. Average normalized partner coreness deviation $\langle dc \rangle$ before and after t_c obtained from the agent-based model, using the calibration from the R&D network.

`before` and `after` t_c for the reshuffled network. We see that the trend is the same as for the empirical network. However, the *differences* between the two curves are much larger for the *empirical* network than for the reshuffled network. This means that the *observed* change `before` and `after` t_c is not random. We performed a two-sided Kolmogorov–Smirnov test to the distributions of the $\langle dc \rangle$ for the empirical and the reshuffled network, and we can reject that they are the same with $p = 0.056$.

3.3. *Validation of the agent-based model*

To validate our agent-based model on the macro level, we have to verify that the dynamics observed in Fig. 4 for firms and for scientists can indeed be replicated by

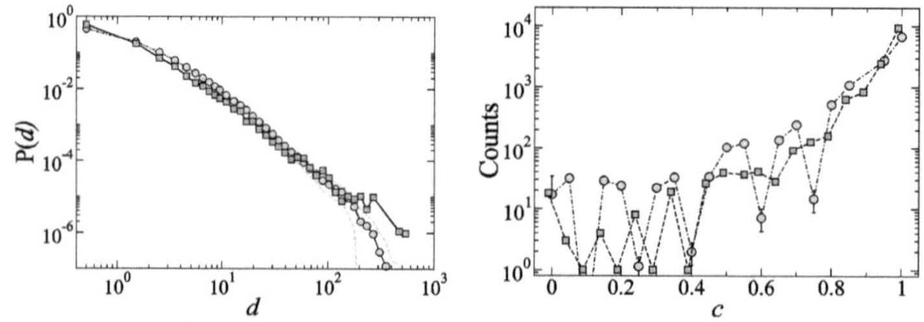

(a) (b)

Fig. 6. (Color online) Model results and validation for R&D collaborations of firms: (a) Degree distribution of the R&D network obtained from the agent-based model (blue line) and from the empirical network (circles). (b) Comparison of the coreness distribution for the R&D network obtained from the agent-based model (orange) and from the empirical network (blue). The results are averaged over 100 model realizations and the error bars (when visible) indicate standard errors.

our model. Our main result is presented in Fig. 5 which should be compared to the empirical findings shown in Fig. 4. It demonstrates that our agent-based model generates the same pattern in partner selection that was observed empirically, namely after reaching their lowest coreness value, agents establish collaborations with more peripheral agents.

To demonstrate that the agent-based model is also able to reproduce other empirical findings without overfitting, we plot in Fig. 6(a) the degree distribution of the network ensembles obtained from the 100 realizations, alongside of the empirical degree distribution for the R&D network, both for the final time. The excellent match of the two distributions should be noted. We further plot in Fig. 6(b) the distribution of the coreness values both from the empirical data and from the computer simulations of the R&D network. Here, we use the normalized coreness C' instead of C_F. While one could argue about some deviations between the two in the range of small coreness values, we note that the core–periphery structure of the network is well captured by the model — without any additional assumptions.

3.4. *Improving network embeddedness: Chance or choice?*

Our observations about the impact of network embedding on the partner selection raise the question whether this impact should be interpreted as a change in the *strategy* of an agent in selecting its partners.

Such a change of strategy could indeed have a rational explanation, as follows. Agents new to the network may have little chances to get connected to core agents. Therefore, in the absence of better alternatives, they may eventually team up with other newcomers or agents from the periphery with comparable coreness. Together with their partners, they then try to improve their network position. However, at the time of maximum network integration, the competition with other agents of similar or lower coreness can become more important than the opportunity to further increase their (already optimal) position. So, while previous partners may have become competitors, successful agents more likely search for, and to team up with, newcomers with fresh ideas.

The question is whether the observed change in partner selection indeed follows a *strategy*, i.e., a deliberative process to become more successful, or whether the "strategy" is still the same but opportunities have changed. Then, contrasting the above explanation, one could argue that differences in partner selection are caused by different opportunities to be involved in a collaboration. To decide between these two alternative explanations, we can use our agent-based model because it allows to disentangle strategic behavior from probabilistic actions. Precisely, in our model agents are assigned constant probabilities for linking to newcomers or incumbents. Hence, differences in observed actions are not expressed by the link probabilities, which are the same for all agents, but by the process to become initiators of collaborations, and to be selected within a partner category "same/different/no label".

Our results have demonstrated that this model is able to reproduce the observed change in partner selection together with other topological features, such as degree distribution and coreness distribution. Therefore, we can conclude that the change in choosing partners can be reproduced *without* assuming changes in the selection rules. This does *not* allow to conclude that agents do not follow strategies in selecting their partners, or change these strategies dependent on the network position. But it demonstrates that agents do not need to *change strategies* to act in a way that is observed in their evolution of network embeddedness.

In fact, the abundance of collaboration opportunities is one of the driving forces behind the process of improving network embeddedness. While it is true that agents with a high activity are more often selected, it is also true that the number of newcomers or peripheral agents is much larger than the number of core agents. Whenever newcomers or peripheral agents are activated to become initiators, they follow the preferential selection rule, i.e., they tend to select partner agents with a larger degree. These are most likely agents from the core, which are well embedded in the network. Because these agents are more often selected for collaboration, their degree increase over time which further increases their probability to be selected, next time. A higher degree, on the other hand, relates to a lower coreness value as outlined above, albeit not in a linear manner. It is this feedback process that eventually helps some agents to further improve their network embeddedness, whereas the majority of agents still stay in the periphery.

4. Discussion

In this paper we have focused on the dynamics of collaboration networks, using two data sets from different domains, about R&D collaborations between firms and about co-authorship collaborations between scientists. To answer our initial question, whether these different collaboration networks can be characterized and modeled from a unifying perspective, we made two contributions.

First, we proposed a new measure for network embeddedness, the (relative) coreness value $c^i(t)$, to compare the topological positions of individual agents. We would like to emphasize that our measure of network embeddedness is also a good predictor of success of individual agents. Looking at the R&D collaboration network of firms and their corresponding patent data [30], we verified that for the successful firms a better coreness comes along with more patents whereas for the "normal" firms both the position and the number of patents is rather level (in comparison to the successful firms).

Monitoring the change of individual coreness values, we found in both data sets the same dynamics (shown in Fig. 4). Some agents over time have improved their network embeddedness by moving from the *periphery* of the collaboration network close to the *core*, i.e., from high to low coreness values. From the data, we obtained a time t_c^i for each agent when the minimal coreness, i.e., the best network embeddedness, is reached. At about t_c^i we observed a *change* in partner selection, from partners of *similar* coreness to partners of *different* (i.e., high) coreness.

To explain this seemingly strategic behavior was one aim of our agent-based model, which is our second contribution. We utilized a stochastic label propagation model with five probabilities to form collaboration links between newcomers and established agents with similar or different labels. These probabilities could be calibrated using aggregated quantities from the respective empirical collaboration networks. As the result, we obtained a *data-driven* agent-based model, where domain specific information is included in interpretable link probabilities. Additionally, we implemented a hypothesis of how agents choose collaboration partners *within* the three partner categories same/different/no label, namely by preferentially choosing agents with a higher degree.

Our model assumes that link formation is unilateral, i.e., mutual consensus is not modeled. This shortcoming is discussed in [27]. Further, our model does not include the termination of collaborations or the exit of agents because the data sets do not contain the respective information. As a consequence, the model may overestimate the number of active agents. To assess the impact on the maximum k-core value, let us assume that the number of active partners in the R&D collaboration network is proportional to the size of a firm. It was observed [5] that firms with a high degree less likely disappear. Hence, we expect that $\langle dc_{\text{after}} \rangle$ is less negative, as links to firms with low degree (which implies low coreness) are more likely to disappear. In other words, the distance between $\langle dc_{\text{before}} \rangle$ and $\langle dc_{\text{after}} \rangle$ could be smaller than estimated. A similar argument applies to the model of scientific collaborations.

Despite these simplifying modeling assumptions we could demonstrate that the model is able to explain the observed impact of network embeddedness on the selection of collaboration partners. In a nutshell, there exists a feedback between an agent's activity on the one hand and its ability to increase the degree by establishing new collaborations and to propagate the own label to newcomers, on the other hand. More collaborations not only lead to higher degrees, but also to lower coreness values, i.e., agents become embedded in the core of the network. The chance to propagate the own label later increases the chance to be selected for new collaborations, because empirics has shown that firms or scientists prefer to collaborate with partners with the same label.

These combined effects eventually explain the observation that core agents tend to collaborate preferably with newcomers or agents from the periphery. In fact, newcomers and peripheral agents choose these core agents with larger probability, once they managed to establish their strong network embeddedness. Thus, in conclusion, what seems to be a deliberative strategy of successful agents, namely to switch their rules of partner selection, can be basically explained without strategic considerations. These cannot be excluded, but the model suggests that the empirical observations do not already imply such considerations. Hence, the emergence of realistic core–periphery structures in collaboration networks can be successfully modeled without deliberative agents.

Acknowledgments

The authors acknowledge financial support from the EU-FET project MULTIPLEX 317532. The calculations of network measures and the statistical treatment of data were performed using the R software for statistical analysis v2.15.2 and the igraph library v0.6.2.

Author Contributions

A.G. and F.S. designed the research, A.G., M.V.T., GV and LV performed the research and analyzed the data. F.S. and A.G. wrote the manuscript.

References

[1] Bala, V. and Goyal, S., A noncooperative model of network formation, *Econometrica* **68**(5) (2000) 1181–1229.

[2] Borgatti, S. P. and Everett, M. G., Models of core/periphery structures, *Soc. Networks* **21**(4) (2000) 375–395.

[3] Couzin, I. D., Krause, J. *et al.*, Self-organization and collective behavior in vertebrates, *Adv. Study Behav.* **32**(1) (2003) 1–75.

[4] Deisboeck, T. S. and Couzin, I. D., Collective behavior in cancer cell populations, *Bioessays* **31**(2) (2009) 190–197.

[5] Dunne, P. and Hughes, A., Age, size, growth and survival: Uk companies in the 1980s. *J. Ind. Econ.* **42**(2) (1994) 115–140.

[6] Ebeling, W. and Schweitzer, F., Self-organization, active brownian dynamics, and biological applications, *Nova Acta Leopold.* **88**(332) (2003) 169–188.

[7] Erban, R. and Othmer, H. G., From individual to collective behavior in bacterial chemotaxis, *SIAM J. Appl. Math.* **65**(2) (2004) 361–391.

[8] Freeman, L. C., Centrality in social networks conceptual clarification, *Soc. Networks* **1**(3) (1979) 215–239.

[9] Garas, A., Schweitzer, F. and Havlin, S., A *k*-shell decomposition method for weighted networks, *New J. Phys.* **14**(8) (2012) 083030.

[10] Goyal, S. and Moraga-González, J. L., R&d networks, *RAND J. Econ.* **32**(4) (2001) 686–707.

[11] Hecht, I., El, Y. B., Balmer, F., Natan, S., Tsarfaty, I., Schweitzer, F. and Jacob, E. B., Tumor invasion optimization by mesenchymal-amoeboid heterogeneity, *Sci. Rep.* **5** (2015) 10622.

[12] Jackson, M. O. and Wolinsky, A., A strategic model of social and economic networks, *J. Econ. Theory* **71**(1) (1996) 44–74.

[13] Jeanson, R. and Deneubourg, J.-L., Conspecific attraction and shelter selection in gregarious insects, *Am. Naturalist* **170**(1) (2007) 47–58.

[14] Kerth, G., Perony, N. and Schweitzer, F., Bats are able to maintain long-term social relationships despite the high fission-fusion dynamics of their groups, *Proc. R. Soc. B: Biological Sci.* **278**(1719) (2011) 2761–2767.

[15] Koenig, M. D., Battiston, S., Napoletano, M. and Schweitzer, F., Recombinant knowledge and the evolution of innovation network, *J. Econ. Behav. Organ.* **79**(3) (2011) 145–164.

[16] Koenig, M. D., Battiston, S., Napoletano, M. and Schweitzer, F., The efficiency and stability of r&d networks, *Games Econ. Behav.* **75**(2) (2012) 694–713.
[17] Koenig, M. D., Tessone, C. J. and Zenou, Y., Nestedness in networks: A theoretical model and some applications, *Theor. Econ.* **9**(3) (2014) 695–752.
[18] Otjes, S. and Rasmussen, A., The collaboration between interest groups and political parties in multi-party democracies: Party system dynamics and the effect of power and ideology, *Party Polit.* **23**(2) (2017) 96–109.
[19] Sarigol, E., Pfitzner, R., Scholtes, I., Garas, A. and Schweitzer, F., Predicting scientific success based on coauthorship networks, *EPJ Data Sci.* **3** (2014) 9.
[20] Schweitzer, F., Fagiolo, G., Sornette, D., Redondo, F. V. and White, D. R., Economic networks: What do we know and what do we need to know? *Adv. Complex Syst.* **12**(4) (2009) 407–422.
[21] Schweitzer, F., Lao, K. and Family, F., Active random walkers simulate trunk trail formation by ants, *Biosystems* **41**(3) (1997) 153–166.
[22] Seidman, S. B., Network structure and minimum degree, *Soc. Networks* **5**(3) (1983) 269–287.
[23] Suzuki, N., Kamiya, T., Umata, I., Ito, S. and Iwasawa, S., Analyzing the structure of the emergent division of labor in multiparty collaboration, in *Proceedings of the ACM 2012 Conference on Computer Supported Cooperative Work* (2012), pp. 1233–1236.
[24] Takeishi, A., Knowledge partitioning in the interfirm division of labor: The case of automotive product development, *Organ. Sci.* **13**(3) (2002) 321–338.
[25] Tomasello, M. V., Napoletano, M., Garas, A. and Schweitzer, F., The rise and fall of R&D networks, *Ind. Corporate Change* **26**(4) (2016) 617–646.
[26] Tomasello, M. V., Perra, N., Tessone, C. J., Karsai, M. and Schweitzer, F., The role of endogenous and exogenous mechanisms in the formation of r&d networks, *Sci. Rep.* **4** (2014) 5679.
[27] Tomasello, M. V., Burkholz, R. and Schweitzer, F., Modeling the formation of r&d alliances: An agent-based model with empirical validation, Economics Discussion Papers 2017-107, Kiel Institute for the World Economy (IfW Kiel) (2017).
[28] Tomasello, M. V., Tessone, C. J. and Schweitzer, F., A model of dynamic rewiring and knowledge exchange in r&d networks, *Adv. Complex Syst.* **19**(1–2) (2016) 1650004.
[29] Tomasello, M. V., Vaccario, G. and Schweitzer, F., Data-driven modeling of collaboration networks: A cross-domain analysis, *EPJ Data Sci.* **6** (2017) 22.
[30] Vaccario, G., Verginer, L., Garas, A., Tomasello, M. V. and Schweitzer, F., Network embeddedness of firms predicts their innovation potential, submitted for publication (2022).
[31] Van Dyke, N. and McCammon, H. J., *Strategic Alliances: Coalition Building and Social Movements* (University of Minnesota Press, 2010).
[32] Zingg, C., Nanumyan, V. and Schweitzer, F., Citations driven by social connections? A multi-layer representation of coauthorship networks, *Quant. Sci. Stud.* **1**(4) (2020) 1493–1509.

PART 2
Network Models

Chapter 9

Overview

Networks on the rise. Networks have been discussed in different scientific disciplines for a long time. Early studies of social networks date back to the 1930s. The seminal random graph models were developed in the late 1950s in mathematics. Boolean networks were introduced for biological applications in the late 1960s. But only in the 1990s did the interest in network science start to rise after fundamental contributions, such as the configuration model, the preferential attachment model, or the small-world model, were proposed.

Advances in Complex Systems has published about network science from its beginning. Already the first two volumes contained papers about communication networks [1] and gene regulatory networks [3, 4]. Every significant development in network science is reflected in the journal's publications, as also the list of topical issues witnesses. Today, networks are the dominant topic across submissions.

Focus on interactions. Complex networks are a specific representation of complex systems. The system elements, i.e., the agents, are represented as *nodes* in a network, and their interactions as *links* between nodes. Network models are primarily focus on the link structure, i.e., on the *topology*. In this respect, network models and agent-based models are complementary.

Applying the network approach requires decomposing all agents' interactions into bilateral interactions between any two agents. This reductionistic view allows one grasping regularities in the distribution of interactions that are hardly visible otherwise. But it makes network models less appropriate to model, e.g., group interactions or indirect interactions. Very recent concepts of *higher-order networks* address this problem using simplicial complexes, a topic *Advances in Complex Systems* has been discussed already in 2012.

Network reconstruction. Many publications about networks assume that the network is given to analyze its topology. They neglect one of the biggest problems in network science, namely reconstructing the network from a given data set, e.g., from sequential data. All topological quantities depend on this process. Different publications address this problem, e.g., for financial data [40], for email data [45], for literary texts [116], or for survey data [148].

The topology is often analyzed for the *aggregated* network that contains all links. The aggregated statistics, in particular the degree distribution, allows to characterize the *heterogeneity* of networks [20], i.e., the variability of node properties like the number of links, the distribution of motives and clustering coefficients [58, 102] or the k-coreness [97] across networks of different origin. A challenge is to generate complex networks for simulations so that specific aggregated empirical properties are preserved [57, 106].

Identifying important nodes. Some problems in network science have been persistently discussed over the last 25 years, as the publications of *Advances in Complex Systems* show. One of these problems is identifying important nodes and quantifying their importance at the topological level. In our publications, new *centrality* measures have been proposed for weighted networks [131], for multilayer networks, [142], and for temporal networks [107]. Further, applications of centrality measures to quantify economic integration [42] or to study financial markets [40] have been published. One of our most cited papers explores the relationship between centrality and traffic flow [13].

Community detection. A second persistent problem regards the identification of communities, i.e., groups of nodes. One of our early published and most cited papers analyzes community structures in Jazz [12], another one in Cricket [86] or in the airport network of the United States [84]. Applications of community detection to software refactoring [98] or to social tagging or blogging systems [36, 52] are also discussed.

A significant share of related publications address the computational problems in identifying communities and suggests new algorithms [63, 67, 76, 100, 102, 123]. Even more, some works aim at a unifying perspective for community detection [90, 134]. Special applications are targeted at heterogeneous multi-relational networks [61, 95] or at small communities in online social networks [64].

Dynamics of networks. After quantifying the network topology, it is of interest how this topology changes over time, for instance, in growing networks or if links are deleted or rewired. *Advances in Complex Systems* has published frameworks to generalize the growth dynamics beyond the known growth model classes [29, 33, 59, 66]. The growth of coupled networks [15] was discussed, but also the robustness of networks against adverse node or link removal [94, 145]. Some papers tried to motivate better the reasons and the rules for link formation dependent on the disciplinary context, e.g., in collaboration networks of firms [117] or social networks [16]. But also growth models for particular networks, like self-organized corona graphs [138], or contact networks in granular matter [9] have been introduced.

Dynamics on networks. Different from the dynamics *of* networks, the dynamics *on* a network assumes that a dynamic process, e.g., a spreading process or a synchronization process [133], is constrained by the underlying network topology [62, 92]. For the dynamics, one has to distinguish between conserved and non-conserved quantities. The transport of packages [10] over a communication network, for instance, would be an example for conserved quantities, also the visitation probabilities of nodes in a random walk [110] are constrained. This conservation impacts dynamic properties such as the transit time or the load distribution. The spreading of viruses, opinions, or information in general [82] refers to non-conserved quantities. The spreading of fads or viruses may further depend on a critical threshold for the respective quantities [19, 69]. Such assumptions require modeling the dynamics inside the nodes, which turns these network models into agent-based model [87].

Network control. The dynamics *on* a network can be utilized to control the network. This approach combines classical control theory and network science [23, 136]. There are also relations to network interventions [109] and to nudging discussed in the social sciences [144]. To apply network control, one first needs to identify which nodes should be targeted with a control signal to steer the network dynamics most efficiently [114, 140, 141, 153]. Further, the energy, i.e., the "cost" for the control signal needed to control the network and the fraction of nodes that can be controlled, are of interest [111, 122, 127, 129, 130]. Applications to cell biology are also discussed [137]. *Advances in Complex Systems* has published a topical issue in 2019, which contains an excellent introduction to the problems [132].

Bipartite and multiplex networks. Recent developments in network science consider that the data contain relations between different types of entities, for instance, between firms and their respective board members [11] or between developers and their projects [99]. From these bipartite networks, different projections can be analyzed. For instance, a link between two firms exists if they are served by the same board member, or a link between two board members exists if they work for the same firm. But instead of such projections, the data can also be represented as a *multi-layer* network. One layer may contain the board members, the other one the firms, and *inter-layer links* connect these two layers. Often, we also study *multiplex* networks with different layers, where each layer contains the same type of nodes, e.g., individuals, and layers reflect the different relations between them, e.g., friendship and work relations. This bears the problem of identifying whether individuals in different levels are the same [56]. Centrality measures must be modified for such networks [142]. Also, spreading processes are affected by the inter-layer and intra-layer topologies, for example, the opinion diffusion in a multi-layer social network [96, 120].

Temporal networks. If links represent interactions, they may not be present all the time but only during particular time intervals. It has consequences, for instance, for spreading information or infectious diseases, which were addressed in *Advances in Complex Systems* already in 2008. It is required that an ordered temporal sequence, or a causal path [35], between nodes a, b, c, d exists to pass on the information or the disease from a to d. An early analysis of such a temporal network for the case of literary texts [116] demonstrates that, for instance, importance measures of fictional characters are considerably impacted. A practical application is the improved trend prediction in online services based on temporal information [88]. Applications to optimal routing strategies are also discussed [124]. More theoretical analysis with helpful background information is found in [107, 135].

Application: Economic networks. Applications of network concepts to socio-economic problems are an important focus of *Advances in Complex Systems* since its launch. In addition to principal discussions about the challenges in economic network research [49], ownership networks [147], networks of direct investments [27, 112] and the world trade network [91, 126] have been analyzed both empirically and with network models. Another important topic is the impact of networks on innovations [30, 34, 149] and knowledge exchange [2, 117].

More abstract models used Boolean networks and scale-free networks to simulate industry relations and markets [37, 51] or tried to understand the transaction network in the housing market [6]. On the other hand, practical insights have been derived from a data-driven study of firms' bankruptcy in production networks [38].

Application: Financial networks. Bankruptcy and, more general, systemic risk is also the main topic in studies about financial networks. Again, these models could be seen as agent-based models because they require understanding of the conditions for the failure of individual nodes from a financial context. On the one hand, these models bear analogies to epidemic spreading. On the other hand, they are more complex because not only the dynamics of nodes but also the dynamics along the links need to be explicitly taken into account.

Balance sheets [73, 108], banking relations [105] and transaction networks [71] have been studied empirically. Based on these insights, cascade models are utilized to study the origin and the spread of *financial contagion*, mainly with an application to interbank networks [68, 77, 113, 118, 139]. Other publications study user networks in electronic auctions [101] or explain market fluctuations through investor networks [121]. Artificial neural networks are utilized to forecast price increments [7].

Application: Infrastructure networks. In transportation or road networks, the stability problem is addressed under the more general term resilience. It denotes the ability of a system to regain its functionality after a shock. These shocks can be caused *exogenously* from the dynamics *of* the network, for instance, from the removal of links or nodes in a transportation network due to an earthquake. More often, these shocks are generated *endogenously* as a result of the dynamic processes *on* the network, e.g., from a fluctuating traffic demand [104, 115]. To model resilience, it is essential to understand the role of capacity constraints [72] and the impact of hierarchical network structures [46] on the emergence of congestion.

These networks' *spatial embedding* is the bigger challenge. Links have a spatial and a temporal dimension that translates to costs and considerably constrains the network [26]. The spatial influence becomes visible in the different topological features of complex networks [53, 129]. Some papers explicitly map spatial models to the geography of countries, for instance, the international networks of innovators [150] or the motorway network of the Netherlands [83]. The latter is a very particular way of modeling spatial networks using the slime mould approach, which links transportation and biology.

Application: Software development. The great versatility of the network approach is demonstrated in software engineering. Both the *product*, i.e., the software, and the *production*, i.e., the collaboration of software developers, can be modeled as networks. *Advances in Complex Systems* has published a topical issue about such studies in 2014.

The network approach allows to analyze the structure and the dependencies of software modules [47, 103, 125], also different software systems can be classified using their topological features [97]. Evidently, software functionality requires some level of modularity [74]. The dynamics of software developments were also studied empirically [5, 31] and problems of collaborative software engineering were reflected from a complex network perspective [89]. This regards the interactions between users and developers [93] but also the project size [99].

Application: Social networks. Social systems are studied as complex networks at various levels of their organization. Popular are online communities because they provide rich data sets from user interactions, for example, from political discussions [79]. Very intriguing is the idea to set up large-scale online games such that specific user interactions are induced and to analyze the respective data [81]. But there are also experiments with small groups [75] to identify their signed relations in virtual social networks. A more formal and lesser known index to quantify signed relations was proposed recently [151].

Other studies focus on collaboration networks [14] or higher-order structures in social networks needed to understand cooperation and conflict [78].

Application: Language and text analysis. Communication is the main process in social networks. Consequently, the research focus in social science is extended from the communicating actors to the communication means, i.e., language, and the artifacts, i.e., literary texts. What can the network approach contribute to such analyses? On the most basic level, the language structure is described by network tools [39]. Further, language acquisition is mediated by the structure of social networks [44].

However, most effort is spent exploring the network structure in literary texts. It can be used to classify specific texts, e.g., the epic poems of Ossian [119], but also to reveal relations between fictional characters in these texts [116, 143]. At the same time, these studies provide more insights into linguistic structures [48, 128].

This type of structural analysis can be also utilized to detect network relations between scientific concepts [152], or Wikipedia categories [80]. Eventually, the structural analysis can also be applied to music [25] to identify network motives.

Application: Computer networks. Networks involving computers and algorithms provide a challenging combination of infrastructure and communication networks. At the communication level, various load balancing protocols and their applications are simulated or implemented [28]. The formation of ad-hoc networks [146] or the performance of unstructured peer-to-peer networks is also studied [65]. A different application domain is algorithms for distributed computing [55] and data mining [85].

Application: Functional networks. Publications in *Advances in Complex Systems* about biological networks mainly focus on biomolecular networks. They have been more frequent in the earlier days, whereas now authors prefer disciplinary journals. In metabolic networks, the nodes are usually proteins or enzymes, and links describe their interaction with other proteins or enzymes. The structure and functioning of such networks can be revealed by studying the network topology or by reverse engineering [21, 24, 60]. Because of feedback processes, the role of cycles is particularly important [8, 17]. Further, network control issues also play a role [50].

In gene regulatory networks, interactions between molecular regulators such as RNA and DNA molecules influence the expression of genes and, hence, morphogenesis. Models of these processes often take an abstract perspective, such as in Boolean networks [22], when studying oscillations [41] or on the role of time delays [54]. Regulatory networks evolve, for instance, because of neutral mutations that may impact their function [18]. Processes generating network motives can also be studied by using artificial regulatory networks [32]. On the other hand, gene regulatory networks have also inspired other disciplines, for instance, artificial intelligence [43].

Comments on the reprinted publications

The role of geography and traffic in networks. This paper is not about road traffic but air traffic, specifically in North America. It starts with constructing a large-scale weighted network from the available data. Nodes represent airports and links direct flight connections with a weight given by the number of maximum passengers on the corresponding connection. These weights are also used to define the "strength" of the nodes. Topological analysis of the aggregated network shows correlations between the heavy-tailed degree distribution and the distribution of betweenness centralities. To study the role of space on these correlations, a network model couples the topological growth with the growth of the weights. Adding a spatial component that reflects typical distances between airports, the growth mechanism is turned into a strength-driven preferential attachment with spatial selection. If new nodes are connected to the network, this also implies an adaptation of existing link weights. The paper explores the role of the two free parameters of the growth model on the topology of possible spatial networks. In particular, the relation of betweenness centrality and spatial constraints is studied. Weights, space, and coupled growth lead to nonlinear correlations between topology and traffic.

Filtered graphs from dynamical financial correlations [40]. Time series data from the stock market contain essential information about the dependencies within an economy. But it is not easy to make this information accessible in an understandable manner. The method utilized in this paper represents it as a special kind of network, which can be subsequently investigated using standard topological measures. Instead of drawing networks from correlations, which is quite common but less informative, the analysis focuses on the level of aggregated centralities to map different economic sectors. Figure 1 provides a fresh look at the economy, with room for further investigations.

Social and opportunistic behavior in email networks [45]. Email exchange is still among the most important means of online communication. From the information about the sender, receiver, and time stamp, one can construct a *directed* and *temporal* social network. But email communication is prone to *spamming*, denoted here as "opportunistic" behavior to contrast it with "social" communication. It raises the question of to what extent the network constructed from spam data differs from the network of social communication. In addition to topological differences, this paper also analyses temporal patterns using information entropy. The findings are not only interesting to further quantify human behavior, they are also applicable to improve spam detection algorithms.

Coupling structure and function in dynamic networks [70]. What can be done if the dynamics *on* the network and the dynamics *of* the network cannot simply be decoupled? That means network topology impacts network function and the other way round. To model such problems requires recasting formally the relation between agent-based and network modeling, which both highlight different aspects of the structure and dynamics of complex systems. This paper provides a framework to address such issues. Because of its explicit agent dynamics, we could have presented it under agent-based models. But its aim is to demonstrate how the approach proposed can generate different types of networks. Examples inspired by biology illustrate the applicability.

Temporal network analysis of literary texts. [116] Temporal networks have become a hot topic in network science. So, why not try to turn a literary text into a temporal network? A link exists if two fictional characters from a novel meet face to face in discrete time. This requires solving the underlying problem of network reconstruction. Before this problem is discussed, the paper gives a nice introductory overview of the relations between network science and literature as of 2016.

Based on the constructed series of temporal networks, a supra-centrality matrix is defined, from which different importance measures for the characters are calculated, as well as derived aggregated measures such as "vitality". One can now study the impact of temporal correlations within a literary text. But one can also try to compare different texts, using quantities such as the Freeman index that measures how much a story is focused on a single character. The authors correctly point to the problem when interpreting such results: We now know the answer, but what was the question? If these issues are taken seriously, it will help to bridge the gap between complexity science and digital humanities, which have a growing interest in applying quantitative methods. *Advances in Complex Systems* has published a topical issue about these studies in 2022.

Polarization in survey-based attitude networks. [148] The close relation between agent-based and network models becomes evident when studying social networks in which communities or groups are identified by their opinions. To model opinions or attitudes requires understanding the internal dynamics of agents. Identifying communities in large networks is a common problem of network science. This paper is part of network models because its focus is clearly on the latter issue.

Surveys are used to determine opinions, and empirical data from different years are analyzed to proxy the dynamics. Multidimensional opinions lead to overlapping communities of individuals. This problem can be overcome by applying different methods to identify the most important issues and basing community detection on these issues. An asset of the paper is the comparison of different community detection algorithms and different empirical data sets. On the downside, polarization is only characterized by non-overlapping opinion-based groups. But it is interesting to see how similarity-based networks can lead to different group structures.

Selected publications

[1] Heusse, M., Snyers, D., Guérin, S., and Kuntz, P., *Adaptive Agent-Driven Routing and Load Balancing in Communication Networks*, Advances in Complex Systems **01** (**1998**) 237–254.

[2] Richards, D., McKay, B. D., and Richards, W. A., *Collective Choice and Mutual Knowledge Structures*, Advances in Complex Systems **01** (**1998**) 221–236.

[3] Kopp, S. and Reidys, C. M., *Neutral Networks: A Combinatorial Perspective*, Advances in Complex Systems **02** (**1999**) 283–301.

[4] SolÉ, R. V., Salazar-Ciudad, I., and Garcia-FernÁNdez, J., *Landscapes, Gene Networks and Pattern Formation: on the Cambrian Explosion*, Advances in Complex Systems **02** (**1999**) 313–337.

[5] Dalle, J.-M. and Jullien, N., *Windows Vs. Linux: Some Explorations Into the Economics of Free Software*, Advances in Complex Systems **03** (**2000**) 399–416.

[6] Rosenthal, L., *Simulating Transaction Networks in Housing Markets*, Advances in Complex Systems **03** (**2000**) 371–384.

[7] Castiglione, F., *Forecasting Price Increments Using an Artificial Neural Network*, Advances in Complex Systems **04** (**2001**) 45–56.

[8] Gleiss, P. M., Stadler, P. F., Wagner, A., and Fell, D. A., *Relevant Cycles in Chemical Reaction Networks*, Advances in Complex Systems **04** (**2001**) 207–226.

[9] Moukarzel, C. F., *Random Multiplicative Response Functions in Granular Contact Networks*, Advances in Complex Systems **04** (**2001**) 523–533.

[10] Tadić, B. and Rodgers, G. J., *Packet Transport on Scale-Free Networks*, Advances in Complex Systems **05** (**2002**) 445–456.

[11] Battiston, S., Weisbuch, G., and Bonabeau, E., *Decision Spread in the Corporate Board Network*, Advances in Complex Systems **06** (**2003**) 631–644.

[12] Gleiser, P. M. and Danon, L., *Community Structure in Jazz*, Advances in Complex Systems **06** (**2003**) 565–573.

[13] Holme, P., *Congestion and Centrality in Traffic Flow on Complex Networks*, Advances in Complex Systems **06** (**2003**) 163–176.

[14] Pütsch, F., *Analysis and Modeling of Science Collaboration Networks*, Advances in Complex Systems **06** (**2003**) 477–485.

[15] Zheng, D. and Ergün, G., *Coupled Growing Networks*, Advances in Complex Systems **06** (**2003**) 507–514.

[16] Carayol, N. and Roux, P., *Behavioral Foundations and Equilibrium Notions for Social Network Formation Processes*, Advances in Complex Systems **07** (**2004**) 77–92.

[17] Wagner, A. and Wright, J., *Compactness and Cycles in Signal Transduction and Transcriptional Regulation Networks: A Signature of Natural Selection?*, Advances in Complex Systems **07** (**2004**) 419–432.

[18] Attolini, C. S. O. and Stadler, P. F., *Neutral Networks of Interacting RNA Secondary Structures*, Advances in Complex Systems **08** (**2005**) 275–283.

[19] Grönlund, A. and Holme, P., *A Network-Based Threshold Model for the Spreading of Fads in Society and Markets*, Advances in Complex Systems **08** (**2005**) 261–273.

[20] Hu, H.-B. and Wang, L., *The Gini Coefficient'S Application to General Complex Networks*, Advances in Complex Systems **08** (**2005**) 159–167.

[21] Lee, P.-H., Huang, C.-H., Fang, J.-F., Liu, H.-C., and Ng, K.-L., *Hierarchical and Topological Study of the Protein–Protein Interaction Networks*, Advances in Complex Systems **08** (**2005**) 383–397.

[22] Richardson, K. A., *Simplifying Boolean Networks*, Advances in Complex Systems **08** (**2005**) 365–381.

[23] Duan, Z., Wang, J., and Huang, L., *Some Special Decentralized Control Problems in Continuous-Time Interconnected Systems*, Advances in Complex Systems **09** (**2006**) 277–286.

[24] Forst, C. V., Cabusora, L., Mawuenyega, K. G., and Chen, X., *Biological Systems Analysis by A Network Proteomics Approach and Subcellular Protein Profiling*, Advances in Complex Systems **09** (**2006**) 299–314.

[25] Itzkovitz, S., Milo, R., Kashtan, N., Levitt, R., Lahav, A., and Alon, U., *Recurring Harmonic Walks and Network Motifs in Western Music*, Advances in Complex Systems **09** (**2006**) 121–132.

[26] Barthélemy, M., Barrat, A., and Vespignani, A., *The Role of Geography and Traffic in the Structure of Complex Networks*, Advances in Complex Systems **10** (**2007**) 5–28.

[27] Battiston, S., Rodrigues, J. F., and Zeytinoglu, H., *The Network of Inter-Regional Direct Investment Stocks Across Europe*, Advances in Complex Systems **10** (**2007**) 29–51.

[28] Eidenbenz, S., Hansson, A. A., Ramaswamy, V., and Reidys, C. M., *on A New Class of Load Balancing Network Protocols*, Advances in Complex Systems **10** (**2007**) 359–377.

[29] Fang, J., Bi, Q., Li, Y., Lu, X.-B., and Liu, Q., *Toward A Harmonious Unifying Hybrid Model for Any Evolving Complex Networks*, Advances in Complex Systems **10** (**2007**) 117–141.

[30] Fleming, L. and Frenken, K., *The Evolution of Inventor Networks in the Silicon Valley and Boston Regions*, Advances in Complex Systems **10** (**2007**) 53–71.

[31] Jenkins, S. and Kirk, S. R., *An Investigation of Merging and Collapsing of Software Networks*, Advances in Complex Systems **10** (**2007**) 379–393.

[32] Leier, A., Kuo, P. D., and Banzhaf, W., *Analysis of Preferential Network Motif Generation in an Artificial Regulatory Network Model Created by Duplication and Divergence*, Advances in Complex Systems **10** (**2007**) 155–172.

[33] Qin, S., Dai, G., Wang, L., and Fan, M., *Scale-Free Evolving Networks With Accelerated Attachment*, Advances in Complex Systems **10** (**2007**) 143–154.

[34] Verspagen, B., *Mapping Technological Trajectories As Patent Citation Networks: A Study on The History of Fuel Cell Research*, Advances in Complex Systems **10** (**2007**) 93–115.

[35] Ay, N. and Polani, D., *Information Flows in Causal Networks*, Advances in Complex Systems **11** (**2008**) 17–41.

[36] Cattuto, C., Baldassarri, A., Servedio, V. D. P., and Loreto, V., *Emergent Community Structure in Social Tagging Systems*, Advances in Complex Systems **11** (**2008**) 597–608.

[37] Easton, G., Brooks, R. J., Georgieva, K., and Wilkinson, I., *Understanding the Dynamics of Industrial Networks Using Kauffman Boolean Networks*, Advances in Complex Systems **11** (**2008**) 139–164.

[38] Fujiwara, Y., *Chain of Firms' Bankruptcy: A Macroscopic Study of Link Effect in A Production Network*, Advances in Complex Systems **11** (**2008**) 703–717.

[39] Mukherjee, A., Choudhury, M., Basu, A., Ganguly, N., and Chowdhury, S. R., *Rediscovering the Co-Occurrence Principles of Vowel Inventories: A Complex Network Approach*, Advances in Complex Systems **11** (**2008**) 371–392.

[40] Pozzi, F., Di Matteo, T., and Aste, T., *Centrality and Peripherality in Filtered Graphs from Dynamical Financial Correlations*, Advances in Complex Systems **11** (**2008**) 927–950.

[41] Radde, N., *The Effect of Time Scale Differences and Time Delays on the Structural Stability of Oscillations in A Two-Gene Network*, Advances in Complex Systems **11** (**2008**) 471–483.

[42] Reyes, J., Schiavo, S., and Fagiolo, G., *Assessing the Evolution of International Economic Integration Using Random Walk Betweenness Centrality: The Cases of East Asia and Latin America*, Advances in Complex Systems **11** (**2008**) 685–702.

[43] Zahadat, P. and Katebi, S. D., *Tartarus and Fractal Gene Regulatory Networks With Inputs*, Advances in Complex Systems **11** (**2008**) 803–829.

[44] Corominas-Murtra, B., Valverde, S., and Solé, R., *The Ontogeny of Scale-Free Syntax Networks: Phase Transitions in Early Language Acquisition*, Advances in Complex Systems **12** (**2009**) 371–392.

[45] Gomes, L. H., Almeida, V. A. F., Almeida, J. M., Castro, F. D. O., and Bettencourt, L. M. A., *Quantifying Social and Opportunistic Behavior in Email Networks*, Advances in Complex Systems **12** (**2009**) 99–112.

[46] Hayashi, Y., *Necessary Backbone of Superhighways for Transport on Geographical Complex Networks*, Advances in Complex Systems **12** (**2009**) 73–86.

[47] Kohring, G. A., *Complex Dependencies in Large Software Systems*, Advances in Complex Systems **12** (**2009**) 565–581.

[48] Masucci, A. P. and Rodgers, G. J., *Differences Between Normal and Shuffled Texts: Structural Properties of Weighted Networks*, Advances in Complex Systems **12** (**2009**) 113–129.

[49] Schweitzer, F., Fagiolo, G., Sornette, D., Vega-Redondo, F., and White, D. R., *Economic Networks: What Do We Know and What Do We Need to Know?*, Advances in Complex Systems **12** (**2009**) 407–422.

[50] Tang, B., He, L., Jing, Q., and Shen, B., *Model-Based Identification and Adaptive Control of the Core Module in A Typical Cell Cycle Pathway Via Network and System Control Theories*, Advances in Complex Systems **12** (**2009**) 21–43.

[51] Tseng, J.-J., Li, S.-P., Chen, S.-H., and Wang, S.-C., *Emergence of Scale-Free Networks in Markets*, Advances in Complex Systems **12** (**2009**) 87–97.

[52] Uchida, M., Shibata, N., Kajikawa, Y., Takeda, Y., Shirayama, S., and Matsushima, K., *Identifying the Large-Scale Structure of the Blogosphere*, Advances in Complex Systems **12** (**2009**) 207–219.

[53] Wang, J. and Provan, G., *Topological Analysis of Specific Spatial Complex Networks*, Advances in Complex Systems **12** (**2009**) 45–71.

[54] Wu, F.-X., *Delay-Independent Stability of Genetic Regulatory Networks With Time Delays*, Advances in Complex Systems **12** (**2009**) 3–19.

[55] Anna Lázár, K., *A Formal Language Theoretic Approach to Distributed Computing on Dynamic Networks*, Advances in Complex Systems **13** (**2010**) 253–280.

[56] Du, F., Xuan, Q., and Wu, T.-J., *One-To-Many Node Matching Between Complex Networks*, Advances in Complex Systems **13** (**2010**) 725–739.

[57] Hakansson, N., Jonsson, A., Lennartsson, J., Lindström, T., and Wennergren, U., *Generating Structure Specific Networks*, Advances in Complex Systems **13** (**2010**) 239–250.

[58] House, T., *Generalized Network Clustering and Its Dynamical Implications*, Advances in Complex Systems **13** (**2010**) 281–291.

[59] König, M. D., Tessone, C. J., and Zenou, Y., *From Assortative to Dissortative Networks: The Role of Capacity Constraints*, Advances in Complex Systems **13** (**2010**) 483–499.

[60] Lai, D., Lu, H., Lauria, M., Di Bernardo, D., and Nardini, C., *Mania: A Gene Network Reverse Algorithm for Compounds Mode-Of-Action and Genes Interactions Inference*, Advances in Complex Systems **13** (**2010**) 83–94.

[61] Murata, T. and Ikeya, T., *A New Modularity for Detecting One-To-Many Correspondence of Communities In Bipartite Networks*, Advances in Complex Systems **13** (**2010**) 19–31.

[62] Scholtes, I., Botev, J., Esch, M., and Sturm, P., *Epidemic Self-Synchronization in Complex Networks of Kuramoto Oscillators*, Advances in Complex Systems **13** (**2010**) 33–58.

[63] Shi, C., Yan, Z., Wang, Y., Cai, Y., and Wu, B., *A Genetic Algorithm for Detecting Communities in Large-Scale Complex Networks*, Advances in Complex Systems **13** (**2010**) 3–17.

[64] Slanina, F. and Konopásek, Z., *Eigenvector Localization As A Tool to Study Small Communities in Online Social Networks*, Advances in Complex Systems **13** (**2010**) 699–723.

[65] Chandra, J. and Ganguly, N., *on Coverage Bounds of Unstructured Peer-To-Peer Networks*, Advances in Complex Systems **14** (**2011**) 611–633.

[66] Giabbanelli, P. J., *The Small-World Property in Networks Growing by Active Edges*, Advances in Complex Systems **14** (**2011**) 853–869.

[67] Jin, D., Liu, D., Yang, B., Liu, J., and He, D., *Ant Colony Optimization With A New Random Walk Model for Community Detection In Complex Networks*, Advances in Complex Systems **14** (**2011**) 795–815.

[68] Li, S., *Contagion Risk in an Evolving Network Model of Banking Systems*, Advances in Complex Systems **14** (**2011**) 673–690.

[69] Lucatero, C. R. and Jaquez, R. B., *Virus and Warning Spread in Dynamical Networks*, Advances in Complex Systems **14** (**2011**) 341–358.

[70] Smith, D. M. D., Onnela, J.-P., Lee, C. F., Fricker, M. D., and Johnson, N. F., *Network Automata: Coupling Structure and Function in Dynamic Networks*, Advances in Complex Systems **14** (**2011**) 317–339.

[71] Wang, H., Van Boven, E., Krishnakumar, A., Hosseini, M., Van Hooff, H., Takema, T., Baken, N., and Van Mieghem, P., *Multi-Weighted Monetary Transaction Network*, Advances in Complex Systems **14** (**2011**) 691–710.

[72] Berche, B., Ferber, C. V., Holovatch, T., and Holovatch, Y., *Transportation Network Stability: A Case Study of City Transit*, Advances in Complex Systems **15** (**2012**) 1250063.

[73] Bezemer, D. J., *The Economy As a Complex System: The Balance Sheet Dimension*, Advances in Complex Systems **15** (**2012**) 1250047.

[74] Geipel, M. M., *Modularity, Dependence and Change*, Advances in Complex Systems **15** (**2012**) 1250083.

[75] Guazzini, A., Vilone, D., Bagnoli, F., Carletti, T., and Grotto, R. L., *Cognitive Network Structure: an Experimental Study*, Advances in Complex Systems **15** (**2012**) 1250084.

[76] He, D., Liu, J., Yang, B., Huang, Y., Liu, D., and Jin, D., *An Ant-Based Algorithm With Local Optimization for Community Detection In Large-Scale Networks*, Advances in Complex Systems **15** (**2012**) 1250036.

[77] Li, S. and He, J., *The Impact of Bank Activities on Contagion Risk in Interbank Networks*, Advances in Complex Systems **15** (**2012**) 1250086.

[78] Maletić, S., Horak, D., and Rajković, M., *Cooperation, Conflict and Higher-Order Structures of Social Networks*, Advances in Complex Systems **15** (**2012**) 1250055.

[79] Sobkowicz, P. and Sobkowicz, A., *Properties of Social Network in an Internet Political Discussion Forum*, Advances in Complex Systems **15** (**2012**) 1250062.

[80] Suchecki, K., Salah, A. A. A., Gao, C., and Scharnhorst, A., *Evolution of Wikipedia'S Category Structure*, Advances in Complex Systems **15** (**2012**) 1250068.

[81] Szell, M. and Thurner, S., *Social Dynamics in A Large-Scale Online Game*, Advances in Complex Systems **15** (**2012**) 1250064.

[82] Tasgin, M. and Bingol, H. O., *Gossip on Weighted Networks*, Advances in Complex Systems **15** (**2012**) 1250061.

[83] Adamatzky, A., Lees, M., and Sloot, P., *Bio-Development of Motorway Network in The Netherlands: A Slime Mould Approach*, Advances in Complex Systems **16** (**2013**) 1250034.

[84] Gegov, E., Postorino, M. N., Atherton, M., and Gobet, F., *Community Structure Detection in the Evolution of the United States Airport Network*, Advances in Complex Systems **16** (**2013**) 1350003.

[85] Hegedűs, I., Ormándi, R., and Jelasity, M., *Massively Distributed Concept Drift Handling in Large Networks*, Advances in Complex Systems **16** (**2013**) 1350021.

[86] Mukherjee, S., *Complex Network Analysis in Cricket: Community Structure, Player'S Role And Performance Index*, Advances in Complex Systems **16** (**2013**) 1350031.

[87] Roth, C., *Socio-Semantic Frameworks*, Advances in Complex Systems **16** (**2013**) 1350013.

[88] Zeng, A., Gualdi, S., Medo, M., and Zhang, Y.-C., *Trend Prediction in Temporal Bipartite Networks: The Case of Movielens, Netflix, and Digg*, Advances in Complex Systems **16** (**2013**) 1350024.

[89] Cataldo, M., Scholtes, I., and Valetto, G., *A Complex Networks Perspective on Collaborative Software Engineering*, Advances in Complex Systems **17** (**2014**) 1430001.

[90] Crampes, M. and Plantié, M., *A Unified Community Detection, Visualization and Analysis Method*, Advances in Complex Systems **17** (**2014**) 1450001.

[91] Dueñas, M. and Fagiolo, G., *Global Trade Imbalances: A Network Approach*, Advances in Complex Systems **17** (**2014**) 1450014.

[92] Faletra, M., Palmer, N., and Marshall, J. S., *Effectiveness of Opinion Influence Approaches in Highly Clustered Online Social Networks*, Advances in Complex Systems **17** (**2014**) 1450008.

[93] Geipel, M. M., Press, K., and Schweitzer, F., *Communication in Innovation Communities: An Analysis of 100 Open Source Software Projects*, Advances in Complex Systems **17** (**2014**) 1550006.

[94] Ghedini, C. G. and Ribeiro, C. H. C., *Improving Resilience of Complex Networks Facing Attacks and Failures Through Adaptive Mechanisms*, Advances in Complex Systems **17** (**2014**) 1450009.

[95] Liu, X., Liu, W., Murata, T., and Wakita, K., *A Framework for Community Detection in Heterogeneous Multi-Relational Networks*, Advances in Complex Systems **17** (**2014**) 1450018.

[96] Luo, S., Xia, H., and Yin, B., *Continuous Opinion Dynamics on an Adaptive Coupled Random Network*, Advances in Complex Systems **17** (**2014**) 1450012.

[97] Meyer, P., Siy, H., and Bhowmick, S., *Identifying Important Classes of Large Software Systems Through K-Core Decomposition*, Advances in Complex Systems **17** (**2014**) 1550004.

[98] Pan, W., Li, B., Jiang, B., and Liu, K., *Recode: Software Package Refactoring Via Community Detection in Bipartite Software Networks*, Advances in Complex Systems **17** (**2014**) 1450006.

[99] Schweitzer, F., Nanumyan, V., Tessone, C. J., and Xia, X., *How Do Oss Projects Change In Number And Size? A Large-Scale Analysis To Test A Model Of Project Growth*, Advances in Complex Systems **17** (**2014**) 1550008.

[100] Shin, J. K., *Opinion Dynamics Approach for Identifying Community Structures in Complex Networks*, Advances in Complex Systems **17** (**2014**) 1450023.

[101] Slanina, F., *Dynamics of User Networks in On-Line Electronic Auctions*, Advances in Complex Systems **17** (**2014**) 1450002.

[102] Zhao, Y., Li, S., and Wang, S., *Agglomerative Clustering Based on Label Propagation for Detecting Overlapping And Hierarchical Communities in Complex Networks*, Advances in Complex Systems **17** (**2014**) 1450021.

[103] Šubelj, L., Žitnik, S., Blagus, N., and Bajec, M., *Node Mixing and Group Structure of Complex Software Networks*, Advances in Complex Systems **17** (**2014**) 1450022.

[104] Andreotti, E., Bazzani, A., Rambaldi, S., Guglielmi, N., and Freguglia, P., *Modeling Traffic Fluctuations and Congestion on A Road Network*, Advances in Complex Systems **18** (**2015**) 1550009.

[105] Aparicio, D. and Fraiman, D., *Banking Networks and Leverage Dependence in Emerging Countries*, Advances in Complex Systems **18** (**2015**) 1550022.

[106] Bo, J., Zhi-Yong, L., and Jian-Mai, S., *The Application of Semidefinite Integer Programming Model for the Simulation Of Complex Networks*, Advances in Complex Systems **18** (**2015**) 1550015.

[107] Costa, E. C., Vieira, A. B., Wehmuth, K., Ziviani, A., and Da Silva, A. P. C., *Time Centrality in Dynamic Complex Networks*, Advances in Complex Systems **18** (**2015**) 1550023.

[108] Garrido, P., Campos, P., and Dias, A., *Balance Sheet Analysis of Credit and Debt Networks*, Advances in Complex Systems **18** (**2015**) 1550025.

[109] Greetham, D. V., Sengupta, A., Hurling, R., and Wilkinson, J., *Interventions in Social Networks: Impact on Mood and Network Dynamics*, Advances in Complex Systems **18** (**2015**) 1550016.

[110] Kosmidis, K., Beber, M., and Hütt, M.-T., *Network Heterogeneity and Node Capacity Lead to Heterogeneous Scaling Of Fluctuations in Random Walks on Graphs*, Advances in Complex Systems **18** (**2015**) 1550007.

[111] Wang, D., Jin, S., Wu, F.-X., and Zou, X., *Estimation of Control Energy and Control Strategies for Complex Networks*, Advances in Complex Systems **18** (**2015**) 1550018.

[112] Garas, A., Lapatinas, A., and Poulios, K., *The Relation Between Migration and FDI in the OECD from A Complex Network Perspective*, Advances in Complex Systems **19** (**2016**) 1650009.

[113] Guan, Y. and Pollak, M., *Contagion in Heterogeneous Financial Networks*, Advances in Complex Systems **19** (**2016**) 1650001.

[114] Hou, W., Tamura, T., Ching, W.-K., and Akutsu, T., *Finding and Analyzing the Minimum Set of Driver Nodes in Control of Boolean Networks*, Advances in Complex Systems **19** (**2016**) 1650006.

[115] Patil, G. R. and Bhavathrathan, B. K., *Effect of Traffic Demand Variation on Road Network Resilience*, Advances in Complex Systems **19** (**2016**) 1650003.

[116] Prado, S. D., Dahmen, S. R., Bazzan, A. L., Carron, P. M., and Kenna, R., *Temporal Network Analysis of Literary Texts*, Advances in Complex Systems **19** (**2016**) 1650005.

[117] Tomasello, M. V., Tessone, C. J., and Schweitzer, F., *A Model of Dynamic Rewiring and Knowledge Exchange in R&D Networks*, Advances in Complex Systems **19** (**2016**) 1650004.

[118] Xu, T., He, J., and Li, S., *Multi-Channel Contagion in Dynamic Interbank Market Network*, Advances in Complex Systems **19** (**2016**) 1650011.

[119] Yose, J., Kenna, R., Maccarron, P., Platini, T., and Tonra, J., *A Networks-Science Investigation Into the Epic Poems of Ossian*, Advances in Complex Systems **19** (**2016**) 1650008.

[120] Hu, H.-B., Li, C.-H., and Miao, Q.-Y., *Opinion Diffusion on Multilayer Social Networks*, Advances in Complex Systems **20** (**2017**) 1750015.

[121] Oldham, M., *Market Fluctuations Explained by Dividends and Investor Networks*, Advances in Complex Systems **20** (**2017**) 1750007.

[122] Wang, D. and Zou, X., *Control Energy and Controllability of Multilayer Networks*, Advances in Complex Systems **20** (**2017**) 1750008.

[123] Zhao, F., Zhang, L., Yang, G., He, L., and Yan, F., *Application of Cut Algorithm Based on Algebraic Connectivity to Community Detection*, Advances in Complex Systems **20** (**2017**) 1750002.

[124] Kumari, S. and Singh, A., *Time-Varying Network Modeling and Its Optimal Routing Strategy*, Advances in Complex Systems **21** (**2018**) 1850006.

[125] Pan, W., Hu, B., Dong, J., Liu, K., and Jiang, B., *Structural Properties of Multilayer Software Networks: A Case Study in Tomcat*, Advances in Complex Systems **21** (**2018**) 1850004.

[126] Pan, Z., *Varieties of Intergovernmental Organization Memberships and Structural Effects In the World Trade Network*, Advances in Complex Systems **21** (**2018**) 1850001.

[127] Tu, C. and Yan, R., *Estimating and Enhancing the Feedbackability of Complex Networks*, Advances in Complex Systems **21** (**2018**) 1850005.

[128] Buk, S., Krynytskyi, Y., and Rovenchak, A., *Properties of Autosemantic Word Networks in Ukrainian Texts*, Advances in Complex Systems **22** (**2019**) 1950016.

[129] Cao, Q., Ramos, G., Bogdan, P., and Pequito, S., *The Actuation Spectrum of Spatiotemporal Networks With Power-Law Time Dependencies*, Advances in Complex Systems **22** (**2019**) 1950023.

[130] Duan, G., Li, A., Meng, T., and Wang, L., *Energy Cost for Target Control of Complex Networks*, Advances in Complex Systems **22** (**2019**) 1950022.

[131] Glattfelder, J. B., *The Bow-Tie Centrality: A Novel Measure for Directed and Weighted Networks With An Intrinsic Node Property*, Advances in Complex Systems **22** (**2019**) 1950018.

[132] Li, A. and Liu, Y.-Y., *Controlling Network Dynamics*, Advances in Complex Systems **22** (**2019**) 1950021.

[133] Lin, X., Zhou, S., Sun, L., and Wu, Y., *Complex Function Projective Synchronization in Fractional-Order Complex Networks and Its Application in Fractal Pattern Recognition*, Advances in Complex Systems **22** (**2019**) 1950010.

[134] Long, H. and Liu, X.-W., *A Unified Community Detection Algorithm in Large-Scale Complex Networks*, Advances in Complex Systems **22** (**2019**) 1950004.

[135] Mellor, A., *Event Graphs: Advances and Applications of Second-Order Time-Unfolded Temporal Network Models*, Advances in Complex Systems **22** (**2019**) 1950006.

[136] Ravandi, B., Ansari, F. S., and Mili, F., *Controllability Analysis of Complex Networks Using Statistical Random Sampling*, Advances in Complex Systems **22** (**2019**) 1950012.

[137] Rozum, J. C. and Albert, R., *Controlling the Cell Cycle Restriction Switch Across the Information Gradient*, Advances in Complex Systems **22** (**2019**) 1950020.

[138] Sharma, R., Adhikari, B., and Krueger, T., *Self-Organized Corona Graphs: A Deterministic Complex Network Model With Hierarchical Structure*, Advances in Complex Systems **22** (**2019**) 1950019.

[139] Walters, N., Zyl, G. V., and Beyers, C., *Financial Contagion in Large, Inhomogeneous Stochastic Interbank Networks*, Advances in Complex Systems **22** (**2019**) 1950002.

[140] Xiang, L. and Chen, G., *Minimal Edge Controllability of Directed Networks*, Advances in Complex Systems **22** (**2019**) 1950017.

[141] Zhang, Y., Garas, A., and Schweitzer, F., *Control Contribution Identifies Top Driver Nodes in Complex Networks*, Advances in Complex Systems **22** (**2019**) 1950014.

[142] Cheriyan, J. and Sajeev, G. P., *M-Pagerank: A Novel Centrality Measure for Multilayer Networks*, Advances in Complex Systems **23** (**2020**) 2050012.

[143] Prado, S. D., Dahmen, S. R., Bazzan, A. L. C., Maccarron, M., and Hillner, J., *Gendered Networks and Communicability in Medieval Historical Narratives*, Advances in Complex Systems **23** (**2020**) 2050006.

[144] Schweitzer, F., *Designing Systems Bottom Up: Facets and Problems*, Advances in Complex Systems **23** (**2020**) 2020001.

[145] Shang, Y., *Multi-Hop Generalized Core Percolation on Complex Networks*, Advances in Complex Systems **23** (**2020**) 2050001.

[146] Varga, I., *A Complex SIS Spreading Model in Ad Hoc Networks With Reduced Communication Efforts*, Advances in Complex Systems **23** (**2020**) 2050009.

[147] Biggiero, L. and Magnuszewski, R., *The General Ownership Structure of the European Aerospace Industry A Statistical and Network Analysis*, Advances in Complex Systems **24** (**2021**).

[148] Dinkelberg, A., O'Sullivan, D. J., Quayle, M., and Maccarron, P., *Detecting Opinion-Based Groups and Polarization in Survey-Based Attitude Networks and Estimating Question Relevance*, Advances in Complex Systems **24** (**2021**) 2150006.

[149] Edet, S., Panzarasa, P., and Riccaboni, M., *Global Cities in International Networks of Innovators*, Advances in Complex Systems **24** (**2021**).

[150] Edet, S., Panzarasa, P., and Riccaboni, M., *Global Cities in International Networks of Innovators*, Advances in Complex Systems **24** (**2021**).

[151] Leal, W., Restrepo, G., Stadler, P. F., and Jost, J., *Forman–Ricci Curvature for Hypergraphs*, Advances in Complex Systems **24** (**2021**).

[152] Palchykov, V., Krasnytska, M., Mryglod, O., and Holovatch, Y., *Network of Scientific Concepts: Empirical Analysis and Modeling*, Advances in Complex Systems **24** (**2021**).

[153] Xiao, S., She, B., Mehta, S., and Kan, Z., *Design of Controllable Leader–Follower Networks Via Memetic Algorithms*, Advances in Complex Systems **24** (**2021**).

Advances in Complex Systems, Vol. 10, No. 1 (2007) 5–28

World Scientific
www.worldscientific.com

THE ROLE OF GEOGRAPHY AND TRAFFIC IN THE STRUCTURE OF COMPLEX NETWORKS

MARC BARTHÉLEMY[*,†,‡], ALAIN BARRAT[§,¶] and ALESSANDRO VESPIGNANI[†,‖]

[†]*School of Informatics and Biocomplexity Center, Indiana University, Bloomington, IN 47406, USA*

[§]*Laboratoire de Physique Théorique (UMR du CNRS 8627), Bâtiment 210, Université de Paris-Sud 91405 Orsay, France*

[‡]*marc.barthelemy@cea.fr*
[¶]*alain.barrat@th.u-psud.fr*
[‖]*alexv@indiana.edu*

Received 25 July 2005
Revised 27 October 2005

We report a study of the correlations among topological, weighted and spatial properties of large infrastructure networks. We review the empirical results obtained for the air-transportation infrastructure that motivates a network modeling approach which integrates the various attributes of this network. In particular, we describe a class of models which include a weight-topology coupling and the introduction of geographical attributes during the network evolution. The inclusion of spatial features is able to capture the appearance of non-trivial correlations between the traffic flows, the connectivity pattern and the actual distances of vertices. The anomalous fluctuations in the betweenness-degree correlation function observed in empirical studies are also recovered in the model. The presented results suggest that the interplay between topology, weights and geographical constraints is a key ingredient in order to understand the structure and evolution of many real-world networks.

Keywords: Networks; traffic; transportation systems.

1. Introduction

The empirical evidence coming from studies on systems belonging to areas as diverse as social sciences, biology and computer science have shown that in a wide range of networks the occurrence of vertices with a very large degree (number of links to other vertices) is very common [2, 4, 17, 44]. The presence of these "hubs" often goes along with very large degree fluctuations. The large topological heterogeneity associated with these features is statistically expressed by the presence of heavy-tailed degree distributions with diverging variance that have a very strong impact

*On leave of absence from CEA — Centre d'Etudes de Bruyères-le-Châtel, Département de Physique Théorique et Appliquée BP12, 91680 Bruyères-Le-Châtel, France.

on the networks' physical properties such as resilience and vulnerability, or the propagation of pathogen agents [1, 15, 38, 43].

The purely topological definition of networks, however, misses important attributes which are frequently encountered in real-world networks. In the first instance, networks are far from a binary structure, and are better represented as weighted graphs with the intensity of links that may vary over many orders of magnitude. Indeed, in many graphs ranging from food-webs to metabolic networks, large variations of the link intensities are empirically observed [3, 7, 19, 24, 29, 31, 31]. Notably, the statistical properties of weights indicate non-trivial correlations and association with topological quantities [7].

Another important element of many real networks is their embedding in real space. For instance, most people have their friends and relatives in their neighborhood, transportation networks depend on distance, and many communication networks include devices with definite transmission ranges [20, 21, 23, 27, 37]. A particularly important example of such a "spatial" network is the Internet, which is a set of routers linked by physical cables with different lengths and latency times [30, 44]. An analogous situation is faced in the air transportation network with routes covering very different distances. The length of the link is a very important quantity usually associated with an intrinsic cost in the establishment of the connection. If the cost of a long-range link is high, most of the connections starting from a given node will go to the closest neighbors in the embedding space. Long-range links, on the other hand, usually correspond to connections towards already well-connected nodes (hubs). This seems natural in the case of the air transportation network, for instance: short connections go to small airports while long distance flights are directed preferentially towards large airports (i.e. well connected nodes). It is therefore natural to find that spatial constraints can have important consequences for the topology of the resulting network [13, 25, 34]. This issue is particularly important in spatial economics, where the evolution of an economic system depends strongly on the geographical distribution of economic activity [5].

In this article, we discuss the interplay of the three aforementioned ingredients (heterogeneous topology, weights and spatial constraints) in a model of a growing network combining these ingredients at the same time. The proposed model is obtained as the embedding of the weighted growing network introduced in Ref. 8 in a two-dimensional geometrical space. Spatial constraints are translated into a preference for short links and combined with the coupling between the evolution of the network and the dynamical rearrangement of the weights. This mechanism naturally leads to the appearance of many features observed in real-world networks, in particular, the non-linear correlations between weights and topology, and the large fluctuations of the betweenness centrality. Let us note that although we were motivated by the airport network, we do not intend here to present a specific model for this network, but rather to investigate the effects of different ingredients.

The paper is organized as follows. In Sec. 2, we briefly review some important empirical results on the North American airline network, highlighting the most salient geographical effects. Sections 3 and 4 are devoted to the presentation and to the numerical study of the spatially weighted model, stressing the effect of the spatial embedding and constraints on the properties of the resulting network. In Sec. 5, we present a summary of the results and conclusions about large network modeling.

2. A Case Study: Space, Topology and Traffic in the North American Airline Network

2.1. *Topological characterization*

The characteristics of the world-wide air-transportation network (WAN) using the International Air Transportation Association (IATA) database [28] have been presented in Ref. 7. The network consists of $N = 3{,}880$ vertices and $E = 18{,}810$ edges and shows both small-world and scale-free properties, as has also been confirmed in different datasets and analysis [25, 26, 31, 32]. In particular, the average shortest path length, measured as the average number of edges separating any two nodes in the network shows the value $\langle \ell \rangle \simeq 4.4$, very small compared to the network size N. The degree distribution takes the form $P(k) = k^{-\gamma} f(k/k_x)$, where $\gamma \simeq 2.0$ and $f(k/k_x)$ is an exponential cut-off function. The degree distribution is therefore heavy-tailed with a cut-off that finds its origin in the physical constraints on the maximum number of connections that a single airport can handle [4,25,26]. The airport connection graph is therefore a clear example of a small-world network showing a heavy-tailed degree distribution and heterogeneous topological properties. This heavy-tailed distribution is the signature of very large degree fluctuations[a] which will have a crucial impact on all dynamical processes that take place on this type of network.

The world-wide airline network necessarily mixes different effects. In particular, there are clearly two different spatial scales: global (intercontinental) and domestic. The intercontinental scale defines two different groups of travel distances, and for statistical consistency we eliminate this specific geographical constraint by focusing on a single continental case. Namely, in the following we will essentially consider the North American network consisting of $N = 935$ vertices with an average degree $\langle k \rangle \approx 8.4$ and an average shortest path $\langle \ell \rangle \simeq 3.9$. The statistical topological properties of the North American network are consistent with the WAN. In particular, the North American network presents a degree distribution statistically consistent with the world-wide airline network. Indeed, we observe (Fig. 1) also in this case a power-law behavior over almost two orders of magnitude, followed by a cut-off

[a]This can also be observed by a very large value of the variance or equivalently by the fact that $\langle k^2 \rangle \gg \langle k \rangle^2$.

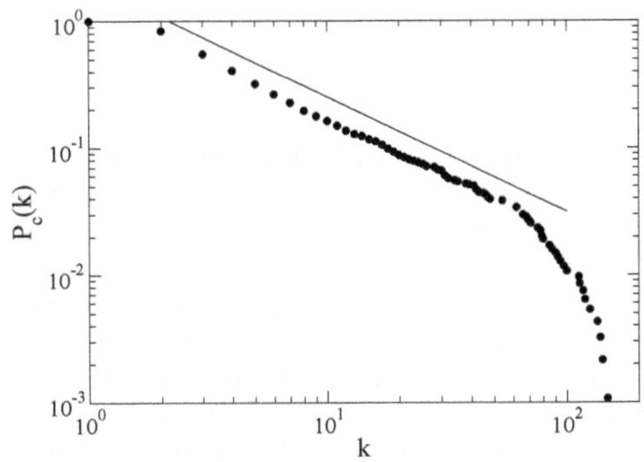

Fig. 1. Cumulative degree distribution $P_c(k)$ for the North American network. The straight line indicates a power-law decay with exponent $\gamma - 1 = 0.9$.

indicating the maximum number of connections possible due to limited airport capacity and to the size of the network considered.

A further characterization of the network is provided by considering quantities that take into account the global topology of the network. For instance, the degree of a vertex is a local measure that gives a first indication of its centrality. However, a more global approach is needed in order to characterize the real importance of various nodes. Indeed, some particular low-degree vertices may be essential because they provide connections between otherwise separated parts of the network. In order to take properly into account such vertices, the betweenness centrality (BC) is commonly used [14, 18, 22, 39, 40]. The betweenness centrality of a node v is defined as

$$g(v) = \sum_{s \neq t} \frac{\sigma_{st}(v)}{\sigma_{st}}, \tag{2.1}$$

where σ_{st} is the number of shortest paths going from s to t and $\sigma_{st}(v)$ is the number of shortest paths going from s to t and passing through v. This definition means that central nodes are part of more shortest paths within the network than peripheral nodes. Moreover, the betweenness centrality in transport networks provides an estimate of the traffic handled by the vertices, assuming that the number of shortest paths is a zero-th order approximation to the frequency of use of a given node. It is generally useful to represent the average betweenness centrality for vertices of the same degree:

$$g(k) = \frac{1}{N(k)} \sum_{v/k_v = k} g(v), \tag{2.2}$$

where $N(k)$ is the number of nodes of degree k. For most networks, $g(k)$ is strongly correlated with the degree k and in general, the larger the degree the larger the centrality. For scale-free networks, it has been shown that the centrality scales with k as

$$g(k) \sim k^{\mu}, \tag{2.3}$$

where μ depends on the network [14, 22, 39, 40]. The value for the North American network is $\mu \approx 1.2$ (see Fig. 2). For some networks, however, the BC fluctuations around the behavior given by Eq. (2.3) can be very large and "anomalies" can occur, in the sense that the variation of the centrality versus degree is not a monotonous function. Guimerà and Amaral [25] have shown that this is indeed the case for the air-transportation network. This is a very relevant observation in that very central cities may have a relatively low degree and *vice versa.* In Fig. 2, we report the average behavior along with the scatter plot of the betweenness centrality versus degree of all airports of the North American network. Also in this case, we find very large fluctuations around the average with a behavior similar to those observed in Ref. 25. Interestingly, Guimerà and Amaral have put forward a network model embedded in real space that considers geopolitical constraints. This model appears to reproduce the betweenness centrality features observed in the real network and highlights the importance of space as a relevant ingredient in the structure of networks. In the following, we focus on the interplay between spatial embedding, topology and weights in a simple general model for weighted networks in order to provide a modeling framework considering all three aspects simultaneously.

2.2. *Traffic properties*

The airline transportation infrastructure is a paramount example of a large scale network that can be represented as a complex weighted graph: the airports are the

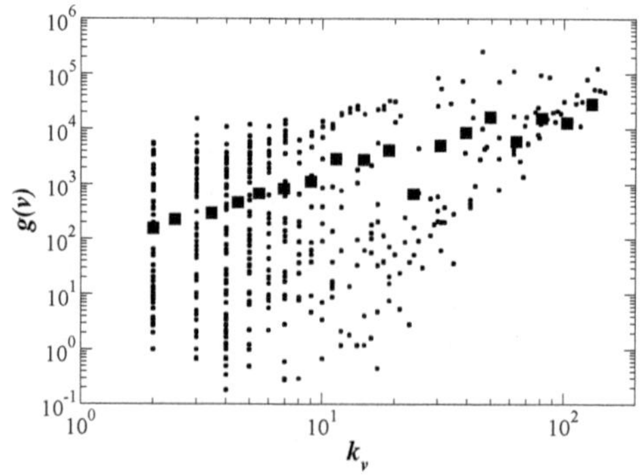

Fig. 2. Scatter-plot of the betweenness centrality versus degree for nodes of the North American air-transportation network. The squares correspond to the average BC versus degree and can be fitted by a power law with exponent $\mu \approx 1.2$.

vertices of the graph and the links represent the presence of direct flight connections among them. The weight on each link is the number of maximum passengers on the corresponding connection.

More generally, the properties of a graph can be expressed via its adjacency matrix a_{ij}, whose elements take the value 1 if an edge connects the vertex i to the vertex j and 0 otherwise (with $i, j = 1, \ldots, N$ where N is the size of the network). Weighted networks are usually described by a matrix w_{ij} specifying the weight on the edge connecting the vertices i and j ($w_{ij} = 0$ if the nodes i and j are not connected). In the following, we will consider only the case of symmetric positive weights $w_{ij} = w_{ji} \geq 0$.

Along with the degree of a node, a very significative measure of the network properties in terms of the actual weights is obtained by looking at the vertex *strength* s_i^w defined as [7]

$$s_i^w = \sum_{j \in \mathcal{V}(i)} w_{ij}, \tag{2.4}$$

where the sum runs over the set $\mathcal{V}(i)$ of neighbors of i. The strength of a node integrates the information both about its connectivity and the importance of the weights of its links and can be considered as the natural generalization of the connectivity. When the weights are independent of the topology, we obtain $s^w \simeq \langle w \rangle k$, where $\langle w \rangle$ is the average weight. In the presence of correlations, we obtain in general $s^w \simeq Ak^{\beta_w}$ with $\beta_w = 1$ and $A \neq \langle w \rangle$ or $\beta_w > 1$.

For the North American air-transportation network, we observe a nonlinear behavior with exponent $\beta_w \simeq 1.7$ (see Fig. 3) while for the global WAN network [7] we obtain $\beta_w(WAN) = 1.5 \pm 0.1$. These values of β_w larger than one imply that the strength of vertices grows faster than their degree, i.e. the weight of edges belonging to highly connected vertices tends to have a value higher than that corresponding to

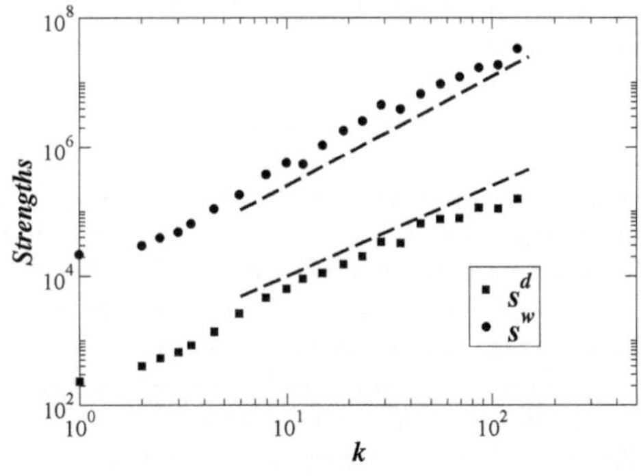

Fig. 3. Weight and distance strengths versus degree for the North American network. The dashed lines correspond to the power-laws $\beta_d \simeq 1.4$ and $\beta_w \simeq 1.7$.

a random assignment of weights. (In the randomized version of the network where the weights are randomly re-distributed on the existing topology of the network, we checked that we obtain $\beta_w = 1$.) There is thus a strong correlation between the weight and the topological properties, which means that the larger the airport, the more traffic it can handle. The fingerprint of these correlations is also observed in the behavior of the average weight as a function of the end points degrees $\langle w_{ij}\rangle \sim (k_i k_j)^\theta$ with an exponent $\theta_{NA} = 0.7 \pm 0.1$ for the North American network and $\theta_{\rm WAN} = 0.5 \pm 0.1$ for the WAN [7] (a simple argument presented in Ref. 7 suggests that in most cases, the exponents θ and β_w are related by $\beta_w = 1 + \theta$).

The topological clustering [46] is defined as the fraction of connected neighbors

$$C(i) = \frac{2E_i}{k_i(k_i - 1)}, \tag{2.5}$$

where E_i is the number of links between the neighbors of node i, and $k_i(k_i - 1)/2$ is the maximum number of such links (k_i is the degree of node i).

The topological clustering, however, does not take into account the fact that some neighbors are more important than others. We thus have to introduce a measure of the clustering that combines the topological information with the weight distribution of the network. The *weighted clustering coefficient* is defined as [7]

$$C^w(i) = \frac{1}{s_i^w(k_i - 1)} \sum_{j,h} \frac{(w_{ij} + w_{ih})}{2} a_{ij} a_{ih} a_{jh}. \tag{2.6}$$

This quantity $C^w(i)$ is counting for each triple formed in the neighborhood of the vertex i the weight of the two participating edges of the vertex i. The normalization factor $s_i^w(k_i - 1)$ ensures that $0 \le C_i^w \le 1$ and that C_i^w recovers the topological clustering coefficient in the case that $w_{ij} = \text{const}$. It is customary to define C^w and $C^w(k)$ as the weighted clustering coefficient averaged over all vertices of the network and over all vertices with degree k, respectively. The ratio C^w/C (and similarly $C^w(k)/C(k)$, which allows an analysis with respect to the degree k) indicates if the interconnected triples are more likely formed by the edges with larger weights.

Another quantity used to probe the networks' architecture is the behavior of the average degree of nearest neighbors, $k_{nn}(k)$, for vertices of degree k [45]. This last quantity is related to the correlations between the degree of connected vertices since it can be expressed as $k_{nn}(k) = \sum_{k'} k' P(k'|k)$, where $P(k'|k)$ is the conditional probability that a given vertex with degree k is connected to a vertex of degree k'. In the absence of degree correlations, $P(k'|k)$ does not depend on k and neither does the average nearest neighbors' degree; i.e. $k_{nn}(k) = \text{const.}$ [45]. In the presence of correlations, the behavior of $k_{nn}(k)$ identifies two general classes of networks. If $k_{nn}(k)$ is an increasing function of k, vertices with high degree have a larger probability to be connected with large degree vertices. This property is referred in physics and in social sciences as *assortative mixing* [41]. Conversely, a decreasing behavior of $k_{nn}(k)$ defines *disassortative mixing* in the sense that high degree vertices have a majority of neighbors with low degree, while the opposite holds for low degree vertices.

Along with the weighted clustering coefficient, we can measure the weighted average nearest neighbors degree, defined as

$$k_{nn}^w(i) = \frac{1}{s_i} \sum_{j=1}^{N} a_{ij} w_{ij} k_j. \tag{2.7}$$

This definition implies that $k_{nn}^w(i) > k_{nn}(i)$ if the edges with the larger weights are pointing to the neighbors with larger degree and $k_{nn}^w(i) < k_{nn}(i)$ in the opposite case. The $k_{nn}^w(i)$ thus measures the effective *affinity* to connect with high or low degree neighbors according to the magnitude of the actual interactions. As well, the behavior of the function $k_{nn}^w(k)$ (defined as the average of $k_{nn}^w(i)$ over all vertices i with degree k) reflects the weighted assortative or disassortative properties considering the actual interactions among the system's elements. Figure 4 displays the behavior of these various quantities as a function of the degree for the North American airport network. An essentially flat $k_{nn}^w(k)$ is obtained and a slight disassortative trend is observed at large k for the $k_{nn}(k)$, due to the fact that large airports have in fact many intercontinental connections to other hubs which are located outside of North America and are not considered in this "regional" network. The clustering is very large and slightly decreasing at large k. This behavior is often observed in complex networks and is here a direct consequence of the role of large airports in providing non-stop connections to different regions which are not interconnected themselves. Moreover, weighted correlations are systematically larger than the topological ones, signaling that large weights are concentrated on links between large airports which form well inter-connected cliques (see also Ref. 7 for more details).

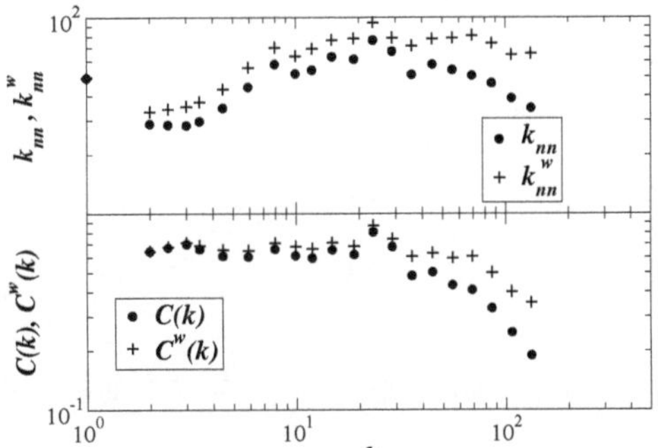

Fig. 4. Assortativity and clustering for the North American network. Circles correspond to topological quantities while squares are for affinity and weighted clustering.

For a given node i with connectivity k_i and strength s_i^w different situations can arise. All weights w_{ij} can be of the same order s_i^w/k_i. In contrast, the most heterogeneous situation is obtained when one weight dominates over all the others. A simple way to measure this "disparity" is given by the quantity Y_2 introduced in other contexts [12, 16]:

$$Y_2(i) = \sum_{j \in \mathcal{V}(i)} \left[\frac{w_{ij}}{s_i}\right]^2. \tag{2.8}$$

If all weights are of the same order then $Y_2 \sim 1/k_i$ (for $k_i \gg 1$), and if a small number of weights dominate then Y_2 is of order $1/n$ with n of order unity. This quantity was recently used for metabolic networks [3] to show that for these networks one can identify dominant reactions. For the WAN and the North American network, we observe a behavior consistent with a decay for large degrees of the form $Y_2 \sim 1/k$, which indicates that for large airports, all connections carry essentially the same number of passengers (even if there are obviously differences between these connections).

2.3. *Spatial analysis*

The spatial attributes of the North American airport network are embodied in the physical spatial distance, measured in kilometers or miles, characterizing each connection. Figure 5 displays the probability distribution $P(d_{ij})$ of the distances d_{ij} of the direct flights. These distances correspond to Euclidean measures of the links between airports and clearly show a fast decaying behavior reasonably fitted by an exponential. The exponential fit gives a value for a typical scale of the order

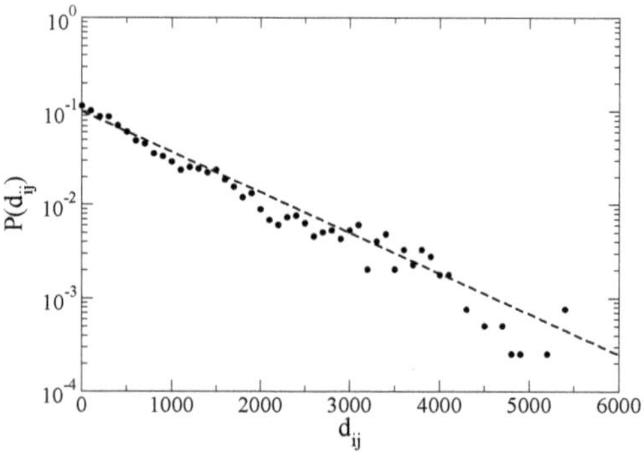

Fig. 5. Distribution of distances (in kms) between airports linked by a direct connection for the North American network. The straight line indicates an exponential decay with scale of order 1,000 kms.

1,000 kms. The origin of the finite scale can be traced back to the existence of physical and economical restrictions on airline planning in a continental setting.

Since space is an important parameter in this network, another interesting quantity is the *distance strength* of i

$$s_i^d = \sum_{j \in \mathcal{V}(i)} d_{ij}, \tag{2.9}$$

where d_{ij} are the *Euclidean* distances of the connections departing from the airport i. This quantity gives the cumulated distances of all the connections from (or to) the considered airport. Similarly to the usual weight strength, uncorrelated random connections would lead to a linear behavior of $s^d(k) \propto k$, while we observe in the North American network a power law behavior

$$s^d(k) \sim k^{\beta_d} \tag{2.10}$$

with $\beta_d \simeq 1.4$ (Fig. 3). This result shows the presence of important correlations between topology and geography. Indeed, the fact that the exponents appearing in the relations (2.4) and (2.10) are larger than one have different meanings. While Eq. (2.4) means that larger airports have connections with larger traffic, (2.10) implies that they also have farther-reaching connections. In other terms, the traffic (and the distance) per connection is not constant but increases with k. As intuitively expected, the airline network is an example of a very heterogeneous network where the hubs have at the same time large connectivities, large weight (traffic) and long-distance connections [7], related by super-linear scaling relations.

3. The Model

Early modeling of weighted networks just considered weight and topology as uncorrelated quantities [49]. This is not the case in real world networks, where a complex interplay between the evolution of weights and topological growth exists as well as spatial constraints. In the following, we discuss the effects of these ingredients on a simple model of an evolving network.

3.1. *The weight-topology coupling ingredient*

Previous approaches to the modeling of weighted networks focused on growing topologies where weights were assigned statically, i.e. once for ever, with different rules related to the underlying topology [49, 51]. These mechanisms, however, overlook the dynamical evolution of weights according to the topological variations. We can illustrate this point in the case of the airline network. If a new airline connection is created between two airports it will generally provoke a modification of the existing traffic of both airports. In the following, we review a model that takes into account the coupled evolution in time of topology and weights. The model dynamics starts from an initial seed of N_0 vertices connected by links with assigned weight w_0. At each time step, a new vertex n is added with m edges (with initial weight

w_0) that are randomly attached to a previously existing vertex i according to the probability

$$\prod_{n\to i} = \frac{s_i}{\sum_j s_j}. \tag{3.1}$$

This rule of "busy get busier" relaxes the usual degree preferential attachment, focusing on a strength-driven attachment in which new vertices connect more likely to vertices handling larger weights and which are more central in terms of the strength of interactions. This weight-driven attachment [Eq. (3.1)] appears to be a plausible mechanism in many networks [8].

The presence of the new edge (n, i) introduces variations of the existing weights across the network. In particular, we consider the local rearrangements of weights between i and its neighbors $j \in \mathcal{V}(i)$ according to the simple rule

$$w_{ij} \to w_{ij} + \Delta w_{ij}, \qquad \text{with} \quad \Delta w_{ij} = \delta \frac{w_{ij}}{s_i}. \tag{3.2}$$

This rule considers that the establishment of a new edge of weight w_0 with the vertex i induces a total increase of traffic δ that is proportionally distributed among the edges departing from the vertex according to their weights (see Fig. 6), yielding $s_i \to s_i + \delta + w_0$. We will focus on the simplest model with $\delta = \text{const.}$, but one can consider different choices [9, 10, 42] of Δw_{ij} depending on the specific properties of each vertex (w_{ij}, k_i, s_i). After the weights have been updated, the growth process is iterated by introducing a new vertex with the corresponding re-arrangement of weights.

The model depends only on the dimensionless parameter δ (rescaled by w_0), that is the fraction of weight which is "induced" by the new edge onto the others. If $\delta \approx 1$, the traffic generated by the new connection will be dispatched in the already existing connections. In the case of $\delta < 1$, we face situations where a new connection is not triggering a more intense activity on existing links. Finally, $\delta > 1$ is an extreme case in which a new edge generates a sort of multiplicative effect that is bursting the weight or traffic on neighbors.

The network's evolution can be inspected analytically by studying the time evolution of the average value of $s_i^w(t)$ and $k_i(t)$ of the ith vertex at time t, and

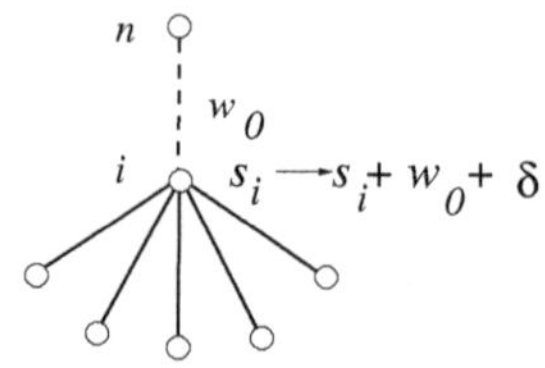

Fig. 6. Illustration of the construction rule. A new node n connects to a node i with probability proportional to $s_i/\sum_j s_j$. The weight of the new edge is w_0 and the total weight on the existing edges connected to i is modified by an amount equal to δ.

by relying on the continuous approximation that treats k, s^w and the time t as continuous variables [2,7,17]. One obtains $s_i^w(t) = (2\delta+1)k_i(t)$, which implies $\beta_w = 1$ and a prefactor different from $\langle w \rangle$, which indicates the existence of correlations between topology and weights. The fact that $s^w \propto k$ is also particularly relevant since it states that the weight-driven dynamics generates in Eq. (3.1) an effective degree preferential attachment that is parameter independent. This highlights an alternative microscopic mechanism accounting for the presence of the preferential attachment dynamics in growing networks. The behavior of the various statistical distribution can be easily computed and one obtains in the large time limit $P(k) \sim k^{-\gamma}$ and $P(s) \sim s^{-\gamma}$ with

$$\gamma = \frac{4\delta + 3}{2\delta + 1}. \tag{3.3}$$

This result shows that the obtained graph is a scale-free network described by an exponent $\gamma \in [2, 3]$ that depends on the value of the parameter δ. In particular, when the addition of a new edge does not affect the existing weights ($\delta = 0$), the model is topologically equivalent to the Barabási–Albert model [6] and the value $\gamma = 3$ is recovered. It is also possible to show analytically [8] that $P(w) \sim w^{-\alpha}$, where $\alpha = 2 + 1/\delta$. The exponent α has large variations as a function of the parameter δ and this feature clearly shows that the weight distribution is extremely sensitive to changes in the microscopic dynamics ruling the network's growth.

3.2. *The spatial ingredient*

Space is an important ingredient in many systems and it is crucial to understand its effect on the formation of networks. We consider this problem in a simple model of a growing weighted network whose nodes are embedded in a two-dimensional space. As in the weighted model above, it is reasonable to think that a newly created node n will establish links towards pre-existing nodes with heavy traffic or strength (hubs). Costs, however, are usually associated with distances, and there is a trade-off between the need to reach a hub in a few hops and the connection costs. The cost naturally increases with the distance, implying that the probability of establishing a connection between the new node n and a given vertex i decays as a function of the increasing Euclidean distance d_{ni}. As in the case of topological preferential attachment (i.e. connecting probability proportional to the degree [6]), this trade-off can be expressed in two different ways: the connecting probability can decrease either as a power-law of the distance [33, 48, 50] or as an exponential with a finite typical scale [13] as it seems more natural for networks such as transportation networks (see Fig. 5) or technological networks [47]. All the effects described here are obtained in the case of an exponential decay $\exp(-d_{ni}/r_c)$ but are also present in the case of a power-law d_{ni}^{-a} (the effect of a decreasing scale r_c is qualitatively the same as the effect of an increasing exponent a). Eventually, the creation of new edges will introduce new traffic that will trigger perturbations in the network. This model therefore consists of two combined mechanisms:

(i) *Growth.* We start with an initial seed of N_0 vertices randomly located (with uniform distribution) on a two-dimensional disk (of radius L) and connected by links with assigned weight w_0. At each time step, a new vertex n is placed on the disk at a randomly assigned position $\mathbf{x}_n$ (still according to a uniform distribution). This new site is connected to m previously existing vertices, choosing preferentially nearest sites with the largest strength. More precisely, a node i is chosen according to the probability

$$\prod_{n \to i} = \frac{s_i^w e^{-d_{ni}/r_c}}{\sum_j s_j^w e^{-d_{nj}/r_c}}, \tag{3.4}$$

where r_c is a typical scale and d_{ni} is the Euclidean distance between n and i. This rule of *strength-driven preferential attachment with spatial selection* generalizes the preferential attachment mechanism driven by the strength to spatial networks. Here, new vertices connect more likely to vertices which correspond to the best interplay between Euclidean distance and strength.

(ii) *Weights dynamics.* The weight of each new edge (n, i) is fixed to a given value w_0 (this value sets a scale so we can take $w_0 = 1$). The creation of this edge will perturb the existing interactions, and we consider *local* perturbations for which only the weights between i and its neighbors $j \in \mathcal{V}(i)$ are modified:

$$w_{ij} \to w_{ij} + \delta \frac{w_{ij}}{s_i^w}. \tag{3.5}$$

After the weights have been updated, the growth process is iterated by introducing a new vertex, i.e. going back to step (i) until the desired size of the network is reached.

The previous rules have simple physical and realistic interpretations. Equation (3.4) corresponds to the fact that new sites try to connect to existing vertices with the largest strength, with the constraint that the connection cannot be too costly. This adaptation of the rule "busy get busier" allows us to take into account physical constraints. The weight dynamics Eq. (3.5) expresses the perturbation created by the addition of the new node and link. It yields a global increase of $w_0 + \delta$ for the strength of i, which will therefore become even more attractive for future nodes.

The model contains two relevant parameters: the ratio between the typical scale and the size of the system $\eta = r_c/L$, and the ability to redistribute weights, δ. Depending on the values of η and δ we obtain different networks whose limiting cases are summarized in Fig. 7. More precisely, we expect:

- For $\eta \gg 1$, the effect of distance is negligible and we recover the properties of the weighted model discussed above. The effect of the redistribution parameter δ is to broaden the various probability distributions and to increase the correlations

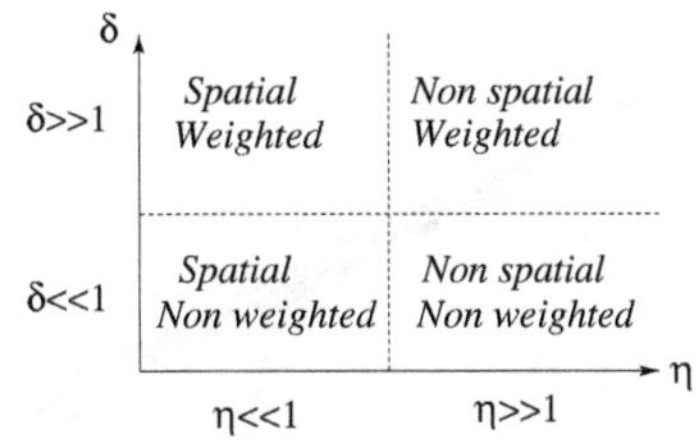

Fig. 7. Different limiting regimes of the model depending on the value of its parameters δ and η.

between topology and weights. Moreover, in this case, no correlations are introduced between the topology of the network and the underlying two-dimensional space, so that the distance strength s^d grows simply linearly with the degree $s^d \sim k$.

- When η decreases, additional constraints appear and have consequences that we will investigate numerically in the following. Unless otherwise specified, the simulations correspond to the parameters $m = 3$ (i.e. an average degree $\langle k \rangle = 6$), and $\delta = 1.0$. We consider networks of size up to N=10,000, and the results are averaged over up to 100 realizations. All the observed dependences in η are essentially the same for other investigated values of δ.

4. Simulation of the Model

4.1. *Topology and weights*

At a purely topological level, the principal effect of a typical finite scale r_c in the creation of new connections is to introduce a cut-off in the scale-free degree distributions [13]. In Fig. 8, we report the degree distribution for a fixed value of δ and decreasing values η. A more pronounced cut-off appears at decreasing value of η, signaling the onset of a trade-off between the number of connections and their cost in terms of Euclidean distance. The small-world properties of the network are also modified [13]: on the one hand, the increasing tendency to establish connections in the geographical neighborhood favors the formation of cliques and leads to an increase in the clustering coefficient (see inset of Fig. 9). On the other hand, this same tendency leads to an increase in the diameter of the graph, measured as the average shortest path distances between pairs of nodes. The diameter, however, still increases logarithmically with the size of the graph, as shown in Fig. 9: the constructed networks do display the small-world property, even if strong geographical constraints are present.

The correlations appearing between space, traffic and topology of the network are affected by the presence of spatial constraints. Strikingly, the effect of the spatial constraint is to increase both exponents β_w and β_d to values larger than 1, and although the redistribution of the weights [Eq. (3.5)] is linear, nonlinear relations

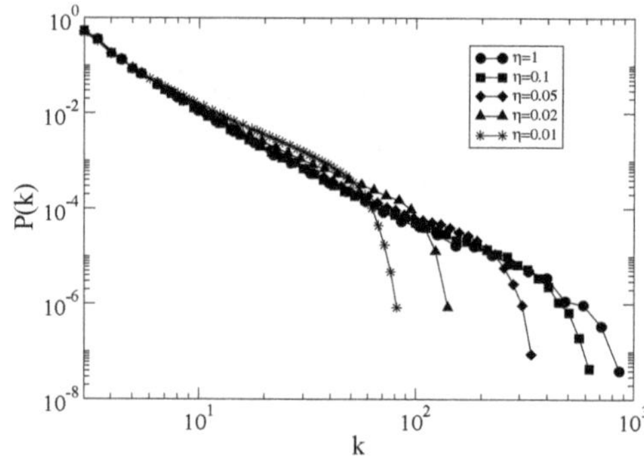

Fig. 8. Degree distribution $P(k)$ for different values of η and $\delta = 1$. The degree distribution is averaged over 50 networks of size $N = 10^4$ and minimum degree $m = 3$.

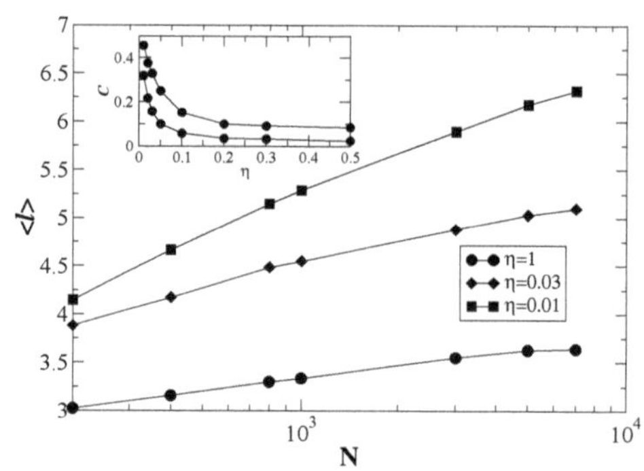

Fig. 9. Average shortest path distances as a function of the network size N for different values of η, with $m = 3$ and $\delta = 1$. Inset: clustering coefficient as a function of η for $N = 10^4$, $\delta = 1$ (circles) and $\delta = 5$ (squares). The data are averaged over 50 networks.

$s^w(k)$ and $s^d(k)$ as a function of k appear. For the weight strength, the effect is not very pronounced, with an exponent of order $\beta_w \approx 1.1$ for $\eta = 0.01$, while for the distance strength the nonlinearity has an exponent of order $\beta_d \approx 1.27$ for $\eta = 0.02$ (the value of the exponents β_w and β_d depend on η; see also Ref. 36 for a spatial model with $\beta_d > 1$). We show in Fig. 10 the distance strength for two extreme situations for which spatial constraints are nonexistent ($\eta = 10.0$) or in contrast very strong ($\eta = 0.02$).

The nonlinearity induced by the spatial structure can be explained by the following mechanism affecting the network growth. The increase of spatial constraints

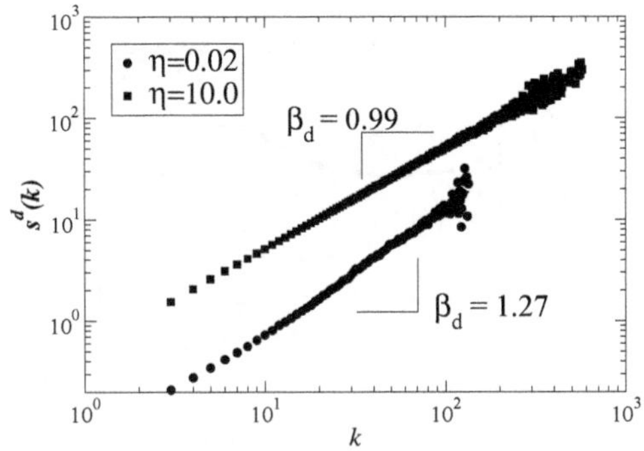

Fig. 10. Distance strength versus k for $\eta = 0.02$ and $\eta = 10.0$ (the networks are obtained for $\delta = 1$, $N = 10^4$, $\langle k \rangle = 6$, and averaged over 100 configurations). When η is not small, space is irrelevant and there are no correlations between degree and space. When spatial effects are important ($\eta = 0.02 \ll 1$), non-linear correlations appear. We observe a crossover for $k \simeq 10$–20 to a power-law behavior and the power-law fit over this range of values of k is shown (full lines).

affects the trend to form global hubs, since long distance connections are less probable, and drives the topology towards the existence of "regional" hubs of smaller degree. The total traffic, however, is not changed with respect to the case $\eta = \infty$, and is in fact directed towards these "regional" hubs. These medium-large degree vertices therefore carry a much larger traffic than they would if global "hubs" were available, leading to a faster increase of the traffic as a function of the degree, eventually resulting in a super-linear behavior. Moreover, as previously mentioned, the increase in distance costs implies that long range connections can be established only towards the hubs of the system: this effect naturally leads to a super-linear accumulation of $s^d(k)$ at larger degree values.

Spatial constraints also have a strong effect on the correlations between neighboring nodes (Fig. 11). At large η, a disassortative network is created, as is the case in most growing networks [11]; as η decreases, k_{nn} decreases, and an increasing range of flat $k_{nn}(k)$ appears: the tendency for small nodes to connect to hubs is contrasted by the need to use small-range links. For small enough η, a nearly neutral behavior more similar to what is actually observed in the airport network is reached. Moreover, the affinity of nodes to establish strong links to large nodes, measured by $k_{nn}^w(k)$, goes from a flat behavior at large η to a slightly assortative one at small η. In all cases, the weighted correlation $k_{nn}^w(k)$ remains clearly larger than the unweighted $k_{nn}(k)$, showing that links to busier nodes are typically stronger.

A non-trivial clustering hierarchy is already displayed by the model without spatial constraints. As previously mentioned, the decrease of η leads to an increase in clustering. Moreover, the weighted clustering is always significantly larger than the unweighted one, showing that the cliques typically carry an important traffic

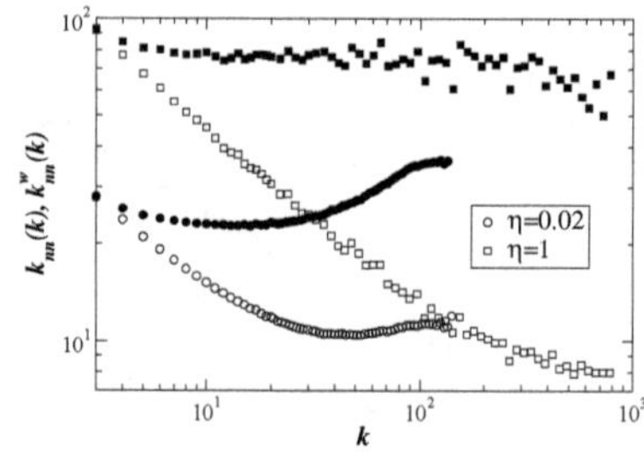

(a)

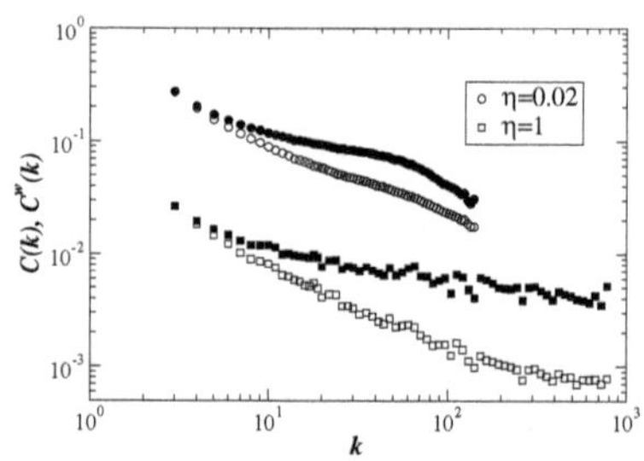

(b)

Fig. 11. (a) Assortativity and (b) clustering versus k obtained for the model for $\delta = 1$, $N = 10^4$, $\langle k \rangle = 6$, $\eta = 0.02$ (circles) and $\eta = 1$ (squares). Data are averaged over 100 configurations. Empty symbols refer to topological correlations while full symbols correspond to the weighted quantities k^w_{nn} and C^w.

load (see Fig. 11). These effects are a general signature of spatial constraints, as also observed in a unweighted network [13].

4.2. *Spatial constraints and betweenness centrality*

The spatial constraints act at both local and global levels of the network structure by introducing a distance cost in the establishment of connections. It is therefore important to look at the effect of space in global topological quantities such as the betweenness centrality. The betweenness centrality of a vertex is determined by its ability to provide a path between separated regions of the network. Hubs are natural crossroads for paths and it is natural to observe a marked correlation between g and k as expressed in the general relation $g(k) \sim k^{\mu}$. The exponent μ depends on the characteristics of the network, and we expect this relation to be altered when spatial constraints become important. In the present model, Fig. 12 clearly shows that this correlation in fact increases when spatial constraints become large (i.e. when η decreases). This can be understood simply by the fact that the probability

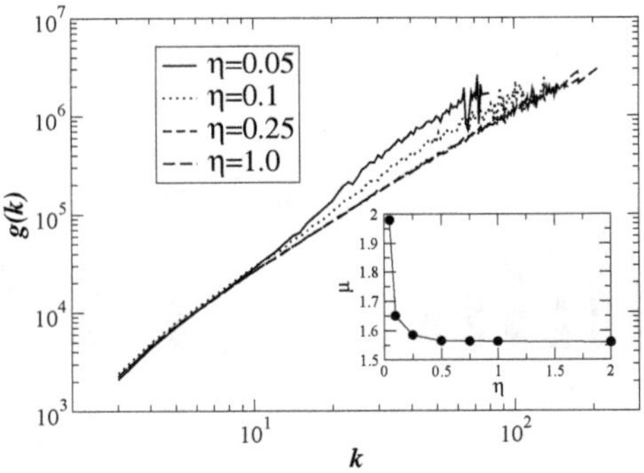

Fig. 12. Betweenness centrality versus degree for different values of η. Inset: Exponent μ (obtained by fitting the data for $k > 10$) of the betweenness centrality versus η (for $N = 5{,}000$, $m = 3$ and averaged over 50 configurations). For strong spatial constraints long-range shortcuts are very rare and hubs connect regions that are otherwise almost disconnected, which in turn implies a larger centrality of the hubs.

of establishing far-reaching shortcuts decreases exponentially in Eq. (3.4) and only the large traffic of hubs can compensate for this decay. Distant geographical regions can thus only be linked by edges connected to large degree vertices, which implies a more central role for these hubs.

In order to better understand the effect of space on the properties of betweenness centrality, we have to explicitly consider the geometry of the network along with the topology. In particular, we need to consider the role of the spatial position by introducing the spatial barycenter of the network. Indeed, in the presence of a spatial structure, the centrality of nodes is correlated with their position with respect to the barycenter G, whose location is given by $\mathbf{x}_G = \sum_i \mathbf{x}_i/N$. For a spatially ordered network — the simplest case being a lattice embedded in a one-dimensional space — the shortest path between two nodes is simply the Euclidean geodesic. In a limited region, for two points lying far away, the probability that the shortest path passes near the barycenter of all nodes is very large. In other words, this implies that the barycenter (and its neighbors) will have a large centrality. In a purely topological network with no underlying geography, this consideration does not apply anymore, and the full randomness and the disordered small-world structure are completely uncorrelated with the spatial position. It is worth remarking that the present argument applies in the absence of periodic boundary conditions that would destroy the geometrical ordering. This point is illustrated in Fig. 13 in the simple case of a one-dimensional lattice.

The present model defines an intermediate situation in that we have a random network with space constraints that introduces a local structure since short distance connections are favored. Shortcuts and long distance hops are present along with a spatial local structure that clusters spatially neighboring vertices. In Fig. 14,

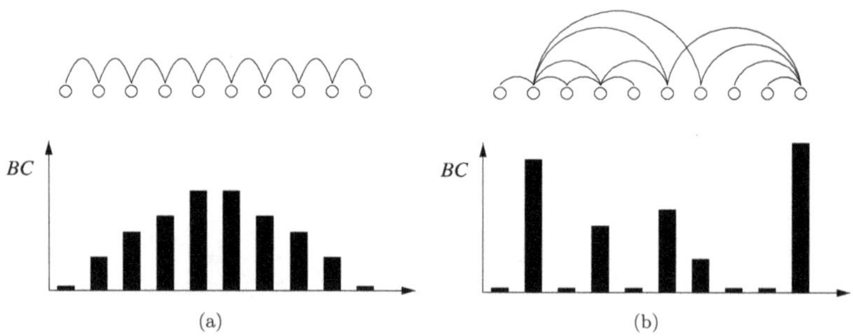

Fig. 13. (a) Betweenness centrality for the (one-dimensional) lattice case. The central nodes are close to the barycenter. (b) For a general graph, the central nodes are usually the ones with large degree.

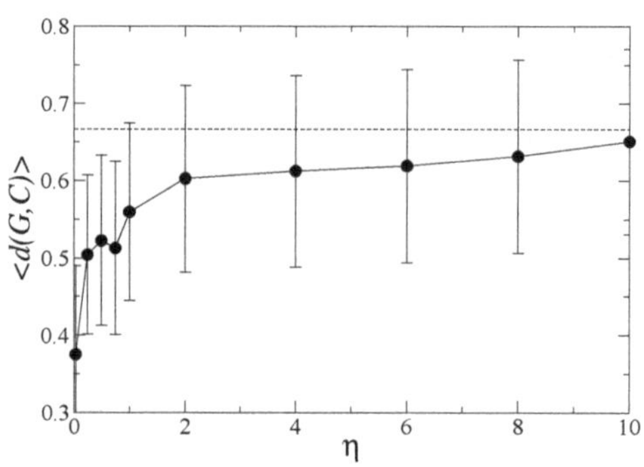

Fig. 14. Average Euclidean distance between the barycenter G of all nodes and the ten most central nodes (C) versus the parameter η (Here $\delta = 0$, $N = 5{,}000$ and the results are averaged over 50 configurations). When space is important (i.e. small η), the central nodes are closer to the gravity center. For large η, space is irrelevant and the average distance tends to the value corresponding to a uniform distribution $\langle r \rangle_{\text{unif}} = 2/3$ (dotted line).

we plot the average distance $d(G, C)$ between the barycenter G and the ten most central nodes. As expected, as spatial constraints become more important, the most central nodes get closer to the spatial barycenter of the network.

Another effect observed when the spatial constraints become important are the large fluctuations of the BC. Figure 15(a) displays the relative fluctuation

$$\delta g(k) = \frac{\sqrt{\langle \delta g^2(k) \rangle}}{\langle g(k) \rangle}, \tag{4.1}$$

where $\langle \delta g^2(k) \rangle$ is the variance of the BC and $\langle g(k) \rangle$ its average (computed for each value of k). The value of η modifies the degree cut-off, and in order to be

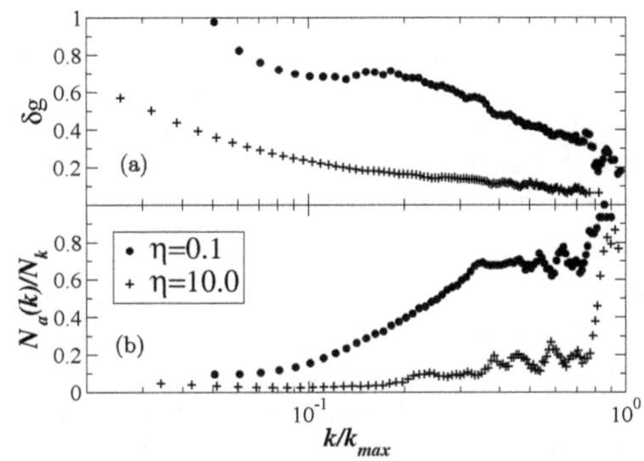

Fig. 15. (a) Relative fluctuations of the betweenness centrality versus $k/k_{\max}$ for two values of η ($N=5{,}000$ and the results are averaged over 50 configurations and binned). The fluctuations increase when η decreases (i.e. when spatial constraints increase). (b) Number of anomalies $N_a(k)$ rescaled by the number $N(k)$ of nodes of degree k versus $k/k_{\max}$ for different values of η. The relative number of anomalies is larger when spatial constraints are large, especially for large k.

able to compare the results for different values of η we rescale the abscissa by its maximum value $k_{\max}$. This plot [Fig. 15(a)] clearly shows that the BC relative fluctuations increase as η decreases and become quite large. This means that nodes with small degree may have a relatively large BC (or the opposite), as observed in the air-transportation network (see Fig. 2 and Ref. 25). In order to quantify these "anomalies" we compute the fluctuations of the betweenness centrality $\Delta_{RN}(k)$ for a randomized network with the same degree distribution as the original network and constructed with the Molloy–Reed algorithm [35]. We consider a node i as being "anomalous" if its betweenness centrality $g(i)$ lies outside the interval $[\langle g(k)\rangle - \alpha\Delta_{RN}(k), \langle g(k)\rangle + \alpha\Delta_{RN}(k)]$, where we choose $\alpha \simeq 1.952$ so that the considered interval would represent 95% of the nodes in the case of Gaussian distributed centralities around the average.

In Fig. 15(b), we show the relative number of anomalies versus $k/k_{\max}$ for different values of η. This plot shows that the relative number of anomalies $N_a(k)/N(k)$ increases when the degree increases and more interestingly strongly increases when η decreases. Note that since for increasing k the number of nodes $N(k)$ is getting small, the results become noisier.

The results of Figs. (12)–(15) can be summarized as follows. In a purely topological growing network, centrality is strongly correlated with degree since hubs have a natural ability to provide connections between otherwise separated regions or neighborhoods [14]. As spatial constraints appear and become more important, two factors compete in determining the most central nodes: (i) on the one hand hubs become even more important in terms of centrality since only a lot of traffic can compensate for the cost of long-range connections, which implies that the correlations between degree and centrality become even stronger; (ii) on the other hand, many paths go through the neighborhood of the barycenter, reinforcing the centrality of less-connected nodes that happen to be in the right place; this yields larger fluctuations of g and a larger number of "anomalies."

We finally note that these effects are not qualitatively affected by the weight structure and we observe the same behavior for $\delta = 0$ or $\delta \neq 0$.

5. Conclusions

In this article, we have reviewed the main empirical characteristics of the airline network. In particular, we have shown that weight–topology correlations as well as geographical effects are essential features of this system. We have also reviewed a simple model of growing weighted networks that introduces the effect of space and geometry in the establishment of new connections. When spatial constraints are important, the effects on the network structure can be summarized as follows:

- *Effect of spatial embedding on topology–traffic correlations.* Spatial constraints induce strong nonlinear correlations between topology and traffic. The reason for this behavior is that spatial constraints favor the formation of regional hubs and locally reinforces the preferential attachment, leading for a given degree to a larger strength than the one observed without spatial constraints. Moreover, long-distance links can connect only to hubs, which yields a value $\beta_d > 1$ for small enough η. The existence of constraints such as spatial distance selection induces some strong correlations between topology (degree) and non-topological quantities such as weights or distances.
- *Effect of space embedding on centrality.* Spatial constraints also induce large betweenness centrality fluctuations. While hubs are usually very central, when space is important central nodes tend to get closer to the gravity center of all points. Correlations between spatial position and centrality compete with the usual correlations between degree and centrality, leading to the observed large fluctuations of centrality at fixed degree.
- *Effect of space embedding on clustering and assortativity.* The existence of spatial constraints implies that the tendency to connect to hubs is limited by the need to use small-range links. This explains the almost flat behavior observed for the assortativity. Connection costs also favor the formation of cliques between spatially close nodes, and thus increase the clustering coefficient.

Including spatial effects in a simple model of weighted networks thus yields a large variety of behaviors and interesting effects. This study sheds some light on the importance and effect of different ingredients, such as the geographical distribution of nodes or the diversity of interaction weights, in the structure of large complex networks. We believe that this attempt at explaining network typology could be useful in understanding and modeling real-world networks. We finally note that the distance used here is a simple measure of the cost of a connection, but that the model we proposed could be easily generalized to the more complex cost functions that can be found in economic systems.

Acknowledgments

We thank L. A. N. Amaral, E. Chow, S. Dimitrov, R. Guimerà, P. de Los Rios, and T. Petermann for interesting discussions at various stages of this work. A. Barrat and A. Vespignani are partially funded by the European Commission (contract 001907, DELIS).

References

[1] Albert, R., Jeong, H. and Barabási, A.-L., Error and attack tolerance of complex networks, *Nature* **406** (2000) 378.
[2] Albert, R. and Barabási, A.-L., Statistical mechanics of complex networks, *Rev. Mod. Phys.* **74** (2002) 47.
[3] Almaas, E., Kovacs, B., Viscek, T., Oltvai, Z. N. and Barabási, A.-L., Global organization of metabolic fluxes in the bacterium, Escherichia coli, *Nature* **427** (2004) 839.
[4] Amaral, L. A. N., Scala, A., Barthélemy, M. and Stanley, H. E., Classes of small-world networks, *Proc. Natl. Acad. Sci.* (*USA*) **97** (2000) 11149.
[5] Andersson, C., Hellervik, A. and Lindgren, K., A spatial network explanation for a hierarchy of urban power laws, *Physica A* **227** (2005) 345.
[6] Barabási, A.-L. and Albert, R., Emergence of scaling in random networks, *Science* **286** (1999) 509.
[7] Barrat, A., Barthélemy, M., Pastor-Satorras, R. and Vespignani, A., The architecture of complex weighted networks, *Proc. Natl. Acad. Sci.* (*USA*) **101** (2004) 3747.
[8] Barrat, A., Barthélemy, M. and Vespignani, A., Weighted evolving networks: Coupling topology and weights dynamics, *Phys. Rev. Lett.* **92** (2004) 228701.
[9] Barrat, A., Barthélemy, M. and Vespignani, A., Modeling the evolution of weighted networks, *Phys. Rev. E* **70** (2004) 066149.
[10] Barrat, A., Barthélemy, M. and Vespignani, A., Traffic-driven model of the World Wide Web graph, in Lecture Notes in Computer Science, Vol. 3243 (Springer, 2004), p. 56.
[11] Barrat, A. and Pastor-Satorras, R., Rate equation approach for correlations in growing network models, *Phys. Rev. E* **71** (2005) 036127.
[12] Barthélemy, M., Gondran, B. and Guichard, E., Spatial structure of the internet traffic, *Physica A* **319** (2003) 633.
[13] Barthélemy, M., Crossover from scale-free to spatial networks, *Europhys. Lett.* **63** (2003) 915.
[14] Barthélemy, M., Betweenness centrality in large complex networks, *Euro. Phys. J. B* **38** (2003) 163.
[15] Cohen, R., Erez, K., ben Avraham, D. and Havlin, S., Resilience of the Internet to random breakdowns, *Phys. Rev. Lett.* **85** (2000) 4626.
[16] Derrida, B. and Flyvbjerg, H., Statistical properties of randomly broken objects and of multivalley structures in disordered systems, *J. Phys. A* **20** (1987) 5273.
[17] Dorogovtsev, S. N. and Mendes, J. F. F., *Evolution of Networks: From Biological Nets to the Internet and WWW* (Oxford University Press, Oxford, 2003).
[18] Freeman, L. C., A set of measures of centrality based upon betweeness, *Sociometry* **40** (1977) 35.

[19] Garlaschelli, D., Battiston, S., Castri, M., Servedio, V. D. P. and Caldarelli, G., The scale-free topology of market investments, *Physica A* **350** (2005) 491.
[20] Gastner, M. T. and Newman, M. E. J., The spatial structure of networks, cond-mat/0407680.
[21] Gastner, M. T. and Newman, M. E. J., Shape and efficiency in spatial distribution networks, condmat/0409702.
[22] Goh, K.-I., Kahng, B. and Kim, D., Universal behavior of load distribution in scale-free networks, *Phys. Rev. Lett.* **87** (2001) 278701.
[23] Gorman, S. P. and Kulkarni, R., Spatial small worlds: New geographic patterns for an information economy, submitted to *Environment and Planning Journal B* (2003).
[24] Granovetter, M., The strength of weak ties, *Am. J. Sociology* **78**(6) (1973) 1360.
[25] Guimerà, R. and Amaral, L. A. N., Modeling the world-wide airport network, *Euro. Phys. J. B* **38** (2004) 381.
[26] Guimerà, R., Mossa, S., Turtschi, A. and Amaral, L. A. N., The worldwide air transportation network: Anomalous centrality, community structure, and cities' global roles, *Proc. Natl. Acad. Sci.* (*USA*) **102** (2005) 7794.
[27] Helmy, A., Small large-scale wireless networks: Mobility-assisted resource discovery, cs.NI/0207069.
[28] International Air Transport Association, http://www.iata.org.
[29] Krause, A. E., Frank, K. A., Mason, D. M., Ulanowicz, R. E. and Taylor, W. W., Compartments revealed in food-web structure, *Nature* **426** (2003) 282.
[30] Lakhina, A., Byers, J. B., Crovella, M. and Matta, I., On the geographic location of Internet resources, technical report, online version available at: http://www.cs.bu.edu/techreports/pdf/2002-015-internet-geography.pdf'
[31] Li, W. and Cai, X., Statistical analysis of airport network of China, *Phys. Rev. E* **69** (2004) 046106.
[32] Li, C. and Chen, G., Network connection strength: Another power-law?, cond-mat/0311333 (2003).
[33] Manna, S. S. and Sen, P., Modulated scale-free network in the euclidean space, *Phys. Rev. E* **66** (2002) 066114.
[34] Masuda, N., Miwa H. and Konno, N., Geographical threshold graphs with small-world and scale-free properties, *Phys. Rev. E* **71** (2005) 036108.
[35] Molloy, M. and Reed, B., A critical point for random graphs with a given degree sequence, *Random Struct. Algorithms* **6** (1995) 161.
[36] Mukherjee, G. and Manna, S. S., Growing spatial scale-free graphs by selecting local edges, cond-mat/0503697.
[37] Nemeth, G. and Vattay, G., Giant clusters in random *ad hoc* networks, *Phys. Rev. E* **67** (2003) 036110.
[38] Callaway, D. S., Newman, M. E. J., Strogatz, S. H. and Watts, D. J., Network robustness and fragility: Percolation on random graphs, *Phys. Rev. Lett.* **85** (2000) 5468.
[39] Newman, M. E. J., Who is the best connected scientist? A study of scientific coauthorship network I, *Phys. Rev. E* **64** (2001) 016131.
[40] Newman, M. E. J. , Who is the best connected scientist? A study of scientific coauthorship network II, *Phys. Rev. E* **64** (2001) 016132.
[41] Newman, M. E. J., Assortative mixing in networks, *Phys. Rev. Lett.* **89** (2002) 208701.
[42] Pandya, R. V. R., A note on "Weighted evolving networks: Coupling topology and weight dynamics", cond-mat/0406644 (2004).
[43] Pastor-Satorras, R. and Vespignani, A., Epidemic spreading in scale-free networks, *Phys. Rev. Lett.* **86** (2001) 3200.

[44] Pastor-Satorras, R. and Vespignani, A., *Evolution and Structure of the Internet: A Statistical Physics Approach* (Cambridge University Press, Cambridge, 2004).
[45] Vázquez, A., Pastor-Satorras, R. and Vespignani, A., Large-scale topological and dynamical properties of Internet, *Phys. Rev. E* **65** (2002) 066130.
[46] Watts, D. J. and Strogatz, S. H., Collective dynamics of "small-world" networks, *Nature* **393** (1998) 440.
[47] Waxman, B. M., Routing of multipoint connections, *IEEE J. Select. Areas. Commun.* **6** (1988) 1617.
[48] Xulvi-Brunet, R. and Sokolov, I. M., Evolving networks with disadvantaged long-range connections, *Phys. Rev. E* **66** (2002) 026118.
[49] Yook, S. H., Jeong, H., Barabási, A.-L. and Tu, Y., Weighted evolving networks, *Phys. Rev. Lett.* **86** (2001) 5835.
[50] Yook, S.-H., Jeong, H. and Barabási, A.-L., Modeling the Internet's large-scale topology, *Proc. Natl. Acad. Sci. (USA)* **99** (2002) 13382.
[51] Zheng, D., Trimper, S., Zheng, B. and Hui, P. M., Weighted scale-free networks with stochastic weight assignments, *Phys. Rev. E* **67** (2003) 040102.

Advances in Complex Systems, Vol. 11, No. 6 (2008) 927–950

CENTRALITY AND PERIPHERALITY IN FILTERED GRAPHS FROM DYNAMICAL FINANCIAL CORRELATIONS

F. POZZI*, T. DI MATTEO† and T. ASTE‡

Department of Applied Mathematics,
The Australian National University,
0200 Canberra, ACT, Australia
**francesco.pozzi@anu.edu.au*
†tiziana.dimatteo@anu.edu.au
‡tomaso.aste@anu.edu.au

Received 22 February 2008
Revised 22 May 2008

Minimum spanning trees and planar maximally filtered graphs are generated from correlations between the 300 most-capitalized NYSE stocks' daily returns, computed dynamically over moving windows of sizes between 1 and 12 months, in the period from 2001 to 2003. We study how different economic sectors differently populate the various regions of these graphs. We find that the financial sector is always at the center whereas the periphery is shared among different sectors. Four extremes are observed: stocks well-connected and central; stocks well-connected but at the same time peripheral; stocks poorly-connected but central; stocks poorly-connected and peripheral. Two principal components of centrality measures are individuated. The economic meaning of this hierarchical disposition is discussed.

Keywords: Econophysics; networks; minimum spanning tree; planar maximally filtered graph; financial data correlations.

1. Introduction

One of the main goals in the field of complex systems is the selection and extraction of relevant and meaningful information about the properties of the underlying system from large data sets. In the last few years different methods have been proposed for filtering financial data by extracting a structure of interactions from cross-correlation matrices where only a subset of relevant entries are selected by means of criteria borrowed from network theory [1–14]. In particular, two methods that have been proven to be very effective are the minimum spanning tree (MST) [1, 15] and the planar maximally filtered graph (PMFG) [7, 8]. They are both based on the principle of iteratively constructing a constrained graph (a tree or a planar graph[a]) which retains the largest correlations between connected nodes.

In this paper we study MST and PMFG graphs generated dynamically from correlations computed over a moving window. Dynamics adds a quantification of stability/variability over time which is very important in systems such as financial markets that are constantly evolving. In a previous paper [16] we have shown that the dynamical PMFG preserves the same hierarchical structure as the dynamical MST, providing in addition a more significant and richer structure, a stronger robustness and dynamical stability in the long run. In this paper we address the question of how these graphs are populated and we investigate the typology of stocks which stay in the center and those that are relegated to the peripheries.

The paper is organized as follows. In Sec. 2 we present the data set and introduce the correlation matrices associated with the complete and the filtered graphs. Central and peripheral nodes are classified in Sec. 3. Results are discussed in Sec. 4. Conclusions are drawn in Sec. 5. Appendices A–D contain definitions and supporting material.

2. From Returns to Graphs

We have analyzed daily time series of the $n = 300$ most-capitalized NYSE stocks from 2001 to 2003, for a total of $T = 748$ days [9]. Returns are computed as logarithmic differences of daily prices, and daily prices are computed as averages of daily quotations. Closing quotations are excluded from the computation. Stocks are classified into 12 economic sectors and 77 economic subsectors, according to the classification of *Forbes* magazine. The names of sectors, the codes used in this paper and the number of stocks in each sector are shown in Table 1.

Table 1. Names of sectors, codes, and corresponding number of stocks.

Sector	Code	Number of stocks
Basic Materials	S01	24
Capital Goods	S02	12
Conglomerates	S03	8
Consumer Cyclical	S04	22
Consumer Noncyclical	S05	25
Energy	S06	17
Financial	S07	53
Healthcare	S08	19
Services	S09	69
Technology	S10	34
Transportation	S11	5
Utilities	S12	12

[a]A planar graph is a graph that can be represented on a Euclidean plane with no intersections between edges.

We have considered moving windows from time t to time $(t + \Delta t - 1)$, where $t = 1, 2, \ldots, T - \Delta t + 1$ and $\Delta t = 21, 42, 63, 84, 126, 251$ market days, corresponding approximately to $1, 2, 3, 4, 6, 12$ months. For each t and Δt, the resulting matrix of returns is denoted as $Y_s(t, \Delta t)$, with $s = 1, 2, \ldots, 300$.

For each t and Δt we have computed the correlation matrix $C(t, \Delta t)$ with coefficients given by the formula

$$c_{i,j}(t, \Delta t) = \frac{\langle Y_i Y_j \rangle - \langle Y_i \rangle \langle Y_j \rangle}{\sqrt{(\langle Y_i^2 \rangle - \langle Y_i \rangle^2)(\langle Y_j^2 \rangle - \langle Y_j \rangle^2)}}, \tag{1}$$

where $\langle f \rangle = \frac{1}{\Delta t} \sum_{\tau=0}^{\Delta t - 1} f(t + \tau)$ is the time average of a given series $f(t)$ over the window Δt. From these correlation coefficients $c_{i,j}$, we compute distances between stocks i and j: $d_{i,j} = \sqrt{2(1 - c_{i,j})}$ [1, 17–19]. The resulting matrix, $D(t, \Delta t) = \sqrt{2(1 - C(t, \Delta t))}$, is the dynamical distance matrix of the weighted complete graphs K_{300}, which has $n(n-1)/2 = 300 \times 299/2 = 44\,850$ edges connecting all pairs of nodes. Different methods exist in the literature in order to filter such a huge amount of data, which is otherwise hardly readable or usable.

A widely used filtering method is the MST, a connected graph with no cycles and $n - 1$ edges. In the MST construction, edges are selected in order to minimize the sum of the distances [20]. An almost linear running time algorithm has recently been developed by Chazelle [21]. Since we have computed almost 4,000 MSTs out of 300 nodes' graphs, the efficiency of the algorithm had to be considered. We have used Prim's algorithm [22, 23] implemented in Matlab and we verified that its efficiency $O(n^2)$ is enough for our purposes.

A filtering method which uses a similar principle, but allows more interactions and builds a more complex and rich structure, is the PMFG [7, 8, 16]. It is built by constructing an ordered list of edges from the smallest to the largest distances and then connecting the shortest edges whilst constraining the condition of planarity[b] [7, 8]. The resulting network is a connected planar graph in which the number of edges is $3(n-2)$, approximately three times the number of edges of the MST. It has been proven in Ref. 8 that the MST is always a subgraph of the PMFG.

3. Analysis: Who Stays in the Center?

3.1. *Central and peripheral nodes*

For each node of both dynamical MSTs and PMFGs and for each Δt, we computed the time average of the following quantities: degree, betweenness, eccentricity and closeness (see definitions in Appendix A). For the list of 300 stocks (the graph's vertices), the time averages of these four measures have been sorted, respectively in *descending* order for degree and betweenness and in *ascending* order for eccentricity and closeness. These two kinds of ranking are due to the opposite meaning of the

[b]The numerical code for constructing the PMFG is provided in Ref. 8 ("Additional Material").

two couples of measures: the first two are properly centrality measures, the second two are peripherality measures. As a consequence of such ordering, nodes at the top of the ranking are the most central while those at the bottom are the most peripheral from the point of view of each of the measures considered. In Tables 2 and b three nodes with the highest averages (at the top of the ranking) and three nodes with the lowest (at the bottom of the ranking) are reported. Tables 2(a) and (b) illustrate the results for degrees in MSTs and PMFGs; Tables 2(c) and (d) illustrate the results for betweennesses; Tables 3(e) and (f) illustrate the results for eccentricities; and Tables 3(g) and (h) illustrate the results for closenesses.

We see that the most central nodes belong predominantly to the Financial Sector (S07). Some stocks are particularly recurrent: Franklin Resources (BEN), Merrill Lynch (MER), Suntrust Banks (STI), Bear Stearns (BSC), Jefferson–Pilot (JP), A. G. Edwards (AGE) and Legg Mason (LM). Two particularly recurrent central stocks that are not financial are PPG Industries (PPG), which belongs to the sector of Basic Materials (S01), and Eaton (ETN) of the Capital Goods sector (S02). Newmont Mining (NEM) and Barrick Gold (ABX), both belonging to Basic Materials (S01), are two clear examples of peripheral nodes — nodes that are always at the bottom of all lists. There are other stocks, though, that are at the bottom of the list only for degrees and betweennesses, like Sociedad Anonima ADS (YPF, Energy–S06), BCE INC (BCE, Services–S09) and Mattel (MAT, Consumer Cyclical–S04). There are, however, others that are at the bottom of the list only for eccentricities and closenesses, like Schlumberger (SLB), Baker Hughes (BHI), BJ Services (BJS), Halliburton (HAL) and Smith International (SII), which interestingly all belong to the Energy sector (S06).

The unambiguous predominance of Financial among central nodes is even more evident if we look only at eccentricity and closeness. In fact, only financial stocks are present in lists e, f, g and h of Table 3 among the three with the smallest eccentricity and closeness, while lists a, b, c and d for degrees and betweennesses in Table 2 show a more variegated situation, with also Basic Materials (S01) and Capital Goods (S02) present.

As Δt increases, the average degrees of central nodes increase remarkably, both in the dynamical MSTs and in the dynamical PMFGs, while the average degrees of poorly-connected nodes slightly decrease to their minimum possible value, i.e. 1 for the MST and 3 for the PMFG. When $\Delta t = 12$ months, there are 28 stocks with time average degrees exactly equal to 1 in the MST, which means that these companies have always been extremely badly connected, with only one connection to the rest of the tree for all of the period considered. The same patterns are followed by time average betweennesses: as Δt increases, central nodes become more central and poorly-connected nodes become less connected. We see that the less central nodes according to the degree and the betweenness are the same: ABX, BCE, MAT, NEM, YPF, which are not sectorially characterizable.

Eccentricity and closeness show a less clear behavior. As Δt increases, both average eccentricities and closenesses of central nodes diminish in the MSTs, i.e.

Table 2. The three lowest and highest averages for degree and betweenness, for the MST and the PMFG, computed at $\Delta t = 1, 2, 3, 4, 6, 12$ months. For each Δt, the first, second and third columns indicate respectively the NYSE code for companies; the code for economic sectors of activity (S01–S12) and the time average values.

$\Delta t = 1$			$\Delta t = 2$			$\Delta t = 3$			$\Delta t = 4$			$\Delta t = 6$			$\Delta t = 12$		
(a) Degree — MST																	
JP	S07	3.67	*JP*	S07	5.42	*JP*	S07	7.10	*JP*	S07	7.84	*JP*	S07	8.78	*ETN*	S02	11.33
STI	S07	3.62	*STI*	S07	4.65	*PPG*	S01	5.63	*PPG*	S01	6.24	*PPG*	S01	7.15	*JP*	S07	11.17
BEN	**S07**	**3.57**	**BEN**	**S07**	**4.59**	*STI*	S07	5.21	**BEN**	**S07**	**5.59**	*ETN*	S02	6.77	*PPG*	S01	9.91
...	...	...	...	...	...	...	...	...	...	...	...	...	...	...	...	...	...
MCD	S09	1.31	*HCP*	S09	1.16	*HCP*	S09	1.08	*LUK*	S03	1.05	*BCE*	S09	1.00	*SVU*	S09	1
BCE	S09	1.30	*BCE*	S09	1.11	*BCE*	S09	1.05	*BCE*	S09	1.02	*MAT*	S04	1	*UST*	S05	1
YPF	**S06**	**1.15**	**YPF**	**S06**	**1.05**	**YPF**	**S06**	**1.02**	**YPF**	**S06**	**1**	**YPF**	**S06**	**1**	**YPF**	**S06**	**1**
(b) Degree — PMFG																	
STI	S07	13.1	**BEN**	**S07**	**17.3**	**BEN**	**S07**	**20.0**	**BEN**	**S07**	**21.5**	**BEN**	**S07**	**24.8**	*ETN*	S02	28.8
BEN	**S07**	**12.6**	*STI*	S07	17.0	*PPG*	S01	18.5	*PPG*	S01	20.3	*PPG*	S01	22.9	*PPG*	S01	26.8
JP	S07	12.0	*PPG*	S01	16.6	*STI*	S07	17.6	*JP*	S07	18.5	*JP*	S07	20.0	**BEN**	**S07**	**24.5**
...	...	...	...	...	...	...	...	...	...	...	...	...	...	...	...	...	...
BCE	S09	3.7	*GDT*	S08	3.3	*HCP*	S09	3.2	*XRX*	S10	3.2	*XRX*	S10	3.1	*SVU*	S09	3.0
NEM	S01	3.6	*BCE*	S09	3.3	*BCE*	S09	3.1	*BCE*	S09	3.1	*SVM*	S09	3.0	*UST*	S05	3.0
YPF	**S06**	**3.3**	**YPF**	**S06**	**3.1**	**YPF**	**S06**	**3.0**	**YPF**	**S06**	**3.0**	**YPF**	**S06**	**3**	**YPF**	**S06**	**3.0**

Table 2. (*Continued*)

$\Delta t = 1$			$\Delta t = 2$			$\Delta t = 3$			$\Delta t = 4$			$\Delta t = 6$			$\Delta t = 12$		
(c) Betweenness — MST																	
STI	S07	15939	*STI*	S07	20860	*STI*	S07	22269	**BEN**	**S07**	**22408**	*STI*	S07	25114	**BEN**	**S07**	**33570**
BEN	**S07**	**14978**	*MER*	S07	18964	*MER*	S07	21411	*UPC*	S07	22112	**BEN**	**S07**	**24431**	*ETN*	S02	29038
BSC	S07	14074	**BEN**	**S07**	**18665**	**BEN**	**S07**	**20597**	*MER*	S07	21740	*PPG*	S01	22375	*STI*	S07	28537
...	...	...	...	...	...	...	...	...	...	...	...	...	...	...	...	...	...
MCD	S09	335	*TSS*	S10	147	*FNF*	S07	73.7	*FNF*	S07	37.6	*BCE*	S09	1.0	*BCE*	S09	0
GDT	S08	273	*BCE*	S09	86.6	*BCE*	S09	57.6	*BCE*	S09	14.3	*MAT*	S04	0	*TSS*	S10	0
YPF	**S06**	**140**	**YPF**	**S06**	**41.1**	**YPF**	**S06**	**15.6**	**YPF**	**S06**	**0**	**YPF**	**S06**	**0**	**YPF**	**S06**	**0**
(d) Betweenness — PMFG																	
STI	S07	5024	*STI*	S07	9969	**BEN**	**S07**	**12526**	**BEN**	**S07**	**13453**	**BEN**	**S07**	**16267**	**BEN**	**S07**	**18577**
BEN	**S07**	**4464**	**BEN**	**S07**	**8852**	*STI*	S07	11012	*PPG*	S01	11372	*PPG*	S01	13524	*STI*	S07	18008
PPG	S01	3583	*PPG*	S01	8109	*PPG*	S01	9647	*STI*	S07	11119	*STI*	S07	12150	*ETN*	S02	17461
...	...	...	...	...	...	...	...	...	...	...	...	...	...	...	...	...	...
ABX	S01	30.5	*NEM*	S01	15.8	*BCE*	S09	13.4	*MAT*	S04	7.50	*TSS*	S10	1.41	*BCE*	S09	0.189
YPF	**S06**	**23.0**	*BCE*	S09	15.1	*TSS*	S10	11.0	*BCE*	S09	2.10	*MAT*	S04	1.33	*TSS*	S10	0.068
NEM	S01	18.6	**YPF**	**S06**	**8.56**	**YPF**	**S06**	**2.28**	**YPF**	**S06**	**0.180**	**YPF**	**S06**	**0**	**YPF**	**S06**	**0.001**

Table 3. The three highest and lowest averages for eccentricity and closeness, for the MST and the PMFG, computed at $\Delta t = 1, 2, 3, 4, 6, 12$ months. For each Δt, the first, second and third columns indicate respectively the NYSE code for companies; the code for the 12 economic sectors of activity and the time average values.

$\Delta t = 1$			$\Delta t = 2$			$\Delta t = 3$			$\Delta t = 4$			$\Delta t = 6$			$\Delta t = 12$		
(e) Eccentricity — MST																	
BEN	**S07**	**19.3**	*BSC*	S07	18.0	*MER*	S07	17.6	**BEN**	**S07**	**16.7**	**BEN**	**S07**	**15.4**	**BEN**	**S07**	**13.6**
AGE	S07	19.4	*MER*	S07	18.0	*BSC*	S07	17.7	*AGE*	S07	16.8	*AGE*	S07	15.5	*AGE*	S07	13.9
BSC	S07	19.4	**BEN**	**S07**	**18.0**	**BEN**	**S07**	**17.7**	*MER*	S07	16.9	*LM*	S07	15.8	*LM*	S07	14.0
...	...	...	...	...	...	...	...	...	...	...	...	...	...	...	...	...	...
BJS	S06	23.6	*BR*	S06	22.9	*ABX*	S01	23.2	*WEC*	S12	22.4	*SLB*	S06	21.6	*BHI*	S06	21.5
ABX	S01	24.2	*ABX*	S01	23.1	*MUR*	S06	23.3	*ABX*	S01	22.7	*ABX*	S01	21.8	*HAL*	S06	21.6
NEM	**S01**	**24.2**	**NEM**	**S01**	**23.2**	**NEM**	**S01**	**23.4**	**NEM**	**S01**	**22.7**	**NEM**	**S01**	**21.9**	*SLB*	S06	21.9
(f) Eccentricity — PMFG																	
BEN	**S07**	**4.90**	**BEN**	**S07**	**6.64**	**BEN**	**S07**	**6.98**	**BEN**	**S07**	**6.88**	**BEN**	**S07**	**6.51**	**BEN**	**S07**	**6.30**
AGE	S07	4.93	*MER*	S07	6.67	*BSC*	S07	7.04	*AGE*	S07	7.00	*AGE*	S07	6.68	*AGE*	S07	6.53
BSC	S07	4.94	*STI*	S07	6.68	*MER*	S07	7.05	*BSC*	S07	7.01	*BSC*	S07	6.69	*LM*	S07	6.57
...	...	...	...	...	...	...	...	...	...	...	...	...	...	...	...	...	...
HCP	S09	6.04	*BJS*	S06	8.50	*WEC*	S12	9.15	*MUR*	S06	9.22	*BJS*	S06	9.10	*BHI*	S06	9.47
ABX	S01	6.15	*APC*	S06	8.50	**NEM**	**S01**	**9.17**	*ABX*	S01	9.29	**NEM**	**S01**	**9.27**	*HAL*	S06	9.76
NEM	**S01**	**6.19**	**NEM**	**S01**	**8.52**	*MUR*	S06	9.18	**NEM**	**S01**	**9.33**	*ABX*	S01	9.27	*SLB*	S06	9.81

Table 3. (*Continued*)

$\Delta t = 1$			$\Delta t = 2$			$\Delta t = 3$			$\Delta t = 4$			$\Delta t = 6$			$\Delta t = 12$		
(g) Closeness — MST																	
BEN	**S07**	**9.40**	*MER*	S07	8.27	*MER*	S07	7.86	**BEN**	**S07**	**7.52**	**BEN**	**S07**	**6.89**	**BEN**	**S07**	**6.24**
BSC	S07	9.43	*BSC*	S07	8.28	**BEN**	**S07**	**7.93**	*AGE*	S07	7.55	*AGE*	S07	7.05	*AGE*	S07	6.61
MER	S07	9.46	**BEN**	**S07**	**8.34**	*BSC*	S07	7.96	*MER*	S07	7.57	*MER*	S07	7.21	*LM*	S07	6.65
...	...	...	...	...	...	...	...	...	...	...	...	...	...	...	...	...	...
BJS	S06	13.55	*BR*	S06	13.26	*APC*	S06	13.53	*MUR*	S06	13.22	*SLB*	S06	12.98	*HAL*	S06	13.63
ABX	S01	13.96	*APC*	S06	13.27	**NEM**	**S01**	**13.58**	*ABX*	S01	13.27	*ABX*	S01	13.03	*BHI*	S06	13.68
NEM	**S01**	**13.98**	**NEM**	**S01**	**13.32**	*MUR*	S06	13.67	**NEM**	**S01**	**13.27**	**NEM**	**S01**	**13.13**	*SLB*	S06	14.10
(h) Closeness — PMFG																	
BEN	**S07**	**2.53**	**BEN**	**S07**	**3.30**	**BEN**	**S07**	**3.42**	**BEN**	**S07**	**3.42**	**BEN**	**S07**	**3.22**	**BEN**	**S07**	**3.12**
AGE	S07	2.57	*STI*	S07	3.34	*MER*	S07	3.47	*AGE*	S07	3.47	*BSC*	S07	3.37	*LM*	S07	3.27
BSC	S07	2.58	*MER*	S07	3.35	*AGE*	S07	3.49	*MER*	S07	3.50	*AGE*	S07	3.37	*AGE*	S07	3.27
...	...	...	...	...	...	...	...	...	...	...	...	...	...	...	...	...	...
SII	S06	3.60	*APC*	S06	5.14	*BJS*	S06	5.59	*ABX*	S01	5.69	*BJS*	S06	5.73	*BHI*	S06	6.09
ABX	S01	3.69	*BR*	S06	5.15	*APC*	S06	5.65	**NEM**	**S01**	**5.73**	**NEM**	**S01**	**5.79**	*HAL*	S06	6.30
NEM	**S01**	**3.73**	*BJS*	S06	5.16	*MUR*	S06	5.71	*MUR*	S06	5.74	*ABX*	S01	5.80	*SLB*	S06	6.42

the central nodes become more central. But a slight decrease can be seen also for eccentric nodes, which means that the eccentric nodes become less eccentric. In the PMFGs, instead, eccentricities and closenesses of central nodes slightly increase, i.e. the central nodes become slightly less central, and a more pronounced increase can be seen for eccentric nodes. However, we have noticed that by dividing each value by the average over all nodes, we retrieve the situation where, as Δt increases, the central nodes become more central and less central nodes become even less central according to all measures. This result confirms the general result that, as Δt increases, the structure of the system becomes more definite and robust. This is in agreement with previous results found in a past work [16].

We note that the most eccentric nodes according to both eccentricity and closeness are the same (SLB, BHI, BJS, HAL, SII) and are all Energy stocks, but they are different from those selected by degree and betweenness.

Summarizing, we find that all the measures provide similar results from the point of view of centrality but different results from the point of view of peripherality. Intuitively, we may think that there could be four possible extremes that need to be considered: nodes well-connected and central; nodes well-connected but eccentric at the same time; nodes poorly-connected but central; nodes poorly-connected and eccentric. In Sec. 5 we will investigate further this reasoning.

3.2. *Central and peripheral sectors*

In order to assess the relevance of sectors from a centrality/peripherality point of view, we have counted, for all the measures considered above and for each sector, the number of stocks present in the top 50 and in the bottom 50 of the rankings, in descending order for degree and betweenness and in ascending order for eccentricity and closeness. Results are reported in Tables 4 and 5.

We find that Financial is always strongly predominant among the central nodes of the system. All the four measures, for all the values of Δt, for both the dynamical MSTs and PMFGs, indicate an extraordinary presence of Financial within the most central nodes of the system, as we can see from column S07 of Tables 4(a), (b), (e), (f) and Tables 5(a), (b), (e), (f). Such predominance slightly decreases as Δt increases. Though proportions are very similar for all the four measures, eccentricity and closeness [Tables 5(a), (b), (e), (f)] show higher figures than degree and betweenness [Tables 4(a), (b), (e), (f)].

In general, for all measures and for both MST and PMFG, a strong presence among the central nodes (compared to the low number of stocks associated with them) can also be attributed to Basic Materials (S01), Capital Goods (S02), Conglomerates (S03) and Consumer Cyclical (S04). Basic Materials shows a very strong presence when $\Delta t = 1$ month but dramatically decreases as Δt increases. The presence of all other sectors is overall negligible or very negligible, as in the cases of Healthcare (S08), Transportation (S11) and Utilities (S12).

Table 4. Number of stocks, for each sector, among the 50 highest and the 50 lowest values for degree and betweenness, for both the MST and the PMFG, for each value of Δt (descending order ranking).

Degree

	(a) MST's top 50											(a)
	S01	S02	S03	S04	S05	S06	**S07**	S08	S09	S10	S11	S12
$\Delta t = 1$	8	3	2	3	1	1	**25**	0	2	5	0	0
$\Delta t = 2$	7	3	3	3	1	3	**21**	0	4	5	0	0
$\Delta t = 3$	5	3	2	3	1	4	**20**	1	6	4	0	1
$\Delta t = 4$	5	3	2	2	2	4	**20**	1	5	4	0	2
$\Delta t = 6$	3	3	2	3	2	4	**18**	2	8	4	0	1
$\Delta t = 12$	3	3	3	2	5	4	**15**	2	8	3	0	2
	(c) MST's bottom 50											(c)
	S01	S02	S03	S04	S05	S06	S07	S08	S09	**S10**	S11	S12
$\Delta t = 1$	3	1	0	1	9	2	3	7	14	**9**	0	1
$\Delta t = 2$	3	1	1	2	8	2	5	5	11	**10**	0	2
$\Delta t = 3$	1	1	1	3	7	2	5	3	14	**11**	0	2
$\Delta t = 4$	1	1	1	2	6	2	6	4	15	**10**	0	2
$\Delta t = 6$	1	0	1	1	6	2	8	4	13	**12**	0	2
$\Delta t = 12$	3	0	1	2	7	1	4	2	14	**14**	0	2

	(b) PMFG's top 50											(b)
	S01	S02	S03	S04	S05	S06	**S07**	S08	S09	S10	S11	S12
$\Delta t = 1$	8	3	3	3	1	2	**26**	0	1	3	0	0
$\Delta t = 2$	8	3	3	3	1	2	**23**	0	1	6	0	0
$\Delta t = 3$	8	3	2	2	1	3	**22**	0	3	6	0	0
$\Delta t = 4$	8	3	2	2	1	3	**21**	0	4	6	0	0
$\Delta t = 6$	5	3	2	2	1	3	**20**	2	6	6	0	0
$\Delta t = 12$	4	3	3	2	5	2	**17**	2	5	6	0	1
	(d) PMFG's bottom 50											(d)
	S01	S02	S03	S04	S05	S06	S07	S08	S09	**S10**	S11	S12
$\Delta t = 1$	3	2	0	1	8	2	2	7	15	**7**	1	2
$\Delta t = 2$	3	2	1	3	6	2	4	5	14	**9**	0	1
$\Delta t = 3$	3	1	1	3	6	2	4	4	14	**10**	1	1
$\Delta t = 4$	3	1	1	4	6	2	4	3	14	**10**	1	1
$\Delta t = 6$	2	1	1	2	7	2	4	3	14	**12**	1	1
$\Delta t = 12$	4	0	0	3	5	2	3	2	14	**15**	1	1

Table 4. (*Continued*)

Betweenness																								
	(e) MST's top 50 (e)												(f) PMFG's top 50 (f)											
	S01	S02	S03	S04	S05	S06	**S07**	S08	S09	S10	S11	S12	S01	S02	S03	S04	S05	S06	**S07**	S08	S09	S10	S11	S12
$\Delta t = 1$	10	3	3	2	0	1	**27**	0	0	4	0	0	8	3	4	2	0	2	**27**	0	1	3	0	0
$\Delta t = 2$	9	3	3	2	0	2	**26**	0	1	4	0	0	9	3	3	3	1	3	**23**	0	1	4	0	0
$\Delta t = 3$	10	3	4	2	1	2	**25**	0	0	3	0	0	9	3	3	2	1	3	**23**	0	2	4	0	0
$\Delta t = 4$	9	3	4	1	1	3	**25**	0	1	3	0	0	10	3	4	2	1	4	**22**	0	2	2	0	0
$\Delta t = 6$	6	3	4	1	1	4	**24**	0	5	2	0	0	8	3	5	1	1	5	**21**	1	3	2	0	0
$\Delta t = 12$	3	3	3	5	2	5	**17**	2	7	3	0	0	4	3	3	4	2	5	**18**	2	5	4	0	0
	(g) MST's top 50 (g)												(h) PMFG's top 50 (h)											
	S01	S02	S03	S04	S05	S06	S07	S08	S09	**S10**	S11	S12	S01	S02	S03	S04	S05	S06	S07	S08	S09	**S10**	S11	S12
$\Delta t = 1$	2	1	0	1	10	2	4	8	17	**4**	1	0	2	1	0	1	9	2	7	7	16	**5**	0	0
$\Delta t = 2$	3	1	1	2	7	2	3	6	16	**9**	0	0	3	1	1	3	9	2	3	5	15	**8**	0	0
$\Delta t = 3$	3	1	1	3	6	2	5	4	13	**11**	0	1	3	3	1	2	5	2	6	4	15	**9**	0	0
$\Delta t = 4$	1	1	1	4	6	2	4	4	16	**11**	0	0	2	2	2	4	5	2	3	5	16	**9**	0	0
$\Delta t = 6$	0	1	1	1	7	2	6	4	16	**11**	1	0	2	1	1	1	8	2	5	3	15	**11**	1	0
$\Delta t = 12$	4	1	1	2	7	1	5	2	11	**15**	0	1	2	0	0	2	6	3	6	2	15	**13**	1	0

Table 5. Number of stocks, for each sector, among the 50 lowest and the 50 highest values for eccentricity and closeness, for both the MST and the PMFG, for each value of Δt (ascending order ranking).

	Eccentricity																							
	(a)					MST's top 50						(a)	(b)					PMFG's top 50						(b)
	S01	S02	S03	S04	S05	S06	**S07**	S08	S09	S10	S11	S12	S01	S02	S03	S04	S05	S06	**S07**	S08	S09	S10	S11	S12
$\Delta t = 1$	11	4	4	3	0	0	**28**	0	0	0	0	0	11	3	4	2	0	0	**29**	0	0	1	0	0
$\Delta t = 2$	10	3	3	3	1	0	**30**	0	0	0	0	0	8	3	4	4	0	0	**30**	0	1	0	0	0
$\Delta t = 3$	6	4	4	4	0	0	**28**	0	3	1	0	0	7	3	3	2	0	0	**30**	0	3	2	0	0
$\Delta t = 4$	2	3	4	2	0	0	**34**	0	5	0	0	0	3	3	4	2	0	0	**31**	0	7	0	0	0
$\Delta t = 6$	2	3	5	3	0	0	**28**	0	6	3	0	0	2	3	4	4	0	0	**26**	0	9	2	0	0
$\Delta t = 12$	2	3	4	4	0	1	**22**	0	8	6	0	0	1	3	2	3	0	2	**21**	0	14	4	0	0
	(c)					MST's bottom 50						(c)	(d)					PMFG's bottom 50						(d)
	S01	S02	S03	S04	**S05**	**S06**	S07	S08	S09	S10	S11	**S12**	S01	S02	S03	S04	**S05**	**S06**	S07	S08	S09	S10	S11	**S12**
$\Delta t = 1$	2	0	0	0	**7**	**13**	2	6	7	1	0	**12**	2	0	0	0	**12**	**11**	1	5	7	1	0	**11**
$\Delta t = 2$	2	0	0	0	**9**	**13**	2	5	7	0	0	**12**	2	0	0	0	**12**	**12**	2	4	6	0	0	**12**
$\Delta t = 3$	2	0	0	0	**13**	**14**	1	3	5	0	0	**12**	2	0	0	0	**10**	**13**	1	5	6	1	0	**12**
$\Delta t = 4$	3	1	0	0	**12**	**13**	1	3	5	0	0	**12**	3	0	0	0	**11**	**13**	0	3	5	3	0	**12**
$\Delta t = 6$	3	2	0	0	**12**	**12**	0	2	7	0	0	**12**	4	0	0	0	**13**	**10**	0	2	5	4	0	**12**
$\Delta t = 12$	4	1	0	1	**14**	**10**	0	2	8	1	0	**9**	5	3	0	2	**13**	**10**	0	2	6	1	0	**8**

Table 5. (*Continued*)

	Closeness																							
	(e)					MST's top 50						(e)	(f)					PMFG's top 50						(f)
	S01	S02	S03	S04	S05	S06	**S07**	S08	S09	S10	S11	S12	S01	S02	S03	S04	S05	S06	**S07**	S08	S09	S10	S11	S12
$\Delta t = 1$	10	3	4	4	0	0	**29**	0	0	0	0	0	9	3	5	3	0	0	**29**	0	0	1	0	0
$\Delta t = 2$	10	3	4	2	0	0	**30**	0	1	0	0	0	8	3	4	5	0	0	**29**	0	0	1	0	0
$\Delta t = 3$	6	4	4	4	0	0	**29**	0	3	0	0	0	7	4	4	4	0	0	**27**	0	3	1	0	0
$\Delta t = 4$	8	3	4	3	0	0	**28**	0	3	1	0	0	7	3	4	2	0	0	**28**	0	6	0	0	0
$\Delta t = 6$	3	3	5	3	0	0	**27**	0	6	3	0	0	3	3	5	4	0	0	**24**	0	9	2	0	0
$\Delta t = 12$	3	3	6	7	0	0	**23**	0	5	3	0	0	2	3	4	6	0	1	**20**	1	10	3	0	0
	(g)					MST's bottom 50						(g)	(h)					PMFG's bottom 50						(h)
	S01	S02	S03	S04	**S05**	**S06**	S07	S08	S09	S10	S11	**S12**	S01	S02	S03	S04	**S05**	**S06**	S07	S08	S09	S10	S11	**S12**
$\Delta t = 1$	2	0	0	0	**8**	**13**	2	6	6	1	0	**12**	2	1	0	0	**9**	**12**	1	5	7	1	0	**12**
$\Delta t = 2$	2	0	0	0	**8**	**15**	2	5	6	0	0	**12**	2	0	0	0	**9**	**15**	2	4	6	0	0	**12**
$\Delta t = 3$	2	0	0	0	**10**	**15**	1	3	7	0	0	**12**	2	0	0	0	**11**	**15**	1	3	5	1	0	**12**
$\Delta t = 4$	2	1	0	0	**8**	**15**	2	4	6	0	0	**12**	3	1	0	0	**8**	**14**	1	5	5	1	0	**12**
$\Delta t = 6$	3	1	0	0	**11**	**12**	1	2	8	0	0	**12**	4	1	0	0	**10**	**11**	0	2	6	4	0	**12**
$\Delta t = 12$	3	1	0	1	**10**	**14**	1	2	7	1	0	**10**	4	1	0	1	**11**	**12**	0	2	6	2	0	**11**

Technology (S10) and Services (S09) occupy many of the lower 50 more peripheral positions, according to degree and betweenness but not to eccentricity and closeness, which, instead, offer surprisingly strong evidence for the role of eccentric nodes belonging to Utilities (S12), Energy (S06) and Consumer Noncyclical (S05). In particular, we find absolutely outstanding the relevance of the Utilities sector among the lower 50 positions of the latter two measures: almost 100% of the 12 Utilities stocks are counted.

From a joint analysis of Tables 2–5, we find particularly noteworthy the similarity of the behaviors of MSTs and PMFGs, which show a remarkable correspondence of sectorial structures.

3.2.1. *Robustness with respect to* Δt

We have noted in the previous analysis that changes in the ranking occur as the time window Δt changes. In order to check and quantify the robustness of the sectorial population in MSTs and PMFGs with respect to Δt, we have computed for all measures the number of common elements between the most central or the less central stocks extracted at different Δt.

If we consider the first 100 nodes with a higher degree for $\Delta t = 1$ month and $\Delta t = 12$ months, we find that they have 70 common elements for the MSTs and 72 common elements for the PMFGs, showing a high level of robustness of both MSTs and PMFGs with respect to changes in Δt. Similar figures, respectively 68 and 72, are found for nodes characterized by a lower degree. Respectively, 71 and 69 common elements are found among the first 100 nodes with higher betweenness, and 60 and 68 among the 100 nodes with a lower betweenness. Similar results have been found for eccentricity and closeness.

Moreover, we have verified that the number of common elements between the 100 most central nodes when $\Delta t = 1$ month and the 100 less central nodes when $\Delta t = 12$ months is always negligible for both MSTs and PMFGs. The same happens for the number of common elements between the 100 most central nodes when $\Delta t = 12$ months and the 100 less central nodes when $\Delta t = 1$ month. This means that central nodes do not become peripheral as Δt increases and, vice versa, peripheral nodes do not become central. Therefore, the graphs are very robust with respect to Δt.

3.2.2. *Similarity between MSTs and PMFGs*

In order to check and quantify the strength of the structural similarity between MSTs and PMFGs, we have computed for all measures the number of common elements between the most central or the less central stocks extracted by the two filtered graphs.

If we consider the first 100 nodes with a higher degree for MSTs and PMFGs, we find that they have 94 common elements when $\Delta t = 1$ month and 83 common elements when $\Delta t = 12$ months, showing therefore a very similar structure. We find respectively 86 and 82 common elements if we consider the 100 nodes with a lower degree. Very high figures are also found for the betweenness: 93 and 93 for the 100 higher values; 91 and 86 for the 100 lower ones. We find 95 and 86 common nodes for the 100 lower eccentricities; 94 and 86 common nodes for the 100 higher ones. We find 72 and 69 for the 100 lower closenesses; 77 and 74 for the 100 higher ones.

Moreover, we have verified that the number of common elements between the 100 most central nodes in the MST and the 100 less central nodes in the PMFG is systematically zero for all Δt. The same happens for the number of common elements between the 100 most central nodes in the PMFG and the 100 less central nodes in the MST. This means that it never happens that a node appears to be central in the MST and peripheral in the PMFG at the same time (or vice versa).

3.3. *Persistence over time*

In order to check the robustness of ranks over time, we have computed the standard deviations of the ranks over time for each node and for each of the eight measures when $\Delta t = 12$. We find parabolic curves in which the highest and the lowest ranks have a smaller standard deviation, while nodes in the middle have a rather larger variability. We verify that most of the stocks in the first 30 positions remain in the first 100 positions during turbulence periods with probabilities between 60% and 99.9%. We find that degree, betweenness and closeness are relatively robust whereas eccentricity is much more variable. Such variability is due to periods of strong market turbulence during which the network is particularly shaken. The last 30 positions are much more stable: they remain in the last 100 positions with probabilities between 70% and 99.9%.

We measure the persistence over time of centrality or peripherality by computing, for each measure, the number of common elements between the first 20 and 50 stocks in the four following cases at the same time: the MST with $\Delta t = 1$ and $\Delta t = 12$; the PMFG with $\Delta t = 1$ and $\Delta t = 12$. We find that a considerable number of central stocks are common elements of all measures and are predominantly financial. The most peripheral stocks are more heterogeneous from a sectorial point of view. Nonetheless, Utilities emerges for closeness and eccentricity. Details are given in Appendix B.

3.4. *Capitalization*

In general, we observe that big companies seem usually to be slightly more connected and central than small ones. Average capitalizations of the most central nodes are systematically higher than those of the most peripheral. In Appendix C a set of statistical tests can be found in which we demonstrate that the association between large capitalization and centrality remains valid — and strong — for all the measures, for all Δt considered, for both the MST and the PMFG.

4. Discussion

The analyses of degree, betweenness, eccentricity and closeness suggest that there could be two different "principal components" which fully explain the differences between degree and betweenness on one side and eccentricity and closeness on the other. Average fractional ranks[c] have been computed for both the MST and the PMFG accordingly with each of the four measures: degree, betweenness (descending order); eccentricity, closeness (ascending order). An analysis of the principal components of these eight variables provides a robust confirmation that all are centrality measures (the first principal factor explains the 74.50% of the total variance and there is an 82–88% correlation with all the measures). But it also provides a strong evidence that a second principal factor, explaining the 23.74% of the total variance, is also statistically significant. We have found that the first principal component is essentially an average of all the eight variables, while the second principal component is approximately equal to the difference between the average of degrees and betweennesses and the average of eccentricities and closenesses.

Projections of each stock for the two components are shown in Fig. 1, where the first factor is on the horizontal axis and the second factor on the vertical one.

All points in Fig. 1 are by construction inscribed in a rhombus whose edges correspond to the four extreme behaviors of nodes that we are about to illustrate. We can see from the figure that central and well-connected nodes are at the center on the right; peripheral and poorly-connected nodes are at the center on the left; central but poorly-connected nodes are at the top; and eccentric but well-connected nodes are at the bottom. We see a continuity of points from left to right but not from top to bottom. Along the sides of the rhombus we have: peripheral nodes from the less to the most connected; poorly-connected nodes from the most peripheral to the most central; central nodes from the less to the most connected; well-connected nodes from the less to the most central.

We can see from the figure some interesting cases:

- Nodes that are central and well-connected: AGE, BEN and MER (S07).
- Eccentric nodes but with high connectivity: APA (Apache Corp., S06), KMG (Kerr–McGee Corp. Holding Company, S06), SII (S06) and CL (Colgate–Palmolive Co., S05).
- Nonperipheral nodes but characterized by low connectivity: EK (Estman Kodak Co., S04), LUK (Leucadia National Corp., S03), GDW (Golden West Financial Corp., S07) and UIS (Unisys Corp., S10).
- Nodes neutral from both points of view: BBY (Best Buy Co. Inc., S09), MDP (Meredith Corp., S09) and JNY (Jones Apparel Group Inc., S04).
- Eccentric and poorly-connected nodes: HCP (Health Care Property Invs. Inc., S09), VLO (Valero Energy Corp. New, S06), YPF (S06) and SLE (Sara Lee Corp., S05).

[c]The fractional ranking consists in assigning the same mean rank to entries with the same score.

Centrality and Peripherality in Filtered Graphs 943

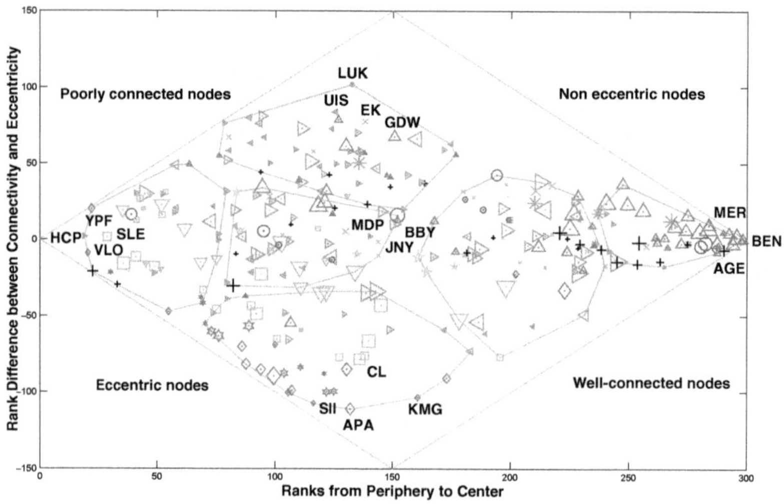

Fig. 1. Map of the nodes with respect to the two "principal components" of the centrality measures. The first axis is an average of the eight rankings of the nodes ordered by degree, betweenness, eccentricity and closeness, each computed for the MSTs and the PMFGs and for all cases, $\Delta t = 1, 2, 3, 4, 6, 12$ months; it is an indicator of the centrality of a node in the graph. The second axis is the difference between the average of degree and betweenness ranks and the average of eccentricity and closeness ranks; as a result, nodes with high connectivity and low and high peripherality are projected toward the bottom of the figure, and nodes with low connectivity and very low eccentricity are projected toward the top. The size of the marker indicates the capitalization of the relative stock: the larger, the more capitalized. The lines delimit polygonal regions enclosing the six clusters. Symbols and colors correspond to economic sectors according to the following: $+$, black — Basic Materials; $\bigcirc$, blue — Capital Goods; $*$, dark green — Conglomerates; $\times$, light green — Consumer Cyclical; $\square$, cyan — Consumer Noncyclical; $\diamond$, brown — Energy; $\triangle$, violet — Financial; $\triangledown$, light brown — Healthcare; $\triangleright$, red — Services; $\triangleleft$, magenta — Technology; $\star$, peach — Transportation; ✡, gray — Utilities.

Qualitatively, we can recognize from Fig. 1 that there are regions dominated by particular sectors. In order to quantify this observation, we have performed a cluster analysis which leads to the six clusters enclosed in lines in the figure. The analysis has been performed by using SPAD software where a hierarchical tree (dendrogram) has been constructed based on all the axes obtained by principal component analysis and the Wards aggregation criterion with a cut performed using a consolidation procedure [24].

In the first cluster we find 44 stocks that are both extremely central and well-connected. This cluster is clearly dominated by 25 financial stocks, with a much lighter presence of stocks belonging to Basic Materials (6), Conglomerates (4) and Capital Goods (3). This group is concentrated on the right side of the figure.

In the second cluster we find 66 stocks that are central and connected, but less than in the first cluster. Some of these stocks tend to be slightly eccentric but still remain well-connected. There are again many stocks belonging to Financial (10). Significant are the stocks belonging to Consumer Cyclical (12), Basic Materials (7), Conglomerates (2) and Capital Goods (4). There are also 15 stocks from Services and 7 from Technology. This is the group that is close to the first in the figure.

The third cluster is characterized by 45 stocks that are eccentric but generally well-connected. They are concentrated at the bottom of the figure. Not surprisingly, we find that this cluster is dominated by Utilities (9), Energy (12) and Consumer Noncyclical (10), with a moderate presence of Services (8).

Scarcely connected, but also not eccentric are the 53 stocks belonging to the fourth cluster. They are mainly from Technology (13), Services (19), and Financial (10). They are concentrated at the top of the figure.

The fifth cluster is characterized by 51 stocks which are generally poorly-connected and peripheral. This cluster is not clearly characterized from a sectorial point of view, apart from 14 stocks belonging to Services. It is concentrated in a left region of the figure.

In the sixth cluster we find 41 stocks that are definitely peripheral and poorly-connected to the system. There are 10 stocks from Consumer Noncyclical, 8 from Healthcare and 9 from Services, all concentrated at the extreme corner of the rhombus on the left side.

In Appendix D the complete composition of each cluster is reported.

We also observe that subsectors are strongly clustered too. For example, in Fig. 1, Electric Utilities, Oil Well Services & Equipment and Oil & Gas Operations lie along the left bottom side, while Food Processing is slightly above them; Semiconductors lies along the right bottom side; Chemical-Plastic & Rubber and Chemical Manufacturing lie along the right side of the diagonal; Investment Services lies along the right top side, while Regional Banks and Conglomerates lie slightly below them; and Computer Services lies slightly below the top vertex.

5. Conclusions

We have mapped the economic sectors onto MST and PMFG filtered graphs, retrieving a complex but rather well-defined hierarchical organization. It turns out that a proper classification follows two significant principal components that divide the system toward four extremes: stocks well-connected and central; stocks well-connected but at the same time peripheral; stocks poorly-connected but central; stocks poorly-connected and peripheral.

In the light of economic theory we can see that Financial must play a central role in the entire system: all companies involved in a production activity need funds before they start their business. Funds are provided directly and primarily by investors (self-financing), then conspicuously by the financial system and only at the end by private lenders for the residual part. It is therefore straightforward that we find banks and stocks belonging to Financial at the center of the networks. But companies also need raw materials (such as steel, aluminum or copper) and other intermediate goods (all those goods used as inputs in the production of other goods), as well as capital goods or physical capital (such as factories, machinery, tools, and various buildings): these are all "specific inputs" of the production. It is no surprise then that, after Financial, we find that Basic Materials, Capital Goods and Conglomerates also share the central part of the filtered graphs. Sectors specialized in final products, such as Consumer Noncyclical and Healthcare, are concentrated on the periphery instead. Similarly, sectors like Transportation, Energy and Utilities, which are general inputs and serve indistinctly all other activities, are rather peripheral. We have also observed that, although this hierarchical organization is rather robust over time, there are periods of huge turbulence during which the system hierarchy is shaken. Future works will be devoted to the investigation of this dynamical mixing during financial turbulence.

Acknowledgments

This work was partially supported by the ARC Discovery Projects DP0344004 (2003) and DP0558183 (2005), and the COST P10 Physics of Risk project.

Appendix A. Definitions

Here we recall the definition of *degree*, *betweenness*, *eccentricity*, *closeness*, *shortest path* and its length [25, 26]. In this paper we consider unweighted, undirected connected graphs only.

The *degree* of a vertex in a graph is the number of edges connected to that vertex. The average degree is equal to $2(n-1)/n$ for the MST and $6(n-2)/n$ for the PMFG.

The *shortest path*, or *geodesic*, between vertex i and vertex j is the shortest chain of vertices connecting vertices i and j. The length of a path from vertex i to vertex j is the number of edges included in the path. Since the MST is embedded in the $PMFG$, the average length of the shortest paths connecting two vertices in the MST needs to be larger than the same average computed for the $PMFG$.

An important centrality measure for each node i is the *betweenness*, which is the total number of shortest paths between all possible pairs of vertices that pass through vertex i. For a graph with n vertices, the betweenness centrality for vertex i is defined as $CB(i) = \sum_{s \neq i \neq t, s \neq t} \frac{\sigma_{st}(i)}{\sigma_{st}}$, where σ_{st} is the number of shortest paths from s to t and $\sigma_{st}(i)$ is the number of shortest paths from s to t that pass through vertex i. If a large percentage of paths from s to t pass through i, then i must be more "between" than others.

The *eccentricity* for each node i is the maximum length of the shortest paths that connect i to any other vertex j. If the eccentricity is high, the node is relatively peripheral; if it is low, the node is relatively central. The eccentricity is then a measure of peripherality. The maximum eccentricity is the graph's *diameter*. The minimum eccentricity is the graph's *radius*.

The *closeness* of vertex i is the average length of all shortest paths that connect i and any other vertex j. If the average is high, vertex i is relatively peripheral: a long path, on average, is needed to reach other vertices. If the average is low, node i is relatively central, and any other vertex is easily reached starting from i. The closeness of a vertex is another measure of peripherality.

We have calculated all these quantities for all dynamical MSTs and PMFGs and for each Δt.

Appendix B. Robust Central and Peripheral Nodes

For each of the four measures [degree, betweenness, (-)eccentricity, (-)closeness], we looked at the set of stocks with the largest values and investigated the corresponding sectors.

In the case of the 20 most central stocks for degree, we have found 12 stocks: 8 Financial (AGE, BEN, JP, LM, MER, NCC, STI, UPC), 1 Basic Materials (PPG), 1 Capital Goods (ETN), 1 Conglomerates (TIN) and 1 Technology (ADI). If we extend the analysis to the first 50 stocks, we find 14 more cases: 5 Financial (BSC, KEY, KRB, LEH, TMK), 2 Basic Materials (PH, ROH), 2 Capital Goods (CAT, ITW), 1 Conglomerates (DOV), 1 Consumer Cyclical (WHR), 1 Consumer Noncyclical (WWY), 1 Energy (KMG) and 1 Technology (NSM). The less central stocks according to the degree are only three in number: 2 Services (BCE, SVU) and 1 Energy (YPF). If we extend the analysis to the worst 50 stocks, we find 18 more cases: 5 Technology (CSC, GTK, SBL, TSS, XRX), 4 Consumer Noncyclical (ADM, CCE, NWL, UST), 4 Services (HCP, HRB, MCD, SVM), 2 Financial (FNF, TRH), 1 Consumer Cyclical (MAT), 1 Healthcare (GDT) and 1 Utilities (WEC).

In the case of the 20 most central stocks for betweenness, we have found 11 stocks: 9 Financial (AGE, BEN, BSC, JP, LEH, LM, MER, NCC, STI), 1 Basic Materials (PPG) and 1 Capital Goods (ETN). If we extend the analysis to the first 50 stocks, we find 15 more cases: 6 Financial (AXP, KEY, MEL, ONE, TMK, UPC), 2 Basic Materials (PH, ROH), 2 Capital Goods (CAT, ITW), 2 Conglomerates (GE, TIN), 1 Consumer Cyclical (WHR), 1 Energy (RD) and 1 Technology (ADI). The less central stocks according to the betweenness are only three in number: 1 Services (BCE), 1 Energy (YPF) and 1 Consumer Cyclical (MAT). If we extend the analysis to the worst 50 stocks, we find 15 more cases: 5 Consumer Noncyclical (CCE, GIS), 5 Services (HRB, KIM, MCD, ODP, SVU), NWL, SLE, UST), 3 Technology (CSC, TSS, XRX), 1 Healthcare (GDT) and 1 Financial (TRH).

In the case of the 20 most central stocks for eccentricity, we have found 11 stocks: 10 Financial (AGE, AXP, BEN, BSC, LEH, LM, MEL, MER, ONE, STI) and 1 Capital Goods (ETN). If we extend the analysis to the most central 50 stocks, we find 12 more cases: 8 Financial (BK, COF, JP, KRB, NCC, PNC, SCH, TMK), 2 Capital Goods (CAT, ITW), 1 Conglomerates (GE) and 1 Basic Materials (PPG).

The less central stocks according to the eccentricity are eight in number: 5 Services (APC, BHI, BJS, SII, VLO), 2 Basic Materials (ABX, NEM) and 1 Utilities (WEC). If we extend the analysis to the higher 50 stocks, we find 24 more cases: 7 Utilities (AEP, CIN, DPL, ETR, PEG, SO, WMB), 5 Energy (APA, BR, HAL, MUR, SLB), 6 Consumer Noncyclical (AVP, BUD, CLX, GIS, HSY, K), 4 Services (GGP, HCP, KIM, SPG) and 2 Healthcare (HMA, MME).

In the case of the 20 most central stocks for closeness, we have found 13 stocks: 11 Financial (AGE, AXP, BEN, BSC, LEH, LM, MEL, MER, NCC, ONE, STI), 1 Capital Goods (ITW) and 1 Basic Materials (PPG). If we extend the analysis to the most central 50 stocks, we find 13 more cases: 8 Financial (BK, JP, KRB, LNC, MMC, PNC, SCH, TMK), 2 Conglomerates (DOV, GE), 2 Capital Goods (CAT, ETN) and 1 Basic Materials (PH). The less central stocks according to the closeness are 11 in number: 8 Services (APC, BHI, BJS, BR, MUR, SII, SLB, VLO), 2 Basic Materials (ABX, NEM) and 1 Utilities (WEC). If we extend the analysis to the higher 50 stocks, we find 21 more cases: 9 Utilities (AEP, CIN, DPL, EQT, ETR, FPL, PEG, SO, WMB), 3 Energy (APA, HAL, YPF), 4 Consumer Noncyclical (AVP, BUD, K, SLE), 3 Services (GGP, HCP, KIM) and 2 Healthcare (HMA, MME).

Appendix C. Does Capitalization Matter?

We have two choices when we consider either the MST or the PMFG; we have two choices when we consider the first and last 50 or the first and last 100 stocks on the lists; we have six choices when we consider either $\Delta t = 1, 2, 3, 4, 6$ or 12 months; and, finally, we have four different measures (degree, betweenness, eccentricity and closeness). We have a total of $2 \times 2 \times 6 \times 4 = 96$ cases.

We first consider at the same time all the 96 cases, assuming that they are extracted from the same independent, normally distributed statistical population, and we calculate the confidence interval with a 1.00% significance level by means of the T-Student. We find that it is (26.38, 31.76) centered on a general sample mean of 29.07. This means that the most central stocks are approximately 29 ranks more capitalized than the less central ones.

A second test was done by using the 48 cases in which only the first and last 50 stocks on the list were considered. The confidence interval was (24.74, 30.96) centered on a sample mean equal to 27.85.

When we use the 48 cases in which only the first and last 100 stocks on the list are considered, we obtain a (25.80, 34.77) confidence interval centered on a sample mean equal to 30.28.

By considering only the 16 cases for each Δt, we obtain the following confidence intervals: (18.45, 28.47) when $\Delta t = 1$ month, centered on a sample average equal to 23.46; (20.51, 34.10) when $\Delta t = 2$ months, centered on a sample average equal to 27.30; (20.61, 36.31) when $\Delta t = 3$ months, centered on a sample average equal to 28.46; (19.89, 38.20) when $\Delta t = 4$ months, centered on a sample average equal to

29.05; (25.25, 38.63) when $\Delta t = 6$ months, centered on a sample average equal to 31.94; and (28.35, 40.03) when $\Delta t = 12$ months, centered on a sample average equal to 34.19. These results clearly show that the presence of the most-capitalized stocks among the central ones tends to increase as Δt increases.

We have also tested the differences considering only the 48 cases when degrees and betweennesses are involved, and we have found the following confidence interval: (35.1112, 39.8122) centered on a sample average equal to 37.46. When we consider only the 48 cases when eccentricities and closenesses are involved, we obtain (18.7999, 22.5405) centered on a sample average equal to 20.67. The difference seems to be definitely more pronounced in the first case than in the second. This is also another confirmation that degree and betweenness behave in a different way with respect to eccentricity and closeness.

Finally, we have tested the 48 cases when only the MST is involved, and we have found a confidence interval of (24.69, 32.52) centered on a sample average equal to 28.61. We have repeated the test when only the PMFG is considered, and we found a confidence interval of (25.6731, 33.3773) centered on a sample average equal to 29.53. This result shows that there is no difference between the MST and the PMFG.

Appendix D. Description of Clusters

In the first cluster we find 44 stocks that are both extremely central and well-connected: BEN, MER, AGE, STI, BSC, PPG, LM, UPC, JP, NCC, ETN, LEH, AXP, CAT, ITW, ONE, DOV, MEL, PH, PNC, KEY, KRB, GE, TMK, ASO, TIN, ROH, CMA, SNV, LNC, JCI, BK, DD, WY, MMC, SCH, IP, WMT, NWS, DJ, BAC, TEF, SNE, EMR. This cluster is clearly dominated by 25 financial stocks, with a much lighter presence of stocks belonging to Basic Materials (6), Conglomerates (4) and Capital Goods (3). This group is concentrated on the right side of the figure.

In the second cluster we find 66 stocks that are central and connected, but less than the first cluster: WHR, APD, TXT, TCB, NFB, SWK, TRB, ADI, MBI, PX, S, COF, EMN, HD, DHR, MAR, KRI, WFC, AIG, EC, CB, RD, DOW, HDI, NAV, MYG, TIF, IBM, DIS, PGR, KSS, GM, UTX, IRF, FD, BDK, TOT, ECL, GR, BC, CCU, PFE, LEG, OMC, WWY, DE, AL, LSI, VMC, LOW, TXN, AT, LEA, GCI, SPC, PD, GWW, LIZ, GPC, MRK, ABK, CSX, LTR, LTD, BNI, UNP. Some of these stocks tend to be slightly eccentric on certain occasions but still remain well-connected. There are again many stocks belonging to Financial (10). Significant are the stocks belonging to Consumer Cyclical (12), Basic Materials (7), Conglomerates (2) and Capital Goods (4). There are also 15 stocks from Services and 7 from Technology. This is the group that is close to the first in the figure.

The third cluster is characterized by 45 stocks that are eccentric but generally well-connected: NSM, UCL, TER, KMG, EQR, KO, SBC, PG, BLS, CPB, CL, APA, OXY, DRE, HNZ, PEG, FPL, BMY, PNW, ABT, SII, BFB, CIN, BHI, WLP,

SPG, BJS, VNO, AEP, HSY, AHC, SLB, BR, G, CAG, DUK, APC, HUM, HAL, AIV, SO, K, GGP, ETR, DPL. They are concentrated at the bottom of the figure. Not surprisingly, we find that this cluster is dominated by Utilities (9), Energy (12) and Consumer Noncyclical (10), with a moderate presence of Services (8).

Scarcely connected but also not eccentric are the 53 stocks belonging to the fourth cluster: HIB, TMO, EFX, AVY, HLT, FDC, SHW, CTL, GDW, UNM, SEE, PVN, PBI, GP, EK, CYN, IPG, TYC, RHI, ROK, RCL, LUK, MAN, MGG, BCR, AFL, AOC, TV, UIS, PLL, NUE, CI, SSP, SFA, DBD, HAR, WAG, IGT, WPO, CNA, DNY, TOY, GTK, HRB, BLL, CSC, CDN, TSS, SBL, MAT, TMX, BCE, TRH. They are mainly from Technology (13), Services (19) and Financial (10), and are concentrated at the top of the figure.

The fifth cluster is characterized by 51 stocks which are generally poorly-connected and peripheral: MTG, BA, BBY, MHP, VFC, JNY, MDP, NSC, LLY, MAS, PCL, ORI, PHM, EMC, TJX, AGN, ALL, FNM, UN, VSH, CAH, MDT, FDO, N, ABS, AMD, RBK, JNJ, CA, CTX, KR, JCP, MCK, NKE, GPS, GD, FRE, PEP, SWY, AZO, SYY, SLR, MU, GLW, MYL, LUV, IFF, BEC, FON, KMB, OCR. This cluster is not clearly characterized from a sectorial point of view, apart from 14 stocks from Services. It is concentrated in a left region of the figure.

In the sixth cluster we find 41 stocks that are definitely always peripheral and poorly-connected to the system: ODP, EAT, WEN, BDX, APH, MOT, CLX, WMB, EQT, BAX, FNF, SGP, HB, XRX, HMT, MUR, NWL, SVU, ADM, BSX, MME, AVP, KIM, VAR, HMA, MCD, HRL, CCE, UST, SVM, GIS, NOC, BUD, GDT, ABX, WEC, SLE, NEM, YPF, VLO, HCP. There are 10 stocks from Consumer Noncyclical, 8 from Healthcare and 9 from Services, all concentrated at the extreme corner of the rhombus on the left side.

References

[1] Mantegna, R. N., Hierarchical structure in financial markets, *Eur. Phys. J. B* **11** (1999) 193–197.

[2] Strogatz, S. H., Exploring complex networks, *Nature* **410** (2001) 268–276.

[3] Onnela, J.-P., Taxonomy of financial assets, M.Sc. thesis, Helsinki University of Technology, 2002.

[4] Onnela, J.-P., Chakraborti, A., Kaski, K. and Kertèsz, J., Dynamic asset trees and portfolio analysis, *Eur. Phys. J. B* **30** (2002) 285–288.

[5] Onnela, J.-P., Chakraborti, A., Kaski, K., Kertèsz, J. and Kanto, A., Dynamics of market correlations: Taxonomy and portfolio analysis, *Phys. Rev. E* **68** (2003) 056110.

[6] Onnela, J.-P., Chakraborti, A., Kaski, K. and Kertèsz, J., Dynamic asset trees and Black Monday, *Physica A* **324** (2003) 247–252.

[7] Aste, T., Di Matteo, T. and Hyde, S. T., Complex networks on hyperbolic surfaces, *Physica A* **346** (2005) 20–26.

[8] Tumminello, M., Aste, T., Di Matteo, T. and Mantegna, R. N., A tool for filtering information in complex systems, *PNAS* **102/30** (2005) 10421–10426.

[9] Tumminello, M., Aste, T., Di Matteo, T. and Mantegna, R. N., Correlation based networks of equity returns sampled at different time horizons, *Eur. Phys. J. B* **55** (2007) 209–217.

[10] Ohlenbusch, H. M., Aste, T., Dubertret, B. and Rivier, N., The topological structure of 2D disordered cellular systems, *Eur. Phys. J. B* **2** (1998) 211–220.
[11] Aste, T. and Sherrington, D., Glass transition in self-organizing cellular patterns, *J. Phys. A* **32** (1999) 7049–7056.
[12] Aste, T. and Di Matteo, T., Tumminello, M. and Mantegna, R. N., Correlation filtering in financial time series, *Proc. SPIE* **5848** (2005) 100–109.
[13] Aste, T. and Di Matteo, T., Dynamical networks from correlations, *Physica A* **370** (2006) 156–161.
[14] Di Matteo, T. and Aste, T., Extracting the correlation structure by means of planar embedding, *Proc. SPIE* **6039** (2006) 60390P-1.
[15] Gower, J. C. and Ross, G. J. S., Minimum spanning trees and single linkage cluster analysis, *Appl. Stat.* **18** (1969) 54–64.
[16] Pozzi, F., Aste, T., Rotundo, G. and Di Matteo, T., Dynamical correlations in financial systems, *Proc. SPIE* **6802** (2008) 68021E.
[17] Di Matteo, T. and Aste, T., How does the eurodollar interest rate behave?, *Int. J. Theor. Appl. Fin.* **5** (2002) 107–122.
[18] Di Matteo, T., Aste, T. and Mantegna, R. N., An interest rates cluster analysis, *Physica A* **339** (2004) 181–188.
[19] Di Matteo, T., Aste, T., Hyde, S. T. and Ramsden, S., Interest rates hierarchical structure, *Physica A* **355** (2005) 21–33.
[20] Eisner, J., State-of-the-art algorithms for MSTs: A tutorial discussion. Unpublished manuscript, University of Pennsylvania (1997). Available online: http://cs.jhu.edu/jason/papers/#ms97
[21] Chazelle, B., A minimum spanning tree algorithm with inverse-Ackermann type complexity, *JACM* **47/6** (2000) 1028–1047.
[22] Jarník, V. O jistém problému minimálním, *Práce Moravské Přírodovědecké Společnosti* **6** (1930) 57–63. (in Czech).
[23] Prim, R. C., Shortest connection networks and some generalisations, *Bell Syst. Tech. J.* **36** (1957) 1389–1401.
[24] Lebart, L., Morineau, A. and Piron, M., *Statistique Exploratoire Multidimensionnelle* (Duond, Paris, 1995), pp. 155–175.
[25] Caldarelli, G., *Scale-Free Networks: Complex Webs in Nature and Technology* (Oxford University Press, 2007).
[26] Dorogovtsev, S. N. and Mendes, J. F. F., *Evolution of Networks: From Biological Nets to the Internet and WWW* (Oxford University Press, 2003).

Advances in Complex Systems, Vol. 12, No. 1 (2009) 99–112

QUANTIFYING SOCIAL AND OPPORTUNISTIC BEHAVIOR IN EMAIL NETWORKS

LUIZ H. GOMES*, VIRGILIO A. F. ALMEIDA†,
JUSSARA M. ALMEIDA‡ and FERNANDO D. O. CASTRO§

Computer Science Department,
Universidade Federal de Minas Gerais,
Belo Horizonte, Brazil
*lhg@dcc.ufmg.br
†virgilio@dcc.ufmg.br
‡jussara@dcc.ufmg.br
§fernando@dcc.ufmg.br

LUÍS M. A. BETTENCOURT

Theoretical Division, MS B284,
Los Alamos National Laboratory,
Los Alamos NM 87545, USA

Santa Fe Institute, 1399 Hyde Park Road,
Santa Fe NM 87501, USA
lmbett@lanl.gov

Received 19 October 2008
Revised 24 January 2009

Email graphs have been used to illustrate the general properties of social networks of communication and collaboration. However, increasingly, the majority of Internet traffic reflects opportunistic rather than symbiotic social relations. Here we use email data drawn from a large university to construct directed graphs of email exchange that quantify the differences between social and opportunistic behavior, represented by legitimate messages and spam, respectively. We show that while structural characteristics typical of other social networks are shared to a large extent by the legitimate component, they are not characteristic of opportunistic traffic. To complement the graph analysis, which suffers from incomplete knowledge of users external to the domain, we study temporal patterns of communication to show dynamical properties of email traffic. The results indicate that social email traffic has lower entropy (higher structural information) than opportunistic traffic for periods covering both working and non-working hours. We see in general that both social and opportunistic traffics are not random, and that social email shows stronger temporal structure with a high probability for long silences and bursts of a few messages. These findings offer insights into the fundamental differences between social and opportunistic behavior in email networks, and may generalize to the structure of opportunistic social relations in other environments.

Keywords: Complex networks; email networks; opportunistic behavior.

1. Introduction

The fast pace of recent progress in the quantitative understanding of complex networks that mediate social interactions has been largely due to new ways of harvesting data, mainly by electronic means. The World Wide Web has seen a significant growth in the scale and variety of online communities through the rise of communication and social networking services [1]. Understanding the structure and dynamics of communication graphs is a natural goal for network analysis, since different types of communication and social relations are embedded within network structures. For this reason, graphs of email communication, where nodes represent email users and links denote messages exchanged between them, have become important example social networks.

The statistical mechanics of these networks makes possible a quantification of aspects of human social behavior and their comparison to the structure of interactions in other complex systems. Our focus here is on network characteristics that distinguish social and opportunistic behavior in email communication networks. A recent study [2] has provided evidence for structural properties that are characteristic of social graphs, but not of other complex networks. These are a nontrivial clustering coefficient (network transitivity) and the presence of positive degree correlations (assortative mixing by degree) between adjacent nodes. Moreover, it has been suggested that social networks can be largely understood in terms of the organization of nodes into communities [2–4, 6, 8], a feature that, to some extent, can explain the observed values for the clustering coefficient and degree correlations. This observation has indeed led to the interesting suggestion that email networks can be used to infer informal communities of practice within organizations [4], as well as their hierarchical structure [3, 4, 7], features that can in principle be useful for the efficient management of human collective behavior. In fact, the nature of such hierarchies can be quantified [3, 9], and may be self-similar [3].

Beyond these characteristics that are, at least at the qualitative level, general to social networks there are features of email graphs that are specific. The most important property of email is the low cost involved in delivering a message to a large group of recipients. This tends to make communication between any two nodes more indiscriminate, as email senders may easily send copies of a message to multiple parties that play no active role in the relationship between sender and recipient. As such, we may expect that networks of email may contain nodes with very high degree, and that degree distributions exhibit less severe or no practical constraints to their high degree tails.

The ease with which messages can be distributed to many recipients is also at the root of most opportunistic behavior involving email. Many services on the Web have been targets for opportunistic strategies with consequences of varying degree including the reduction of quality of information, fraud and increasing infrastructure costs. In fact, there has been growing interest in uncovering evidence of opportunistic behavior in communication and online social networks such as social phishing, self-promotion, unsolicited advertising, hostile flaming, and spurious traffic [10, 12, 14, 15]. Email as a means of potential mass distribution is particularly associated with the dissemination of computer viruses as well as spam traffic that flood the Internet with unwanted messages usually containing commercial propositions or, more recently, a variety of other scams. Estimates [11] suggest that in 2006 spam messages comprised over 80% of all Internet email traffic. This behavior, which we call generically opportunistic, displays different characteristics from other types of social relations for which social networks have been constructed and analyzed. Here we present a comparative analysis of structural and temporal features of these networks, comparing those created by the exchange of email between legitimate users to networks of spam.

In all previous characterizations of email communications as networks, the problem that these networks also mediate opportunistic relations has not been addressed. In order to attempt to eliminate such behaviors, as well as to deal with incomplete network reconstruction, authors have used several strategies such as restricting the analysis of email traffic to within the organization's domain [3, 4, 6, 7, 16], taking into account only links that display communication in both directions [3, 6, 7], eliminating nodes associated with very high message volumes [3, 6, 7], and setting minimal message thresholds for a link to exist [4]. Here we provide a more complete study of email networks by lifting most of these restrictions. Then email networks become directed, and the number of users and links in our dataset is dominated by spam traffic. The remainder of this manuscript is organized as follows. In Sec. 2 we give details about our data and the several networks of social and opportunistic behavior constructed. We then proceed to analyze them via standard network measures for which we expect opportunistic behavior to differ from social. In Sec. 3 we give an additional characterization of the temporal structure of time series of email and show that social and opportunistic traffics differ in several characteristic ways. Finally we present our conclusions.

2. Data Set and Network Structural Analysis

To construct networks of email communication we consider the email traffic from a department of a large university. Email messages arriving at the departmental server are classified either as spam or legitimate by SpamAssassin, a standard and widely used filtering software [17]. We construct four graphs representing different email networks. A *social network* is built from the legitimate (as classified by the filtering system) messages exchanged between all users, including those external to the department that send/receive emails to/from internal users. Similarly, a *spam network* is built from the messages classified as spam, exchanged between all users. An *internal social network* is built by considering internal users exclusively involved in legitimate internal email communication. Finally, the internal spam traffic[a] is used to build an *internal spam network*. In general these networks are

[a]Originating from and addressed to an internal user. These are usually the result of forged identifiers.

directed. We note that messages exchanged through legitimate mailing lists, which also involve bulk email traffic, may exhibit some characteristics similar to spam. As in [18], aiming at minimizing the impact of such communication patterns in our analysis, we remove users who exchange emails with fifty or more other users from our internal social network.

Our four networks are built from a thirty day log including 562,664 messages, of which 270,491 are spam. The set consists of 19,504 internal and 259,069 external users. Of these, 164,998 external users are senders of spam, while that number is only 721 for those internal to the domain, most of them under fabricated identifiers. Also note that the number of users in our log is orders of magnitude larger that those included in several previously analyzed datasets [19, 21].

Ebel, Mielsh and Bornholdt [19] analyzed a similarly constructed email network, although without drawing the distinction between spam and legitimate traffic. They characterized the degree k distributions for the entire graph as a power law $P(k) \propto 1/k^{\alpha}$, with exponent $\alpha = 1.81$. For the network composed exclusively of internal users they found a smaller exponent $\alpha = 1.32$. Similarly we find power law degree distributions for the undirected versions of our four networks, with exponents $\alpha = 1.82$ ($R^2 = 0.942$) for the full social network, $\alpha = 2.03$ ($R^2 = 0.925$) for the entire spam network, and $\alpha = 1.22$ ($R^2 = 0.958$) and $\alpha = 1.79$ ($R^2 = 0.831$) for the internal social and spam networks, respectively. It is remarkable that our results are broadly consistent with those of [19], for entirely different data. Because the spam network contains nodes with very high degree, we find a tendency for the exponent to be larger for opportunistic behavior, which suggests that the true social exponent may be overestimated if the two traffics are not separated. The degree distribution is though a weak discriminator between social and opportunistic behavior and is clearly affected by incomplete knowledge of parts of the network, which is a limitation of real data sets whenever external users are considered. Such lack of knowledge results in the incorrect shift of external users to lower degree, and consequently leads to larger estimates of the exponent α. Thus, both the failure to exclude spam traffic and the incomplete knowledge of links between external users contribute to an overestimation of the exponent α.

Next, we recall that according to Newman and Park [2], high clustering coefficient and positive assortative mixing are two graph theoretical quantities typical of social networks. Therefore, we investigate whether these two structural properties of email graphs can distinguish the social imprint of legitimate email communication from the opportunistic characteristics of spam. In order to do so we compare the average values of these network measures determined for the four networks constructed from actual data with corresponding values obtained for networks with randomized links with the same degree sequence.

Indeed, considering the undirected versions of our networks, the average clustering coefficient over the internal social network is $C = 0.241 \pm 0.008$, whereas the clustering coefficient in the internal opportunistic network is much lower, at $C = 0.052 \pm 0.006$. Both values are much larger than the average clustering in random networks with the same size and degree distribution, $C = 0.0188$ and $C = 0.0103$, for the social and spam networks, respectively. These results compare to the clustering coefficient of internal domain users of $C = 0.154$, found by Ebel *et al.* [19]. Considering the networks that include external users, whose neighbors are only known incompletely, we find $C = 0.137 \pm 0.003$ for the social network and $C = 0.026 \pm 0.001$ for the opportunistic network, in contrast with a $C = 0.003$ for the entire network of Ebel *et al.* [19]. Figure 1 shows the distribution of the clustering coefficient for social, spam and their corresponding randomized networks.

All four networks contain a significant fraction of their nodes with vanishing clustering coefficient, but this proportion is much higher for graphs that include external users and/or opportunistic components. Specifically, 61% of all nodes in the entire social network have $C = 0$, while this becomes more than 81% for the entire opportunistic component. The internal social network has only 25% of its nodes with

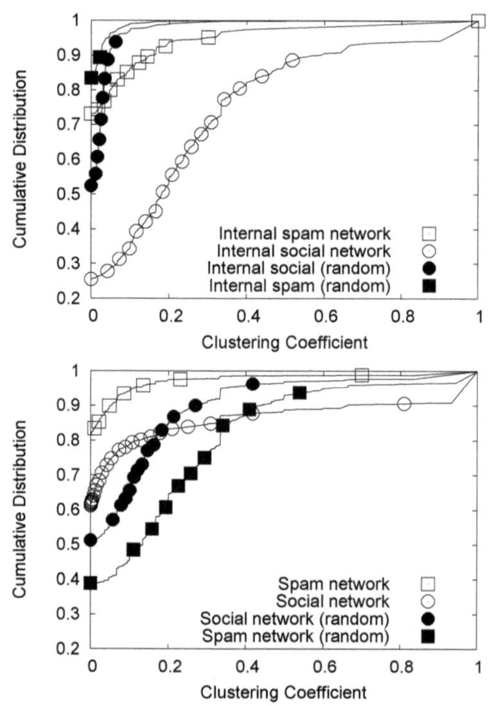

Fig. 1. Distribution of the clustering coefficient for social, opportunistic networks and their corresponding randomized networks with preserved degree sequence for the internal networks (*top*) and complete networks (*bottom*) built from messages exchanged between all users, internal and external.

$C = 0$, compared to 73% for the internal spam network. These features indicate that there are clear differences on average between clustering in a social versus clustering in a spam component of email networks, and also that low clustering is not a sufficient condition for a node to be associated with opportunistic behavior. Similar to the analysis of the degree distribution, these results also indicate that the separation of the two traffics is important in order to identify the truly social component. Failure to do so will result in the underestimation of the average social network clustering.

We now analyze the nature of degree correlations between nodes by computing the corresponding Pearson correlation coefficient [22] r

$$r = \frac{\sum_i j_i k_i - M^{-1} \sum_i j_i \sum_{i'} k_{i'}}{\sqrt{[\sum_i j_i^2 - M^{-1}(\sum_i j_i)^2][\sum_i k_i^2 - M^{-1}(\sum_i k_i)^2]}}, \tag{1}$$

where j_i and k_i are the excess in-degree and out-degree of the vertices that the ith edge leads into and out of, respectively, and M is the total number of edges in the graph.

The expectation of assortative mixing by degree in a social network of email is not obvious. In fact as we argued above, a user's degree is a very variable property that can be easily changed drastically by the inclusion of the user's address in, or by the use of, distribution lists. This common use of email can create huge imbalances of degree between senders and recipients and may generate negative values for the Pearson coefficient even for groups of legitimate users. If this can be expected of the degree correlation in the social network, then such an effect should be even more pronounced in the opportunistic graph. There, spam senders follow the strategy of increasing their degree indiscriminately and maximally, and consequently reach on average a population of recipients with much lower degree than their own, which are statistically much more abundant for a scale free distribution.

These qualitative expectations are borne out by estimation of r. Using (1) we computed the Pearson coefficient r for each of the four *directed* networks, and obtained $r = -0.135$ for the entire social network (with $r = -0.082$ for its corresponding randomized network), $r = -0.139$ (-0.111) for the entire opportunistic network, and $r = 0.232$ (0.095) and $r = 0.049$ (0.073) for the social and opportunistic internal networks, respectively. Standard errors are smaller than 1%. For the internal social network, we also observed that the positive value of r is the result of an approximately linear correlation between the out-degree of the sender and the in-degree of the recipient. Such systematic correlation across degree is absent for the other three networks, with the difference that for networks containing external users, there is an average imbalance between the degrees of senders and recipients that leads to a negative r. Both social networks show significantly stronger assortativity (internal social network) and disassortativity (social network) than their corresponding randomized networks. The dissortative nature of the entire social network probably stems from incomplete knowledge. The fact is that almost all external users are only connected to internal users, for which all connections are

Table 1. Summary results for structural measures applied to the Social (legitimate email) and Opportunistic (spam exchange) total networks and to those restricted to internal traffic within the domain.

Network measure	Internal social	Internal opportunistic	Social	Opportunistic
Degree distribution (α)	1.22	1.79	1.82	2.03
Clustering coefficient (real/random)	0.2409/ 0.0188	0.0521/ 0.0103	0.1374/ 0.0089	0.0261/ 0.0124
Assortative mixing (real/random)	0.2324/ 0.0946	0.0493/ 0.0727	−0.1347/ −0.0824	−0.1387/ −0.1110
Preferential exchange ($\langle E \rangle$)	0.27568	0.06246	0.03288	0.00007

known, and as a consequence have typically higher average degree than actually measured.

In summary, we see that the consideration of this set of standard network measures places networks of email communication in a unique position. On the one hand, the legitimate component of a completely known email network shares its transitivity and positive degree correlation properties with other social networks. Unlike some other social networks however its degree distribution is scale free and characterized by a small exponent ($2 > \alpha > 1$), which implies that although the distribution remains normalizable, no finite moments exists as the network size goes to infinity.

Despite these considerations, the opportunistic network built from the exchange of spam messages shows definite properties, specifically negligible transitivity and assortative mixing near their corresponding random network with preserved degree sequence. Moreover, our analysis shows that, in contrast to previous expectations [2], social email networks involving users that are external to the local domain may present a negative degree correlation, presumably reflecting in part the incomplete knowledge of external links, but also resulting from message exchanges characteristic of email, such as the widespread subscription to legitimate distribution lists.

These differences suggest mechanisms to differentiate legitimate human communication from opportunistic behavior on the basis of network structure, and have indeed been proposed as the basis for spam detection algorithms [23–25]. In fact, the structural relationships between email senders and recipients is the basis for a spam detection algorithm previously proposed by the authors [24]. However, much remains unsatisfactory about the transitivity and assortative mixing measures as means to characterize patterns of human communication. The most serious flaw is that their estimation relies on the knowledge of all neighbors of each node. This is not possible beyond a small subset, corresponding to users in the local domain; a general problem of the construction of any email network. A solution to this problem may follow from the consideration of quantities that characterize the dynamics of communication links between senders and recipients to which we now turn.

3. Temporal Patterns of Email Communication

We start with the simplest measure of communication between two users: reciprocity [30]. This measure aims at capturing the social behavior of human users that usually provide responses to other users within their social groups. Contrarily, a spam network has a strong structural communication imbalance among senders and receivers. We build a simple coefficient of preferential exchange E_i for user i this way:

$$E_i = 1 - \left| \frac{\sum_{j \in C_i} [k(j \to i) - k(i \to j)]}{\sum_{j \in C_i} [k(j \to i) + k(i \to j)]} \right| \tag{2}$$

where C_i is the set of all users that have contact with user i within a given time period, and $k(j \to i)$ is the number of messages sent by user j to i. Therefore, $0 \leq E_i \leq 1$, with the lower end corresponding to no message being replied to, and the upper end to every message obtaining a response. This can be further averaged over all users to generate network expectation values $\langle E \rangle$. Considering internal as well as external users, we find that $\langle E \rangle = 0.0329 \pm 0.0005$ in the social network, whereas a significantly lower $\langle E \rangle = 0.00007 \pm 0.00002$ is observed in the spam network. Values of $\langle E \rangle = 0.2757 \pm 0.0083$ and $\langle E \rangle = 0.0625 \pm 0.0056$ are found in the internal social and spam networks, respectively. Therefore, opportunistic networks are naturally associated with small (but potentially non-zero) reciprocity, whereas social networks, particularly those containing legitimate users whose behavior we know completely, are associated with the highest reciprocity.

Up to this point we have concentrated on the structure of the network of interactions mediated by email messages. In its construction as a graph we have not paid attention to the detailed temporal structure of message exchanges. An interesting question then is whether the dynamical properties of email traffic can distinguish different types of social relations. This question has recently become a subject of interest. Eckmann, Moses and Sergi [18] have shown that coherent structures emerge from the temporal correlations between time series expressing short periods of intense message exchange between groups of users. Barabasi [26], on the other hand, has suggested that the distribution of time intervals between email messages is bursty and may be well described by a power law distribution $P(\tau) \sim \tau^{-\gamma}$ with $\gamma \simeq 1$. Subsequently it has been shown that lognormal statistics gives a better description of the same data [27]. Recently, Malmgren, Stouffer, Motter and Amaral have shown that the power law scaling of the distribution of time intervals between email messages is a consequence of the circadian and weekly cycles of human activity [28]. These characterizations identify properties of legitimate email traffic — temporal correlations between users and inter-message time statistics — that are

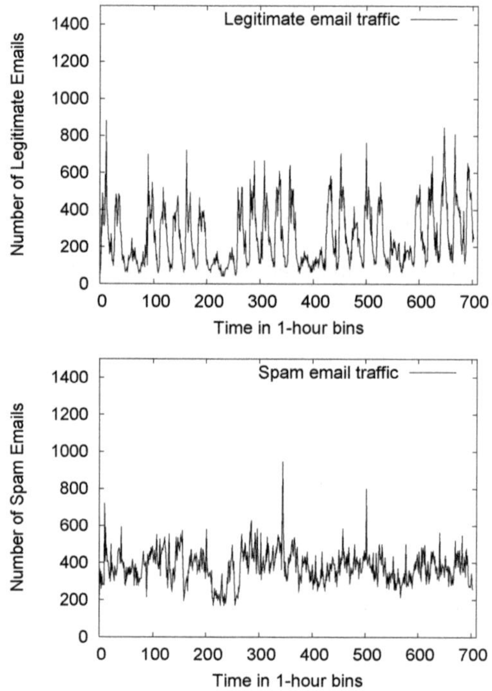

Fig. 2. Hourly load variations in the number of emails for social and spam traffic. Traditional social email traffic presents two distinct and roughly stable regions: a high load diurnal period, typically during working hours. On the other hand, the intensity of spam traffic is roughly insensitive to the time of the day.

thought to be exclusively social and thus not shared by the opportunistic traffic component.

To evade effects of variability associated with individual users, we chose to investigate the statistics of our social and spam aggregate traffics for each class. Figure 2 shows hourly load variations in the number of emails for social and spam traffic. Figure 2 (top) shows that hourly load variations in the social email traffic exhibit the traditional bell-shape behavior, typically observed in other types of web traffic [20, 33], with load peaks during weekdays and a noticeable decrease in load intensity over the weekend. While social email traffic shows large temporal variations, from night to day, working days to weekends, opportunistic traffic is roughly insensitive to the time of the day, see Fig. 2 (bottom). This stability in the daily spam traffic was previously observed in [13, 20] for different data sets. In [20], in particular, the authors extensively analyzed the statistical properties of characteristics of spam and social email aggregate traffics. In addition to results on various other characteristics, the inter-arrival time distributions were found to follow Poisson processes, for both aggregate social email traffic and aggregate spam traffic, with significantly different rate parameters.

To capture the different levels of variation observed in email traffic, we attempted to identify statistical temporal patterns of communication that are characteristic of the social versus opportunistic aggregated traffics. In so doing we average over the behaviors of many users. Specifically, we represent temporal patterns of message arrival through the definition of a state in terms of a communication *word* of size L. The dimension L is the number of time intervals, or letters, in the communication word, which is written as a vector $W = [i_1, i_2, \ldots, i_L]$. The simplest representation of the traffic is through a binary assignment, where the value of i_j is set to 1 if one or more messages were exchanged in the corresponding time interval, or $i_j = 0$ otherwise. In other words,

$$W = [01001\ldots01], \tag{3}$$

where there are L Boolean variables, each corresponding to the exchange, or not, of at least one message in consecutive time periods Δt. For stationary processes the probability of a message exchange occurs with a fixed probability per time unit. The representation of time series in terms of binary words is familiar from other contexts in physics and information theory [29, 31], from the analysis of the time evolution of dynamical systems, to time series of action potential in neuronal activity [32] or bit streams in noisy communication channels. The entropy of the distribution and its variation with the word size L gives us in fact some of the essential properties of the dynamical rules that generate these dynamical patterns [29, 31].

To illustrate these statements, consider the simplest statistical model that generates a binary time series subject to a given message arrival rate p. Then p can be written as the probability to obtain a 1 at each letter. If we further assume that bits corresponding to different letters are uncorrelated, then the bit value at each letter can be regarded as the result of an independent Bernoulli trial.

Under these assumptions the probability of a given number of events k in L trials is given by the binomial distribution

$$f(k; L, p) = \binom{L}{k} p^k (1-p)^{L-k}. \tag{4}$$

Moreover the probability of a sequence with the same number of events is the same regardless of their order, as each occurrence is independent for different bins. Thus to obtain the probability for a particular sequence of k events in L bins we must divide by the number of possible arrangements $\binom{L}{k}$. Then the probability for a particular sequence or binary word with k ones and length L is

$$p_W = \frac{p^k (1-p)^{L-k}}{\sum_{k'=0}^{L} \binom{L}{k'} p^{k'} (1-p)^{L-k'}} = p^k (1-p)^{L-k}. \tag{5}$$

Because all words with a given number n of 1s are equally likely, their probability is $p_W(n; L, p) = p^n\ (1-p)^{L-n}$. This implies that the Shannon entropy of the time series can be written as

$$H = -\sum_W p_W \log_2(p_W) = -\langle k \rangle \log_2\left(\frac{p}{1-p}\right) - L\log_2(1-p)$$
$$= m\,L, \qquad (6)$$

with $\langle k \rangle = Lp$. Thus, in the absence of temporal correlations, the Shannon entropy is a strict linearly growing function of the word length L, with slope $m = -(1-p)\log_2(1-p) - p\log_2 p > 0$.

These expressions become especially simple if the temporal bin for each letter is chosen such that $p = 1/2$, in which case $m = 1$ is maximal. This independent message model (IMM) is the maximal entropy distribution for a traffic characterized by an average message arrival probability p. Real traffics, which show temporal structure, must therefore display lower entropy relative to the idealized IMM message stream. We refer to the difference of the traffic entropy to that of the corresponding $H_{\text{IMM}}(L)$, measured with the same average choice of p as the traffic's structural information for a given L.

Figure 3 shows the difference between the entropy of the independent message model and the real traffics, legitimate and spam. We aggregated the data into two distinct sets based on temporal periods: work hours (i.e. the period from 8AM to 8PM of the weekdays, except holidays, in the log) and remaining times which we refer to as non-work hours, essentially nights and weekends.

The results show that the social email traffic has lower entropy (higher structural information) than the opportunistic traffic for both work and non-work periods. This difference becomes more noticeable the larger the word, thus capturing longer patterns of communication and the presence of time correlations. The difference between the independent message model, where for $p = 1/2$ all words are equally likely, and the real traffics is that in the latter, words with many 1s (0s) are suppressed while the probability of words with two to three 1s separated by one to three 0s is enhanced. The difference between social and opportunistic traffics is more subtle, with social email traffic displaying a greater probability for words with an isolated message in a long stream of silence. Nevertheless, we see that both social and opportunistic traffics are far from random. In quantitative terms, the deviations from the independent message model observed in the social email traffic are from 19.7% to 21.8% larger than the deviations in the spam traffic during work periods. During non-work periods, the deviations are from 61.5% to 65.6% larger in the social email traffic. Moreover, for given real traffic and temporal period, we found little variation on the entropy computed for multiple periods, with the half-length of a 95% confidence interval under 1% of the mean. Thus, social email shows richer (less random) temporal structure with a high probability for long silences and bursts of a few messages.

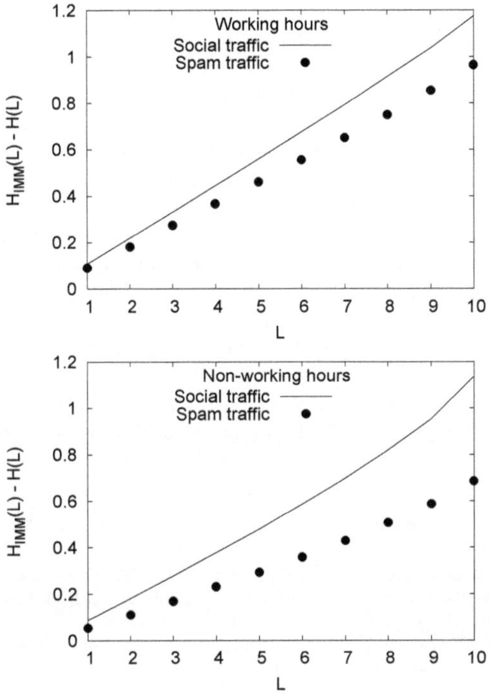

Fig. 3. The variation of the difference between the independent message model entropy $H_{\mathrm{IMM}}(L)$ and the entropy of the legitimate and spam traffics $H(L)$, with word size L, during work (top) and non-work (bottom) periods. All word probability distributions were constructed by normalizing the time bin for each letter word so that $p = 1/2$. As a result the time bin for each letter of the social traffic during work hours was set to 4s, and 11s for the corresponding non-work period. Time bins for the opportunistic traffic were set at 4s during work hours and 5s otherwise. The slight excess curvature for large L is the result of poorer estimation of rare long words.

4. Conclusions

We have shown a number of empirical results and theoretical arguments that underscore the differences between social and opportunistic behavior in networks of email communication. Opportunistic nodes in the email network display a behavior that can be captured graphically through the absence of definite metrics present in other social networks. However, incomplete knowledge of parts of email networks demand a combination of metrics to distinguish between dissimilar types of email traffic. The coefficient of preferential exchange minimizes the incomplete knowledge problem and expresses the lack of reciprocity in opportunistic communication networks. Perhaps even more directly, opportunistic email traffic can be identified by a greater statistical simplicity (higher randomness) in temporal patterns of communication, typical of the fact that each sender/recipient relationship is not a unique idiosyncratic agent and instead the same schemes are used to reach many recipients indiscriminately. Moreover, the ease in exchanging email messages that leads to these opportunistic behaviors also has consequences for estimates of the structural properties of the truly social component of the network, which exhibits a power law degree distribution with a small exponent and, in some cases, small or negative assortative mixing by degree. We believe that the quantitative characteristics of opportunistic communication patterns observed here for email networks may generalize to other social environments. The ever-changing nature of opportunistic actions in digital domain leads to an arsenal of techniques and strategies to distribute spam and other types of unwanted content, as has been shown by the growth of illegitimate Internet traffic [11]. The analysis of structural network properties and dynamic temporal patterns presented here offers insights that can be useful to differentiate social from opportunistic traffic in the Internet and are currently being implemented to complement standard spam identification strategies [24, 25].

References

[1] Backstrom, L., Huttenlocher, D., Kleinberg, J. and Lan, X., *Proc. 12th ACM SIGKDD Intl. Conf. on Knowledge Discovery and Data Mining*, 2006.

[2] Newman, M. E. J. and Park, J., *Phys. Rev. E* **68** (2003) 036122.

[3] Guimerà, R., Danon, L., Díaz-Guilera, A., Giralt, F. and Arenas, A., *Phys. Rev. E* **68** (2003) 065103(R).

[4] Tyler, J. R., Wilkinson, D. M. and Huberman, B. A., *Communities and Technologies* (Kluwer, B.V., The Netherlands, 2003).

[5] Newman, M., Watts, D. and Strogatz, S., *Proc. Natl. Acad. Sci. USA* **99** (2002) 2566–2572.

[6] Arenas, A., Danon, L., Díaz-Guilera, A., Gleiser, P. M. and Guimera, R., *Eur. Phys. J. B* **38** (2004) 373.

[7] Guimerà, R., Danon, L., Díaz-Guilera, A., Giralt, F. and Arenas, A., *J. Economic Behaviour* **61** (2006) 4.

[8] Casado, J. M., Garfinkel, T., Cui, W., Paxson, V. and Savage, S., *Proc. of the 4th Workshop on Hot Topics in Networks*, 2005.

[9] Trusina, A., Maslov, S., Minnhagen, P. and Sneppen, K., *Phys. Rev. Lett.* **92** (2004) 178702.

[10] Jagatic, T., Johnson, N., Jakobsson, M. and Menczer, F., *Commun. ACM* **50** (2007) 10.

[11] Anderson, D., Fleizach, C., Savage, S. and Voelker, G., *Proc. of the USENIX Security Symposium*, 2007.

[12] Heymann, P., Koutrika, G. and Garcia-Molina, H., *IEEE Internet Computing*, **11** (2007) 6.

[13] Cranor, L. and LaMacchia, B., *Communications of the ACM* **41** (1998) 8.

[14] Wellman, B., Salaff, J., Dimitrova, D., Garton, L., Gulia, M. and Haythornthwaite, C., *Annual Review of Sociology* **22** (1996) 213.

[15] Kossinets, J. G. and Watts, D. J., *Science* **311** (2006) 5757.

[16] Newman, M. E. J., Forrest, S. and Balthrop, J., *Phys. Rev. E* **66** (2002) 035101(R).

[17] SpamAssassin Home Page: http://www.spamassassin.org

[18] Eckmann, J. P., Moses, E. and Sergi, D., *Proc. Natl. Acad. Sci. USA* **101** (2004) 14333.
[19] Ebel, H., Mielsch, L. I. and Bornholdt, S., *Phys. Rev. E* **66** (2002) 035103(R).
[20] Gomes, L., Cazita, C., Almeida, J., Almeida, V. and Meira, W., *Performance Evaluation*, **64** (2007) 7–8.
[21] Shetty, J. and Adibi, J., *Proc. of The 11th ACM SIGKDD Int. Conf. Knowledge Discovery and Data Mining*, 2005.
[22] Newman, M. E. J., *Phys. Rev. Lett.* **89**(20) (2002) 208701.
[23] Gomes, L. H., Almeida, R. B., Bettencourt, L. M. A., Almeida, V. A. F. and Almeida, J. M., Second Conference on Email and Anti-Spam (CEAS 2005), CA, 2005.
[24] Gomes, L., Castro, F., Bettencourt, L., Almeida, V., Almeida, J. and Almeida, R., Steps to Reducing Unwanted Traffic on the Internet (SRUTI 2005), USENIX, 2005.
[25] Boykin, P. O. and Roychowdhury, V., *IEEE Computer* **38-4** (2005) 61.
[26] Barabási, A. L., *Nature* **435** (2005) 207.
[27] Stouffer, D. B., Malmgren, R. D. and Amaral, L. A. N., *Nature* **435** (2005) 207.
[28] Malmgren, R. D., Stouffer, D. B., Motter, A. E. and Amaral, L. A. N., *Proc. Natl. Acad. Sci. USA* **105** (2008) 18152–18158.
[29] Bialek, W., Nemenman, I. and Tishby, N., *Neural Computation* **13** (2001) 2409–2463.
[30] Garlaschelli, D. and Loffredo, M. I., *Phys. Rev. Letters* **93** (2004) 268701.
[31] Crutchfield, J. P. and Feldman, D. P., Santa Fe Institute technical report 01-02-012 (2001).
[32] Rieke, F., Warland, D., de Ruyter Van Steveninck R. R. and Bialek, W., *Spikes: Exploring the Neural Code* (MIT Press, Cambridge, MA, 1997).
[33] Veloso, E., Almeida, V., Wagner Meira, Bestavros, A. and S. Jin, *IEEE/ACM Trans. on Networking* **14** (2006) 1.

Advances in Complex Systems, Vol. 14, No. 3 (2011) 317–339

DOI: 10.1142/S0219525911003050

World Scientific
www.worldscientific.com

NETWORK AUTOMATA: COUPLING STRUCTURE AND FUNCTION IN DYNAMIC NETWORKS

DAVID M. D. SMITH

Centre for Mathematical Biology, Oxford University, Oxford OX1 3LB, UK

Oxford Centre for Integrative Systems Biology, Department of Biochemistry, Oxford University, South Parks Road, Oxford, OX1 3QU, UK
d.smith3@physics.ox.ac.uk

JUKKA-PEKKA ONNELA

Department of Biomedical Engineering and Computational Science, Helsinki University of Technology, P.O. Box 9203, FIN-02015 HUT, Finland

CABDyN Complexity Centre, Oxford University, Oxford, OX1 1HP, UK

CHIU FAN LEE

Max Planck Institute for the Physics of Complex Systems, Noethnitzer Str. 38, 01187 Dresden, Germany

MARK D. FRICKER

Department of Plant Sciences, University of Oxford, Oxford, OX1 3RB, UK

NEIL F. JOHNSON

Physics Department, University of Miami, Coral Gables, Florida FL 33124, USA

Received 26 May 2010
Revised 12 February 2011

We introduce Network Automata, a framework which couples the topological evolution of a network to its structure. To demonstrate its implementation we describe a simple model which exhibits behavior similar to the "Game of Life" before recasting some simple, well-known network models as Network Automata. We then introduce Functional Network Automata which are useful for dealing with networks in which the topology evolves according to some specified microscopic rules and, simultaneously, there is a dynamic process taking place on the network that both depends on its structure but is also capable of modifying it. It is a generic framework for modeling systems in which network structure, dynamics, and function are interrelated. At the practical level, this framework allows for precise, unambiguous implementation of the microscopic rules involved in such systems. To demonstrate the approach, we develop a family of simple biologically inspired models of fungal growth.

Keywords: Complex system; functional, dynamic networks; computer simulation; integrative biology.

PACS Number(s): 89.75.Fb, 87.15.A-, 87.85.Xd

1. Introduction

The network framework has proved very successful in the study of various complex interacting systems. In the network description, the interacting elements are depicted as nodes and the interactions between the elements are represented by links connecting the corresponding nodes. The science of complex networks has progressed very quickly in the last few years, and some excellent reviews have been written covering both the methodology and key results [1, 6, 21]. The strength of the complex network paradigm lies in its ability to capture some of the essential structural characteristics of interacting systems while reducing the details of both the elements and their interactions. Consequently, the early complex network literature was almost exclusively focused on structural properties of networks.

In many out-of-equilibrium growing networks, the evolution at a given time is dependent on the configuration of the network at that time, as exemplified in the preferential-attachment model of Barabási and Albert [4]. In this paper, we develop Network Automata (NA), which can be seen as a natural extension of Cellular Automata (CA) [18, 32]. In contrast to Cellular Automata *on* a network, in which the states of nodes evolve and whose neighborhoods are defined by the network [7, 19, 20, 29], in the NA the network topology itself changes in time. We describe some different variants of NA in Secs. 2 and 3 and to demonstrate the versatility, we show how some familiar network models can be recast as NA in Sec. 4.

Whilst structural properties remain important in constraining the behavior of a system [19], there is also significant interest in understanding dynamical processes taking place on networks [16]. Indeed, the marriage between structure and dynamic processes is so strong that the behavior of dynamic processes can be used to detect structure [14]. While a network's topology constrains the type of dynamics that may unfold on it, in many scenarios the dynamical process may influence the subsequent evolution of the topology — meaning that the structural properties of the network are coupled to its function. A real-world example might be the evolution of transport links within a city. The dynamics of the human population using this network in turn affect the reinforcement or removal of those transport links and the feedback process is apparent. A similar situation may arise in the context of social networks, where one's current social opportunities and dynamics are limited by the existing network structure, but they can be widened by extending the network.

There have been several specific attempts in the literature to inter-relate a network's structure, dynamics, and function [7, 11, 15, 22, 29, 33]. While many network generating algorithms and Cellular Automata models can be easily recast within the Network Automata framework [24], its usefulness is exemplified when it is applied to systems in which topological evolution is coupled to a dynamic process occurring upon the network. We call these systems Functional Network Automata (FNA) and introduce the concept in detail in Sec. 5. Unlike certain CA models on networks in which the topology is evolved according to a global fitness function such as the ability of the system to perform a specific function (e.g. the density classification problem) [7, 29], the topological update in FNA is performed simultaneously with the functional process resulting in configurations which are emergent phenomena of the system [13]. FNA are conceptually similar to Structurally Dynamic Cellular Automata (SDCA) in which topological and functional evolution are coupled [2, 13]. Whilst topological updates in SDCA are governed by the CA states of the nodes, in the FNA the network evolution can be coupled to any functional process occurring on the network. Its contribution to the network field lies in the fact that it enables a *precise* specification of the microscopic rules underlying the structural *and* functional evolution of a given network-based system. This is a desirable feature which circumnavigates issues commonly associated with simulating complex systems whereby microscopic implementation biases can have emergent macroscopic consequences [23] and any comparison of the macroscopic behavior of the system might be rendered redundant. The (evolving) topology inherent in FNA removes the constraint of spacial homogeneity associated with CA models [23], further bridging the gap between CA and Multi-Agent Systems. As a demonstration of such an application, we introduce a family of progressively more realistic, biologically-inspired models of woodland fungal growth in Sec. 6.

2. Network Automata

Consider an arbitrary weighted or unweighted, directed or undirected network at some time t which is to be grown to some size N_{tot}. As nodes might be added to the system at each timestep, the network would conventionally be considered to be growing. An alternative representation is to treat the system as being of size N_{tot} at all times where at some time t many of the nodes have no links. Information regarding the network's topology is entirely encompassed within the adjacency matrix $\mathbf{A}(t)$ which is of dimension $N_{tot} \times N_{tot}$. The matrix holds information about which links exist, their direction and, perhaps, weights. One might consider the evolution of the network as a process that alters the elements within this adjacency matrix, updating the attributes of any of the possible links which could exist in the system. We base a framework of network growth around this concept. If the microscopic ruleset governing the network's evolution is solely related to quantities which can be derived from the network's current topology (and hence from $\mathbf{A}(t)$) then the evolution of the network might be expressed in terms of some operation F acting upon the adjacency matrix:

$$\mathbf{A}(t+1) = F(\mathbf{A}(t)). \tag{1}$$

The ruleset employed could relate to any property of the nodes (their degree, betweenness, clustering and so on) or the links (weights or direction) and might be deterministic or stochastic.

This concept can be achieved in practice by visiting *all* possible links within the adjacency matrix $\mathbf{A}(t)$ for a network comprising N_{tot} nodes (whether part of a component or not) once every timestep.[a] The update as to the nature of the link at the next time step (its existence or its weight or direction) is then prescribed by the ruleset. Because each link can see the requisite information at the start of the time step, the system is synchronously updated. The ruleset can take the form of a lookup table in which the *state* of a link is evaluated and the *color* (existence/direction/weight) is prescribed for the next timestep. This update process is analogous to the update of a cell within a conventional Cellular Automaton [18, 32] *except* that it acts upon the connectivity of an node, thereby generating $\mathbf{A}(t+1)$. This link-orientated update is a generic description of a dynamic network and all the essential features of that network's evolution are then contained within the exhaustive ruleset. We describe such a system as a *Network Automaton.*[b] No restriction has yet been made as to the directionality or weight of links. Unlike Structurally Dynamic Cellular Automata [13], in the simplest instance of NA the nodes do not carry values, the network update is purely governed by current topology.[c]

3. Restricted Network Automata

Having defined Network Automata we shall now explore a simple example that can encapsulate some familiar Cellular Automata (CA) behavior [30]. We will impose some constraints such that this example is a small subclass of NA. First, we shall look at the undirected graph with unweighted links such that the existence of a link at time t can be described as the element $A_{i,j}(t)$. The ruleset governing the system's evolution determines the existence of this link at the next time step $A_{i,j}(t+1)$. At the start of the update process, we let the information that the ruleset can act upon be simply the current existence of the link and the degrees of the two nodes which it could possibly connect namely $k_i(t), k_j(t)$. We impose some simple rules to govern the evolution of the system, namely that the existence of a link at time $t+1$ is a function of the combined degree of the adjacent nodes $k_i(t)+k_j(t)$ at the beginning of the timestep and its own existence $A_{i,j}(t)$.

We now restrict the network to an underlying lattice, $\mathbf{U}$ such that only those links that exist within the underlying structure can be formed in $\mathbf{A}$. Note that the

[a]The process of running through all links might be streamlined to include only those eligible for updating. This might be particularly useful in scenarios where the subset of links whose states might change is significantly less than total corresponding to all pairs of nodes or in models in which many nodes might remain unconnected indefinitely.
[b]The concept of a Network Automaton has been alluded to before although not formally defined [32].
[c]This was proposed as a possible extension to SDCA in [13].

underlying lattice is undirected such that $U_{i,j} = U_{j,i}$. For visual clarity, we shall make this an undirected, degree 4 lattice with cyclic boundary conditions. Clearly this construction fulfills the criterion of being a Network Automaton, but since it is restricted to an underlying static network $\mathbf{U}$, we call it a Restricted Network Automaton (RNA). The evolution of the state of a specific link can be described as some operation

$$A_{i,j}(t+1) = F\left(A_{i,j}(t), U_{i,j}, \sum_j A_{i,j}(t) + \sum_i A_{i,j}(t)\right). \tag{2}$$

As an example of the RNA framework, let us construct the rules of the game in the nomenclature of the "life-like" CA models [30]. The survival and birth of a link is determined according to the value of $k_i(t) + k_j(t)$. We can express a ruleset for the survival process of the link in terms of a number (or a set of numbers) x_s such that if $k_i(t) + k_j(t) = x_s$ and the state of the link $A_{i,j}(t) = 1$, then $A_{i,j}(t+1) = 1$ and is zero otherwise. Likewise, for link birth, we have a number x_b such that if $k_i(t) + k_j(t) = x_b$ and $A_{i,j}(t) = 0$, then $A_{i,j}(t+1) = 1$. These rules are conventionally described in terms of a *fertility interval* and *survival interval* [30] or *birth set/survival set* [8], and here we follow the latter convention. The rules in this particular example relate the number of neighboring links to the future existence of a link. For example, according to rule $B2/S3$ a non-existent link will be "born" if the combined degrees of the two nodes between which it might exist is 2 and a link will survive if the combined degree of the two nodes it connects is 3. This ruleset is given explicitly in Table 1.

The top three rows of Table 1 refer to the birth of links and the next two refer to link survival. Clearly, if a link exists then the degree of the nodes at each end must be greater than zero. As such, the fourth and fifth lines of Table 1 cover all eventualities of link survival for this ruleset. The explicit inclusion of the (symmetric) underlying matrix $\mathbf{U}$ reflects the restricted nature of the automaton. Naturally, $k_i(t)$ and $k_j(t)$ can be expressed in terms of the network's adjacency matrix $\mathbf{A}(t)$ as the ith and jth element of $\mathbf{A}(t)\ \underline{1}$ where $\underline{1}$ is a vector with all elements equal to one and of

Table 1. The rules of game $B2/S3$ on an arbitrary underlying network $\mathbf{U}$. A link will be born if it has 2 neighboring links and it must have only 1 neighbor to survive.

Time t				Time $t+1$	
$k_i(t)$	$k_j(t)$	$A_{i,j}(t)$	$A_{j,i}(t)$	$A_{i,j}(t+1)$	$A_{j,i}(t+1)$
1	1	0	0	$U_{i,j}$	$U_{i,j}$
0	2	0	0	$U_{i,j}$	$U_{i,j}$
2	0	0	0	$U_{i,j}$	$U_{i,j}$
2	1	1	1	1	1
1	2	1	1	1	1
Otherwise				0	0

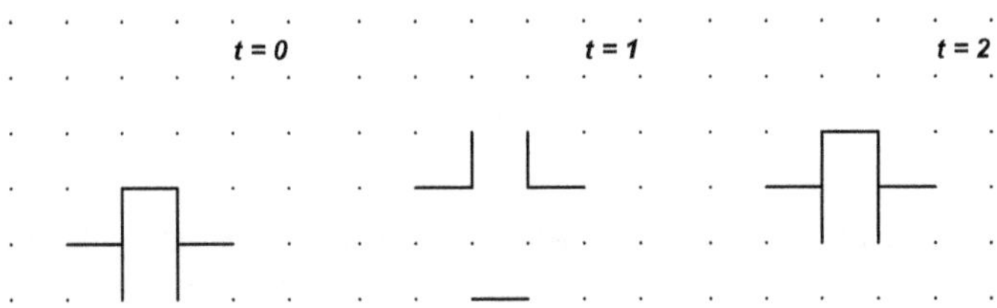

Fig. 1. A simple spaceship of period two in the $B2/S3$ game with underlying lattice of degree 4. Repeated application of the automaton rules will clearly perpetuate its motion.

dimension N_{tot}. More concisely, in terms of some operation F

$$\mathbf{A}(t+1) = F(\mathbf{A}(t), \mathbf{U}). \tag{3}$$

We can now observe the evolution of the automaton on some initial configurations of $\mathbf{A}(0)$ using the rule set $B2/S3$ and underlying lattice of degree 4. We can observe "blinkers" which are motifs which return to their original position after some period. There are also motifs that replicate themselves after a number of timesteps but are spatially translated as shown in Fig. 1. These are known as "spaceships" in the CA nomenclature because they propagate through the space. There are many other interesting configurations and many rulesets to explore, even with the degree 4 underlying lattice. There are a number of "still lifes" (objects that remain unchanged), blinkers of long periods, and "puffers" (debris leaving spaceships) which have been found [24].

It is interesting to note that this particular Network Automata example could be interpreted as a conventional Cellular Automata *on* a network which is the *line graph* [3] of the underlying lattice used in this example. Nodes in the line graph represent edges in the original network and are inter-connected if the edges they represent in the original network share a common node. An implementation of the equivalent Cellular Automata on this line graph would comprise of each node being assigned a binary state $\{0, 1\}$ and the ruleset acting on the neighborhood of a given node in the new network, i.e. birth/survival is related to the number of neighboring nodes in state 1.

4. Stochastic Network Automata

We can also use Network Automata to construct more conventional evolving networks. To do this we augment the NA, which consists of purely deterministic rules, by adding one or more stochastic rules and, thus, arrive at Stochastic Network Automata (SNA). To implement SNA, we need two additional definitions. We denote the outcome of a Bernoulli trial with $\mathcal{B}(p)$ defined as

$$\begin{cases} P(\mathcal{B}(p) = 0) = 1 - p \\ P(\mathcal{B}(p) = 1) = p. \end{cases}$$

We also define the Heaviside-like step function $\phi(x)$ as

$$\phi(x) = \begin{cases} 1 & \text{if } x > 0 \\ 0 & \text{if } x \leq 0. \end{cases} \tag{4}$$

Note that SNA can be restricted to a fixed underlying lattice (network) **U**, resulting in Restricted Stochastic Network Automata (RSNA). We follow this approach in developing the biologically-inspired model in Sec. 6, but emphasize that in this section there is no constraining underlying lattice structure imposed.

4.1. *Random attachment model*

The random attachment algorithm for building a single component network is straightforward. We consider a simple case in which at each time step a new node is connected to the existing network with *one* undirected link. Growing networks such as this one are known as non-equilibrium networks to distinguish them from equilibrium networks in which the number of nodes is constant. The conventional, master-equation analysis [6] provides the degree distribution of such a system, $P(k) = c_k = 2^{-k}$.

We now emulate this process as an SNA. Suppose we wish to grow the network to N_{tot} nodes, so that the adjacency matrix is of dimension N_{tot}. At each time step, we consider the update of all possible links in the network but only wish to update the links from nodes within the connected network component to nodes outside of it. Consequently, there are a total of $N(t)(N_{tot} - N(t))$ links that may be added, of which we wish that, on average, only one link will be added. Having identified those links which may be added, the required probability associated with one of them being added is $P(t) = [N(t)(N_{tot} - N(t)]^{-1}$.

The explicit rules for this particular SNA which replicates random attachment are expressed in Table 2. They state that the link can only be born if the node i is already part of the component and node j is not in the component (or vice-versa) *and* $\mathcal{B}(P(t)) = 1$. If the link already exists, it stays. We seed the automaton with initial configuration of $A_{1,2}(0) = A_{2,1}(0) = 1$ reflecting a single component of two nodes. All links are only considered once in the update stage and both $A_{i,j}(t+1)$ and

Table 2. The stochastic ruleset to replicate random-attachment network growth. The resulting distribution is shown in Fig. 2.

Time t				Time $t+1$	
$A_{i,j}(t)$	$A_{j,i}(t)$	$\phi(k_i(t))$	$\phi(k_j(t))$	$A_{i,j}(t+1)$	$A_{j,i}(t+1)$
0	0	0	0	0	0
0	0	0	1	$\mathcal{B}(P(t))$	$A_{i,j}(t+1)$
0	0	1	0	$\mathcal{B}(P(t))$	$A_{i,j}(t+1)$
0	0	1	1	0	0
1	1	1	1	1	1

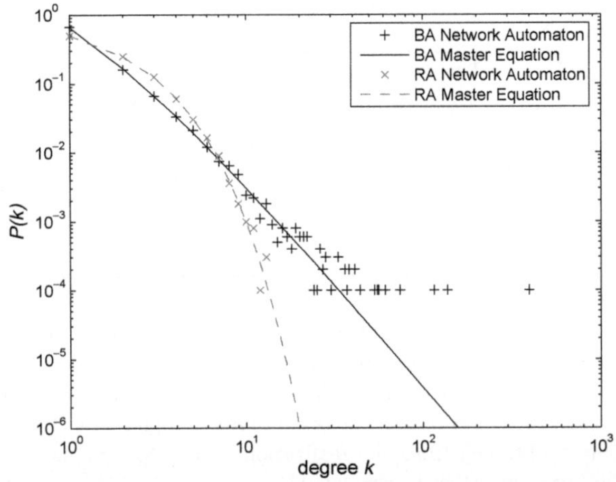

Fig. 2. The resulting degree distributions $P(k)$ for the random-attachment (RA) model and the preferential-attachment (BA) model. The analytically-obtained degree distributions for these models are $P(k) = 2^{-k}$ and $P(k) = 2m(m+1)/[k(k+1)(k+2)]$ respectively [1, 6], with $m = 1$ with here. These are plotted with lines. Superimposed, are the corresponding distributions for one realization of the SNA, grown to only 10,000 nodes. Clearly the two approaches are consistent.

$A_{j,i}(t+1)$ are simultaneously updated.[d] The comparison of the degree distribution of the network generated by the SNA to that of the master-equation analysis is shown in Fig. 2.

It is clear from the binomial process governing the addition of new links (and nodes) to the existing component of the network that, on average, one new link and one new node are added, although clearly more than one new node could be attached with more than one new link allowing cycles to be formed.

4.2. *Barabási–Albert model*

SNA can also emulate a preferential-attachment model such as that by Barabási and Albert [4]. In the BA model the probability for a new node to attach to an existing node is proportional to the degree of the existing node, i.e., the attachment probability is a linear function of the node degree. This attachment mechanism is achieved in the SNA by simply modifying the probability of link birth, and the ruleset for this process is presented in Table 3, where $K(t) = \sum_i k_i(t)$ denotes the sum of degrees over all nodes in the network.

[d] This might be achieved in practice by running sequentially through all nodes $i \in \{1, \ldots, N_{tot}\}$ and then potential neighbors $j \in \{1, \ldots, i\}$ (the lower triangular portion of the adjacency matrix).

Table 3. The stochastic ruleset to generate preferential-attachment network growth. A small simulation distribution is shown in Fig 2.

Time t				Time $t+1$	
$A_{i,j}(t)$	$A_{j,i}(t)$	$\phi(k_i(t))$	$\phi(k_j(t))$	$A_{i,j}(t+1)$	$A_{j,i}(t+1)$
0	0	0	0	0	0
0	0	0	1	$\mathcal{B}\left(\frac{k_j(t)}{K(t)(N_{tot}-N(t))}\right)$	$A_{i,j}(t+1)$
0	0	1	0	$\mathcal{B}\left(\frac{k_i(t)}{K(t)(N_{tot}-N(t))}\right)$	$A_{i,j}(t+1)$
0	0	1	1	0	0
1	1	1	1	1	1

The analytical result obtained using the master-equation approach is given by $P(k) = 2m(m+1)/[k(k+1)(k+2)]$ where m is the (fixed) degree of the new node entering the network [1]. Asymptotically this leads to $P(k) \sim k^{\gamma}$ with $\gamma = 3$. The analytically-obtained distribution is plotted in Fig. 2 together with the corresponding distribution obtained from one realization of an SNA simulation. The match between the two approaches is very good.

4.3. *Watts–Strogatz model*

We have demonstrated how non-equilibrium growing networks can be generated as an SNA. Here we turn to equilibrium networks and discuss the rules necessary to generate small-world networks emulating those of Watts and Strogatz [31]. The WS model starts from considering a one-dimensional lattice comprising N nodes with all nodes having the same degree k (through connections to nearest neighbors, then next nearest neighbors, etc.) and cyclic boundary conditions.[e] An initial configuration for $k = 4$ is shown in Fig. 3, as an example. Each link in the network is visited and it is rewired with probability p. The original rewiring mechanism was

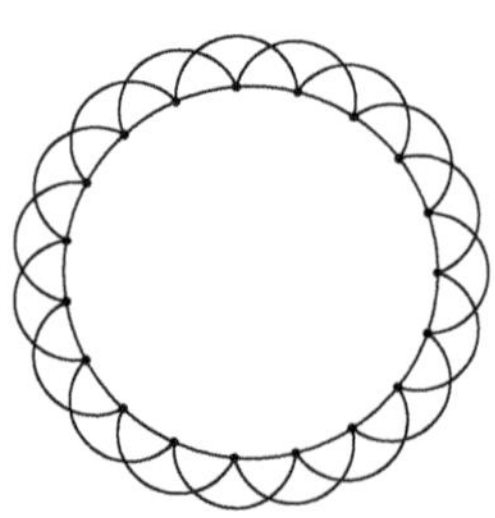

Fig. 3. The initial network configuration of the Watts-Strogatz model for $k = 4$.

[e]Note that a two-dimensional lattice of degree 4 would be inappropriate in that it has zero clustering.

such that one end of the link remained where it was and the other vertex was chosen at random from the rest of the network. In practice, the addition of shortcuts is the important aspect of this model so we choose a slightly simpler mechanism such that both ends of the rewired link are chosen at random. We implement this modified process as an SNA because it results in a somewhat simpler ruleset than the original model whilst retaining the salient features of the original model.

Initially, there are $Nk/2$ links within the system. The expected number which are to be rewired is $pNk/2$. This process might be considered "link death". The number of nodes remains constant, and we wish the number of links to remain constant too. The expected number of links to be born is therefore set equal to the expected number of links which are removed. This is comparible to the notion of link rewiring in the original model of Watts and Strogatz although we note that the number of links removed and those added are not necessarily equal. Assuming no loops of length one (self-connected nodes) such that $A_{i,i} = 0$, the total number of links in the system that are not alive (and therefore capable of being born) is

$$\frac{N(N-1)}{2} - \frac{Nk}{2}. \tag{5}$$

We can then describe the time-independent birth probability of links by

$$P(t) = p_b = \frac{kp}{(N-k-1)}. \tag{6}$$

The ruleset for this system, which is is given in terms of the link-removal probability p and the link-birth probability p_b, is shown in Table 4. In an actual implementation of the model, one runs through all possible links, whether they exist or not, and updates their state according to the rules. The result of applying this ruleset is shown in Fig. 4, depicting the normalized clustering coefficient and mean shortest path for the networks generated. It is interesting to note that this implementation of the SNA requires only one time step. After sufficient successive applications of the automata, all the original links will have be removed and any remaining links will exist between random pairs of nodes. The result will be a network comparable to the classic random graphs of Erdös and Rényi [9].

Table 4. The simple, stochastic ruleset for emulation of the Watts and Strogatz small-world network model. The top row represents the removal of a link with probability p and the bottom, the rewiring of a link with probability p_b as defined in Eq. (6). The network is undirected, yielding a symmetric adjacency matrix $\mathbf{A}$.

Time $t=0$		Time $t=1$	
$A_{i,j}(0)$	$A_{j,i}(0)$	$A_{i,j}(1)$	$A_{j,i}(1)$
1	1	$1-\mathcal{B}(p)$	$A_{i,j}(1)$
0	0	$\mathcal{B}(p_b)$	$A_{i,j}(1)$

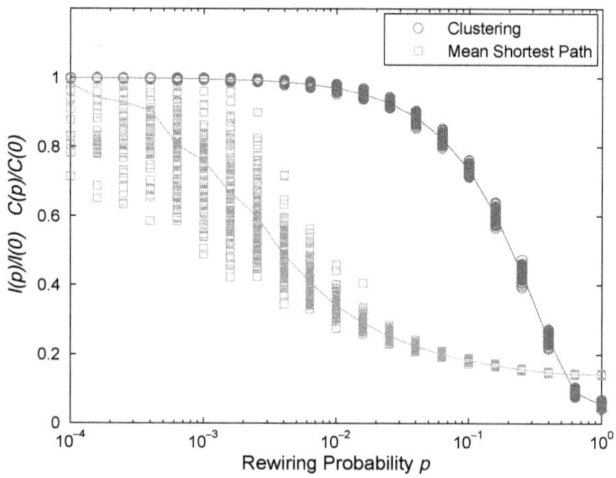

Fig. 4. The normalized clustering coefficient and average shortest path of the networks generated using the Network Automata mechanism outlined in Table 4 to emulate the Watts–Strogatz small-world model. Here, $k = 10$ and $N = 400$. Each marker represents a simulated network, with the line being the mean values over 100 simulations. The normalizing coefficients $C(0)$ and $l(0)$ are the clustering coefficient and mean shortest path of the network prior to any rewiring.

5. Functional Network Automata

We now consider a situation in which the topology of a network evolves while there is simultaneously some process taking place on the network. At any given time the topology of the network constrains the type of dynamics that may unfold on it. However, the dynamical process may influence the subsequent topological evolution of the network, so that its structural properties are coupled to its function and vice-versa. The ruleset governing the topological update process relates not only to network-derived quantities but also functional aspects of the nodes and/or links. Since the functional process requires a network on which to perform, we decouple the evolution of the network into two distinct phases, namely, that affecting its topology and that governing the functional process. Writing the functional information (relating to nodes and/or links) at some time t as some matrix $\mathbf{S}(t)$, the formal description of the evolution can be expressed in terms of some operations F and G as

$$\mathbf{A}(t+1) = F(\mathbf{A}(t), \mathbf{S}(t)), \tag{7}$$

$$\mathbf{S}(t+1) = G(\mathbf{A}(t+1), \mathbf{S}(t)). \tag{8}$$

This expression states that the network evolves according to some process, which is determined by its own current topology $\mathbf{A}(t)$, *and also* by some attributes of its nodes and links that includes function-based information $\mathbf{S}(t)$. The functional process then occurs on this network to generate the new set of information $\mathbf{S}(t+1)$. The global state of the system is encompassed by the matrices **A** and **S**.[f] The system is synchronously updated by running through all possible links between all pairs of nodes and updating the state of each link in accordance to the ruleset employed.

6. Biologically Inspired Model

We now construct a series of simple models of woodland fungi [5, 10, 26, 27] to demonstrate the versatility of Functional Network Automata. Although the models, which describes the growth of the fungi and distribution of resources within it, are biologically inspired, their aim is not to incorporate a large number of biological details. Instead, we adopt a minimalist approach to emulate fungal growth and internal nutrient transport from a small set of microscopic rules. These models are, nevertheless, capable of producing structures qualitatively similar to mycelial network development (e.g. Fig. 5). This example is *Phanerochaete velutina*, a foraging saprotrophic woodland fungus that builds an adaptive network to translocate

Fig. 5. (Color online) Transport of non-metabolized, radioactively-labeled amino acid within *Phanerochaete velutina*, a foraging, woodland fungi. The fungi was grown in a Petri dish from a 12 mm agar inoculum plug across a scintillation screen. The image is of an area approximately 10 × 10 cm and depicts relative intensities of photon counts derived from movement of a radiolabeled amino acid integrated over a 60 minute interval with dark blue indicating the lowest and dark red indicating the highest relative intensity [26–28].

[f]Here, we have defined the topological and functional updates as being sequential to each other. In general, a topologically and functionally coupled system system could also be defined synchronously, $\mathbf{A}(t+1) = F(\mathbf{A}(t), \mathbf{S}(t))$ and $\mathbf{S}(t+1) = G(\mathbf{A}(t), \mathbf{S}(t))$ although any (dis)advantage from this formulation is unclear. The suitability of either formulation is likely to be situation (model) dependent and related to the relative timescales of the functional and topological evolution.

resource [5, 27, 28]. The transport flux is oscillatory and still not fully understood [10, 25, 27, 28].

We start by specifying biologically-naive and mathematically-simple update rules (Ruleset **a**) which govern the topological part of the system's evolution as in Eq. (7). This is paired to a simple functional update process (Process **1**) representing Eq. (8). The ruleset and the process fully define the model. We then modify both the topological rules (Ruleset **b** and **c**) and functional process (Process **2**) according to some basic physical and biological considerations providing six distinct models. The end product may serve as a platform for more elaborate future models of fungal growth, demonstrating the effectiveness of using the framework.

Consider a system of agents who might each be interpreted as a cell in a two-dimensional lattice. The connectivity between agents is North, South, East and West reflecting a maximum possible connectivity of $d = 4$ for all agents who are subsequently restricted to local information as long-range communication is assumed unlikely in the biological system. The agent layer is superimposed on a resource layer as in Fig. 6. The rules of the system are very simple. If an agent is above a resource, it absorbs that resource at some rate R_E. The objective of each agent is two-fold: to grow into available space in the search for more resource, and to redistribute excess internal resource to support further exploration. We allow each agent to grow only one new neighbor (in a random direction) at a particular time, but only if the agent has resources to do so. To mimic active transport of resources to the growing tip, an agent passes resources to its neighbor cell provided that the neighbor does not pass resources to it. We now endeavor to categorize this simple multi-agent system as a Restricted Stochastic Network Automaton. This

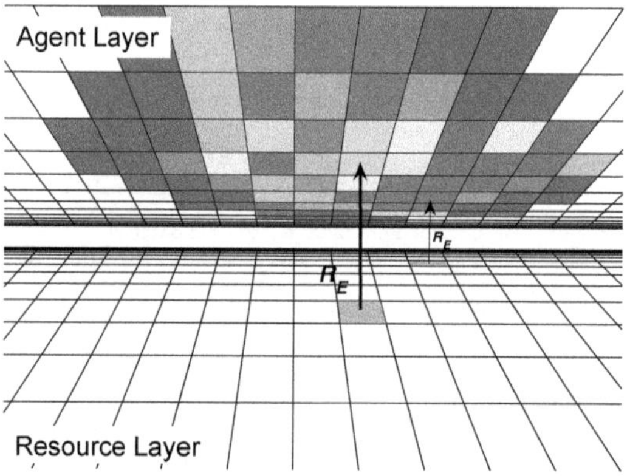

Fig. 6. (Color online) A biologically-inspired multi-agent model whereby the agent layer is superimposed upon a resource layer. Agents above a resource can accumulate resources at some rate R_E.

serves not only to clarify any ambiguities that arise in the programming of a multi-agent system [23], but also as a potential aid to improving efficiency in that the required iteration and information storage/retrieval aspects are clearly defined by the rulesets imposed.

Let us first look at the growth (topological) stage. Each cell or agent represents a node and the boundary between two cells through which resource is passed is represented by a directed link. Consider that the information upon which the topological ruleset will act to update the attributes of a link in the network is simply the amount of resource that each of the two nodes has at each end of the link and their in and out-degrees. We can simply write the functional information as a vector such that $S_i(t)$ refers to the resource (functional variable) that agent (node) i has at time t. For clarity, we can express some topological information as vectors too, such that the element $k_i(t)$ represents the total degree of a node, ${}_{\text{out}}k_i(t)$ its out-degree which, in turn, provides its in-degree ${}_{\text{in}}k_i(t)$, each of which is obtainable from the adjacency matrix $\mathbf{A}(t)$.

We will grow this Network Automaton in an unweighted but directed adjacency matrix $\mathbf{A}$ so that if $A_{i,j} = 1$ the link exists and is directed from i to j, whereas if $A_{j,i} = 1$ the link exists and is directed from j to i. If neither $A_{i,j} = 1$ nor $A_{j,i} = 1$ then the link does not exist. Here $A_{i,j} = 1$ and $A_{j,i} = 1$ are mutually exclusive. As each node has limited possible connectivity (here 4), we only consider the subset of links in the system which could possibly exist. The structural update process runs through all of these possible links and each pair of nodes which could be connected is considered once. The link attributes $A_{i,j}$ and $A_{j,i}$ are then updated at the same time.

We can now write the network update procedure for this topological update as Ruleset **a** in terms of an exhaustive truth table as in Table 5. Note that if $A_{i,j}(t) = A_{j,i}(t) = 0$, the total degree of both nodes i and j is less than the connectivity d of the lattice in which the system evolves. Again, we denote the outcome of a Bernoulli trial as $\mathcal{B}(p)$ such that $P(\mathcal{B}(p) = 1) = p$ and conversely $P(\mathcal{B}(p) = 0) = 1 - p$. Again, we employ a step function $\phi(x)$ defined as $\phi(x) = 1$ for $x > 0$ and $\phi(x) = 0$ for $x \leq 0$ to distinguish active or "alive" agents. That is, an agent i is active if it has some amount of resource, $\phi(S_i(t)) = 1$. The first rule in Ruleset **a** states that if there does not exist a link between two inactive nodes at the present timestep, no link shall be formed at the next timestep. The second and third rules reflect an agent growing by establishing a link towards (on average) one neighbor in the lattice. Once fully connected ($k_i(t) = d$), an agent cannot generate any further links. The fourth rule reflects two active agents who are adjacent in the lattice but not connected. A link is established between them with its direction determined via a coin toss. The remaining eight rules in Ruleset **a** simply state that once a link is established, it remains indefinitely: there is no rewiring.

We now describe the resource distribution (functional) stage and start by mapping the adjacency matrix $\mathbf{A}(t+1)$ to a normalized transition matrix $\mathbf{T}(t+1)$

Table 5. Different rulesets (the columns labeled 'time $t+1$') for three biologically inspired models: (a) the simplest scenario, (b) incorporating conservation of resources, and (c) implementing a delay factor (see text for details). As $A_{i,j=1}$ and $A_{j,i}=1$ are mutually exclusive, there are only 12 possible states of a link at time t.

Time t				Model **a** Time $t+1$		Model **b** Time $t+1$		Model **c** Time $t+1$	
$A_{i,j}(t)$	$A_{j,i}(t)$	$\phi(S_i(t))$	$\phi(S_j(t))$	$A_{i,j}(t+1)$	$A_{j,i}(t+1)$	$A_{i,j}(t+1)$	$A_{j,i}(t+1)$	$A_{i,j}(t+1)$	$A_{j,i}(t+1)$
0	0	0	0	0	0	0	0	0	0
0	0	0	1	0	$\mathcal{B}\left(\frac{1}{d-k_j(t)}\right)$	0	$\mathcal{B}\left(\frac{1}{d-k_j(t)}\right)$	0	$\mathcal{B}\left(\frac{g}{d-k_j(t)}\right)$
0	0	1	0	$\mathcal{B}\left(\frac{1}{d-k_i(t)}\right)$	0	$\mathcal{B}\left(\frac{1}{d-k_i(t)}\right)$	0	$\mathcal{B}\left(\frac{g}{d-k_i(t)}\right)$	0
0	0	1	1	$\mathcal{B}\left(\frac{1}{2}\right)$	$1-A_{i,j}(t+1)$	$\mathcal{B}\left(\frac{1}{2}\right)$	$1-A_{i,j}(t+1)$	$\mathcal{B}\left(\frac{1}{2}\right)$	$1-A_{i,j}(t+1)$
0	1	0	0	0	1	0	1	0	1
0	1	0	1	0	1	0	1	0	1
0	1	1	0	0	1	0	1	0	1
0	1	1	1	0	1	$\delta({}_{\text{in}}k_i, d)\mathcal{B}\left(\frac{1}{d}\right)$	$1-A_{i,j}(t+1)$	$\delta({}_{\text{in}}k_i, d)\mathcal{B}\left(\frac{1}{d}\right)$	$1-A_{i,j}(t+1)$
1	0	0	0	1	0	1	0	1	0
1	0	0	1	1	0	1	0	1	0
1	0	1	0	1	0	1	0	1	0
1	0	1	1	1	0	$1-A_{j,i}(t+1)$	$\delta({}_{\text{in}}k_i, d)\mathcal{B}\left(\frac{1}{d}\right)$	$1-A_{j,i}(t+1)$	$\delta({}_{\text{in}}k_i, d)\mathcal{B}\left(\frac{1}{d}\right)$

describing the flow of resource between adjacent cells:

$$T_{i,j}(t+1) = \begin{cases} A_{i,j}(t+1)/{}_{\text{out}}k_i(t+1) & \text{for } {}_{\text{out}}k_i(t+1) > 0 \\ 0 & \text{for } {}_{\text{out}}k_i(t+1) = 0 \end{cases}$$
$$T_{i,i}(t+1) = \begin{cases} 0 & \text{for } {}_{\text{out}}k_i(t+1) > 0 \\ 1 & \text{for } {}_{\text{out}}k_i(t+1) = 0. \end{cases} \tag{9}$$

Through this process, an agent distributes resource equally amongst those neighbors which are not transmitting resource to it.

We can then write the update for the resource distribution process as

$$\underline{S}(t+1) = \mathbf{T}^{\top}(t+1)\underline{S}(t) + \underline{\xi}(t), \tag{10}$$

where the vector $\underline{\xi}$ corresponds to the accumulation of resource by agents from the resource (substrate) layer and $\top$ denotes the matrix transpose. We impose the constraint that only "alive" (i.e. active) agents can accumulate resource through this process so

$$\xi_\eta(t) = R_E\ \phi(S_\eta(t))\ L_\eta, \tag{11}$$

where the vector $\underline{L}$ denotes the (binary) existence of resource at position of node (agent) η in the resource layer, R_E is the rate at which an agent accumulates the resource and $\phi(x)$ is the step function defined earlier to identify active agents. This resource accumulation from the substrate could be made time dependent (i.e. finite resources such that $\underline{L}$ represents quantities of resource at specific locations) and the accumulation rate could also be made site specific reflecting resource "quality" although here we will not consider these effects. Equations (9)–(11) constitute Process **1** and represent the functional update stage of Eq. (8).

For the example of Fig. 7, the amount of resources that node i has at time $t+1$ is

$$\begin{aligned} S_i(t+1) &= S_j(t)\frac{A_{j,i}(t+1)}{{}_{\text{out}}k_j(t+1)} + S_m(t)\frac{A_{m,i}(t+1)}{{}_{\text{out}}k_m(t+1)} \\ &\quad + S_r(t)\frac{A_{r,i}(t+1)}{{}_{\text{out}}k_r(t+1)} + S_q(t)\frac{A_{q,i}(t+1)}{{}_{\text{out}}k_q(t+1)} \\ &= S_j(t)T_{j,i}(t+1) + S_r(t)T_{r,i}(t+1) + S_q(t)T_{q,i}(t+1). \end{aligned} \tag{12}$$

In this example node i is not accumulating resource from the substrate layer. The resource it receives is from neighbors to whom it is connected and whose links are directed towards node i. The amount it receives from a given neighbor is simply determined by the out-degree of that neighbor.

We can now observe the Network Automaton in operation as shown in Fig. 8. We start with a single node η above a single food source so that at time $t = 0$ the agent has some resource. In the initial configuration the adjacency matrix is all zeros $A_{i,j}(0) = 0\ \forall\ i, j$, and the resource information vector is all zeros except

NA: Coupling Structure and Function in Real-World Networks 333

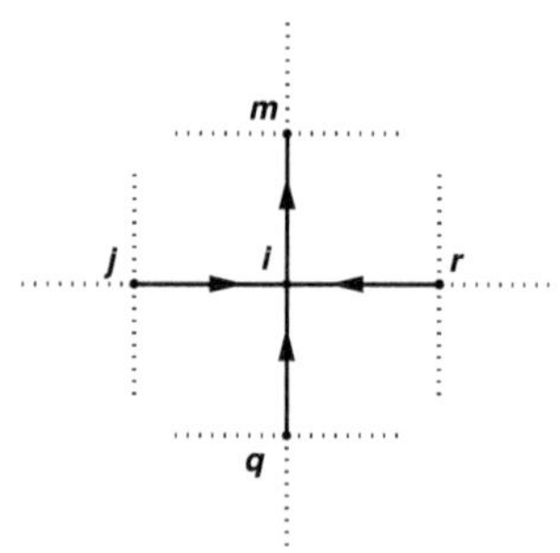

Fig. 7. The influx of resource into node labeled i. The amount this node receives from node j is related to the out-degree of node j. This amount would be expressed as $S_j(t)/_{\text{out}}k_j(t+1)$. Similarly, node i receives an amount $S_q(t)/_{\text{out}}k_q(t+1)$ of resource from node q and $S_r(t)/_{\text{out}}k_r(t+1)$ from node r whereas it receives nothing from node m.

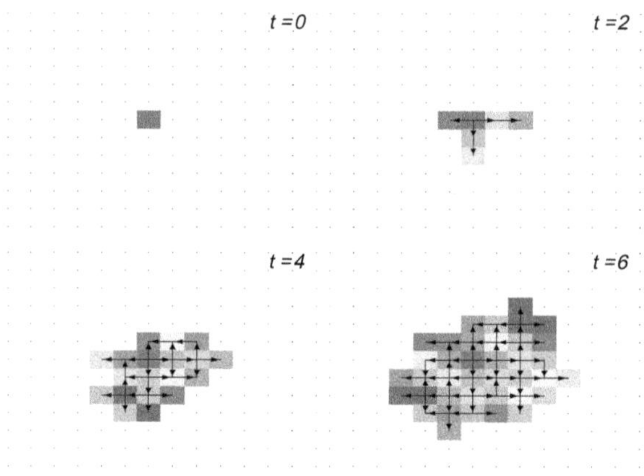

Fig. 8. (Color online) The evolution of the biologically-inspired NA over 6 timesteps. The colors represent the amount of resource $S_i(t)$ a node has (the functional aspect) superimposed on the evolving, directed network (the structural aspect). Dark blue denotes the agent with the lowest amount of resource and dark red the highest.

$S_\eta(0) = R_E$ such that the initial agent has amount R_E. For this example, the resource accumulation vector $\underline{\xi}(t)$ is also all zeros except $\xi_\eta(t) = R_E \,\forall\, t$. We observe both the network and functional aspect of the system. The nodes (agents) are superimposed on the directed network, and the amount of resources a node has is indicated by its color ranging from blue (low concentration) to dark red (high concentration) [10, 24]. Only nodes that have resources are included and the result is independent of the choice of R_E. A longer simulation is shown in Fig. 9.

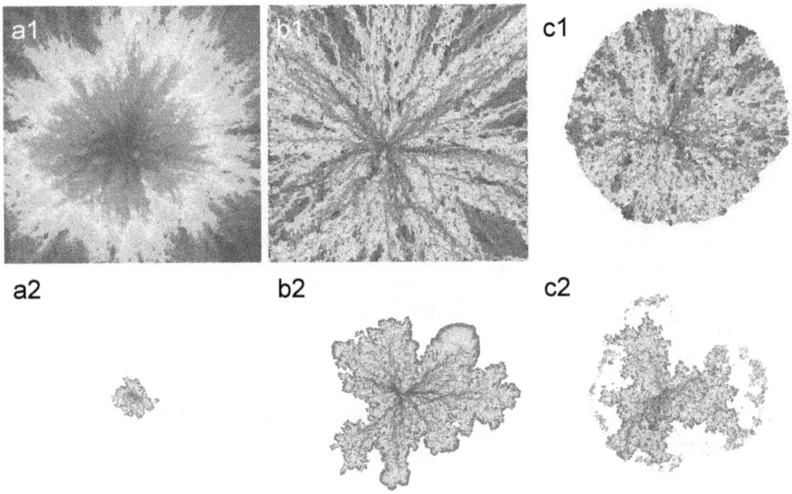

Fig. 9. (Color online) Simulated output for the six models pairing each of the topological updates, Rulesets **a**, **b** and **c**, and functional updates of Processes **1** and **2**. For all models, the lattice is of dimension 400 × 400 with periodic boundary conditions. Each was seeded with a single initial agent above a single (infinite) resource. The rate of resource accumulation for that agent is $R_E = 80,000$ and the simulation was run for 2000 time steps for each model. For Process **2** the residual consumption rate of each agent is set at $R_C = 1$. For Ruleset **c**, the delay factor is $g = 0.1$. Again, dark blue denotes the agent with lowest amount of resource and dark red the highest.

Note that under this ruleset and functional update stage an agent might accumulate resource indefinitely. By allowing a node which has in-degree equal to the maximum connectivity (${}_{\rm in}k_i = d$) to randomly flip a direction of one of its links overcomes this biologically-unfeasible behavior and results in the topological update of (Ruleset **b**) where the Kronecker delta function is defined as $\delta(x, y) = 0$ for $x \neq y$ and $\delta(x, y) = 1$ for $x = y$. A simulation of this ruleset coupled with the functional update of Process **1** is shown in Fig. 9 and emergent canalized flux channels are clear. Such channels have been observed experimentally in a wide class of real biological fungi [27].

In Rulesets **a** and **b**, the physical transport of nutrient is of comparable speed to that of the growth, which is clearly not reasonable for most biological systems. A further development can be made to incorporate two different time scales in the model so that the growth and redistribution of resources take place at different rates, leading to Ruleset **c**. To establish this feature within the model, we can delay the growth process by introducing a parameter $g \in (0, 1]$ to the stochastic growth terms in the topological ruleset, so that an agent grows, on average, one neighbor every $1/g$ time steps. This is depicted explicitly in Table 5. Whilst this additional development slows the rate of growth, the automaton will grow indefinitely until the domain is fully occupied (the system illustrated in plot c1 of Fig. 9 is still growing).

We can also incorporate consumption of resources by agents in the model by making further modifications to the functional update stage, resulting in Process **2**. If an agent has more than some residual consumption amount, R_C, then this rate of consumption is deducted. If the agent has less than this value, all of that agent's resources are removed and that agent might be considered dead. We introduce an intermediate stage in the functional update process which can be described in a similar manner to Eq. (10):

$$\underline{S}^*(t+1) = \mathbf{T}^\top(t+1)\underline{S}(t) + \underline{\xi}(t), \tag{13}$$

where the resource distribution is described by the transition matrix $\mathbf{T}(t+1)$ defined in Eq. (9) and the accumulation of resource from the substrate is described by the vector $\underline{\xi}(t)$ defined in Eq. (11). We can then write the update for $\underline{S}(t+1)$, including consumption of resource as

$$S_i(t+1) = \phi(S_i^*(t+1) - R_C)(S_i^*(t+1) - R_C), \tag{14}$$

which makes use of the step function $\phi(x)$. Only agents (nodes) active in the network can accumulate resources from the resource layer. The effect of this "cost" of living clearly limits the potential size of the system. Equations (9), (11), (13) and (14) constitute Process **2** and the effect of using this functional update are illustrated in Fig. 9.

Whilst the aim of development of these models of fungal growth is to serve as an illustration of an application FNA, we can qualitatively observe the effects of the different rulesets and functional update processes, a pairing of which constitute a fully-defined model. We observe from the upper row of plots in Fig. 9 that the functional update process, Process **2** is robust to the imposed variations in the topological updates (**a**, **b** and **c**) with respect to coverage of the domain. The model **c1** will eventually occupy the full space and is still growing at the point illustrated in the plot c1 of Fig. 9. Interstingly, comparison of plots a1 and a2 demonstrates that the incorporation of resource conservation in ruleset **b** results in stronger canalization of flow than the simplest ruleset **a**. The lower plots in Fig. 9 depict the use of Process **2** incorporating the consumption of resource. Models **a2** and **b2** have reached a steady state in which no more growth occurs whereas in model **c2**, the peripheral islands unconnected to the main population body will eventually die out.

7. Concluding Remarks and Discussion

In this paper we have developed the concepts of Network Automata (NA) together with its restricted (RNA) and stochastic (SNA) variants. This generic framework can encompass the behavior of many familiar models. We have demonstrated its

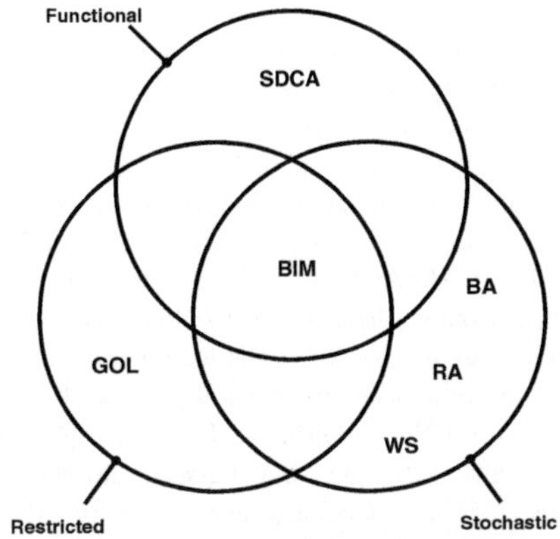

Fig. 10. The functional, stochastic and restricted behaviors which can be encompassed by (Functional) Network Automata. Also illustrated is the position of the Barabási–Albert (BA) network growth model, the random attachment (RA) network growth model, the Watts–Strogatz small-world model (WS), the Game of Life Cellular Automata (GOL), the biologically inspired models (BIM) introduced in Sec. 5 and Structurally Dynamic Cellular Automata [13].

ability to mirror Cellular Automata behavior in Sec. 3, and also non-equilibrium growing network models (random attachment, Barabási-Albert) and equilibrium (Watts–Strogatz) network models in Sec. 4 which are illustrated schematically in Fig. 10. Whilst these pedagogical cases (which might be implemented directly more efficiently) have comprised undirected, unweighted networks, these features can be easily incorporated within the NA framework.

The development of NA naturally leads to the concept of Functional Network Automata which couples the evolution and function of complex networks by using simple microscopic rules at the level of nodes and links. We have demonstrated the practicality of the framework by applying it to a family of biologically inspired models, which produce qualitatively similar canalized flow patterns to those observed in real woodland fungi [10, 25–27]. This suggests that for organisms which have to adapt their morphology to a variable environment, function may play a crucial role in determining structure. The well-defined and simple rulesets not only make replication straightforward but also aid implementation at the programming level. In the study of emergent phenomena involving networks, use of an FNA implementation enables concise and clear establishment of microscopic rules such that the observed differences in resulting behavior are known to be resultant from the rules as opposed to any emergence from implementation biases. Speculatively, we might infer that the growth of "limbs" in models **b2** and **c2** evident in Fig. 9 might constitute a primitive emergent search strategy in that a larger distance from the initial resource has been reached than would have been if the same number of agents had grown with isotropic radial coverage. One can easily think of more complex rulesets to more accurately model a real system, such as adding transport costs or finite resources, both of which can easily be accomplished, or even a time-dependent ruleset. It is then interesting to ask what level of complexity is required to more accurately model real biological systems.

We would expect certain limitations of FNA consistent with the advantages of Multi-Agent Systems over conventional CA approaches. Specifically, MAS are well suited to mixed populations of agents [23] and implementation of such models as FNA could prove difficult. However, we also expect there to be many application domains for Network Automata from social to biological systems such as modeling the evolution of genetic regulatory networks where selection pressure at the functional (phenotypic) level influences evolution at the topological (genetic) level [12]. One can envisage applying FNA to Diffusion Limited Aggregation (DLA) systems. That DLA has been recently recast as a CA suggests that the FNA might be suitable for this type of problem [23] or, indeed, any system in which the dynamics of network topology is related to the function performed thereon. There are certain biological systems which can quasi-solve increasingly complex problems in a constant time. An example is *Physarum polycephalum*, the true slime mold, which approximately identifies the Steiner points when placed upon multiple food sources [17]. Given that it might be possible to model these systems within the NA framework, it might suggest how to design a hardware-based implementation to perform similar calculations in constant time. It is interesting to pose the question as to what kind of problems could be solved by such a system and how complex the microscopic rules would be for a given problem. This would reflect the minimum length of ruleset that would have to be employed by the system in both the network and functional update stages.

Acknowledgments

D.M.D.S. acknowledges funding by EPSRC and BBSRC and J.-P.O. by a Wolfson College Junior Research Fellowship (Oxford). D.M.D.S acknowledges funding from the European Union (MMCOMNET) for part of this research. We thank Felix Reed-Tsochas for comments and suggestions.

References

[1] Albert, R. and Barabási, A.-L., Statistical mechanics of complex networks, *Rev. Mod. Phys.* **74** (2002) 47–97.

[2] Alonso-Sanz, R., A structurally dynamic cellular automaton with memory, *Chaos Soliton. Fract.* **32** (2006) 1285–1295.

[3] Balakrishnan, V. K., *Schaums Outline of Graph Theory* (Mcgraw-Hill Publishing Company, New York, 1997).

[4] Barabási, A.-L. and Albert, R., Emergence of scaling in random networks, *Science* **286** (1999) 509–512.

[5] Bebber, D. P., Hynes, J., Darrah, P. R., Boddy, L. and Fricker, M. D., Biological solutions to transport network design, *Proc. R. Soc. B* **274** (2007) 2307–2315.

[6] Dorogovtsev, S. N. and Mendes, J. F. F., Evolution of networks, *Adv. Phys.* **51** (2002) 1079–1187.
[7] Darabos, C., Giacobini, M., Tomassini, M., Performance and robustness of cellular automata computation on irregular networks, *Adv. Complex Syst.* **10** (2007) 85–110.
[8] Eppstein, D., Growth and decay in life-like cellular automata, in Adamatzky, A. (ed.), *Game of Life Cellular Automata* (London: Springer, 2010).
[9] Erdös, P. and Rényi, A., On random graphs, *Publ. Math. Debrecen* **6** (1959) 290.
[10] Fricker, M. D., Lee, J. A., Bebber, D. P., Tlalka, M., Hynes, J., Darrah, P. R., Watkinson, S. C. and Boddy, L., Imaging complex nutrient dynamics in mycelial networks, *J. Microscopy* **231** (2008) 317–331.
[11] Fronczak, P., Fronczak, A. and Hołyst, J. A., Self-organized criticality and coevolution of network structure and dynamics, *Phys. Rev. E* **73** (2006) 046117.
[12] Foster, D. V., Kaufmann, S.A. and Socolar, J. E. S., Network growth models and genetic regulatory networks, *Phys. Rev. E* **73** (2006) 031912.
[13] Ilachinski, A. and Harpern, P., Structurally dynamic cellular automata, *Complex Systems* **1** (1987) 503–527.
[14] Lambiotte, R., Delvenne, J.-C. and Barahona, M., Laplacian dynamics and multiscale modular structure in networks, arXiv:0812.1770v3 (2009).
[15] Meisel C. and Gross, T., Self-organized criticality in a realistic model of adaptive neural networks, arXiv:0903.2987v1 (2009).
[16] Motter, A. E., Matías, M. A., Kürths, J. and Ott, E., Dynamics on Complex Networks and Applications, *Physica D* **224** (2006) 7–8.
[17] Nakagaki, T., Kobayashi, R., Nishiura, Y. and Ueda, T., Obtaining multiple separate food sources: Behavioural intelligence in the physarum plasmodium, *Proc. Biol. Sci.* **271** (2004) 2305–2310.
[18] von Neumann, J., *The Theory of Self-Reproducing Automata* (University of Illinois Press, Urbana, 1966).
[19] Marr, C. and Hütt, M.-T., Topology regulates pattern formation capacity of binary cellular automata on graphs, *Physics A* **354** (2005) 641–662.
[20] Marr, C. and Hütt, M.-T., Outer-totalistic cellular automata on graphs, *Phys. Lett. A* **373** (2009) 546–549.
[21] Newman, M. E. J., The structure and function of complex networks, *SIAM Rev.* **45** (2003) 167–256.
[22] Rohlf, T. and Bornholdt, S., Self-organized criticality and adaptation in discrete dynamical networks, arXiv:0811.0980v1 (2008).
[23] Spicher, A., Fatés, N. and Simonin, O., Translating discrete multi-agents systems into cellular automata: application to diffusion-limited aggregation, Filipe, J., Fred, A. and Sharp, B. (eds.), *CCIS* **67** *Proceedings of ICAART09–Revised Selected Papers*, 270–282 (Berlin: Springer, 2010).
[24] Smith, D. M. D., *Agents, Games and Networks*, D. Phil. thesis, University of Oxford, UK (2007).
[25] Tlalka, M., Bebber, D. P., Darrah, P. R., Watkinson, S. C. and Fricker, M. D., Emergence of self-organised oscillatory domains in fungal mycelia, *Fungal Genet. Biol.* **44** (2007) 1085–1095.
[26] Tlalka, M., Bebber, D. P., Darrah, P. R., Watkinson, S. C. and Fricker, M. D., Quantifying dynamic resource allocation illuminates foraging strategy in *Phanerochaete velutina*, *Fungal Genet. Biol.* **45** (2008) 1111–1121.
[27] Tlalka, M., Hensman, D., Darrah, P. R., Watkinson, S. C. and Fricker, M. D., Noncircadian oscillations in amino acid transport have complementary profiles in assimilatory and foraging hyphae of *Phanerochaete velutina*, *New Phytol.* **158** (2003) 325–335.

[28] Tlalka, M., Watkinson, S. C., Darrah, P. R. and Fricker, M. D., Continuous imaging of amino-acid translocation in intact mycelia of *Phanerochaete velutina* reveals rapid, pulsatile fluxes, *New Phytol.* **153** (2002) 173–184.
[29] Tomassini, M., Giacobini, M. and Darabos, C., Evolution and dynamics of small-world cellular automata, *Complex Systems* **15** (2005) 261–284.
[30] de La Torre, A. C. and Mártin, H. O., A survey of cellular automata like the "game of life", *Physica A* **240** (1997) 560–570.
[31] Watts, D. J. and Strogatz, S. H., Collective dynamics of 'small-world' networks, *Nature* **393** (1998) 440.
[32] Wolfram, S., *A New Kind of Science* (Wolfram Media, 2002).
[33] Zimmerman, M. G., Eguíluz, V. M. and San Miguel, M., Coevolution of dynamical states and interactions in dynamic networks, *Phys. Rev. E* **69** (2004) 065102(R).

Advances in Complex Systems
Vol. 19, No. 3 (2016) 1650005 (19 pages)

DOI: 10.1142/S0219525916500053

TEMPORAL NETWORK ANALYSIS OF LITERARY TEXTS

SANDRA D. PRADO* and SILVIO R. DAHMEN†
Instituto de Física da UFRGS
91501–970, Porto Alegre, Brazil
**sandra.prado@ufrgs.br*
†silvio.dahmen@ufrgs.br

ANA L.C. BAZZAN
Instituto de Informática da UFRGS, 91501–970, Porto Alegre, Brazil
bazzan@inf.ufrgs.br

PADRAIG MAC CARRON
Department of Experimental Psychology, University of Oxford
Oxford, OX1 3UD United Kingdom
padraig.maccarron@psy.ox.ac.uk

RALPH KENNA
Applied Mathematics Research Centre, Coventry University
Coventry CV1 5FB United Kingdom
r.kenna@cov.ac.uk

Received 23 February 2016
Revised 30 April 2016
Accepted 8 June 2016
Published 5 September 2016

We study temporal networks of characters in literature focussing on *Alice's Adventures in Wonderland* (1865) by Lewis Carroll and the anonymous *La Chanson de Roland* (around 1100). The former, one of the most influential pieces of nonsense literature ever written, describes the adventures of Alice in a fantasy world with logic plays interspersed along the narrative. The latter, a song of heroic deeds, depicts the Battle of Roncevaux in 778 A.D. during Charlemagne's campaign on the Iberian Peninsula. We apply methods recently developed by Taylor *et al.* [Taylor, D., Myers, S. A., Clauset, A., Porter, M. A. and Mucha, P. J., Eigenvector-based centrality measures for temporal networks, *CoRR* (2015).] to find time-averaged eigenvector centralities, Freeman indices and vitalities of characters. We show that temporal networks are more appropriate than static ones for studying *stories*, as they capture features that the time-independent approaches fail to yield.

Keywords: Structure and dynamics of complex networks; graph theory; networks and literature.

†Corresponding author.

1. Introduction

This work was motivated by a simple question: Is there something in the structure of a narrative or in the way characters interact throughout, that is similar across genre and time barriers? Any approach one may contemplate when trying to 'understand' a piece of good literature is of course contingent on the question being asked. Finding an answer — if there is one — to what constitutes 'good' literature requires much more than a naive analysis that disregards aspects like the underlying theme, the historical, cultural and sociological contexts at the time of writing and how the readership relates to these. Therefore, the purpose of the authors is *not* to find answers to questions which can only be tackled with an approach that contemplates all these aspects of literary analysis. Our main objective is to explore some aspects of the structure of the time-evolving network of characters of a novel and see if network theory can be used as a viable tool in literature. It is known, for instance, that renowned authors construct a character's *persona* as it interacts with other characters and settings [27]. Since network theory is the epitome of a theory of interconnectedness, can it be used to tell us something about how the connections define one character's importance? Is there a correlation between the mathematical results it yields and our perception, as readers, of the relevance of this or that character? These are some of the questions we will try to address in the present work.

Even when one confines oneself to the exclusive use of networks without resorting to other forms of literary analysis [11], there remain several questions that need to be tackled. The first main challenge is the question of what exactly constitutes the network in a narrative. Should nodes represent characters or groups of characters, objects or places? What do edges exactly represent? This question has no simple answer as interactions between characters can be complex and play a fundamental role in the story, as do objects and locations. In the case of historical texts, narratives depend to a great extent on the narrator and how he/she interprets historical events [13]. In spite of having a well-defined theoretical framework of 'character networks' of Woloch and Moretti [19, 20, 30], choices still have to be made and there is always more than one possible network for a given narrative [21].

The second challenge one faces when treating networks of actors in a novel or historical narrative is that they are not fixed in time. Most of the studies on networks have been conducted for systems frozen on a time scale large enough to allow one to completely disregard any dynamics. All nodes, irrespective of any time ordering regarding the events they depict, are connected into a single, time-independent network. Following the current terminology, we can call these *aggregate* networks. This may be reasonable in some context [3, 7, 8, 12, 15, 16, 22, 28, 29] but networks involving interpersonal relations and events — either historical or fictional — are intrinsically dynamic [13, 14, 18]. Edges may come and go, as well as characters. Even though one may correctly argue that in a certain sense they are immutable — history cannot be changed and once a piece of fiction is written, it remains so — the stories they depict happen over time, which may typically span a period of a few days to several decades. If network theory is to have any predictive power, the first question to come to mind is if measures which make sense for time-independent networks will still make sense for dynamical ones and the insights gained from studying the former can still be applied to the latter [14, 25].

The main goal of the present paper is the application of temporal networks to literature and to describe the kinds of questions one can answer with the methods at hand. To do this, we combine the theoretical framework of Woloch [30] and Moretti [19–21] for literary studies with techniques developed by Taylor *et al.* [26] for temporal networks on the mathematical side. To the best of our knowledge, the present paper is the first one where these methods are applied to literature. We chose two texts as study objects: Lewis Carroll's *Alice's Adventures in Wonderland* [6] of 1865 and the anonymous *La Chanson de Roland* from *ca.* 1100 AD [2]. Carroll's *Alice* was chosen for several reasons: it has a story with a well defined number of chapters (12), which we took as being our time sequence. The number of characters is small, which makes it easier to follow an individual character and see how measures associated with it change over time. It has also another advantage: some of its characters are by now household names in many of the 174 languages the book has been translated into. So it is possible to check if some more 'popular' characters, like the Cheshire Cat or the Dodo really stand the test of a mathematical analysis as to their relevance for the plot. We also looked for a text with a larger set of characters and which would be as far removed in time and *genre* from *Alice* as possible: the choice fell on *La Chanson de Roland*, a *chanson de geste* (song of heroic deeds) describing the slaughtering of Roland and his knights. The story is based on the real Battle of Roncevaux of 778, when the retreating army of Charlemagne was ambushed and the rearguard, to which Roland belonged, was decimated. The fact that there is an interval of approximately 8 centuries between the two texts guarantees that any results we obtain is genre-independent and not influenced by the styles characteristics of their time of composition.

This paper is organized as follows: In the first section, we briefly review the theoretical framework of character networks in literature while addressing the first question: how to choose a network. This is followed by an exposition of the method propounded in [26] to deal with time-dependent networks, thus addressing the second problem: to incorporate the dynamics in the system. We then present the results for the two books in question, *Alice's Adventures in Wonderland* and *La Chanson de Roland.* To better show the differences when dynamics is considered, we compare these results with the ones obtained from aggregate, time-independent networks of all characters of each text. We close the paper with some conclusions and perspectives on the use of network theory to literary texts.

2. Networks and Literature: Some Background

In this section, we discuss the theoretical aspects of networks in literature while addressing the problem of how to construct one. The theory of networks of characters, or *character networks* as it is called in the specialized literature was introduced by Moretti [19–21] and Woloch [30]. An extensive discussion of this theory is presented by Rochat in [23], who applied their ideas to study Jean-Jacques Rousseau autobiography *Les Confessions* from the viewpoint of networks. The interested reader should refer to this work for more details. Here, we concentrate on those aspects which are relevant to the present paper. According to Rochat, a character network represents '*relations between characters from a text. The relations are based on text proximity, shared scenes/events, quoted speech, etc.*' From a literary perspective, one starts by defining the concepts of *character-space* and *character-system* [30]. The character-space would be narrative environment of a given character and the character-system the union of all these spaces. When one picks out one particular network from the character-system, one ends up with character network, which can be represented as a simple or multiplex, time-dependent or dynamic network. This implies, of course, making choices as to what relations are important. One key point in the approach is the realization of the fact that *characters in any narrative are defined not by themselves, but as part of a web where each one helps define the others.* In this sense, a story really is a complex system. This is a matter-of-course fact in literary studies but here lies the link to the application of network theory to narratives: taking networks as the natural theoretical framework for studying the interdependence between agents that act through space and time, the main task one faces is how to disregard, in a principled way, aspects of narratives which might not render themselves amenable to a mathematical treatment. Then comes the question of how data should be collected, by actually reading the book or finding ways of automatically obtaining information via an adequate software.

Automated network extraction in a context-free way is one option [10, 23, 24]: one collects characters which are mentioned in a page and connect them all into one complete network, even if they do not interact. The procedure is repeated for the next page and then pages are fused: if edges between two characters appear in both pages, they are reinforced. New edges are added. This procedure is repeated for all pages, and at the end of the data collecting one is left with a network which resembles a complete network (everyone connected to everyone else). This is the approach favored by [23]. Such context-free method has its obvious limitations, as for instance placing edges where there might be none. Tolstoy's *Anna Karenina* provides us with a good example of the problems one might run into: the character Levin and the philosopher Schopenhauer appear repeatedly together, but to the reader it is clear that Levin speaks *of* Schopenhauer but not *to* Schopenhauer. Another problem is that this method, in our opinion, tend to attribute more importance to characters which are peripheral, or keep characters in a plot which at some point of the narrative disappear. Some ways of avoiding these difficulties or minimizing them have been discussed in [23]. This approach has its merits, for instance if one is interested in statistics, when questions regarding not one particular character of a book but literary styles or types of characters are the main focus. As one does not have the time or resources to read hundreds of books, automated extraction of networks have to be considered a serious alternative. Another approach is that of Kenna and Mac Carron [7, 8]. These authors improved on these ideas by connecting only those characters who actually meet at some point of the narrative. Of course, this requires reading the books and consequently the number of works that can be tackled is reduced. The character networks of their studies were drawn from a corpora of myths and sagas as the Irish *Táin Bó Cúailnge*, the Icelandic *Sagas* and the Greek *Iliad*. Moreover, being mainly narratives of conflicts, links were also given the attributes *friendly* and *hostile*. Without pretending to be exhaustive, these authors calculated a series of measures for the complete, friendly and hostile graphs and showed that some of the networks in these texts bear a striking resemblance to social networks from the real world, an indication that the stories they depict could have some elements drawn from real events [8]. A more detailed analysis of the *Iliad* beyond that of [7] can be found in [9]. In all the aforementioned studies, one does not differentiate the time in the narrative where the link is made, so there is no dynamics involved.

The importance of dynamics has been discussed by Agarwal *et al.* [1]. These authors studied *Alice's Adventures in Wonderland*, albeit in an online-accessible, abridged version (10 chapters instead of 12). The data extraction was automatic, but checked later by hand. Their main motivation is in line with multiplex networks, where different attributes of links changes the relevance of characters according to what is being observed. This can be accomplished by discriminating between uni/bi-directional and categories: interaction and observation. One character may observe another character but is not observed by them, as when Alice sees the Rabbit but is not seen by it. The way they introduced dynamics is by looking at some relevant measures chapter by chapter and comparing them. Taylor *et al.* devised in [26] a way of generalizing eigenvector-based measures usually studied in time-independent scenarios to temporal networks. These authors look at snapshots of the network at different time layers and treating inter-layer connections (same time) and intra-layer connections (different times) as being essentially different. For a system with N nodes and T time steps they construct an $N \times T$ block diagonal supra-centrality matrix $\mathbb{M}$ where each $N \times N$ diagonal block is the adjacency matrix (or any given function thereof) at a given time step $t = t_1, t_2, \ldots, t_T$. A parameter ε controls how strongly time layers are connected to each other. One may then use all the results from spectral analysis already known to work for aggregate networks. This ideas were tested with 3 networks: the exchange of Ph.D's in mathematics in the United States, the costarring networks of top-billed actors during the Golden Age of Hollywood and citations of decisions of the American Supreme Court [26]. The results presented in [1] represent a valuable approach towards incorporating some sort of dynamics into networks as well as in the way they attribute values to edges, but we believe that the methods of Taylor and coworkers capture the dynamics in a more consistent and rigorous way. We discuss in more detail the differences between these approaches in the next sections.

Since our main goal is to combine character networks with temporal networks, our choices of network building and time reckoning were the following. As for network extraction, our option follows closely that of Kenna and Mac Carron. As we are interested in multiplex networks in a *temporal* sense and not in a *attribute* sense, we consider an edge to exist whenever characters meet face to face but do not give them any attribute. This allows us to concentrate on the dynamical aspects of the plot without complicating to much on our analysis. Our choice of time layers was that of chapters, except in the case of *La Chanson de Roland*, as we explain below. This choice of chapters seem a reasonable one, as it was made by the author of the narrative and usually contains what one could describe as 'scenes' of a play. For most stories, the scene depicted in a chapter happens *after* the scenes of previous chapters, particularly in the case of *Alice*. As for *La Chanson de Roland*, one should bear in mind that it was actually recited for an audience and it was thus divided in stanzas or irregular sizes, called *laisses*.[a] There is a total of 247 to 272 stanzas, depending on the paper consulted. As the emphasis of the story is on the actions rather than on the characters and their introspection, we grouped some stanzas which described a particular scene — a gathering of Charlemagne and his knights or a battle, for example — into one single chapter. Stanzas describing the scenery were not considered. With this we ended up with 44 chapters, each one describing a scene.

Going back to the question we discussed in the introduction, we are interested in seeing whether temporal networks can serve as a tool to select characters according to their relevance to the plot. According to [16, 21] the narrative importance of a character can be evaluated by deleting it from the network and comparing to the network before deletion. As one character is removed, so are the links connected to it. The network becomes more sparse and some minor characters might become completely disconnected from the main narrative. The quantity which measures this difference is the so-called vitality and it will be discussed in the following section.

3. Temporal Networks: Some Definitions

In the theory of static networks, a number of measures have been developed to account for the importance of nodes [22, 29]. Many of these centrality measures can be expressed in terms of the eigenvector related to the largest eigenvalue of the adjacency matrix [4, 17, 26]. The adjacency matrix $\mathbf{A}$, for a network with N nodes is simply defined as the $N \times N$ matrix whose element ij is 1 if node i and node j are connected or zero otherwise. For a given network, one may simply count all existing edges, irrespective of when they actually were formed, and define a aggregate adjacency matrix. However, if this matrix is to more realistically represent some network which evolves in time, so that for instance a given edge between i

[a]It is generally believed that this division has to do with the time required for the bard to rest or play some instrument in between stanzas. For further discussion, on the composition of the song, see the introduction in [2].

and j appears at a latter instant of time when, say, nodes k and l are no longer extant, one has to find ways to represent how the adjacency of a given time changes from time layer to time layer. One way of doing this is to define a supra-adjacency matrix $\mathbb{M}$ that can be constructed in the following way [26]: let us assume time is discretized in T steps. Following [26], we define $\mathbb{M}$ as the $(NT \times NT)$-matrix

$$\mathbb{M} = \begin{bmatrix} \varepsilon \mathbf{M}^{(1)} & \mathbf{I} & 0 & \cdots \\ \mathbf{I} & \varepsilon \mathbf{M}^{(2)} & \mathbf{I} & \ddots \\ 0 & \mathbf{I} & \varepsilon \mathbf{M}^{(3)} & \ddots \\ \vdots & \ddots & \ddots & \ddots \end{bmatrix}, \qquad (1)$$

where each block diagonal $(N \times N)$-matrix $\mathbf{M}^{(t)}$ represents the adjacency matrix $\mathbf{A}(t)$ at a given time layer t $(t = 1, 2, \ldots, T)$. The choice $\mathbf{M}^{(t)} = \mathbf{A}(t)$ is appropriate for our purposes, but $\mathbf{M}^{(t)}$ can be a written as a more general function of $\mathbf{A}(t)$, as for example the hub $(\mathbf{A}\mathbf{A}^{\mathbf{T}})$ and authority $(\mathbf{A}^{\mathbf{T}}\mathbf{A})$ scores, where $\mathbf{T}$ denotes the transpose matrix [26]. The upper (and lower) diagonal identity matrices are introduced to guarantee that node i at a given time t is identified with itself at the next time step $t + 1$. The parameter ε controls how strongly a given node is coupled to itself between neighboring time layers. A value of $\varepsilon \to 0^+$ implies a strong correlation between time layers, meaning order-preserving aggregation. A value of $\varepsilon \to \infty$ implies the decoupling of layers. As shown in [26], the limits of small and large values of ε are well understood. The centrality trajectories depend on the choices for ε and there is not a clear interpretation for the intermediate regime. However, as pointed out by the authors, the limit $\varepsilon \to 0^+$ can still give valuable information. In this paper, only this latter limit has been explored.

Nodes (characters) are connected by an edge if they, at some point in the story, actually meet face to face. If one character talks about another character or thinks of him/her/it, no edge exists. No value or classification is given to links. Time is measured in chapters. For each chapter (time layer t), we calculate an adjacency matrix $\mathbf{M}^{(t)}$: if, say, characters i and j were connected in previous times but do not contact each other at time step t, the corresponding entry $M_{ij}(t)$ is set to zero. With these partial matrices, we build the supra-adjacency matrix $\mathbb{M}$ from which our analysis follows.

In fact, the leading eigenvector of $\mathbb{M}$ yields what one calls a joint node-layer centrality, since it reflects the centrality of both node (character) i and time layer (chapter) t. Since, in general, these quantities are ε-dependent and not easy to interpret, the trajectory of node centrality in time is given by the conditional node-layer centrality of the corresponding node-layer (i, t). The conditional node-layer centrality is the joint node-layer centrality normalized by the MLC of layer t. Here, MLC stands for *Marginal Layer Centrality*, obtained by summing the centrality of all nodes i in a given time layer t.

In the limit of $\varepsilon \to 0^+$ one expects a slow variation of the conditional node-layer centrality as function of time. This is what the authors in [26] call time-averaged centrality and it ranks nodes so that their centralities are constant in time. In all the analyses that follow in this paper, what we call centrality is exactly this time-averaged centrality. We should briefly comment on computational complexity here. The supra-centrality matrix given by Eq. (1) whose dominant eigenvector gives the joint node-layer centralities, has size $NT \times NT$, and that can be problematic for large networks with many time layers. Conversely, Taylor *et al.* have proved the time-averaged node centralities are given by the solution of an eigenvalue equation of rank N.

We are interested in this paper in the Freeman index of the network. Suppose an unweighted network of N nodes, one particular node j having the highest degree centrality $C_{\max}$ amongst nodes. The degree centrality is the number of edges connected to a given node. The Freeman index C_F is defined as

$$C_F = \frac{\sum_{i=1}^{N} [C_{\max} - C_i]}{(N-1)(N-2)}, \tag{2}$$

where the sum runs over all nodes with degree centrality C_i. The denominator can be explained as follows: the highest centrality a node j can have is in the so-called star graph, where all nodes are connected to j and to no other node. If this were indeed the case, the centrality of j would be $C_{\max} = (N-1)$ while $C_i = 1$ for $i \neq j$ and the sum in the numerator of the expression above would be

$$\begin{aligned} \sum_{i=1}^{N} [C_{\max} - C_i] &= 0 + (N-1) \times (N-1-1) \\ &= (N-1)(N-2). \end{aligned} \tag{3}$$

In our approach, we replace C_i by the eigenvector centrality c_i obtained by diagonalizing $\mathbb{M}$ in the limit $\varepsilon \to 0^+$ and multiply (2) by a factor $\sqrt{N(N-1)}$ which comes from the eigenvector normalization. The values of C_F vary between 0 and 1, where 1 corresponds to a star graph while 0 corresponds to a complete graph where all nodes have the same number of connections. Geometrically, one may interpret the Freeman index as an indicator of how close a graph is to a star graph. In a literary, context this means how 'centered' a story is on one particular character or more 'distributed' amongst characters. For example, we might expect a biography to have a Freeman index closer to 1, as the different persons would be mentioned mostly in their connection to the person being biographed and not in their relation to each other. Another interesting measure of the relative importance of a given node or its vitality. The vitality is a concept which measures how some structural properties of a network depend on a node and can be evaluated by deleting it from the network. In short, by removing a node i, another node j emerges as more prominent or is completely disconnected from what is left of the network. To the best of our knowledge, Moretti was the first to use it in character network analysis.

In a case study of Hamlet [21], he demonstrates the cohesive narrative role of a central actor in that play. Vitalities can be defined with respect to any real-valued measure $\mathcal{G}$ defined over a graph G. The vitality of node j is given by the difference between G calculated for the whole set of nodes and G for the set of nodes without j, namely

$$V(\mathcal{G}) = \mathcal{G}(G) - \mathcal{G}(G\backslash j). \tag{4}$$

By deleting a character, we mean that all edges attached to it are removed. In our particular case, we measure the vitality with respect to the eigenvector centralities of nodes.

Recently, Fenu and Higham [5] have argued that the way the supra-centrality matrix is constructed in Eq. (1) can be problematic when one considers directed graphs. This is particularly true in the case of centrality measures based on the concept of traversals through a network. In their study, they consider an alternative formulation where only the identity matrices of the upper diagonal are present. Since in our case when characters A and B meet, it is irrelevant who addresses whom, we proceed with the matrix to be given by (1).

4. The Chapter-by-Chapter Case

In what follows, whenever we refer to the books we are studying we will denote them by italics (*Alice*, *Roland*) whereas characters will be denoted by normal type (Alice, Roland). Before we present our results for temporal networks, it is interesting to ask whether they are really necessary. Most studies so far have combined all characters into one single (time-independent) network but according to [14], this can lead to inaccurate results. One could for instance think of a situation where one particular character, highly connected, dies at the beginning of the story, a fact that an aggregate network does not show at all. One option would be to go to the opposite extreme, by taking snapshots of the network at different instants of time and considering them as independent from each other (in the language of [26], this corresponds to the limit of $\epsilon \to \infty$). In order to better illustrate this idea, we study *Alice* as being actually 12 networks, one for each chapter. We calculate each character's eigenvector centrality at a given chapter, regardless of what happened before or happens after that given chapter. This is the approach also used by Agarwal *et al.* [1] in their study of *Alice*.

The result can be seen in Fig. 1 in the form a temperature map of characters' degrees. In spite of having calculated different centralities, the graphs of [1] are equivalent to this temperature map. For each chapter, an adjacency matrix is built and diagonalized. The leading eigenvector of this matrix carries the centralities of each character. The centralities are normalized chapter by chapter. The result is easy to interpret, as it is equivalent to the impression one gets from actually reading the book, but trying to imagine that each chapter is a different story. Note that chapter 1 has only three characters: Alice, her sister and the Rabbit, Alice having

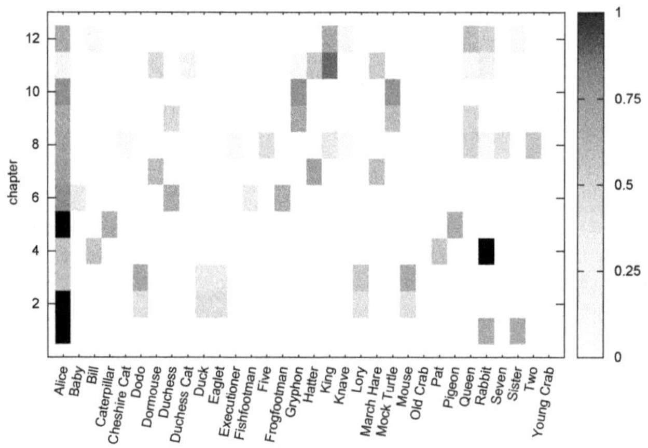

Fig. 1. Heat map of characters eigenvector centrality for each of the 12 chapters of *Alice's Adventures in Wonderland*. Centralities are normalized by chapter.

the highest degree centrality. In chapter 2, new characters appear while the Rabbit and Alice's sister disappear. The Rabbit shows up again in chapter 4, while Alice's sister reappears only at the end of the book, when Alice wakes from her dream. One may clearly see that chapter 8 has the largest number of characters and correspond to the point when Alice arrives at the Queen's croquet-ground. By looking at this image, one could conclude that the Rabbit appears to be more important than, say, the Queen, as he appears in more chapters. However, one should keep in mind that each centrality is relative to one particular chapter of the book. This means that in chapter 1, for instance, the Rabbit is being compared to only two other characters, namely Alice and her sister. So its centrality there cannot be compared to that of the Queen on chapter 8, which has many more characters. One should also bear in mind that subjectively one identifies the Rabbit in chapter 1 with the same Rabbit in chapter 8 and it seems natural to define an overall centrality by some sort of weighted or unweighted average. However, from a mathematical point of view, the characters appearing in each chapter have nothing to do with previous or posterior 'reincarnations' of their selves: rigorously they are completely different actors. What one needs is to find a mathematical way of identifying characters from chapter to chapter, thus giving the narrative an inner consistency which does not depend on subjective identification of characters and respects, at the same time, causality.

Previous studies on character networks [7–9, 21, 23] have used what we call the aggregate network, that is all connections made along a plot are considered together, irrespective of whether they are, at some point of the narrative, no longer extant. For the sake of completeness we show what the aggregate networks of *Alice* and *Roland*

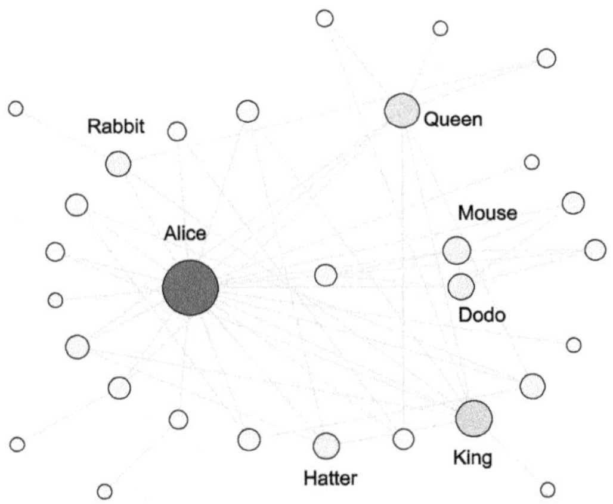

Fig. 2. The aggregate network of characters in *Alice*. Only some well-known characters are named. Two characters are not depicted (the young and old crab) as they do not relate to any other character.

look like in Figs. 2 and 7, respectively. All these studies however emphasize the importance of incorporating the dynamics of the stories in their studies, something that an aggregate network does not provide since the causal timeline is lost with the sum of partial results [14]. If one expects to get meaningful results, one must therefore consider the networks in their full time-dependence and identify a given character with itself as the story evolves. This is exactly what the method of [26] does, as we show in our case-studies of *Alice* and *Roland* in what follows.

5. The Strong Time-Coupling Limit

We consider now the dynamic case. As mentioned previously the limit of $\varepsilon \to 0^+$ corresponds to the strong coupling between time layers. In this limit, the method presented in [26] yields a centrality which is time-independent. To better see the difference in both cases, we plot in Figs. 3 and 4 the Freeman index calculated for *Alice* in the (a) aggregate and (b) dynamic case. The red horizontal line represents the Freeman index of the network. The dots above the name of each character (horizontal axis) represent the new value of this index when that particular character is taken out of the network. From these two images, it is clear that Alice, the Queen and the King are the characters which most affect the structure of the network. In the aggregate network, the King and Queen's importance surpasses that of Alice, as the relative change they cause in the Freeman index of the network is larger. Not only this changes in the dynamic case (Alice becomes more influential) but also the direction of change is different by the removal of Alice. To understand these differences, we first recall that a smaller Freeman index means that the network looks more like a complete graph while a higher Freeman index implies that the network

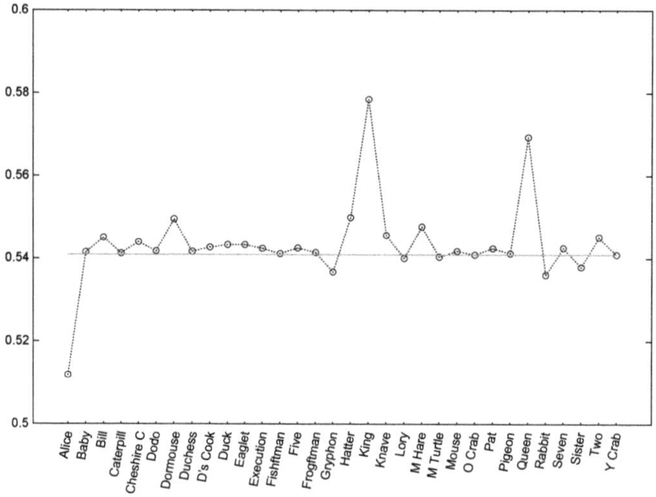

Fig. 3. (Color online) Freeman index calculated for the aggregate case (red line) of *Alice*. Dots represent the same quantity when a particular character is deleted (named on horizontal axis). The vitality is given by the deviates from the red line.

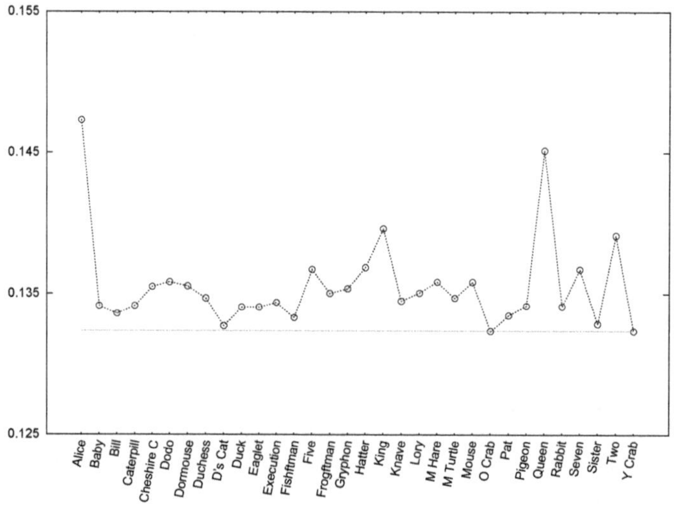

Fig. 4. (Color online) Freeman index calculated for the dynamic case (red line) of *Alice*. Dots represent the same quantity when a particular character is deleted (named on horizontal axis). The vitality is given by the deviates from the red line.

is more star-like. If one compares the aggregate network of Fig. 2 with the result depicted in Fig. 3, by removing the node Alice the network becomes less star-like, hence the drop in the Freeman index. In the dynamic case, depicted in Fig. 4, the change in the index is larger, when compared to the removal of the node Queen or node King. This means that the node associated with Alice is more relevant as her actions are distributed along the text and not concentrated on a few chapters at the end, as is the case of the Queen and King. Moreover, for the dynamic case, her removal makes the *average* network more star-like as the remaining networks keep shifting from being more centered on the Queen or the King.

Another way of seeing these results is depicted in Figs. 5 and 6, where we plot what the eigenvector centrality of all characters would be if some (the most central ones) were to be removed. One can read two things from this graph. On the one hand, it gives the centrality when everybody is in the story (red bars) and how these change as a certain character is removed from the story (cyan for a network without Alice, blue without the King and yellow without the Queen). By removing Alice, all characters have the largest gain compared to their previous values, which implies that Alice has the highest vitality (in both cases, aggregate and dynamic). However, in the dynamic case, the Queen once again assumes a more prominent role than the King, a fact verified above when considering the Freeman index of the network.

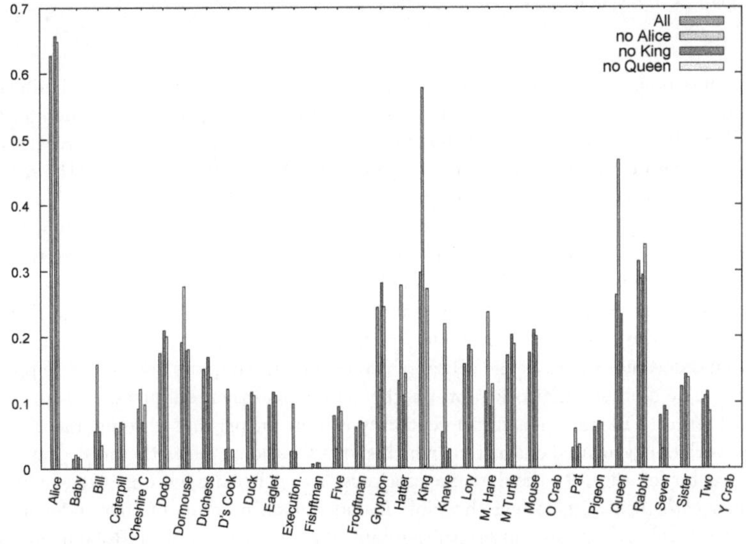

Fig. 5. Centrality of characters in *Alice* without some of its characters, as indicated by the colors of the bars. Aggregate case.

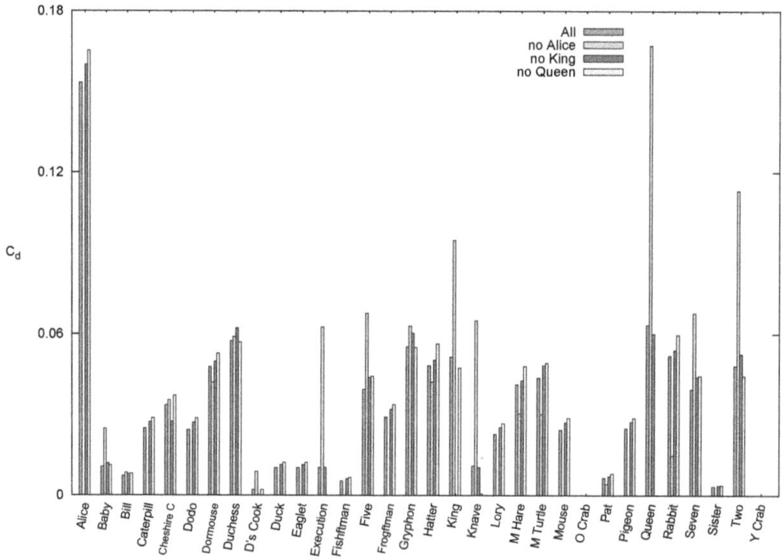

Fig. 6. Centrality of characters in *Alice* without some of its characters, as indicated by the colors of the bars. Dynamical case.

One could argue that *Alice*'s network is too small and changes are not significant. In order to validate the method in a larger data set, we applied the same ideas to *Roland*. In this case, the difference between aggregate (Fig. 8) and dynamic networks (Fig. 9) become more pronounced, as depicted in the figures below. For the sake of clarity, we also plot the whole aggregate network of Roland in Fig. 7.

For the aggregate network, if one deletes the node associated with Roland the graph becomes more star-like; one can see in Fig. 7 that the action shifts to well-defined star structure of Charlemagne, since Roland is not as well connected as Charlemagne. However, in the dynamic case the opposite is actually true: for the narrative, the absence of the node Roland makes the network more distributed. Roland does not figure as the center of attention, but subplots become more prominent. The dynamic network captures the actions and their subplots.

Figure 11 for the eigenvector centrality is more surprising: in comparison to the network of aggregated characters, which shows Charlemagne as the most connected character (cf. Fig. 10), the dynamic case reveals that Roland has on average a higher eigenvector centrality along the plot. If one thinks in terms of time-independent network this result is counterintuitive but what it means is that Roland, in spite of having less connections, maintains more of them active during the plot, thus playing a more prominent role as Charlemagne does.

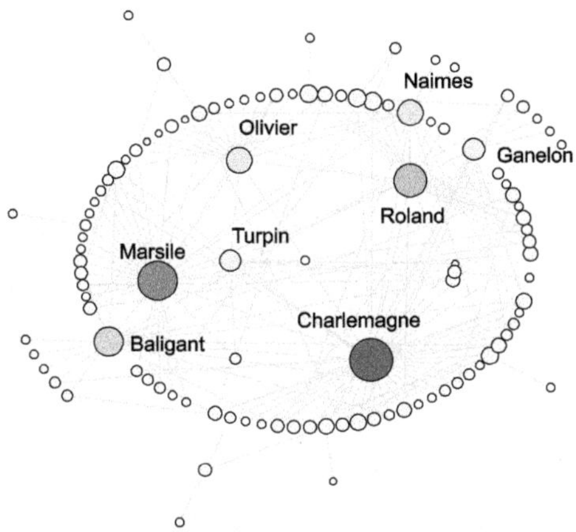

Fig. 7. The aggregate network of *Roland*. Only better known characters are indicated in the figure.

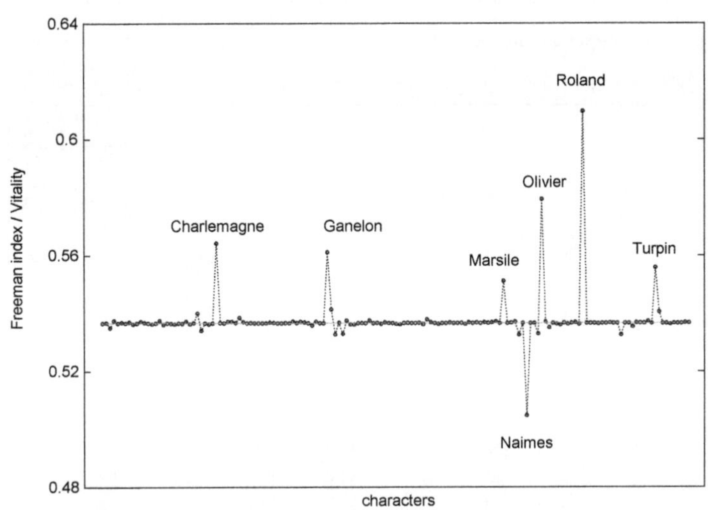

Fig. 8. (Color online) Freeman index of the aggregate network (red line). Dots represent the same quantity when a particular character is taken out (named on horizontal axis). The vitality is given by the deviates from the red line.

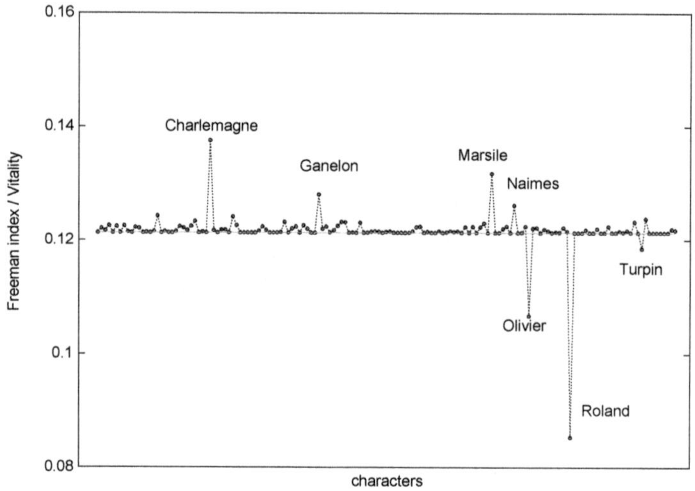

Fig. 9. (Color online) Freeman index of the dynamical network (red line). Dots represent the same quantity when a particular character is taken out (named on horizontal axis). The vitality is given by the deviates from the red line.

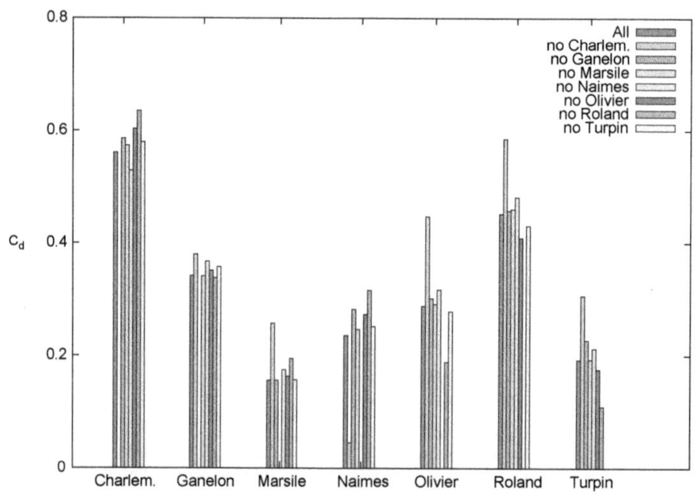

Fig. 10. (Color online) Eigenvector centrality of selected characters of *Roland* in the aggregate network. The blue bar represents the values obtained when all characters are present. The other bars are the values when some characters are removed from the network (color coded according to character).

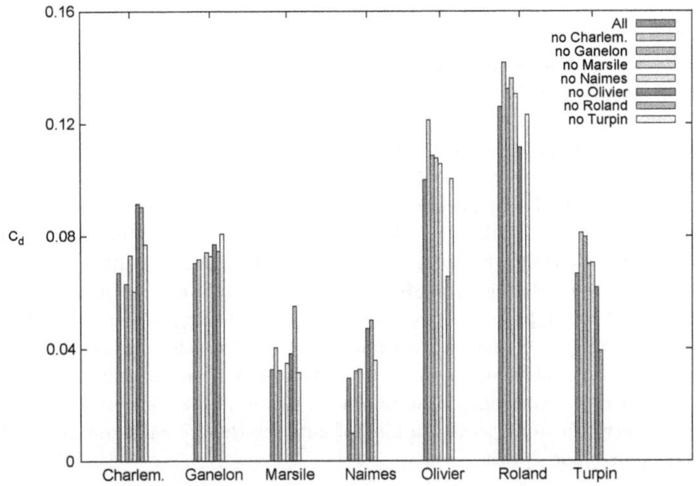

Fig. 11. Same figure as above but using [26] for the dynamic case. Note that in this case Roland becomes the character with highest centrality.

In short, by removing a node x, another node y emerges as more prominent for what is left of the network.

6. Conclusion

In this work, we applied the methods developed in [26] for finding eigenvector-based centralities of temporal networks and from these to determine the vitality and Freeman index of characters of two selected literary texts: Lewis Carroll's *Alice Adventures in Wonderland* and the anonymous epic poem *La Chanson de Roland.* These two quantities are calculated using the vector-based centrality of a character, which is given by eigenvector associated with the highest eigenvalue of a supra-centrality matrix $\mathbb{M}$ defined in [26]. It is worth emphasizing that this method can be applied to any centrality that can be expressed as a function of the adjacency matrix of the network [4].

Our results confirm the utility of this new approach, with results becoming more pronounced as the network becomes bigger. The application of $\mathbb{M}$ offers new insights into evolving stories, picking out the most important characters in a way that static networks do not. As for *Alice in Wonderland,* the differences between static and dynamic network are there, but not in such a pronounced way, as the story is rather short (12 chapters) and the number of characters small (32 in total). For *Roland,* however, the dynamic method selects Roland as the most relevant character, in contrast to the static case, where Charlemagne is the character with highest vitality and Freeman index. Extrapolating these ideas and techniques, we are left to wonder what kind of interpretations, answers or even predictions one would be able to make when applying this approach to model temporal networks in other stories. Any such approach would of course demand knowing what the 'right' questions to ask might be.

As for the amount of subjectivity associated with the construction of a network in literary context, there is no straightforward solution nor *the* right network. As pointed along the text, there are several networks which can be constructed from a set of nodes. The connections are contingent on the question one is trying to answer. This does not make network theory *a priori* an invalid tool for the analysis of complex relations between actors in a given setting. On the contrary, comparison between what network theory predicts and reality should be used as a criterium for the validity of this approach, as we hoped to have shown in the results we presented.

Finally, there exists an enormous amount of texts which could be analyzed. It is of course an impossible task to read all of them, but if one is able to devise a way of systematically gathering information from classics, network theory, in particular temporal networks, may provide a kind of network-theoretical signature to classify authors, genres and epochs.

Acknowledgments

A.L.C.B, S.R.D. and S.D.P. were supported by IRSES Grant Project PIRSES-GA-2011-295302. The hospitality of the AMRC in Coventry is gratefully acknowledged. PMC was supported by a European Research Council Advanced grant to R.I.M. Dunbar.d

References

[1] Agarwal, A., Corvalan, A., Jensen, J. and Rambow, O., Social network analysis of alice in wonderland, in *Proc. NAACL-HLT 2012 Workshop on Computational Linguistics for Literature* (Association for Computational Linguistics, Montréal, Canada, 2012), pp. 88–96, Available at http://www.aclweb.org/anthology/W12-2513.

[2] Anonymous, *The Song of Roland* (Hackett Publishing Company, Indianapolis and Cambridge, 2012).

[3] Barrat, A., Barthélemy, M., Pastor-Satorras, R. and Vespignani, A., The architecture of complex weighted networks, *Proc. Natl. Acad. Sci.* **101** (2004) 3747–3752.

[4] Benzi, M. and Klymko, C., On the limiting behavior of parameter-dependent network centrality measures, *SIAM J. Matrix Anal. Appl.* **36** (2015) 686–706.

[5] C. Fenu and Higham, D. J., Block matrix formulations for evolving networks, arXiv:1511.07305v1.

[6] Carroll, L., *The Annotated Alice: Alice's Adventures in Wonderland & Through the Looking Glass* (Clarkson N. Potter, New York, 1960).

[7] Mac Carron, P. and Kenna, R., Universal properties of mythological networks, *Europhys. Lett.* **99** (2012) 28002.

[8] Mac Carron, P. and Kenna, R., Network analysis of the íslendiga sögur — The sagas of the icelanders, *Eur. Phys. J. B* **86** (2013) 407.

[9] Kydros, D., Georgios, E. and Notopoulos, P., Homer's iliad — A social network analytic approach, *Int. J. Humanit. Arts Comput.* **9** (2015) 115–132.

[10] Elson, D. K., Dames, N. and McKeown, K., Extracting social networks from literary fiction, in *Proc. 48th Annual Meeting of the Association for Computational Linguistics (ACL 2010)* (Association for Computational Linguistics, Uppsala, Sweden, 2010), Available at http://aclanthology.info/events/acl-2010.
[11] Eagleton, T., Literary Theory. An Introduction (The University of Minnesota Press, Minneapolis, 2003).
[12] Freeman, L. C., Centrality in social networks: Conceptual clarification, *Social Netw.* **1** (1979) 215–239.
[13] Gramsch, R., Das Reich als Netzwerk der Fursten (Jan Thorbecke Verlag, Ostfildern, 2012).
[14] Holme, P., Modern temporal network theory: A colloquium, arXiv: 1508.01303.
[15] Holme, P. and Saramäki, J., Temporal networks, *Phys. Rep.* **519** (2012) 97–125.
[16] Koschützki, D., Lehmann, K., Peeters, L., Richter, S., Tenfelde-Podehl, D. and Zlotowski, O., Centrality indices, in *Network Analysis*, Brandes, U. and Erlebach, T. (eds.), *Lecture Notes in Computer Science*, Vol. 3418 (Springer, Berlin/Heidelberg, 2005), pp. 16–61.
[17] Solá, L., Romance, M., Criado, R., Flores, J., del Amo, A. G. and Boccaletti, S., Eigenvector centralities of nodes in multiplex networks, *Chaos* **23** (2013) 033131.
[18] Düring, M. and Stark, M., Historical network analysis, in *Encyclopedia of Social Networks*, ed. Barnett, G. (Sage Publishing, London, 2011).
[19] Moretti, F., Conjectures on world literature, *New Left Rev.* **1** (2000) 54–68.
[20] Moretti, F., *Graphs, Maps, Trees: Abstract Models for a Literary History* (Verso, 2005).
[21] Moretti, F., Network theory, plot analysis, *New Left Rev.* **80** (2011).
[22] Newman, M., *Networks: An Introduction* (Oxford University Press, New York, 2010).
[23] Rochat, Y., *Character Networks and Centrality*, Ph.D. thesis, Université de Lausanne (2014).
[24] Sack, G., haracter networks for narrative generation: Structural balance theory and the emergence of proto-narratives, in *2013 Workshop on Computational Models of Narrative, OASICS*, Vol. 32 (Leibniz-Zentrum für Informatik, Dagstuhl, 2013).
[25] Saramäki, P. H. P. J., *Temporal Networks: Understanding Complex Systems* (Springer Verlag, Berlin and Heidelberg, 2013).
[26] Taylor, D., Myers, S. A., Clauset, A., Porter, M. A. and Mucha, P. J., Eigenvector-based centrality measures for temporal networks, *CoRR* (2015), arXiv: 1507.01266v2.
[27] Truby, J., *The Anatomy of Story* (Faber and Faber, New York, 2007).
[28] Waerzeggers, C., Social network analysis of cuneiform archives: A new approach, in *Documentary Sources in Ancient Near Eastern and Greco-Roman Economic History: Methodology and Practice*, Baker, H. and Jursa, M. (eds.) (Oxbow, Oxford, 2014).
[29] Wasserman, S. and Faust, K., *Social Network Analysis: Methods and Applications*, Vol. 8 (Cambridge University Press, 1994).
[30] Woloch, A., *The One vs. The Many: Minor Characters and the Space of the Protagonist in the Novel* (Princeton University Press, 2003).

OPEN ACCESS
Advances in Complex Systems
Vol. 24, No. 2 (2021) 2150006 (37 pages)

DOI: 10.1142/S0219525921500065

World Scientific
www.worldscientific.com

DETECTING OPINION-BASED GROUPS AND POLARIZATION IN SURVEY-BASED ATTITUDE NETWORKS AND ESTIMATING QUESTION RELEVANCE

ALEJANDRO DINKELBERG*
MACSI, Department of Mathematics and Statistics,
University of Limerick, Limerick, V94 T9PX, Ireland
Centre for Social Issues Research,
University of Limerick, Limerick, V94 T9PX, Ireland
alejandro.dinkelberg@ul.ie

DAVID JP O'SULLIVAN
MACSI, Department of Mathematics and Statistics,
University of Limerick, Limerick, V94 T9PX, Ireland
david.osullivan@ul.ie

MICHAEL QUAYLE
Centre for Social Issues Research,
University of Limerick, Limerick, V94 T9PX, Ireland
Department of Psychology, School of Applied Human Sciences,
University of KwaZulu-Natal, Pietermaritzburg, South Africa
mike.quayle@ul.ie

PÁDRAIG MACCARRON
MACSI, Department of Mathematics and Statistics,
University of Limerick, Limerick, V94 T9PX, Ireland
Centre for Social Issues Research,
University of Limerick, Limerick, V94 T9PX, Ireland
padraig.maccarron@ul.ie

Received 20 May 2021
Revised 16 August 2021
Accepted 14 September 2021
Published 29 November 2021

*Corresponding author.

Networks, representing attitudinal survey data, expose the structure of opinion-based groups. We make use of these network projections to identify the groups reliably through community detection algorithms and to examine social-identity-based groups. Our goal is to present a method for revealing polarization and opinion-based groups in attitudinal surveys. This method can be broken down into the following steps: data preparation, construction of similarity-based networks, algorithmic identification of opinion-based groups, and identification of important items for community structure. We assess the method's performance and possible scope for applying it to empirical data and to a broad range of synthetic data sets. The empirical data application points out possible conclusions (i.e. social-identity polarization), whereas the synthetic data sets mark out the method's boundaries. Next to an application example on political attitude survey, our results suggest that the method works for various surveys but is also moderated by the efficacy of the community detection algorithms. Concerning the identification of opinion-based groups, we provide a solid method to rank the item's influence on group formation and as a group identifier. We discuss how this network approach for identifying polarization can classify non-overlapping opinion-based groups even in the absence of extreme opinions.

Keywords: Attitude networks; opinion-based groups; community detection; survey analysis; polarization; data mining.

1. Introduction

Shared opinions are an important feature in the formation of social groups [34]. It has been shown that clusters of opinions become signifiers of group identity [11]. In recent studies, public health opinion groups have been shown to coalesce around a growing trust/distrust in science, with those having distrust being less compliant with regards to hand-washing and maintaining distance [39]. This sort of behavior has major consequences for public health compliance [47]. As a result, it is important to be able to identify such groups accurately, and to reveal if different opinion-based groups are, or will become, polarized on the clusters of topics they share.

In online communities, such as Facebook groups or subreddit memberships, mutual interests in a subject, or attitudes, are often the primary shared commonality, rather than prior acquaintanceship or geographical proximity. It has been found that in many online communities, users tend to share media aligned with their own values and dismiss alternative views [7]. These groups tend to be driven by homophily [41]. In this study, we will use this idea of shared attitudes to uncover opinion-based groups from surveys. In a survey, a participant provides responses on many topics with only a small number of possible response options. These responses are typically on an ordinal scale (e.g. a Likert scale) [17]. Often the scale has a limited range of discrete response options, for example five-point and seven-point scales are commonly employed [27]. We use a distance metric, akin to the Manhattan distance, on these scales across all the survey questions, referred to as items, to identify participants with similar opinions. Thus, we construct a network of participants linked by shared opinions. This method is discussed in more detail in Ref. 38.

There are methods for investigating the social structure and belief networks, see for example [12, 13]. However, in these methods the network is constructed by placing edges between pairs of nodes with a correlation above a certain threshold. This makes the edges difficult to interpret and the choice of threshold is arbitrary. In our approach, we introduce a cut-off when a giant component is formed containing almost all participants. An edge represents shared agreement; the stronger the weight of the edge, the more agreement between these participants. We use community detection techniques to identify clusters of participants with similar opinions, i.e. opinion-based groups. We compare this to statistical methods, such as hierarchical clustering on the refined survey data and show they give consistent results with each other and, hence, this is a viable method for detecting opinion-based groups and polarization.

Surveys can contain hundreds of items, many of which are not expressing attitudes but answering trivial questions leading up to an attitudinal item. We aim to select attitudes, which are closely linked to attitudes of the identified clusters. To do this, we apply two feature selection methods to either identify or rank the most relevant items. Ranking the items allows us to reduce the number of items and to highlight influential items.

In this paper, we lay out a novel approach on how to remodel attitudinal survey data in order to identify opinion-based groups, applying three different community detection algorithms. We present a new item rank method, which ranks and identifies the items' importance for an opinion-based group structure. Finally, we demonstrate how to apply this approach to existing data sets. The paper is laid out as follows, in Sec. 2, we outline the method for forming the networks, identifying the clusters and the feature selection methods for picking the relevant questions. In Sec. 3, we show the results and identify the community detection algorithms as robust methods for detecting the opinion-based groups in similarity networks. Finally, in Sec. 4, we discuss the results, give concluding remarks and discuss further research avenues.

2. Methods

The detection of opinion-based groups using survey data is conducted in a multi-step procedure. This can be broadly broken down into data restructuring, construction of similarity-based networks [38], community detection [25, 33, 42] and item importance. Based on survey data, the method creates links between individuals by constructing a similarity-based network. The emergent structure can reveal opinion-based groups and predict social-group formation [6, 14]. Once we detect these opinion-based groups, our approach provides a method to rank each item's influence on the opinion-based group structure.

Survey data often cover multiple contexts with a large number of items. Hence, a subset of items has to be chosen depending on the subject matter of interest. For example, if we focus on political polarization, then we are interested in identifying politically relevant items which cover attitudes related to party alignment. To uncover these attitude connections, we employ a method to project survey data as a similarity network based on the answers of participants. In the resulting network, nodes are participants and weighted links are the similarity scores between participants.

Reference 38 shows that the network visualization provides information about groups of individuals that share similar opinions. However, the visualization and the distinction of groups in the network is highly dependent on layout algorithms, chosen by the user (in [38], the Kamada–Kawai layout algorithm [32] was used). A common way in complex networks to partition a graph is the application of community detection algorithms [22]. The use of community detection algorithms in our context has the benefit that they do not rely on the visual inspection of the network and that it takes the approach one step further: to reliably uncover opinion-based groups.

Over the last two decades, a range of different community detection algorithms have evolved (see, for example, [8]). Based on the high complexity of this challenge, there exists no generally applicable algorithm [22]. The choice of using three distinct algorithms ensures the performance and robustness of the community detection. Initially, we apply the Girvan–Newman algorithm [25], which uses the edge betweenness centrality to minimise the cross-cutting links between communities. We then run the statistical-driven Hierarchical Clustering algorithm [42] and finally the Stochastic Block Model used for community detection [33].

In the following sections, we explain our approach step by step. Although we will later show, using extensive simulations, that our method provides robust results, in order to illustrate the application we first run through a specific example: the American National Election Study (ANES) from 2016 [4]. This large data set captures a broad range of general and political attitudes from the American people and includes over 4000 participants and more than 650 items. We aim to detect opinion-based groups and polarization in the data set. As an example the ANES data set delivers an ideal candidate to reconfirm polarization. Although the ANES data set is not intended to reveal opinion-based groups or polarization, it captures the particular structure of the American two-party system, which is perceived as bipolar [3, 21, 30]. We take this party alignment as a reference group for community detection and polarization in this data set. For our method, the ANES data set is suitable to investigate polarization [9]. The first step is to project the survey data as a network.

2.1. *Identifying opinion-based groups from survey data: A score-based linking method*

Attitudinal survey data provides the basis for a network, using the individuals as nodes and their similarity score as links. The scales of the items are reformatted into a range between −1 and 1. For instance, a seven-point scale will then be defined as a scale with values of −1, −2/3, −1/3, 0, 1/3, 2/3 and 1. The scale represents a clear ordinal structure. The reformatting is applied to the whole data set.

The similarity measure S_{ij} between the individuals, i and j, is the sum of differences between all n_f answers to the items q_n (i.e. the Manhattan).

$$S_{ij} = n_f - \sum_{f=1}^{n_f} |q_{if} - q_{jf}|. \tag{1}$$

The similarity measure ranges between $-n_f$ and n_f. This is to support comprehension, so that an edge with a value of n_f displays full agreement between two nodes on all items. The similarity measure S is at its maximum n_f, if two individuals have identical responses to all items. Links are drawn, where the similarity exceeds a threshold θ, which is chosen when a giant component is formed. Its success criterion is fulfilled if there are enough links in the network to build a giant component, where a path can be drawn to at least 80% of the individuals. To achieve this, the threshold will successively be lowered until the network matches the success criterion. This means that the procedure includes stepwise links with a lower similarity score. While the integration of the threshold reduces the number of included individuals, it also reduces the total number of additional links. After these three steps, the data can be shown as a network in order to identify opinion-based groups.

For the ANES data set, we identified eight items based on a study from [40] to measure political attitudes. Malka and colleagues' evaluations rely on cultural, economic and self-reported political ideology attitudes. We then run the data refinement and the network construction on these eight selected items (see Table 1).

We removed individuals who did not answer all eight items and so our maximum network size here is 3,081 nodes. With a threshold of 7.0, we get 50,143 links between 2,714 individuals, forming a giant component (88.1% of individuals), where all individuals are connected (see Fig. 1). In our next step, we introduce the community detection for identifying possible opinion-based groups in our network.

2.2. *Detecting opinion-based groups*

Community detection in graphs is an active and ongoing field of research in and of itself, see for example [22, 25, 31]. Currently there exist a large variety of algorithms to detect group structure from network characteristics [22, 23].

In our analysis, we have chosen three different approaches: Girvan–Newman community detection, Hierarchical Clustering and the Stochastic Block Model. The Girvan–Newman algorithm is a network-based method, which is directly applied to our constructed network. In general, Hierarchical Clustering is applied on the refined

Table 1. American National Election Study 2016 — Selected item and their answer range. For a detailed description of the items, see Appendix D.

Item	Label	Answer range
Abortion	V161232	1–4
Race relations	V161198	1–7
Immigration	V161192	1–4
Welfare	V161209	1–3
Homosexuality	V161231	1–3
Business	V161201	1–7
Guns	V161187	1–3
Income	V161189	1–7

Fig. 1. (Color online) American National Election Study data 2016, constructed similarity network from the refined data set. The nodes' color marks the self-identified party affiliation: Republican (red), Democrat (blue) or unknown/independent (yellow).

data set. The Stochastic Block Model is an inference algorithm which detects communities by model fitting. Detailed descriptions of these can be found in the Supplementary Information.

2.2.1. *Within sum of squares*

The Within Sum of Squares (WSS) forms a building block of multiple parts of this analysis, for example, comparing the identified communities of the community detection methods. It is the sum of the squared distance of each individual from their assigned cluster centres. We can calculate the WSS as follows:

$$\text{WSS} := \sum_{k=1}^{n_k} \sum_{i \in C_k} \sum_{f=1}^{n_f} (q_{if} - \bar{q}_{kf})^2, \tag{2}$$

where the number of clusters is n_k. C_k is the set of individuals in cluster k, q_{if} is individual i's response to item f and $\bar{q}_{kf} = \sum_{i \in C_k} q_{if}/|C_k|$ is the average answer to item f in cluster k. The goal of our three community detection methods is to reduce WSS substantially while using the least number of communities possible. For two different community assignments, but with the same number of communities, the community with the lower WSS fits better to the data. The preferred partition contains communities, where, on average, the distance between individuals to others in their community is smaller.

With the WSS, we can generate an elbow plot (see Appendix C) for the communities, determined by our community detection methods. An elbow plot displays the WSS versus the number of communities and gives information about the ideal number of communities in the data [49]. We could use the number of parties that people self-identify as our optimal number of communities. However, the optimal community structure could contain sub-groups within these partisan groups; for instance, we might observe that the community structure is well explained by Republicans or Democrats that are "centralist", or "left" and "right" of the centre, for instance. The elbow plot provides a method for exploring this optimal number of communities. The "elbow" in the plot indicates a striking mark for the curve. Successively adding clusters to the data will reduce the total WSS. If the reduction is exceptionally high for an additional cluster, it gives the hint that this might be the ideal number of clusters for the data [10]. In this way adding more clusters to the data will lead to comparatively small changes in the curve (see Fig. C.1).

2.2.2. *Girvan–Newman algorithm*

The Girvan–Newman algorithm was one of the first community detection algorithms in complex networks [25]. This top-down approach divides the network into communities by successively removing links with the highest edge betweenness centrality. Using the edge betweenness centrality, the algorithm intends to identify the community bridging links. It is based on the assumption that links between communities have a higher edge betweenness centrality, caused by their linking ability.

Once the cross-cutting links are identified, they relate two opinion-based groups and mark an attitudinal intersect. The nodes holding these links are positioned at the border of the opinion spaces. Hence, links with a high edge betweenness centrality mark regions, where the participants are in the middle of two opinion spaces. The Girvan–Newman algorithm detects and removes cross-cutting links, conceptually dividing the opinion space into smaller, more internally intertwined opinion-based groups. Also, the edge betweenness centrality is used to measure within community polarization [24].

Our goal is to detect polarization in the ANES data set from 2016. We run the Girvan–Newman algorithm on the constructed network until it splits into two communities. In order to obtain a statement about the overall structure, we recompute the Girvan–Newman community detection on the biggest community if the first division results in one very small community (less than 5% the size of the larger community). Applying the Girvan–Newman algorithm to the ANES data set resulted in communities of 41 nodes and 2673 nodes. Running the algorithm again on the larger community, only 190 edges needed to be removed in order to split this community into two communities of 1818 and 855 nodes (see Fig. 2). It can be seen that the larger of these components corresponds mostly to democrats (blue nodes) and the smaller to republicans (red nodes).

2.2.3. *Stochastic block model for community detection*

An algorithm for broad range of applications to produce a model to generate networks with community (block) structure is called the Stochastic Block Model [29].

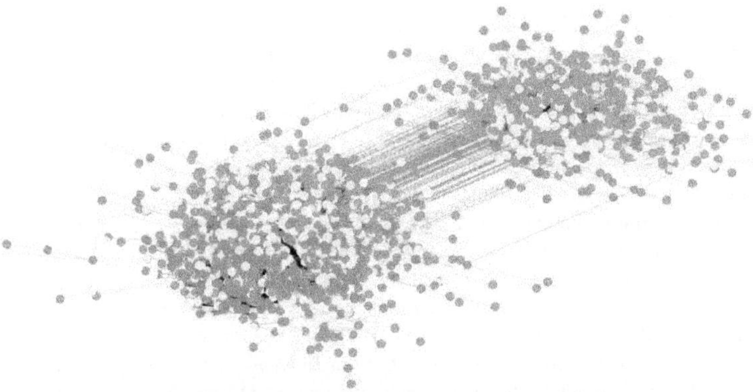

Fig. 2. (Color online) American National Election Study data 2016, constructed similarity network from the refined data set. With help of the Girvan–Newman algorithm the network is separated into two communities. The purple links are the eliminated links between the communities, and are not part of the network anymore.

The model, based on statistical inference, describes the link formation as a process that takes place more often within than between communities. The community detection is viewed as a challenge of fitting the Stochastic Block Model to a network in order to reveal a probability-based community structure. Based on this, through an integrated optimization process a suitable Stochastic Block Model candidate is selected. The flexibility of the Stochastic Block Model means that there exist a variety of approaches for applying and configuring it [23]. Besides the flexibility, another advantage is the computational complexity in $O(N \ln N)$ [23], and therefore the speed of execution is fast compared to the Girvan–Newman algorithm. One drawback of this method, in comparison to the Girvan–Newman algorithm and the Hierarchical Clustering method, is that is built on stochastic computation. Multiple runs of this method may yield different communities for the same network. It is also not guaranteed that the result is the optimal solution. Nonetheless, Fortunato and Hric [23] assess the Stochastic Block Model as a strong candidate for community detection.

In our approach, we use an algorithm in the Python module *graph-tool* [44]. The algorithm provides a degree-corrected or a non-corrected version, which, in our case, will be run according to the minimal description length criterion as described in [45]. This function uses an agglomerative heuristic, the Markov Chain Monte Carlo algorithm, for optimization [43]. The core of the function is a one-dimension minimization based on the golden section search. More details about the algorithm and its variants can be found in Refs. 33, 43, 45.

2.2.4. *Hierarchical clustering*

The Hierarchical Clustering method is applied directly to the data set, i.e. without constructing a similarity network. The core of analysis is a distance matrix which contains every distance between the individuals. The distance projects the dissimilarity in their answers over all items. In an iterative process the Hierarchical Clustering merges individuals by aggregating the most similar clusters together, which is decided by a linkage function. In this case, we use a "group average" linkage function [20]. This computes the average of the distance between people in different clusters and aggregates the closest. This leads to the interpretation that the people's clusters who are, on average, close in their opinion are aggregated together initially. Additionally, it is considered to be a intermediate version of the single and compete linkage methods and is relatively robust to outliers. The more common Ward linkage function, which aggregates clusters together that increase the within cluster sum of squares the least, tends to find spherical clusters and is sensitive to outliers [20]. However, other linkages functions produce similar clustering results.

The comparison of the three community detection methods arises from the need to choose the ideal number of communities. One approach is to compute a measurement which takes the distances of the answers in each community, the WSS, into account. The WSS makes it possible to quantify the variability between individuals for a given community assignment. With it, we are able to compare the three methods and, additionally, decide which is the ideal number of communities.

2.3. *Selecting relevant items*

Selecting relevant items from large data sets is an important component of our method. Often, to reduce complexity and to include only relevant items, a selection step for the items must be made *a priori*. Therefore, a tool for distinguishing between influential and noisy items would be beneficial to assess the item selection and moreover, rank them in relation to their influence on opinion-based group structure.

The responses of the survey data constitutes a corresponding vector of opinions for every individual. The differences in their responses form our network structure and opinion-based groups. After the determination of community assignments, we introduce a measurement to locate the relevant items for this particular community structure. Thus, every item is ranked by its meaningfulness.

The basic concept consists of randomly selecting an item from a data set, shuffling the responses and reallocating them to the individuals. Through this, we break possible correlations to other items and influence on the community assignment if one exists. The method is build up as follows:

(1) The WSS is calculated to obtain a reference value. The calculation is based on the Girvan–Newman, Stochastic Block Model or Hierarchical Clustering community assignments.

(2) At random the method chooses one item and modifies the data set. Consequently, all features are like in the original data set but answers of the selected item are now shuffled.
(3) On the basis of the community assignment, a new WSS is computed. In an additional step, we calculate the ratio of the difference between the old and the new WSS.
(4) To make a reliable statement about the item ranking, the procedures in (2) and (3) is repeated M times per item. In the end, the mean of all WSS-differences is taken to assess each item.
(5) Finally, a value for each item determines the average percentage change of the WSS. Whereas, a higher value means higher influence on the community assignment and values near zero suggest no influence on community assignments.

The results of the method can be used to produce a violin plot[a] (see Fig. 3).

Following our example, we computed the item rank method to evaluate the items influence on the community assignments. We ran our method on the eight items from the ANES data set 2016 and simulated it 1000 times, so that, on average, each WSS-distance distribution is based on 125 shuffles of that item. It shows that the item *Welfare* ($V161209$) had the highest and the item *Immigration* ($V161192$) the lowest influence. As a comparison for our item rank method we test it against two other methods of feature selection, the *Random Forest* classifier [15], and *Boruta* [36].

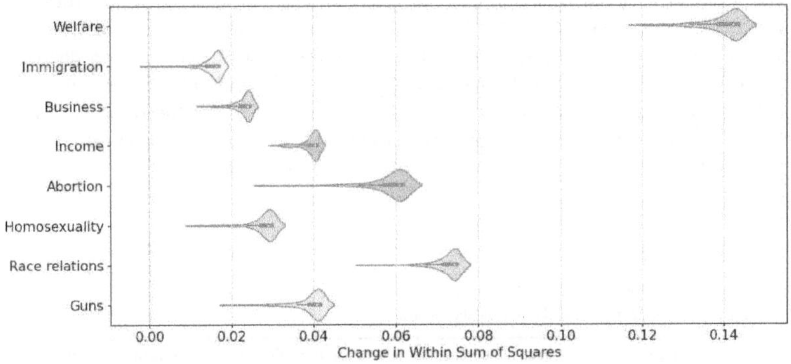

Fig. 3. Item selection: Violin plot for the eight items from the ANES data set 2016. It shows the distribution of the average percentage change between the original WSS and the recalculated WSS, in the case of shuffling the items.

[a] It works similar to a common box plot: it marks the median for the WSS-difference for each item, displays the interquartile range and it draws the distribution for WSS-differences using a kernel density estimation.

Random Forests are a substantial modification of the classification trees method that attains near state-of-the-art performance for classification across a wide range of data sets [2, 16]. A Random Forest model is formed from an ensemble of classification trees, where the trees are constructed so they are uncorrelated with each other. A new data point is classified in the model by checking the class that each of the classification trees gives and taking the majority vote of these. The Random Forest model also natively provides item importance measures that can be used to rank the importance of items to the opinion group classification. Please refer to Appendix B.1 for further details along with useful references to consider when implementing Random Forests.

Boruta is another feature selection method that builds on the Random Forests classifier. It is noted for tackling the "all-relevant" problem, where, as the name suggests, we seek to find all features that are relevant for the model's ability to classify the opinion-based groups. Several studies have used it successfully as a feature selection tool in a wide range of areas from Fisheries' management [18] to gene expression [35]. It is a wrapper for the Random Forest algorithm, where it uses a statistical test to identify items that are confirmed to be important, unimportant or undetermined. We are concerned with those items that are deemed to be important to the opinion-based groups under study here. Please refer to Appendix B.2 for further details.

The results of the feature selection for the eight items are shown in Table 2. They are also used in the violin plot (see Fig. 3) and represent here the average change in the WSS for every item. The second column (Random Forest) shows the values to assess the rank of each item. Evidently, it also ranks *Welfare* and *Race relations* as the two most important items but differs in the rest. The Boruta method defines 7 out of 8 items as important for the community split-up, and validates therefore the selected items for the community detection. Additionally, like the Random Forest method it ranks the item *Gay marriage* as the least important item, whereas the item rank method evaluates *Immigration* as the least important one.

Table 2. Results for the feature selection by the item rank method, random forest classification and Boruta. The methods were applied on the selected features from the ANES data set 2016, and based on the community detection from the Girvan–Newman algorithm.

Item	Variable	Item rank	Random Forest	Boruta
Welfare	V161209	0.140	0.279	Important
Race relations	V161198	0.072	0.202	Important
Abortion	V161232	0.058	0.134	Important
Gun control	V161187	0.039	0.062	Important
Income	V161189	0.039	0.162	Important
Gay marriage	V161231	0.028	0.030	Undetermined
Business	V161201	0.023	0.084	Important
Immigration	V161192	0.016	0.046	Important

3. Results

In this section, we validate the previously described methods on synthetic data, expand the analysis to new data sets and discuss how to apply it to consecutive data sets.

3.1. *Data sets*

3.1.1. *Synthetic data sets*

In this section, we compare the three different community detection algorithms and under which circumstances they can be applied. The results will show that the detection of the opinion-based groups is not an artefact of just one community detection method. To test our approach, we use simulated survey data, where we specify the ground truth for who belongs to each opinion-based group. Additionally, by building in items that are stronger, weaker or no predictors of group membership, we test an item rank method to evaluate the items' influence on the community structure. This will provide sound footing for its performance when applying it to real-world data sets. The application to synthetic data sets will reveal, due to gradual variations of their parameter, the effectiveness of the presented approach. It will also show that the performance of the community detection aligns with our determination of the predefined groups (see Appendix E).

In the simulated data sets we fix the size of the network, the number of items and the group membership of each individual. An individual's answer to each item is simulated by drawing from a normal distribution with mean μ_a if they are in group a and μ_b if they are in group b. The standard deviation is the same for the sake of simplicity. The μ-distance, $|\mu_a - \mu_b|$, is a measure of the maximal difference the two groups are on an item (see Fig. 4).

Community detection

The synthetic data sets have 100 individuals with responses to 7 items on a scale from 1 to 7. The community structure is an equal division into two groups of size 50. The items are ranked in four different categories of information content about the group structure, determined by an increasing standard deviation. We consider values of the μ-distance from 0.6 to 6.0 (y-axis) and the values of standard deviation from 0.3 to 3.0 (x-axis), both with a step size of 0.3.

For every parameter combination we simulate 30 data sets to which we apply the community detection algorithms. The heatmaps in Fig. 5 show the percentage of correctly allocated nodes from the network by the community detection algorithms, averaged over the 30 simulations. The heatmaps delineate two regions. The dark blue region is where the community detection works reliably. In this region, there is a small overlap in the distributions of item responses for each group due to the high μ-distance and the low standard deviation. In the light blue region a lower μ-distance and a higher *standard deviation* leads to a larger overlap of the responses in the two communities.

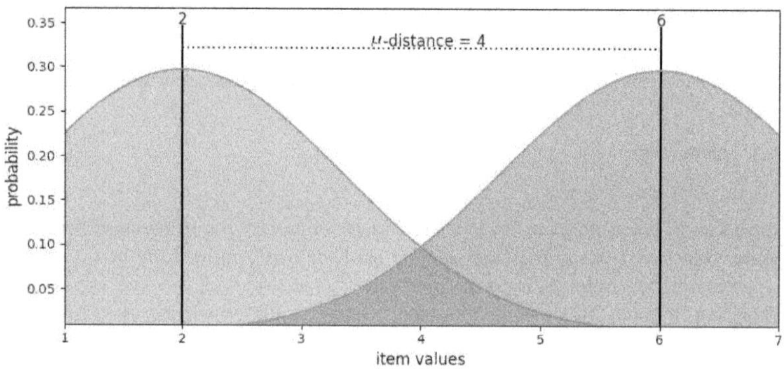

Fig. 4. The μ-distance for an item. Group a has mean $\mu_a = 2$ and group b has mean $\mu_b = 6$. The standard deviation in both cases is $\sigma = 1.4$.

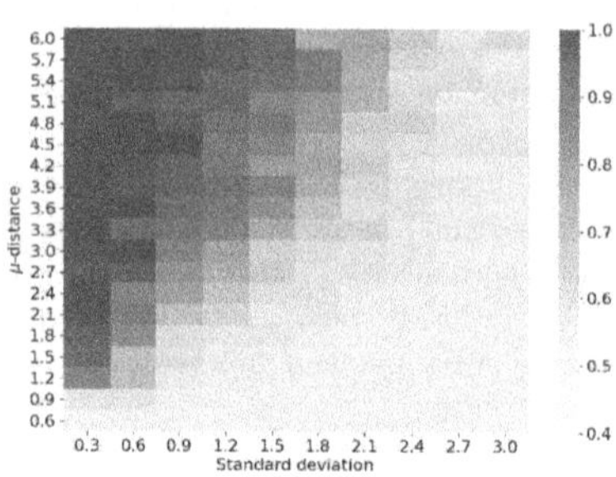

(a) Girvan–Newman

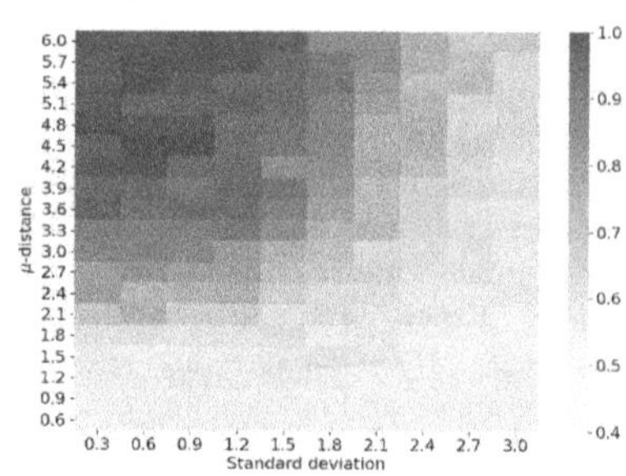

(b) Hierarchical Clustering

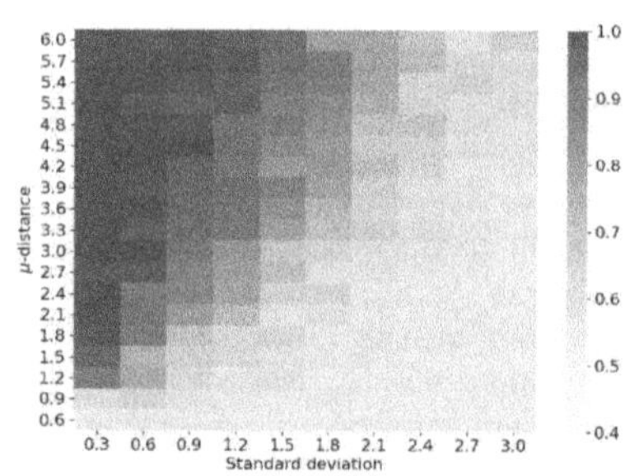

(c) Stochastic Block Model

Fig. 5. Heatmaps for the mean correct allocation of the community detection algorithms for synthetic data sets, based on 30 runs per parameter constellation.

We also simulated data sets with additional items, which we describe in Appendix E. If we add additional items there is more information on the group structure, leading to an improved performance of the community detection algorithms (see Figs. E.1, E.2 and E.3). For the shown data set, the difference between the community detection algorithms is minor. A notable difference is only a lack of performance for the Hierarchical Clustering where the μ-distance is between 1.2 and 3.3 and the standard deviation is 0.3. The data sets with additional items show similar results.

Item rank

To generate the synthetic data sets, items with different levels of information are included. In this way, we provide an order of items, concerning their information about the built-in group structure. The items split up into highly informative, less informative and uniformly distributed noise questions. Likewise the community detection methods, we assess the item rank method by means of synthetic data sets. We test the method's ability to rank the items correctly by their informational value. In the following, we will see that the item rank method's performance depends on the ability of the community detection methods to reveal the group structure in the attitude network.

The bar chart (Fig. 6), representing a cross-section of the heatmaps (see Appendix E), shows an equivalent performance of the community detection algorithms. The bar chart captures: (a) the proportion of successful community detection in comparison to the ground truth, (b) the proportion of correctly detected importance of items, and (c) the performance of the *Random Forest* classification

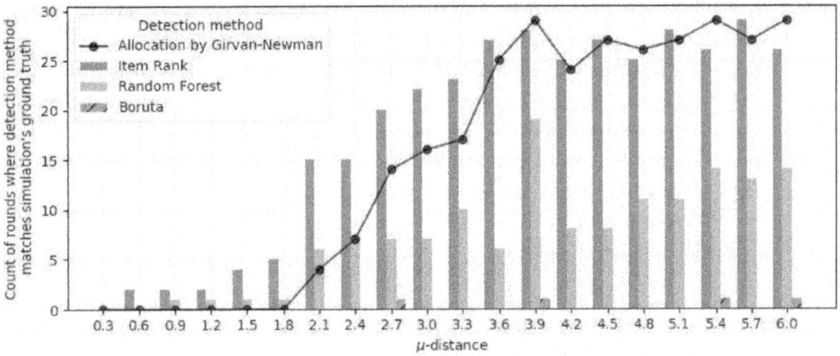

Fig. 6. Results of the correct item ranking, depending on the μ-difference. For each of the item ranking methods (Item Rank, Random Forest, Boruta), each bar shows how often each method detects the correct item ranking for each configuration of the synthetic data set. For every μ-distance, we run 30 simulations with 30 different synthetic data sets. We added the number that the community detection algorithm (here: Girvan–Newman algorithm) allocated the individuals correctly in relation to the ground truth from the synthetic data set. The graph describes the development of the number of correct allocations for each μ-distance.

algorithm and the *Boruta* method for feature selection. By this, it shows what happens in the transition phase, when moving from a dark blue to a light blue region (see Appendix E, heatmaps). The bar charts show as expected a similar number of correct allocations for the Girvan–Newman algorithm, Hierarchical Clustering and Stochastic Block Model. The number of completely correct ranked items by the item rank method is around 25 out of 30 for the simulations with a maximal μ-distance between 3.3 and 6.0. Below 3.0, the proportion of completely correct rankings of the items decreases. The lack of predictive power of the Random Forest model is of note. It is only able to pick out the correct ranking in a small number of simulations, even when there is a clear community split. The data shows that it performs better in allocating correctly the higher ranked questions but worse in determining the overall ranking. For the same reason, the *Boruta* method does not determine the rankings accurately. However, it is effective at distinguishing between important and unimportant items.

Applying the approaches on synthetic data sets was beneficial for exploring and comparing the methods under artificial conditions to ensure their robustness in various scenarios. Nevertheless, synthetic data is no substitute for real-world empirical data sets. Furthermore, only by analysing real data sets, inferences can be made and results can be interpreted.

3.1.2. *Wellcome trust data*

Here, we present a data set from the Wellcome Global Monitor 2018 which has not been previously studied to reveal opinion-based groups. The Wellcome Global Monitor conducted a survey in 2018 in over 140 countries with over 140,000 participants [48]. The survey encompasses public attitudes to science and health. We select attitude-related items from the data set; 10 items deal with trust in organizations, institutions and science, and 3 are attitudes towards vaccines.

We apply our approach to each listed country to detect opinion-based groups and, if applicable, relevant items for community structure. We refine and normalise the data to construct country-specific networks. We apply the Girvan–Newman algorithm to detect opinion-based groups, and Hierarchical Clustering to confirm these results. Our approach detects polarization on health and science attitudes in five countries: Singapore, Venezuela, Cameroon, Congo and Nicaragua (see Table 3 and Table S1 for the complete table).

Table 3. Results from Wellcome Global monitor in the five countries where we identified polarized opinion-based groups by Girvan–Newman algorithm.

Country	Size	Split-up (GN)	Links	Removed links	Threshold	Split-up (HC)	Overlap
Singapore	456	[327, 129]	5015	12	11.5	[326, 130]	0.985
Venezuela	575	[380, 195]	3757	109	11	[452, 123]	0.854
Cameroon	493	[318, 175]	4767	161	10	[401, 92]	0.542
Congo	356	[191, 165]	2180	47	10	[209, 147]	0.933
Nicaragua	614	[433, 181]	5466	105	11	[423, 191]	0.925

The most noteworthy result is Singapore. Even though the threshold is very large (11.5), over 5015 links were added to the network, indicating that the individuals have a high consensus within the items. A link between the individuals in the network with a threshold of 11.5 means that in over 90% their item responses are very close or identical. Next to the high agreement, it was possible to separate the network into two communities by only erasing 12 links. The Hierarchical Clustering provides similar results, with an overlap of over 98.5% in community allocation. In the other four countries two large opinion-based groups are shown for both community detection methods, with the exception of Cameroon where the overlap of the two methods is only about 54%.

The examination through the item rank method reveals three items as the most important for the communities that we detect via the Girvan–Newman: trust in charity workers, trust in traditional healers and trust in scientists. We showed how to analyze large data sets to uncover the existence of opinion groups. For countries with large opinion-based groups, we are also able to uncover and rank the important items for the community structure.

3.1.3. *Consecutive data sets: ANES 2012 & 2016*

Polarization is often seen as an intrasocietal process of moving toward the extremes on political attitudes, e.g. being further away from each others' opinion on a scale. Our method identifies polarization — even in the absence of extreme opinions — by classifying non-overlapping opinion-based groups. In the previous section, the ANES data set from 2016 was examined with an item selection based on [40]. Here, we investigate the ANES data set from 2012 to demonstrate an approach to consecutive data. Instead of relying on a predetermined selection, we apply the Boruta method to reveal the important items for our opinion-based groups. To apply Boruta to our data set, we reduced the amount of items from the ANES data set 2016 and 2012 to

Table 4. American National Election Studies 2012 and 2016 — Item labels and their answer range. The Boruta method selected the item as important in at least one of the data sets. Only the selected item *Birthright Citizenship* from 2016 is not mentioned due to the fact that there was no corresponding item in 2012. The table shows the items that are used for the network projection method and later for the community detection.

Item	Label 2012	Label 2016	Scale	Boruta
Abortion	abortpre_4point	V161232	1–4	2012
Environment-jobs	envjob_self	V161201	1–7	2012/2016
Race relations	aidblack_self	V161198	1–7	2012/2016
Immigration	immig_policy	V161192	1–4	2016
Govt. guaranteed income	guarpr_self	V161189	1–7	2012/2016
Death penalty	penalty_favopp_x	V161233x	1–4	2016
Defence-spending	defsppr_self	V161181	1–7	2012/2016
Govt. spending & services	spsrvpr_ssself	V161178	1–7	2012/2016
Medical insurance	inspre_self	V161184	1–7	2012/2016
Gay marriage	gayrt_marry	V161227x	1–3/1–6	2016

each 34 items based on relevance (see Table S2 for the reduced item list), selecting those items obviously related to a personal, political attitude position. Further, we only included participants who self-identified as republicans or democrats as the Boruta method requires group categorization for the item selection. Table 4 shows the items which the Boruta method identifies as important in distinguishing between Democrats and Republicans. After the normalization process, the selection of the important items allows us to construct the score-based similarity networks for the ANES data 2012 and 2016. The question is whether the opinion-based clusters are getting more separated, and so easier to detect, or is the opinion-scored network closer together, and therefore it is more difficult to distinguish between communities.

The network for the ANES data set 2012 consists of 2,039 nodes and 31,619 links with a threshold of 8.0 (see Fig. 7). The two communities detected by the Girvan–Newman algorithm have 2,493 and 546 nodes. The first community includes 1,004 democrats, 547 republicans and 942 unknown, while the second community consists of 308 republicans, 71 democrats and 167 unknown.

The network constructed from the ANES data set 2016 includes 2,274 participants and 27,326 links with a threshold of 7.8 (see Fig. 8). Applying the Girvan–Newman algorithm to the network results in a community with 596 republicans, 151 democrats and 424 unknown (total: 1171) and a second community with 612 democrats, 119 republicans and 372 unknown (total: 1103). To reveal the

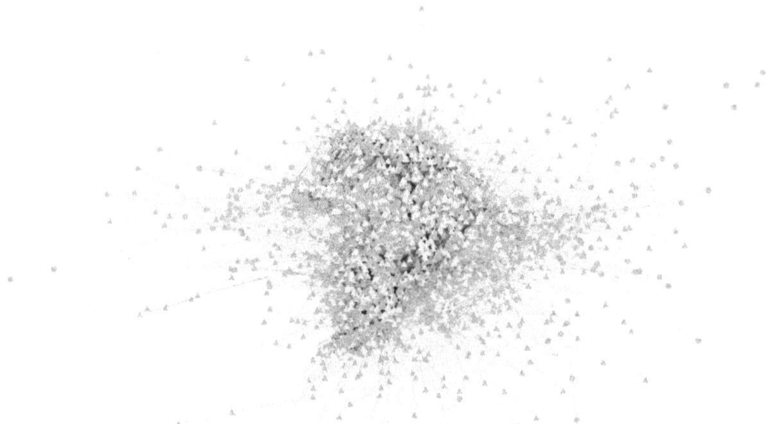

Fig. 7. (Color online) American National Election Survey data 2012, constructed similarity network from the refined data set, with 2 communities, detected by the Girvan–Newman algorithm. The position and shape of the nodes is used to distinguish between the communities. The color of the nodes represents their party affiliation: republican (red), democrat (blue) and unknown (yellow).

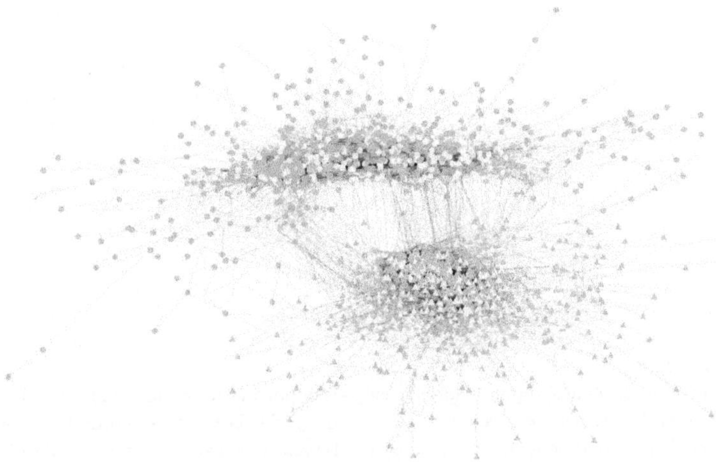

Fig. 8. (Color online) American National Election Studies data 2016, constructed similarity network from the refined data set, with 2 communities, detected by the Girvan–Newman algorithm. The Boruta method provides the item selection. The network projection uses 10 items from the data set. The position and shape of the nodes is used to distinguish between the communities. The color of the nodes represents their party affiliation: Republican (red), Democrat (blue) and unknown (yellow).

opinion-based groups considerably more links had to be removed in 2012 in comparison to 2016 and the graph had to be re-split several times as it did not fulfil the minimum community size criterion.

The application of our opinion-based group detection leads to the conclusion that, based on the ten important items, the ANES data is getting more polarized over time (from 2012 to 2016). The ANES data set from 2016 can be split up by erasing less cross-cutting links than 2012, and the groups are visually easier to distinguish.

The results show what is already observed: survey participants become increasingly polarized along party lines on several key opinions [30] in the ANES data set from 2012 and 2016. The communities are formed around the party affiliation, i.e. each community includes a majority of either Republicans or Democrats. Nevertheless, some participants are more aligned in their attitudes with the other group, contrary to their self-reported partisanship.

4. Conclusions

In this paper, we created a network of individuals linked by similar responses from a survey. We used three different clustering algorithms and show that all are consistent with each other at identifying communities of opinion-based groups on both empirical and simulated data. Further to this, we developed a method to identify the rank and importance of the items in a survey. We compared this to the Random Forest and Boruta method to validate it on simulated survey data. All methods can identify important items, but the method introduced here is more robust at ranking the survey items most important to the identified opinion-based groups. This allowed us to identify which items are most important to the opinion-based group that we found in the ANES and Wellcome Trust data. Additionally, instead of a post-validation, the Boruta method, given a predefined group categorization, is able to reduce the items of a survey to a subset of group-relevant items. The exploration of our approach on simulated data also points out limitations for our methods (i.e. opinion-based group detection and item rank). They rely on the performance of the community detection algorithms and, therefore, on the detected communities' meaningfulness.

Being able to identify opinion-based groups is important for understanding a wide range of social issues that can only be solved by the large-scale coordination of opinions (e.g. climate change; public health interventions; vaccination etc.). This is particularly important in understanding online social media interactions, which provide clear affordances for opinion exchange (e.g. via "likes" and "shares"). While identity has been shown to be central to social opinion processes (e.g. [19, 26]), until now it has been difficult to clearly identify links between bundles of opinions and social identities.

The value of this approach is demonstrated in [39] which shows opinion-based groups emerging at the start of the COVID crisis, progressively polarizing on the dimension of distrust in science; and leading to identity-based differences in compliance with public health guidance. We presented a secondary analysis of Wellcome Trust data, identifying countries like Singapore that are highly divided on trust in charity workers and science. Similarly, when we analyse the ANES 2012 and 2016 survey data, we identify items in the US that Democrats and Republicans are becoming increasingly polarized on. This phenomenon is widely observed in political and social sciences (see e.g. [30]).

While we identify separate opinion-based groups here, we do not quantify the level of polarization, which we aim to address in the future. Network measures to quantify polarization exist, including using edge betweenness [24]. However, these methods all rely on identifying hubs to detect polarized groups. As we construct similarity-based networks, which are dense and weighted, our topology is different. We tend not to have hubs as every node in an opinion-based group will be directly or indirectly linked to every other node in that group. In order to bring methods like this to bear we will need to modify them.

This method for detecting polarization in opinion-based groups paves the way to investigate the co-constitutive relationship between attitudes and social identity and related phenomena using a network approach.

Acknowledgments

This project has received funding from the European Research Council (ERC) under the European Union's Horizon 2020 research and innovation programme (grant agreement No. 802421). The authors would like to thank Susan C. Fennell for fruitful conversations and feedback.

Appendix A. Community Detection Algorithms

A.1. *Girvan–Newman algorithm*

The community detection algorithm by [25] is a divisive approach which successively separates the network into communities. Contrary to other algorithms like Louvain (modularity optimization method) [8], it is based on edge betweenness. In our case, we make use of the unweighted edge betweenness centrality in which the edge weights have no influence on the community detection. Using this measure to calculate the centrality of links, it intends to identify through it the community bridging links. It is based on the assumption that links between communities have a higher edge betweenness centrality, caused by their linking ability. Given this, a high amount of shortest paths go through the links to connect nodes between the communities. The Girvan–Newman algorithm is structured as follows [25]:

(1) The edge betweenness centrality ranks each link.
(2) The link with the highest edge betweenness centrality is selected and removed from the graph.
(3) All links which were influenced by the removal are selected and their edge betweenness is recalculated.
(4) Steps 2 and 3 are repeated until every link has been removed from the graph. In our case, we repeat the steps until we split the graph into the desired number of components and terminate the algorithm then.

The community detection algorithms are not only assessed due to their ability to select communities, but as well by their computational complexity [8]. The Girvan–Newman algorithm's bottleneck is the repeated calculation of the edge betweenness centrality for every link in the network. Its algorithmic complexity is $O(m^2n)$, where the input involves m, the number of links, and n, the amount of nodes. This shows that the performance time is exponentially increasing in relation to the input. Due to the very input-sensitive behavior, the computational limits the usage of this method to networks with a maximum of a couple of thousand nodes.

A.2. *Hierarchical clustering*

The Hierarchical Clustering method is applied directly to the data set, thus without constructing a network. The core of analysis is a distance matrix which contains every distance between the individuals i and j. There are various ways of calculating the distance between individuals. Here, we choose the euclidean distance:

$$d(i,j) = \sum_{f=1}^{n_f} (q_{if} - q_{jf})^2, \tag{A.1}$$

where $d(i,j)$ is the distance square distance between individuals (nodes) i and j; and n_f is the number of items and q_{if} is individuals i's response to item f. The distance between two individuals demonstrates the similarity in their answers over all their items. The calculation of distance and the distance matrix are the essential components for the application of the Hierarchical Clustering.

The agglomerative character defines the starting point of the Hierarchical Clustering, each individual is defined as a single, separated cluster. The number of clusters is correspondingly as high as the number of individuals, N. From there, the algorithm works as follows:

(1) Considering the distance matrix, select the pair of closest clusters (minimal distance).
(2) Merge the clusters together and recalculate the distance matrix with the new cluster. The merging of clusters uses the group average linkage function.
(3) Steps 1 and 2 are repeated until there is only a giant component left that contains all individuals.

Appendix B. Feature Selection Methods

B.1. *Random forest*

As mentioned in the main text, the Random Forest model is a classification method that has attained near state-of-the-art performance for classification [2, 16]. Making it an attractive option for classification not only due to its accuracy but its in built variable importance measures and extensions like Bourta (which is discussed in the following section). The ethos of Random Forests is to build a series of Classification Trees using randomized data and items and then apply the majority vote of this ensemble of trees to classify data. The process to construct a Random Forest is as follows. For each tree, we draw a bootstrap sample, sampling with replacement. Any individuals that are not used to construct the tree are held as validation data (referred to as an "out-of-bag" sample). Using the bootstrapped sample, we grow a Classification Tree. This process is identical to normal Classification Trees except for one notable difference. At each split, we randomly select p items from the total set of predictors. Using these p variables, we then choose the best variable and split-point. The randomization of the training set helps to avoid over-fitting. The randomization of the predictors selected at each split ensures the trees are uncorrelated; otherwise powerful predictors would be likely to be selected, resulting in each tree in the ensemble providing the same information [28]. We repeat this process until we have grown the desired number F of trees. Thanks to the randomization of the data and predictors, we do not need to prune any of the trees that make up the forest, as would be the case for Classification Trees. The predicted probability for any class is the class's proportion that each member of the forest voted for.

A valuable side-effect of randomly selected variables for each split in the ensemble of Classification Trees is that it gives us native access to variable importance measures [15, 37]. When a variable is used in a split, the decrease in the Gini node impurity is recorded. We can then rank variables by the average of all the Gini impurity reductions, allowing us to observe which questions are most important in forming an opinion-based group. Care must be taken when interpreting variable importance scores as shown in [46]. They noted that issues can occur when many categorical variables are included with a diverse range in the number of levels. Impurity measures can favor those with many levels. This is not an issue for us as the number of levels in the categorical variables from any survey remains relatively small and similar to each other. Also, if multiple continuous predictors are used with varying scales this can also make the importance measures unreliable. It was noted in [5] that for continuous predictors that, although the most powerful predictors were not always given the highest importance, these variables were consistency ranked among the top on a ranking of variable importance. In [1], it was noted that multicollinarity is not an issue in Random Forest when ranking item importance. For example, if we added highly correlated predictors the importance is not diluted between them, the ranking is rather preserved. Both [1] and [37] provide useful information on the choice of hyper-parameters; the former noting that *ntree* increases the stability of the importance measures and *mtry* increases the distance between the improvement measures (and the latter reference providing useful implementation notes for practitioners, such as the number of trees used should grow with the number of predictor variables used). Additionally, we note similar in our simulation of synthetic survey data, where for categorical variables, the most powerful predictors were ranked highest. Also of note, though we obtain a rank for the importance of estimates we do not know where the cut off for where an item becomes *unimportant* occurs, of even if it does. The Boruta algorithm addresses this in the Sec. (B.2).

B.2. *Boruta*

As mentioned in Sec. 3, Boruta is a feature selection method where we are concerned with teasing out all relevant features that are predictive, in our case, of the opinion groups that we have found. Boruta is a wrapper for the Random Forest model, that builds on the easy access to the variable importance measures. Providing a means of identifying which items (variables) are not predictive for classification. Please refer to [36] for a more extensive discussion of the algorithm's implementation but we will provide the broads strokes here.

An iteration of Boruta is as follows: It begins by adding a copy of each item to the data set, where each of these are shuffled randomly. These randomized items are called *shadow features*. The shadow features hold no correlation with the classification that the random forest is trying to build and will provide the benchmark for when an item can be declared important. Variable importance is calculated for each item (including shadow features). We note the shadow feature with the largest variable importance and compare all items importance to it. If an item has a variable importance larger than it, they are declared a "hit", if not they are declared a "miss". This process is repeated multiple times so we get, for each items, the fraction of times it was at hit, p_h.

To find when an item is important or not, we take this fraction, p_h, and perform a two-tailed statistical test based on the binomial distribution.[b] By default this significance level is set to 5%. For an item, if we fail to reject the null hypothesis, then the item's importance is indistinguishable from that of the shadow features. As a result we can not say it is better or worse than the shadow features. If we can reject the null in favor of the alternative hypothesis, then the item's importance is different from that of the shadow features. Interpreting the sign of the test statistic yields whether the item is important or unimportant to the formation of the observed opinion based group. This process provides a method of isolating which of the features are important to the formation of opinion based groups that we wish to study. As Boruta is an extension of the Random Forest variable importance measure, providing a way of finding a cut-off for which opinions are important, it inherits the same caveats for application to data. Thus, the points and references from Sec. B.1 that deal with the reliability of the Random Forest variable importance measures should be kept in mind when using Bourta (scale of continuous predictors, large difference in number of categories between variables, etc.).

Appendix C. Example Elbow Plot

For each of the community detection methods that we have used we wish to isolate (1) the number of communities that we have and (2) the relative performance of each method. To do this we can construct an elbow plot.

This plots the Within-cluster Sum of Squares (WSS) we obtain for a specified community assignment, which we sweep successively through.

If we see an "elbow" in the plot, that point defines the ideal number of communities in the data [49]. The elbow defines a point where before it, adding more communities provides large reductions in the WSS; however, after this point, additional clusters only provide marginal reductions in the WSS [10]. Beyond the elbow, for a community detection method, additional communities are unlikely to provide increased explanatory power in the community assignment.

We generate three curves for both synthetic data (Fig. C.1) and the ANES data (Fig. C.2) using the Hierarchical Clustering, Girvan–Newman Algorithm and

[b] In fact, thanks to the number of iterations of Boruta we can use the t-test based on population proportions.

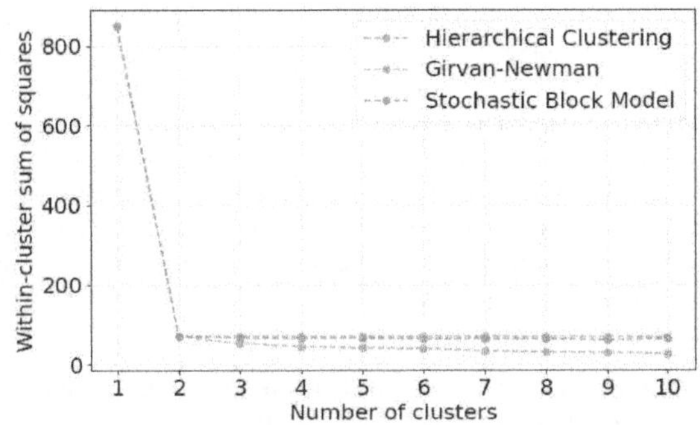

Fig. C.1. Example of a successful elbow plot. We generated the plot for a synthetic data set that has two communities by definition. We apply the three community detection algorithms Hierarchical Clustering, Girvan–Newman algorithm and Stochastic Block Model and calculate for each number of community the WSS.

Stochastic Block Model outlined previously for a range of 1 to 10 communities. Figure C.1 shows the results for the synthetic data, in which all community detection algorithms behave the same and select similar network communities. The "elbow" in the plot yields the correct number of communities as two communities.

When we apply this to the 2016 ANES data set using the eight variables discussed in the main text, we find that the picture is similar, if not naturally a little noisier than the synthetic data (see Fig. C.2). As you can see, the Hierarchical Clustering provides the lowest WSS than the other methods, and all three confirm that three is the optimal number of clusters. Even though the Hierarchical clustering method offers the lowest number of communities, each method may provide a different but

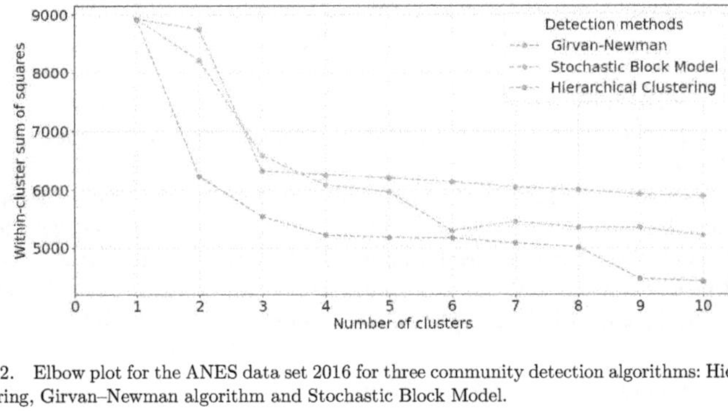

Fig. C.2. Elbow plot for the ANES data set 2016 for three community detection algorithms: Hierarchical Clustering, Girvan–Newman algorithm and Stochastic Block Model.

complementary community partition that may be worth exploring. How both the Girvan–Newman and Stochastic Block Model arrive at their community partition leads to a valuable interpretation of these identity-based groups being separated by a small number of tenuous links to the other groups. As such, differences in the beliefs held between these groups might be of interest in-of-themselves. It is also worth noting that using the WSS as a criterion for model selection may be overly harsh on the network-based community detection methods. The Hierarchical Clustering algorithm is built to minimise the distance matrix between communities, so if performing adequately, it should have the lowest WSS.

Appendix D. Description of Selected Items From the American National Election Study 2016

The selected items are assigned to eight different categories. The ANES 2016 provides a full description of the items (see ANES 2016 user guide).[c] Here, we describe the items and the extracted information from the user guide to allow a quick access and understanding of the used items.

Abortion (V161232). The item focuses on the position of the legal regulations of abortion. The participant positions themself and chooses between (1) "By law, abortion should never be permitted"; (2) "By law, only in case of rape, incest, or woman's life in danger"; (3) "By law, for reasons other than rape, incest, or woman's life in danger if need established"; (4) By law, abortion as a matter of personal choice".
Race relations (V161198). The participant is asked to evaluate the government's racial re-distributive policies. The participants can choose from an answer categories which range from *Government should help the black people* to *The black people should help themselves.*
Immigration (V161192). The item is about the government's policy toward unauthorized immigrants. The participant positions themself and chooses between (1) "Make all unauthorized immigrants felons and send them back to their home country"; (2) "Have a guest worker program in order to work"; (3) "Allow to remain and eventually qualify for U.S. citizenship, if they meet"; and (4) "Allow to remain and eventually qualify for U.S. citizenship without penalties".
Welfare (V161209). The item's topic is the amount of government spending on welfare programs. The question is whether it should be "increased", "decreased" or "kept the same".
Homosexuality (V161231). The item addresses the position and legal acceptance of gay marriage and allows the participant to choose between "Gay and lesbian couples should be allowed to legally marry", "Gay and lesbian couples should be allowed to form civil unions but not legally marry" and "there should be no legal recognition of a gay or lesbian couple's relationship".

[c]https://electionstudies.org/wp-content/uploads/2018/12/anes_timeseries_2016_userguidecodebook.pdf.

Business (V161201). The item rates the position of the participants towards the potential cost of jobs through implementing environmental regulations. The answers can range from "regulate business to protect the environment and create jobs" to "no regulation because it will not work and will cost jobs"
Guns (V161187). The question is whether the should it make more difficult to to buy a gun. The participant's answers are "increased", "decreased" or "kept the same".
Income (V161189). The item is to determine the participants' attitude towards the governmental effort to job and income. The item's scale ranges from Government should see to jobs and standard of living to Government should let each person get ahead on own.

Appendix E. Method to Create Our Synthetic Data Sets

The idea of generating a synthetic data set is to systematically vary properties of the groups in a data set and apply our method to this data to access its performance.

For this method, we can define the following variables required to create simulated data for testing:

- *n_agents* = number of individuals in the data set.
- *n_items* = number of questions of the created survey.
- *scale_steps* = size of scale for every question. It will be the same for every n questions.
- *mu_max* = maximal μ-difference, which defines the highest difference available in the questions. The questions can have a smaller μ-difference, if their ranking is lower.
- *number_ranks* = number of differently ranked questions in the data set.
- *n_comp* = number of predefined communities in the data.
- *sd* = lowest standard deviation for the highest ranked questions.
- *split_up* = percentages to define the size of the community in relation to the overall number of the individuals.

The number of questions per ranking depends on the number of questions and on the number of ranks and is then normally distributed around an expected value to allow variation. The importance of the questions is set due to a higher or lower μ-difference. The higher the importance of the question, the higher the μ-difference. The *mu_max* defines the μ-difference for the questions with the highest ranking, all the other questions will have a lower μ-difference or are noise questions. The method constructs a data set for the number of requested individuals and questions. It is structured like the data used in from the ANES 2016 but without missing data points, and therefore meets our requirements of replicating attitudinal survey data.

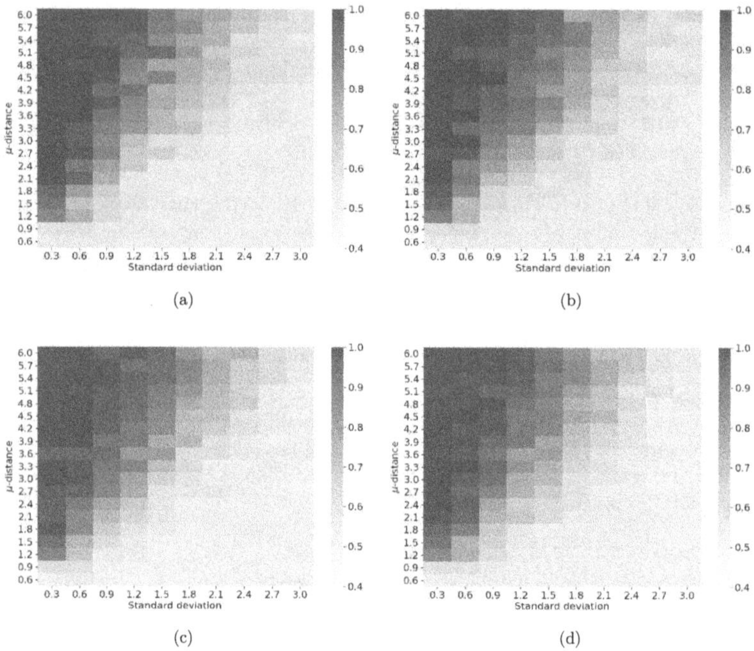

Fig. E.1. Heatmaps for the mean correct allocation of Girvan–Newman algorithm for synthetic data sets, based on 30 runs per parameter constellation. The four heatmaps only differ in the number of integrated items: (a) 6, (b) 7, (c) 8, (d) 9.

E.1. *Simulations based on synthetic data*

In order to examine the performance of the three community detection algorithms, we introduced the synthetic data set construction process. This allows us to to explore the limits of each community detection algorithm. We simulate a large number of data sets and run the algorithms on them.

The synthetic data sets are constructed on artificial results from 100 individuals, with answers to 6, 7, 8 or 9 questions on a scale from 1 to 7. The questions are ranked in four different categories of influence, determined by an increasing mean. The community structure is an equal division into two groups of 50. The μ-distance ranges from 0.6 to 6.0 with a step size of 0.3 (y-axis). The corresponding *standard deviation* ranges from 0.3 to 3.0 (x-axis).

The heatmaps in Figs. E.1, E.2 and E.3 indicate the mean percentage of correctly allocated individuals by the community detection algorithms. Each square of the

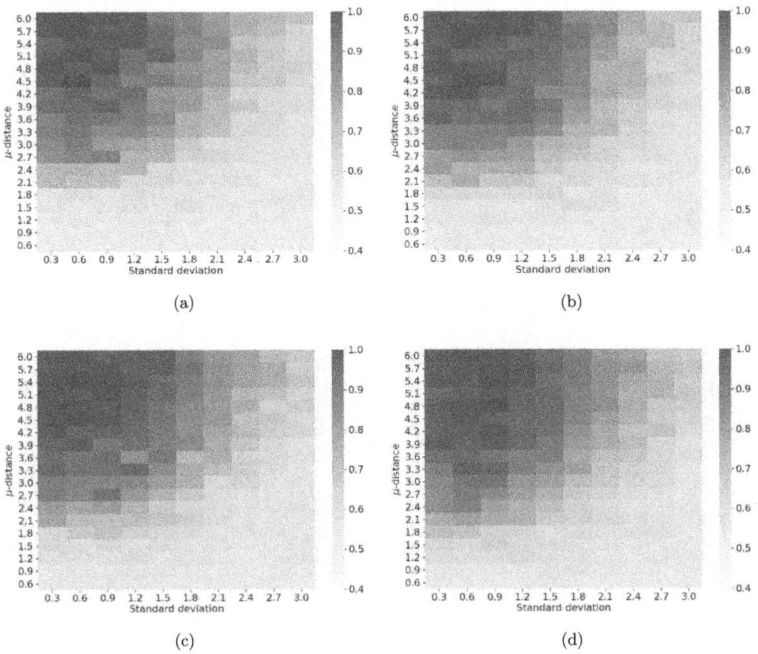

Fig. E.2. Heatmaps for the mean correct allocation of hierarchical clustering for synthetic data sets, based on 30 runs per parameter constellation. The four heatmaps only differ in the number of integrated items: (a) 6, (b) 7, (c) 8, (d) 9.

heatmap represents a mean of 30 simulation runs. For example, the value of 1.0 reports that in 30 simulation runs the algorithm allocated all individuals to the correct community.

The results for the Girvan–Newman show the best performance for relatively high μ-distance and a low standard deviation. With an increasing standard deviation and a decreasing μ-distance, the algorithm is not capable of allocating correctly. The values around 0.5 mean that for example a random allocation algorithm would perform likewise. Adding additional items (questions) that are informative to the community structure with-in the simulated data provides slight improvements when trying to recover the communities via the community detection methods (see Figs. E.1(a)–E.1(d), dark regions). Similar behavior and results can be seen in the heatmaps for the hierarchical clustering and the stochastic block model (see Figs. E.2 and E.3). However, it is striking that the hierarchical clustering method shows a lack of competitiveness, for a low μ-distance (between 0.3 and 3.3) and a

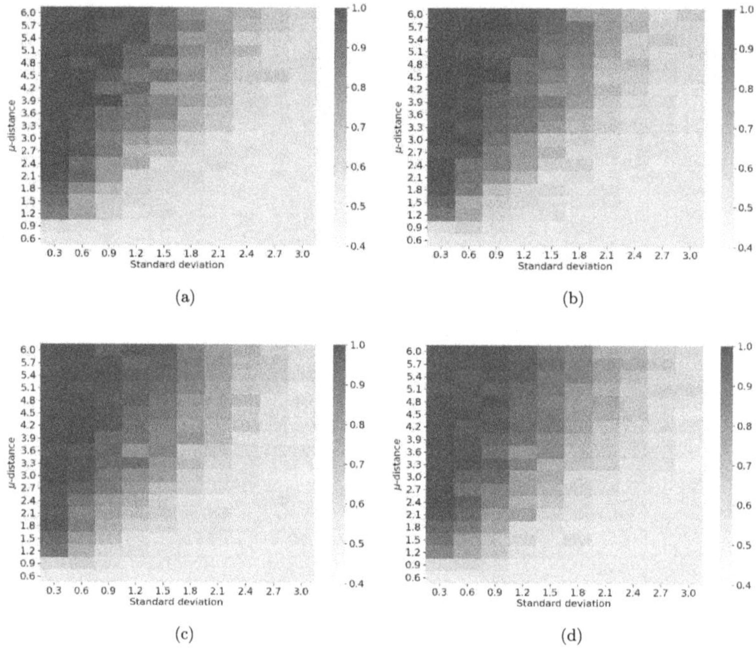

Fig. E.3. Heatmaps for the mean correct allocation of stochastic block model for synthetic data sets, based on 30 runs per parameter constellation. The four heatmaps only differ in the number of integrated items: (a) 6, (b) 7, (c) 8, (d) 9.

standard deviation of 0.3 or 0.6. For that parameter group, the results of the Girvan–Newman algorithm and the stochastic block are more convincing.

A subsequent step to the analysis of synthetic data and the determination of communities is the evaluation of the questions and their influence on the community structure.

Within the synthetic data set, we determine the importance of the questions by their overlap of the answer distributions of distinct communities. A higher overlap means less information concerning the community structure. The question selection method is therefore able to rank the questions by their influence.

Figure E.4 shows a detailed section of the heatmaps. The construction of the synthetic data sets is the same as in the heatmaps, solely the standard deviation is fixed to 0.7. The figure is separated into the results from the three community detection algorithms. Moreover, the ranking results of the question selection method, the Random forest method and the Boruta package are shown. The bars

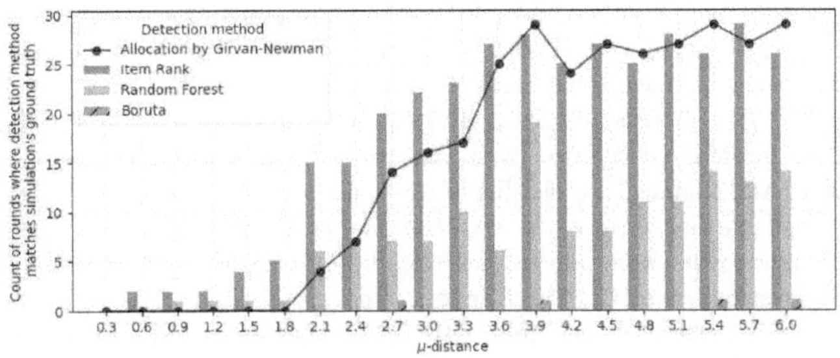

(a) Girvan-Newman algorithm

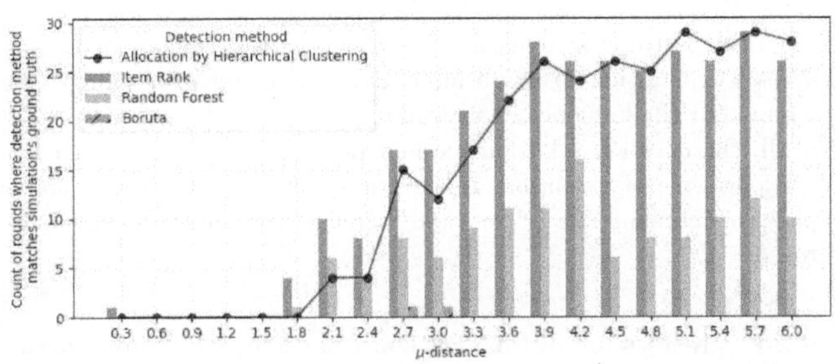

(b) Hierarchical clustering

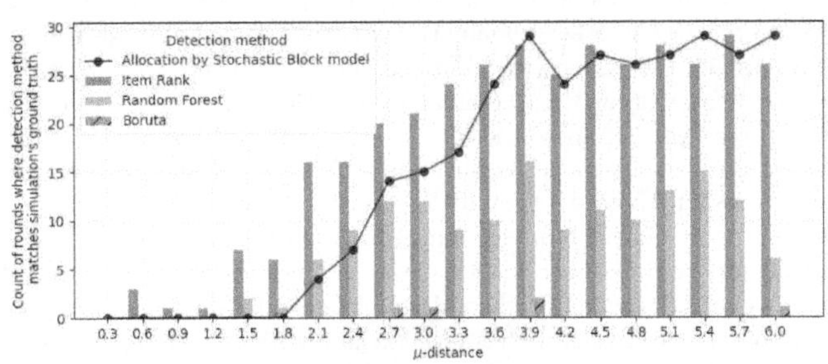

(c) Stochastic block model

Fig. E.4. Relation between feature selection and community detection algorithms. Each bar represents the simulation results of 30 simulations with the same parameters, the μ-distance is the variable displayed on the x-axis. The lowest standard deviation is 0.7, but it increases for less important questions. The results are based on the communities from: (a) Girvan–Newman algorithm, (b) Hierarchical clustering and (c) Stochastic block model. The bars show the performance of the questions selection method, Random forest classifier and Boruta (with Random forest classifier). The bar reflects the correct ranking of the questions for the 30 per μ-distance.

report the number of successful rankings of all 7 questions for 30 runs. The curve points out the frequency in which the community detection algorithm allocates all individuals of a simulation run correctly. The maximal possible count is 30 for each μ-distance. Over all, it is notable that the curves of all three community detection algorithms represent the same dynamics, drawing parallels to the results of the heatmaps. Additionally, the ranking of the questions is related to the performance of the algorithm as it is based on their community allocation. For large μ-distance (3.6–6), the question selection method classifies in more than 25 cases the ranking of the 7 questions correctly. The Random forest method generally does not exceed 15. In the cases where the community detection does not work, 1.8 and below, the question selection method and Random forest hardly work. The results for ranking the questions in the case of the Boruta method show that it is not working. It has to be mentioned that the Boruta algorithm focuses on the determination of important and unimportant features or items, and not on the correct ranking of the questions. Nevertheless, the question selection method is able to uncover a high amount of information about the influence of each item.

All in all, the question selection method performs very well, which may then justify the long time of execution. However, the Random forest method and the Boruta package are many times faster and therefore applicable on a much larger set of features.

Appendix F. Results for the ANES Data Set From 2012 and 2016

The analysis of the ANES data set from 2012 and 2016 was run for the Girvan-Newman algorithm, the hierarchical clustering and the stochastic block model. Only the networks for the Girvan–Newman algorithm were displayed in the main-section. In order to provide the reader with additional information and to be able to compare the network division of the three community detection algorithms, the networks are shown here (see Figs. F.1 and F.2).

Reduced results for ANES data set 2016

The analysis of the ANES data set from 2012 and 2016 was run for the Girvan–Newman algorithm, the hierarchical clustering and the stochastic block model. Only the networks for the Girvan–Newman algorithm were displayed in the main section. In order to provide the reader with additional information and to be able to compare the network division of the three community detection algorithms, the networks are shown here.

To confirm the communities from the ANES data set 2016 in Fig. 2 and to test its robustness, we reduced the used variables to construct the network and the communities. We rerun the analysis for the same set of variables but without the variable *Immigration*. We selected the variable *Immigration* to be eliminated because according to the item rank method it has the least impact on the community structure.

(a) Girvan-Newman partition

(b) Hierarchical clustering partition

(c) Stochastic block model partition

Fig. F.1. (Color online) American National Election Study data 2012, constructed similarity network with 2 communities, detected by: (a) Girvan–Newman algorithm, (b) Hierarchical clustering, (c) Stochastic block model. The position and shape of the nodes is used to distinguish between the communities. The color of the nodes represents their party affiliation: republican (red), democrat (blue) and unknown (yellow).

Figure F.3 displays two communities with similar sizes and similar partisan sorting like in Fig. 2. The sizes of the communities of the Girvan–Newman algorithm are 1879 and 1966. The reduction from eight to seven variables did not change the overall structure of the network.

Appendix G. Additional Results for Singapore

From the Wellcome Global Monitor, we used all items which express positions to several institutions and vaccines. Here, we reapply the method and only use the available data without the items on vaccines. We show that it is possible to receive similar communities without the vaccine questions as they are not listed at the top of item ranking.

The reduction of the data leads to more individuals who answered all items. Therefore, the size of the constructed network increased to 531. Nonetheless, the overall structure remains very similar to the results with the vaccine items. The community split up is 386 and 145 individuals and they are only connected by 5

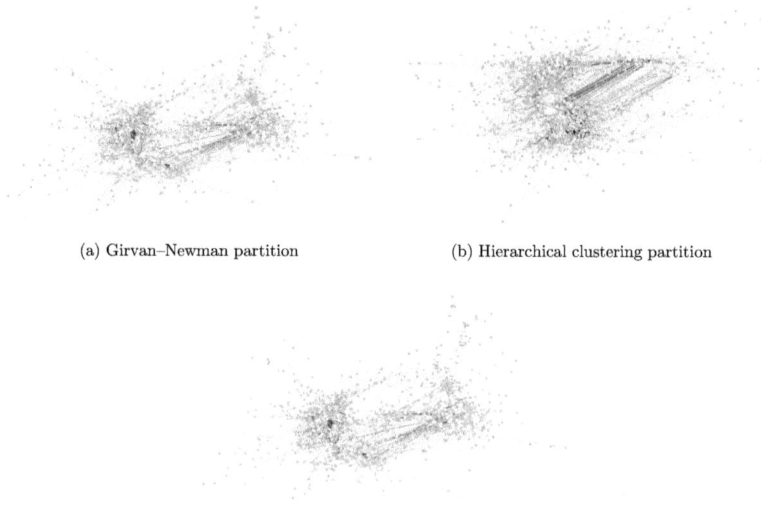

(a) Girvan–Newman partition

(b) Hierarchical clustering partition

(c) Stochastic block model partition

Fig. F.2. (Color online) American National Election Survey data 2016, constructed similarity network with 2 communities, detected by: (a) Girvan–Newman algorithm, (b) Hierarchical clustering, (c) Stochastic block model. The position and shape of the nodes is used to distinguish between the communities. The color of the nodes represents their party affiliation: republican (red), democrat (blue) and unknown (yellow).

cross-cutting links. The threshold is 9 and the total amount of links is 7136. The method detects two opinion-based groups that are connected within the groups and have nearly no connection between groups.

The item ranking provides us with the information about the influence on the community structure. For an additional analysis, we investigated how many of the most relevant items do we need to get a similar network structure and communities. We successively added an item until we received a similar network structure. We integrated the first 9 out of 13 items to get to a similar network structure. With a threshold of 8.2, the method constructed a network with a size of 597, with two communities (492, 105) and a total number of 8270 links where only 2 of them were cross-cutting links.

Appendix H. Threshold and the Giant Component

To decide the threshold for agreement between participants θ we plot the fraction of participants in the giant component against θ in Fig. H.1 for the ANES data in 2012

(a) Girvan–Newman partition

(b) Hierarchical clustering partition

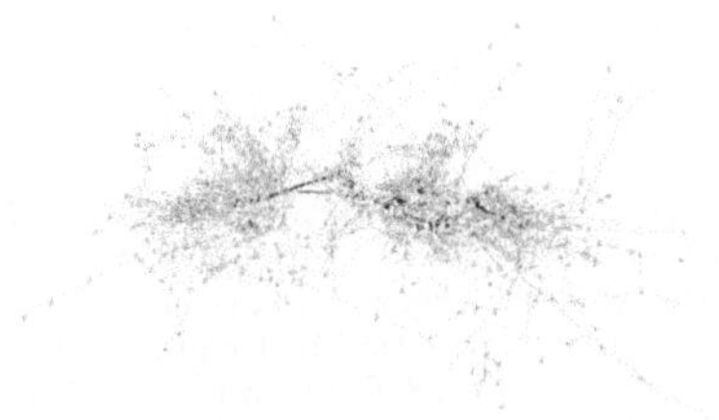

(c) Stochastic block model partition

Fig. F.3. (Color online) American National Election Survey data 2016, constructed similarity network with 2 communities, detected by: (a) Girvan–Newman algorithm, (b) Hierarchical clustering, (c) Stochastic block model. We reduced the number of variables from 8 to 7. The shape of the nodes is used to distinguish between the communities. The color of the nodes represents their party affiliation: republican (red), democrat (blue) and unknown (yellow).

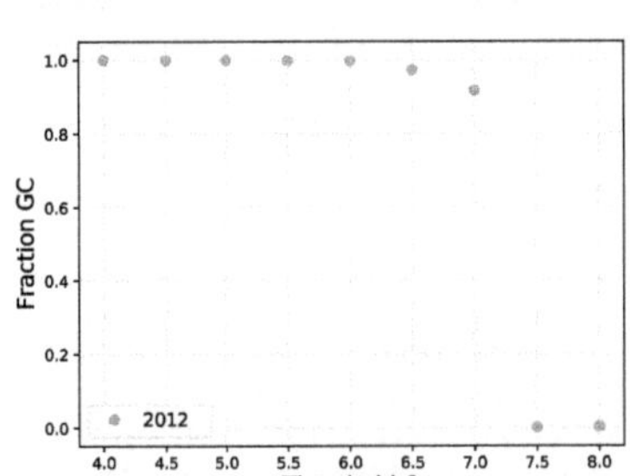

(a)

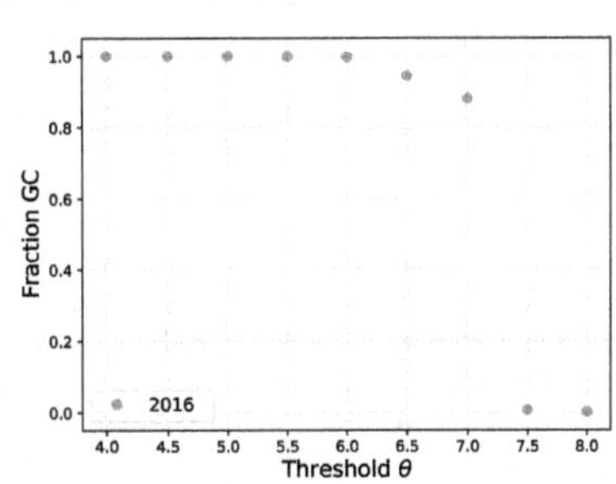

(b)

Fig. H.1. The fraction of the giant component versus the agreement threshold θ increasing in intervals of 0.5 for the ANES data in (a) 2012 and (b) 2016. As can be seen, there is a sudden jump from $\theta = 7.5$ to $\theta = 7$.

and 2016. In each case when $\theta = 7$ there is a jump in the number of nodes in the giant component and then it slowly increases to include all participants. Hence, we use this value of θ as the network is not too dense yet includes the majority of the nodes. Further decreasing θ vastly increases the number of links. We find similar communities in this case but it is less efficient due to the high density of links.

References

[1] Genuer, R., Poggi, J.-M. and Tuleau-Malot, C., Variable selection using random forests, *Pattern Recognit. Lett.* **31** (2010) 2225–2236.

[2] Zhang, C., Liu, C., Zhang, X. and Almpanidis, G., An up-to-date comparison of state-of-the-art classification algorithms, *Expert Syst. Appl.* **82** (2017) 128–150.

[3] Abramowitz, A. and McCoy, J., United States: Racial resentment, negative partisanship, and polarization in Trump's America, *Ann. Am. Acad. Polit. Soc. Sci.* **681** (2018) 137–156.

[4] American National Election Studies (ANES), University of Michigan, and Stanford University, Anes 2016 time series study (2017), doi:10.3886/ICPSR36824.V2.

[5] Archer, K. J. and Kimes, R. V., Empirical characterization of random forest variable importance measures, *Comput. Stat. Data Anal.* **52** (2008) 2249–2260.

[6] Băbeanu, A.-I., van de Vis, J. and Garlaschelli, D., Ultrametricity increases the predictability of cultural dynamics, *New J. Phys.* **20** (2018) 103026.

[7] Bakshy, E., Messing, S. and Adamic, L. A., Exposure to ideologically diverse news and opinion on facebook, *Science* **348** (2015) 1130–1132.

[8] Barabasi, A.-L., *Network Science (Online)* (Cambridge University Press, Cambridge, UK), http://networksciencebook.com/chapter/9.

[9] Baumann, F., Lorenz-Spreen, P., Sokolov, I. M. and Starnini, M., Emergence of polarized ideological opinions in multidimensional topic spaces, *Phys. Rev. X* **11** (2021) 011012.

[10] Bholowalia, P. and Kumar, A., EBK-means: A clustering technique based on elbow method and k-means in WSN, *Int. J. Comput. Appl.* **105** (2014) 17–24.

[11] Bliuc, A.-M., McGarty, C., Reynolds, K. and Muntele, D., Opinion-based group membership as a predictor of commitment to political action, *Eur. J. Soc. Psychol.* **37** (2007) 19–32.

[12] Boutyline, A. and Vaisey, S., Belief network analysis: A relational approach to understanding the structure of attitudes, *Am. J. Sociol.* **122** (2017) 1371–1447.

[13] Brandt M. J., Sibley, C. G. and Osborne, D., What is central to political belief system networks? *Pers. Soc. Psychol. Bull.* **45** (2019) 1352–1364.

[14] Breiger, R. L., Schoon, E., Melamed, D., Asal, V., and Rethemeyer, R. K., Comparative configurational analysis as a two-mode network problem: A study of terrorist group engagement in the drug trade, *Soc. Netw.* **36** (2014) 23–39.

[15] Breiman, L., Random forests, *Mach. Learn.* **45** (2001) 5–32.

[16] Caruana, R. and Niculescu-Mizil, A., An empirical comparison of supervised learning algorithms, in *ACM Int. Conf. Proc. Series*, Vol. 148 (ACM Press, New York, New York, USA, 2006), ISBN 1595933832, pp. 161–168, doi: 10.1145/1143844.1143865, www.cs.cornell.edu http://portal.acm.org/citation.cfm?doid=1143844.1143865.

[17] DeVellis, R., *Scale Development: Theory and Applications*, Applied Social Research Methods Series (Sage Publications, Inc., Thousand Oaks, 2003).

[18] Di Franco, A., Thiriet, P., Di Carlo, G., Dimitriadis, C., Francour, P., Gutiérrez, N. L., Jeudy De Grissac, A., Koutsoubas, D., Milazzo, M., Otero, M. D. M., Piante, C., Plass-

Johnson, J., Sainz-Trapaga, S., Santarossa, L., Tudela, S. and Guidetti, P., Five key attributes can increase marine protected areas performance for small-scale fisheries management, *Sci. Rep.* **6** (2016) 1–9.
[19] Doell, K. C., Prnamets, P., Harris, E. A., Hackel, L. M. and Bavel, J. J. V., Understanding the effects of partisan identity on climate change, *Curr. Opin. Behav. Sci.* **42** (2021) 54–59.
[20] Everitt, B. S., Landau, S., Leese, M. and Stahl, D., *Cluster Analysis: Fifth Edition* (John Wiley & Sons, 2011).
[21] Fiorina, M. P. and Abrams, S. J., Political polarization in the american public, *Ann. Rev. Polit. Sci.* **11** (2008) 563–588.
[22] Fortunato, S., Community detection in graphs, *Phys. Rep.* **486** (2010) 75–174.
[23] Fortunato, S. and Hric, D., Community detection in networks: A user guide, *Phys. Rep.* **659** (2016) 1–44.
[24] Garimella, K., Morales, G. D. F., Gionis, A., and Mathioudakis, M., Quantifying controversy on social media, *ACM Trans. Soc. Comput.* **1** (2018) 1–27.
[25] Girvan, M. and Newman, M. E. J., Community structure in social and biological networks, *Proc. Natl. Acad. Sci.* **99** (2002) 7821–7826.
[26] Gollwitzer, A., Martel, C., Brady, W. J., Prnamets, P., Freedman, I. G., Knowles, E. D. and Bavel, J. J. V., Partisan differences in physical distancing are linked to health outcomes during the COVID-19 pandemic, *Nat. Human Behav.* **4** (2020) 1186–1197.
[27] Groves, R. M., FowlerJrF. J., Couper, M. P., Lepkowski, J. M., Singer, E. and Tourangeau, R., *Survey Methodology*, Vol. 561 (John Wiley & Sons, 2011).
[28] Ho, T. K., A data complexity analysis of comparative advantages of decision forest constructors, *Pattern Anal. Appl.* **5** (2002) 102–112.
[29] Holland, P. W., Laskey, K. B. and Leinhardt, S., Stochastic blockmodels: First steps, *Soc. Netw.* **5** (1983) 109–137.
[30] Iyengar, S., Lelkes, Y., Levendusky, M., Malhotra, N. and Westwood, S. J., The origins and consequences of affective polarization in the united states, *Ann. Rev. Polit. Sci.* **22** (2019) 129–146.
[31] Javed, M. A., Younis, M. S., Latif, S., Qadir, J., and Baig, A., Community detection in networks: A multidisciplinary review, *J. Netw. Comput. Appl.* **108** (2018) 87–111.
[32] Kamada, T. and Kawai, S., An algorithm for drawing general undirected graphs, *Inf. Process. Lett.* **31** (1989) 7–15.
[33] Karrer, B. and Newman, M. E. J., Stochastic blockmodels and community structure in networks, *Phys. Rev. E* **83** (2011) 016107.
[34] Kruglanski, A. W., Pierro, A., Mannetti, L., and De Grada, E., Groups as epistemic providers: Need for closure and the unfolding of group-centrism., *Psychol. Rev.* **113** (2006) 84.
[35] Kursa, M. B., Robustness of random forest-based gene selection methods, *BMC Bioinform.* **15** (2014) 8.
[36] Kursa, M. B. and Rudnicki, W. R., Feature selection with the Boruta Package, *J. Stat. Softw.* **36** (2010) 1–13.
[37] Liaw, A., Wiener, M. and Others, Classification and regression by randomForest, *R News* **2** (2002) 18–22.
[38] MacCarron, P., Maher, P. J. and Quayle, M., Identifying opinion-based groups from survey data: A bipartite network approach, *preprint* (2020).
[39] Maher, P. J., MacCarron, P. and Quayle, M., Mapping public health responses with attitude networks: The emergence of opinion-based groups in the uk's early covid-19 response phase, *Br. J. Soc. Psychol.* **59** (2020) 641–652.

[40] Malka, A., Soto, C. J., Inzlicht, M. and Lelkes, Y., Do needs for security and certainty predict cultural and economic conservatism? A cross-national analysis, *J. Pers. Soc. Psychol.* **106** (2014) 1031–1051.
[41] McPherson, M., Smith-Lovin, L. and Cook, J. M., Birds of a feather: Homophily in social networks, *Ann. Rev. Sociol.* **27** (2001) 415–444.
[42] Murtagh, F. and Contreras, P., Algorithms for hierarchical clustering: An overview, *WIREs Data Min. Knowl. Discov.* **2** (2011) 86–97.
[43] Peixoto, T. P., Efficient monte carlo and greedy heuristic for the inference of stochastic block models, *Phys. Rev. E* **89** (2014) 012804.
[44] Peixoto, T. P., The graph-tool python library, figshare (2014).
[45] Peixoto, T. P., Nonparametric bayesian inference of the microcanonical stochastic block model, *Phys. Rev. E* **95** (2017) 012317.
[46] Strobl, C., Boulesteix, A. L., Zeileis, A. and Hothorn, T., Bias in random forest variable importance measures: Illustrations, sources and a solution, *BMC Bioinform.* **8** (2007) 25.
[47] Vaughan, E. and Tinker, T., Effective health risk communication about pandemic influenza for vulnerable populations, *Am. J. Public Health* **99** (2009) S324–S332.
[48] Wellcome Trust, The Gallup Organization Ltd., *Wellcome Global Monitor*, 2018. [data collection], 2nd Edition, (UK Data Service, 2019) SN: 8466, http://doi.org/10.5255/UKDA-SN-8466-2.
[49] Yuan, C. and Yang, H., Research on k-value selection method of k-means clustering algorithm, *J* **2** (2019) 226–235.

PART 3
System Dynamics Models

Chapter 16

Overview

Focus on the system. System dynamics is the oldest among the four modeling approaches discussed in this book. It has roots in control theory from engineering, in the cybernetic concepts of system thinking in the 1950s, in chemical reaction kinetics, and population dynamics. Early models of population growth date back to Verhulst (1838) or Lotka and Volterra in the 1920s. Different from, e.g., agent-based modeling, where the focus is on emerging systemic properties, in system dynamics, the structure and dynamics of a system are given. The system elements are not (many) individual agents but (few) aggregated quantities. For instance, populations are described as spatio-temporal densities. The system elements can also be regarded as *representative agents.* For example, macroeconomic models of the output market describe the dependencies between *the* household and *the* firm.

Nonlinearity and feedback loops. It is not the *topic* that distinguishes system dynamics from other modeling approaches for studying complex systems, but the *method.* We recall that there are *two* primary sources of systemic complexity: (i) the collective interactions of a large number of system elements, which are addressed in agent-based modeling, and (ii) the nonlinear feedback between system elements. The scientific interest in system dynamics is in studying this nonlinear feedback which can be positive, i.e., amplifying, or negative, i.e., damping [1, 12, 52, 59, 130].

Different feedback processes are combined in feedback *loops.* Nested feedback loops characterize most real systems, i.e., many concurrent loops influence the system dynamics in different directions simultaneously. Importantly, these influences are *directed* and not symmetric, resulting in rather complicated interdependencies. For instance, A *directly* influences B which directly influences C. But C directly influences A. Thus, B also influences A, but *indirectly.* If these influences can amplify and dampen, it becomes difficult to anticipate the resulting system dynamics.

Variables and rates. The directed influences, sometimes denoted as causal influences, depend on *parameters* quantifying their *strength.* To characterize their meaning, stock variables, e.g., the sizes or densities of populations X and Y, are distinguished from flow variables which describe the influence of X on Y. In a predator-prey system, for instance, X reduces Y at a given rate via consumption, but both X and Y can also have positive feedback via reproduction.

That means flow variables usually have a *time dimension*, e.g., number of offspring per individual per time unit. In contrast, stock variables are numbers or densities that *change* in time. Starting from a diagram with nested feedback loops, we can write down dynamic equations for the stock variables straightforwardly. The system dynamics is then given as a set of *coupled* differential equations in continuous time or difference equations in discrete time [3, 4, 95]. They can be interpreted as input-output relations, where the strengths of the flow variables determine the *rates* of change [67]. Such systems have close relations to kinetic reaction networks [68, 94] or gene regulatory networks, discussed under Network Models and Models of Evolution.

Equilibria, stability, and control parameters. The coupled differential equations are deterministic, which means they do not consider fluctuations or random influences. Setting the time derivatives to zero then defines possible equilibrium states. The most important question regards the *number* and the *stability* of these equilibrium states [30, 47, 54, 74]. These are classified in applied mathematics, in particular in nonlinear dynamics. Equilibria can be *fixed points*, i.e., have one value, or *periodic solutions*, i.e., have a repeating sequence of values [3, 34, 35, 48, 49, 72, 125]. They can further be *stable* or *unstable*, i.e., in the case of small perturbations, the system either returns to the stable state or moves away from the unstable state.

The number and the stability of these equilibrium states crucially depend on the parameters characterizing the dynamic equations and the boundary conditions. If these parameters cross particular critical values, stable equilibria can turn into unstable equilibria or the other way round. That means the behavior of the macroscopic system can change drastically, known as "phase transition" in physico-chemical or as "regime shift" in bio-social systems. The respective parameters are called *control parameters* because they control to a large degree the systemic stability.

The micro-macro link. System dynamics aims at forecasting the development of aggregated quantities. Some agent-based models propose methods to *derive* the system dynamics from the agent dynamics, which becomes possible only under particular assumptions about the underlying interactions [11, 80, 91, 120, 123]. Popular is the mean-field approach, which assumes that every agent interacts with all other agents in the same manner. Hence, the influence of all other agents is summarized in a mean field acting on each agent. It neglects spatial heterogeneity or path dependence in interactions but is a first and valuable step toward justified system dynamics.

Order parameters. In addition to *control parameters*, system dynamics models use the concept of *order parameters* to capture the aggregated behavior succinctly.

Whereas control parameters specify the conditions for a phase transition or a regime shift, order parameters describe in a coarse-grained manner what happens when a system approaches these critical conditions. Continuous or discontinuous changes of the order parameter, critical slowing down or speed up of its dynamics, or scaling laws with particular exponents allow for classifying the possible transitions around such critical points. Thus, knowing the approximate dynamics of the order parameter helps forecast the systemic dynamics without paying attention to all details.

In most cases, the order parameter is not a parameter, but an aggregated variable, e.g., the net magnetization in a ferromagnetic system. Regarding non-standard applications, the order parameter can also have more than one dimension (see [134] below for an epidemic model). Then the problem arises with obtaining an appropriate order parameter and how to derive its approximate dynamics [38]. Mathematical tools, such as the elimination of fast variables, and empirical studies of critical transitions, help to solve these problems. Sometimes, a mean-field approximation of the interaction dynamics allows deriving an equation for the order parameter.

Long-term predictability. Even if the dynamics are deterministic, their long-term predictability is not always guaranteed. Under specific conditions which also depend on the control parameters, small differences between neighboring trajectories can grow exponentially. Thus, after finite time, two similar system states can become completely different. In time discrete systems, so-called maps, this is known as deterministic chaos [97]. But chaos also exists in higher dimensional continuous-time dynamics.

This problem involves two scientific questions: (i) how to detect whether and at what time horizon a dynamic system lacks long-term predictability, (ii) how to push the system with targeted interventions back to a predictable regime. One possible answer to the first question is the Ljapunov exponent, which can be calculated from a time series of the system dynamics. A positive Lyapunov exponent is a necessary but not a sufficient condition for chaos.

To answer the second question, different mechanism of chaos control have been proposed. In essence, they provide an additional feedback cycle to stabilize the system dynamics [2, 44, 66]. But the right moment and the right amount of the control signal need to be determined. These problems are closely related to network control, discussed under Network Models.

Focus on distributions. In various cases, system dynamics models the distribution of an underlying variable. If empirical data is available, such distributions can be obtained from the observed frequencies [25, 90]. Hence, many data studies can be attributed to the broad range of system dynamics models, as long as they focus on the systemic properties, notably the distributions [100, 111].

To describe their dynamics, one studies the change of the distribution's characteristic parameters, i.e., the mean and the standard deviation. If the dynamics of the underlying variable are not known or too challenging to formalize, we can consider a statistical model instead. That means a random process generates the respective distribution. Unlike an agent-based model, the random process does not capture microscopic interactions of system elements. Instead, it "simulates" dynamics that lead to the same probability distribution [56]. The best example is a random walker model to simulate a diffusion process. Also, the Metropolis-Hasting algorithm to simulate the relaxation of a system into an equilibrium state follows a similar idea. Ultimately, the interest is in the resulting distribution that characterizes the *system*.

Universality. One remarkable insight from system dynamics is the fact that the same frequency distribution can describe data from very different systems. For instance, in an election, each candidate receives a specific number of votes, or in science, each publication gets a particular number of citations. These numbers depend on the country, the scientific disciplines, etc. However, suppose we scale these numbers appropriately, e.g., by the mean number of votes or citations. In that case, we find that the distribution of votes and citations have the same form, namely a log-normal distribution. Hence, different systems seem to follow the same statistical laws from an abstract and aggregated perspective. So, the laws are, in some sense, universal and apply to very different phenomena.

Such findings raise several issues: (i) to find these universal laws and classify them, (ii) to detect their limits, (iii) to understand their meaning, e.g., by using a generative model for the underlying variables, It should help clarify the meaning of, e.g., "social" if social and physical systems follow the same universal laws.

Information, inequality, and ranking. In many cases, a distribution reflects the underlying heterogeneity of agents or system elements. We can then use this distribution to derive aggregated measures that characterize the system as a whole. For instance, to quantify *inequality*, we can calculate the well-known Gini coefficient.

The degree of the order in a system is captured be the *information entropy* as an aggregated measure. It can be interpreted as the number of bit needed to capture the state of a system comprised of many elements. If the system is in an ordered state, we need less information because of the underlying order. For disordered or unpredictable systems, however, we would need a large amount of information to describe the system state correctly.

We can also use the underlying distribution to obtain a ranking of elements against the value of their respective variable. Hence, order, information, inequality, or rankings are *systemic properties*, and we use a system dynamics perspective based on distributions to quantify them.

Scaling. It is essential to know how systemic properties, e.g., conductivity in a physical system or consensus in a social system, depend on the size of the system or other input variables [53, 64, 85, 90]. Technically speaking, how do these properties scale with size? The term "scale" already implies that we do not need to know an exact closed-form expression for this relation, but a coarse-grained argument of the form $Y(X) \sim X^{\alpha}$, where α is the scaling exponent. Such a relationship makes sense only if it can be applied to a large range of X values, ideally over several orders of magnitude.

In macroeconomic theory, such relations are known as returns to scale. The input variables are often capital and labor, and the output describes a production. If we double capital or labor, will we obtain twice the output (constant returns to scale), or will we get even more than that (increasing returns to scale) or less (decreasing returns to scale)? The law of diminishing returns states that we will obtain more output if we increase the input, but with every new unit of input the gain in output becomes smaller. It sets natural limits for extending economic activities.

Scaling laws imply that the respective system quantity changes with size or other input variables continuously [32, 42]. That means we do not expect disruptive regime shifts or first-order phase transitions. But such sharp transitions usually only occur in very large systems. In finite systems, we instead observe a smoothed-out transition to the new state, which itself could be described by a scaling law.

Application: Economic dynamics. Macroeconomic models bear many similarities with system dynamics models. In particular, they exhibit spontaneous oscillations known as business cycles which need to be explained [33, 102, 125]. Other types of fluctuations are less regular, e.g., in financial markets, but can be captured based on dynamic information flows [137] or through scaling laws [27, 71, 73, 89, 101].

Economic innovation diffusion models often adopt a system dynamics perspective [98, 106, 131]. Other economic models focus on the role of feedback cycles that, in combination with shocks, lead to instability [12, 105, 107] or to stability of unwanted equilibrium states [103].

Eventually, scaling laws in macro economics [69], but also financial systems [24, 78, 87, 99, 101] are studied. Generalized Lotka-Volterra models are shown to reproduce some of the observed power-law distributions in economic systems [13].

Application: Environmental dynamics. Papers about ecological or environmental systems published in *Advances in Complex Systems* are often related to the problem of long-term prediction. Interestingly, the meaning of the term "prediction" has changed during the development of complexity science [61]. If environmental system dynamics exhibit spatial or temporal chaos, prediction is hampered. Therefore the conditions for chaos need to be identified [19, 28, 43]. On the other hand, long-term prediction is also restricted by systemic instability [23, 74, 118, 124]. In general, we note that evolutionary processes require some periods of instability for the system to evolve into a new stable state. Related papers are discussed under "Models of Evolution".

Other models of ecological systems use the arsenal of nonlinear dynamics to analyze critical parameters for adaptation processes in insects [79, 121, 136], infectious diseases [40, 45, 55, 65], spatial plankton distributions [51], single species [36], or predator-prey systems [39, 52, 60, 75, 76] and food webs [50, 93]. Human populations are modeled as well [104]. Also, polluted environmental systems are studied [57, 63]. Eventually, scaling laws for species in real-world habitats are investigated [5].

Application: Social dynamics. Studies of social phenomena apply the system perspective to understand better inequality and rankings, often concerning measures of performance and success. *Advances in Complex Systems* has published a related topical issue in 2018. Case studies investigate the success of scientists [86], in soccer teams [128] and of pop songs [129]. One particular issue is gender disparity, e.g., among top rated computer scientists [127, 138].

Other studies pay particular attention to the system dynamics, using, for example, data from the Eurovision Song Contest [110] or from popular album charts [7]. It is interesting whether such data show signatures of self-organized criticality found in other systems [8, 10]. A population dynamics model was also adapted to describe the activities of religious movements [108, 109] or political systems [68]. And, of course, system dynamics models of diffusion [30, 36, 106] or epidemic spreading [40, 41, 92, 122, 134] are enriched by considering social behavior [88, 141].

Application: Flow dynamics. The dynamics of transport, mobility, or flow can be simulated using agent-based models as long as the laws of motion are known. But in many cases, it is more convenient to model the resulting flow as a macroscopic phenomenon using continuum descriptions [9, 26, 29, 82]. It can ease the computation and also allows for analyzing critical conditions. For instance, hydrodynamic phenomena can be modeled by solving partial differential equations or simulating particle movement using molecular dynamics techniques.

One important application area of flow dynamics is *granular matter* [15, 16, 17, 18, 20, 58], and *Advances in Complex Systems* has published a topical issue about that. Another application area is the flow of *pedestrians* which is important to understand crowd dynamics [70, 77, 114, 140]. A third application area is *traffic*. It mostly concerns car traffic [116, 119], but *Advances in Complex Systems* has also published flow models of Internet traffic [96] and ant traffic, which was additionally supported by multi-agent simulations [46].

Application: Information measures. Most papers published in *Advances in Complex Systems* measure information by information entropy, an aggregated variable characterizing the system. But instead of the plain entropy, usually derived aggregated measures, such as the mutual information [83, 115] or multiscale complexity measures [31, 37] are evaluated.

An information-theoretic analysis allows to quantify the volatility of financial markets [133], the semantics of written language [23], word recognition [135] or music [84, 139]. It can also be used to describe transient regimes in synchronization phenomena [14, 34]. A particular method based on information bottlenecks [22, 126] helps to identify optimal system models.

Information entropy can be further employed to optimize the description of partially ordered system states [81] or to model the transition toward ordered or optimized states [112, 113], sometimes denoted as guided self-organization [62]. The principle of minimal information is also used to define the most accurate distribution under the minimal available information of the considered system [117].

Comments on the reprinted publications

Specialization and generalization within biological populations [6]. At a first glimpse, the modeling approach of this paper could be seen as an agent-based model. Agents have to solve tasks. There are n different tasks, and each task is present x times. Solved tasks disappear, i.e., there is a depleting "resource". Agents increase their propensity to perform specific tasks once they solve them. On the other hand, they decrease their propensity for all other tasks not solved, which is a negative feedback process.

But a closer look shows that agents do not interact. They are mainly used to introduce the respective (unknown) distribution of tasks solved, which depends on two control parameters, the rates of learning and forgetting. Thus, the paper sells itself correctly as a *population* dynamics paper. But instead of analyzing the (unknown) system dynamics on the aggregated level, it resorts to *simulating* an underlying process of learning and forgetting on the micro level. What matters, in the end, is the distribution of aggregated values; in this case, the average number of solved tasks depending on the two control parameters (see Figure 2). Differences in this distribution allow to characterize specialists and generalists.

The paper points out the important relationship between system dynamics and agent-based models. If the system dynamics is not explicitly given, we can still obtain it by simulating an underlying generative process.

Inference versus imprint in climate modeling [21]. Climate models are the prototype of complex system dynamics models. They are complex because of the large number of dynamic variables coupled with nonlinear relations and the large number of control parameters involved. A system dynamics approach is needed because the generative processes for the emergence of the systemic properties are mostly not accessible. The problem complexity is exacerbated by the need to model systemic quantities at a specific spatial and temporal resolution. Therefore, the model needs to offer an interface to calibration with empirical data, and the nonlinear dynamics itself are hardly predictable in a turbulent or chaotic regime. But most of all, the precise dynamics are often not even known. In this situation, we are reminded of the mathematician Georges Polya, who concluded that if the problem is too difficult to solve, we should turn to a simpler one. But given the importance of climate change, that is not an alternative.

To better understand the advancements in climate modeling, it is worth looking back at early attempts to approach system dynamics. One such attempt is the ϵ-machine, a particular type of stochastic cellular automata. The machine resembles a network where the nodes are different states of a machine, and the links indicate possible transitions between states. In the paper at hand, the ϵ-machine tries to learn from the available spatio-temporal data a dynamic model of turbulent transport processes that applies to smaller spatio-temporal scales. Hence, the stochastic and deterministic structure of the model is inferred from observations rather than imprinted a priori.

It could be enlightening for those familiar with today's advanced tools of deep learning to see the differences in the methodological approach. Unlike an artificial neural network, the model is not hidden behind layers. However, when this paper was published in 2002, the limited memory for the computation and the choice of a reference model to assess computational effort and model accuracy were rendered problematic.

Quantifying emergence with mutual information [83]. This paper is about predicting the future, which is, according to Nils Bohr, a difficult task. For most systems, we have some data available about their history. How much information can we derive from this data to extrapolate future behavior? It depends on the system dynamics. In the case of simple deterministic systems, we may have enough information, but if the system, for instance, exhibits chaos, long-term prediction may be impossible.

To scrutinize the analysis, a distinction between strong and weak emergence is picked up in the paper. This discussion played a role in the sciences of complexity around the year 2000 and also involved philosophers. Weak emergence refers to a scenario where an expectation of the dynamic outcome can be formed, e.g., from an ensemble approach or averaging different computer simulation runs. On the other hand, strong emergence cannot be described from an ensemble, but relies on the specific history.

An information-theoretic approach is used to distinguish between these two cases. The paper defines permanently persistent mutual information, roughly speaking, the part of the information that makes a system evolution unique. i.e., it is the amount of information about the future which is determined by the past, and hence the extent to which the future can be forecast from past observations (of the same realization). This information is hidden in the system dynamics and cannot be anticipated by studying similar systems. It is worth noting that there are close relationships to other complexity measures, e.g., Excess Entropy, Efficient Measure

Complexity, or Statistical Complexity. Thus, the paper reminds us of the ongoing discussion about aggregated complexity measures that can reflect the dynamics' complexity and the uniqueness of its outcome.

Coevolution of road expansion and urban traffic growth [116]. This paper would also fit the part about models of evolution. But a closer look reveals that two coupled nonlinear growth processes are modeled on their way toward equilibrium, defined by the saturation level. These two equations are supposed to reflect the influence of the transportation system, characterized by the number of travelers, vehicles, and road mileage. The paper focuses on the control parameters' role as in other system dynamics models. The study identifies three periods of different lengths in the dynamics. To apply the model to a real-world scenario, Tokyo and Beijing are used as two different examples. The model is calibrated against Beijing data to determine the four relevant model parameters. The simulated aggregated outcome then shows the expected behavior.

The paper illustrates the typical system dynamics approach and its limitations. For instance, no spatial effects are considered, the transportation system is not modeled as a network, but as a number, and growth is assumed to follow a power law. From a more general perspective, one should not simply criticize a model for what it does not reflect but also praise it for what it can achieve. In the case at hand, we learn how to adapt a simple saturation dynamics to an urban problem and how measurable control parameters impact the model outcome.

Dynamics of exploited fish populations [124]. The role of long-term memory, or correlations, in the dynamics of complex systems is inherently related to the problem of predictability. It was discussed above from the perspective of weak versus strong emergence. This paper pays particular attention to higher than second order correlations. These are more difficult to detect in time series data and to model analytically. The framework of fractional dynamic processes is often used, with fractional Brownian motion as the main example. The framework assumes higher correlations in the increments, usually reflected in the integration kernel of the dynamic equation.

The paper applies this framework to the time series of the growth rate of three fish populations. If this growth rate were purely random, it would imply limited controllability. Thus, it is important to know if higher-order correlations exist for implementing sustainability measures. The paper demonstrates how to separate pure random features from the non-trivial dynamics of amplitude fluctuations in the dynamics of the three populations. It, of course, leaves open the question of what biological or ecological reasons impact the stochastic growth dynamics. But the discussion points to human exploitation and environmental change as the main influences causing long-range dependencies.

In conclusion, we should seek to apply fractional differential equations as a modeling tool to other complex processes in natural or socio-economic systems. We should also watch out for higher-order correlations.

Economic complexity measures from Lotka–Volterra equations [132]. How to define appropriate macroscopic complexity measures is one of the persistent questions in complexity science from its very beginning. As mentioned above, entropy-based measures are often considered, which can be recast as information measures. This general discussion challenges disciplinary approaches, notably in economics, where aggregated indices have long been used to quantify the status of economies or the performance of financial markets.

This paper enters a recent and partially emotional discussion about quantifying the economic complexity of countries using the diversity of their production. The novel idea of the approach proposed in this paper is the ecological perspective. The model explicitly states that technological progress drives product diversity, in addition to geographical positions and economic relations. Instead of calculating an economic complexity index from the data, a dynamic model of three coupled quantities, countries, technologies, and products, is proposed. Potentially this dynamics reaches an equilibrium state, but one can also study the convergence process.

The paper takes data from 41 countries to compare its index with two other indices from the literature. To measure technological knowledge, it uses patenting activities of a country. The calculations of the indices result in three respective rankings of countries regarding their economic fitness, which partly deviate, and partly match. So, the dilemma is similar to other studies of this kind. Without ground truth, we are left to discuss these measures' differences and correlations. That is why the authors make a point here also regarding a theoretical justification of their ecological approach.

Simple amplitude dynamics of Covid-19 outbreaks [134]. This paper uses a system dynamics model of epidemic spreading. It considers susceptible, infected, and recovered subpopulations in the basic scenario and further an exposed subpopulation in the extended scenario. A standard analysis gives the equilibrium states of the coupled differential equations conditional on the control parameters, and a stability analysis further explains how the system responds to perturbations of these equilibrium states. This dynamics can be expressed in terms of an order parameter, a time-dependent three-component vector comprised of the fixed points, the basis vectors, and time-dependent amplitudes. Focusing on their interpretation allows more concise insights into the driving forces behind the epidemic dynamics and a better comparison across systems.

The paper demonstrates in a clear way how tools of nonlinear dynamics can be applied to system dynamics models of four coupled equations. It highlights the advantage of this approach. But even more, the paper applies the order parameter dynamics to describe the real-world spreading of COVID-19 in China. So, it gives a didactic example of how to theoretically and empirically understand system dynamics.

Selected publications

[1] Anderson, C., *Simulation of the Feedbacks and Regulation of Recruitment Dancing in Honey Bees*, Advances in Complex Systems **01** (**1998**) 267–282.

[2] Bienenstock, E. and Lehmann, D., *Regulated Criticality in the Brain?*, Advances in Complex Systems **01** (**1998**) 361–384.

[3] Laugesen, J., Mosekilde, E., Maistrenko, Y. L., and Maistrenko, V. L., *On-Off Intermittency and Riddled Basins of Attraction in a Coupled Map System*, Advances in Complex Systems **01** (**1998**) 161–180.

[4] Postnov, D. E., Balanov, A. G., and Mosekilde, E., *Synchronization Phenomena in an Array of Population Dynamic Systems*, Advances in Complex Systems **01** (**1998**) 181–202.

[5] Solé, R. V. and Alonso, D., *Random Walks, Fractals and the Origins of Rainforest Diversity*, Advances in Complex Systems **01** (**1998**) 203–220.

[6] Spencer, A. J., Couzin, I. D., and Franks, N. R., *The Dynamics of Specialization and Generalization Within Biological Populations*, Advances in Complex Systems **01** (**1998**) 115–127.

[7] Bentley, R. A. and Maschner, H. D. G., *Subtle Nonlinearity in Popular Album Charts*, Advances in Complex Systems **02** (**1999**) 197–208.

[8] Jan, N., Moseley, L., Ray, T., and Stauffer, D., *Is the Fossil Record Indicative of a Critical System?*, Advances in Complex Systems **02** (**1999**) 137–141.

[9] Tabony, J., Vuillard, L., and Papaseit, C., *Biological Self-Organisation and Pattern Formation By Way of Microtubule Reaction-Diffusion Processes*, Advances in Complex Systems **02** (**1999**) 221–276.

[10] Vázquez, A., Sotolongo-Costa, O., and Brouers, F., *Avalanche Dynamics and Long Time Relaxation in Complex Systems*, Advances in Complex Systems **02** (**1999**) 173–183.

[11] Ballot, G. and Taymaz, E., *Competition, Training, Heterogeneity Persistence, and Aggregate Growth in A Multi-Agent Evolutionary Model*, Advances in Complex Systems **03** (**2000**) 335–351.

[12] Mazzucato, M., *Firm Size, Innovation, and Market Share Instability: the Role of Negative Feedback and Idiosyncratic Events*, Advances in Complex Systems **03** (**2000**) 417–431.

[13] Solomon, S., *Generalized Lotka-Volterra (Glv) Models of Stock Markets*, Advances in Complex Systems **03** (**2000**) 301–322.

[14] Crutchfield, J. P. and Feldman, D. P., *Synchronizing to the Environment: Information-Theoretic Constraints on Agent Learning*, Advances in Complex Systems **04** (**2001**) 251–264.

[15] Dufty, J. W., *Kinetic Theory and Hydrodynamics For a Low Density Granular Gas*, Advances in Complex Systems **04** (**2001**) 397–406.

[16] Khakhar, D. V., Orpe, A. V., and Ottino, J. M., *Surface Granular Flows: Two Related Examples*, Advances in Complex Systems **04** (**2001**) 407–417.

[17] Losert, W. and Kwon, G., *Transient and Steady-State Dynamics of Granular Shear Flows*, Advances in Complex Systems **04** (**2001**) 369–377.

[18] Makse, H. A., *Nonlinear Elasticity and Thermodynamics of Granular Materials*, Advances in Complex Systems **04** (**2001**) 491–501.

[19] Petrovskii, S. V. and Malchow, H., *Spatio-Temporal Chaos in an Ecological Community as a Response to Unfavourable Environmental Changes*, Advances in Complex Systems **04** (**2001**) 227–249.

[20] Del Rio, J. A., Kostoff, R. N., Garcia, E. O., Ramirez, A. M., and Humenik, J. A., *Phenomenological Approach To Profile Impact of Scientific Research: Citation Mining*, Advances in Complex Systems **05** (**2002**) 19–42.

[21] Palmer, A. J., Schneider, T. L., and Benjamin, L. A., *Inference Versus Imprint in Climate Modeling*, Advances in Complex Systems **05** (**2002**) 73–89.

[22] Shalizi, C. R. and Crutchfield, J. P., *Information Bottlenecks, Causal States, and Statistical Relevance Bases: How To Represent Relevant Information in Memoryless Transduction*, Advances in Complex Systems **05** (**2002**) 91–95.

[23] Bouzat, S. and Zanette, D. H., *Survival and Extinction in the Replicator Model: Dynamics and Statistics*, Advances in Complex Systems **06** (**2003**) 223–240.

[24] Capobianco, E., *Functional Approximation in Multiscale Complex Systems*, Advances in Complex Systems **06** (**2003**) 177–204.

[25] Kaizoji, T. and Kaizoji, M., *Empirical Laws of a Stock Price Index and a Stochastic Model*, Advances in Complex Systems **06** (**2003**) 303–312.

[26] Maselko, J., *Pattern Formations in Chemical Systems*, Advances in Complex Systems **06** (**2003**) 3–14.

[27] Staliunas, K., *Anticorrelations and Subdiffusion in Financial Systems*, Advances in Complex Systems **06** (**2003**) 251–262.

[28] Strizhak, P. E., *Application of Chemical Chaos to Analytical Chemistry*, Advances in Complex Systems **06** (**2003**) 137–153.

[29] Taylor, A., *Chemical Patterns in Simple Flow Systems*, Advances in Complex Systems **06** (**2003**) 155–162.

[30] Ahmed, E., Hegazi, A. S., and Elgazzar, A. S., *On Persistence and Stability of Some Biological Systems With Cross Diffusion*, Advances in Complex Systems **07** (**2004**) 65–76.

[31] Bar-Yam, Y., *Multiscale Complexity/Entropy*, Advances in Complex Systems **07** (**2004**) 47–63.

[32] Campos, P. R. A., De Oliveira, V. M., and Maia, L. P., *Emergence of Allometric Scaling in Genealogical Trees*, Advances in Complex Systems **07** (**2004**) 39–46.

[33] Delli Gatti, D., Di Guilmi, C., Gaffeo, E., Giulioni, G., Gallegati, M., and Palestrini, A., *Business Cycle Fluctuations and Firms' Size Distribution Dynamics*, Advances in Complex Systems **07** (**2004**) 223–240.

[34] Feldman, D. P. and Crutchfield, J. P., *Synchronizing to Periodicity: the Transient Information and Synchronization Time of Periodic Sequences*, Advances in Complex Systems **07** (**2004**) 329–355.

[35] Zeng, G., Chen, L., Sun, L., and Liu, Y., *Permanence and the Existence of the Periodic Solution of the Non-Autonomous Two-Species Competitive Model With Stage Structure*, Advances in Complex Systems **07** (**2004**) 385–393.

[36] Chen, F., Chen, X., and Shi, J., *Dynamic Behavior of a Nonlinear Single Species Diffusive System*, Advances in Complex Systems **08** (**2005**) 399–417.

[37] Davison, M. and Shiner, J. S., *Extended Entropies and Disorder*, Advances in Complex Systems **08** (**2005**) 125–158.

[38] Livadiotis, G., *Numerical Approximation of the Percentage of Order For One-Dimensional Maps*, Advances in Complex Systems **08** (**2005**) 15–32.

[39] Pei, Y., Li, C., Chen, L., and Wang, C., *Complex Dynamics of One-Prey Multi-Predator System With Defensive Ability of Prey and Impulsive Biological Control On Predators*, Advances in Complex Systems **08** (**2005**) 483–495.

[40] Schinaia, G., *Empirical Data and Mathematical Structures in the Epidemic Modeling of Parenteral Hepatitis in Italy*, Advances in Complex Systems **08** (**2005**) 33–58.

[41] Zeng, G.-Z. and Chen, L.-S., *Complexity and Asymptotical Behavior of a Sirs Epidemic Model With Proportional Impulsive Vaccination*, Advances in Complex Systems **08** (**2005**) 419–431.

[42] Łaszkiewicz, A., Cebrat, S., and Stauffer, D., *Scaling Effects in the Penna Ageing Model*, Advances in Complex Systems **08** (**2005**) 7–14.

[43] Chaudhuri, S., *Predictability of Chaos Inherent in the Occurrence of Severe Thunderstorms*, Advances in Complex Systems **09** (**2006**) 77–85.

[44] Duan, Z., Wang, J., and Huang, L., *Some Special Decentralized Control Problems in Continuous-Time Interconnected Systems*, Advances in Complex Systems **09** (**2006**) 277–286.

[45] Liu, H., Yu, J., and Zhu, G., *Analysis of a Vector-Host Malaria Model With Impulsive Effect and Infection-Age*, Advances in Complex Systems **09** (**2006**) 237–248.

[46] Peters, K., Johansson, A., Dussutour, A., and Helbing, D., *Analytical and Numerical Investigation of Ant Behavior Under Crowded Conditions*, Advances in Complex Systems **09** (**2006**) 337–352.

[47] Xiaoxing, C., *Permanence and Global Stability For Nonlinear Discrete Model*, Advances in Complex Systems **09** (**2006**) 31–40.

[48] Yang, X.-S. and Wang, L., *Emergent Periodic Behavior in Coupled Chaotic Systems*, Advances in Complex Systems **09** (**2006**) 249–261.

[49] Zhang, H. and Chen, L., *Oscillation Criteria For a Class of Second-Order Impulsive Delay Difference Equations*, Advances in Complex Systems **09** (**2006**) 69–76.

[50] Zhang, S., Wang, F., and Chen, L., *A Food Chain System With Density-Dependent Birth Rate and Impulsive Perturbations*, Advances in Complex Systems **09** (**2006**) 223–236.

[51] Bhattacharyya, R. and Mukhopadhyay, B., *Modeling Fluctuations in a Minimal Plankton Model: Role of Spatial Heterogeneity and Stochasticity*, Advances in Complex Systems **10** (**2007**) 197–216.

[52] Chen, X., *Permanence in a Discrete N-Species Non-Autonomous Lotka–Volterra Competitive Systems With Infinite Delays and Feedback Control*, Advances in Complex Systems **10** (**2007**) 463–477.

[53] Ciamarra, M. P. and Coniglio, A., *Universality in City Morphology and the Morphology of a City and Its Implications For City Evacuation Plans*, Advances in Complex Systems **10** (**2007**) 373–377.

[54] Fang, N. and Chen, X., *Global Stability of a Nonlinear Discrete Population Model With Time Delays*, Advances in Complex Systems **10** (**2007**) 315–333.

[55] Jiao, J. and Chen, L., *Global Attractivity and Permanence of a Stage-Structured Pest Management Si Model With Time Delay and Diseased Pests Impulsive Transmission*, Advances in Complex Systems **10** (**2007**) 479–494.

[56] Kennedy, J., *The Particle Swarm As Collaborative Sampling of the Search Space*, Advances in Complex Systems **10** (**2007**) 191–213.

[57] Liu, B. and Tao, F., *The Optimal Pulse Harvesting Policy on a Single-Species Population Model With Birth Pulses in a Polluted Environment*, Advances in Complex Systems **10** (**2007**) 173–196.

[58] Loguinova, N. and Vlasov, Y., *Oscillations in Granular Dynamics*, Advances in Complex Systems **10** (**2007**) 287–299.

[59] Ma, S., Lu, Q., and Hogan, S. J., *Double Hopf Bifurcation For Stuart–Landau System With Nonlinear Delay Feedback And Delay-Dependent Parameters*, Advances in Complex Systems **10** (**2007**) 423–448.

[60] Pang, G. and Chen, L., *Complexity of An Ivlev'S Predator-Prey Model With Pulse*, Advances in Complex Systems **10** (**2007**) 217–231.

[61] Parisi, G., *Physics Complexity and Biology*, Advances in Complex Systems **10** (**2007**) 223–232.

[62] Rodríguez, A., Grushin, A., and Reggia, J. A., *Swarm Intelligence Systems Using Guided Self-Organization For Collective Problem Solving*, Advances in Complex Systems **10** (**2007**) 5–34.

[63] Sivakumar, B., Wallender, W. W., Horwath, W. R., and Mitchell, J. P., *Nonlinear Deterministic Analysis of Air Pollution Dynamics in a Rural And Agricultural Setting*, Advances in Complex Systems **10** (**2007**) 581–597.

[64] Valdez-Cepeda, R. D., Delgadillo-Ruiz, O., Magallanes-Quintanar, R., Miramontes-De León, G., García-Hernández, J. L., Enciso-Muñoz, A., and Mendoza, B., *Scale-Invariance of Normalized Yearly Mean Grain Yield Anomaly Series*, Advances in Complex Systems **10** (**2007**) 395–412.

[65] Wang, X. and Song, X., *Global Properties of a Model of Immune Effector Responses To Viral Infections*, Advances in Complex Systems **10** (**2007**) 495–503.

[66] Yang, X.-S., Yuan, Q., and Wang, L., *What Connection Topology Prohibit Chaos in Continuous Time Networks?*, Advances in Complex Systems **10** (**2007**) 449–461.

[67] Bosse, T. and Treur, J., *A Philosophical Foundation For Unification of Dynamic Modeling Methods Based On Higher-Order Potentialities and Their Reducers*, Advances in Complex Systems **11** (**2008**) 831–860.

[68] Dittrich, P. and Winter, L., *Chemical Organizations in a Toy Model of the Political System*, Advances in Complex Systems **11** (**2008**) 609–627.

[69] Gatti, D. D., Di Guilmi, C., Gallegati, M., Gaffeo, E., Giulioni, G., and Palestrini, A., *Scaling Laws in the Macroeconomy*, Advances in Complex Systems **11** (**2008**) 131–138.

[70] Johansson, A., Helbing, D., Al-Abideen, H. Z., and Al-Bosta, S., *From Crowd Dynamics to Crowd Safety: A Video-Based Analysis*, Advances in Complex Systems **11** (**2008**) 497–527.

[71] Liu, R., Di Matteo, T., and Lux, T., *Multifractality and Long-Range Dependence of Asset Returns: The Scaling Behavior of the Markov-Switching Multifractal Model With Lognormal Volatility Components*, Advances in Complex Systems **11** (**2008**) 669–684.

[72] Radde, N., *The Effect of Time Scale Differences and Time Delays on the Structural Stability of Oscillations in a Two-Gene Network*, Advances in Complex Systems **11** (**2008**) 471–483.

[73] Rosales, F., Posadas, A., and Quiroz, R., *Multifractal Characterization of Spatial Income Curdling: Theory and Applications*, Advances in Complex Systems **11** (**2008**) 861–874.

[74] Voorhees, B., Senez, J., Keeler, T., and Connors, M., *A Population Model of the Stability–Flexibility Tradeoff*, Advances in Complex Systems **11** (**2008**) 443–470.

[75] Zeng, G., Wang, F., and Nieto, J. J., *Complexity of a Delayed Predator–Prey Model With Impulsive Harvest and Holling Type II Functional Response*, Advances in Complex Systems **11** (**2008**) 77–97.

[76] Zhang, S., Tan, D., and Gu, H., *Dynamic Complexities of a Chemostat Model With Pulsed Input and Washout At Different Times*, Advances in Complex Systems **11** (**2008**) 65–76.

[77] Chattaraj, U., Seyfried, A., and Chakroborty, P., *Comparison of Pedestrian Fundamental Diagram Across Cultures*, Advances in Complex Systems **12** (**2009**) 393–405.

[78] Conlon, T., Ruskin, H. J., and Crane, M., *Multiscaled Cross-Correlation Dynamics in Financial Time-Series*, Advances in Complex Systems **12** (**2009**) 439–454.

[79] Diwold, K., Merkle, D., and Middendorf, M., *Adapting to Dynamic Environments: Polyethism in Response Threshold Models For Social Insects*, Advances in Complex Systems **12** (**2009**) 327–346.

[80] Jacobi, M. N. and Görnerup, O., *A Spectral Method for Aggregating Variables in Linear Dynamical Systems With Application to Cellular Automata Renormalization*, Advances in Complex Systems **12** (**2009**) 131–155.

[81] Salge, C. and Polani, D., *Information-Driven Organization of Visual Receptive Fields*, Advances in Complex Systems **12** (**2009**) 311–326.

[82] Ali, A. and Grosskinsky, S., *Pattern Formation Through Genetic Drift At Expanding Population Fronts*, Advances in Complex Systems **13** (**2010**) 349–366.

[83] Ball, R. C., Diakonova, M., and Mackay, R. S., *Quantifying Emergence in Terms of Persistent Mutual Information*, Advances in Complex Systems **13** (**2010**) 327–338.

[84] Boon, J. P., *Complexity, Time and Music*, Advances in Complex Systems **13** (**2010**) 155–164.

[85] Gotts, N. M. and Polhill, J. G., *Size Matters: Large-Scale Replications of Experiments With Fearlus*, Advances in Complex Systems **13** (**2010**) 453–467.

[86] Krapivin, M., Marchese, M., and Casati, F., *Exploring and Understanding Citation-Based Scientific Metrics*, Advances in Complex Systems **13** (**2010**) 59–81.

[87] Zhang, J., Wang, J., and Shao, J., *Finite-Range Contact Process on the Market Return Intervals Distributions*, Advances in Complex Systems **13** (**2010**) 643–657.

[88] Apolloni, A. and Gargiulo, F., *Diffusion Processes Through Social Groups' Dynamics*, Advances in Complex Systems **14** (**2011**) 151–167.

[89] Deng, W., Li, W., Cai, X., and Wang, Q. A., *On the Application of the Cross-Correlations in the Chinese Fund Market: Descriptive Properties and Scaling Behaviors*, Advances in Complex Systems **14** (**2011**) 97–109.

[90] Guéret, C., Wang, S., Groth, P., and Schlobach, S., *Multi-Scale Analysis of the Web of Data: a Challenge To the Complex System'S Community*, Advances in Complex Systems **14** (**2011**) 587–609.

[91] Irle, A., Kauschke, J., Lux, T., and Milaković, M., *Switching Rates and the Asymptotic Behavior of Herding Models*, Advances in Complex Systems **14** (**2011**) 359–376.

[92] Kamp, C., *Mixing Patterns Among Epidemic Groups*, Advances in Complex Systems **14** (**2011**) 537–547.

[93] Palamara, G. M., Zlatić, V., Scala, A., and Caldarelli, G., *Population Dynamics on Complex Food Webs*, Advances in Complex Systems **14** (**2011**) 635–647.

[94] Peter, S. and Dittrich, P., *On the Relation Between Organizations and Limit Sets in Chemical Reaction Systems*, Advances in Complex Systems **14** (**2011**) 77–96.

[95] Schulman, L. S., Bagrow, J. P., and Gaveau, B., *Visualizing Relations Using the "Observable Representation"*, Advances in Complex Systems **14** (**2011**) 829–851.

[96] Smith, R. D., *The Dynamics of Internet Traffic: Self-Similarity, Self-Organization, and Complex Phenomena*, Advances in Complex Systems **14** (**2011**) 905–949.

[97] Zeraoulia, E. and Sprott, J. C., *Robustification of Chaos in 2d Maps*, Advances in Complex Systems **14** (**2011**) 817–827.

[98] Bentley, R. A. and Ormerod, P., *Accelerated Innovation and Increased Spatial Diversity of Us Popular Culture*, Advances in Complex Systems **15** (**2012**) 1150011.

[99] Caccioli, F., Catanach, T. A., and Farmer, J. D., *Heterogeneity, Correlations and Financial Contagion*, Advances in Complex Systems **15** (**2012**) 1250058.

[100] Jäger, G., *Power Laws and Other Heavy-Tailed Distributions in Linguistic Typology*, Advances in Complex Systems **15** (**2012**) 1150019.

[101] Kristoufek, L., *Fractal Markets Hypothesis and the Global Financial Crisis: Scaling, Investment Horizons and Liquidity*, Advances in Complex Systems **15** (**2012**) 1250065.

[102] Mcnelis, P. D. and Yoshino, N., *Macroeconomic Volatility Under High Accumulation of Government Debt: Lessons From Japan*, Advances in Complex Systems **15** (**2012**) 1250057.

[103] Mueller, G. P., *The Dynamics and Evolutionary Stability of Cultures of Corruption: Theoretical And Empirical Analyses*, Advances in Complex Systems **15** (**2012**) 1250082.

[104] Premo, L. S., *Local Extinctions, Connectedness, and Cultural Evolution in Structured Populations*, Advances in Complex Systems **15** (**2012**) 1150002.

[105] Raberto, M., *Managing Financial Instability in Capitalist Economies*, Advances in Complex Systems **15** (**2012**) 1203005.

[106] Scholnick, J. B., *The Spatial and Temporal Diffusion of Stylistic Innovations in Material Culture*, Advances in Complex Systems **15** (**2012**) 1150010.

[107] Semmler, W. and Chappe, R., *Ponzi Finance and the Hedge Fund Industry*, Advances in Complex Systems **15** (**2012**) 1250037.

[108] Vitanov, N. K., Ausloos, M., and Rotundo, G., *Discrete Model of Ideological Struggle Accounting For Migration*, Advances in Complex Systems **15** (**2012**) 1250049.

[109] Ausloos, M., *Another Analytic View About Quantifying Social Forces*, Advances in Complex Systems **16** (**2013**) 1250088.

[110] García, D. and Tanase, D., *Measuring Cultural Dynamics Through the Eurovision Song Contest*, Advances in Complex Systems **16** (**2013**) 1350037.

[111] Kostanjčar, Z. and Jeren, B., *Emergence of Power-Law and Two-Phase Behavior in Financial Market Fluctuations*, Advances in Complex Systems **16** (**2013**) 1350008.

[112] Polani, D., Prokopenko, M., and Yaeger, L. S., *Information and Self-Organization of Behavior*, Advances in Complex Systems **16** (**2013**) 1303001.

[113] Van Dijk, S. G. and Polani, D., *Informational Constraints-Driven Organization in Goal-Directed Behavior*, Advances in Complex Systems **16** (**2013**) 1350016.

[114] Maiti, R. R., Mallya, A., Mukherjee, A., and Ganguly, N., *Understanding the Correlation of the Properties of Human Movement Patterns*, Advances in Complex Systems **17** (**2014**) 1450019.

[115] Pfante, O., Bertschinger, N., Olbrich, E., Ay, N., and Jost, J., *Comparison Between Different Methods of Level Identification*, Advances in Complex Systems **17** (**2014**) 1450007.

[116] Wu, J., Xu, M., and Gao, Z., *Modeling the Coevolution of Road Expansion and Urban Traffic Growth*, Advances in Complex Systems **17** (**2014**) 1450005.

[117] Yukalov, V. I. and Sornette, D., *Self-Organization in Complex Systems As Decision Making*, Advances in Complex Systems **17** (**2014**) 1450016.

[118] Zgonnikov, A. and Lubashevsky, I., *Unstable Dynamics of Adaptation in Unknown Environment Due to Novelty Seeking*, Advances in Complex Systems **17** (**2014**) 1450013.

[119] Andreotti, E., Bazzani, A., Rambaldi, S., Guglielmi, N., and Freguglia, P., *Modeling Traffic Fluctuations and Congestion on a Road Network*, Advances in Complex Systems **18** (**2015**) 1550009.

[120] Banisch, S. and Lima, R., *Markov Chain Aggregation for Simple Agent-Based Models on Symmetric Networks: The Voter Model*, Advances in Complex Systems **18** (**2015**) 1550011.

[121] Cabrera, O. and Zanette, D. H., *Avoiding Extinction By Migration: The Case of the Head Louse*, Advances in Complex Systems **18** (**2015**) 1550010.

[122] Eckalbar, J. C., Tsournos, P., and Eckalbar, W. L., *Dynamics in An Sir Model When Vaccination Demand Follows Prior Levels of Disease Prevalence*, Advances in Complex Systems **18** (**2015**) 1550021.

[123] Lindgren, K., Jonson, E., and Lundberg, L., *Projection of a Heterogenous Agent-Based Production Economy Model to a Closed Dynamics of Aggregate Variables*, Advances in Complex Systems **18** (**2015**) 1550012.

[124] Mendes, H. C., Murta, A., and Mendes, R. V., *Long Range Dependence and the Dynamics of Exploited Fish Populations*, Advances in Complex Systems **18** (**2015**) 1550017.

[125] Bashkirtseva, I., Pisarchik, A., Ryashko, L., and Ryazanova, T., *Excitability and Complex Mixed-Mode Oscillations in Stochastic Business Cycle Model*, Advances in Complex Systems **19** (**2016**) 1550027.

[126] Lamarche-Perrin, R., Banisch, S., and Olbrich, E., *The Information Bottleneck Method For Optimal Prediction of Multilevel Agent-Based Systems*, Advances in Complex Systems **19** (**2016**) 1650002.

[127] Jadidi, M., Karimi, F., Lietz, H., and Wagner, C., *Gender Disparities in Science? Dropout, Productivity, Collaborations and Success of Male and Female Computer Scientists*, Advances in Complex Systems **21** (**2018**) 1750011.

[128] Pappalardo, L. and Cintia, P., *Quantifying the Relation Between Performance and Success in Soccer*, Advances in Complex Systems **21** (**2018**) 1750014.

[129] Shin, S. and Park, J., *On-Chart Success Dynamics of Popular Songs*, Advances in Complex Systems **21** (**2018**) 1850008.

[130] Burgos, C., Cortés, J., Martínez-Rodríguez, D., and Villanueva, R., *Computational Modeling With Uncertainty of Frequent Users of E-Commerce in Spain Using An Age-Group Dynamic*

Nonlinear Model With Varying Size Population, Advances in Complex Systems **22** (**2019**) 1950009.

[131] Gallay, O., Hashemi, F., and Hongler, M.-O., *Imitation, Proximity, and Growth — a Collective Swarm Dynamics Approach*, Advances in Complex Systems **22** (**2019**) 1950011.

[132] Ivanova, I., Strand, O., and Leydesdorff, L., *An Eco-Systems Approach to Constructing Economic Complexity Measures: Endogenization of the Technological Dimension Using Lotka–Volterra Equations*, Advances in Complex Systems **22** (**2019**) 1850023.

[133] Pfante, O. and Bertschinger, N., *Information-Theoretic Analysis of Stochastic Volatility Models*, Advances in Complex Systems **22** (**2019**) 1850025.

[134] Frank, T. D., *Simplicity From Complexity: on the Simple Amplitude Dynamics Underlying Covid-19 Outbreaks in China*, Advances in Complex Systems **23** (**2020**) 2050022.

[135] Mehta, A. and Luck, J.-M., *Hearings and Mishearings: Decrypting the Spoken Word*, Advances in Complex Systems **23** (**2020**) 2050008.

[136] Tereshko, V., *Kinetic Phase Transition in Honeybee Foraging Dynamics: Synergy of Individual And Collective*, Advances in Complex Systems **23** (**2020**) 2050019.

[137] Zheng, J. and Nie, C.-X., *The Dynamics of Price–Volume Information Transfer in the Cryptocurrency Markets*, Advances in Complex Systems **23** (**2020**) 2050014.

[138] Jaramillo, A. M., Macedo, M., and Menezes, R., *Reaching to the Top: the Gender Effect in Highly-Ranked Academics in Computer Science*, Advances in Complex Systems **24** (**2021**).

[139] Su, J. and Zhou, P., *Use of Gaussian Process To Model, Predict and Explain Human Emotional Response To Chinese Traditional Music*, Advances in Complex Systems **24** (**2021**).

[140] Subramanian, G. H., Karthika, P., and Verma, A., *Macroscopic Fundamental Flow Diagrams of a Spiritually Motivated Crowd*, Advances in Complex Systems **24** (**2021**).

[141] Xu, W.-J., Zhong, C.-Y., Ye, H.-F., Chen, R.-D., Qiu, T., Ren, F., and Zhong, L.-X., *Risk Awareness to Epidemic Information and Self-Restricted Travel Behavior On Contagion*, Advances in Complex Systems **24** (**2021**).

J. Complex Systems (1998) **1**, 115–127

The Dynamics of Specialization and Generalization within Biological Populations

Andrew J. Spencer[†] **Iain D. Couzin**

Nigel R. Franks

Centre for Mathematical Biology and
Department of Biology and Biochemistry
University of Bath
Bath BA2 7AY, UK

(Received 29 May 1998)

ABSTRACT. We develop an abstract model to explore specialization and generalization in task performance by individuals within biological populations. Individuals follow simple rules of increasing and decreasing task propensities that could, for example, be based on learning and forgetting. The model does not explore efficiency per se, but makes the prediction that where behavioural specialization occurs in nature, organisms are likely to be reaping sufficient benefits from improved handling efficiency to offset the costs of increased search time. A second prediction is that among specialists, there will be a trade-off between stability and responsiveness. The model reveals potential similarities between a wide range of complex biological systems.

KEYWORDS: Specialists, generalists, simulation model, predation, pollination, switching, task allocation, search image, ants, bumblebees, social insects.

† Email: A.J.Spencer@bath.ac.uk.

1. Introduction

One of the greatest contributions of evolutionary biology is recognizing the importance of the uniqueness of the individual (Medawar, 1957). Indeed much of our understanding of evolutionary change through natural selection is associated with genetic biodiversity within populations (Ridley, 1993). However, both nature and nurture contribute to the uniqueness of individuals. Individuals within a population have different genotypes which encounter different environmental influences and such so-called genotype by environment interactions (Falconer, 1981) amplify phenotypic diversity. The most labile and hence arguably the most interesting aspect of phenotypes is behaviour. Here we will focus on behavioural diversity within populations by considering how learning may lead to generalization and specialization and how specialists may switch. These are important considerations over a surprisingly broad range of biological examples.

In ecology, for example, the stability of communities may be greatly influenced by individual predators specializing on one or a subset of many possible prey types (MacArthur, 1955; May, 1973; Pimm and Lawton, 1977). For example, if predators form search images for the more abundant of two (cryptic) prey types they may continue to hunt that prey type disproportionately even when it has become less abundant than the alternative (Begon *et al.*, 1990). Here population dynamics would be linked with the dynamics of individuals learning and forgetting certain search images. Similar reasoning can be applied to the behaviour of pollinators in which members of the same population may specialize in visiting only a subset of accessible flowers (Heinrich, 1979).

The advantage of specialization by individuals within groups is also considered to be of overwhelming importance in many of the major transitions in the evolution of life (Maynard Smith and Szathmáry, 1995). One such transition is from single-celled to multicellular organisms; another major evolutionary transition is from solitary organisms to highly social ones. The selective advantage of multicellular organisms over single-celled organisms is probably associated in part with cellular specialization leading to an efficient division of labour (Maynard Smith and Szathmáry, 1995). (Implicit in the division of labour is that individuals become more efficient as they specialize (Smith, 1776).) The evolutionary transition from solitary organisms to highly integrated societies of individual organisms (e.g. colonies of ants, termites and certain bees and wasps) is also associated with efficiencies that accrue from a division of labour and task specialization. Social insect colonies have been compared to factories within fortresses (Oster and Wilson, 1978, p. 21-23) and there are many different tasks that workers must perform, from building the nest and guarding the colony to tending the queen, rearing many different stages of brood, and feeding and grooming one another.

Several authors have modelled the association between learning and task allocation in social insects. Deneubourg *et al.* (1987) look at the influence of learning on the spatial fidelity of ants during foraging bouts, and discuss its implications for the division of labour. Using a conceptually similar model to our own, Plowright and Plowright (1988) model the emergence of "elitism" (i.e. that a small number of individuals do a disproportionate amount of the colony's work). More recently, Theraulaz *et al.* (1998) investigate the effects of variable response thresholds of individuals to tasks (see Discussion).

In diverse areas of biology, transitions from generalized behaviour to specialized behaviour are of major evolutionary importance. Clearly, there are fundamental differences, that must not be overlooked, between the systems we have just described. For example, individual predators will specialize on particular types of prey for their own immediate benefit, whereas cells within an organism, or ants within a colony, may specialize for mutual benefit (favouring the selfish genes they have in common). In other words, specialization within organisms or societies occurs because the entities involved belong to a community of mutual interest (Cosmides and Tooby, 1981; see also Bourke and Franks, 1995) and co-operate to favour their self interest indirectly, whereas specialization within ecological populations of distantly related individuals occurs due to direct self interest. Recognizing such fundamental differences should not, however, obscure key similarities in the dynamics of specialization and switching. It is these similarities that we explore here, by considering a model that investigates the dynamics of specialization within populations.

2. The Model

In outline:

1 Agents encounter one or more tasks in their environment.
2 At each time step each agent may perform one task.
3 If it performs a particular task its propensity to perform that task increases.
4 If it does not perform a task its propensity for performing the task decreases.

In detail:

Tasks are abstracted as discrete items, one task item being defined as the amount of task that one agent can complete in one unit of time. Hence we have not modelled the effects of changing task efficiency. For simplicity and generality, we have not specified time scales. All the parameters of the model scale with the time step. The reason for this is that time scales for different organisms are likely to differ over several orders of magnitude, and the time period represented by one time step must reflect the behaviour under consideration.

Space is not modelled explicitly; instead, each agent experiences a probability, P, of encountering a task item during each time step. This probability is defined as

$$P = \frac{T}{(T + \alpha)} \tag{2.1}$$

where T is the total number of task items within the simulation, and α is a parameter relating the size of the arena within which the agents and tasks exist to the physical size of a task item and the area searched by an agent within one timestep. This provides a reasonable approximation for the situation where individuals and task items are distributed randomly within an arena of fixed size. At each time step, each agent has a probability P_n of encountering an item of each task, n, such that

$$P_n = \frac{T_n}{(T + \alpha)} \tag{2.2}$$

where T_n is the number of items of task n within the simulation.

At the start of the simulation, each individual has the same propensities for carrying out each specific task. These propensities represent the probability that, on encountering an item of task x, the individual will work on that task item. If an individual's propensity for carrying out task x is defined as π_x, then when the individual encounters an item of task x it has probability π_x of working on that item.

If a task item is worked on by an agent, the task item disappears from the simulation for all future time steps, and the individual's propensity for that task is increased by

$$\Delta\pi_x = \pi_x(1 - \pi_x)\lambda \tag{2.3}$$

where λ is the parameter governing the individual's "learning rate". Simultaneously, for all tasks n other than x, the individual's propensity decreases by

$$\Delta\pi_n = -(\pi_n(1 - \pi_n)\phi + \pi_n\phi^2) \tag{2.4}$$

where ϕ is the parameter governing the individual's "forgetting rate".

Thus if an individual performs a task, it becomes $\Delta\pi_x$ more likely to perform that task should it encounter it in the next time step, and it becomes less likely to perform every other task by $\Delta\pi_n$ where π_n is its previous propensity for that task.

Task choice is determined purely by individual task propensities that are adjusted according to task experience. However, it is essential that a mechanism should exist to prevent task propensities from becoming trapped at zero. For the types of situation we model here, stochastic effects mean that, in time, individuals will always fail to encounter any given task for a long enough period of time that their propensity for that task reaches zero. Once a propensity has reached zero, the individual will never perform that task and so there will be no potential for its propensity to increase again. Unless individuals can be assured of a constant supply of a task for which they have high propensity, they must become prepared to do tasks which they would at first refuse if faced with a persistent dearth of their 'preferred' task(s).

We incorporate such a mechanism into this simulation by allowing individuals' propensities for all tasks to approach a low but non-zero 'resting level', R, on each occasion that a task item is refused. On refusal of a task item, for all tasks n,

$$\Delta\pi_n = -\left(\frac{(\pi_n - R)(1 - \pi_n)\phi}{1 - R} + \frac{(\pi_n - R)\phi^2}{1 - R}\right) \tag{2.5}$$

if $\pi_n > R$, or

$$\Delta\pi_n = \frac{\pi_n(R - \pi_n)\phi}{R} + \frac{(R - \pi_n)\phi^2}{R} \tag{2.6}$$

if $\pi_n < R$.

In the model, the relationship between propensities and time is sigmoidal, given successive iterations of the equations 2.3, 2.4, 2.5 or 2.6. As a result, propensities change slowly when they are close to 1 or close to 0, but change more rapidly at intermediate values. We believe that this is biologically plausible, but other relationships might exist and are currently being explored.

3. Results

The principal factors that affect the behaviour of the simulation are the rates of learning λ(increasing propensity) and forgetting ϕ (decreasing propensity). To simplify this initial analysis we studied the dynamics of task allocation where only two tasks are present, and the learning and forgetting rates, λ and ϕ, are manipulated. Each simulation involves twenty initially identical individuals, and the following parameters were universal: resting propensity level, R, 0.2; rate

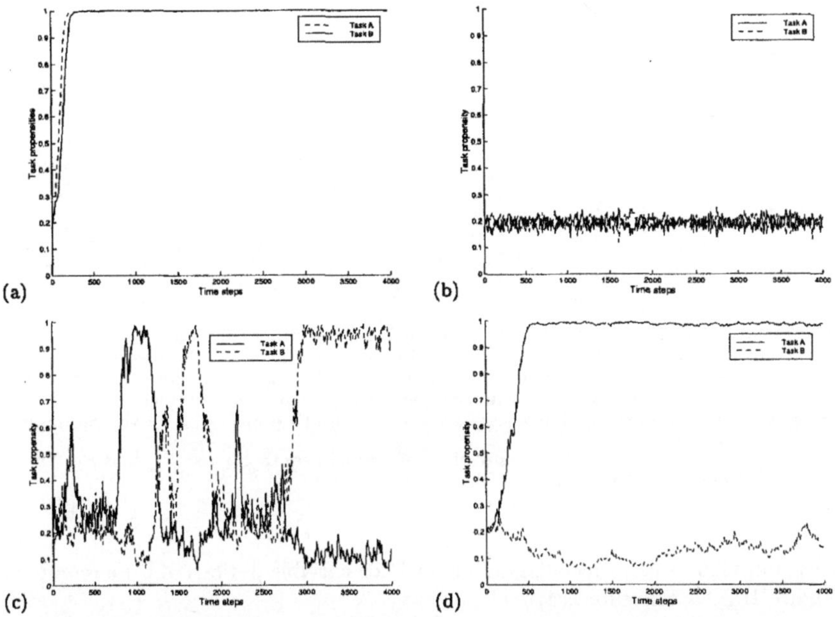

Figure 1. Individual task propensities against time for a range of learning (λ) and forgetting (ϕ) rates: (a) $\lambda = 0.199$, $\phi = 0.0066$; (b) $\lambda = 0.073$, $\phi = 0.1004$; (c) $\lambda = 0.199$, $\phi = 0.0752$; (d) $\lambda = 0.073$, $\phi = 0.0206$.

of task accumulation, 5 items per task per timestep; starting propensities, 0.2; density parameter, α, 50.

3.1. Individual behaviour

A range of individual responses to these conditions may be exhibited according to the values of learning and forgetting. When learning is rapid relative to forgetting, individuals exhibit behaviour such as that shown in Figure 1(a). The individual's propensities for both tasks rise rapidly and stabilise at close to 1, implying that the individual will perform almost every task it encounters. When individuals forget at a greater rate, as in Figure 1(b), their propensities for both tasks never increase significantly above the resting level of 0.2. These individuals rarely perform tasks of either type. In certain areas of parameter space, how-

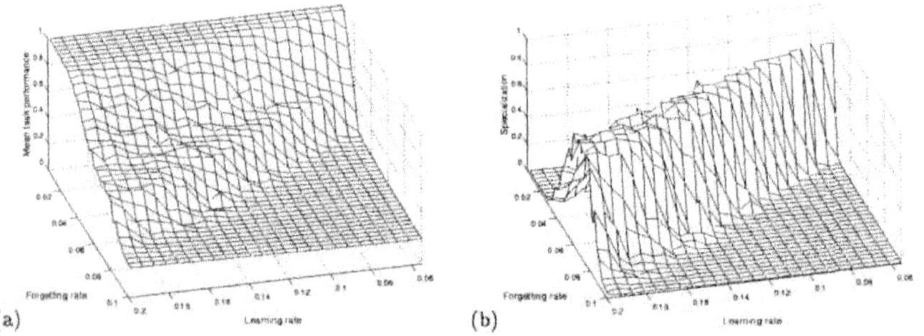

(a) (b)

Figure 2. Collective responses across learning (λ) and forgetting (ϕ) parameter space. In both cases the z-axis represents mean values taken over the course of 500 timesteps, after the simulation had been allowed to run for 3500 timesteps. (a) Mean task performance, measured as the mean number of task items dealt with per agent per timestep. (b) Specialization, measured as the mean of the differences between each agent's propensities for the two tasks, $\sum_{i=1}^{n} \frac{|P_{i_A} - P_{i_B}|}{n}$ where n is the number of agents (equal here to 20) and P_{i_A} and P_{i_B} are the agent propensities for tasks A and B.

ever, individuals can specialise on one task for variable periods of time (compare Figure 1(c) with Figure 1(d)).

3.2. Population behaviour

Figure 2(a) shows the population average propensity for task performance as a function of the rates of learning and forgetting by individuals. The upper plane is characterised by learning rates that are high compared with forgetting rates; the general trend in this region is for agents to perform any task that they encounter (Figure 1(a) exemplifies individual behaviour under these conditions). The lower plane is characterised by learning rates that are lower or only moderately higher than forgetting rates; the general trend in this region is for agents to perform very few of the tasks that they encounter (Figure 1(b) exemplifies individual behaviour under these conditions). In the transition zone between these two regions, intermediate states are found.

Figure 2(b) shows specialization (i.e. the tendency to perform one task more than the other) as a function of the rates of learning and forgetting by individuals (the parameter range for learning and forgetting is identical to that shown in Figure 2(a)). At one end of the specialization zone, that is when learning is slow, specialization tends to be stable (as exemplified by Figure 1(d)). At the other

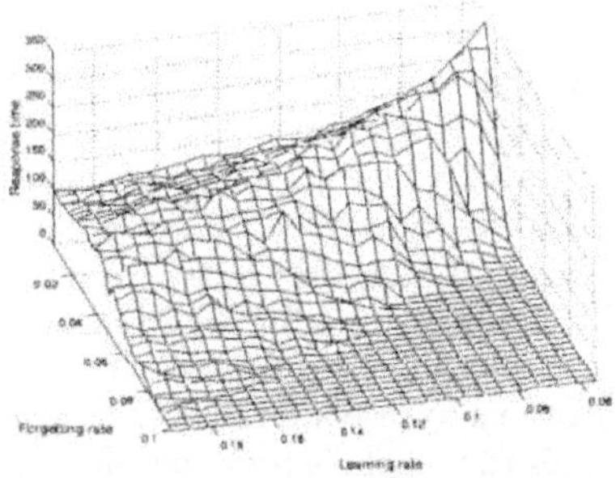

Figure 3. Collective responses across learning (λ) and forgetting (ϕ) parameter space. The simulation was run for 4000 timesteps with only task A present, at which point task B was introduced. The z-axis indicates the time taken before the rate (per agent per timestep) at which task B was performed approached (to within 0.2) the rate at which task A was performed.

end of the zone, where learning is fast, individual specializations are transient (see Figure 1(c)).

Figure 3 explores the ability of individuals within a population to respond to the introduction of a second task after a period in which they have encountered only one task. When the second task is initially introduced, none of the individuals in the group have encountered it previously and consequently they all have extremely low propensities for performing it. As they begin to encounter the second task, however, their propensities gradually increase until eventually the second task is performed at an equal rate to the first. The time taken to reach this point depends upon their learning and forgetting rates, and Figure 3 shows the time elapsed between the introduction of the second task and the point at which the second task is performed at a rate close to that of the first, across the same range of learning and forgetting rates used in Figure 2. This value is a measure of responsiveness: the shorter the elapsed time, the more rapidly individuals are responding to the change in conditions. However, it should be noted that the extremely low area of the graph associated with high forgetting rates is not indicative of a rapid response. This is because when individuals have a high forgetting rate they are incapable of learning even a single task (i.e. their propensities for the first task remain close to the resting level of 0.2); propensities for the second task were at the same resting level of 0.2 before its introduction and consequently little or no time elapsed before both tasks began to be performed at the same (very low) level. This graph shows that the ability to respond to change is associated with high values of both learning and forgetting.

4. Discussion

Our model shows that both generalization and specialization may occur in biological populations as a result of learning and forgetting, and that specialization is associated with a restricted range of parameter space. It further reveals a classic trade-off between stability and responsiveness.

Specialization is common in nature, both in the form of genetic predisposition or adaptation to a task, and in the form of behavioural specialization amongst potential alternative tasks. In this paper we examine the latter, which is much more labile allowing organisms to respond rapidly to their environment. Examples of behavioural specialization include prey specificity in predators (Begon *et al.*, 1990), flower specificity in pollinators (Heinrich, 1979), and task specificity in social insects (Hölldobler and Wilson, 1990; Bourke and Franks, 1995).

There are, however, costs to specialization due for example to increased search time (Krebs and Davies, 1993). Our model incorporates the effects of search time, since specialists 'waste' time steps rejecting task items. Figure 2 shows this cost of increased search time (the area of specialization in Figure 2(b) corresponds to the area of decreasing task performance in Figure 2(a)).

The costs of specialization beg the question why it is found so frequently in nature. One answer is that these costs may be offset by efficiency benefits through improved handling of tasks (Krebs and Davies, 1993). Since we do not model efficiency of task performance, this cost is not offset in our model. However, we can predict from our results that where behavioural specialization occurs in nature, organisms are likely to be reaping sufficient benefits from improved handling efficiency to offset the costs of increased search time.

The most likely cause of improved handling efficiency is learning (in the sense of skill refinement). Specialization will be favoured if individuals are not able to learn tasks concurrently, as can occur due to cognitive limitations. Specialization will also be favoured if the costs of learning a second task (incurred because time spent learning a second task is time that could be used to carry out a first task) are greater than the benefits. For simplicity, we have looked only at two tasks here, but the same principle applies wherever organisms are more efficient if they learn only a subset of possible tasks. A well-studied example of this is specialization among flower types by bumblebee foragers. Heinrich (1979) describes the way in which bumble bees learn to cope with different flower morphologies in the search for nectar and pollen. An interesting case, in which specialization is favoured by physiological as well as by behavioural adaptation to a task, is that of the digestive physiology of mallard ducks. It has been shown that these ducks have an increased digestive efficiency when they specialize on a particular food type (Miller, 1975, cited in Begon *et al.*, 1990).

Classical behavioural ecology models have shown, with respect to dietary breadth, that generalists should be expected to occur when the ratio of energy gained per unit handling time is the same for all items, or when search times are extremely long (Krebs and Davies, 1993). Such generalization may also be favoured when there is no significant cognitive limitation to learning tasks concurrently, or when there are no efficiency gains to be made through skill refinement.

Where specialization occurs, our model predicts a trade-off between stability and responsiveness. The population of agents (workers, predators or pollinators) can respond rapidly to the introduction of a new opportunity (task, prey species, flower type), but only if they have high learning and high forgetting rates (see Figure 3). The penalty of such behaviour is that such individuals will rarely specialize on one task for long. Typically they will flip stochastically from one task to another (see Figure 1(c)). In real biological situations this inconsistency would be likely to incur costs: efficiency gained from specializing and developing skills with certain types of item might be thrown away in too rapid switching to alternatives. The penalty of the opposite strategy (that is, stable specializations associated with lower learning and forgetting rates) is that in a changing environment, individuals will adjust their behaviour only slowly.

It is also notable that the area of parameter space where specialization occurs (Figure 2(b)) is also associated with slower responsiveness (Figure 3). This suggests that jacks of all trades, although they may be masters of none, are more likely to be successful opportunists.

There is very good evidence from both vertebrate and invertebrate predators for switching between prey types according to their relative density (Lawton *et al.*, 1974; Murdoch and Oaten, 1975; Murton, 1971; for review see Begon *et al.*, 1990, chapter 9). One traditional hypothesis for such behaviour is that predators form search images (Tinbergen, 1960; but see also Guilford and Dawkins, 1987; Giraldeau, 1997). The notion of a search image is that an organism forms a mental image of a cryptic prey type upon successive encounters and as a consequence tends to be less aware of other prey types. This implies that if the prey for which a predator has a search image becomes very rare, the search image is progressively forgotten, and also that search images are exclusive (a cognitive limitation).

There is also good evidence of switching behaviour in pollinators. Individual bees switch their preferences in response to changing relative abundance of flower types (Heinrich, 1979). For more recent work on bumble bee decision making during foraging, see Cartar (1992), Dukas and Real (1993), Dukas and Waser (1994). Rissing (1981) has shown that individual workers in seed harvesting ants also show preferences for particular seed types and switching behaviour. For an example of switching behaviour in a non-social pollinator, see Goulson *et al.* (1997).

All of these examples emphasize the advantages of maintaining a specialization whilst a task remains available, and of being able to switch when circumstances change. The importance of individuals being able to respond to a changing environment suggests why in social insect colonies with workers that are physically polymorphic, the least specialized workers are always in the majority. In such cases, extreme physical castes are hard-wired for particular roles and therefore the ability of a colony to respond to a changing environment depends on having large numbers of generalists who can specialize behaviourally according to the needs of their colony (Oster and Wilson, 1978; Tofts and Franks, 1992; Bourke and Franks, 1995).

In solitary organisms, members of a population switch according to their own needs and local circumstances; in a population of workers in a eusocial colony switching can occur that benefits the entire community. If one viewed a colony of social insects exhibiting this trait, decisions might appear to be taken at a global level, but in reality this global behaviour is more likely to arise from independent, local decisions by individuals. An alternative model for social insects has been to consider individuals as having thresholds, which may be fixed (i.e. no learning—Bonabeau *et al.*, 1996) or variable (due to learning—Theraulaz *et al.*, 1998), and that there are global stimuli that emanate from each of the tasks. If the stimulus is greater than this threshold point, then every individual with a threshold below the stimulus performs the behaviour. By contrast, the model presented here has been based purely on local stimuli, and this has, we believe, given it an increased generality.

Our goal in this paper has been to explore in very general terms the dynamics of specialization and generalization in biological populations. In order to achieve some generality, we have deliberately kept our model simple. More specifically, we have excluded explicit spatial considerations in this first version of the model. (Clearly if different tasks, resources or prey types were spatially segregated and agents had restricted patterns of movement then tendencies to specialization could be greatly enhanced.) In the model, learning by the agents is a form of positive feedback and forgetting is a form of negative feedback, and hence our modelling can be seen as an exploration of stability versus responsiveness in terms of the relative magnitude of positive and negative feedback. The model is highly abstract, but for this very reason has revealed potential similarities among a wide range of complex biological systems.

5. Acknowledgments

The authors thank their colleagues in the Ant Lab for many helpful discussions, Nick Britton for advice, and an anonymous referee for useful comments. This work was supported by grants to A.J.S. from the Department of Biology and Biochemistry and to I.D.C. from NERC. N.R.F. wishes to acknowledge the financial support of NATO and the Leverhulme Trust.

References

Begon, M., Harper, J.L. and Townsend, C.R. (1990). *Ecology: Individuals, Populations and Communities*. Blackwell Scientific Publications, Boston.

Bonabeau, E., Theraulaz, G. and Deneubourg, J.-L. (1996). Quantitative study of the fixed threshold model for the regulation of division of labor in insect societies. *Proc. R. Soc. Lond.* B. **263**, 1565–1569.

Bourke, A.F.G. and Franks, N.R. (1995). *Social Evolution in Ants*. Princeton University Press, Princeton.

Cartar, R.V. (1992). Adjustment of foraging effort and task switching in energy-manipulated wild bumblebee colonies. *Anim. Behav.* **44**, 75–87.

Cosmides, L.M. and Tooby, J. (1981). Cytoplasmic inheritance and intragenomic conflict. *J. theor. Biol.* **89**, 83–129.

Deneubourg, J.-L., Goss, S., Pasteels, J.M., Fresneau, D. and Lachaud, J.P. (1987). Self-organization mechanisms in ant societies (II): Learning in foraging and division of labor. In: (Pasteels, J.M. and Deneubourg, J.-L. eds) *From Individual to Collective Behaviour in Social Insects*, pp. 177–196. *Experientia* Supplementum **54**. Birkhäuser Verlag, Basel.

Dukas, R. and Real, L.A. (1993). Learning constraints and floral choice behaviour in bumble bees. *Anim. Behav.* **46**, 637–644.

Dukas, R. and Waser, N.M. (1994). Categorization of food types enhances foraging performance in bumblebees. *Anim. Behav.* **48**, 1001–1006.

Falconer, D.S. (1981). *Introduction to Quantitative Genetics*, 2nd Edition. Longman, Harlow U.K.

Giraldeau, L-A. (1997). The Ecology of Information Use. In: (Krebs, J.R. and Davies, N.B. eds.) *Behavioural Ecology: An Evolutionary Approach*, 4th edition, pp. 42–68. Blackwell Science, Oxford.

Goulson, D., Ollerton, J. and Sluman, C. (1997). Foraging strategies in the small skipper butterfly, *Thymelicus flavus*: when to switch? *Anim. Behav.* **53**, 1009–1016.

Guilford, T. and Dawkins, M.S. (1987). Search images not proven: a reappraisal of recent evidence. *Anim. Behav.* **35**, 1838–45.

Heinrich, B (1979). *Bumblebee Economics*. Harvard University Press. Cambridge, Massachusetts.

Hölldobler, B. and Wilson, E.O. (1990). *The Ants*. Springer-Verlag, Berlin

Krebs, J.R. and Davies, N.B. (1993). *An Introduction to Behavioural Ecology*, 3rd edition. Blackwell Scientific Publications, Oxford.

Lawton, J.H., Beddington, J.R. and Bonser, R. (1974). Switching in invertebrate predators. In: (Usher, M.B. and Williamson, M.H. eds.) *Ecological Stability*. Chapman and Hall, London.

MacArthur, R.H. (1955). Fluctuations of animal populations and a measure of community stability. *Ecology* **36**, 533–536.

May, R.M. (1973). *Stability and complexity in model ecosystems*. Princeton University Press, Princeton.

Maynard Smith, J. and Szathmáry (1995). ***The Major Transitions in Evolution.*** Oxford University Press, Oxford.
Medawar, P.B. (1957). ***The Uniqueness of the Individual.*** Methuen, London.
Miller, M.R. (1975). Gut morphology of mallards in relation to diet quality. ***Journal of Wildlife Management* 39**, 168–173.
Murdoch, W.W. and Oaten, A. (1975). Predation and population stability. ***Advances in Ecological Research* 9**, 1–131.
Murton, R.K. (1971). The significance of a specific search image in the feeding behaviour of the wood pigeon. ***Behaviour* 40**, 10–42.
Oster, G.F. and Wilson, E.O. (1978). ***Caste and Ecology in the Social Insects.*** Princeton University Press, Princeton.
Pimm, S.L. and Lawton, J.H. (1977). The number of trophic levels in ecological communities. ***Nature* 268**, pp.329–331.
Plowright, R.C. and Plowright, C.M.S. (1988). Elitism in social insects : a positive feedback model. In: (Jeanne R.L. ed.) ***Interindividual Behavioral Variability in Social Insects***, pp. 419–431. Westview Press, Boulder CO.
Ridley, M. (1993). ***Evolution.*** Blackwell Scientific Publications, Boston.
Rissing, S.W. (1981). Foraging specializations of individual seed-harvester ants. ***Behav. Ecol. Sociobiol.* 9**, 149–152.
Smith, A. (1776). ***The Wealth of Nations***, Books I–III. Reprinted 1986 (A. Skinner, ed.). Penguin, Harmondsworth, U.K.
Theraulaz, G., Bonabeau, E. and Deneubourg, J.-L. (1998). Response threshold reinforcement and division of labour in insect societies. ***Proc. R. Soc. Lond.*** B. **265**, 327–332.
Tinbergen, L. (1960). The natural control of insects in pinewoods. 1: Factors influencing the intensity of predation by songbirds. ***Archives Néerlandaises de Zoologie* 13**, 266–336.
Tofts, C. and Franks, N.R. (1992). Doing the right thing: ants, honeybees and naked mole-rats. ***Trends in Ecology and Evolution* 7** (10), 346–349.

Advances in Complex Systems, Vol. 5, No. 1 (2002) 73–89

INFERENCE VERSUS IMPRINT IN CLIMATE MODELING

A. J. PALMER,[†] T. L. SCHNEIDER*[,†] and L. A. BENJAMIN[‡]
[†]*NOAA Environmental Technology Laboratory,*
[‡]*NOAA Forecast Systems Laboratory,*
325 Broadway, Boulder, CO 80305, USA

Received 7 April 2001
Revised 23 January 2002
Accepted 2 February 2002

A statistical inference method known as ε-machine reconstruction is introduced as a modeling procedure for turbulent transport processes in a climate model. Observational data on the atmospheric boundary layer obtained with a radar wind profiler, a radio-acoustic sounding system, and a Raman lidar system was assembled to construct this type of model for use within the unresolved (sub-grid) scales of a numerical climate model. An ensemble of 500 single-column model runs using the inferred sub-grid turbulent transport models demonstrated comparable performance to an identical ensemble of runs using the standard, eddy-diffusivity parametrizations for the turbulent transport. The primary advantages of the ε-machine models are that they are a less biased modeling framework for complex processes such as turbulent transport, and that they are more memory efficient.

Keywords: Climate modeling; statistical inference; sub-grid parametrization; ε-machine.

1. Introduction

Atmospheric general circulation models (AGCMs), also referred to as 'climate models,' are the principle tools used to understand and predict human influences on the Earth's climate system. The resolution of AGCM's is typically quite crude, with horizontal resolutions of the order of hundreds of kilometers, vertical resolutions of the order of kilometers and temporal resolutions of tens of minutes. Even so, the complexity of these models and the need for very long simulations and/or ensembles of simulations, make them among the most computationally demanding of computer applications. Also, many critical physical processes such as thermodynamics and clouds, radiative transfer and turbulent transport, operate on much finer spatial and temporal scales than is utilized in these models. These processes cannot be represented explicitly and must be approximated, hence the term sub-grid parametrization. In general, AGCM's do a reasonably good job of simulating

*Corresponding author.

the global mean state of the atmosphere, but have difficulties in getting the correct amount of variability and in simulating specific aspects of the atmospheric general circulation, such as the El Niño-Southern Oscillation (e.g. Ref. 5). Thus, if we are to have full confidence in our predictions of climate change, and more importantly the regional impacts of climate change, we must do a better job representing the physical processes occurring on sub-grid scales and make optimal use of the computational resources applied to this task.

1.1. *The problem: Sub-grid parametrizations*

As noted above, the climate system is modeled with a combination of dynamical and empirical modeling frameworks, and the available computational resources are simply allotted to meet the requirements of this modeling framework. Specifically, climate models today allocate most of the active computational memory to the symbols of the modeling grammar (e.g. the 32-bit or 64-bit floating point representations of the physical fields) rather than to the syntax of the modeling grammar. This is because the modeling syntax is relatively simple and static throughout the model run. On the grid points of the model, the principle syntax is the set of Navier–Stokes equations of fluid dynamics. On the sub-grid scale of the model, the syntax is a set of empirical models known as physical parametrizations.

Few will argue that the set of Navier–Stokes equations is not the correct modeling syntax for the resolved scale of the model. However, the syntax used for modeling sub-grid processes is generally subject to argument. While these parametrizations are often guided by sound physical principles such as similarity scaling, these same guidelines can become biases when the physical symmetries that they are based on become spontaneously broken (a symmetry is said to be spontaneously broken if it is consistent with the equations of motion and the boundary conditions but is not present in the solution). As a flow parameter is changed, such as increasing the Reynolds number, the various symmetries permitted by the equations of motion (and the boundary conditions) are often successively broken. In some flow regimes, such as at very high Reynolds number, there is a tendency for the symmetries to be restored in a statisitcal sense far from the boundaries, and statistical parametrizations generally work well in this case. However, in other flow regimes, such as near bifurcations of the dynamical attractor, the symmetries are broken in both an instantaneous and statistical sense and the statistical parametrizations will fail. In this case, a new modeling syntax for the sub-grid process is needed.

Unfortunately, in gathering observational data on the sub-grid processes, there is a tendency to immediately filter the data into statistical means and moments or other linear measures of variability in order to fit a favored parametrization, and the memory content of the data (i.e. the memory allocation required to model the variability) is lost. A more scientific approach to the problem would be to let observational data on the sub-grid process tell us what active memory is required for a given observational window on the data. A potential outcome of this type of approach to sub-grid modeling is that a more efficient modeling syntax for the sub-processes will be found that requires less active memory than do the existing parametrizations. If this were to occur for enough of the sub-grid parametrizations, it is conceivable that memory resources previously tied up in sub-grid modeling could be made available for increasing the resolution and/or accuracy of the resolved-scale portion of the model.

1.2. *The solution: ε-machines*

Such a modeling approach is proposed and tested here for the first time for sub-grid turbulent transport processes. The method is based on rigorous statistical inference of causal pattern in observational data. It is a symbolic dynamic modeling procedure known as "ε-machine reconstruction" [1]–[3]. In general, the causal patterns found in the data will contain both stochastic and deterministic structure, and the ε-machine statistical inference method is designed to capture this dual nature. The value of adding stochastic structure to an approximate model of a nonlinear dynamic system has been clearly illustrated in Ref. 13. The underlying true dynamical attractor of the system is more thoroughly explored with such a model structure. The stochastic component of the ε-machine brings this value to the model, but with the added advantage that deterministic as well as stochastic structure is inferred from observational data rather than imprinted onto the model *a priori*.

The platform we have selected for our initial evaluation of this new, empirical, sub-grid, modeling methodology is the single column version of the National Center for Atmospheric Research (NCAR) Community Climate Model (or SCCM). The sub-grid processes we model are the vertical transport of momentum, heat and moisture. The standard physical parametrization for these processes is based on the assumption of local eddy diffusivity supplemented with an algorithm that accounts for "non-local" transport by large scale convective turbulence in the boundary layer [10]. For the five lowest vertical levels in the model, we replaced this entire parametrization with an algorithm built from observational data using the ε-machine statistical inference method. The momentum transport algorithm was constructed from wind-profiler data, the heat transport algorithm from radio acoustic sounding (RASS) data, and the moisture transport from Raman lidar data.

Comparison of the model performance using the standard parametrizations and the ε-machine algorithms was then made. Using an ensemble of 500 model runs for each comparison case, the model performance was measured by the ensemble mean of the model predicted temperature minus the observed temperature at a fixed level of the model. On the basis of this performance measure, the models with the ε-machine algorithms performed as well as those with the standard parametrizations.

In the following sections we review the ε-machine statistical inference method, propose a recipe for applying the method to the sub-grid modeling problem, and report results of the SCCM experiments.

2. The ε-Machine Statistical Inference Method

ε-machine reconstruction refers to a new complex systems modeling procedure that builds a model, called an ε-machine [1]–[3], directly from observational data. "Machine" refers to a finite state machine as the basic framework of the model. A finite state machine is simply a set of transition rules between a finite number of states. Examples are Markov chains and regular grammars in formal language theory [9]. These can be thought of as discrete analogues to the transition rules for continuous systems encoded as differential equations. "Reconstruction" refers to the manner in which the machines are built from observational data. The first two reconstruction steps shift the representation of the data from a string (e.g. a time series, or spatial sequence) to a tree representation, and then to a finite state machine [1].

Two features of the machine reconstruction program make it a compelling new method for sub-grid modeling of complex processes in the climate system. The first is the nature of the constructed ε-machine. The ε-machine is a particular type of finite state machine known as a stochastic automata [9]. Such machines have both deterministic and probabilistic structure. This is exactly what is needed to remove the arbitrary selection of one or other of these modeling frameworks in sub-grid climate modeling. The second feature is that machine reconstruction builds models directly from observational data. This is similar to training an artificial neural network to find a causal pattern in data, except that in the case of the neural network, once it is trained on the data it is a fixed, the deterministic algorithm and the discovered causal pattern is hidden from the modeler. On the contrary, the reconstructed ε-machine represents an active and synergistic combination of deterministic and stochastic structure, and this computational structure is displayed to the modeler [1]–[3].

Once the machines are constructed, we may compute a fundamental measure of complexity for the underlying dynamical system that produced the data. This measure of complexity is termed "statistical complexity" and is defined as the entropy of the recurrent causal states of the machine [1]–[3]:

$$C = -\sum p(v) \log_2(p(v)), \tag{1}$$

where $p(v)$ is the probability of occurrence of the recurrent states v of the machine. This measure of complexity can be shown to be equal to the computational memory required to run the ε-machine model. The statistical complexity is zero for both a pure random process, and for a pure deterministic process, both of which, as illustrated below, can be represented as a single-state machine.

The machines themselves can be represented as digraphs where the vertices are the machine states and the edges represent the transitions between the states [1]. The edges are labeled with the symbol that is emitted on the transition and with the probability of the transition. For example, 0|0.5 indicates the transition is made on a symbol 0 with 0.5 probability. Examples of machines for two simple processes,

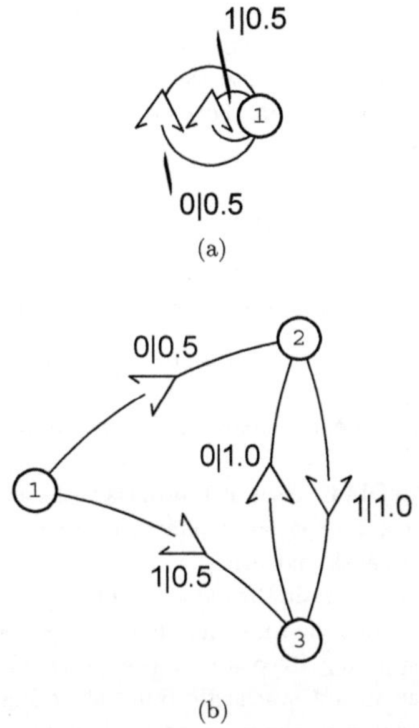

Fig. 1. Example of finite-state machine representation of (a) an unbiased random string of zeros and ones, and (b) a periodic string of zeros and ones with a period of two symbols. The edge labels identify the transition symbols and transition probabilities, e.g. 0|0.5 indicates the transition is made on a symbol 0 with probability 0.5.

the random coin-flip process and the periodic process with a period of two symbols, are shown in Fig. 1. Machine representations of these processes and many more are discussed at length in Refs. 1–3. The machines represent probabilistic structure in a process as a branching of edges, and deterministic structure as non-branching edges. The branching in the machine representation of the periodic process shown in Fig. 1(b) represents the uncertainty in phase when the machine is in the start state (here the start state is labeled state-1).

The "machine reconstruction" procedure for producing a stochastic automaton representation of a string generated by a more general process is a straightforward computer programming problem, and is outlined in Ref. 1. Briefly, it consists of first producing a parse tree representation of the string by sliding a window of fixed length through the string and laying down a branch of the tree at each window position. The states of the automaton are simply invariant subtrees found within the parse tree and the state transition probabilities and transition symbols are determined by the total number of state transitions found in the construction of the parse tree.

The importance of representing the complexity of a process in this way rests primarily in its lack of bias; probabilistic and deterministic structure is discovered in the data rather than imprinted onto the data. The ε-machine representations captures pattern and regularities in the data in a way that reflects the causal structure of the process [1]–[3].

3. ε-Machines as Sub-Grid Models

We propose the following six steps for using an ε-machine as a model of a sub-grid process. We elaborate further on these steps later in the text.

(i) Process Selection: Identify a sub-grid process whose standard parametrization in the model is suspect.

(ii) Dynamical Variables: Identify relevant dynamical variables governed by the sub-grid process.

(iii) Observational Data: Obtain a set of appropriate observational data on the sub-grid process, and select a spatial or temporal order and data sequence length in which to look for causal pattern.

(iv) Coarse-Graining: Coarse-grain the data into a finite set of symbols with respect to a partition that yields the maximum entropy for the chosen sequences.

(v) ε-Machine Reconstruction: Reconstruct the coarse grained data, first into a parse tree and then into a stochastic finite state machine whose states are invariant subtrees found in the parse tree. Compute the transition rates between the machine states. This is the basic ε-machine reconstruction procedure documented in the literature [1].

(vi) ε-Machine Sub-grid Model: Replace the suspect parametrization with the ε-machine transition matrix. The transition symbol encodes the coarse-grained value of the sub-grid process dynamical variable given to the resolved scale portion of the model. The machine is returned to the start state if the machine and resolved scale portions of the model fail to satisfy any imposed boundary conditions.

- If one is using the machine to model the temporal order of the sub-grid process, a single machine transition is made at each model time step beginning with the machine in the start state.
- If one is using the machine to model the spatial order of a sub-grid process, then at each model time step the machine begins in the start state and makes all of the transitions needed to span the modeled sub-grid spatial domain.

These steps establish a general procedure for unbiased modeling of a sub-grid process based on coarse-grained observational data for the process. An expanded discussion of each step is provided in the next section which describes our particular application of the steps to sub-grid turbulent transport processes.

The ε-machine model of a sub-grid process captures the same deterministic and stochastic structure found in the data under a given observational window on the process. In practice, the observational window is a narrow view of the process which can distort the model of the process [2]. The coarse graining of the data and the model output, the use of relatively short sequences as candidate causal structure imposed by limited data records, and even the choice of casual states as invariant sub-trees found in the parse tree all represent ε-machine model constraints. However, ε-machine modeling is designed to allow all of these constraints to be systematically relaxed as increased amounts of observational data and computational memory become available. Keeping these resources in focus is an important modeling discipline that allows conditioned performance comparisons to be made based on equal allocation of a given resource. In particular, by being able to specify the computational memory of an ε-machine, the performance comparison of an ε-machine model and parametrization model of a sub-grid process should ultimately be conditioned on equal *total* memory allocation (symbols plus syntax).

4. Application to a Climate Model

Below we describe our application of the above six steps to modeling of sub-grid, vertical transport of momentum, heat, and moisture in a single-column climate model. The single column model we use is a one-dimensional version of the National Center for Atmospheric Research (NCAR) Community Climate Model [10] known as the SCCM [7]. The SCCM was designed as a platform for testing various parametrizations in the climate model without the need to run the entire climate model over all of the horizontal grid points of the globe. Only a single vertical column is modeled by the SCCM. The horizontal advection of fields that would normally be computed over the horizontal grid points of the full climate model is replaced by time-dependent boundary conditions specified on the single column and assimilated into the SCCM during a model run. These boundary conditions are obtained from measured fields during an intensive observation period (IOP) conducted at the location of the modeled column.

The period we chose for running the SCCM is known as the summer 1995 SCM IOP [7]. This data set is derived from observations made at the Atmospheric Radiation Measurement (ARM) site in Oklahoma from July 18 to August 4, 1995. The data we used for building our ε-machine models was obtained in the same general area as this IOP and during the same season, but were not part of the assimilation or validation data used for the SCCM run for this IOP.

4.1. *Process selection (Step 1)*

We identified the sub-grid parametrizations for vertical transport of momentum, heat and moisture in the NCAR single column climate model (SCCM) as our "suspect" parametrizations. The standard parametrization for these processes in the SCCM model as well as in the complete NCAR global climate model is a model based on the diffusion equation. In these parametrizations an "eddy diffusivity" constant is computed which is designed to represent local as well as nonlocal vertical transport.

The eddy diffusivity parametrization is a reasonable model of turbulent transport when the time and spatial scales over which transport is modeled take place in the dissipation range or even in the inertial sub-range of turbulence. However, in climate models such as the SCCM, the temporal and spatial scales of the sub-grid process to be modeled are typically much larger than these scales, so the diffusivity parametrization is suspect primarily because of its failure to properly model the temporal and spatial variability of the transport process, in particular, the stochastic component of the variability.

One way to demonstrate this failure is to plot the diffusivity parametrization at each 20-minute time step that is invoked in the model for a particular IOP, together with a similar plot of observational data for the same sampling time for a climatologically similar location and time. Such a plot is shown in Fig. 2 for

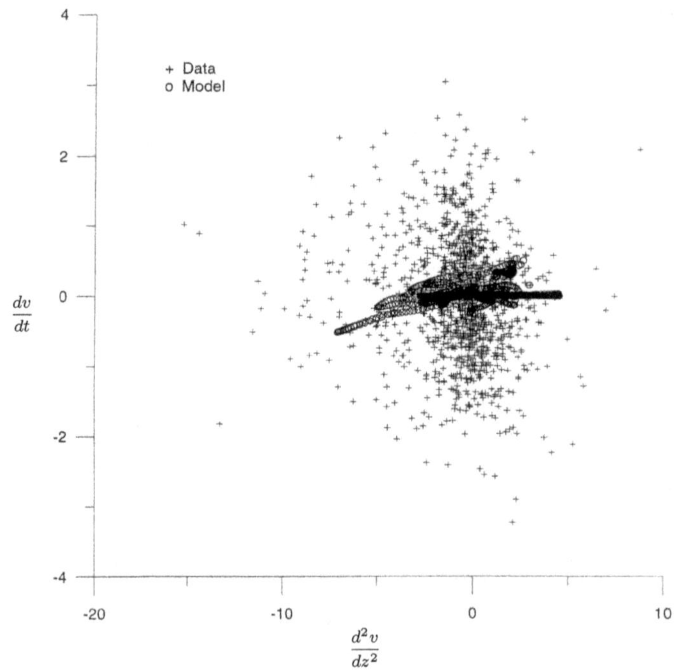

Fig. 2. Plot of diffusivity parametrization implemented every 20 minutes during the SCCM run described in the text compared to observation data from radar wind profilers for 18-minute time steps.

momentum transport. The figure plots the time rate of the change of horizontal wind speed during the model time step versus the second derivative of the wind speed with respect to height for the four lowest levels used in the SCCM. The model run was for the IOP mentioned above using the standard SCCM eddy diffusivity parametrizations, and the observational data were radar wind profiler data obtained at Lamont, Oklahoma, during the same 18-day period for which the model was run. For the chosen graph coordinates, the diffusivity parametrization, with a fixed diffusivity constant, would produce a single straight line. The 20-minute sampled, model generated points on the plot are seen, as expected, to lie mostly along several straight lines corresponding to different model-selected diffusivity constants including zero for night-time periods when the boundary layer is below the top of the lowest level of the model. The observational data that were sampled at roughly the same interval (18 minutes) is seen to show no support for the linear relationship generated by the diffusivity equation used in the parametrization. The observational data for dv/dt appear randomly distributed in relation to d^2v/dz^2.

If the sampling time were reduced sufficiently to enter the inertial sub-range of the boundary layer turbulence or the dissipation range, a diffusivity equation-based relationship would ultimately appear in the observational data. However, for current climate model resolutions and time steps, the parametrizations for vertical transport based on the diffusivity equation is simply not validated by the observations.

4.2. *Dynamical variables (Step 2)*

The dynamical variable that is given to the resolved scale portion of the SCCM by the diffusivity parametrization for sub-grid vertical transport of a field is the change in field that occurs at each time step of the model at a given vertical level. These field differences are the dynamical variables chosen for ε-machine modeling. Only the field differences themselves are used to build the ε-machine models; i.e. the models are univariate models. Multivariate ε-machines can also be built. For example, if the plot in Fig. 2 had shown evidence of a residual dependence of dv/dt on d^2v/dz^2, one might have chosen to construct a bivariate model for these two dynamical variables. Appendix A presents a brief description of how a multivariate ε-machine model could be built and used as a sub-grid model.

4.3. *Observational data (Step 3)*

The observational data that were used to build ε-machine models for sub-grid vertical transport are measured vertical profiles for winds, temperature and humidity. These data were obtained from a wind profiler radar, a radio acoustic sounding system (RASS) and a Raman lidar system, respectively. The wind and RASS profilers are Doppler radar systems that utilize a back-scattered signal from Bragg structures in the atmospheric boundary layers caused by natural turbulent variability in the first case and a co-directed sound wave in the second case [12]. The Raman lidar determines the profile of the water vapor mixing ratio based on the principle of Raman scattering by molecules. It determines the range resolved ratio of the back-scatter by water vapor (at 408 nm) to back-scatter by nitrogen (at 387 nm) [4].

The wind and temperature data were obtained with a 404 MHz radar wind profiler and RASS system operated by the NOAA Forecast Systems Laboratory at Lamont, Oklahoma and Purcell, Oklahoma, respectively, during the same 18-day period of the SCCM model run (19 July–5 August, 1995). As mentioned above, these data were not used as part of the validation measures for the model output, nor as part of the boundary condition data. The Raman lidar is operated at the central facility of the Department of Energy's Southern Great Plains (SGP) Cloud and Radiation Testbed (CART) near Lamont, OK (see http://www.arm.gov). The moisture data were obtained over 26 days from July (10, 14, 17–23, 25) and August (6, 7, 13-20, 25, 27-31) of 1998. Thus, these data correspond to the same season and location as for the model run, but are from a different year. Quality control checks were made on all of the observational data to eliminate spurious signals from aircraft, birds and other random interference.

All of the data are processed into samples of field differences that occur over an approximate 20 minute time step as required by the resolved scale portion of the SCCM. We chose to build the ε-machine using these samples in a spatial (height) domain rather than the temporal domain because of the many data gaps that existed in the temporal domain. Under this condition, time can be considered as merely an exemplar index of the height-ordered data sequences in which we searched for causal pattern. The sequence length chosen for our search for pattern was two height-ordered samples. The total number of height levels chosen to parse the data into was the five lowest levels used in the model. These levels are approximately 500 m, 750 m, 1000 m, 1750 m and 2750 m above the ground.

The choice of sequence length is constrained by total number of exemplars in the data record as follows. For maximum entropy patterns, one expects to see a statistical fluctuation of the occurrence of a given height ordered sequence of D symbols in the data given by

$$\delta = \left[\frac{S^D}{N}\right]^{1/2}, \tag{2}$$

where S is the number of symbols chosen to coarse grain the data into, and N is the total number of parse tree branch exemplars in the data [8]. Below, we will chose binary coarse graining of the data, i.e. $S = 2$. For our data records, N is near 400 for the wind profiler and RASS data, and near 3000 for the Raman lidar data. Thus, a choice of $D = 5$ keeps the expected statistical fluctuation of pattern down to about 28% for the wind profiler and RASS data and about 10% for the lidar data. The two-symbol word length is chosen at $D/2$ or half the parse tree depth as the recommended optimal length for finding causal structure in the parse tree [1].

4.4. *Coarse-graining* (*Step 4*)

For these first experiments we choose the simplest of the possible coarse graining symbol sets, the two symbols 0, 1. This symbol set has been used for all of the ε-machine models presented thus far in the literature. However, the ε-machine reconstruction method allows the use of an arbitrary set of symbols.

Once a symbol set is selected, the partition chosen for coarse-graining the data into the symbol set must be chosen. The ideal partition is what is known as a generating partition. A generating partition defines a partition for which the resulting coarse-grained symbols can be shown to encode the entire dynamics generated by the original algorithm that produced the real-valued data. For example, the logistic map, which is a commonly used chaotic analytical model, has a binary generating partition, at $x = 1/2$ [2, 3]. For experimental data where the real-valued algorithm is not known, one cannot prove the existence of a generating partition. However, a theorem of Ref. 11 states that if a data series does have a generating partition, then that partition will also be a maximum entropy partition. This result suggests the use of a maximum entropy partition for symbolic dynamic modeling of experimental data [14]. In practice, once our sequence length of two symbols was selected in accordance with the discussion below Eq. (2), we chose a binary partition to be near the median value for the data set and confirmed that it was a local maximum entropy partition by explicit computation of the entropy of the sequences by varying the partition value about the median.

4.5. *ε-machine reconstruction* (*Step 5*)

Using the coarse-grained data format described above, ε-machines were constructed from the vertical profile data in accordance with the procedure described in Ref. 1. Recall that the ε-machines are constructed to locate the causal structure between length-2 subsequences of symbols ordered in height, where the symbols encode the field difference that occurs in the 20-minute SCCM time step. For all four field difference data sets (horizontal wind speed components, temperature and moisture) the reconstructed ε-machine was found to be the single state, purely random process machine shown in Fig. 1(a); i.e. no deterministic structure was found in the vertical profiles for the described observation window onto the data.

4.6. *Replacement of SCCM parametrizations* (*Step 6*)

The standard diffusivity parametrizations for vertical transport of momentum, heat, and moisture that are invoked at each time step in the SCCM were replaced by the ε-machine algorithm at the vertical levels that the ε-machine was designed to model (the four lowest levels in the SCCM.) Like the diffusivity equation, the ε-machine algorithm provides a value for the change of the field during a time step, at each of the four levels. Thus, at each time step of the model, the ε-machine was set to the start state and then was run through four state transitions producing four symbols that encode the change of the field at the four lowest levels of the model. The symbols $0, 1$ encode one of two distinct values of the modeled dynamical variable (field differences over the 20-minute time step). In practice, we chose the two encoded values to be equal to the median value of the modeled field difference plus or minus its standard deviation. The standard diffusivity parametrization was retained for the levels higher than the lowest four levels chosen for ε-machine modeling.

5. SCCM Performance Comparisons

The SCCM used here employs the standard "physics packages" of the NCAR-CCM Version 3.6. As described in the SCCM, Version 1.2 users' guide [7], on the resolved scale of the model, the vertical advection of water vapor is evaluated explicitly using a semi-Lagrangian procedure, whereas the temperature and momentum are advected with an Eulerian finite difference scheme. However, the large-scale forcing (including the horizontal flux divergence) is prescribed from input data sets. The forcing data were derived from the 1995 ARM SCM IOP described earlier (one of the standard input options of the NCAR-SCCM).

In executing our numerical experiments we adopted the approach of Ref. 6. These authors have shown that in order to minimize uncertainties inherent in the single-column model framework, ensemble averages are required. The initial conditions were perturbed randomly over 500 realizations. Maximum temperature perturbations were $|T_{\text{pert}}|\max < 0.9°\text{C}$, and the water vapor mixing ratio perturbations (in the boundary layer) were of the order $|q_{\text{pert}}|\max < 6\%$ of the local values.

In the present study we examine five cases, each averaged over 500 member ensembles: (i) a control run using the standard eddy diffusivity parametrizations; (ii) replacing the momentum term with our inferred model; (iii) replacing the thermal diffusivity parametrization with our inferred model; (iv) replacing the moisture diffusivity parametrization with our inferred model; and finally, (v) a run in which all of the eddy diffusivity parametrizations (momentum, thermal and moisture) were replaced with our corresponding inferred models. For each case and each realization, the modeled temperature is subtracted from the observed temperature (truth) to form a model error. Time series of the ensemble averaged model errors are presented in Fig. 3.

For this case (the summer 1995 IOP) it has been shown that the model solutions bifurcate beyond about 200 hours [6], reflecting the strong nonlinearities in the model, and this is seen also in our model runs shown in Fig. 3. The variations between the ensemble mean time series shown in Fig. 3, is within the mean variations between the individual members of the ensemble of time series (not shown). Thus, there is no statistically significant difference between the performance of the inferred, "coin-flip" models of vertical transport and the conventional eddy diffusivity parametrization. As is expected, the difference in performance between

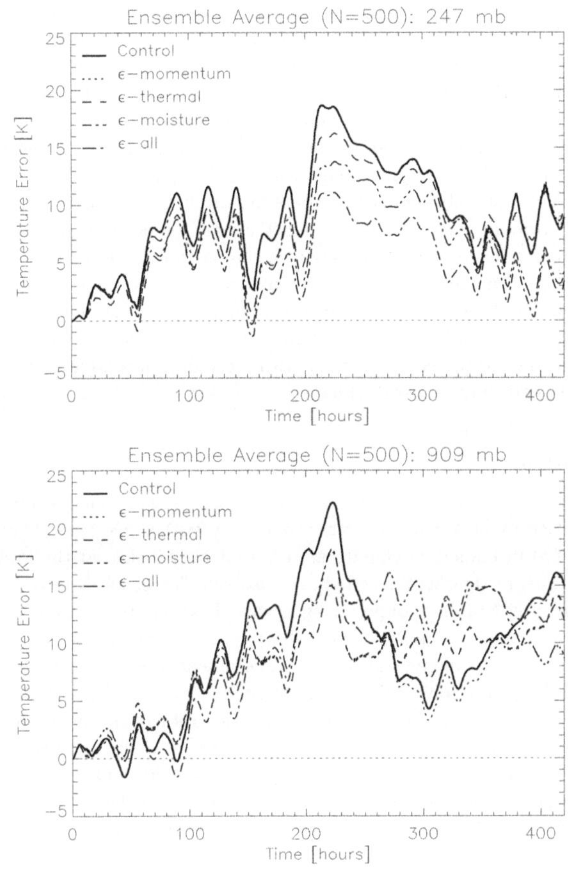

Fig. 3. Time series of the ensemble mean temperature error for the control run, and the four numerical experiments: respectively, the ε-machine representation of the vertical diffusion in the ABL of momentum; sensible heat; water vapor; and lastly momentum, heat and vapor combined. (a) For a model layer in the upper troposphere (247 mb) and (b) for one in the lower troposphere (909 mb).

the model runs with different momentum transport models is small, compared to the difference seen for the more strongly coupled thermal and moisture sub-grid models.

Other stochastic parametrizations of turbulent transport, alternatives to the one-bit, random process inferred by the ε-machine approach, would undoubtedly perform equally well in the SCCM model runs. For example, as pointed out by a reviewer of this work, the data on momentum transport shown in Fig. 2 appears normally distributed, and there is no *a priori* reason why a Gaussian representation of the modeled fluxes would not perform as well as the inferred, one-bit "fair coin" model. There are two reasons why such models are not evaluated here. First, according to Eq. (2), the amount of data utilized here for the inference of a transport model involving minimal two-symbol sequences, is simply insufficient to produce a statistically significant model for anything beyond the binary symbol set ($S = 2$). More importantly, as mentioned above, the performance comparison of the transport models should ultimately be a comparison conditioned on the amount of computational memory required of the model. Gaussian distributed symbol sets consume, by definition, far greater memory than a binary symbol set, and would likely offer no statistically significant performance advantage over the binary symbol choice when so conditioned. In a study with more data available, a demonstration of this would certainly be worthwhile.

6. Conclusions

The climate system, with its plethora of nonlinear dynamic modes and couplings, is characterized by broken symmetries, even for statistical averages. The dynamical equations used in numerical climate models that are invoked at the grid points of the model capture this nonlinear behavior as well as can be expected, i.e. within the limits of predictability imposed by chaos. However, the parametrizations invoked in the models to capture the behavior of sub-grid processes often fail to perform reliably in modeling such complex, nonlinear processes as turbulent transport, especially at the longer time scales used in climate models. A new modeling framework is needed for these sub-grid processes that does not assume that the usual symmetries are satisfied by the flow, either in an instantaneous or statistical sense. A modeling framework that accomplishes this will, in general, contain both deterministic and stochastic structure. Ideally, the stochastic and deterministic structure of the model should be inferred from observations rather than imprinted *a priori*.

We have put forth a six-step approach to such sub-grid process modeling, and have applied the approach to modeling of sub-grid turbulent transport of momentum, heat and moisture in a single column climate model. We inferred our sub-grid process models from observational data using the ε-machine statistical inference method put forth in Ref. 3. For the observational window used, this inference procedure resulted in a random two-alternative "coin-flip" model for the sub-grid turbulent transport of momentum, heat and moisture at the lowest four vertical levels of the model. When the standard diffusivity parametrizations for these processes were replaced by the inferred, purely random model, the model runs showed comparable performance as measured by the ensemble mean of 500 model realizations of temperature at two model levels.

Sub-grid models based on ε-machine and on parametrizations have fundamentally different structures and philosophies. For a meaningful comparison to be made of these two methods, it is essential that we define a conditional comparison, the most appropriate condition for comparison being that of equal computational memory. In the SCCM experiments described above, the inferred sub-grid model for turbulent transport uses (in principle) only one bit of active memory as opposed to many hundreds of bits used by the standard parametrizations. This suggests that the inferred sub-grid model structures can be used to better optimize the application of computational resources in climate models. For example, the released memory allocation accompanying the shift to this type of inferred model structure for the many other sub-grid processes may ultimately enable model resolution to be improved for the same total memory allocation and computational speed.

Acknowledgment

The Raman lidar data and the IOP data for the SCCM run were obtained from the Atmospheric Radiation Measurement (ARM) Program sponsored by the U.S. Department of Energy, Office of Science, Office of Biological and Environmental Research, Environmental Sciences Division.

Appendix A. Multivariate ε-Machine Modeling

In the general, "multivariate" case, there will be several "diagnostic" variables for the sub-grid process, as well as several associated "prognostic" variables from the resolved scale portion of the climate model. In the most direct ε-machine approach to sub-grid modeling, a set of simultaneously measured values for all of these variables is first represented as a single string. If there is an underlying nonlinear dynamic system governing the sub-grid process that involves these variables, then a theorem in Ref. 11 implies that the most efficient partition for the data for describing this underlying dynamical system is the maximum entropy partition. Finding this partition generally requires a global search process, and is the first step in building the ε-machine model for the sub-grid process [14]. For n-dimensional, multivariate data, the maximum entropy partition is found (approximately) by coarse graining the data with respect to a trial hyperplane partition of dimension $n - 1$. In other words, the data is binned into a finite set of symbols, each of which corresponds to the position of the n-dimensional data point relative to the partition. The simplest coarse-graining is to assign a 0 or 1 to the data point, depending on whether a particular data point is on either side of the partition [14].

The constructed ε-machine can now be used as a multivariate sub-grid model in the following way. Say the dynamical portion of the larger model has reached a point where it requires a value for a sub-grid process variable. Consider, for example, that the sub-grid variable is surface stress. The resolved scale portion of the large model has computed, among other variables, a value for the horizontal wind speed at a grid point. (The standard sub-grid parametrization for surface stress simply multiplies this wind speed with a drag coefficient to obtain the needed sub-grid stress value.) In order for the ε-machine model to provide a stress value, the current state of the sub-grid, wind-stress system, as described by the machine, must be known. If the ε-machine model is being interrogated for the first time, then the current machine state is set to the start state. Next, the machine makes a transition from the start state in accordance with the transition probabilities specified for the machine. The ε-machine is left in a new state and has emitted a specified symbol in accordance with the transition matrix defined for the machine. This symbol encodes a sector of wind and stress data relative to the maximum entropy partition. The current value of the wind speed from the resolved-scale portion of the model is then used to define a subset of stress values within this sector by referring to the original wind-stress data set. A median value from this subset of stress values is then given to the dynamic model.

If there are no data points in the identified sector at the current wind-speed value, then the ε-machine is returned to the start state (state of total ignorance). This action is an unbiased representation of the fact that the ε-machine model of the wind-stress system and the resolved scale portion of the model have lost synchronism. Otherwise, the machine simply makes the next transition in synchronism with the time step taken by the dynamic portion of the model. The implementation of this cooperative operation of the sub-grid model with the dynamical model is not possible with the standard sub-grid parametrization schemes.

References

[1] Crutchfield, J. P., Knowledge and meaning: Chaos and complexity, in *Modeling Complex Systems*, Lam, L. and Morris, H. C., eds. (Springer Verlag, Berlin, 1992), pp. 66–101.

[2] Crutchfield, J. P., The calculi of emergence: Computation, dynamics and induction, *Physica* **D75**, 11–54 (1994).

[3] Crutchfield, J. P. and Young, K., Inferring statistical complexity, *Phys. Rev. Lett.* **63**, 105–108 (1989).

[4] Goldsmith, J. E. M., Blair, F. H., Bison, S. E. and Turner, D. D., Turn-key Raman lidar for profiling atmospheric water vapor, clouds and aerosols, *Applied Optics* **37**, 4979–4990 (1998).

[5] Hack, J. J., Kiehl, J. T. and Hurrell, J. W., The hydrologic and thermodynamic characteristics of the NCAR CCM3, *Journal of Climate* **11**, 1179–1206 (1998).

[6] Hack, J. J. and Pedretti, J. A., Assessment of solution uncertainties in single-column modeling frameworks, *Journal of Climate* **13**, 352–365 (2000).

[7] Hack, J. J., Pedretti, J. A. and Petch, J. C., SCCM user's guide (1999), available from www.cgd.ucar.edu/cms/sccm/userguide.html.

[8] Hanson, J. E., *Computational Mechanics of Cellular Automata*, Ph.D. thesis (University of California, Berkeley, 1993).

[9] Hopcroft, J. E. and Ullman, J. D., *Introduction to Automata Theory, Languages and Computation* (Addison-Wesley, Reading, Mass., 1979).

[10] Kiehl, J. T., Hack, J. J., Bonan, G. B., Boville, B. A., Briegleb, B. P., Williamson, D. L. and Rasch, P. J., *Description of the NCAR Community Climate Model (CCM3)*, Technical Note TN-420+STR (National Center for Atmospheric Research, 1996).

[11] Kolmogorov, A. N., A new metric invariant of transient dynamical systems and automorphisms in Lebesque spaces, *Doklady Akademii nauk SSSR* **119**, 861 (1958).
[12] Martner, B. E., Wuertz, D. B., Stanov, B. B., Strauch, R. G., Westwater, E. R., Gage, K. S., Ecklund, W. L., Martin, C. L. and Dabberdt, W. F., An evaluation of wind profiler, RASS and microwave radiometer performance, *Bulletin of the American Meteorological Society* **74**, 599–613 (1993).
[13] Palmer, T. N., A nonlinear dynamic perspective on model error, private communication (2000).
[14] Young, K., private communication (1999).

Advances in Complex Systems, Vol. 13, No. 3 (2010) 327–338

DOI: 10.1142/S021952591000258X

World Scientific
www.worldscientific.com

QUANTIFYING EMERGENCE IN TERMS OF PERSISTENT MUTUAL INFORMATION

ROBIN C. BALL* and MARINA DIAKONOVA†

Centre for Complexity Science and Department of Physics,
University of Warwick, Coventry, CV4 7AL, UK
**R.C.Ball@warwick.ac.uk*
†M.Diakonova@warwick.ac.uk

ROBERT S. MACKAY

Centre for Complexity Science and Department of Mathematics,
University of Warwick, Coventry, CV4 7AL, UK
R.S.MacKay@warwick.ac.uk

Received 11 December 2009
Revised 24 April 2010

We define Persistent Mutual Information (PMI) as the Mutual (Shannon) Information between the past history of a system and its evolution significantly later in the future. This quantifies how much past observations enable long-term prediction, which we propose as the primary signature of (Strong) Emergent Behavior. The key feature of our definition of PMI is the omission of an interval of "present" time, so that the mutual information between close times is excluded: this renders PMI robust to superposed noise or chaotic behavior or graininess of data, distinguishing it from a range of established Complexity Measures. For the logistic map, we compare predicted with measured long-time PMI data. We show that measured PMI data captures not just the period doubling cascade but also the associated cascade of banded chaos, without confusion by the overlayer of chaotic decoration. We find that the standard map has apparently infinite PMI, but with well-defined fractal scaling which we can interpret in terms of the relative information codimension. Whilst our main focus is in terms of PMI over time, we can also apply the idea to PMI across space in spatially-extended systems as a generalization of the notion of ordered phases.

Keywords: Emergence; persistent mutual information; chaotic dynamical systems; complexity measure; logistic map.

1. Introduction

Our starting point is the desire to discover and quantify the extent to which the future evolution of a dynamical system can be predicted from its past, and from the standpoint of Complexity Theory we are interested in assessing this from observed data alone without any prior parametric model.

We should nevertheless admit prior classification as to the general nature of the system, and simple constraining parameters such as its size, composition and local laws of motion. Given that information, there may prove to be entirely reproducible features of its subsequent evolution which inevitably emerge over time, such as eventual steady state behavior (including probability distributions). This we follow [1, 3] and others in terming *weak emergence.* The emergence is weak in terms of there being no choice of outcome, it can be anticipated without detailed inspection of the particular instance.

We focus on *Strong Emergence* by which we mean features of behavior significantly into the future which can only be predicted with knowledge of prior history. The implication is that the system has made conserved choices not determined by obvious conservation laws, or at least nearly conserved choices which imply the existence of associated slow variables.

More formally, we must conceive of an ensemble comprising a probability distribution of realizations of the system (and its history), from which observed histories are drawn independently. The behavior of a particular realization which can be anticipated from observing other realizations is weakly emergent, whilst that which can only be forecast from the observation of the past of each particular instance is strong emergence. A related distinction between weak and strong emergence is given in [19], but with quantification based on a metric on the underlying space, rather than purely measure-theoretic.

2. Persistent Mutual Information

Within an ensemble of histories of the system, we can quantify strong emergence in terms of mutual information between past and future history which persists across an interval of time τ. This *Persistent Mutual Information* is given by

$$I(\tau) = \int \log\left(\frac{P[x_{-0}, x_{\tau+}]}{P[x_{-0}]P[x_{\tau+}]}\right) P[x_{-0}, x_{\tau+}]dx_{-0}dx_{\tau+} \tag{1}$$

where x_{-0} designates a history of the system from far past up to present time 0; $x_{\tau+}$ is the corresponding history of the system from later time τ onwards; $P[x_{-0}, x_{\tau+}]$ is their joint probability density within the ensemble of histories; and $P[x_{-0}]P[x_{\tau+}]$ is the product of corresponding marginal probability densities for past and future taken separately. If the history variables $x(t)$ are discrete-valued, then the integration over histories is interpreted as summation; in the continuous case, $I(\tau)$ has the merit of being independent of continuous changes of variable, so long as they preserve time labelling.

Quantitatively, $I(\tau)$ measures the deficit of (differential) Shannon Entropy in the joint history compared to that of past and future taken independently, that is

$$I(\tau) = H[P[x_{-0}]] + H[P[x_{\tau+}]] - H[P[x_{-0}, x_{\tau+}]] \tag{2}$$

where the separate (differential) Shannon Entropies for a probability density P of a set of variables y are given generically by

$$H\,[P] = -\int \log(P[y])P[y]dy. \tag{3}$$

For continuously distributed variables, the differential Shannon Entropies carry an infinite offset from absolute information content, but those offsets can be shown to cancel from Eq. (2) above. Thus $I(\tau)$ is precisely the amount of information (in Shannon Entropy) about the future which is determined by the past, and hence *the extent to which the future can be forecast from past observations* (of the same realization).

The key distinguishing feature of our definition above is the exclusion of information on $x_{0\tau}$, that is the intervening time interval of length τ. This ensures that $I(\tau)$ is only sensitive to system memory which persists right across time τ; any features of shorter term correlation do not contribute. The choice of τ must inevitably be informed by observation, but the extreme cases have sharp interpretation.

$I(0)$ corresponds directly to the Excess Entropy as introduced in [12] where it was called "Effective Measure Complexity", and makes no distinction of timescale in the transmission of information. This quantity has been discussed in many guises: as effective measure complexity in [12, 17], Predictive Information in [2] and as Excess Entropy in [7, 11], to name but a few. See also [22, 13, 11] for measurements of Excess Entropy and the related Entropy Rate on a variety of systems, including the logistic map.

Our sharpest measure of Strong Emergence is the *Permanently Persistent Mutual Information* (PPMI), that is the PMI $I(\infty)$ which persists to infinite time. This quantifies the degree of permanent choice spontaneously made by the system, which cannot be anticipated without observation but which persists for all time. A prominent class of example is spontaneous symmetry breaking by ordered phases of matter: here a physical system is destined to order in a state of lower symmetry than the probability distribution of initial conditions, and hence must make a choice (such as direction of magnetization) which (on the scale of microscopic times) endures forever. As a result, Strong Emergence can only be diagnosed by observing multiple independent realizations of the system, not just one long-time history. An interesting though anomalous case is presented by clock phase, where time shift leads to different phases. This is exploited in measuring PPMI for the logistic map in the following section (see Fig. 1).

PPMI corresponds to some partitioning of the attracting dynamics of the system into negligibly communicating (and negligibly overlapping) subdistributions. If the dynamics evolves into partition i with probability p_i, then the PPMI is simply given by

$$I(\infty) = -\sum_i p_i \log(p_i) \tag{4}$$

which is the entropy of the discrete distribution p_i. For deterministic dynamics, each p_i is simply determined by the sampling of its associated basin of attraction in the distribution of initial conditions, so in this case, the PPMI is sensitive to the

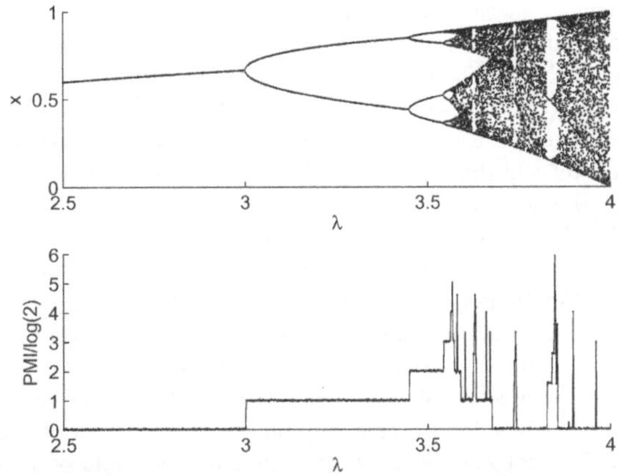

Fig. 1. Bifurcation diagram (above) and measured Persistent Mutual Information (below) for the logistic map, as a function of the map control parameter λ. For each value of λ, the map was allowed a minimum of 10^5 iterations to settle and then the Mutual Information measured across a "time" interval of 10^5 iterations. Each MI measurement used the distance to 4th nearest neighbor to estimate probability density ($k = 4$), based on a sample of $N = 5000$ iterate pairs. Before chaos sets in, PMI increases stepwise in jumps of $\log 2$, reflecting the doubling of the resolved period. It is also seen to pick up the band periodicity after the onset of chaos (resolving some more fine periods within the band hopping), as well as a nonchaotic period three regime.

latter. However, for stochastic dynamics, it is possible for the p_i to be predominantly determined by the distribution of early time fluctuations.

3. PPMI in the Logistic Map

We consider time series from the logistic map $x_{n+1} = \lambda x_n(1 - x_n)$ as a simplest non-trivial example which brings out some non-trivial points. Depending on the control parameter λ, its attracting dynamics can be a limit cycle, fully chaotic, or the combined case of banded chaos where there is a strictly periodic sequence of bands visited but chaotic evolution within the bands [5].

In the case of a periodic attracting orbit with period T, the choice of phase of the cycle is a permanent choice leading directly to positive PPMI. For this case, the phase separation is in the time domain, so the attractor can be fully sampled by shifting the start time of observation. If we assume the latter is uniformly distributed over a range of time large enough compared to the period, then the observed phase is a result of a uniformly selected and symmetry breaking choice in which each phase has $p_i = 1/T$, and this leads to PPMI

$$I(\infty) = \log(T). \tag{5}$$

This is a generic result and not special to the logistic map. Because there is just an attracting orbit, the Excess Entropy gives the same value.

For fully chaotic attracting dynamics such as at $\lambda = 4$, we have to be careful in principle about limits. Provided the probability densities are measured with only limited resolution δx in x, then we expect past and future to appear effectively independent for $\tau > \tau(\delta x)$ and hence $I(\tau) \to 0$ and there is zero PPMI. Thus, for chaotic motion, the associated values are quite different: the Excess Entropy is positive and reflects the complexity of its dynamics, whereas the PPMI is zero reflecting the absence of long-time correlations.

Both of the above results can be seen in measured numerical data for the logistic map in Fig. 1. What is more pleasing still is the behavior of PMI for banded chaos, where a T-periodic sequence of bands shows through to give $I(\infty) \geq \log(T)$ (assuming random initiation phase as before) with equality when the Tth iterate restricted to one band is mixing. The fact that the numerical results overshoot $\log(T)$ for many parameter values can be attributed to the presence of a finer partition than the T bands, for example, around an attracting periodic orbit of period a multiple of T or into sub-bands with a period a multiple of T. Even in cases where the dynamics really is T-periodic mixing (meaning it cyclically permutes T bands and the Tth iterate restricted to one is mixing), and hence PPMI is precisely $\log(T)$, the numerics might pick up some long-time correlation that does not decay until after the computational time interval. One can see this in the more detailed data of Fig. 2 where there are background plateaux corresponding to the periodicity of the observed major bands, decorated by narrower peaks corresponding to higher

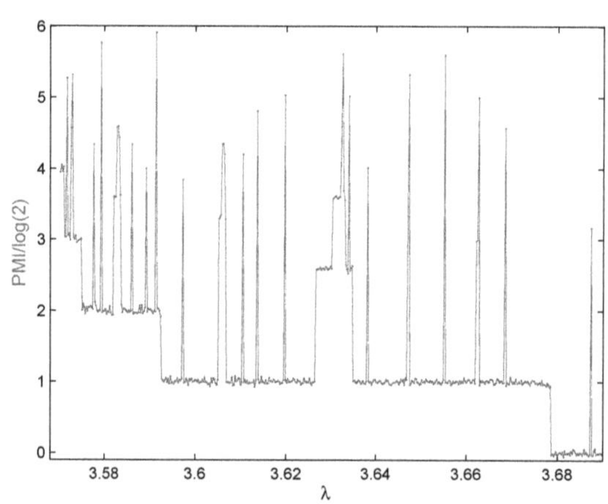

Fig. 2. PMI for a chaotic region of the control parameter (sample size $N = 5000$, $k = 4$, settle time 10^{10}, time separation 10^6). In this range, chaotic bands are known to merge (see bifurcation diagram in Fig. 1). PMI picks out the relevant decrease in overall periodicity as 16 bands merge pairwise into eight, four, and finally two at λ slightly less than 3.68. PMI also detects a period tripling regime, which can be seen around $\lambda = 3.63$.

periodicities. The steps in PMI are particularly clear where bands merge because these special points have strong mixing dynamics within each band cycle [20]. Figures 1 and 2 for the PPMI can be usefully contrasted with the Excess Entropy graphed in Fig. 1 of [11]. On the period-doubling side, they display the same values (log of the period), whereas in the regions of banded chaos the PPMI picks out the number of bands whilst the Excess Entropy is complicated by its sensitivity to short-time correlations in the chaotic decoration.

4. Issues Measuring PMI

Measuring Mutual Information and in particular the implied measurement of the entropy of the joint distribution suffers from standard challenges in measuring the entropy of high dimensional data. The naive "histogram" method, in which probability densities are estimated directly from frequency counts in pre-selected (multidimensional) intervals, is easy to apply but can require very large sample sizes in order to ensure that the significant frequencies are estimated from multiple (rather than single) counts. In practice, we found the kth neighbor approach of Kraskov *et al.* [16] a more effective tool (from now on referred to as the k-NN method). It is more limited in sample size due to unfavorable order of algorithm, but this was outweighed by its automatic adjustment of spatial resolution to the actual density of points.

The basis of the k-NN method is to estimate the entropy of a distribution from the following estimate of the logarithm of local probability density about each sampled point:

$$\log(p) \simeq \log\left(\frac{k/N}{\epsilon_k^d}\right) + [\Psi(k) - \Psi(N) - \log(k/N)] = -\log \epsilon_k^d + \Psi(k) - \Psi(N) \quad (6)$$

where N is the total number of sample points and ϵ_k^d is the volume of space out to the location of the kth nearest neighbor of the sample point in question. The combination $\frac{k/N}{\epsilon_k^d}$ in the first logarithm is simply interpretable as an amount of sampled probability in the neighborhood divided by corresponding volume. In the remaining term $\Psi(z) = \Gamma'(z)/\Gamma(z) \simeq \log(z) - \frac{1}{2z} \text{as} |z| \to \infty$ is the digamma function and the whole of this term is only significant for small k, where it corrects a slight bias associated with finite sampling of the neighborhood [16].

We interpret $\log(N/k) \simeq \Psi(N) - \Psi(k)$ as the (logarithmic) probability resolution in these measurements. When N and either k is large or held fixed, the variation of these two forms is equivalent and we generally show the first for simplicity of exposition. The main exception is for entropy data on the standard map below, where taking data down to $k = 1$ significantly enhances our appreciation of scaling and the use of the more accurately scaled second form is important.

Because PMI is invariant under changes of variable, there is considerable scope for choice of how to parametrize past and future before feeding into the PMI measurement. For the logistic map, we exploited its deterministic nature, by which one value of x_n provides all information about the past up to iterate n which is relevant to the future, and similarly all influence of the past on the future beyond iterate n' is captured by the value of $x_{n'}$. Note however that we did not require to identify minimalist causal states in the sense discussed in Sec. 5 below.

For systems without known causal coordinates, the practical measurement of PMI has a rich time parametrization. In principle, what we can directly measure is the mutual information $I(t_1, t_2; t_3, t_4)$ between time intervals $[t_1, t_2]$ and $[t_3, t_4]$. If we assume stationarity, then this is more naturally parametrized as $I(\tau; T_-, T_+)$ in terms of the intervals $T_- = t_2 - t_1$ and $T_+ = t_4 - t_3$ of past and future respectively as well as the intervening interval $\tau = t_3 - t_2$. Then the full PMI is defined as

$$I(\tau) = \lim_{T_-, T_+ \to \infty} I(\tau; T_-, T_+). \tag{7}$$

If the PPMI is desired, it is computationally efficient to set $\tau \to \infty$ before taking the limits above, because the dimension of space in which entropy must be measured is set by $T_- + T_+$ alone. By contrast with PMI, the Predictive Information developed at length in [2] is in the present notation $I(0; T, T)$.

Practical measurement of PMI entails some limited resolution, whether explicitly by histogram methods or implicitly through the depth of sampling in k-NN and other adaptive methods. This inevitably leads to long periodic orbits being capped in their apparent period and hence their measured $I(\tau)$. We can be fairly concrete in the case of measurement by the k-NN method, which looks out across a neighborhood whose aggregate measure is k/N. The longest period one can thereby detect is of the order N/k so we are led to expect $I(\tau) \simeq \log(\min(T, N/k))$. One point

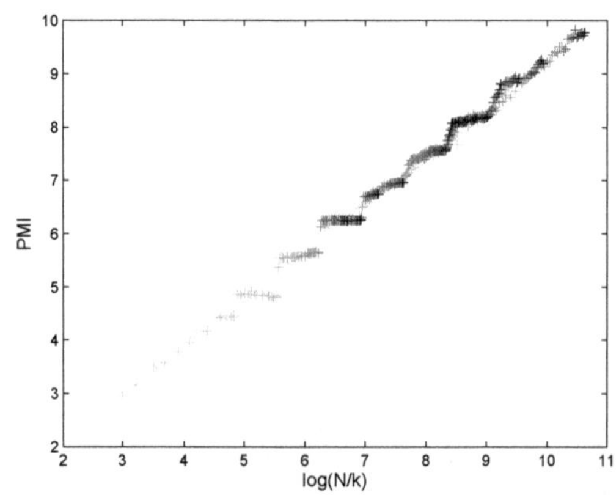

Fig. 3. PMI of the logistic map at the period-doubling accumulation point plotted against the effective resolution with which probability density has been measured, given by $\log(N/k)$. ($N = 1000, 2000, \ldots, 41000$ where runs with higher N correspond to darker points; $k = 1, 2, 3, 4, 5, 10, \ldots, 50$, time separation 10^8, settle time 10^{10}.) The periodicity measured is resolution limited, with apparent overall slope of this plot ca. 0.9 in fair agreement with slope unity predicted on the basis of resolvable period $\propto N/k$ (see text).

where we can check this quantitatively is the accumulation point of the period-doubling sequence [6, 10, 15]: Fig. 3 shows the measured results agreeing with the expectation that $I(\tau) \simeq \log(N/k)$.

5. Relationship with Statistical Complexity

Statistical Complexity (in certain contexts, it is equivalent to "True Measure Complexity" first introduced in [12]; see also [8, 21]) is built on the projection of past and future down to optimal causal states $S_-(t)[x_{t-}]$ such that $P[x_{t+}|x_{t-}] = P[x_{t+}|S_-(t)]$. In terms of these, one readily obtains that the PMI is given by the Mutual Information between time-separated forward and reverse causal states, that is

$$I(\tau) = H[P[S_-(t)]] + H[P[S_+(t+\tau)]] - H[P[S_-(t), S_+(t+\tau)]] \tag{8}$$

as a straightforward generalization of the corresponding result for the Excess Entropy $I(0)$ [12].

In general, one cannot simplify the above formula (I_{-+} in a natural extension of the notation) to use other combinations of choice between S_- or S_+. However, we conjecture that under fairly general conditions they are all equivalent for the PPMI, i.e., $\tau \to \infty$. It is an interesting open question whether for general time gap τ it can be proved that $I_{-+} \geq (I_{--}, I_{++}) \geq I_{+-}$, and perhaps also $I_{--} = I_{++}$. For $\tau = 0$, similar forward and reverse time dependencies for ϵ-machines have been considered in [9], where it was noted that in general $I_{--} \neq I_{++}$, and bidirectional machines were defined that incorporate this time asymmetry.

6. Fractal and Multifractal PMI: Example of the Standard Map

As we already observed for the accumulation points of the standard map, where the attractor of the dynamics is a Cantor fractal adding machine, the measured PMI may go to infinity as the resolution increases. The archetypal case of this is where the probability measures themselves have fractal support or more generally exhibit multifractal scaling. The general phenomenology is that dividing space of dimension d into cells c of linear width ϵ of integrated measure $\mu_c = \int_c P(x)d^dx$, the density in a cell is estimated as μ_c/ϵ^d and hence to leading order as $\epsilon \to 0^+$ one expects

$$H[P] = -\sum_c \mu_c \log\left[\frac{\mu_c}{\epsilon^d}\right] = d\log\epsilon - D\log\epsilon + \text{constant} \tag{9}$$

where D is the information dimension of the integrated (natural) measure μ, defined to be

$$D = \lim_{\epsilon\to 0^+} \frac{\sum_c \mu_c \log \mu_c}{\log \epsilon}. \tag{10}$$

Applying this to the PMI through Eq. (2) then leads to

$$I(\tau) = I_0(\tau) - (D_- + D_+ - D_{-+})\log\epsilon \tag{11}$$

where D_{-+} is the information dimension of the joint distribution of past and future, D_- and D_+ the information dimensions of the respective marginal distributions, and $I_0(\tau)$ the extrapolated resolution-free PMI (note the dimensions of the underlying spaces cancel from $I(\tau)$ because $d_{-+} = d_- + d_+$).

Applying the equivalent analysis to the k-NN method, we have to be careful to insist that Eq. (1) is used, meaning in particular that it is a neighborhood of k neighbors in the *joint* distribution which is taken to determine the ratio of joint and marginal probability densities within the logarithm. With this understanding, $\log \epsilon$ in the above expression for PMI can be written in terms of probability distribution $\frac{1}{D_{-+}} \log (k/N)$ leading to

$$I(\tau) = I_0(\tau) + \Gamma \log(N/k) \tag{12}$$

where

$$\Gamma = \frac{D_- + D_+ - D_{-+}}{D_{-+}} \tag{13}$$

is the relative information codimension.

Our first check is the accumulation point of period doubling of the logistic map, at which the dynamics causally orbits its Cantor set attractor. In this case all the information dimensions above are each equal to the fractal dimension of the Cantor set, leading to $\Gamma = 1$ in agreement with our earlier observations and interpretation based on resolution limited period. Figure 3 shows the directly measured PMI with apparent $\Gamma \simeq 0.9$ in fair agreement, where sampling over the attractor set is achieved by using different times of measurement. A uniform distribution of measurement times approximates uniform sampling of the unique invariant probability measure of the attractor.

The standard map provides a much more subtle test of this phenomenology. This two-dimensional map

$$p' = (p + K \sin x)_{\text{modulo } 2\pi}, \quad x' = (x + p')_{\text{modulo } 2\pi} \tag{14}$$

is strictly area-preserving in the x, p plane (reduced modulo 2π) and has uniform invariant measure. Thus if we launch this dynamics with random initial conditions, the marginal distributions (joint of x, p at fixed iteration) remain strictly uniform forever. The joint distribution between distant iterations of the standard map is far from simple and uniform, at least for moderate values of the map parameter K. Figure 4 shows the measured PMI as a function of the probability resolution k/N, displaying clear fractal behavior for values of $0 \le K < 6$.

The corresponding estimates of the relative information codimension are shown in Fig. 5. For the left end point we can anticipate $\lim_{K\to 0} \Gamma(K) = 1/3$ on theoretical grounds, because this limit corresponds to a continuous time Hamiltonian dynamics which is energy (Hamiltonian) conserving. The joint distribution can therefore only explore $D_{-+} = 4 - 1$ degrees of freedom leading to $\Gamma = 1/3$. Note this result depends on the assumption that the shear in the dynamics between close energy shells is sufficient to destroy correlation of their in-shell coordinates, and we did

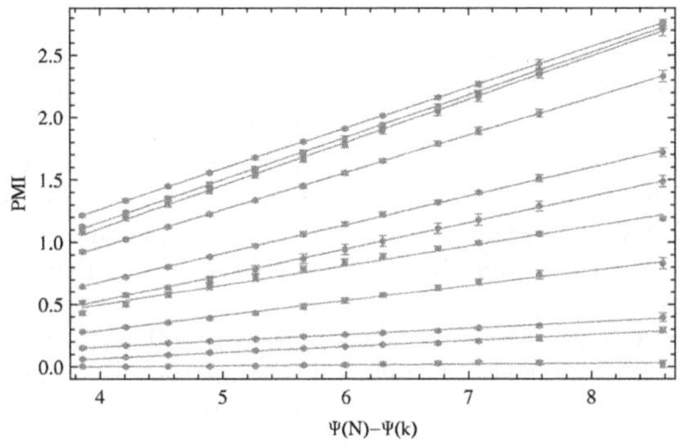

Fig. 4. PMI for the standard map at 3000 iterations, as a function of the probability resolution used to measure it, for map parameters $K = 0.1, 0.5, 0.9, 1.0, 1.1, 1.2, 1.5, 2, 3, 4, 6$ (top to bottom). The resolution of probability is plotted as $\Psi(N) - \Psi(k) \simeq Ln(N/k)$, where $N = 3000$ is the number of sample points used, and k is the rank of neighbor used in the measurement of Mutual Information (see text) which ranges from 1 to 64 across each plot. For each map parameter, there is a clear linear dependence on this logarithmic resolution, consistent with fractal phenomenology and the interpretation of the slope as a relative information codimension. The data have been averaged over five independent sets of such measurements with the error bars showing the 2σ error in each mean, and the drawn lines correspond to the slopes plotted in Fig. 5.

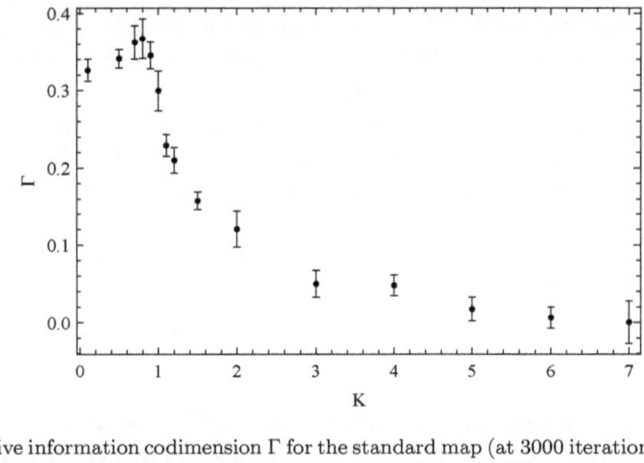

Fig. 5. Relative information codimension Γ for the standard map (at 3000 iterations) as a function of map parameter K. These are the slopes of the data shown in Fig. 4, and the error bars are estimated as 2σ based on separate best fit to the five independent runs of data. The intercept at $K = 0$ matches theoretical expectation of 1/3 (see text) and the fall to zero at large K is consistent with dominance by chaotic dynamics. It is interesting that Γ peaks in the vicinity of $K_g = 0.97$ where the golden KAM curve breaks up, but anomalously slow dynamical relaxation in this region (see [4] and references therein) means that the peak may not reflect the limit of infinite iterations and hence true PPMI.

correspondingly find that we had to use a large number of map iterations for the expected behavior to emerge.

The apparent peak in Γ around $K = 1$ is particularly interesting because this is the vicinity of K_c where momentum becomes diffusive, closely associated (and often identified) with the breakup of the golden KAM (Kolmogorov–Arnold–Moser) curve at $K_g = 0.971635\ldots$ [14, 18]. Dynamical anomalies have been observed around this critical value of K which might underlie the peak we observe in PPMI. On the other hand, corresponding long-time dynamical correlations pose a threat to whether our results are adequately converged.

For larger K, our measurements are consistent with $\Gamma(K) = 0$ for $K > K_1$ where $6 < K_1 < 7$, and indeed the full PMI is within the uncertainty of zero in this regime, meaning the map appears fully chaotic to the level we have resolved.

7. Conclusions

We have shown that Persistent Mutual Information is a discriminating diagnostic of hidden information in model dynamical systems, and the Permanent PMI is a successful indicator of Strong Emergence.

The detailed behavior of the logistic map is sufficiently re-entrant, with periodicity and cascades of period multiplication intermingled amidst chaos, that we are unlikely to have the last word on the full quantitative behavior of the PPMI as a function of map parameter λ beyond the first cascade.

For the standard map, PPMI reveals some of the subtlety only otherwise accessible through dynamical properties such as explicit orbits. Precise relationships remain an open issue, particularly around critical map parameter K_c. The observed fractal behavior with a deficit between the joint information dimension and those of the marginals is we suggest a general phenomenology. Whether it reflects truly fractal and multifractal behavior in any particular case should rest on a wider multifractal analysis of the joint probability measure, which we intend to address in later work on a wider range of non-trivial dynamical systems.

Application to intrinsically stochastic systems and real-world data are outstanding challenges. We can however readily invoke a wide variety of examples associated with ordering phenomena in statistical physics where a dynamically persistent and spatially coherent order parameter emerges. In these cases, there is clearly PPMI in time, but we can also consider just a time slice and let one spatial coordinate take over the role of time in our PMI analysis. Spin Glasses are a key instance where the two viewpoints are not equivalent: these have order and hence Mutual Information persists in time but not in space.

Acknowledgments

We acknowledge separate communications from P. Grassberger, D. Feldman and W. Bialek which were particularly helpful in terms of references. This research was supported by the EPSRC funding of Warwick Complexity Science Doctoral Training Centre, which fully funded MD, under grant EP/E501311/1.

References

[1] Bar-Yam, Y., A mathematical theory of strong emergence using multiscale variety, *Complexity* **9** (2004) 15–24.

[2] Bialek, W., Nemenman, I. and Tishby, N., Predictability, complexity, and learning, *Neural Comput.* **13** (2001) 2409–2463.

[3] Chalmers, D. J., *Strong and Weak Emergence*, The Re-Emergence of Emergence. (Oxford University Press., 2002).

[4] Chirikov, B. V. and Shepelyansky, D. L., Asymptotic statistics of Poincaré recurrences in Hamiltonian systems with divided phase space, *Phys. Rev. Lett.* **82** (1999) 528–531.

[5] Collet, J.-P. and Eckmann, P., *Iterated Maps on the Interval as Dynamical System*, Modern Birkhäuser Classics (Birkhäuser, 1980).

[6] Coullet, P. and Tresser, C., Iterations d'endomorphismes et groupe de renormalisation, *J. Phys. Colloque C* **539** (1978) C5 – 25.

[7] Crutchfield, J. and Packard, N., Symbolic dynamics of noisy chaos, *Physica D* **7** (1983) 201–223.

[8] Crutchfield, J. and Young, K., Inferring statistical complexity, *Phys. Rev. Lett.* **63** (1989) 105–108.

[9] Ellison, C. J., Mahoney, J. R. and Crutchfield, J. P., Prediction, retrodiction, and the amount of information stored in the present, *J. Stat. Phys.* **136** (2009) 1005–1034.

[10] Feigenbaum, M. J., Quantitative universality for a class of nonlinear transformations., *J. Stat. Phys.* **19** (1978) 25–52.

[11] Feldman, D. P., McTague, C. S. and Crutchfield, J. P., The organization of intrinsic computation: Complexity-entropy diagrams and the diversity of natural information processing, *CHAOS* **18** (2008).

[12] Grassberger, P., Toward a quantitative theory of self-generated complexity, *International Journal of Theoretical Physics* **25** (1986) 907–938.

[13] Grassberger, P., On symbolic dynamics of one-humped maps of the interval, *Z. Naturforschung* **43a** (1988) 671–680.

[14] Greene, J. M., A method for determining a stochastic transition, *Journal of Mathematical Physics* **20** (1979) 1183–1201.

[15] Grossman, S. and Thomae, S., *Z. Naturforschung* **32a** (1977) 1353.

[16] Kraskov, A., Stogbauer, H., and Grassberger, P., Estimating mutual information, *Phys. Rev. E* **69** (2004).

[17] Lindgren, K. and Nordahl, M. G., Complexity measures and cellular automata, *Complex Systems* **2** (1988) 409–440.

[18] MacKay, R. and Percival, I., Converse KAM: Theory and Practice, *Commun. Math. Phys.* **98** (1985) 469–512.

[19] MacKay, R. S., Nonlinearity in complexity science, *Nonlinearity* **21** (2008) T273–T281.

[20] Misiurewicz, M., Absolutely continuous measures for certain maps of an interval, *Publications Mathematiques* (1981) 17–51.

[21] Shalizi, C., Shalizi, K. and Haslinger, R., Quantifying self-organization with optimal predictors, *Phys. Rev. Lett.* **93** (2004).

[22] Zambella, D. and Grassberger, P., Complexity of forcasting in a class of simple models, *Complex Systems* **2** (1988) 269–303.

Advances in Complex Systems
Vol. 17, No. 1 (2014) 1450005 (18 pages)

DOI: 10.1142/S0219525914500052

World Scientific
www.worldscientific.com

MODELING THE COEVOLUTION OF ROAD EXPANSION AND URBAN TRAFFIC GROWTH

JIANJUN WU*,‡, MINGTAO XU† and ZIYOU GAO†

*State Key Laboratory of Rail Traffic Control and Safety,
Beijing Jiaotong University,
Beijing 100044, P. R. China

†MOE Key Laboratory for Urban Transportation
Complex Systems Theory and Technology,
Beijing Jiaotong University,
Beijing 100044, P. R. China
‡jjwu1@bjtu.edu.cn

Received 24 May 2013
Revised 23 October 2013
Accepted 15 January 2014
Published 20 March 2014

The evolution of road expansion and traffic growth (motor vehicle) of urban system is a quite complex process. To investigate the interaction between them, a coevolution dynamics model is proposed in this paper to capture the relationships among traveler, vehicle and road. Then stability analysis and numerical simulation are conducted. The results show that the coevolution model can be stable under certain conditions and there exists a dynamic equilibrium in the evolution process. In order to verify the proposed model, Beijing as a case study is analyzed. Results show that the proposed model provides a new perspective for capturing the characteristic of road and urban traffic and contributes to the coordinated development of the whole transportation system.

Keywords: Coevolution; dynamics model; road expansion; urban traffic growth; simulation.

1. Introduction

With the rapid development of urbanization and urban transport motorization, the supply of traffic facilities cannot satisfy the rapidly growing traffic demand. As a result, many traffic problems become more and more serious, e.g., traffic congestion, air pollution and traffic noise, which have become the bottleneck of a city's further development. At present, the conventional method of solving traffic problems is to build more traffic infrastructures or renovation or adjust the structure of urban road which will increase road network density and its capacity. But the capital and urban space is limited. In addition, the improvement of traffic condition will stimulate the

‡Corresponding author.

rapid increase of vehicle. Therefore, the traffic congestion cannot be alleviated but become more serious and step into so-called "Downs Law" (the peak-hour congested highway law) problems [2, 30].

Improving the traffic supply simply cannot alleviate traffic congestion, a number of countries turn to study the integration of traffic supply and demand. At first, the macroscopic characteristic between road expansion and urban traffic growth is studied. As for the microcosmic structure characteristics, it is difficult to analyze the relationship because the interaction of signalized arterials, freeways and local roads is very complex. In addition, only can we get the interaction between road and urban traffic, we can study the relationship of their internal structure further. Therefore, we only study the macroscopic relationship between road expansion and urban traffic growth. For simplification, urban traffic growth means the evolution of motor vehicle in this paper.

The key relationship between road surface and urban traffic can be regarded as the problem of road resource supply and traffic demand. Nagureny [19] and Dafemos *et al.* [5] were among the earliest to establish the correlation models to study the equilibrium problem between them. Li *et al.* [17] proposed a calculation model of urban road network capacity and traffic demand in which they used road supply and demand matching index to evaluate the balance between road supply and traffic demand. Shao *et al.* [23] developed a mathematics programming model using the difference of network capacity and traffic demand and got the optimal matching of urban road structure and traffic demand mode. Meanwhile, from the perspective of optimization, Schweitzer *et al.* [22] introduced an optimization model to determine and evaluate whether the existing road network matched with the residents' trip demand or not. In addition, the relationship between carrying capacity of road network and traffic demand has been given much attention. Some works indicated that the road network which holds a high-density and higher connectivity will be a better solution to traffic problems. Wu *et al.* [29] studied the coevolution of road surface and urban traffic structure and indicated that ignoring the relations between road surface and urban traffic structure will lead to the disequilibrium development, and eventually cause the instability of the urban transport system in the chaotic state. Bruno [1] and Kulash [11] demonstrated that road network with higher density can amplify the capability of traffic in each lane and then increase road capacity. Other interesting results [3, 10, 13, 27, 24, 26, 31, 32, 33] can be found. Furthermore, with the development and extensive application of computer technology, many scholars expected to obtain the exact relationship between road and urban traffic by analyzing a city's evolution data for each year. Some papers [4, 12, 21] studied the evolution data of Atlanta, King County and San Francisco and indicated that the road density can not only affect the supplying capability, but also influence the travel modes of the inhabitants and their travel distance. Recently, the topological complexity of transportation network has attracted many interests. Many works showed that the topology has great effects on the traffic system performance and its capacity. Holme [9] studied the relationship between centrality measures and traffic density for simple particle hopping models on networks with emerging scale-free degree distributions and how the speed of the dynamics was affected by the underlying network structure. Porta *et al.* [20] analyzed the topological structure of road networks of six cities with different form and historical background and showed that these networks belonged to scale free networks and exhibited some characteristic of small world. Wu *et al.* [28] investigated the congestion problem of three typical networks, e.g., scale-free networks, small-world networks and random networks. The results showed that the scale-free network is more sensitive to the traffic congestion than other two networks at the beginning. But compared with other two topologies, the scale-free network can support much more volume of traffic.

To summarize these works, efforts to display the correlation of road surface and urban traffic have ranged between three levels: economic studies that study the relationship of traffic supply and demand aiming to replicate the essence of traffic problems; geographical studies that aim to replicate road geometries and traffic structures based on intuitive and heuristic rules, and complexity studies that exhibit transportation system network owning some characteristics of complex networks. These efforts, however, have been limited to model and capture the relationship between road surface and urban traffic which is ignored in previous works.

To fill this gap, in contrast to previous work, we pay much attention to the following aspects. First, based on the correlation in their evolution process, a dynamics coevolution model of road expansion and urban traffic growth is established, and the stability of this model is provided. Another main contribution of this study is that we validate the model against empirical facts which can be used to predict the quantification of evolution scale during the coevolution process and to give the suggestion in the urban development planning.

The rest of this paper is organized as follows. In Sec. 2, the coevolution dynamics model is given, and its mathematical expression is provided. In Sec. 3, we analyze the model stability. This is followed by the simulation of our model and the analysis of simulation results in Sec. 4. Meanwhile, a validation of the model using the data from Beijing of China is carried out in Sec. 5. Finally, conclusion and future directions are given in Sec. 6.

2. Model

The coevolution relationship between road surface and urban traffic growth is an interactional process which ranges from coordination to incoordination, and then from incoordination to a new higher coordination continuously. The urban traffic system is regarded as a self-organization system which is contained in the dissipative structure system by Levinson *et al.* [15] and Wu *et al.* [29]. Moreover, Haken [7] discussed that under certain controlled circumstance, this kind of self-organization system will possess a stable ordered structure under the internal interaction. A coevolution dynamics model is established in this paper which is based on logistic equation. At present, this equation has been widely used to model the growth process [6, 14, 16] to describe the interaction of road surface and urban traffic in city system.

2.1. *Assumptions*

To facilitate the model formulation, some assumptions are made throughout the paper explained in this section.

- The evolution scale of roads mileage R stands for the general state of the process of road expansion while the quantity of urban motor vehicle C and population N represent the process of urban traffic growth.
- The number of vehicles C, population N and road mileage R are the continuous differentiable functions of time t.
- The evolution of vehicle C, population N and road mileage R are relatively independent, all of them accord with the evolution discipline of Logistic.

2.2. *Coevolution model*

Luís *et al.* [18] indicated that many diverse properties of cities from road mileage and vehicle quantity follows a power-law function of population size with scaling exponent, β and $\beta > 0$. The power-law scaling takes the form as

$$Y(t) = Y(0)N(t)^{\beta}, \tag{1}$$

where Y denotes the road miles. Y_0 is a normalization constant and $N(t)$ is the population at time t. Thereby, we can get the following equations:

$$C(t) = C(0)N(t)^{\beta_1}, \tag{2}$$

$$R(t) = R(0)N(t)^{\beta_2}, \tag{3}$$

where $C(0)$ and $R(0)$ are the initial value of vehicle quantity and road mileage. From Eq. (1), we can see that the interaction among traveler, road mileage and vehicle is not equivalent. Equations (2) and (3) describe that the scaling exponent β_1 and β_2 can be applied to reflect the effect of population to the coevolution of urban road surface and vehicle mileage. Moreover, let $C_m = C(0)N_m^{\beta_1}$ and $R_m = R(0)N_m^{\beta_2}$, where C_m, N_m and R_m are the maximum level of vehicle amount, population and road mileage respectively under nature growth. Then, the coevolution model can be given as follows:

$$\begin{cases} \dfrac{dC}{dt} = k_1 C\left(1 - \dfrac{C}{C_m} + \delta_1 \dfrac{R}{R_m}\right) \\ \dfrac{dR}{dt} = k_2 R\left(1 - \dfrac{R}{R_m} + \delta_2 \dfrac{C}{C_m}\right) \\ C_m = C(0)N_m^{\beta_1} \\ R_m = R(0)N_m^{\beta_2} \end{cases}, \tag{4}$$

where k_1 and k_2 are severally the natural growth rate of vehicle and road mileage, while δ_1 and δ_2 represent the feedback coefficient between the evolution of urban traffic and road surface.

3. Stability Analysis

In order to inspect whether the interactional process of road expansion and urban traffic can reach a stable equilibrium state or not, the stability analysis of the coevolution model is discussed in the following.

3.1. *Equilibrium points*

Based on the stability theory of differential equation [8], the first and the second sub-equations at the right side of Eq. (4) do not contain the parameter t, therefore, these two expressions are autonomous equations. Moreover, the stability analysis of the coevolution model can be realized by investigating the equilibrium points of Eq. (4). Let $dC/dt = dR/dt = 0$, four equilibrium points are obtained which are as follows:

$$P_1 = (0, 0), \quad P_2 = (C(0)N_m^{\beta_1}, 0), \quad P_3 = (0, R(0)N_m^{\beta_2}),$$

$$P_4 = \left(\frac{(1+\delta_1)C(0)}{1-\delta_1\delta_2}N_m^{\beta_1}, \frac{(1+\delta_2)R(0)}{1-\delta_1\delta_2}N_m^{\beta_2}\right).$$

This model will be practical if and only if the equilibrium points locate in the first quadrant, namely $C > 0$ and $R > 0$. As to these equilibrium points, only the point P_4 can satisfy the condition. Consequently, to assure the practicability of our model, we assume $\delta_1\delta_2 < 1$, $\beta_1 > 0$ and $\beta_2 > 0$. Then we will carry out stability analysis to these equilibrium points.

3.2. *Stability analysis*

According to the judgment method of equilibrium point of nonlinear differential equation [8], the first two expressions in Eq. (4) are Taylor expanded in their equilibrium points. Meanwhile, the first term are reserved only and the coefficient matrix of Eq. (4) can be given as follows:

$$J = \begin{bmatrix} a_{11} & a_{12} \\ a_{21} & a_{22} \end{bmatrix} = \begin{bmatrix} k_1 - \dfrac{2k_1C}{C(0)N_m^{\beta_1}} + \dfrac{k_1\delta_1R}{R(0)N_m^{\beta_2}} & \dfrac{k_1\delta_1C}{R(0)N_m^{\beta_2}} \\ \dfrac{k_2\delta_2R}{C(0)N_m^{\beta_1}} & k_2 - \dfrac{2k_2R}{R(0)N_m^{\beta_2}} + \dfrac{k_2\delta_2C}{C(0)N_m^{\beta_1}} \end{bmatrix}. \tag{5}$$

The coefficients of characteristic equation are in the following:

$$p = -(a_{11} + a_{22})|_{P_i}, \quad i = 1, 2, 3, 4, \tag{6}$$

$$q = \det J|_{P_i}, \quad i = 1, 2, 3, 4. \tag{7}$$

Furthermore, when $p > 0$ and $q > 0$, the equilibrium point is steady. Then taking the value of four equilibrium points into Eqs. (6) and (7), the stability analysis process is as follows:

(i) As for the point $P_1 = (0, 0)$, its corresponding coefficient matrix is $J = \begin{bmatrix} k_1 & 0 \\ 0 & k_2 \end{bmatrix}$. Then plugging this matrix into Eqs. (6) and (7), one can get $p = -k_1 - k_2$ and $q = k_1 k_2$. Clearly, $p < 0$ and $q > 0$ so the point P_1 is an instable point.

(ii) Similarly, when locating in the point of $P_2 = (C(0)N_m^{\beta_1}, 0)$ and $P_3 = (0, R(0)N_m^{\beta_2})$, this model is still unsteady. In other words, the points P_2 and P_3 are unsteady points as well.

(iii) Eventually, P_4 is analyzed. Analogously, we can get the following expression:

$$p = -\frac{-k_1(1+\delta_1)}{1-\delta_1\delta_2} - \frac{-k_2(1+\delta_2)}{1-\delta_1\delta_2} = \frac{k_1(1+\delta_1) + k_2(1+\delta_2)}{1-\delta_1\delta_2}, \tag{8}$$

$$\begin{aligned} q &= \frac{k_1(1+\delta_1)}{1-\delta_1\delta_2} \cdot \frac{k_2(1+\delta_2)}{1-\delta_1\delta_2} - \frac{k_2\delta_2(1+\delta_2)}{1-\delta_1\delta_2} \cdot \frac{k_1\delta_1(1+\delta_1)}{1-\delta_1\delta_2} \\ &= \frac{k_1k_2(1+\delta_1)(1+\delta_2)}{1-\delta_1\delta_2}. \end{aligned} \tag{9}$$

In Sec. 3.1, we point out that, for the point P_4, there exists $\delta_1\delta_2 < 1$, $\beta_1 > 0$ and $\beta_2 > 0$. Therefore, P_4 is the steady point of this dynamics coevolution model.

It is obvious that, if $t \to \infty$, under the situation of $\delta_1\delta_2 < 1$, $\beta_1 > 0$ and $\beta_2 > 0$, the evolution scale of road and urban traffic in the coevolution model will tend to an stability value whatever the original values of C and R will be. Therefore, the coevolution of road expansion and urban traffic growth can arrive at a balance state under their interdependent feedback. Besides, the equilibrium value of the dynamics model reads

$$C_{\text{equ}} = \lim_{t\to\infty} C(t) = \frac{(1+\delta_1)C(0)}{1-\delta_1\delta_2} N_m^{\beta_1}, \tag{10}$$

$$R_{\text{equ}} = \lim_{t\to\infty} R(t) = \frac{(1+\delta_2)R(0)}{1-\delta_1\delta_2} N_m^{\beta_2}. \tag{11}$$

Equations (8) and (9) suggest that the stability of our coevolution dynamics model depends simply on parameters δ_1 and δ_2. Moreover, Eqs. (10) and (11) reveal that the equilibrium points of our model is not only impacted by δ_1 and δ_2 but also by β_1 and β_2. In view of this, we will discuss the effects of δ_1 and δ_2, and β_1 and β_2 respectively. Following that, the steady condition $\delta_1\delta_2 < 1$, $\beta_1 > 0$ and $\beta_2 > 0$ can be divided into two large categories and seven small subtypes:

- While δ_1 and δ_2 vary, there are three cases: $\delta_1 < 1$ and $\delta_2 < 1$, $\delta_1 < 1$ and $\delta_2 > 1$, $\delta_1 > 1$ and $\delta_2 < 1$.
- Considering the effects of β_1 and β_2, we classify the condition with four cases: $\beta_1 < 1$ and $\beta_2 < 1$, $\beta_1 < 1$ and $\beta_2 > 1$, $\beta_1 > 1$ and $\beta_2 < 1$, $\beta_1 > 1$ and $\beta_2 > 1$.

4. Simulation and Results Analysis

In the simulations, except for the values of δ_1, δ_2, β_1 and β_2, other variables are given as follows: $k_1 = 0.08$, $k_2 = 0.06$, $C(0) = 0.2$, $R(0) = 9.6$ and $N_m = 2.5$.

4.1. *Effects of δ_1 and δ_2*

Because parameters δ_1 and δ_2 have a great influence on the stability and the equilibrium point of dynamics model, it is necessary to investigate the role of them in the coevolution process. Here $\beta_1 = 1$ and $\beta_2 = 1$ is assumed. Other parameter values are given in Table 1.

The simulation results are shown in Fig. 1. Figures 1(a) and 1(b) show that as time progresses, under different condition, the coevolution of road expansion and urban traffic growth is brought into different steady state separately. The evolution curves of vehicle and road mileage of Case 1 are flat and eventually tend to 1.2 million units and 5000 km, respectively.

The feedback coefficients δ_2 in Case 2 and δ_1 in Case 3 are larger than the corresponding values in Case 1 which is applied to study the influence on the coevolution process while the feedback parameters are changed. From Fig. 1, one obtains that owing to the feedback between urban traffic and road, the number of vehicles will increase due to the improvement of road condition. In turn, the increase of vehicle will stimulate the construction of road infrastructures. For Case 2, the road mileage evolution speed and scale are improved rapidly in a short term which can cause the supply of road infrastructure exceeding urban traffic demand. As to Case 3, its evolution speed and scale of vehicle increase sharply in a short period which may result in traffic demand exceeding the traffic supply and arouse a serious traffic problems, such as pollution aggravation, more serious traffic congestion and so on.

Figure 2 shows that due to the feedback between the urban traffic growth and road expansion, they are promoted each other. The conclusion is similar with the discussion of Fig. 1.

To sum up, the coevolution of urban traffic growth and road expansion can arrive at a steady state under the effect of their mutual feedback. Furthermore, one can control the coevolution process through adjusting the value of δ_1 and δ_2.

4.2. *Effects of δ and β*

Next, we discuss the effect of β_1 and β_2. Some cases will be analyzed as listed in Table 2.

Table 1. Correlated parameters.

Variable	Case 1 $\delta_1 < 1$, $\delta_2 < 1$	Case 2 $\delta_1 < 1$, $\delta_2 > 1$	Case 3 $\delta_1 > 1$, $\delta_2 < 1$
δ_1	0.7	0.7	1.5
δ_2	0.4	1.1	0.4

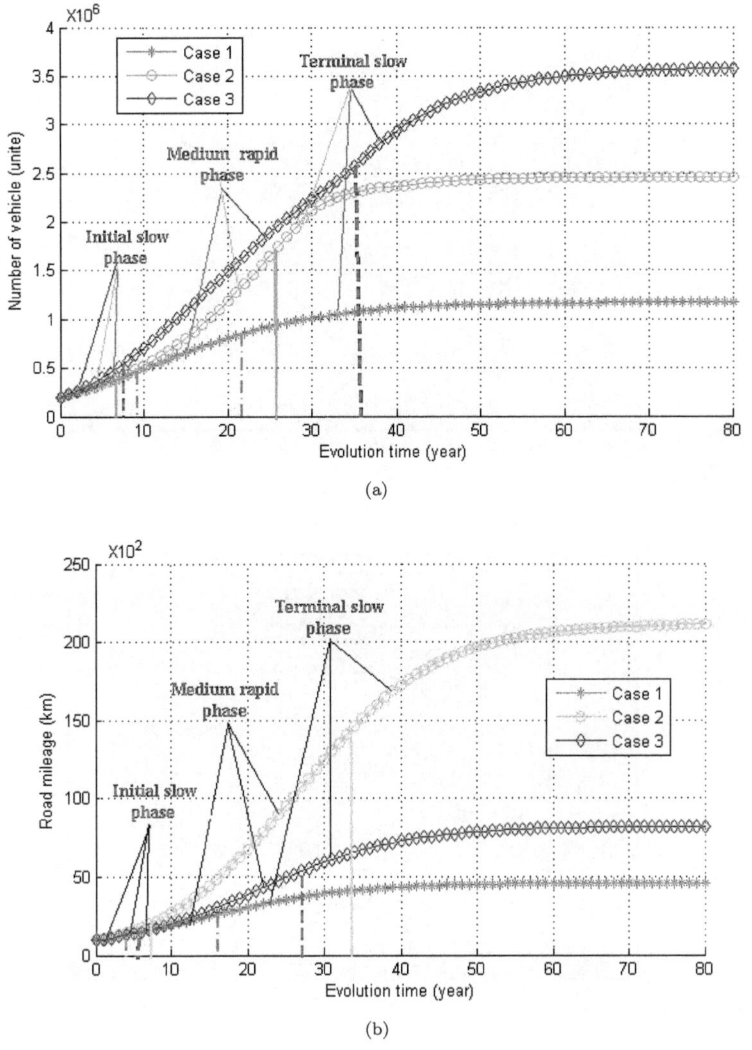

Fig. 1. (Color online) Coevolution of road expansion and urban traffic growth for different δ. (a) The evolution of vehicle quantity. (b) The evolution process of urban road mileage.

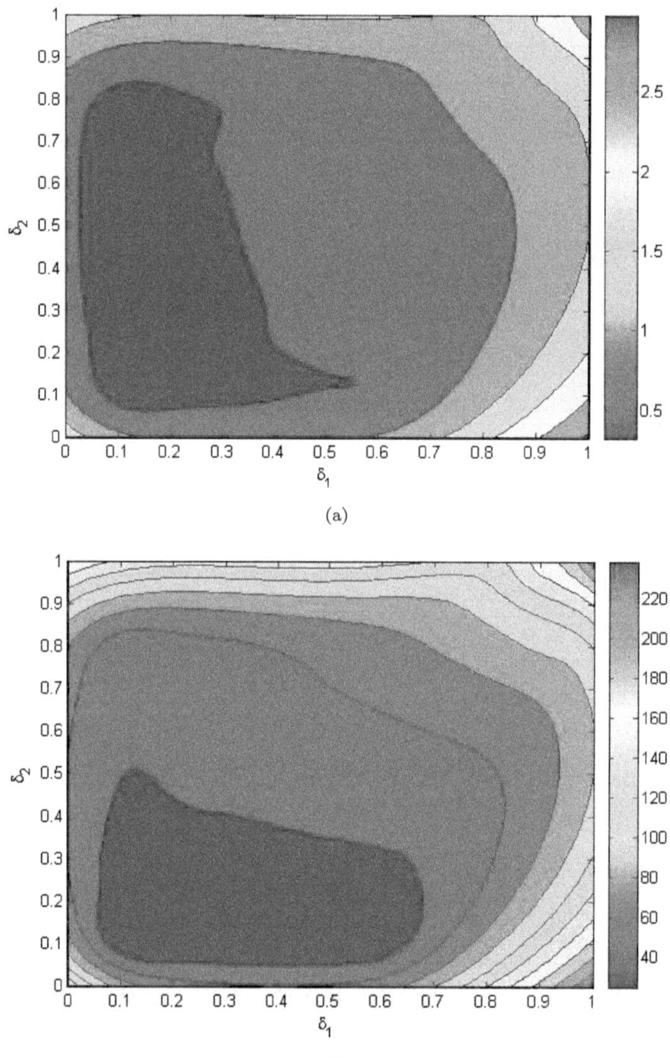

Fig. 2. (Color online) The contour graphs of the coevolution of road expansion and urban traffic growth while δ is different (the value is larger as the color ranges from blue to red). (a) the evolution of vehicle; (b) the evolution of road mileage in the coevolution process.

Table 2. Coefficients under different conditions.

Variable	Case 4 $\beta_1 < 1,\ \beta_2 < 1$	Case 5 $\beta_1 < 1,\ \beta_2 > 1$	Case 6 $\beta_1 > 1,\ \beta_2 < 1$	Case 7 $\beta_1 > 1,\ \beta_2 > 1$
β_1	0.8	0.8	1.2	1.2
β_2	0.7	1.1	0.7	1.1

Figure 3 indicates that population parameters β_1 and β_2 have little effect on the attribute of the coevolution model. Namely, the coevolution process can reach balance on different steady condition. As to Fig. 3(a), the vehicle evolution curves of Case 4 and Case 5 will arrive at the same equilibrium while the curves of Cases 6 and 7 eventually approach to the identical steady status. This demonstrates that the evolution of vehicle quantity is mainly affected by β_1. Similarly, from Fig. 3(b), it is clear that the evolution of road mileage is mainly affected by β_2.

To sum up, the coevolution of urban road expansion together with urban traffic growth can get to a steady status for different β_1 and β_2. However, β_1 primarily adjusts the evolution of urban traffic while β_2 mainly manages the evolution process of road surface. So, to insure the coordination of this coevolution process, β_1 and β_2 should be controlled together.

4.3. *Discussion*

It shows that the coevolution of road expansion and urban traffic growth is a constant development process and that their evolution speed and level do not vary evenly but show obvious phases. This is in agreement with the paper of Wu *et al.* [29]. This phase shows that the coevolution process of road expansion and urban traffic growth consists of three periods: initial slow period, medium rapid period and terminal slow period (showed in Figs. 1 and 3). The first two equations in Eq. (4) show that there exists feedback relationship between road expansion and urban traffic growth and that the rapid development of one party will bring an increase in the evolution speed of the other party. Therefore, the evolution speed of road expansion and urban traffic growth show approximate synchronization.

As mentioned above, this coevolution process includes three phases and the span of each phase is different. Due to lots of limitations, such as the immaturity of technology, the low-level economy and so on, the evolution speed at earlier time is lower and the interaction strength between road expansion and urban traffic growth is weak relatively. However, this phase continues for a short time. With the maturity and innovation of technology, the development of economics, etc., the evolution speed accelerates rapidly while their interaction also becomes greater and greater. According to the limitation of land, energy and other nature resources, population and the length of road will arrive at a stable value eventually. After that, the traffic demand and the road capacity will be stable. If the traffic supply cannot satisfy the travel demand, traffic management, the use of intelligent transport tools and the optimization of traffic internal structure, especially the emergence of new

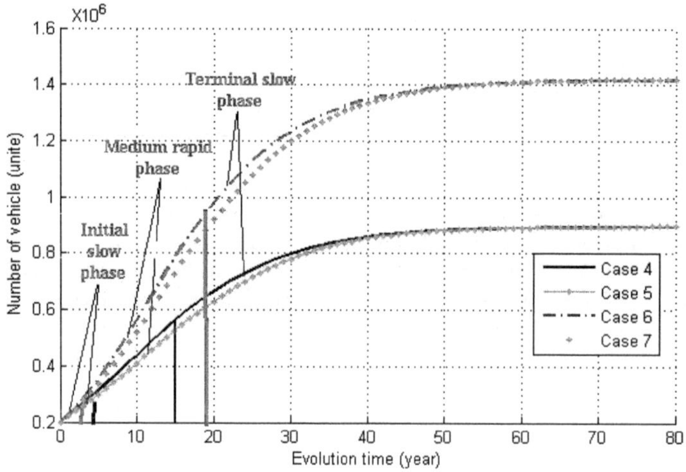

(a)

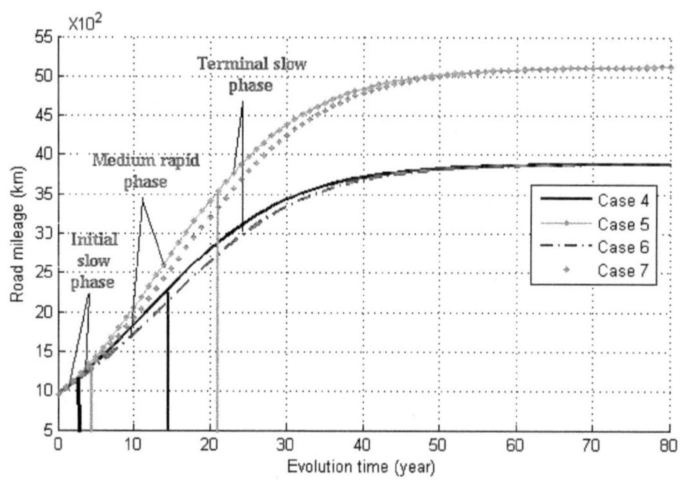

(b)

Fig. 3. (Color online) Coevolution curves of road expansion and urban traffic growth while β is different. (a) The evolution of vehicle. (b) The evolution of road mileage.

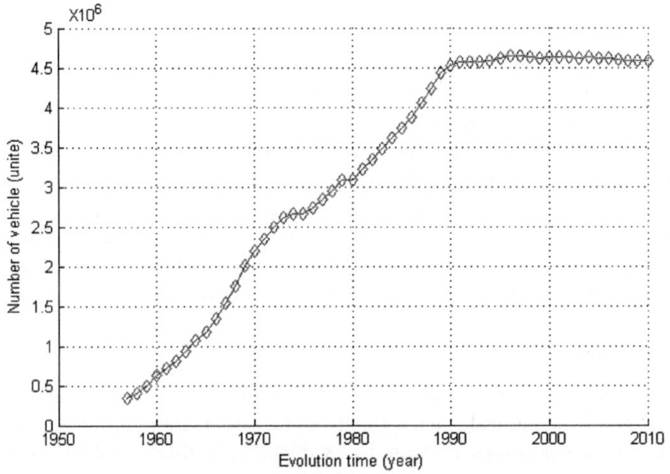

(a)

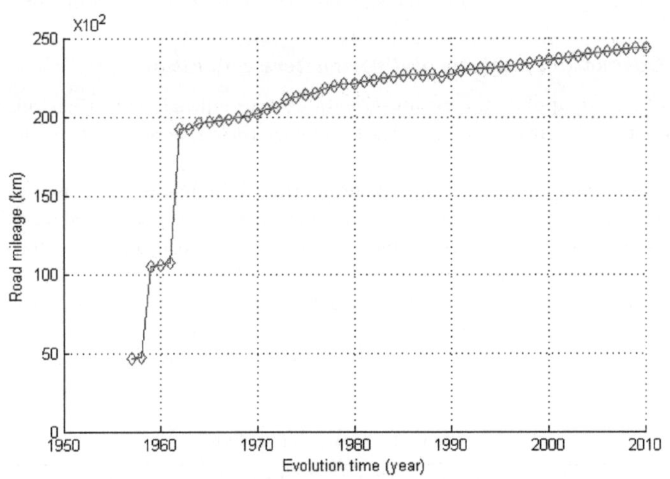

(b)

Fig. 4. (Color online) Coevolution of road expansion and urban traffic growth of Tokyo, Japan. (a) The evolution of vehicle. (b) The evolution of road.

traffic modes, such as subway, the number of vehicle will also get to a dynamic stable value. When approaching to the stable equilibrium values, their coevolution speed slows down again. The coevolution process does not only exhibit dynamics and stage, but also have the characteristics of continuity.

Take the evolution of road expansion and urban traffic growth in Tokyo, Japan as an example to illustrate our model. In Fig. 4, while the evolution of vehicle arrives at equilibrium, the evolution of road mileage still increases. That is reasonable because the dynamic equilibrium of the coevolution of road expansion and urban traffic growth is not strictly synchronous but has a certain lag. Meantime, Fig. 4 shows that the degree of variability in the road mileage becomes smaller and smaller and the evolution process is tending to a stationary value slowly.

5. Case Study

Several sections described the coevolution model from a theoretical perspective and carried out numerical simulation. To some extent, these results show that the coevolution model owns some practical significance and can explain the evolution process of road expansion and urban traffic growth. In order to prove the validity and practicability of our model more closely, we take Beijing as an example to study her evolution process of road expansion and urban traffic growth from 1980 to 2012.

5.1. *Correlation analysis and parameters calibration*

First of all, we applied the evolution data of road mileage and urban vehicle possession of Beijing from 1980 to 2012 to analyze their relevance. The results are as follows.

Here N signifies the number of evolution time of Beijing which is used to analyze the correlation of road mileage and vehicle possession. Table 3 shows that the correlation coefficient between vehicle possession and road mileage is 0.962. They are significant correlation at the 0.01 level.

In order to apply the evolution data of Beijing to validate our coevolution model proposed in this paper, we need to calibrate the four parameters in Eq. (4) next. Based on the evolution data of people, vehicle possession and road mileage of Beijing from 1980 to 2012, we applied the logistic regression in SPSS to calibrate the values

Table 3. Correlation analysis.

Parameters		Vehicle possession	Road mileage
Vehicle possession	Pearson Correlation	1	0.962**
	Sig. (2-tailed)		0.000
	N	33	33
Road mileage	Pearson Correlation	0.962**	1
	Sig. (2-tailed)	0.000	
	N	33	33

Note: **. Correlation is significant at the 0.01 level (2-tailed).

Table 4. Parameters calibration.

Parameters	Coefficient values	Stand. deviation	t Stat	P-value	Lower 95%	Upper 95%
C	0.143	2.406	5.199	0.000	51.504	118.475
R	0.027	0.719	4.343	0.000	10.434	13.319
CR/R_m	0.715	1.119	1.965	0.052	30.516	62.886
CR/C_m	0.143	0.563	2.143	0.039	4.382	5.927

of k_1, k_2, δ_1 and δ_2 in the coevolution model. The calibration results are showed in Table 4.

Therefore, the values of four parameters in Eq. (4) are as follows: $k_1 = 0.143, k_2 = 0.027, \delta_1 = 0.715$ and $\delta_2 = 0.143$.

5.2. *Evolution analysis*

This section is to validate the coevolution model. First of all, we give the value of vehicle possession and road mileage of Beijing in 1980 to $C(0)$ and $R(0)$, namely, $C(0) = 7.90$ and $R(0) = 21.85$. Eventually, we obtain the evolution curves of road mileage and vehicle possession in simulation.

In order to compare with the reality, we describe the evolution of road expansion and urban traffic growth of Beijing from 1980 to 2012 with two evolution curves.

From Fig. 5, we can see that the simulation results of the coevolution of road expansion and urban traffic growth are in accord with the reality of the evolution of road mileage and urban vehicle possession in Beijing. To a large extent, Fig. 5 can demonstrate the practicality and validity of the coevolution model proposed in this paper. In addition, according to the simulation results, the evolution of vehicle of Beijing increases gradually from 2025 and will tend to a dynamic equilibrium state in 2045. At the same time, the evolution of road mileage in Beijing will experience slower growth for a long time and will tend to steady in 2050. Of course, considering that some traffic measures and strategies can be taken to impact the development of road and vehicle in the real management, the changing of the evolution maximum values of road and vehicle and their evolution process may be changed in the future development.

6. Conclusion and Perspective

Few studies involve the dynamic evolution of road expansion and urban traffic growth, this paper proposes a coevolution dynamics model between them which takes three basic factors of transportation system, e.g., travelers, road mileage and vehicle into consideration. Then we prove the stability of the coevolution dynamics model. Furthermore, the data analysis result of Beijing validates the coevolution model proposed in this paper. Of course, this model cannot study all aspects in the coevolution process of road surface and urban traffic. It can at least reasonably

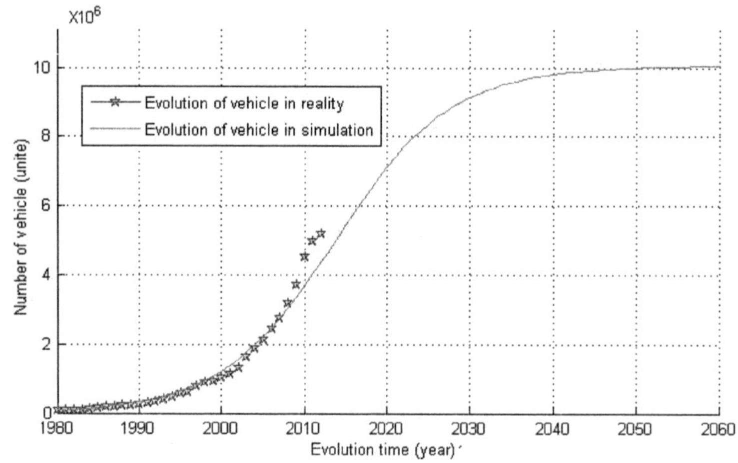

(a)

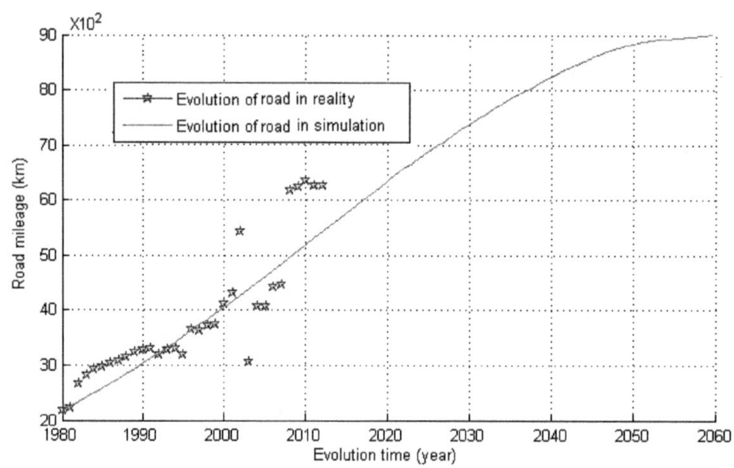

(b)

Data Source: Beijing Municipal Bureau of Statistics.

Fig. 5. Evolution graphs of road expansion and urban traffic growth in simulation based on Beijing and the real evolution process of them of Beijing. (a) The evolution of vehicle possession. (b) The evolution of road mileage.

explain the following aspects:

(i) Due to various constrains including economy, technology and institution, etc., the evolution scale and level of road surface and traffic is limited. While satisfying some requirements, their coevolution will approach a dynamics balance under their interaction.

(ii) Without any constrains, the coevolution of road expansion and urban traffic has a tendency to grow to a larger scale than the maximum level which results in their incoordination development. For instance, traffic congestion is an outcome of the evolution scale of traffic exceeding the development of road and even exceeding its own limitation.

(iii) There exists dynamics interaction between the evolution of road surface and urban traffic. Completed urban road system triggers a larger scale evolution of traffic. In turn, the expansion of the scale of urban traffic brings out a higher request on road. The emergence of rail transit is an inevitable outcome of the evolution scale attaining a certain degree in lots of metropolitans of the world. This correspondence and dynamics coupling balance are identical with the conclusions of our coevolution model.

However, the coevolution of road surface and urban traffic is a complex dynamics process which is influenced by plenty of factors, such as human control, geographical location, economy, etc. The coevolution results are the fruits of combination of multiple elements. How to express their effect and their weight in our model still needs to be discussed in the future work.

In addition, this paper describes only the coevolution of road expansion and urban traffic growth and does not involve in the internal structure of road and traffic. In this paper, the dynamic feedback relationship between road expansion and urban traffic growth and their evolution discipline are obtained which lay a good foundation for the study on the coevolution of their internal structure. Next, referring to the study of Tsekeris and Geroliminis [25], we will further study the coevolution of road structure and urban traffic structure.

Acknowledgments

This paper is partly supported by the National Basic Research Program of China (2012CB725400), NSFC (71271024), Program for New Century Excellent Talents in University (NCET-12-0764), the Fundamental Research Funds for the Central Universities (2012JBZ005) and FANEDD (201170).

References

[1] Bruno, F. S., Antnio, P. A. and Miller, E. J., Interurban road network planning model with accessibility and robustness objectives, *Transport. Plan. Technol.* **33**(3) (2010) 297–313.

[2] Cascetta, E., Pagliara, F. and Papola, A., Governance of urban mobility: Complex systems and intergrated policies, *Adv. Complex Syst.* **10**(2) (2007) 339–354.

[3] Cervero, R. and Duncan, M., Walking, bicycling, and urban landscapes: Evidence from the San Francisco bay area of American, *J. Public Health* **93**(9) (2003) 1478–1483.

[4] Chapman J. and Frank, L., SMARTRAQ: Integrating travel behavior and urban form data to address transportation and air quality problems in Atlanta, *Research Paper*, Regional Transportation Authority and Department of Transportation, Atlanta, Georgia, 2004.

[5] Dafemos, S. and Nagureny, A., Supply and demand equilibrium algorithms for a class of market equilibrium problem, *Transport. Sci.* **23**(2) (1989) 119–124.

[6] Gu, E. G. and Tian, F., Complex dynamics analysis for a duopoly model of common fishery resource, *Nonlinear Dyn.* **61**(4) (2010) 579–590.

[7] Haken, H., *Information and Self-Organization: A Macroscopic Approach to Complex System* (Springer-Verlag, New York, USA, 1988).

[8] Hirsch, M. W. and Smale, S., *Differential Equations, Dynamical Systems and Linear Algebra* (Academic Press, Pittsburgh, USA, 1974).

[9] Holme, P., Congestion and centrality in traffic flow on complex networks, *Adv. Complex Syst.* **6**(2) (2003) 163–176.

[10] ITE Smart Growth Task Force, *Smart Growth Transportation Guidelines: An ITE Proposed Recommended Practice.* Institution of Transportation Engineers, Washington D.C., US, 2003.

[11] Kulash, W., Anglin, J. and Marks, D., Traditional neighborhood development: Will the traffic work, *Proc. 11th Annual Pedestrian Conf.*, Bellevue, US, 1990.

[12] Lawrence Frank and Company, Inc., LUTAQH: A study of land use, transportation, air quality and health in King County, WA, *Research Paper*, Regional Research Institute, King County, WA, 2005.

[13] Leccese, M. and McCormick, K., Congress for the new urbanism, *Charter of the New Urbanism* (McGraw Hill, New York, US, 2000).

[14] Levin, S. A., Hallam, T. G. and Gross, L. J., *Applied Mathematical Ecology* (Springer-Verlag, New York, USA, 1989).

[15] Levinson, D. and Yerra, B., Self organization of surface transportation networks, *Transport. Sci.* **40**(2) (2006) 179–188.

[16] Li, H. C., A biological meaning in asymptotic stability of the delay logistic differential equations, *Int. J. Biomath.* **4**(1) (2011) 119–134.

[17] Li, X. H., Tian, F. and Gu, Z. H., Analytic techniques about the supply and demand of urban road network, *J. Traffic Transport. Eng.* **2**(2) (2002) 88–90.

[18] Luís, M. A. B., José, L., Dirk, H., Christian, K. and Geoffrey, B. W., Growth, innovation, scaling and the pace of life in cities, *Proc. Natl. Acad. Sci.* **104**(17) (2007) 7301–7306.

[19] Nagureny, A., Computational comparisons of spatial price equilibrium methods, *J. Regional Sci.* **27**(1) (1987) 55–76.

[20] Porta S., Crucitti, P. and Latora, V., The network analysis of urban streets: A dual approach, *Phys. A* **369** (2005) 853–866.

[21] Reilly, M. and Landis, J., The influence of built form and land use on mode choice, *Research Paper*, University of California Transportation Center, Berkeley, California, 2003.

[22] Schweitzer, F., Ebeling, W., Rosé, H. and Weiss, O. (1998), Optimization of road networks using evolutionary strategies, *Evol. Comput.* **5**(4) (1998) 419–438.

[23] Shao, Z. Y., Zheng, A. W., Zhu, X. H. and Guo, J. Z., Optimum matching of the city road and mode of traffic based on balance between supply and demand, *J. Basic Sci. Eng.* **11**(1) (2003) 106–111.

[24] Talen, E., Measuring urbanism: Issues in smart growth research, *J. Urban Des.* **8**(3) (2003) 195–215.
[25] Tsekeris, T. and Geroliminis, N., City size, network structure and traffic congestion, *J. Urban Econ.* **76** (2013) 1–14.
[26] Thomas, R., *Sustainable Urban Design: An Environmental Approach* (Spon Press, London, UK, 2003).
[27] Wen, G. W., *The Urban Road Transport System Planning* (Tsinghua University Press, Beijing, China, 2001).
[28] Wu, J. J., Gao, Z. Y., Sun, H. J. and Huang, H. J., Congestion in different topologies of traffic networks, *Europhys. Lett.* **74**(3) (2006) 560–566.
[29] Wu, J. J., Xu, M. T. and Gao, Z. Y., Coevolution dynamics model of road surface and urban traffic structure, *Nonlinear Dyn.* **73**(3) (2013) 1327–1334.
[30] Yang, H., Bell, M. G. H. and Meng, Q., Modeling the capacity and level of service of urban transportation network, *Transport. Res. B* **34**(4) (2000) 255–275.
[31] Yang, P. K., Reconsidering the density of urban trunk road network — The revise proposal for the design standard of urban road planning, *Urban Traffic* **1**(2) (2003) 52–55.
[32] Zhao, J. F., *Urban Road and Aesthetics* (Southeast University Press, Nanjing, China, 1998).
[33] Zhao, P. J., Car use, communiting and urban form in a rapidly growing city: Evidence from Beijing, *Transport. Plan. Technol.* **34**(6) (2011) 509–527.

Advances in Complex Systems
Vol. 18, Nos. 7 & 8 (2015) 1550017 (14 pages)

DOI: 10.1142/S0219525915500174

LONG RANGE DEPENDENCE AND THE DYNAMICS OF EXPLOITED FISH POPULATIONS

HUGO C. MENDES*,‡, ALBERTO MURTA* and R. VILELA MENDES†,§
*Instituto Português do Mar e da Atmosfera,
Avenida Brasília, 1300-598 Lisboa, Portugal
†CMAF and IPFN, University Lisbon,
Av. Prof. Gama Pinto 2, 1649-003 Lisboa, Portugal
‡hugart@gmail.com
§rvilela.mendes@gmail.com
§vilela.mendes@ist.utl.pt

Received 14 June 2014
Revised 6 July 2015
Accepted 28 July 2015
Published 28 August 2015

Long range dependence or long memory is a feature of many processes in the natural world, which provides important insights on the underlying mechanisms that generate the observed data. The usual tools available to characterize the phenomenon are mostly based on second-order correlations. However, the long memory effects may not be evident at the level of second-order correlations and may require a deeper analysis of the nature of the stochastic processes.

After a short review of the notions and tools used to characterize long range dependence, we analyze data related to the abundance of exploited fish populations which provides an example of higher order long range dependence. In particular, we find that fish population time series were thought to have short term memory only because previous studies used averages over species instead of modeling each species individually.

Keywords: Long range dependence; fractional processes; populations.

PACS Number(s): 05.45.Tp, 02.50.Ey, 87.23Cc

1. Introduction

Long range dependence or long memory is an important notion in many processes in the natural world. Studies involving this notion pervade fields from biology to econometrics, linguistics, hydrology, climate, DNA sequencing, etc. Although, at times, it has been considered a nuisance in the study of these processes, the existence of long memory is in fact a bonus, in the sense that it provides further insight in the nature of the process. Whereas short or no memory just points to the essentially

§Corresponding author.

random nature of the phenomena, long memory, by contrast, may provide a window on the underlying mechanisms that generate the observed data.

The most popular definitions of long range dependence are based on the second-order properties of the processes and relate to the asymptotic behavior of covariances, spectral densities and variances of partial sums. However, there are other different points of view, some of which are not equivalent to the characterization of second-order properties. They include ergodic theory notions, limiting behavior, large deviations, fractional differentiation, etc. [1–3].

When looking for or extracting long range dependence from a time series, two important warnings should be taken into account. First, long range dependence may be mimicked by lack of stationarity or by a change of regime. Several methods have been developed [4–6] to distinguish between the two phenomena, detrending and rescaling being probably the simplest one [7]. Second, long range dependence is a feature which might only appear associated to the higher order characteristics of the process. A process that looks short range when looked at through second-order properties, may in fact have an underlying long range dependence of higher order properties.

In Sec. 2, we collect a few definitions of the most usual parameters, used to characterize long range dependence, that will be useful later on and in Sec. 3 analyze some data related to the abundance of exploited fish populations which provides an example of higher order long range dependence.

2. Notions and Tools for Long Range Dependence

When based on second-order properties, long range dependence in a stationary time series $X(t)$ occurs when the covariances

$$\gamma(\tau) = \mathbb{E}\{X(t)X(t+\tau)\}, \tag{1}$$

tend to zero so slowly that their sum

$$\sum_{\tau=0}^{\infty} \gamma(\tau), \tag{2}$$

diverges. Alternative definitions of long range dependence are based on the power-law behavior of the covariances, namely

$$\begin{aligned} \sum_{\tau=-n}^{n} \gamma(\tau) &\backsim n^{\alpha} L_1(n) && \text{when } n \to \infty; 0 < \alpha < 1 \\ \gamma(\tau) &\backsim \tau^{-\beta} L_2(\tau) && \text{when } \tau \to \infty; 0 < \beta < 1 \\ f(\nu) &\backsim |\nu|^{-\xi} L_3(|\nu|) && \text{when } \nu \to 0; 0 < \xi < 1, \end{aligned} \tag{3}$$

$f(\nu)$ being the spectral density

$$f(\nu) = \frac{1}{2\pi} \sum_{\tau=-\infty}^{\infty} e^{-i\nu\tau} \gamma(\tau). \tag{4}$$

L_1, L_2 being slowly varying functions at infinity and L_3 slowly varying at zero.

If $\gamma(\tau)$ is monotone as $\tau \to \infty$, the definitions (3) are equivalent to the divergence of the sum in (2) with $\alpha = 1 - \beta$ and $\xi = 1 - \beta$.

In science, one of the main purposes when observing natural phenomena is the construction of models. A useful approach in this endeavor is the comparison of natural time series with the behavior of well-studied mathematical structures. In the context of long range dependence, a central role is played by the theory of self-similar stochastic processes

$$X(at) \stackrel{d}{=} a^H X(t), \tag{5}$$

with stationary increments

$$X(t+h) - X(t) \stackrel{d}{=} X(t) - X(0), \tag{6}$$

$\stackrel{d}{=}$ meaning equality in distribution. These processes are denoted as $H - sssi$ processes and H is called the Hurst exponent. Notice that there is a close relation between self-similarity and stationarity. If $X(t)$ is H−self-similar then $Y(t) = e^{-tH} X(e^t)$ is stationary and conversely if $Y(t)$ is stationary $X(t) = t^H Y(\ln t)$ is self-similar.

A finite σ^2 variance $H - sssi$ process has a covariance

$$\mathbb{E}\{X(s)X(t)\} = \frac{\sigma^2}{2}\{|t|^{2H} + |s|^{2H} - |t-s|^{2H}\}. \tag{7}$$

Throughout this paper $\mathbb{E}\{\cdots\}$ will denote the expected value which, for all the data examples, one approximates by the empirical average $\langle \cdots \rangle$.

There are non-Gaussian $H - sssi$ processes as well as $H - sssi$ processes with infinite covariance [8]. However, the simplest example of a $H - sssi$ process is a Gaussian process uniquely defined by the covariance (7) and normalized to have $\sigma^2 = 1$. It is called *fractional Brownian motion* (fBm) $B_H(t)$ and the increment process

$$Z(t) = B_H(t+1) - B_H(t), \tag{8}$$

is called *fractional Gaussian noise* (fGn). For $H = \frac{1}{2}, B_{1/2}(t)$ is Brownian motion. fGn noise has covariance

$$\gamma(\tau) = \frac{1}{2}\{|\tau+1|^{2H} - 2|\tau|^{2H} + |\tau-1|^{2H}\}, \tag{9}$$

hence, if $H = \frac{1}{2}$ it has $\gamma(\tau) = 0$ (no memory) and for $H \neq \frac{1}{2}$ and large τ

$$\gamma(\tau) \backsim H(2H-1)|\tau|^{2H-2} \quad \tau \to \infty. \tag{10}$$

For $\frac{1}{2} < H < 1$ the process has long range dependence. Because $\gamma(\tau) > 0$ for $H > \frac{1}{2}$ and $\gamma(\tau) < 0$ for $H < \frac{1}{2}$, the process is called *persistent* in the first case and *anti-persistent* in the second. For the spectral function at $H \neq \frac{1}{2}$

$$f(\nu) \backsim \nu^{1-2H} \quad \nu \to 0, \tag{11}$$

which for $H > \frac{1}{2}$ blows up at the origin.

The Hurst exponent (H) as an index of long range dependence quantifies the tendency of a time series either to regress strongly to the mean or to persist in a deviation from the mean. An H value between 0.5 and 1 signals a time series with long-term positive autocorrelation, meaning that a high (or low) value in the series will probably be followed by another high (or low) value. A value in the range 0 to 0.5 signals long-term switching between high and low values, meaning that an high value will probably be followed by a low value, with this tendency to switch between high and low values lasting into the future.

Although the correlation behavior of the fractional processes is very different from simple Brownian motion, they may be represented as integrals of Brownian motion with the appropriate integration kernel. For example

$$B_H(t) = \int_{-\infty}^{0} \{(t-u)^{H-\frac{1}{2}} - (-u)^{H-\frac{1}{2}}\} dB(u) + \int_{0}^{t} (t-u)^{H-\frac{1}{2}} dB(u), \quad (12)$$

where $B(u) = B_{\frac{1}{2}}(u)$. One practical implication is that to extract an eventual fractional behavior from the data, it is not sufficient an observation of short time intervals where the process may easily be confused with an uncorrelated process.

Another way in which Brownian motion intervenes in modeling processes which are neither uncorrelated nor simple Brownian motion is through the following fundamental result of stochastic analysis [9]: If $X(t)$ is a random variable that is square-integrable in the measure generated by Brownian motion, then

$$dX(t) = \mu(t)dt + \sigma(t)dB(t), \quad (13)$$

where $\mu(t)$ and $\sigma(t)$ are well defined stochastic processes. Therefore, although the increments of $X(t)$ have a representation in terms of the increments of Brownian motion, the process may be very different, depending on the nature of the processes $\mu(t)$ and $\sigma(t)$.

Whenever long range dependence is modeled by fGn one benefits from the extensive theoretical and computational framework that is available for this process. However, fGn is quite rigid in the sense that it specifies the correlations at all time lags, not only at $\tau \to \infty$. It may therefore not be suitable for modeling long range dependent phenomena where the covariance at short time lags differs from fGn. This motivated the development of other models through *Gaussian linear sequences*

$$X(t) = \sum_{j=-\infty}^{\infty} c_{t-j}\epsilon_j, \quad (14)$$

where $\sum_{j=-\infty}^{\infty} c_j^2 < \infty$ and $\{\epsilon_j\}_{j\in\mathbb{Z}}$ are independent identically distributed (i.i.d.) normal random variables, called *innovations*. A Gaussian linear sequence is stationary and for the convergence of the sum in (14) one requires

$$\text{If } \epsilon_j = N(\mu, \sigma^2) \quad \text{with } \mu \neq 0, \quad \sum_j |c_j| < \infty,$$

$$\text{If } \epsilon_j = N(\mu, \sigma^2) \quad \text{with } \mu = 0, \quad \sum_j |c_j|^2 < \infty.$$

An example is the FARIMA (p, d, q) process (fractional autoregressive integrated moving average) [10, 11]

$$X(t) = \Phi_p^{-1}(S)\Theta_q(S)\Delta^{-d}\epsilon_t \quad t \in \mathbb{Z}, \tag{15}$$

$\{\epsilon_t\}$ is an i.i.d. $N(0, \sigma^2)$ sequence and $\Phi_p(S), \Theta_q(S)$ are polynomials on the shift operator

$$S\epsilon_i = \epsilon_{i-1}, \tag{16}$$

Δ^{-d} being

$$\Delta^{-d} = (\mathbf{1} - \mathbf{S})^{-d} = \sum_{i=0}^{\infty} \frac{\Gamma(i+d)}{\Gamma(d)\Gamma(i+1)} S^i. \tag{17}$$

The fractional differencing Δ^{-d} for $0 < d < \frac{1}{2}$ models long range dependence, whereas the auto regressive $\Phi_p(S)$ and the moving average $\Theta_q(S)$ polynomials provide flexibility in modeling the short range dependence.

Finally, as mentioned on the introduction, there are other ways to deal with long range dependence for which the behavior of covariances does not play the main role. A potentially promising way is based on ergodic theory because the notion of memory is related to the connection between a process and its shifts. Then, a possible definition of long range dependent process would be one that is ergodic but non-mixing. However, the mixing property is probably not sufficiently strong to imply that a mixing stationary process has short memory. Stronger requirements may be needed. These notions will not be used here and we refer to [1, 12] for a discussion.

3. The Dynamics of Exploited Fish Populations

Long range dependence has been rarely documented in marine ecology, presumably because of the scarcity of long time series. This lack of extended time series has limited research on long memory in fish stock sizes, whose fluctuations are more often attributed to human exploitation, because most studies focus on highly exploited populations (such as the North Atlantic stocks) and over relatively short time periods. However, for a few fish populations, studies on long-term fluctuations have found long ranging trends related to human activity, mostly through overexploitation and pollution of spawning and nursery areas, environmental changes that affect the recruitment period inducing natural fluctuations in stock size and biotic processes, such as predation, cannibalism and competition [13–15].

A few years ago Niwa [16] studying the time series of 27 commercial fish stocks in the North Atlantic concluded that the variability in the population growth (the annual changes in the logarithm of population abundance $S(t)$)

$$r(t) = \ln\left(\frac{S(t+1)}{S(t)}\right), \tag{18}$$

is described by a Gaussian distribution. That is, the population variability process would be a geometric random walk

$$r(t) = \frac{dS(t)}{S(t)} = \sigma_r dB(t), \tag{19}$$

for some constant σ_r depending on the species. The independence of the increments of Brownian motion would then imply that $r(t)$ is a purely random process.

If completely accurate this would be a sobering conclusion. Natural processes that look purely random, are processes that depend on some many uncontrollable variables that any attempt to handle them is outside our reach. This would be a serious blow to, for example, the implementation of sustainability measures.

In this paper, we reanalyze some of the same type of data to confirm or sharpen the conclusions in [16]. To explore the variability in fish population growth, we extracted information on the Spawning-stock biomass (SSB) data on commercial fish stocks in the North Atlantic. The available SSB time-series data are derived from age-based analytical assessments estimated by the 2013 working groups of the International Council for the Exploration of the Sea (ICES), based on the compilation of relevant data from sampling of fisheries (e.g., commercial catch-at-age) and from scientific research surveys. From the collection of available assessment data, we selected three North Atlantic stocks for which the annual time-series of SSB covers at least 60 years, namely Northeast Arctic cod (Gadus morhua), Arctic haddock (Melanogrammus aeglefinus) and the North Sea autumn-spawning herring (Clupea harengus). At present, ICES classifies these stocks as having above average biomass levels with full reproductive capacity and being harvested sustainably under active management plans. The stock assessment detailed information is available at the ICES webpage [17].

For these three species, we analyze the autocorrelation functions for $r(t)$ and $|r(t)|$

$$C(r,\tau) = \frac{\mathbb{E}\{r(t)r(t+\tau)\}}{\sigma^2}, \tag{20}$$

$$C(|r|,\tau) = \frac{\mathbb{E}\{|r(t)||r(t+\tau)|\}}{\sigma^2}. \tag{21}$$

The results are shown in the Fig. 1.

One sees that, already for time lags of one year, autocorrelations are at noise level, suggestive of uncorrelated processes. However, if $S(t)$ is indeed a geometrical Brownian motion, to rely on correlations or fitting of probability distribution functions is not sufficient. The scaling properties of $r(t)$ should be checked. As Niwa [16] points out, defining

$$r_\Delta(t) = \ln\left\{\frac{S(t+\Delta)}{S(t)}\right\} = \sum_{i=1}^{\Delta} r(t+i), \tag{22}$$

the geometrical Brownian motion hypothesis would imply

$$(\mathbb{E}\{r_\Delta^2\})^{1/2} \backsim \Delta^{1/2}. \tag{23}$$

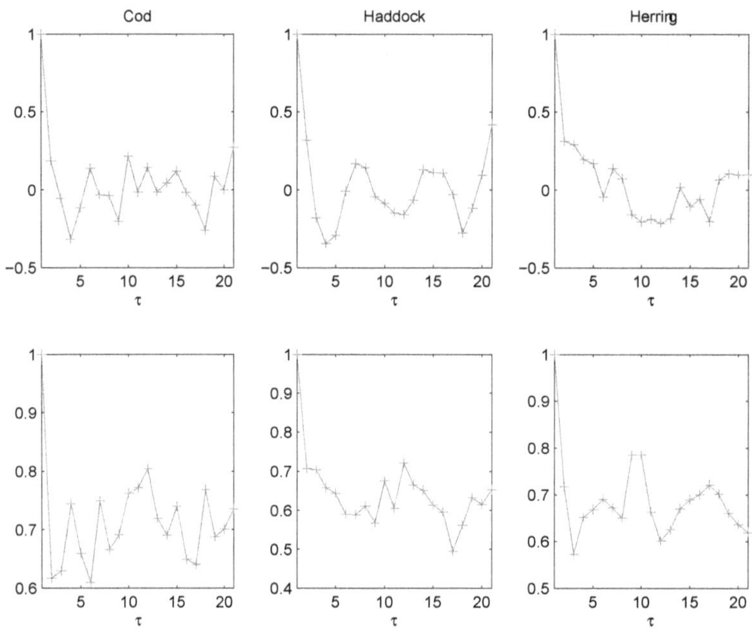

Fig. 1. Autocorrelation functions for $r(t)$ (upper plot) and $|r(t)|$ (lower plot).

Normalizing $\mathbb{E}\{r_\Delta^2\}$ by the covariance σ_r^2 for each species and taking the average over all species, Niwa has obtained a behavior roughly consistent with (23). However, when we analyzed each species separately, the hypothesis does not hold. In Fig. 2 we have plotted in log–log scale the computed $(\mathbb{E}\{r_\Delta^2\})^{1/2}$ as a function of Δ for the three species analyzed in this paper.

One sees that at the species level the geometrical Brownian motion is not a good hypothesis. Even for Herring, where the data seems to follow a scaling law, the slope at large Δ is closer to 0.7 than to 0.5. The conclusion is that whatever is actually determining the stochastic process for each species is somehow washed out when averaging over all the 27 species as Niwa did. Actually this is no surprise. Recall the stochastic analysis result (13). To find a process $r(t)$ that has features close to Brownian motion, but is not exactly Brownian motion only means that the process is square-integrable with respect to the (Wiener) measure generated by Brownian motion. All the interesting features actually lie on the dynamics of the process $\sigma(t)$. That is, on the dynamics of the amplitude of the fluctuations. In fact, this makes sense in biological terms because it is known [18, 19] that fishing magnifies fluctuations in exploited species.

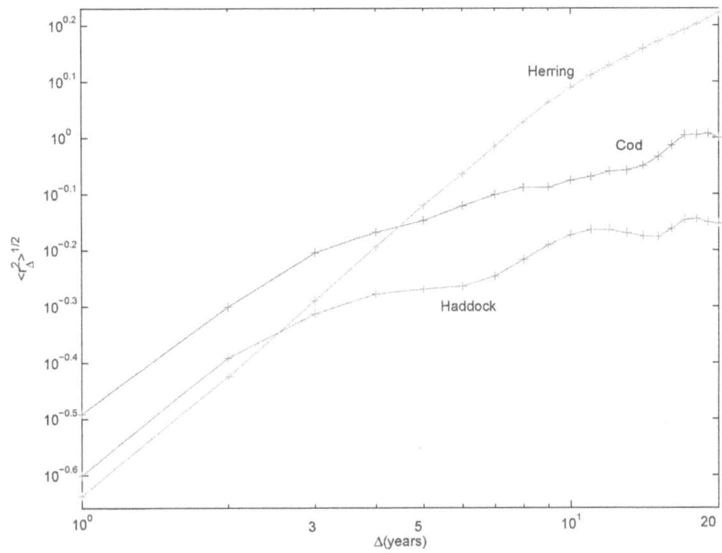

Fig. 2. $(\mathbb{E}\{r_\Delta^2\})^{1/2}$ as a function of Δ.

To reconstruct the dynamics of $\sigma(t)$ from the data we use a standard technique. Given a process

$$\frac{dS(t)}{S(t)} = \mu_t dt + \sigma_t dB(t), \tag{24}$$

the processes μ_t and σ_t would be obtained from

$$\begin{aligned} \mu_t &= \lim_{\varepsilon \to 0} \frac{1}{\varepsilon} \{\mathbb{E}(\log S_{t+\varepsilon} - \log S_t)\}, \\ \sigma_t^2 &= \lim_{\varepsilon \to 0} \frac{1}{\varepsilon} \{\mathbb{E}(\log S_{t+\varepsilon} - \log S_t)^2\}. \end{aligned} \tag{25}$$

However, in practice, we cannot use Eq. (25) to reconstruct μ_t and σ_t, because when the time interval ε is very small, the empirical evaluation of these quantities becomes unreliable. Instead, we estimate σ_t from

$$\sigma_t^2 = \frac{1}{|T_0 - T_1|} var(\log S_t), \tag{26}$$

with a time window $|T_0 - T_1|$ sufficiently small to give a reasonably local characterization of the volatility, but also sufficiently large to allow for a reliable estimate of the local variance of $\log S_t$. Here a window of 6 years has been used. Window sizes from 4 to 7 were also tried which lead to qualitatively similar results (for smaller window sizes, estimates of the local variance becomes meaningless).

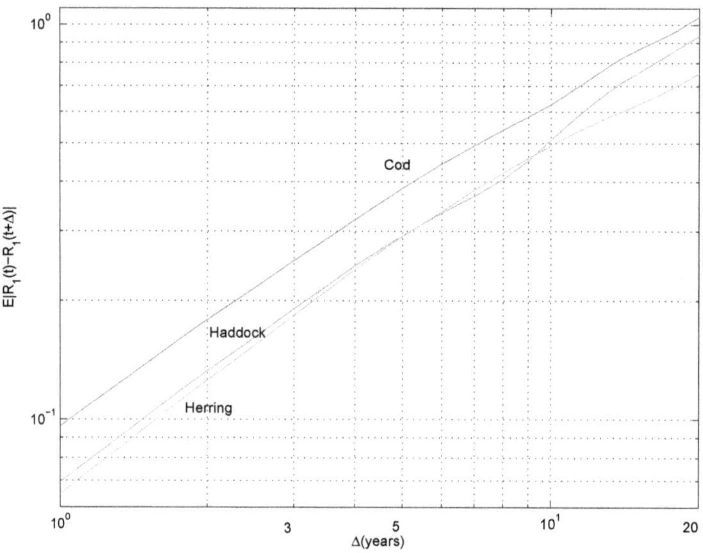

Fig. 3. The scaling behavior of R_1.

As for an estimate of μ_t we notice that, for the species that were analyzed, the mean of $r(t)$ is less than 10^{-2} and the local average with a 6 years window fluctuates in the range $[-0.2, 0.2]$ which is consistent with $\mu_t \simeq 0$.

Once $\sigma(t)$ is computed one forms the cumulative processes

$$\begin{aligned}\sum_{i=1}^{t} \sigma(i) &= \beta_1 t + R_1(t), \\ \sum_{i=1}^{t} \ln \sigma(i) &= \beta_2 t + R_2(t).\end{aligned} \tag{27}$$

β_1 and β_2 being the average values of σ and $\ln \sigma$ and R_1 and R_2 the cumulative processes of the fluctuations about the average. To obtain a model for the fluctuations one looks for the scaling properties of R_1 and R_2, namely the behavior of $\mathbb{E}|R_1(t) - R_1(t+\Delta)|$ and $\mathbb{E}|R_2(t) - R_2(t+\Delta)|$ as a function of the time lag Δ (Figs. 3 and 4).

The conclusion is that although not perfect, which would not be expected with data covering at most 68 points, we may assume that R_1 and R_2 obey an approximate scaling law with exponents H in the range 0.8–0.9. Therefore, R_1 and R_2 may be modeled by fBm implying that the fluctuations of σ and $\ln \sigma$, away from an average value, are modeled by fractional Gaussian noise. We have looked for scaling both for the cumulative σ and $\ln \sigma$ to decide which one would provide a simpler

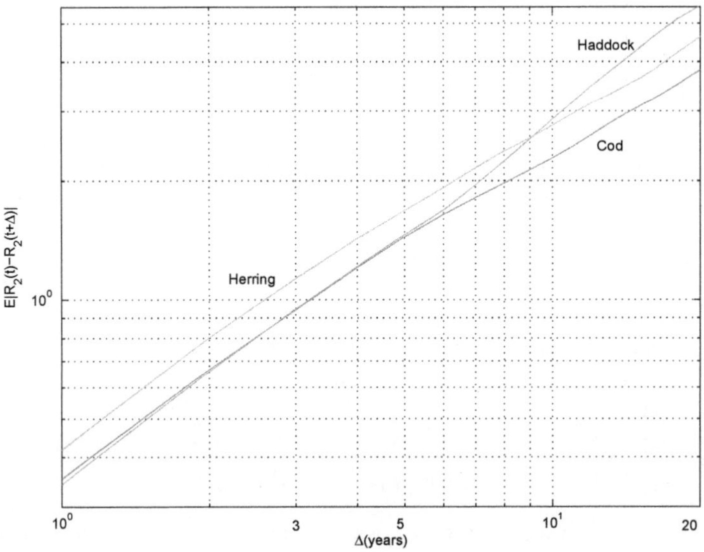

Fig. 4. The scaling behavior of R_2.

model for the amplitude fluctuations. However, with the data available, there is no clear decision. Therefore, two alternative models are proposed for the population fluctuations

$$\begin{aligned} dS(t) &= \sigma(t)S(t)dB_t, \\ \sigma(t) &= \beta_1 + \alpha_1(B_{H_1}(t) - B_{H_1}(t-1)), \end{aligned} \tag{28}$$

or

$$\begin{aligned} dS(t) &= \sigma(t)S(t)dB_t, \\ \ln\sigma(t) &= \beta_2 + \alpha_2(B_{H_2}(t) - B_{H_2}(t-1)). \end{aligned} \tag{29}$$

From the data the following values are obtained for the Hurst coefficients H_1 and H_2

	H_1	H_2
Cod	0.81 ± 0.04	0.80 ± 0.05,
Haddock	0.84 ± 0.04	0.90 ± 0.04,
Herring	0.88 ± 0.04	0.81 ± 0.04.

The ICES data is a complex mixture of actual sampling observations and model extrapolations for which error bars are not usually reported. Therefore, the error estimates that are reported in this table only reflect uncertainties in the estimate of the log–log slopes, not the uncertainties of the original data itself. Nevertheless,

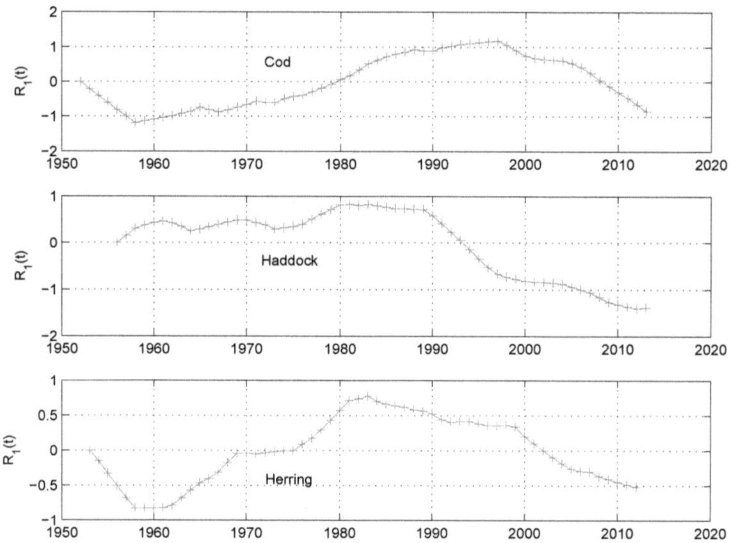

Fig. 5. Dynamics of the cumulative fluctuations of $\sigma(t)$.

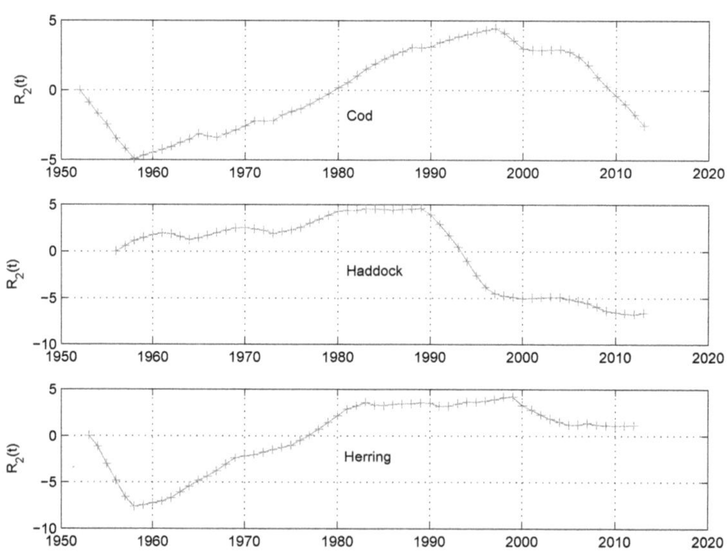

Fig. 6. Dynamics of the cumulative fluctuations of $\ln \sigma(t)$.

the important point to retain is that from the obtained H-values, one sees that the dynamics of the fluctuations is a long range memory process. In addition, we have found out that the processes seem to be species-dependent. For illustration, we plot the cumulative amplitude fluctuations R_1, R_2 in the Figs. 5 and 6.

Processes with such high H-values are almost deterministic processes. This is an important outcome because they may provide clues on the causes of that particular dynamics.

In conclusion, this methodology of separation of the pure random features from the non-trivial dynamics of amplitude fluctuations may be useful for the analysis of other natural processes. It has a solid mathematical basis as a consequence of the stochastic analysis result mentioned in (13). The dynamics of exploited fish populations provides a good example of a phenomenon where the long range dependence features appear not at the level of second-order correlations but only through the analysis of higher order effects.

4. Discussion

(i) Our paper was motivated by concerns on Niwa's [16] assertion that population variability corresponds to a geometric random walk and, consequently, the exploited population trajectory is a series of random uncorrelated abundances over time. The annual change of population abundance is the result of many "shocks", including recruitment variability, natural mortality (e.g., predation and competition) and varying fishing pressure. Despite these collection of independent drivers our results suggests that each fish stock dynamics exhibit long range dependence. Moreover, from the empirically found H-values associated to the fluctuations, one sees that the long range memory process seems to be species dependent. This is the reason behind Niwa's random walk conclusion because, when averaging over species, the $\sigma(t)$ process in Eqs. (28) and (29) would be simply replaced by a fixed number $\overline{\sigma(t)}$ (the average of $\sigma(t)$).

(ii) A full discussion of whatever is actually determining the stochastic process for each species is beyond the scope of this study but the question why fish populations fluctuate has generated much attention from fishery scientists and marine ecologists over the past century. Three general hypotheses have been proposed to answer this question: (i) species interactions generate fluctuating and cyclic population dynamics; (ii) nonlinearity in single-species dynamics generates deterministic fluctuations; and (iii) changes in the environment determines variation in vital rates and recruitment, which in turn drive variation in abundance [21, 19, 24]. These are not mutually exclusive hypotheses as all three could act together to increase variability. For exploited species, fishing can also vary from year to year and translate directly into population variability or could interact with the other drivers to enhance fluctuations in fish abundance [19, 22].

Long range trends are frequently related to external forcing on the populations and are usually derived from human exploitation or environmental change. The dependence in the population growth observed for the haddock, herring and cod stocks could be derived from the different and varying exploitation regimes and/or from large-scale environmental changes as the North Atlantic Oscillation, that can induce low (or high) productivity regimes in fish recruitment (e.g., [20, 26, 27]).

Bjørnstad *et al.* [14] have shown that short-term variability in recruitment caused by environmental change, combined with intercohort interactions can be echoed through the population age structure inducing persistent cycles and long-term fluctuations. It is also known [25] that noise in recruitment combined with a large number of classes of spawners could lead to long-term variations in SSB and yields, as well as to regular cycles, depending on the lifespan of the species. Furthermore, the stocks analyzed here showed species specific stochastic processes that mirror the contrasting life history traits. The distinct growth rates, age at maturity, spawning duration and lifespan are characteristics that make some fish stocks more or less vulnerable to exploitation and environmental conditions [19, 23]. The small bodied and younger herrings population should be less able to smooth out environmental fluctuations and more prone to exhibit unstable dynamics due to changing demographic parameters.

(iii) Long range dependence may also be obtained from fractional differential equations, which however lead to processes different from fBm. Nevertheless, fBm is the simplest process available to model long range dependence and also the one more suitable for reconstruction from the data without too many model-dependent assumptions. Also, the proposal of a realistic deterministic equation model would require a detailed and precise analysis of the physical and ecological mechanisms generating the data. Rather than proposing a specific model, our point of view was to find the correct statistical properties of the data.

(iv) There is no doubt that fisheries management would profit from a clearer understanding of the mechanisms determining the dynamics of the amplitude of the fluctuations in exploited fish stocks. The failure of many fish stocks, despite the implemented management measures, to recover rapidly to former levels of abundance, might arguably be related to the long range memory observed in this study.

References

[1] Samorodnitsky, G., Long range dependence, *Found. Trends Stoch. Syst.* **1** (2006) 163–257.
[2] Dominique, G., How can we define the concept of long memory? An econometric survey, *Economet. Rev.* **24** (2005) 113–149.
[3] Doukhan, P., Oppenheim, G. and Taqqu, M. S., *Theory and Applications of Long-Range Dependence* (Springer, Berlin, 2003).
[4] Berkes, I., Horváth, L., Kokoszka, P. and Shao Q.-M., On discriminating between long-range dependence and changes in mean, *Ann. Stat.* **34** (2006) 1140–1165.
[5] Preuß, P. and Vetter, M., Discriminating between long-range dependence and non-stationarity, *Electron. J. Stat.* **7** (2013) 2241–2297.

[6] Kokoska, P. and Leipus, R., Detection and estimation of changes in regime, in *Theory and Applications of Long-Range Dependence* (Springer, Berlin, 2003), pp. 325–337.
[7] Mendes, R. V., Lima, R. and Araújo, T., A process-reconstruction analysis of market fluctuations, *Int. J. Theor. Appl. Finance* **5** (2002) 797–821.
[8] Samorodnitsky, G. and Taqqu, M. S., *Stable Non-Gaussian Processes: Stochastic Models with Infinite Variance* (Chapman and Hall, New York, 1994).
[9] Nualart, D., *The Malliavin Calculus and Related Topics* (Springer, Berlin, 2006).
[10] Granger, C. and Joyeux, R., An introduction to long-memory time series and fractional differencing, *J. Time Ser. Anal.* **1** (1980) 15–30.
[11] Hosking, J., Fractional differencing, *Biometrika* **68** (1981) 165–176.
[12] Bradley, R., Basic properties of strong mixing conditions. A survey and some open questions, *Probab. Surv.* **2** (2005) 107–144.
[13] Dickson, R. and Brander, K., Effects of a changing windfield on cod stocks of the North Atlantic, *Fish. Oceanogr.* **2** (1993) 124–153.
[14] Bjørnstad, O., Fromentin, J. M., Stenseth, N. C. and Gjøsæter, J., Cycles and trends in cod populations, *Proc. Natl. Acad. Sci. USA* **96** (1999) 5066–5071.
[15] Fromentin, J. M., Myers, R. M., Bjørnstad, O., Stenseth, N. C., Gjøsæter, J. and Christie, H., Effects of density-dependent and stochastic processes on the stabilization of cod populations, *Ecology* **82** (2001) 567–579.
[16] Niwa, H.-S., Random-walk dynamics of exploited fish populations, *ICES J. Mar. Sci.* **64** (2007) 496–502.
[17] http://www.ices.dk/community/advisory-process/Pages/Latest-Advice.aspx.
[18] Hsieh, C.-H., Reiss, C. S., Hunter, J. R., Beddington, J. R., May, R. M. and Sugihara, G., Fishing elevates variability in the abundance of exploited species, *Nature* **443** (2006) 859–862.
[19] Anderson, C. N. K., Hsieh, C.-H., Sandin, S. A., Hewitt, R., Hollowed, A., Beddington, J., May, R. S. and Sugihara, G., Why fishing magnifies fluctuations in fish abundance, *Nature* **452** (2008) 835–839.
[20] Leif, C. S., Ottersen, G., Brander, K., Chan, K. and Stenseth, N. C., Cod and climate: Effect of the North Atlantic Oscillation on recruitment in the North Atlantic, *Mar. Ecol. Prog. Ser.* **325** (2006) 227–241.
[21] Shelton, A. O. and Mangel, M., Fluctuations of fish populations and the magnifying effects of fishing, *Proc. Natl. Acad. Sci. USA* **108** (2011) 7075–7080.
[22] Turchin, P. and Taylor, A. D., Complex dynamics in ecological time series, *Ecology* **73** (1992) 289–305.
[23] Reynolds, J., Jennings, S. and Dulvy, N. K., Life histories of fishes and population responses to exploitation, in *Conservation of Exploited Species*, Reynolds, J. D., Mace, G. M., Redford, K. H. and Robinson, J. G. (eds.) (Cambridge University Press, Cambridge, 2001), pp. 147–168.
[24] Overland, J. E., Alheit, J., Bakun, A., Hurrell, J. W., Mackas, D. L. and Miller, A. J., Climate controls on marine ecosystems and fish populations, *J. Mar. Syst.* **79** (2010) 305–315.
[25] Fromentin, J. and Fonteneau, A., Fishing effects and life history traits: A case study comparing tropical versus temperate tunas, *Fish. Res.* **53** (2001) 133–150.
[26] Borges, M. F., Mendes, H. C. and Santos, A. M. P., Sardine (Sardina pilchardus) recruitment is strongly affected by climate even at high spawning biomass in West Iberia/Canary upwelling system, in *Science and Management of Small Pelagics*, Garcia, S., Tandstad, M. and Caramelo, A. M. (eds.), FAO Fisheries and Aquaculture Proceedings, Vol. 18 (2011), pp. 237–244, FAO, Rome 2012.
[27] Brander, K., Cod recruitment is strongly affected by climate when stock biomass is low, *ICES J. Mar. Sci.* **62** (2005) 339–343.

Advances in Complex Systems
Vol. 22, No. 1 (2019) 1850023 (21 pages)

DOI: 10.1142/S0219525918500236

AN ECO-SYSTEMS APPROACH TO CONSTRUCTING ECONOMIC COMPLEXITY MEASURES: ENDOGENIZATION OF THE TECHNOLOGICAL DIMENSION USING LOTKA–VOLTERRA EQUATIONS

INGA IVANOVA*

*Institute for Statistical Studies and Economics of Knowledge,
National Research University Higher School of Economics (NRU HSE),
20 Myasnitskaya St., Moscow 101000, Russia
inga.iva@mail.ru*

ØIVIND STRAND

*Norwegian University of Science and Technology (NTNU) Aalesund,
Department of International Business,
PO Box 1517, 6025 Aalesund, Norway
oivind.strand@ntnu.no*

LOET LEYDESDORFF

*Amsterdam School of Communication Research (ASCoR),
University of Amsterdam, PO Box 15793,
1001 NG Amsterdam, The Netherlands
loet@leydesdorff.net*

Received 11 January 2018
Revised 28 July 2018
Accepted 11 October 2018
Published 24 December 2018

Economic complexity measures have been constructed on the basis of bipartite country-product network data, but without paying attention to the technological dimension or manufacturing capabilities. In this study, we submit a Ternary Complexity Index (TCI), which explicitly incorporates technological knowledge as a third dimension, measured in terms of patents. Different from a complexity indicator based on the Triple Helix model (THCI) or a measure based on patents and countries (PatCI), TCI — products, countries, and patents — can be modeled in terms of Lotka–Volterra equations and thus the further evolution of an innovation eco-system can be specified. We test the model using empirical data. The results of a regression analysis show that TCI improves on Hidalgo and Hausmann's [The building blocks of economic complexity, *Proc. Natl. Acad. Sci.* **106** (26) (2009) 10570–10575] and Tacchella *et al.*'s [A new metrics for countries fitness and products complexity, *Sci. Rep.* **2** (2012)] complexity measures with respect to both the ranking of countries in terms of their complexity and in terms of the correlation with GDP per capita.

Keywords: Economic complexity; metrics; triple helix; eco-system; simulation.

*Corresponding author.

1. Introduction

The quantitative assessment of the competitive advantages of nations in terms of complexity measures has hitherto not focused on the knowledge intensity of the economy. Assuming that the products in the export portfolio of a country are related to the capabilities needed for manufacturing these products, Hidalgo and Haussmann [24] developed an iterative procedure — the Method of Reflections (MR) — for measuring the complexity of a country's economy. The technological capabilities drive the iteration, but this dimension is otherwise not specified. According to these authors (HH), the merit of the method is that the value of the Economic Complexity Index (ECI) is correlated with a country's GDP per capita [24, Fig. 3, p. 10573]. As a consequence, the deviation of the indicator's value from a country's level of income might be useful for predicting future growth.

Considering that HH's index did not account for empirically observed correlations between a country's competitive advantages and the diversity of its exports, Tacchella *et al.* [46, 47] proposed an alternative — nonlinear — iterative approach: the so-called "Fitness and Product Complexity index" (FCI). These authors noted that their approach resembles models of biological systems in which diversification provides an evolutionary advantage over specialization.

Both FCI and ECI organize the data in terms of bipartite networks of countries versus products. HH noted that they "interpret data connecting countries to the products they export as a bipartite network and assume that this network is the result of a larger, tripartite network, connecting countries to the capabilities they have and products to the capabilities they require" [24, p. 10570]. However, neither HH nor Tacchella *et al.* [46] provide an explicit definition of these intermediating capabilities. Consequently, the capabilities have remained implicit. Cristelli *et al.* [12] submitted that from the perspective of a data-driven approach capabilities can be modeled as a hidden layer of "intangibles" between countries and products.

The endowments of a nation can be considered as one of the factors of manufacturing capabilities. However, these endowments are relatively stable over time. Yet, economic theory has pointed to the importance of technology for explaining economic growth [41]. Long-run economic growth is largely based on the primacy of technological progress [43, 44]. Consequently technological knowledge is the cause of economic growth provided by improvements in manufacturing capabilities.

The capabilities can be considered as the ability to manufacture certain products. HH mention that empirical research "emphasized the accumulation of a few highly aggregated factors of production, such as physical and human capital or general institutional measures" [24, p. 10575]. These factors may also refer to geography, climate, and other regional/national production possibilities and competitive advantages which cannot be exported, or easily acquired from another nation. Furthermore, HH emphasized the importance of new capabilities including the ones originating from technological progress. Considering the present state of technological and economic development, technological knowledge can be expected to play a major role in creating additional value.

Another reason for introducing the technological dimension is that countries exporting the same products do not necessarily have the same capabilities. Some sectors in the economy of developed countries can be offshored to emerging economies. For example, China is the manufacturer and exporter of computers though it does not have the capabilities to produce some key computer components, such as processors. In other words, the degree of localization of offshored technology is also important. One way to account for this problem is to develop and make use of value-added trade data. However, this data are subject to different factors, such as labor costs, taxes, etc.

The major input to the added value is made by technology. Hausmann and Hidalgo [23] provided a more accurate definition of capabilities by accounting the structure of output in the countries-products network. Utkovski *et al.* [51] implemented clustering methods in order to reveal capabilities. Boschma *et al.* [5] used patent classes instead of product groups to measure the complexity in the technology base of US cities, and Balland and Rigby [3] used this method for mapping the diffusion and evolution of knowledge complexity in US cities. However, in terms of the complexity approach, these further studies did not combine the three dimensions of products, countries, and technologies into a single model.

We argue in this study that technological capabilities can be explicitly endogenized into the model of complexity as a third dimension in addition to geographical positions and economic relations. The advancement of technological knowledge can be expected to change the system or, in other words, to disturb the tendencies toward equilibrium Nelson and Winter [35] and Schumpeter [42]. We bring together the ideas of product and knowledge complexity by extending HH's MR to the technological domain and present a nonlinear generalization of ECI. Our model is based on the tripartite network of countries, technologies, and products.

2. Operationalization

We follow Boschma *et al.* [5] and Balland and Rigby [3] in considering patent portfolios as indicators of technological complexity. Patents are analytically independent from products since they are indicators of invention and not innovation. One can consider patents to be a proxy of technological knowledge and the technological knowledge base can hence be measured in terms of patent portfolios [1, 51]. The manufacturing capabilities of a country can be expected to largely overlap with its technological knowledge base [14, 19, 20, 36].

The three interacting dimensions provide a reference to the Triple Helix (TH) model of innovations [16] in which the constituent actors — university, industry, and government — interact among themselves and drive a process of self-organization within the system. In this context, Ivanova *et al.* [28] proposed the Triple-Helix Complexity Index (THCI). In this study, we elaborate the THCI to its nonlinear version, which we designate as TCI. TCI can be evaluated, in terms of Lotka–Voltera (L–V) equations.

L–V equations can be used to model the evolutionary dynamics of eco-systems and thus we can bring the complexity model into the mainstream of evolutionary theorizing (Hodgson & Knudsen [25]). Eco-system approaches have also been used for modeling manufacturing systems [21], business systems [34], from the platform-management perspective [22], and from a multi-actor network perspective [48]. In most of these studies, an eco-system is understood as a number of actors and their relationships [8, 40] (Storper [45]; Mazzucato and Robinson [33]) with an emphasis on relationships. However, there is no precise and agreed definition of an "innovation eco-system" in the innovation-studies literature (Ritala & Almpanopoulon, in press).

Using generalized L–V equations, we are able to show that the complexity measure TCI follows general mechanisms for modeling dynamically evolving ecosystem [7]. As noted, we build on Ivanova *et al.*'s [28] THCI — which remained a linear model — and extend Hidalgo and Hausmann's (HH) Method of Reflections (MR) from two to three dimensions in order to elaborate this Ternary Complexity Index (TCI). We perform model calculations on the basis of empirical data in order to compare the results obtained with HH's MR, Tacchella *et al.*'s [46] FCI, and TCI. The results show that the correlation between TCI and ln(GDP per capita) is improved when compared with ECI. Using this criterion, the complexity ranking of the countries is modified. Since ECI, Fitness, and TCI demonstrate approximately similar results with respect to the prediction of economic growth, this question needs further investigation with extended sets of data.

3. Method

3.1. *HH's method of reflections*

HH's Method of Reflections begins with a country-product export matrix $\{X_{c,p}\}$, where $X_{c,p}$ is the value of product p manufactured by country c. Product p represents a product class. A matrix $M_{c,p}$ is constructed in which the index c refers to a country and p refers to a product group measured as an amount of output. The corresponding matrix elements are set to one if Balassa's [2] RCA is larger than or equal to unity; otherwise the element is equal to zero (Eq. (1)):

$$\mathrm{RCA}_{c,p} = \frac{X_{c,p}/\sum_p X_{c,p}}{\sum_c X_{c,p}/\sum_{c,p} X_{c,p}}. \tag{1}$$

In other words, a country is assumed to export a product if it produces this product proportionally more than the average of the group of countries under consideration.

Summing the elements of matrix $M_{c,p}$ by rows (countries), one obtains a vector with components referring to the corresponding products and indicating a measure of product ubiquity relative to the world market. The sum of matrix elements over the columns (products) provides another vector defining the diversity of a country's exports.

$$\begin{aligned} k_{p,0} &= \sum_{c=1}^{N_c} M_{c,p}, \\ k_{c,0} &= \sum_{p=1}^{N_p} M_{c,p}, \end{aligned} \tag{2}$$

where N_c is the number of countries and N_p is the number of product groups.

More accurate measures of diversity and ubiquity can be obtained by adding the following iterations:

$$\begin{aligned} k_{p,n} &= \frac{1}{k_{p,0}} \sum_{c=1}^{N_c} M_{c,p} k_{c,n-1}, \\ k_{c,n} &= \frac{1}{k_{c,0}} \sum_{p=1}^{N_p} M_{c,p} k_{p,n-1}, \end{aligned} \tag{3}$$

that is, each product is weighted proportionally to its ubiquity on the market, and each country is weighted proportionally to the country's diversity. Substituting the first equation of the system (3) into the second, one obtains

$$k_{c,n} = \frac{1}{k_{c,0}} \sum_{c'=1}^{N_c} \sum_{p=1}^{N_p} M_{c,p} \frac{1}{k_{p,0}} M_{c',p} k_{c',n-2}. \tag{4}$$

Equation (4) can be formulated as a matrix equation

$$\vec{k} = W \cdot \vec{k}, \tag{5}$$

where the vector $\vec{k}$ is a limit of iterations, as follows:

$$\vec{k} = \lim_{n \to \infty} k_{c,n}. \tag{6}$$

HH use the eigenvector $\vec{k}$ of the matrix $W_{c,c'}$

$$W_{c,c'} = \sum_{p} \frac{M_{c,p} M_{c',p}}{k_{c,0} k_{p,0}} \tag{7}$$

associated with the second largest eigenvalue since this eigenvector captures most of the variation[10] for introducing ECI. ECI is defined according to the formula

$$\text{ECI} = \frac{\vec{k} - \langle \vec{k} \rangle}{\text{stdev}(\vec{k})}. \tag{8}$$

In sum, ECI is a vector of which the components refer to the respective countries.

3.2. *Tacchella' et al.'s FCI*

The methods of HH and Tacchella *et al.* have in common that they begin with a binary country–product matrix which is the result of cross-tabling a country's product diversity and product ubiquity as a first step in the iteration. However, Tacchella *et al.* [46] note that the exports of less developed countries require lower levels of sophistication. In their opinion and based on empirical observations, countries produce and export the whole specter of products for which they have production capabilities. Due to uneven development stages of the economies, however, there are a few developed countries producing all products and many less developed ones which produce and export only a limited number of products. Therefore, the binary matrix connecting countries to products has a triangular shape.

In order to more correctly measure a country's manufacturing sophistication, the initial diversity measure is consequently modified in a nonlinear iterative sequence. The newly obtained variable — Fitness — is assumed to measure the level of sophistication of manufacturing capabilities in the respective countries. Tacchella *et al.* [46] augment the weight to different products proportionally to the ubiquity of products when iterating the diversity score.

Since the value of $k_{c,n}$, used to calculate ECI, would deviate with each iteration increasingly from the initial diversity of a country's export $k_{c,0}$, as defined in Eq. (2), these authors propose to iterate a country's product diversity *inversely* proportional to the ubiquity of the products, so that the correlation between initial diversity and Fitness is preserved at each step of the iterations. This modification changes the method from a linear into a nonlinear one. The authors introduce the fitness of countries $F_c^{(n)}$ and the complexity of products $Q_p^{(n)}$, connected by the following iterative sequences:

$$\begin{aligned} \tilde{F}_c^{(n)} &= \sum_p M_{cp} Q_p^{(n-1)}, \\ \tilde{Q}_p^{(n)} &= \frac{1}{\sum_c M_{cp}\left(1/F_c^{(n-1)}\right)}. \end{aligned} \tag{9}$$

At each step of the iteration intermediate values are first computed and are then normalized as follows:

$$\begin{aligned} F_c^{(n)} &= \frac{\tilde{F}_c^{(n)}}{\langle \tilde{F}_c^{(n)} \rangle_c}, \\ Q_p^{(n)} &= \frac{\tilde{Q}_p^{(n)}}{\langle \tilde{Q}_p^{(n)} \rangle_p}. \end{aligned} \tag{10}$$

The initial conditions are: $\tilde{F}_c^{(0)} = 1$ and $\tilde{Q}_p^{(0)} = 1$; the denominators in the system of equation (10) correspond to the average values for each country and product. The following correspondence between first-order values of Tacchella's Fitness and

Product Complexity and HH's diversity/ubiquity holds

$$\tilde{F}_c^{(1)} = k_{c,0},$$
$$\tilde{Q}_p^{(1)} = \frac{1}{k_{p,0}}. \tag{11}$$

The authors note furthermore that in their model the initial meaning of the variables does not change during the iterations.

Although Tacchella *et al.*'s level of development of manufacturing capabilities and HH's level of competitiveness are constructed in similar terms, they are not identical measures. The models are different: the Fitness measure preserves and enhances initial zero-order diversity in Tacchella's models, while the Complexity Index is orthogonally developed to the initial diversity in HH's models. Moreover, ECI is correlated with ln(GDP per capita) and can, according to the claim of the authors, be used as a predictive indicator of long-term growth [24, Fig. 3, p. 10573], whereas as it will be shown below empirically FCI does not correlate with ln(GDP per capita).

3.3. *The ternary complexity index*

Using HH's MR, a country's diversity score is modified directly proportional to the ubiquity of its products. But using FCI the diversity score is modified inversely proportional to the product's ubiquity, that is, more specialized products contribute more to the countries' capabilities. The inverse proportionality is legitimated by the triangular shape of the binary country–product matrix, i.e., the more diversified capabilities a country possesses, the wider the range of products it can produce.

While exporting simple products, developed countries compete with less-developed ones on the market. Due to the number of parties the competition on simple product markets should be especially intensive. Countries with advanced capabilities can concentrate on the manufacturing of technologically advanced products with higher margin profits and a lower number of competitors. The reason for the developed countries to export simple products is that their advanced level of technology makes the manufacturing and export of such products more profitable than for less-developed countries. For example, the achievements of genetic engineering can be applied in agricultural production, where they allow for increased yields. Similarly, sophisticated technologies of shale oil production make the export of shale oil more profitable. In other words, there is a possible effect of technology influence on the range of manufacturing sectors.

The notion of economic complexity can be extended to the technological domain by substituting product values by patent values in Eq. (1) and introducing a country–patent matrix $M_{c,t}$ instead of a country–product matrix $M_{c,p}$. This way, as with ECI and following HH's method, an indicator for technological complexity or Patent Complexity Index (PatCI) was defined by Boschma *et al.* [6] and Ivanova *et al.* [29]. Taking this a step further, one can envisage an additional product-patent

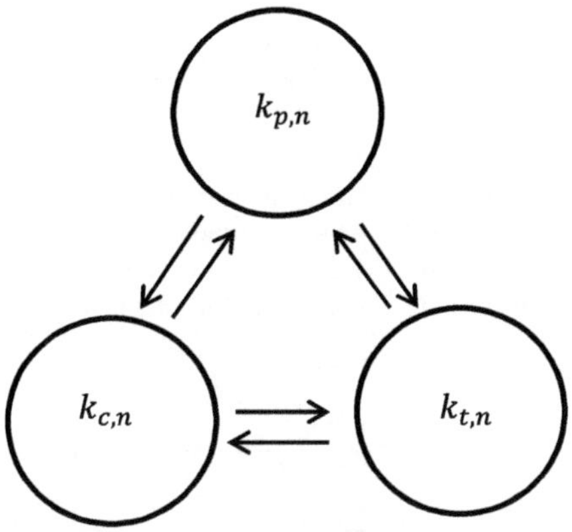

Fig. 1. Interrelations of the complexity measures.

matrix $M_{p,t}$ and a ternary country–product–technology complexity indicator based on the three-dimensional array $M_{p,t,c}$. The latter can be introduced as country–product–patent or patent–product–country cycles with clockwise or counter-clockwise interdependencies [28].

Figure 1 defines these complexity indices in terms of the ecosystems approach which can be extended diachronically.[a]

Instead of the country-product matrix $X_{c,p}$, we use the three-dimensional country–product–technology array $\mathfrak{M}_{c,p,t}$. Initial diversity, product ubiquity and patent ubiquity coefficients are defined as

$$k_{c,0} = \sum_{p=1}^{N_p} \sum_{t=1}^{N_t} \mathfrak{M}_{c,p,t},$$
$$k_{p,0} = \sum_{c=1}^{N_c} \sum_{t=1}^{N_t} \mathfrak{M}_{c,p,t},$$

[a] Ecosystems in biology are defined through the network of interactions among living organisms and the environment. Ecosystems sustain the creation of order against the Second Law of Thermodynamics which is maintained by autocatalysis. Ulanowicz [49, p. 1888] provides the following illustration of the autocatalytic cycle which essentially resembles Fig. 1

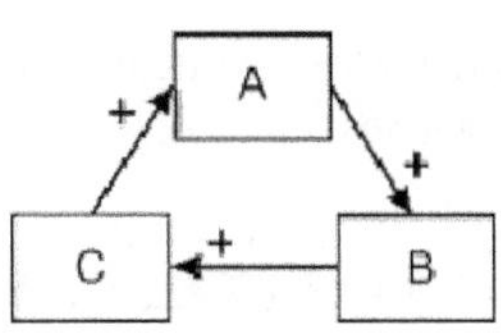

$$k_{t,0} = \sum_{c=1}^{N_c} \sum_{p=1}^{N_p} \mathfrak{M}_{c,p,t}. \tag{12}$$

One can consider iterating across these three dimensions simultaneously. That is, country and product complexity create technology complexity, which goes into calculating product complexity, which goes into country complexity, which goes into technology complexity, etc. That is, instead of the system of equation (3) one can write

$$\begin{aligned} k_{c,n} &= \frac{1}{k_{c,0}} \sum_{p=1}^{N_p} \sum_{t=1}^{N_t} \mathfrak{M}_{c,p,t} k_{p,n-1} k_{t,n-1}, \\ k_{p,n} &= \frac{1}{k_{p,0}} \sum_{c=1}^{N_c} \sum_{t=1}^{N_t} \mathfrak{M}_{c,p,t} k_{c,n-1} k_{t,n-1}, \\ k_{t,n} &= \frac{1}{k_{t,0}} \sum_{c=1}^{N_c} \sum_{p=1}^{N_p} \mathfrak{M}_{c,p,t} k_{p,n-1} k_{c,n-1}. \end{aligned} \tag{13}$$

The three complexity indices: $k_{c,n}$, $k_{p,n}$, $k_{t,n}$ correspond to the geographical, manufacturing, and technological dimensions. The value of each index is determined by the simultaneous action of the other two indices.

The advantage of extending the Method of Reflections to three complexity indices helps to settle the convergence problem. Caldarelli [11, p. 6] formulates that "the major problem in the HH algorithm is that it is a case of consensus dynamics, i.e., the state of a node at iteration t is just the average of the state of its neighbors at iteration $t-1$...... such iterations have the uniform state as the natural fix point..." However, this criterion is not applicable to the case of nonlinear iterative sequence as defined by the set of equation (13).

Figure 1 can also be considered a schematic representation of an autocatalytic cycle with three components. This model is also used for describing the evolution of biological ecosystems [49]. The interplay of indices provides a reference to the interplay of the three actors — university, industry, and government — in a Triple Helix model of innovations. Feed-forward and feed-back cycles may strengthen or weaken a corresponding index in the process of iterations as in the case of an autocatalytic system.

By adding the same terms to the left- and right-hand side of each of equations (13) one can write this system as follows:

$$k_{c,n} - k_{c,n-1} = -k_{c,n-1} + \frac{1}{k_{c,0}} \sum_{p=1}^{N_p} \sum_{t=1}^{N_t} \mathfrak{M}_{c,p,t} k_{p,n-1} k_{t,n-1},$$

$$
\begin{aligned}
k_{p,n} - k_{p,n-1} &= -k_{p,n-1} + \frac{1}{k_{p,0}} \sum_{c=1}^{N_c} \sum_{t=1}^{N_t} \mathfrak{M}_{c,p,t} k_{c,n-1} k_{t,n-1}, \\
k_{t,n} - k_{t,n-1} &= -k_{t,n-1} + \frac{1}{k_{t,0}} \sum_{c=1}^{N_c} \sum_{p=1}^{N_p} \mathfrak{M}_{c,p,t} k_{p,n-1} k_{c,n-1}.
\end{aligned} \tag{14}
$$

Ternary country, product, and technology complexity indices are defined in accordance with the definition of HH's MR as follows:

$$
\begin{aligned}
\mathrm{TCI}_c &= \frac{k_c - \langle k_c \rangle}{\mathrm{stdev}(k_c)}, \\
\mathrm{TCI}_p &= \frac{k_p - \langle k_p \rangle}{\mathrm{stdev}(k_p)}, \\
\mathrm{TCI}_t &= \frac{k_t - \langle k_t \rangle}{\mathrm{stdev}(k_t)},
\end{aligned} \tag{15}
$$

where k_c, k_c, k_c are the limits of iterations

$$
\begin{aligned}
k_c &= \lim_{n \to \infty} k_{c,n}, \\
k_p &= \lim_{n \to \infty} k_{p,n}, \\
k_t &= \lim_{n \to \infty} k_{t,n}.
\end{aligned} \tag{16}
$$

Although the values of the non-normalized indices $k_{c,n}$ grow infinitely, the values of the Ternary Complexity Indicator (TCI) empirically converge to a limit which can be conceptualized as the state of equilibrium obtained through interactions of three different dimensions

3.4. *Constructing the three-dimensional array*

In order to define the array $\{\mathfrak{M}_{c,p,t}\}$ we have to build the three-dimensional array $\{x_{c,p,t}\}$ with respect to c, p, t dimensions in which c refers to countries (or other geographical units), p refers to product classes, and t refers to technology (patent) classes, and then binarize it. We define the matrix elements of a three-dimensional matrix $\{x_{c,p,t}\}$ as follows:

$$
y_{c,p,t} = x_{c,p} Z_{p,t} a_{c,t}. \tag{17}
$$

Here, $x_{c,p}$ is a country–product matrix, $a_{c,t}$ is a country-patent matrix, and $Z_{p,t}$ is a binary matrix which relates product groups to patent classes (hereafter referred to as a concordance matrix) derived from a patent–product concordance table [15] with elements that are assigned the value one if the patent class t relates to product group p and zero otherwise.

Following HH, one can reduce matrix $\{y_{c,p,t}\}$ to a binary form by extending Balassa's RCA index to three dimensions as

$$\mathrm{RCA}_{c,p,t} = \frac{y_{c,p,t}/\sum_{p,t} y_{c,p,t}}{\sum_c y_{c,p,t}/\sum_{c,p,t} y_{c,p,t}} = \frac{y_{c,p,t}/Y_c}{y_{c,p,t}/\sum_c Y_c}. \quad (18)$$

The corresponding array elements $\{\mathfrak{M}_{c,p,t}\}$ are assumed to be one if this extended Balassa's RCA is larger than or equal to unity and otherwise zero. $\mathrm{RCA}_{c,p,t}$ defines the weight of partial patent–product output $y_{c,p,t}$ in a country's production function relative to the weight of the same patent–product output for all the countries in the set.

4. Data

We performed test calculations with the three indicators — HH's ECI, Tacchella's *et al.* Fitness, and TCI for a set of 41 countries, which includes 29 of the 35 OECD member states, three BRICS countries — Brazil, China, Russia, plus nine smaller economies: Croatia, Egypt Georgia, Lithuania, Malaysia, Morocco, Moldova, Romania, and Ukraine. The initial conditions are empirically defined:

(1) Data for the products exported by the 41 countries are harvested from https://comtrade.un.org/data/_in the format of the Standard International Trade Classification (SITC) revision 3 at the 2-digit level.
(2) The patent data organized in terms of 35 technology groups were retrieved from the WIPO statistics database at http://ipstats.wipo.int/ipstatv2/index.htm. We used resident count by filing office since domestic patents fully comply with technology development of the country.
(3) Technology classifications based on the codes of the International Patent Classification (IPC) were obtained from http://www.wipo.int/export/sites/www/ipstats/en/statistics/patents/pdf/wipo_ ipc_technology.pdf.

Correspondence tables connecting SITC Rev.3 and NACE rev. 2 classifications through the sequence: NACE Rev. 2 — ISIC Rev. 4, ISIC Rev. 4 — ISIC Rev. 3.1, ISIC Rev. 3.1 — ISIC Rev. 3, ISIC Rev. 3 — SITC Rev. 3, SITC Rev. 3 — NACE Rev.2 are found at Eurostat Reference and Management of Nomenclatures (RAMON) Index of correspondence tables http://ec.europa.eu/eurostat/ramon/relations/index.cfm?TargetUrl=LST_REL_A concordance table between IPC 8 and NACE Rev.2 concordance table was obtained from http://ec.europa.eu/eurostat/ramon/relations/index.cfm?TargetUrl=LST_REL.

5. Results

The values of TCI for the first 20 iterations in the set of 41 countries are provided in Appendix A. The first 20 iterations for seven major economies (for 2015) selected

I. Ivanova, Ø. Strand and L. Leydesdorff

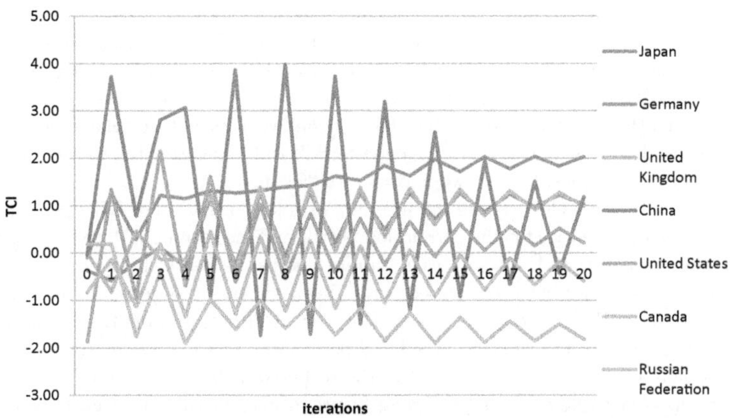

Fig. 2. The first 20 successive iterations of the Ternary Complexity Index for seven selected countries (for 2015).

Table 1. Pearson correlations between the values of TCI, Fitness, ECI, initial diversity score, and ln(GDP per capita) in current USD (for 2015).

	TCI	ECI	Fitness	Diversity	LN (GDP/capita)
TCI					
ECI	−0.728**				
Fitness	−0.165	0.192			
Diversity	0.038	−0.098	0.882**		
LN(GDP/capita	−0.541	0.516	−0.112	−0.078	

Notes: **Correlation is significant at the 0.01 level (2-tailed).
*Correlation is significant at the 0.05 level (2-tailed).

from the set are shown for illustrative purposes in Fig. 2. Using real data, the iterative sequence for all countries in the set empirically converged to a limit. Whereas the interpretation of different iterations in the ECI calculation is difficult, in our approach these iterations can be considered as steps of the autocatalytic cycles which bring the system increasingly into the equilibrium.

Table 1 shows the Pearson correlations between the values of TCI, Fitness,[b] ECI, initial diversity scores, total export values, and the logarithm of nominal GDP per capita in current US$ (for 2015). There is a significant correlation between TCI and ECI and a weak correlation between the pairs of TCI, ECI and Fitness, which can be attributed to the fact that TCI and ECI, on the one side, and Fitness, on the other

[b]FCI measures the Fitness of countries and product complexity; here we use Fitness as analog to ECI.

side, capture different kinds of information[c]. The three indices correlate to different extents with the value of total export. Furthermore, there is no significant relationship between total exports and income. Fitness is significantly correlated with the initial diversity score. ECI and TCI do not correlate with diversity. Note that it can be mathematically shown that HH's complexity is not correlated with the countries' diversity [29], so that one expects ECI to be uncorrelated to the product diversity of countries. Indeed, we find $r = -0.098$ (n.s.).

Both ECI and TCI correlate significantly with the logarithm of GDP per capita, but this is not the case for the correlation between Fitness and GDP. (Applying the Ln function to Fitness [13] also does not improve the situation.) This is not surprising since Fitness is strongly correlated with diversity and the correlation between diversity and GDP per capita is weak. However, TCI outperforms ECI in terms of the correlation with the logarithm of GDP per capita. This may be attributed to the additional accounting of the variety in the technology dimension in TCI. ECI has been reported to be good at predicting future growth in the long run, but not so reliable in short-term predictions This may indicate that the advantage of complexity is more likely to be realized over time [38].

Using an OLS linear regression growth model for a 10-year time period, we tested our data by regressing the rate of growth on the initial level of a country's income and complexity index, according to the Equation provided by Hidalgo and Hausmann [23, p. 10574], as follows:

$$\text{Growth}(t + \Delta t) = A + B \cdot \text{LN}(\text{GDP}(t)) + C \cdot \text{CI}. \tag{19}$$

Here, *Growth* stands for GDP per capita growth (% for the period), CI can stand for ECI, Fitness or TCI. Table 2 shows the results for the three indicators (t-values are

Table 2. OLS 10-year linear regression growth model.

Predicted variable	Growth (2004, 2014)	Growth (2004, 2014)	Growth (2004, 2014)	Growth (2004, 2014) (null model)
Predictors				
LN(GDP/capita)	−53.1	−51.318	−52.564	−53.225
(current USD)	(−7.534)	(−6.148)	(−7.096)	(−7.87)
TCI	−0.656			
	(−0.073)			
ECI		−4.23		
		(−0.398)		
Fitness			−0.371	
			(−0.234)	
Constant	595.089	578.145	595.969	596.236
	(8.826)	(7.258)	(9.094)	(9.212)
Observations	41	41	41	41
R^2	0.613	0.615	0.614	0.613

[c]Cristelli *et al.* [12] mentioned the correlation between Fitness and GDP around 0.45 for 2015 data of 148 countries. One should consider that the correlation with GDP may also depend on the number of countries included in the set (e.g., Ivanova *et al.* [28]).

provided in parentheses). All three indices demonstrate approximately similar results with respect of adjusted R^2 value of the regression.

The last column of Table 2 refers to the model which accounts only for the initial value of GDP per capita (null model):

$$\text{Growth}(t + \Delta t) = \beta_0 + \beta_1 \text{LN}(\text{GDP}(t)). \quad (20)$$

It can be seen from Table 2 that adding complexity dimension slightly improves the situation. The best improvement is provided by the Fitness index which accounts for 34% of the variations. To obtain a better fit one can further introduce additional factors used to explain economic growth.

Traditional growth models account for three factors of growth — increase in labor and labor quality, increase in capital, and increase in technology [41, 43]. We introduced additional country-specific factors to more completely account growth variations, such as gross capital formation, population growth, exchange rate, life expectancy, and unemployment rate. Here population growth, life expectancy, and unemployment rate refer to labor and labor quality, domestic investments refers to capital, trade openness and exchange rate relate to institutional quality, and complexity index refers to an increase in technology. Income per capita serves as a base level of economic development.

$$\begin{aligned}\text{Growth}(t + \Delta t) = \beta_0 &+ \beta_1 \text{LN}(\text{GDP}(t)) + \beta_2 \text{Ln}(\text{GCF}(t)) + \beta_3 \text{Ln}(\text{Pop}(t)) \\ &+ \beta_4 \text{Ln}(\text{ER}(t)) + \beta_5 \text{Ln}(\text{LE}(t)) + \beta_6 \text{Ln}(\text{UE}(t)) + \beta_7 \text{CI}(t), \quad (21)\end{aligned}$$

where *Growth* — GDP per capita growth 10-year period (2004–2014), *GDP* — income per capita; *GCF* — Gross capital formation (% GDP) current USD; *Pop* — population (annual %) growth; *ER* — exchange rate (local currency unit per USD, period averaged); *LE* — life expectancy (years); *UE* — unemployment rate (% of total labor force); *CI* — complexity indicator (ECI, Fitness or TCI). The results of calculations are presented in Table 3. One can mention that all three measures give approximately the same fit.

According to Table 3, the ECI and Fitness coefficient in the regression have a negative sign, meaning that higher values imply lower growth. This may be due to an error term. When regressing growth on independent variables some of the elements may be endogenous. The residual term then can comprise time-invariant component which can be attributed to country-specific fixed effects. To get rid of it and get the better fit for the OLS coefficients one can subtract individual means from the equation for each country in the set. It can be shown that this approach is, in effect, to treat individual effects as coefficients on dummy variables and run least square. OLS 10-year linear regression growth model with country fixed effects removed using panel data for 2003–2005 and 2013–2015 is presented in Table 4.

Note that this time TCI, ECI, Fitness coefficients are all positive. OLS regression with TCI measure substantially improves over regression with ECI or Fitness measures. Also Growth is more sensitive to the change in TCI in comparison with

Table 3. OLS 10-year linear regression growth model with additional growth related variables.

Predicted variable Predictors	Growth (2004–2014)	Growth (2004–2014)	Growth (2004–2014)
LN(GDP per capita)	−55.292	−55.24	−55.966
(current USD)	(−3.523)	(−3.51)	(−3.546)
TCI	5.9		
	(0.652)		
ECI		−5.54	
		(−0.512)	
Fitness			−0.259
			(−0.153)
LN(Gross capital formation)	39.43	32.277	35,357
(% GDP) current USD	(0.772)	(0.621)	(0.672)
LN(Population growth)	−33.332	−39.251	0.074
(annual %)	(−0.489)	(−0.575)	(−0.534)
LN(Exchange rate)	−9.055	−8.575	−8.97
	(−1.741)	(−1.64)	(−1.692)
LN(Life expectancy)	−136.325	−92.182	−121.652
	(−0.381)	(−0.251)	(−0.33)
LN(Unemployment rate)	−59.267	−57.576	−55.70
(% of total labor force)	(−2.417)	(−2.373)	(−2.3)
Constant	1265.775	1102.908	1224.129
	(0.938)	(0,793)	(0.882)
Observations	41	41	41
R^2	0.684	0.682	0.680

Table 4. OLS 10-year linear regression growth model with country fixed effects removed.

Predicted variable Predictors	Growth (2004–2014)	Growth (2004–2014)	Growth (2004–2014)
$LN(GDP_{pc}) - \langle LN(GDP_{pc})\rangle$	−469.139	−442.814	−441.612
	(−4.481)	(−4.043)	(−4.054)
$TCI - \langle TCI\rangle$	5.542		
	(1.956)		
$ECI - \langle ECI\rangle$		2.224	
		(0.309)	
$Fitness - \langle Fitness\rangle$			0.644
			(0.621)
$LN(GCF)-\langle LN(GCF)\rangle$	170.515	167.794	171,564
	(3.547)	(3.304)	(3.361)
$LN(Pop)-\langle LN(Pop)\rangle$	57.08	80.385	82.516
	(0.637)	(0.85)	(0.878)
$LN(ER)-\langle LN(ER)\rangle$	−301.474	−280.335	−282.162
	(−2.82)	(−2.5)	(−2.525)
$LN(LE)-\langle LN(LE)\rangle$	−205.19	−70.712	−42.013
	(−0.385)	(−0.125)	(−0.075)
$LN(UE)-\langle LN(UE)\rangle$	14.794	14.458	16.883
	(0.848)	(0.816)	(0.924)
constant-⟨constant⟩	5.631	5.394	5.596
Observations	(5.006)	(4.574)	(4.582)
	41	41	41
R^2	0.587	0.540	0.544

ECI and Fitness. Fixed effects panel data models offer a solution to endogeneity problem by absorbing time-invariant regressors. The model can consistently be estimated as long as the residual term is uncorrelated with the used regressors.

The ranking of countries (provided in the Appendix) is different for the three indices. The Fitness index, which relies on the diversity score, places countries as China, Germany, Austria, and the UK lower than Poland, Moldova, Egypt, and Croatia, though the countries in the first group have more diversified portfolio of manufacturing. ECI places Ireland, Poland, and Egypt above Canada and China. In our opinion, the ranking of countries according TCI is realistic since manufacturing capabilities are additionally weighted according the respective knowledge bases of the countries.

6. Extension to Continuous Time

In addition to improving the prediction, TCI can be considered as an ecosystem's approach to constructing a complexity indicator for comparative statics (e.g., time series). Let us apply the eco-systems metaphor to model the structure of economic complexity indicators. Assuming that the set of equation (14) present a discrete time form of the continuous equations in which $k_n = k(t+1)$ and $k_{n-1} = k(t)$ and denoting $k_{c,n}$, $k_{p,n}$, and $k_{t,n}$ as vectors $\boldsymbol{x}, \boldsymbol{y}, \boldsymbol{z}$, respectively, and the array $\{\mathfrak{M}_{c,p,t}\}$ as $\mathfrak{M}$, one can write the set of equations (14) in continuous form as follows:

$$\begin{aligned} \frac{dx_i}{dt} &= -x_i + \alpha\,\mathfrak{M}\boldsymbol{y}\boldsymbol{z}, \\ \frac{dy_j}{dt} &= -y_j + \beta\,\mathfrak{M}\boldsymbol{x}\boldsymbol{z}, \\ \frac{dz_k}{dt} &= -z_k + \gamma\,\mathfrak{M}\boldsymbol{x}\boldsymbol{y}. \end{aligned} \tag{22}$$

Here, $i = 1, \ldots, N_c$, $j = 1, \ldots, N_p$, $k = 1, \ldots, N_t$. A negative sign at linear terms in the right-hand side of equation (22) means that the corresponding increment of the left-hand side value (country product diversity, and product and technology ubiquity) will decline unless the appropriate nonlinear term is present.

The set of equation (22) presents a modification of the generalized L–V equation:

$$\frac{dx_i}{dt} = x_i f_i(\boldsymbol{x}), \tag{23}$$

where the vector $\boldsymbol{f}$ is defined as

$$\boldsymbol{f} = -\boldsymbol{I} + \mathbf{A}\mathrm{x}. \tag{24}$$

Here, $\boldsymbol{I}$ is a unity matrix and $\mathbf{A}$ is a community matrix.

Generalized L–V equations can exhibit various kinds of dynamics, including attractors, chaos, and limit cycles [26]. The same kinds of dynamics can be also be expected for equation (22). Whereas interaction of two dimensions shape each other in a coevolution that may lead to relatively stable trajectories, the addition of a third dimension can make these trajectories unstable (hyper-stable, meta-stable, etc.; [32]).

7. Conclusion

A major advantage of the eco-systems approach in constructing complexity measures is the possibility of entertaining models of systems dynamics and self-organization. Hitherto, this approach has not often been applied in innovation studies. An important feature of an eco-system is co-evolution. With respect to constructing economic complexity measures one can account for the co-evolution among different dimensions of innovation systems. In other words, complexity measures can be constructed by following the evolutionary dynamics of innovation eco-systems resulting from interactions among independent dimensions. We show that complexity indicators can be constructed in analogy to eco-systems. Iterative sequences can be approximated by generalized L–V equations in which the three dimensions — countries, products, and technologies — interact and reach a dynamical equilibrium.

TCI can shed light on the meaning and limitations of the complexity approach. While the interpretation of different iteration terms in ECI has remained vague [29], and FCI is defined bottom-up from the data, TCI can straightforwardly be interpreted as a discrete version of generalized L–V equations. This changes the status of the data since we can test a model using this data, and the model guides the interpretation of the different terms in the iterations. The dynamics can be considered as evolutionary stages of interaction dynamics among the three variables: geography, product, and technology (Storper, 1997). Sequential iterations can be considered as a series of successive communications among the variables. During a fixed period, there can be only a limited number of communications. If the equation has an asymptotically stable solution, the specification of this solution may serve as a limit to which iterative communications converge over time.

The introduction of this ecosystems approach in the domain of complexity measures raises further questions. Generalized L–V equations can comprise different dynamics, including limit cycles, chaos, and point attractors. The question of the relations of these dynamics to economic phenomena needs further investigation. A comparison of the results of model simulations of the three complexity measures suggests that TCI can be used for the prediction of the economic growth of countries. This may help decision makers to shape their policy. We believe that elaboration of a paradigm based on the systems-evolutionary dynamics of L–V can bring more predictive power to both the field of developing complexity indicators and innovation studies.

Acknowledgments

Inga Ivanova acknowledges support within the framework of the Basic Research Program at the National Research University Higher School of Economics

(NRU HSE) and within the framework of a subsidy by the Russian Academic Excellence Project '5-100'.

Appendix A

Table A.1. Comparison of the country among TCI, ECI, and Fitness measures (2015).

Rank	*Country*	*TCI*	*Country*	*ECI*	*Country*	*Fitness*
1	Japan	2.124	Switzerland	2.969	United States	27,075
2	Korea, Rep.	1.857	Japan	2.243	Netherlands	24,975
3	Switzerland	1.570	Ireland	2.023	Poland	23,919
4	Germany	1.511	United Kingdom	1.716	Lithuania	22,239
5	United Kingdom	1.443	Korea, Rep.	1.578	Canada	22,039
6	Austria	1.237	Germany	1.415	Spain	20,919
7	China	1.133	United States	0.636	Egypt	20,897
8	Finland	1.102	France	0.599	Japan	20,629
9	Sweden	1.095	Austria	0.561	Denmark	20,510
10	Slovenia	0.901	Hungary	0.502	Korea, Rep.	20,393
11	United States	0.761	Sweden	0.470	Croatia	20,262
12	Luxemburg	0.583	Finland	0.447	Sweden	20,181
13	Slovak Republic	0.503	Luxemburg	0.436	France	19,361
14	France	0.414	Czech Republic	0.420	Latvia	18,898
15	Ireland	0.375	Slovenia	0.242	Moldova	18,492
16	Czech Republic	0.115	Denmark	0.123	Brazil	18,187
17	Romania	0.072	Poland	0.015	China	18,123
18	Morocco	−0.017	Netherlands	−0.117	Slovenia	18,009
19	Netherlands	−0.038	Malaysia	−0.257	Portugal	18,007
20	Denmark	−0.051	Spain	−0.285	Germany	17,884
21	Portugal	−0.067	Greece	−0.388	Czech Republic	17,313
22	Croatia	−0.115	Slovak Republic	−0.412	Hungary	16,519
23	Spain	−0.214	Croatia	−0.436	Austria	16,091
24	Iceland	−0.276	Egypt	−0.464	United Kingdom	15,737
25	Poland	−0.389	New Zealand	−0.509	Estonia	15,644
26	Norway	−0.411	Estonia	−0.563	Ireland	15,59
27	Latvia	−0.473	China	−0.586	Malaysia	14,643
28	Canada	−0.510	Romania	−0.654	Greece	14,631
29	Greece	−0.577	Portugal	−0.676	Luxembourg	14,474
30	Malaysia	−0.650	Georgia	−0.689	Morocco	14,176
31	Brazil	−0.686	Lithuania	−0.696	Australia	13,812
32	Estonia	−0.716	Brazil	−0.711	Finland	13,678
33	Lithuania	−0.876	Canada	−0.731	Slovak Republic	13,604
34	Hungary	−1.039	Moldova	−0.746	New Zealand	13,164
35	Australia	−1.085	Australia	−0.815	Russian Federation	13,027
36	New Zealand	−1.131	Latvia	−0.900	Romania	12,246
37	Russian Federation	−1.198	Iceland	−1.010	Georgia	12,012
38	Moldova	−1.350	Ukraine	−1.029	Ukraine	11,634
39	Egypt	−1.410	Norway	−1.170	Switzerland	10,952
40	Ukraine	−1.410	Russian Federation	−1.181	Norway	5,217
41	Georgia	−2.108	Morocco	−1.370	Iceland	4,335

References

[1] Alkemade, F., Heimeriks, G., Schoen, A., Villard, L. and Laurens, P., Tracking the internationalization of multinational corporate inventive activity: National and sectoral characteristics, *Res. Policy* **44**(9) (2015) 1763–1772.

[2] Balassa, B., Trade liberalization and "revealed" comparative advantage, *Manch. Sch.* **33**(2) (1965) 99–123. doi: 10.1111/j.1467-9957.1965.tb00050.x.

[3] Balland, P. and Rigby, D., The geography of complex knowledge, *Econ. Gepgr.* **93**(1) (2016) 1–23.

[4] Barro, R. J. and Sala-i-Martin, X., *One-Sector Models of Endogenous Growth, in Economic Growth*, 2nd edn. (MIT Press, 2004), pp. 205–237.

[5] Boschma, R., Balland, P. and Cogler, D., Relatedness and technological change in cities: The rise and fall of technological knowledge in US metropolitan areas from 1981 to 2010, *Ind. Corp. Change* **24**(1) (2015) 233–250.

[6] Boschma, R., Balland, P.-A. and Kogler, D. F., Relatedness and technological change in cities: The rise and fall of technological knowledge in US metropolitan areas from 1981 to 2010, *Industrial and Corporate Change* **24**(1) (2014) 223–250.

[7] Brauer, F. and Castillo-Chavez, C., *Mathematical Models in Population Biology and Epidemiology* (Springer-Verlag, 2000).

[8] Chesbrough, H., Sohyeong, K. and Agogino, A., Chez panisse: Building an open innovation ecosystem, *Calif. Manag. Rev.* **56**(4) (2014) 144–171.

[9] Cobb, C. W. and Douglas, P. H., A theory of production, *Am. Econ. Rev.* **18** (1928) 139–165.

[10] Comin, D., Hobijn, B. and Rovit, E., Technology usage lags, *J. Econ. Growth* **13** (2008) 237–256. doi: 10.1007/s10887-008-9035-5

[11] Caldarelli, G., Cristelli, M., Gabrielli, A., Pietronero, L., Scala, A. and Tacchella, A., A network analysis of countries' export flows: Firm grounds for the building blocks of the economy, *PloS One* **7**(10) (2012) e47278.

[12] Cristelli, M., Gabrielli, A., Tacchella, A., Caldarelli, G. and Pietronero, L., Measuring the intangibles: A metrics for the economic complexity of countries and products, *PLoS ONE* **8**(8) (2013) e70726.

[13] Cristelli, M., Gabrielli, A., Tacchella, G. and Pietronero, L., The heterogeneous dynamics of economic complexity, *PLoS One* **10**(2) (2015) e17174.

[14] European Competitiveness Report 2013, towards knowledge driven reindustrialization, http://www.qren.pt/np4/np4/?newsId=3752&fileName=eu_2013_eur_comp_rep_en.pdf.

[15] Eurostat, Patent Statistics: Concordance IPC V8 –NACE REV.2 (2014), https://circabc.europa.eu/sd/a/d1475596-1568-408a-9191-426629047e31/2014-10-16-Final%20IPC_ NACE2_2014.pdf.

[16] Etzkowitz, H. and Leydesdorff, L., The dynamics of innovation: From national systems and "mode 2" to a triple helix of university–industry–government relations, *Res. Policy* **29**(2) (2000) 109–123.

[17] Etzkowitz, H. and Ranga, M., Spaces: A triple helix governance strategy for regional innovation, in: *Innovation Governance in an Open Economy: Shaping Regional Nodes in a Globalized World*, Rickne A., Laestadius S. and Etzkowitz H. (eds.) (Routledge, London, 2012), pp. 51–68.

[18] Feldman, M. and Kogler, D., Stylized facts in the geography of innovations, In *Handbook of the Economics of Innovation*, Hall B. and Rosenberg, N. (eds.) (Oxford, 2010).

[19] Foray, D., *The Economics of Knowledge* (CMIT Press, 2004).

[20] Foray, D. and Lundvall, B.-A., The knowledge-based economy: From the economics of knowledge to the learning economy, *Employment and Growth in the Knowledge-Based Economy*, Lund Vall, B. A. (ed.) (POECD, 1996), pp. 11–32.
[21] Frosch, R. A. and Gallopoulos, N. E., Strategies for manufacturing, *Sci. Am* **261**(3) (1989) 144–152.
[22] Gaver, A. and Cusumano, M., Industry platforms and ecosystem innovation, *J. Prod. Innov. Manag.* **31**(3) (2014) 417–433.
[23] Hausmann, R. and Hidalgo, C., The network structure of economic output, *J. Econ. Growth* **16**(4) (2011) 309–342. doi: 10.1007/s 10887-011-9071-4.
[24] Hidalgo, C. and Hausmann, R., The building blocks of economic complexity, *Proc. Nat. Acade. Sci.* **106**(26) (2009) 10570—10575.
[25] Hodgson, G. and Knudsen, T., *Darwin's Conjecture: The Search for General Principles of Social and Economic Evolution* (University of Chicago Press, Chicago, London, 2011).
[26] Hofbauer, J. and Sigmund, K., *Evolutionary Games and Population Dynamics* (Cambridge University Press, 1998).
[27] Ivanova, I. and Leydesdorff, L., Rotational symmetry and the transformation of innovation systems in a triple helix of university-industry-government relations, *Technol. Forecast. Soc. Change* **86** (2014) 143–156.
[28] Ivanova, I., Strand, Ø., Kushnir, D. and Leydesdorff, L., Economic and technological complexity: A model study of indicators of knowledge-based innovation systems, *Technol. Forecas. Soc. Change* **120** (2017) 77–89.
[29] Kemp-Benedict, E., An interpretation and critique of the method of reflections. MPRA Paper No. 60705, https://mpra.ub.uni-muenchen.de/60705/1/MPRA_paper_60705.pdf.
[30] Kline, S. J. and Rosenberg, N., An overview of innovation, In *The Positive Sum Strategy: Harnessing Technology for Economic Growth*, Landau, R. and Rosenberg, N. (eds.) (National Academy Press, 1986), pp. 275–305.
[31] Leydesdorff, L., Dolfsma, W. and van der Panne, G., Measuring the knowledge base of an economy in terms of triple-helix relations among "technology, organization, and territory", *Res. Policy* **35**(2) (2006) 181–199.
[32] Leydesdorff, L. and Van den Besselaar, P., Technological development and factor substitution in a non-linear model, *J. Soc. Evol. Syst.* **21** (1998) 173–192.
[33] Mazzucato, M. and Robinson, D.K.R., Co-creating and directing innovation ecosystems? NASAs changing approach to public-private partnerships in low-earth orbit, *Technol. Forecast. Soc. Change*, http://dx.doi.org/10.1016/j.techfore.2017.03.034. (in press).
[34] Moore, J. F., Predators and prey: A new ecology of competition, *Harvard Bus. Rev.* **71**(3) (1993) 75–86.
[35] Nelson, R. R. and Winter, S. G., *An Evolutionary Theory of Economic Change* (Cambridge, MA: Belknap Press of Harvard University Press, 1982).
[36] OECD, *The Knowledge-Based Economy* (OECD, 1996), http://www.oecd.org/officialdocuments/publicdisplaydocumentpdf/?cote=OCDE/GD%2896%29102&docLanguage=En.
[37] Oh, D.-S., Phillips, F., Park, S. and Lee, E., Innovation ecosystems: A critical examination, *Technovation* **54** (2016) 1–6.
[38] Ourens, G., Can the Method of Reflections predict future growth? Université catholique de Louvain, Institut de Recherches Economiques et Sociales (IRES). Retrieved from, http://econpapers.repec.org/paper/ctllouvir/2013008.htm; June 12, 2018.
[39] Ritala, P. and Almpanopoulou, A., In defense of 'eco' in innovation ecosystems, *Technovation* **60** (2017) 39–42.

[40] Rohrbeck, R., Hölzle, K. and Gemünden, H. G., Opening up for competitive advantage–How Deutsche Telekom creates an open innovation ecosystem, *R&D Manag.* **39**(4) (2009) 420–430.
[41] Romer, P., Increasing returns and long-run growth, *J. Polit. Economy* **94**(5) (1986) 1002–1037.
[42] Schumpeter, J., *Business Cycles: A Theoretical, Historical and Statistical Analysis of Capitalist Process* (McGraw-Hill, New York, [1939], 1964).
[43] Solow, R. M., A contribution to the theory of economic growth, *Qu. J. Econ.* **70**(1) (1956) 65–94.
[44] Swan, T. W., Economic growth and capital accumulation, *Econ. Rec.* **32**(2) (1956) 334–361.
[45] Storper, M., *The Regional World — Territorial Development in a Global Economy* (Guilford Press, 1997).
[46] Tacchella, A., Cristelli, M., Caldarelli, G., Gabrielli, A. and Pietronero, L., A new metrics for countries' fitness and products' complexity, *Nature: Scientific Reports* **2** (2012) 723.
[47] Tacchella, A., Cristelli, M., Caldarelli, G., Gabrielli, A. and Pietronero, L., Economic complexity: Conceptual grounding of a new metrics for global competitiveness, *J. Econ. Dyn. Control* **37** (2013) 1683–1691.
[48] Tsujimoto, M., Kajikawa, Y., Tomita, J. and Matsumoto, Y., Designing the coherent ecosystem: Review of the ecosystem concept in strategic management, in *Management of Engineering and Technology (PICMET), Portland Int. Conf.* IEEE (2015), pp. 53–63.
[49] Ulanowicz, R. E., The dual nature of ecosystem dynamics, *Ecol. Modell.* **220**(16) (2009) 1886–1892.
[50] Utkovski, Z., Pradier, F., Stojkoski, V., Perez-Cruz, F. and Kocarev, L., Economic complexity unfolded: Interpretable model for the productive structure of economies, *PLoS One* **13**(8) (2017) e0200822.
[51] Verspagen, B., Mapping technological trajectories as patent citation networks: A study on the history of fuel cell research, *Adv. Complex Syst.* **10**(1) (2007) 93–115.
[52] Wang, D., Zhao, X. and Zhang, Z., The time lags effects of innovation input on output in national innovation systems: The case of China, *Discrete Dyn. Nat. Soc.* **2016**(2016), Article ID 1963815, 12 pages.

Advances in Complex Systems
Vol. 23, No. 8 (2020) 2050022 (24 pages)

DOI: 10.1142/S0219525920500228

SIMPLICITY FROM COMPLEXITY: ON THE SIMPLE AMPLITUDE DYNAMICS UNDERLYING COVID-19 OUTBREAKS IN CHINA

T. D. FRANK
Psychological Sciences, University of Connecticut
406 Babbidge Road, Storrs, CT 06269, USA
Physics Department, University of Connecticut
196 Auditorium Road, Storrs, CT 06269, USA
till.frank@uconn.edu

Received 11 February 2021
Revised 30 May 2021
Accepted 2 June 2021
Published 15 July 2021

COVID-19 confronts societies and individuals with unprecedented challenges. It is advocated that complex systems theory, in general, and synergetics, in particular, provide a valuable and comprehensive repertoire of tools and concepts such as the concept of amplitude equations and order parameters to study the spread of COVID-19 in human populations. Specifically, within the framework of SIR and SEIR compartment models COVID-19 trajectories are described in terms of amplitude equations and order parameters. By plotting simulated and semi-empirical COVID-19 case trajectories it is shown that the initial epidemics in China, in general, and Wuhan city, in particular, during the first quarter of the year 2020 followed relatively simple amplitude dynamics in SIR and SEIR model state spaces describing interaction classes of individuals. The amplitudes evolve along certain paths or directions determined by order parameters that are well known to exist in complex systems. In summary, the present work highlights that COVID-19 outbreaks are constrained by general principles that hold for a broad class of phenomena in living and non-living systems.

Keywords: COVID-19; amplitude dynamics; synergetics.

1. Introduction

A thorough understanding of outbreaks of coronavirus disease 2019 (COVID-19) in countries around the world is desirable in view of the gravity of the pandemic that started in December 2019 and is still unfolding in 2021 [47]. In this context, at issue is to take an interdisciplinary perspective such as complex systems theory, in general, and synergetics, in particular [13, 22] in order to identify COVID-19 outbreaks as phenomena belonging to the broad class of bifurcation phenomena determined by amplitude dynamics and order parameters that characterize the interacting components of the systems at hand. By uncovering the general nature of COVID-19 outbreaks, a comprehensive theoretical repertoire becomes available to fight the spread of COVID-19 in addition to the currently used tools.

According to the World Health Organization (WHO), the first case of COVID-19 was confirmed on December 31, 2019 in China [46]. Since then, the disease has spread around the globe and has been reported from all 8 geographic regions of the world: Africa, Asia, the Caribbean, Central America, Europe, North America, Oceania, and South America [45]. In January 2021 the death toll passed the tragic number of 2,000,000 people [47].

Traditionally, the local outbreak of an epidemic is described on the population level with the help of the basic reproduction number R_0. The basic reproduction number is defined as the expected number of secondary cases produced by a typical infected individual in a completely susceptible population [5, 6, 36, 42]. Accordingly, a novel virus will (will not) spread out in a population for $R_0 > 1$ ($R_0 < 1$). Much research has been devoted to determine R_0 directly from COVID-19 case data (e.g. by curve fitting) or indirectly by fitting data to epidemiological compartment models and deriving R_0 from the estimated model parameters. For example, R_0 values between 1.7 and 4.7 have been reported for China for various different scenarios and locations using the aforementioned direct, curve fitting method [51, 52] or the indirect, model-based method [24, 35, 44, 48, 50].

Using epidemiological compartment models [36], the outbreak of an infectious disease, in general, and COVID-19, in particular, is understood as a dynamical system that evolves away from an unstable fixed point characterized by zero-infected individuals [5, 42, 23]. Taking the perspective of nonlinear physics [33] and synergetics [13, 22], instabilities and self-organization in complex systems give rise to a broad class of phenomena [4, 9–11, 18–20, 31] that can be treated on an equal footing. In doing so, the general nature of such phenomena can be explored and a comprehensive repertoire of tools and methods can be applied. A key method in this regard is the description in terms of amplitude equations, eigenvectors (and eigenfunctions) and order parameters [4, 9–11, 13, 14, 18, 19, 22, 33, 31]. Order parameters (or unstable eigenvectors) describe relevant directions in the original state spaces under consideration in which the dominant dynamics away from an unstable fixed point takes place. Amplitude equations describe the dynamics along those directions. In summary, under certain circumstances, simplicity emerges from complexity in the sense that the interacting components of a complex system give rise to a relatively simple dynamics — at least when taking the amplitude dynamics perspective.

Frank [17] worked out these ideas for a SEIR compartment model describing COVID-19 trajectories in China. However, the study [17] has left some crucial issues open for debate. First, the SEIR compartment model is a generalization of SIS and SIR models that serve as benchmark models of infection dynamics [36, 38, 41]. In particular, SIR models have been powerful tools to address COVID-19 epidemics in various countries and regions (e.g. [12, 37, 41, 44]). Consequently, it should be shown how the amplitude equation and order parameter concept applies within a SIR modeling framework. Second, the study [17] did not provide any explicit demonstrations about how actual COVID-19 case data and inferred/simulated COVID-19 case trajectories relate to such order parameters. Both limitations of the previous study will be addressed in this study. In Sec. 2.1, amplitude equations and order parameters will be derived for the SIR model. The COVID-19 outbreak nationwide in China in 2020 will be considered and model-based inferred trajectories will be plotted that demonstrate the COVID-19 outbreak followed a simple amplitude dynamics along the order parameter of the relevant SIR model. Likewise, in Sec. 2.2, a SEIR modeling framework will be evaluated and the amplitude dynamics along the order parameter will be demonstrated for the COVID-19 outbreak in Wuhan city, China, in 2020. Further potential applications will be briefly sketched in Sec. 2.3.

2. COVID-19 Outbreaks in Interacting SIR and SEIR Compartments and Their Amplitude Dynamics

2.1. *SIR framework and the COVID-19 outbreak in China 2020*

The SIR model is a compartment model introduced by Kermack and McKendrick [25, 36] defined by

$$\frac{\mathrm{d}}{\mathrm{d}t}S = -\frac{\beta}{N}IS, \quad \frac{\mathrm{d}}{\mathrm{d}t}I = \frac{\beta}{N}IS - \gamma I, \quad \frac{\mathrm{d}}{\mathrm{d}t}R = \gamma I, \tag{1}$$

where t denotes time and S, I, and R denote the number of susceptible, infectious, and recovered individuals, respectively. The model involves the semi-positive definite parameters β and γ, which quantify the effective contact rate (β) and recovery rate (γ), respectively [36]. The total number of individuals in the population under consideration is constant and given by N with $N = S + I + R$. Using $R(t) = N - S(t) - I(t)$, the model effectively becomes a two-variable model with state vector $\mathbf{X} = (S, I)$. The disease-free state is a fixed point $\mathbf{X}_{\mathrm{st}} = (N, 0)$ in the space spanned by S and I. For this fixed point, we also have $R_{\mathrm{st}} = 0$. Introducing the perturbation vector $\mathbf{Y} = (\delta, I)$ with $S = N + \delta \Longleftrightarrow \delta = S - N$, the linearized dynamics at this fixed point reads $\mathrm{d}\mathbf{Y}/\mathrm{d}t = L_{\mathrm{SIR}}\mathbf{Y}$ with the matrix [29]

$$L_{\mathrm{SIR}} = \begin{pmatrix} 0 & -\beta \\ 0 & \beta - \gamma \end{pmatrix}. \tag{2}$$

The matrix is singular and exhibits the eigenvalues $\lambda_1 = 0$ and $\lambda_2 = \beta - \gamma$. Consequently, the fixed point is stable for $\beta < \gamma$ and unstable for $\beta > \gamma$. The basic reproduction number, which also indicates stability or instability of a fixed point of a population described by a compartment model [5, 29], reads [29, 36]

$$R_{0,\mathrm{SIR}} = \frac{\beta}{\gamma}. \tag{3}$$

Since we have $R_{0,\mathrm{SIR}} < 1 \Longleftrightarrow \beta < \gamma$ and $R_{0,\mathrm{SIR}} > 1 \Longleftrightarrow \beta > \gamma$, the inequality $R_{0,\mathrm{SIR}} < 1$ implies stability and $R_{0,\mathrm{SIR}} > 1$ implies instability of the disease-free state

$\mathbf{X}_{st} = (N, 0)$, which is consistent with the definition (see Sec. 1) of the basic reproduction number [29].

Let us consider the infection dynamics from the perspective of dynamical systems [33] and synergetics [13, 22]. Accordingly, we consider the eigenvectors $\mathbf{v}_j$ with $j = 1, 2$ of the matrix L_{SIR}. Let $v_{j,S}$ and $v_{j,I}$ denote the eigenvector components in S and I directions, respectively, such that $\mathbf{v}_j = (v_{j,S}, v_{j,I})$. From Eq. (2) it follows that

$$\mathbf{v}_1 = \begin{pmatrix} 1 \\ 0 \end{pmatrix}, \quad \mathbf{v}_2 = \frac{1}{\sqrt{1+g^2}} \begin{pmatrix} -1 \\ g \end{pmatrix} \tag{4}$$

with $g = (\beta - \gamma)/\beta$ assuming $\beta > 0$. Note that $\mathbf{v}_1$ and $\mathbf{v}_2$ are normalized vectors but they are not orthogonal to each other. Rather, they are linearly independent of each other. For the following analysis, it is sufficient to require linear independency. The reason for this is that the aim is to describe any state in the two-variable space (S, I) with the help of vectors $\mathbf{v}_1$ and $\mathbf{v}_2$. If they are linearly independent, they can be used to achieve this aim. To this end, we follow the standard fixed point analysis in the theory of pattern formation and self-organization [33], in general, and synergetics [13, 14, 22], in particular. Accordingly, we express the state $\mathbf{X}$ in terms of the fixed point $\mathbf{X}_{st}$, its eigenvectors $\mathbf{v}_j$, and the corresponding amplitudes A_j (measuring distances along the eigenvectors) like

$$\mathbf{X}(t) = \mathbf{X}_{st} + A_1(t)\mathbf{v}_1 + A_2(t)\mathbf{v}_2. \tag{5}$$

Equation (5) describes a mapping $(A_1, A_2) \to (S, I)$ from the amplitude space [13] spanned by A_1 and A_2 to the original state space $\mathbf{X} = (S, I)$. In components, the mapping reads

$$S = N + A_1 - \frac{A_2}{\sqrt{1+g^2}}, \quad I = \frac{g}{\sqrt{1+g^2}} A_2. \tag{6}$$

For $\beta \neq \gamma$ (i.e. $g \neq 0$) the inverse mapping $(S, I) \to (A_1, A_2)$ reads

$$A_1 = S + \frac{I}{g} - N, \quad A_2 = \frac{\sqrt{1+g^2}}{g} I. \tag{7}$$

Differentiating the two relations in Eq. (7) with respect to time, substituting into the right-hand sides the differential equations (1) of the SIR model, and eliminating all thus obtained variables S and I using the mapping (6), we obtain the amplitude equations

$$\frac{\mathrm{d}}{\mathrm{d}t} A_1 = \frac{1-g}{\sqrt{1+g^2}} A_2 p_{\text{SIR}}(A_1, A_2), \quad \frac{\mathrm{d}}{\mathrm{d}t} A_2 = \lambda_2 A_2 + A_2 p_{\text{SIR}}(A_1, A_2) \tag{8}$$

with

$$p_{\text{SIR}} = \frac{\beta}{N} \left(A_1 - \frac{A_2}{\sqrt{1+g^2}} \right). \tag{9}$$

Let us briefly discuss the dynamics of the amplitudes A_1 and A_2 for the case of interest, which is $R_0 > 1 \iff \beta > \gamma$ and $\mathbf{X}(t) \approx \mathbf{X}_{st}$. At the fixed point $\mathbf{X}_{st} = (N, 0)$ we have $A_1 = A_2 = 0$. For $\beta > \gamma > 0$ it follows $g \in (0, 1)$. This implies that $A_2 \geq 0$, see Eq. (7). However, A_1 can assume positive as well as negative values, see Eq. (7) again. For example, for $I = 1$ and $\beta \to \gamma \Rightarrow g \to 0$, we have $A_1 > 0$ (irrespective of S and N). In contrast, for $I = 1$ and $\beta \to \infty \Rightarrow g \to 1$ we have $A_1 \leq 0$ (irrespective of S and N assuming only $N > 0$). The COVID-19 initial stage epidemic in a given population is characterized by a perturbation of the state $\mathbf{X}$ out of $\mathbf{X}_{st}$. Mathematically speaking, we introduce a small parameter ϵ with $\epsilon \ll N$ and assume that $I \propto O(\epsilon)$, $N - S \propto O(\epsilon)$. This implies $A_1 \propto O(\epsilon)$, $A_2 \propto O(\epsilon)$, and $p_{SIR} \propto O(\epsilon)\beta/N$, where β/N is typically a relatively small quantity. In particular, in applications typically $\beta/N \ll \lambda_2$ holds (see below). In summary, from Eq. (8) and the aforementioned considerations, we conclude that $\mathrm{d}A_1/\mathrm{d}t \propto O(\epsilon^2)\beta/N$. In contrast, $\mathrm{d}A_2/\mathrm{d}t \propto \lambda_2 O(\epsilon)$ and we assume that $\lambda_2 \gg \beta/N$. That is, the amplitude A_1 varies initially to a small amount as compared to the amplitude A_2. Consequently, the dynamics along $\mathbf{v}_2$ as described by A_2 dominates initially the overall infection dynamics. From the perspective of synergetics, in general, the eigenvector of an unstable amplitude and the corresponding unstable amplitude play the leading role in a bifurcation away from a fixed point [13, 22]. In this context, the eigenvector is called order parameter and the amplitude is called the order parameter amplitude. Accordingly, $\mathbf{v}_2$ is the SIR order parameter and A_2 the order parameter amplitude.

Finally, we would like to point out that Eq. (1) for $\mathbf{X} = (S, I)$ and Eq. (8) for the amplitudes A_1 and A_2 describe the same dynamical system. They yield identical solutions if the mappings $(A_1, A_2) \to (S, I)$ and $(S, I) \to (A_1, A_2)$ defined by Eqs. (6) and (7) are used.

As part of their comprehensive study, Fanelli and Piazza [12] evaluated COVID-19 data [8] from Italy during the period from February 11 to March 15, 2020. The SIR model parameters were estimated. Note that Fanelli and Piazza [12] estimated the parameter β/N rather than β. Furthermore, in the study by Fanelli and Piazza [12] a fourth compartment was considered: individuals deceased due to COVID-19. In this context, the parameter γ was composed of two contributions (recovery rate and death rate). Therefore, in this application γ is interpreted as removal rate. In this work, the forth compartment will be ignored since (i) it does not affect the order parameter analysis presented above and (ii) in line with Fanelli and Piazza [12], who kept β/N constant, N will be considered as a constant. That is, the drop of N due to COVID-19 associated deaths will be neglected. Finally note that in [12] the parameters $S(0)$ and $I(0)$ were estimated (and $R(0) = 0$) such that $N = S(0) + I(0)$. Explicitly, they found $\beta/N = 3.33 \cdot 10^{-6}/\mathrm{d}$, $\gamma = 0.0210/\mathrm{d}$, $S(0) = 7.92 \cdot 10^4$, and $I(0) = 999$. Based on those estimates, we computed the parameters β, $v_{2,S}$, $v_{2,I}$, λ_2, and the time constant $\tau = 1/\lambda_2$ relevant for this study, see Table 1. Moreover, the angle θ of $\mathbf{v}_2$ with respect to the S-axis was computed from $v_{2,S} = \cos(\theta)$. Accordingly, an angle of $\theta = 0°$ describes an order parameter $\mathbf{v}_2$

Table 1. Parameters of the SIR order parameter approach applied to the COVID-19 disease outbreak in China during the period from January 22 to February 19, 2020.

	Order parameter vector			Speed parameters	
β	$v_{2,S}$	$v_{2,I}$	θ	λ_2	τ
0.26/d	−0.74	0.68	137.3°	0.24/d	4.1d

pointing in the direction of the S-axis. $\theta = 90°$ describes $\mathbf{v}_2$ pointing in the direction of the I-axis. Finally, $\theta = 180°$ describes $\mathbf{v}_2$ pointing in the opposite direction of the S-axis (i.e. towards the negative values).

We found that the contact rate $\beta = 0.26/\text{d}$ was larger than the removal rate γ indicating that the disease-free state in China during the period from January 22 to February 19, 2020 was unstable. The order parameter component $v_{2,I}$ was positive supporting the simple relationship $I \propto A_2$ between the order parameter amplitude A_2 and the number of infectious individuals I. The eigenvalue $\lambda_2 = 0.24/\text{d}$ was positive ($\Rightarrow$ instability of $\mathbf{X}_{\text{st}} = (N, 0)$) and the time constant or e-folding time τ was 4.1 days. Accordingly, during the outbreak in China in the period from January 22 to February 19 the order parameter amplitude A_2 increased (at least initially) by a factor $e \approx 2.72$ in 4.1 days. Finally, comparing λ_2 with β/N (i.e. $\lambda_2 = 0.24/\text{d}$ with $\beta/N = 3.33 \cdot 10^{-6}/\text{d}$), we see that the aforementioned assumption $\lambda_2 \gg \beta/N$ holds.

For the COVID-19 outbreak in China we solved the SIR model (1) numerically. Panel (A) of Fig. 1 shows the thus obtained solutions S (dashed black line) and I (solid black line) as well as the active cases I (gray circles) as reported in the John Hopkins University data repository [2, 8]. Note that the jump of the cases I between

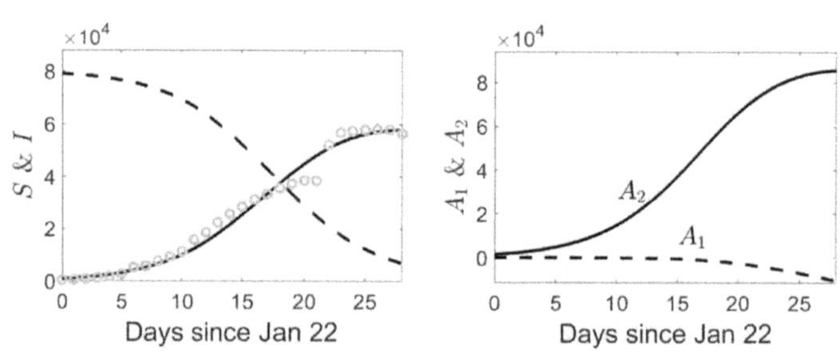

(A) (B)

Fig. 1. State space and amplitude space description of the COVID-19 outbreak in China within a SIR modeling framework (part 1). Panel (A): State space description given by infectious individuals I (data: gray circles; model: solid black line) and susceptible individuals S (dashed black line) versus time. Panel (B): Amplitude space description given by amplitudes A_1 (dashed line) and order parameter amplitude A_2 (solid line) over time.

February 12 and February 13 was due to a change of the reporting procedure of Chinese authorities [12, 24]. By visual inspection, the simulation results approximated well the reported cases and the simulation results presented in panel (A) reproduced the results previously derived by Fanelli and Piazza [12].

Panel (B) of Fig. 1 and panels (A) and (B) of Fig. 2 present the result of the amplitude dynamics analysis. Panel (B) of Fig. 1 shows A_1 (dashed line) and A_2 (solid line) computed numerically from the amplitude equations (8) for initial values $A_1(0)$ and $A_2(0)$ obtained from Eq. (7), $S(0)$, and $I(0)$. We found that the order parameter amplitude A_2 varied by a large amount as compared to A_1 illustrating the leading role of A_2 for the bifurcation that took in January/February 2020 the Chinese population away from the disease-free fixed point $\mathbf{X}_{\rm st} = (N, 0)$. Panel (A) of Fig. 2 sketches the eigenvectors $\mathbf{v}_1$ and $\mathbf{v}_2$ (magnified by a factor 10,000) in the $S - I$ plane. Accordingly, the eigenvectors constitute a skewed coordinate system centered at the unstable fixed point $\mathbf{X}_{\rm st} = (N, 0)$ (open circle). Panel (B) of Fig. 2 shows the axes of this skewed coordinate system as dotted lines. In panel (B) the graphs $S(t)$ and $I(t)$ as shown in panel (A) of Fig. 1 are plotted as a phase space trajectory (solid thick black line). We found that the trajectory followed the $\mathbf{v}_2$ axis. Consequently, panel (B) of Fig. 2 illustrates that the COVID-19 outbreak in China — when described from the perspective of a SIR model — followed the SIR order parameter $\mathbf{v}_2$.

A semi-empirical phase space curve was constructed composed of the reported active COVID-19 cases I between January 22 and February 19 shown in panel (A) of Fig. 1 (gray circles) and the simulated values S shown in panel (A) of Fig. 1 (dashed line). This semi-empirical phase curve is shown in panel (B) of Fig. 2 as gray squares.

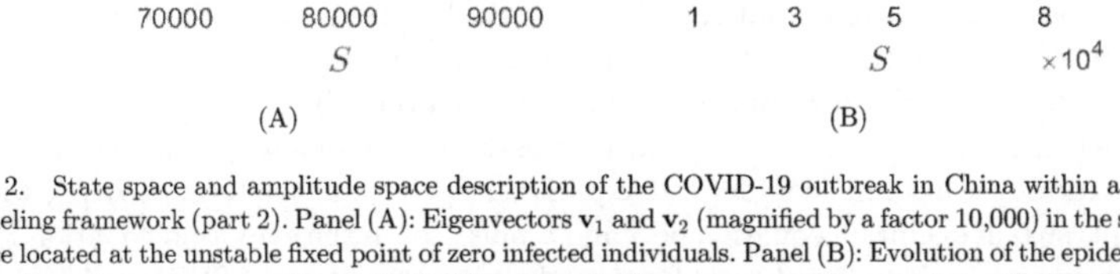

(A) (B)

Fig. 2. State space and amplitude space description of the COVID-19 outbreak in China within a SIR modeling framework (part 2). Panel (A): Eigenvectors $\mathbf{v}_1$ and $\mathbf{v}_2$ (magnified by a factor 10,000) in the state space located at the unstable fixed point of zero infected individuals. Panel (B): Evolution of the epidemics along the order parameter vector $\mathbf{v}_2$. Dotted lines describe the axes spanned by $\mathbf{v}_1$ and $\mathbf{v}_2$ with coordinates A_1 and A_2, respectively. The simulated phase space trajectory $I(t)$ versus $S(t)$ is shown (thick black line) and a semi-empirical trajectory is presented (gray squares). Both evolve in time along the axis defined by the order parameter $\mathbf{v}_2$.

In order to avoid to overload the presentation, not all daily values are presented in panel (B). Only time series values for the days $0, 9, 11, 13, \ldots, 25, 27$ since January 22 are shown.

In closing this section, we would like to return briefly to the fact that Fanelli and Piazza [12] consistent with other research groups [3, 40] estimated the initial size of the susceptible population, i.e. $S(0)$. When naively assuming that all individuals of a population under consideration belong at the beginning of the COVID-19 epidemic in China to the population of susceptible individuals, then $S(0)$ and, consequently, N should not be estimated but taken from demographic data. Therefore, future work may be devoted to conduct the SIR modeling approach with a more realistic value for $S(0)$ and N (as for example in [37]). Having said that, note that from the linearized dynamics involving the matrix $L_{\rm SIR}$ defined by Eq. (2) it follows that the initial dynamics of I does not depend on the population size N. Likewise, the order parameter $\mathbf{v}_2$ and the eigenvalue λ_2 do not depend on the size variable N. Therefore, scaling up $S(0)$ and N to more realistic numbers might not change much of the analyses results presented above. However, a more detailed discussion of this issue is beyond the scope of this study.

2.2. *SEIR framework and the COVID-19 outbreak in Wuhan city, China, 2020*

Following Pang *et al.* [35] and Frank [17], the following SEIR model is considered:

$$\begin{aligned}
\frac{\rm d}{{\rm d}t}S &= -\frac{\beta_1}{N}SI - \frac{\beta_2}{N}SE,\\
\frac{\rm d}{{\rm d}t}E &= \frac{\beta_1}{N}SI + \frac{\beta_2}{N}SE - \alpha E,\\
\frac{\rm d}{{\rm d}t}I &= \alpha E - \gamma I,\\
\frac{\rm d}{{\rm d}t}R &= \gamma I,
\end{aligned} \tag{10}$$

which has extensively been used in the COVID-19 literature [7, 28, 43, 49] and in the context of other virus diseases [26, 30]. For $\beta_2 = 0$ the model reduces to the standard SEIR model [36] with S, E, I, and R describing the number of susceptible, exposed, infectious, and recovered individuals, respectively. However, in the context of the SEIR modeling of the spread of COVID-19, some studies have taken into consideration that exposed individuals can be infectious although they do not show symptoms [28, 34, 35]. One way to account for this transmission mechanism of COVID-19 is to introduce a second contact rate parameter $\beta_2 > 0$ and the term $\beta_2 SE/N$ with a minus sign in evolution equation of S and a plus sign in the evolution equation for E as shown in Eq. (10). For $\beta_2 > 0$, Eq. (10) describes an extended SEIR model with S, E, and I describing susceptible, exposed, and infected individuals, where both exposed and infected individuals are infectious [7, 26, 28, 30, 35, 43, 49]. Moreover, R is a lump compartment that describes removed individuals that cannot infect others. This compartment of removed individuals includes hospitalized individuals, recovered individuals (assumed to be immune on the relative short time period that will be considered in the application below), and individuals deceased due to COVID-19. The parameters in Eq. (10) are the effective contact rate β_1 between infected and susceptible individuals, the effective contact rate β_2 between exposed and susceptible individuals, the rate of progression from being exposed to being infected α, and the removal rate γ. The parameter γ includes the diagnosis rate and the death rate due to COVID-19 (for more details, see [17, 35]). The variable $N = S + E + I + R$ is constant. In the application to the spread of COVID-19 in Wuhan city, China (see below) N can be interpreted as the total population because variations in population size are assumed to be negligibly small on the time scales of interest [17, 23]. Eliminating R with the help of $R(t) = N - S(t) - E(t) - I(t)$ the four-variable model (10) reduces to a three-variable model for the state vector $\mathbf{X}(t) = (S, E, I)$.

The disease-free state $\mathbf{X}_{\rm st} = (S_{\rm st}, E_{\rm st}, I_{\rm st}) = (N, 0, 0)$ is a fixed point of the extended SEIR model (10). Following the procedure from Sec. 2.1, the linearized dynamics at the fixed point $\mathbf{X}_{\rm st}$ is considered. Let $\mathbf{Y}$ denote the perturbation vector defined by $\mathbf{Y} = (\delta S, \delta E, \delta I)$, with $\delta S = S - S_{\rm st} = S - N$, $\delta E = E - E_{\rm st} = E$, and $\delta I = I - I_{\rm st} = I$ are assumed to describe small quantities. Then, $\mathbf{Y}$ satisfies $d\mathbf{Y}/dt = L_1 \mathbf{Y}$, where L_1 is the three-by-three matrix

$$L_1 = \begin{pmatrix} 0 & -\beta_2 & -\beta_1 \\ 0 & -(\alpha - \beta_2) & \beta_1 \\ 0 & \alpha & -\gamma \end{pmatrix}. \tag{11}$$

The matrix L_1 is singular and exhibits the eigenvalues $\lambda_1 = 0$ and [15, 17]

$$\lambda_{2,3} = \frac{\rm Tr}{2} \pm \sqrt{\frac{\rm Tr^2}{4} + \rm DET_{neg}}. \tag{12}$$

In Eq. (12), the trace $\rm Tr = -(\alpha_r + \gamma) < 0$ with $\alpha_r = \alpha - \beta_2$ is computed from the non-singular submatrix L_2 of L_1 defined by

$$L_2 = \begin{pmatrix} -\alpha_r & \beta_1 \\ \alpha & -\gamma \end{pmatrix}. \tag{13}$$

Moreover, $\rm DET_{neg} = \alpha\beta_1 - \alpha_r\gamma$ is the negative of the determinant of L_2. Note that the eigenvalues $\lambda_{2,3}$ are real because we have $\rm Tr^2/4 + DET_{neg} = (\alpha_r - \gamma)^2/4 + \alpha\beta_1 \geq 0$. That is, for the model under consideration there are no complex eigenvalues. For (i) $\rm DET_{neg} < 0$, (ii) $\rm DET_{neg} = 0$, and (iii) $\rm DET_{neg} > 0$ we have (i) $\lambda_2 < 0 \wedge \lambda_3 < 0$, (ii) $\lambda_2 = 0 \wedge \lambda_3 < 0$, and (iii) $\lambda_2 > 0 \wedge \lambda_3 < 0$, respectively. Introducing the weighted contact rate

$$\beta_w = \beta_1 + \frac{\gamma}{\alpha}\beta_2 \tag{14}$$

the cases (i), (ii), (iii) correspond to $\beta_w < \gamma$, $\beta_w = \gamma$, and $\beta_w > \gamma$, respectively. The parameter β_w may be interpreted as bifurcation parameter. At the critical value $\beta_{w,\text{crit}} = \gamma$ the disease-free state becomes an unstable fixed point.

For $\beta_2 = 0$ (standard SEIR model) we simply have $\beta_w = \beta_1$ and the three cases (i), (ii), (iii) correspond to $\beta_1 < \gamma$, $\beta_1 = \gamma$, and $\beta_1 > \gamma$. The basic reproduction ratio for the standard SEIR model ($\beta_2 = 0$) reads [36] $R_{0,\text{SEIR}} = \beta_1/\gamma$. Therefore, for the standard SEIR model the cases (i), (ii), and (iii) refer to $R_{0,\text{SEIR}} < 1$, $R_{0,\text{SEIR}} = 1$, and $R_{0,\text{SEIR}} > 1$ consistent with the definition of the basic reproduction number given in Sec. 1. For the more general case with $\beta_2 > 0$, we have [35]

$$R_{0,\text{SEIR}} = \frac{\beta_2}{\alpha} + \frac{\beta_1}{\gamma} \Rightarrow R_{0,\text{SEIR}} = \frac{\beta_w}{\gamma}. \tag{15}$$

Therefore, the cases (i), (ii), and (iii) with $\beta_w < \gamma$, $\beta_w = \gamma$, and $\beta_w > \gamma$, correspond again to $R_{0,\text{SEIR}} < 1$, $R_{0,\text{SEIR}} = 1$, and $R_{0,\text{SEIR}} > 1$.

The matrix L_1 exhibits the eigenvector $\mathbf{v}_1 = (1, 0, 0)$ with $\lambda_1 = 0$. In the orthogonal subspace spanned by E and I we construct two basis vectors related to the remaining eigenvalues $\lambda_{2,3}$. These two basis vectors can be computed from L_2. Explicitly, all three basis vectors $\mathbf{v}_j$ read [17]

$$\mathbf{v}_1 = \begin{pmatrix} 1 \\ 0 \\ 0 \end{pmatrix}, \quad \mathbf{v}_2 = \frac{1}{W_2}\begin{pmatrix} 0 \\ \beta_1 \\ \lambda_2 + \alpha_r \end{pmatrix}, \quad \mathbf{v}_3 = \frac{1}{W_3}\begin{pmatrix} 0 \\ \beta_1 \\ \lambda_3 + \alpha_r \end{pmatrix} \tag{16}$$

with $W_j = \sqrt{\beta_1^2 + (\lambda_j + \alpha_r)^2}$. The vectors are linearly independent. Therefore, we can proceed with the next step. As in Sec. 2.1, following the concepts of synergetics [13, 14, 22], the state $\mathbf{X}(t)$ is expressed in terms of the fixed point $\mathbf{X}_{\text{st}}$, the basis vectors $\mathbf{v}_j$ and the amplitudes $A_j(t)$ with $j = 1, 2, 3$ like

$$\mathbf{X}(t) = \mathbf{X}_{\text{st}} + A_1(t)\mathbf{v}_1 + A_2(t)\mathbf{v}_2 + A_3(t)\mathbf{v}_3. \tag{17}$$

Equation (17) describes for the SEIR model the mapping $(A_1, A_2, A_3) \to (S, E, I)$ from the amplitude space to the original state space. In components, the mapping reads

$$S = N + A_1, \quad E = A_2 v_{2,E} + A_3 v_{3,E}, \quad I = A_2 v_{2,I} + A_3 v_{3,I}. \tag{18}$$

The inverse mapping $(S, E, I) \to (A_1, A_2, A_3)$ reads

$$A_1 = S - N, \quad A_2 = (v_{3,I}E - v_{3,E}I)/B, \quad A_3 = (v_{2,E}I - v_{2,I}E)/B \tag{19}$$

with $\mathbf{v}_j = (0, v_{j,E}, v_{j,I})$ for $j = 2, 3$ defined by Eq. (16). In Eq. (19), the parameter B is given by the determinant of the basis vector matrix $M = (\mathbf{v}_1\ \mathbf{v}_2\ \mathbf{v}_3)$ and reads $B = \beta_1(\lambda_3 - \lambda_2)/(W_2 W_3) < 0$ for $\alpha > 0, \beta_1 > 0$. The vectors $\mathbf{v}_2$ and $\mathbf{v}_3$ specify directions in the $E - I$ space, while the amplitudes A_2 and A_3 describe how the infection dynamics evolves along those directions. The amplitude A_1 with $S = N + A_1 \Longleftrightarrow A_1 = S - N$, see Eqs. (18) and (19), measures the deviation of

S from the fixed point value N. Using the evolution equations of the SEIR model (i.e. Eq. (10)) and the mappings (18) and (19), the evolution equations for the amplitudes A_1, A_2, and A_3 can be derived in the same way as in Sec. 2. In particular, using Eq. (18) the nonlinear term $S(\beta_1 I + \beta_2 E)/N$ occurring in Eq. (10) can be expressed in terms of the amplitudes like

$$\frac{S}{N}(\beta_1 I + \beta_2 E) = \frac{N + A_1}{N}([\beta_1 v_{2,I} + \beta_2 v_{2,E}]A_2 + [\beta_1 v_{3,I} + \beta_2 v_{3,E}]A_3), \tag{20}$$

such that

$$\frac{S}{N}(\beta_1 I + \beta_2 E) = (N + A_1)p_{\text{SEIR}}(A_2, A_3) \tag{21}$$

with

$$\begin{aligned} p_{\text{SEIR}} &= h_2 A_2 + h_3 A_3, \quad h_2 = (\beta_1 v_{2,I} + \beta_2 v_{2,E})/N, \\ h_3 &= (\beta_1 v_{3,I} + \beta_2 v_{3,E})/N. \end{aligned} \tag{22}$$

Moreover, from $A_1 = S - N$ and $\mathrm{d}S/\mathrm{d}t = -S(\beta_1 I + \beta_2 E)/N$ it follows that $\mathrm{d}A_1/\mathrm{d}t = \mathrm{d}S/\mathrm{d}t = -S(\beta_1 I + \beta_2 E)/N$. Using Eq. (21), this yields the evolution equation $\mathrm{d}A_1/\mathrm{d}t = -(N + A_1)p_{\text{SEIR}}(A_2, A_3)$ for the amplitude A_1. Likewise, the relations for A_2 and A_3 in Eq. (19) can be differentiated with respect to time t and the expressions $\mathrm{d}E/\mathrm{d}t$ and $\mathrm{d}I/\mathrm{d}t$ can be eliminated with the help of Eq. (10). In doing so, evolution equations for A_2 and A_3 can be obtained that depend on the state variables S, E, I. The linear parts must satisfy the eigenvalue equations $\mathrm{d}A_j/\mathrm{d}t = \lambda_j A_j$ for $j = 2, 3$. The nonlinear parts of those evolution equations can be transformed using Eq. (21) again. In summary, a detailed calculation yields [17]:

$$\begin{aligned} \frac{\mathrm{d}}{\mathrm{d}t}A_1 &= -(N + A_1)p_{\text{SEIR}}(A_2, A_3), \\ \frac{\mathrm{d}}{\mathrm{d}t}A_2 &= \lambda_2 A_2 + \frac{v_{3,I}}{B}A_1 p_{\text{SEIR}}(A_2, A_3), \\ \frac{\mathrm{d}}{\mathrm{d}t}A_3 &= \lambda_3 A_3 - \frac{v_{2,I}}{B}A_1 p_{\text{SEIR}}(A_2, A_3). \end{aligned} \tag{23}$$

We would like to point out that the evolution equation for A_2 does not contain a term linear in A_3. Likewise, the evolution equation for A_3 does not contain a term linear in A_2. This is due to the aforementioned feature that the linear dynamics in the $E - I$ subplane satisfies the eigenvalue equations $\mathrm{d}A_j/\mathrm{d}t = \lambda_j A_j$ for $j = 2, 3$. A more detailed discussion of this issue can be found in Appendix A.

A detailed discussion of the dynamics of the amplitude equation model for $\beta_2 = 0$ can be found in [17]. Let us briefly discuss the signs of the amplitudes for the case $\beta_2 > 0$. From $S = N + A_1$ it follows that $A_1 \leq 0$. Furthermore, from Eq. (19) we have $A_2 = (v_{3,I}E - v_{3,E}I)/B$. Note that $v_{3,I} = (\lambda_3 + \alpha_r)/W_3 < 0$, $v_{3,E} = \beta_1/W_3 > 0$, and $B < 0$. Therefore, for any $E, I \geq 0$ we have $A_2 \geq 0$. In contrast, numerical simulations show that A_3 can assume both positive and negative values.

Note that the SEIR model given Eq. (10) for the triplet (S, E, I) and the amplitude equation model (23) for the amplitude triplet (A_1, A_2, A_3) describe the same dynamical system. Both models yield identical solutions if the mappings $(A_1, A_2, A_3) \to (S, E, I)$ and $(S, E, I) \to (A_1, A_2, A_3)$ defined by Eqs. (17) and (19) are used. However, the amplitude equations in combination with the order parameter allow us to illustrate that the interacting epidemiological compartments during the COVID-19 produce a simple amplitude dynamics along the order parameter.

Pang *et al.* [35] studied the infection dynamics in Wuhan city, China, during the period from December 2019 to March 2020 with the help of the SEIR model (10). To this end, they evaluated COVID-19 case data from December 31, 2019 to March 25, 2020. In line with the COVID-19 situation report of WHO [46], Pang *et al.* [35] considered December 31, 2020, as the date on which the first COVID-19 case in Wuhan city was reported. Pang *et al.* [35] distinguished between three stages. Stage 1 covered the period from December 2019 (the month in which presumably COVID-19 emerged in Wuhan city) to January 23, 2020, and included the aforementioned December 31 marker. On January 23, the city of Wuhan was put on lockdown [1, 28, 32, 35]. Stage 2 was given by the period from January 24 to February 11, 2020. During that period certain government interventions were put in place such as the aforementioned lockdown (i.e. individuals could not enter or leave the city). However, a strict quarantine policy was not established during that period. According to Pang *et al.* [35], around February 11, more rigorous intervention measures were put in place. In particular, individuals with COVID-19 symptoms and contact persons of such individuals were quarantined in hospitals and other facilities (see also [21]). Therefore, Pang *et al.* [35] introduced as a third stage the period from February 12 to March 25, 2020.

In this study, only the second stage for which COVID-19 case data can be found in the English literature [37] will be considered. Although in the second stage, the COVID-19 epidemic was already underway, the stage can still be regarded as an initial stage in which the infection dynamics converged away from the disease-free fixed point $\mathbf{X}_{st} = (N, 0, 0)$ (see also [17]). In this case, the considerations made above apply. In contrast, in stage 3 the number of daily new infections in the population of Wuhan city decayed [35], which is consistent with the notion that in stage 3 the infection dynamics converged towards a newly established stable fixed point with zero infected cases [17, 35]. For stage 2, Pang *et al.* [35] obtained $\beta_1 = 0.26/\text{d}$, $\beta_2 = 0.13/\text{d}$, $\alpha = 0.17/\text{d}$, $\gamma = 0.26/\text{d}$, and $N = 9 \cdot 10^6$.

The SEIR model used by Pang *et al.* [35] differed from Eq. (22) by including demographic terms reflecting birth processes and death processes due to causes other than COVID-19. On the one hand, as argued by Frank [17] those terms should have a negligibly small impact on the infection dynamics on the relatively short period of stage 2. On the other hand, when simulating the SEIR model defined by Eq. (23) that does not take into account for demographic variations in population size using the

Table 2. Adjusted contact rate parameters β_1 and β_2 and parameters of the SEIR order parameter approach applied to the COVID-19 disease outbreak in Wuhan city, China, during the period from January 24 to February 11, 2020. The weighted contact rate is listed as well.

Adjusted parameters			Order parameter vector			Speed parameters	
β_1	β_2	β_w	$v_{2,E}$	$v_{2,I}$	θ	λ_2	τ
0.25/d	0.12/d	0.45/d	0.90	0.45	26.4°	0.08/d	12.4d

aforementioned parameters reported by Pang *et al.* [35], we found that the numerical solution systematically overestimated the reported case data. Since for the purpose of this study, it is sufficient to find some reasonably good fit of the model solution to the data, we adjusted (or fine-tuned) only the two contact rates β_1 and β_2. We multiplied both rates with the same factor f and varied f in the range from 90% to 100%. The predicted trajectory $R(t)$ was compared with the reported cumulative COVID-19 cases and the root-mean-square error was computed. The error was minimal for $f = 97\%$. The adjusted parameters β_1 and β_2 are shown in Table 2. Eventually, the following quantities were computed: the weighted contact rate β_w, the order parameter characteristics $v_{2,E}$, $v_{2,I}$, θ, and the speed parameters λ_2 and τ, see Table 2. The order parameter angle θ was defined with respect to the E-axis like $v_{2,E} = \cos(\theta)$. Accordingly, an angle of $\theta = 0°$ describes an order parameter $\mathbf{v}_2$ pointing in the direction of the E-axis. $\theta = 90°$ describes $\mathbf{v}_2$ pointing in the direction of the I-axis. For $0° < \theta < 90°$ the vector $\mathbf{v}_2$ points in some direction in the first quadrant of the $E - I$ plane.

We found that $\beta_w = 0.45/\text{d}$ was larger than γ, which provides evidence that the infection-free state of the population of Wuhan city was unstable during the period of interest. The component $v_{2,E}$ was larger than $v_{2,I}$ suggesting that the increase of exposed individuals was faster as compared to the increase of the infected individuals. The relation $v_{2,E} > v_{2,I} > 0$ implies an order parameter angle $\theta < 45°$. In fact, our analysis suggests that the SEIR order parameter of Wuhan city had an angle of 26.4°. As expected, the eigenvalue $\lambda_2 = 0.08/\text{d}$ was positive ($\Rightarrow$ instability of $\mathbf{X}_{\text{st}} = (N, 0, 0)$). The time constant τ was 12.4 days. That is, A_2 increased (at least initially) by a factor $e \approx 2.72$ in 12.4 days. For λ_3 we found $\lambda_3 = -0.39/\text{d}$. Comparing $|\lambda_3|$ with λ_2 we found that they differed by a factor 5. Consequently, the notion that A_2 and A_3 evolved on slow and fast time scales, respectively, was supported.

We solved the SEIR model (10) numerically using the adjusted β_1 and β_2 parameters shown in Table 2 for the period from January 23 to February 11, 2020 (with January 23 the last day of stage 1 used as initial time point, see [35]). Initial values for the simulation were taken from Pang *et al.* [35]: $E(0) = 3251$, $I(0) = 2731$, $R(0) = 354$, and $S(0) = N - E(0) - I(0) - R(0)$. Panel (A) of Fig. 3 shows the trajectories E (dashed line) and I (solid line) thus obtained. Panel (B) shows the number of individuals in the compartment R over time as obtained from the simulation (solid black line). The number is supposed to reflect the reported cumulative

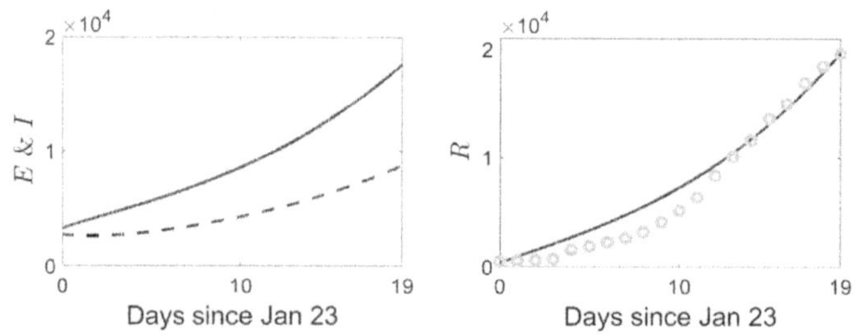

(A) (B)

Fig. 3. State space and amplitude space description of the COVID-19 outbreak in Wuhan city, China (part 1). Panels (A) and (B): State space description in terms of exposed individuals E (solid line; panel (A)), infected individuals I (dashed line; panel (A)), and removed (i.e. COVID-19 diagnosed) individuals R (solid black line; panel (B)) as predicted by the SEIR model (10) over time. Actual cumulative COVID-19 diagnosed cases R are shown as well (gray circles, panel (B)).

COVID-19 cases. The gray circles show reported cumulative COVID-19 cases taken from [37] starting with the 496 case number on January 23 discussed by Pang *et al.* [35]. Our numerical solution $R(t)$ fitted well the reported data. In particular, the SEIR model prediction $R(t)$ shown in panel (B) reproduced overall the prediction obtained by Pang *et al.* [35].

The amplitude equation model (23) was solved numerically for initial values $A_1(0)$, $A_2(0)$, and $A_3(0)$ obtained from Eq. (19) and $S(0)$, $E(0)$, and $I(0)$. Panel (A) of Fig. 4 shows A_2 (solid line) and A_3 (dashed line). As expected, A_2 increased and varied to a large degree as compared to A_3. Panel (B) sketches the eigenvectors $\mathbf{v}_2$ and $\mathbf{v}_3$ in the $E - I$ plane centered at the fixed point coordinates $E = I = 0$ (compare with Table 2 for the vector components $\mathbf{v}_2$). In particular, panel (B) demonstrates that $\mathbf{v}_2$ points at an angle of about $26.4°$ in the $E - I$ plane. The dotted line is the unit circle. Note that $\mathbf{v}_1$ is not shown in panel (B). The eigenvector $\mathbf{v}_1$ would be orthogonal to the $E - I$ plane (i.e. would point out of the plane). The eigenvectors $\mathbf{v}_2$ and $\mathbf{v}_3$ constitute a skewed coordinate system in the $E - I$ plane as shown in panel (C) (dotted lines). Distances along the $\mathbf{v}_2$ and $\mathbf{v}_3$ axes are measured by A_2 and A_3, respectively. Panel (C) shows the exposed $E(t)$ and infected $I(t)$ curves shown in panel (A) as a single phase space trajectory (solid thick black line). The square indicates the initial value of the simulation. Clearly, the phase space trajectory follows the axis $\mathbf{v}_2$. Consequently, panel (C) illustrates that the COVID-19 outbreak in Wuhan city — when described from the perspective of a SEIR model — followed the SEIR order parameter $\mathbf{v}_2$ during the period from January 23 to February 11.

In closing this section, let us briefly compare the present results with the results derived earlier by Frank [17] for the COVID-19 outbreak in Wuhan city. In the earlier study by Frank [17], all three stages introduced by Pang *et al.* [35] were addressed in the sense that the eigenvector $\mathbf{v}_2$ and eigenvalue λ_2 were determined for

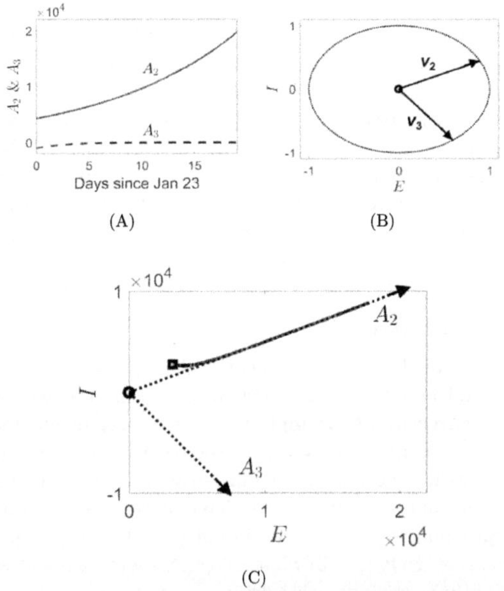

Fig. 4. State space and amplitude space description of the COVID-19 outbreak in Wuhan city, China (part 2). Panel (A): Amplitude equations description in terms of A_2 (solid line) and A_3 (dashed line) as functions of time. Panel (B): Eigenvectors $\mathbf{v}_2$ and $\mathbf{v}_3$ in the $E - I$ space located at the infection-free fixed point. Panel (C): Evolution of the epidemics along the order parameter vector $\mathbf{v}_2$. Dotted lines describe the axes spanned by $\mathbf{v}_2$ and $\mathbf{v}_3$ with coordinates A_2 and A_3, respectively. The simulated phase space trajectory $E(t)$ versus $I(t)$ is shown (thick black line) and evolves along the order parameter vector $\mathbf{v}_2$.

all three stages. In contrast, in this study only stage 2 was analyzed. This study focused on the description of stage 2 because while for stage 2, data from Wuhan city can be found in the English literature, for stages 1 and 3 such data are more difficult to retrieve. Moreover, as reviewed above, stage 3 reflects an epidemiological dynamics that converges towards a newly established fixed point with zero daily new infections. In contrast, the primary focus of the present work is on COVID-19 outbreaks that are characterized by an infection dynamics that diverges away from a fixed point of zero infected individuals. As far as the results presented in Table 2 are concerned, some of those results have in fact been computed in the earlier work by Frank [17]. However, in Frank [17] the contact rate parameters β_1 and β_2 reported by Pang *et al.* [35] rather than the adjusted parameters β_1 and β_2 presented in Table 2 were used. Consequently, the order parameter coefficients shown in Table 2 differ slightly from those reported in Frank [17]. Frank [17] obtained (when rounding to 2 digits) $v_{2,E} = 0.90$ as in this study but $v_{2,I}$ was slightly smaller namely $v_{2,I} = 0.44$ (as compared to $v_{2,I} = 0.45$ in this study). Consequently, the order parameter angle was found to be slightly smaller than the one reported here: Frank [17] obtained an angle of $\theta = 26.0°$ (as compared to $\theta = 26.4°$ in this study). Since the eigenvalue λ_2 also depends on the contact rate parameters, the value reported in Table 2 slightly differs from the one reported earlier in [17], namely, $\lambda_2 = 0.09/\mathrm{d}$ (as compared to $\lambda = 0.08/\mathrm{d}$ in this study). Related to that in [17] the time constant was $\tau = 11.6\mathrm{d}$ (as compared to $\tau = 12.4\mathrm{d}$ in this study). Overall, the quantitative differences by using the adjusted contact rate parameters are relatively small. Importantly, they do not qualitatively change the interpretation of COVID-19 dynamics that took place in Wuhan city in terms of a dynamics that evolved along an appropriately defined order parameter.

2.3. *Further applications*

The amplitude equation description of COVID-19 outbreaks may be used to compare outbreaks across different countries or regions or within a given country or region across different occurrences. For example, the first-wave epidemics may be compared across a selection of countries of interest or the second-wave epidemic of a country may be compared with the corresponding first-wave epidemic. In fact, a comparison of the positive eigenvalues that putatively have driven the first-wave epidemics in several European countries during the Spring period of the year 2020 has been presented in an earlier study [16]. In what follows, a short comment will be presented on two subsequently occurring outbreaks in a particular Chinese province during Spring 2020. The outbreaks took place in the Chinese province Heilongjiang with a population of 38 million and has been study by Sun and Wang [39]. Accordingly, the first outbreak occurred from January 23 to March 11, 2020. The second outbreak took place on April 9, that is, almost one month after the aforementioned March 11 date. Sun and Wang [39] performed a model-based analysis of both outbreaks using a SEIR-like model. The model involves as a 5th variable the compartment of asymptomatic infectious individuals. Their model-based analysis suggests that the first outbreak was induced by a relatively large number of infectious individuals who had entered the providence from other provinces (i.e. imported cases). The intervention measures implemented in Heilongjiang province were sufficient to prevent a major outbreak such that in the 38 million province only about 500 confirmed COVID-19 cases were reported during the January-to-March period. In contrast, the model-based analysis by Sun and Wang [39] suggests that the second outbreak on April 9 took place in an epidemiological situation with relatively weak interventions measures at the time of the outbreak. Sun and Wang [39] concluded that the contact rates were 7 times higher during the second outbreak presumably due to the relaxation of preventative regulations. Nevertheless, as compared to the first outbreak, the second outbreak involved a relatively small number of infected people (less than 50) [39].

Let us briefly comment on that study using the concepts of eigenvalues and eigenvectors. Eliminating the variable of the recovered individuals (just as in

Sec. 2.2), the five-variable model reduces to a four-variable model. Just as for the SIR and SEIR models presented above, the disease-free fixed point exhibits an eigenvalue equal to zero. At issue is to determine the three remaining eigenvalues for the two aforementioned outbreaks. Linearizing the model presented by Sun and Wang [39] at the disease-free fixed point and using the parameter values presented in their study, the eigenvalue spectrums for the two outbreaks can be obtained. The result is shown in Fig. 5 (panel (A)). As expected, for the first outbreak all eigenvalues were negative, indicating that the outbreak took place under relatively strong intervention measures. In contrast, the eigenvalue spectrum of the second outbreak involves a positive eigenvalue. Accordingly, the disease-free fixed point was unstable consistent with the assumed relatively high contact rates. When computing the unstable eigenvector, we found that the vector coefficient of the asymptomatic infectious

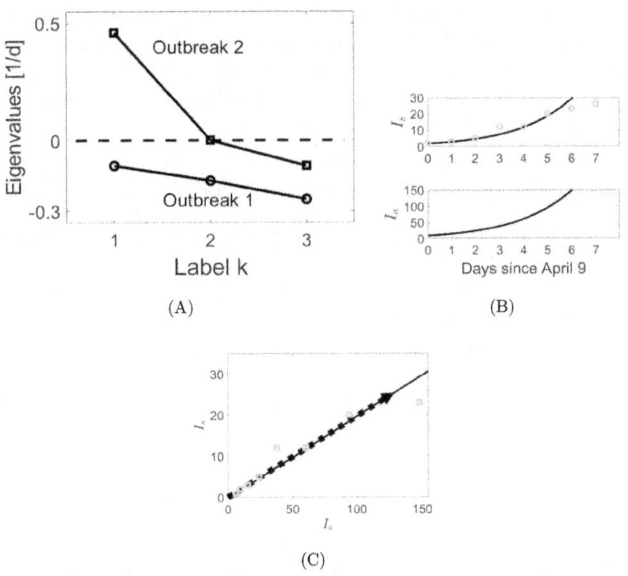

Fig. 5. Characterization of two subsequent COVID-19 outbreaks in Heilongjiang providence, China, using a generalized SEIR model developed by Sun and Wang [39]. Panels (A): The three relevant eigenvalues of the model for the first (circles connected by a solid line) and second (squares connected by a solid line) outbreaks. Panel (B): Evolution of the number of symptomatic, diagnosed individuals I_s and asymptomatic individuals I_a during the second outbreak (solid lines) as predicted by the model. Reported, active COVID-19 cases are shown as well (gray circles). Panel (C): Dynamics of the second outbreak along the unstable eigenvector related to the positive eigenvalue shown in panel (A). The dotted thick black line denotes the unstable eigenvector in the plane spanned by I_a and I_s (the eigenvector was magnified to improve visibility). The simulated phase space trajectory I_s versus I_a (thin black line) and a semi-empirical trajectory (gray squares) are shown as well.

compartment I_a was about 5 times larger than the coefficient of the symptomatic infectious compartment I_s. Panel (B) shows a simulation of the model by Sun and Wang [39] for the aforementioned assumed 7 times higher contact rates. The simulation covers the first week of the second outbreak. By visual inspection, the I_a cases are about 5 times higher than the I_s cases, as expected in view of the aforementioned eigenvector coefficients. In the model by Sun and Wang [39] the diagnosed, active COVID-19 cases correspond to the variable I_s. These COVID-19 cases (as reported in [39]) are shown in panel (B) as well (gray circles). Panel (C) present the unstable eigenvector (dotted black thick line) in the two-dimensional space spanned by the variables I_a and I_s. The phase trajectory (solid black thin line) given by the graphs I_s versus I_a (as presented in panel (B)) is shown there as well and follows the eigenvector. Finally, in panel (C), a semi-empirical phase trajectory is shown (gray squares) that is composed of I_s values given in terms of the empirical data (gray circles shown in panel (B)) and I_a values taken from the model simulation (i.e. the graph I_a shown in the bottom subpanel of panel (B)). While a detailed analysis of the two COVID-19 outbreaks in the province Heilongjiang province as studied by Sun and Wang [39] is beyond the scope of this paper, the comments made so far on that subject and Fig. 5 illustrate that comparing COVID-19 outbreaks is a promising application of the method discussed in Secs. 2.1 and 2.2.

3. Conclusions

This study confirms that COVID-19 outbreaks (and outbreaks of infectious diseases in general) belong to the broad class of bifurcation phenomena and can be treated on an equal footing with bifurcation phenomena in other disciplines. The study supports previously reported results [17] by showing that not only within a SEIR modeling framework but also within the more fundamental SIR modeling framework order parameters can be identified that characterize COVID-19 outbreaks in given populations. Importantly, this work supports explicitly with the help of the applications presented in Sec. 2 the claim that COVID-19 outbreaks evolve along SIR and SEIR order parameters. We conclude that the comprehensive repertoire of tools and methods developed to study self-organization phenomena can readily be applied to study infectious diseases, in general, and COVID-19 outbreaks, in particular. We speculate that the aforementioned repertoire of tools and methods borrowed from the theory of self-organization can also be used to study and improve the effectiveness of intervention measures that aim to reduce the spread of COVID-19. In fact, the impact of intervention measures on the spread of COVID-19 has been extensively discussed in the literature. To this end, frequently, effective reproduction numbers [27] as continuous measures to quantify the success and effectiveness of interventions have been determined. Quantitatively, if they drop below the value of the basic reproduction number this can be taken as evidence for some partial success. Qualitatively, a drop of an effective reproduction number below the critical value 1 is desirable, which indicates that the COVID-19 epidemic in the respective population is about to subside (see also Sec. 1). For example, You *et al.* [50] reported for several Chinese provinces that the effective reproduction number dropped below the basic production number presumable as a result of intervention measures. The studies by Ivorra *et al.* [24] and Pang *et al.* [35] suggest that for China as a whole and for Wuhan city specifically government interventions eventually resulted in a drop of the effective reproduction number below the critical value of 1. Frank [17] suggested that the drop of the effective reproduction number below 1 was related to the stabilization of the fixed point with zero infected individuals in terms of a switch from a positive eigenvalue λ_2 to a negative value. In this context, a recent study on 20 European countries has supported the idea that from the interdisciplinary perspective of synergetics and, in general, from the bird-eye perspective of self-organization, COVID-19 interventions switch eigenvalues of human (physical) systems from positive to negative values [16]. On the contrary, relaxing intervention measures can destabilize the disease-free state (fixed point) as suggested in Sec. 2.3 such that at least one eigenvalue becomes positive. Such a destabilization, in turn, may lead to a relapse (i.e. second outbreak).

The complex systems perspective worked out explicitly in Secs. 2.1 and 2.2 for SIR and SEIR models, respectively, is not limited in its application to those two models. In the epidemiological literature various models that involve more than 4 compartments and go beyond the SIR and SEIR modeling framework are available [36]. In fact, the aforementioned study by Ivorra *et al.* [24] and the study by Sun and Wang [39] (see Sec. 2.3) were based on such more comprehensive models. In this context, the question of model selection arises. Since a general discussion of this topic is beyond the scope of this paper and would be worth to be pursued in its own merit, at this stage, only a few comments will be made. As such, several models may be applied to a given COVID-19 data set in order to determine the model that produces the best fit. However, in the context of the research on COVID-19 epidemics, frequently, models have been selected to meet certain study objectives. For example, the study by Pang *et al.* [35] addressed in Sec. 2.2 used the SEIR model to describe the epidemics in Wuhan in order to determine (among other objectives) the role of the class of exposed individuals during the outbreak. In contrast, for the goals of the study by Fanelli and Piazza [12] discussed in Sec. 2.1 a more coarse grained perspective was sufficient such that the SIR model could be used rather than the SEIR model. Ivorra *et al.* [24] supplemented the SEIR model with several compartments that described different types of diagnosed, quarantined, and hospitalized individuals. With the help of those compartments, Ivorra *et al.* [24] could estimate the effective contact rates β of those individuals in those compartments of diagnosed COVID-19 cases. They could compare those rates with the effective contact rates of non-diagnosed individuals. These examples illustrate cases in which models have been used and developed in line with study goals.

Overall, the results presented in Sec. 2 suggest that in order to fight COVID-19 it is not only important to understand the peculiarities of the COVID-19 pandemic but it is also important to acknowledge the general mechanism that determines the outbreaks of infectious diseases, in general, and COVID-19, in particular. Accordingly, out of the various interactions between individuals of different classes, a relatively simple amplitude dynamics forms that underlies such disease outbreaks. This kind of "reduction of complexity" should occur irrespective of the explicit modeling framework (SIR, SEIR, etc.) used to study a given COVID-19 outbreak.

Data Availability

Data used in Secs. 2.1 and 2.2 are publicly available from the website [2, 8]. Data used in Sec. 2.3 can be found in [39].

Appendix A. Diagonal form of Eigenvalue Equations

In order to derive the linear parts of the evolution equations for A_2 and A_3 occurring in Eq. (23), it is sufficient to study the dynamics in the $E-I$ subplane. In line with the discussion of the linearized dynamics characterized by the matrices L_1 and L_2, see Eqs. (11) and (13), the linearized dynamics in the $E-I$ subplane is determined by

$$\frac{\mathrm{d}}{\mathrm{d}t}\begin{pmatrix} E \\ I \end{pmatrix} = L_2 \begin{pmatrix} E \\ I \end{pmatrix}. \tag{A.1}$$

The eigenvalues $\lambda_{2,3}$ defined by Eq. (12) satisfy the eigenvalue equations

$$L_2 \bar{\mathbf{v}}_j = \lambda_j \bar{\mathbf{v}}_j \tag{A.2}$$

for $j=2,3$, where $\bar{\mathbf{v}}_j = (v_{j,E}, v_{j,I})$ denote the two-dimensional projections of $\mathbf{v}_j$ defined by Eq. (16) into the $E-I$ plane. From Eq. (18) it follows that the mapping

$$\begin{pmatrix} E \\ I \end{pmatrix} = \sum_{j=2,3} A_j \bar{\mathbf{v}}_j \tag{A.3}$$

holds that can be expressed with the help of the matrix M like

$$\begin{pmatrix} E \\ I \end{pmatrix} = M \begin{pmatrix} A_2 \\ A_3 \end{pmatrix}, \quad M = (\,\bar{\mathbf{v}}_2 \;\; \bar{\mathbf{v}}_3\,). \tag{A.4}$$

In this context, $\bar{\mathbf{v}}_j$ act as column vectors that constitute the columns of the matrix M. The matrix M exhibits an inverse matrix M^{-1} composed of the row vectors $\mathbf{w}_2$ and $\mathbf{w}_3$ such that

$$M^{-1} = \begin{pmatrix} \mathbf{w}_2 \\ \mathbf{w}_3 \end{pmatrix}, \quad M^{-1}M = \begin{pmatrix} 1 & 0 \\ 0 & 1 \end{pmatrix} \Rightarrow \mathbf{w}_i\,\bar{\mathbf{v}}_k = \delta_{ik}, \tag{A.5}$$

where δ_{ik} is the Kronecker symbol. Using M^{-1}, the mapping (A.4) can be inverted and reads

$$\begin{pmatrix} A_2 \\ A_3 \end{pmatrix} = M^{-1} \begin{pmatrix} E \\ I \end{pmatrix}. \tag{A.6}$$

In fact, Eq. (A.6) just corresponds to the two-dimensional version of Eq. (19). This also implies that the matrix M^{-1} and its coefficients (or row vectors $\mathbf{w}_j$) are defined explicitly in Eq. (19). Applying M^{-1} to the left-hand side of Eq. (A.1), we obtain

$$\frac{\mathrm{d}}{\mathrm{d}t} M^{-1} \begin{pmatrix} E \\ I \end{pmatrix} = \frac{\mathrm{d}}{\mathrm{d}t} \begin{pmatrix} A_2 \\ A_3 \end{pmatrix}. \tag{A.7}$$

Applying M^{-1} to the right-hand side of Eq. (A.1) and using Eq. (A.4), we obtain

$$M^{-1} L_2 \begin{pmatrix} E \\ I \end{pmatrix} = \underbrace{M^{-1} L_2 M}_{D} \begin{pmatrix} A_2 \\ A_3 \end{pmatrix}. \tag{A.8}$$

As indicated, the product of the three matrices yields a diagonal matrix D. Explicitly, D reads

$$\begin{aligned} L_2 M &= L_2 (\, \bar{\mathbf{v}}_2 \quad \bar{\mathbf{v}}_3 \,) = (\, \lambda_2 \bar{\mathbf{v}}_2 \quad \lambda_3 \bar{\mathbf{v}}_3 \,) \Rightarrow D = M^{-1} L_2 M \\ &= \begin{pmatrix} \mathbf{w}_2 \\ \mathbf{w}_3 \end{pmatrix} (\, \lambda_2 \bar{\mathbf{v}}_2 \quad \lambda_3 \bar{\mathbf{v}}_3 \,) = \begin{pmatrix} \lambda_2 & 0 \\ 0 & \lambda_3 \end{pmatrix}, \end{aligned} \tag{A.9}$$

as expected. That is, the matrix M is the diagonalization matrix that transforms L_2 into diagonal form. Finally, from Eqs. (A.7)–(A.9) it follows that

$$\frac{\mathrm{d}}{\mathrm{d}t} \begin{pmatrix} A_2 \\ A_3 \end{pmatrix} = \begin{pmatrix} \lambda_2 & 0 \\ 0 & \lambda_3 \end{pmatrix} \begin{pmatrix} A_2 \\ A_3 \end{pmatrix}. \tag{A.10}$$

As can be seen from Eq. (A.10), the evolution equation for A_2 does not contain an A_3 term. Likewise, the evolution equation for A_3 does not contain an A_2 term. Consequently, in components, the linear parts of the amplitude equations read $\mathrm{d}A_2/\mathrm{d}t = \lambda_2 A_2$ and $\mathrm{d}A_3/\mathrm{d}t = \lambda_3 A_3$.

References

[1] CNA Asia, China halts flights and trains out of Wuhan as WHO extends talks, Available at http://www.channelnewsasia.com/news/asia/wuhan-virus-quarantine-city-flights-trains-china-12306684.

[2] COVID-19 Tracker, Timeline data from Johns Hopkins Center for Systems Science and Engineering, Available at https://vac-lshtml.shinyapps.io/ncov_tracker.

[3] Croccolo, F. and Roman, H. E., Spreading of infections on random graphs: A percolation-type model for COVID-19, *Chaos Solitons Fractals* **139** (2020) 110077.

[4] Cross, M. C. and Hohenberg, P. C., Pattern formation outside of equilibrium, *Rev. Mod. Phys.* **65** (1993) 851–1112.

[5] Diekmann, O. and Heesterbeek, J. A. P., *Mathematical Epidemiology of Infectious Diseases* (John Wiley and Son, Chichester, 2000).

[6] Diekmann, O., Heesterbeek, J. A. P. and Metz, J. A. J., On the definition and the computation of the basic reproduction ratio R_0 in models for infectious diseases in heterogeneous populations, *Math. Biol.* **28** (1990) 365–382.

[7] Ding, Y. and Gao, L., An evaluation of COVID-19 in Italy: A data-driven modeling analysis, *Infect. Dis. Model.* **5** (2020) 495–501.

[8] Dong, E., Du, H. and Gardner, L., An interactive web-based dashboard to track COVID-19 in real time, *Lancet Infect. Dis.* **20** (2020) 533–534.
[9] Dutt, A. K., Turing pattern amplitude equations for a model glycolytic reaction–diffusion system, *J. Math. Chem.* **48** (2010) 841–855.
[10] Dutt, A. K., Chloride ion inhibition, stirring and temperature effects in an ethylacetoacetat Briggs–Rauscher-oscillator, *J. Phys. Chem. B* **123** (2019) 3525–3534.
[11] Eguiluz, V. M., Herandez-Garcia, E. and Piro, O., Complex Ginzburg–Landau equation in the presence of walls and corners, *Phys. Rev. E* **64** (2001) 036205.
[12] Fanelli, D. and Piazza, F., Analysis and forecast of COVID-19 spreading in China, Italy, and France, *Chaos Solitons Fractals* **134** (2020) 109761.
[13] Frank, T., *Determinism and Self-organization of Human Perception and Performance* (Springer, Berlin, 2019).
[14] Frank, T. D., Formal derivation of Lotka–Volterra–Haken amplitude equations of task-related brain activity in multiple, consecutively preformed tasks, *Int. J. Bifurcation Chaos* **10** (2016) 1650164.
[15] Frank, T. D., Unstable modes and order parameters of bistable signaling pathways at saddle-node bifurcations: A theoretical study based on synergetics, *Adv. Math. Phys.* **2016** (2016) 8938970.
[16] Frank, T. D., COVID-19 interventions in some European countries induced bifurcations stabilizing low death states against high death states: An eigenvalue analysis based on the order parameter concept of synergetics, *Chaos Solitons Fractals* **140** (2020) 110194.
[17] Frank, T. D., COVID-19 order parameters and order parameter time constants of Italy and China: A modeling approach based on synergetics, *J. Biol. Syst.* **28** (2020) 589–608.
[18] Gambino, G., Lombardo, M. C., Rubino, G. and Sammartino, M., Pattern selection in the 2D FitzHugh–Nagumo model, *Ric. Mat.* **68** (2019) 535–549.
[19] Gambino, G., Lombardo, M. C., Sammartino, M. and Sciacca, V., Turing pattern formation in the Brusselator system with nonlinear diffusion, *Phys. Rev. E* **88** (2013) 042925.
[20] Gershenson, C., Trianni, V., Werfel, J. and Sayama, H., Self-organization and artificial life, *Artif. Life* **26** (2020) 391–408.
[21] Globaltimes, Wuhan imposes strict quarantine measures amid fight against the novel coronavirus, Available at http://www.globaltimes.cn/content/1178250.shtml.
[22] Haken, H., *Synergetics: An Introduction* (Springer, Berlin, 1977).
[23] Hethcote, H. W., The mathematics of infectious diseases, *SIAM Rev.* **42** (2000) 599–653.
[24] Ivorra, B., Ferrandez, M. R., Vela-Perez, M. and Ramos, A. M., Mathematical modeling of the spread of coronavirus disease 2019 (COVID-19) taking into account the undetected infections. the case of China, *Commun. Nonlinear Sci. Numer. Simul.* **88** (2020) 105303.
[25] Kermack, W. O. and McKendrick, A. G., A contribution to the mathematical theory of epidemics, *Proc. R. Soc. A* **115** (1927) 700–721.
[26] Li, J. and Cui, N., Dynamic analysis of an SEIR model with distinct incidence for exposed and infectives, *Sci. World J.* **2013** (2013) 871393.
[27] Lipsitch, M., Cohen, T., Cooper, B., Robins, J. M., Ma, S., James, L., Gopalakrishna, G., Chew, S. K., Tan, C. C., Samore, M. H., Fisman, D. and Murray, M., Transmission dynamics and control of severe acute respiration syndrome, *Science* **300** (2003) 1966–1970.
[28] Liu, M., Ning, J., Du, Y., Cao, J., Zhang, D., Wang, J. and Chen, M., Modelling the evolution trajectory of COVID-19 in Wuhan, China: Experience and suggestions, *Public Health* **183** (2020) 76–80.

[29] Ma, J., Estimating epidemic exponential growth rate and basic reproduction number, *Infect. Dis. Model.* **5** (2020) 129–141.
[30] Mukhopadhyay, B. and Bhattacharyya, R., Analysis of a spatially extended nonlinear SEIS epidemic model with distinct incidences for exposed and infectives, *Nonlinear Anal., Real World Appl.* **9** (2008) 585–598.
[31] Murray, J. D., *Mathematical Biology* (Springer, Berlin, 1993).
[32] New York Times, Scale of China's Wuhan shutdown is believed to be without precedent, Available at http://www.nytimes.com/202001/22/world/asia/coronavirus–quarantines–history.html.
[33] Nicolis, G., *Introduction to Nonlinear Sciences* (Cambridge University Press, Cambridge, 1995).
[34] Pan, Y., Zhang, D., Yang, P., Poon, L. M. and Wang, Q., Viral load of SARS-CoV-2 in clinical samples, *Lancet* **20** (2020) 411–412.
[35] Pang, L., Liu, S., Zhang, X., Tian, T. and Zhao, Z., Transmission dynamics and control strategies of COVID-19 in Wuhan, China, *J. Biol. Syst.* **28** (2020) 543–561.
[36] Rock, K., Brand, S., Moir, J. and Keeling, M. J., Dynamics of infectious diseases, *Rep. Prog. Phys.* **77** (2014) 026602.
[37] Shi, P., Dong, Y., Yan, H., Zhao, C., Li, X., Liu, W., He, M., Tang, S. and Xi, S., Impact of temperature on the dynamics of the COVID-19 outbreak in China, *Sci. Total Environ.* **728** (2020) 138890.
[38] Spiedel, L., Klemm, K., Eguiluz, V. M. and Masuda, N., Temporal interactions facilitate endemicity in the susceptible–infected–susceptible epidemic model, *New J. Phys.* **18** (2016) 073013.
[39] Sun, T. and Wang, Y., Modeling COVID-19 epidemic in Heilongjiang province, China, *Chaos Solitons Fractals* **138** (2020) 109949.
[40] Taboe, H. B., Salako, K. V., Tison, J. M., Ngonghala, C. N. and Kakai, R. G., Predicting COVID-19 spread in the face of control measures in West Africa, *Math. Biosci.* **328** (2020) 108431.
[41] Thurner, S., Klimek, P., and Hanel, R., A network-based explanation of why most COVID-19 infection curves are linear, *Proc. Natl. Acad. Sci. USA* **117** (2020) 22684–22689.
[42] van den Driessche, P. and Watmough, J., Reproduction numbers and sub-threshold endemic equilibria for compartment models of disease transmission, *Math. Biosci.* **180** (2002) 29–48.
[43] Wang, Z., Yu, Z., Tian, C., Rodriguez, L. C., Batista, L. S. and Zhao, B., Differential equation analysis of COVID-19, *Novel Tech. Nutr. Food Sci.* **5** (2020) 422–430.
[44] Wangping, J., Ke, H., Yang, S., Wenzhe, C., Shengshu, W., Shanshan, Y., Jianwei, W., Fuyin, K., Penggang, T., Jing, L., Miao, L. and Yao, H., Extended SIR prediction of the epidemics trend of COVID-19 in Italy and compared with Hunan, China, *Front. Med.* **7** (2020) 169.
[45] World Health Organization, WHO COVID-19 Weekly epidemiological update-17, Available at https://www.who.int/emergencies/diseases/novel-coronavirus-2019 (2020).
[46] World Health Organization, WHO novel coronavirus (2019-nCoV) situation report-1, Available at https://www.who.int/emergencies/diseases/novel-coronavirus-2019 (2020).
[47] World Health Organization, WHO COVID-19 weekly epidemiological update-19 January 2021, Available at https://www.who.int/emergencies/diseases/novel-coronavirus-2019 (2021).

[48] Wu, J. T., Leung, K. and Leung, G. M., Nowcasting and forecasting the potential domestic and international spread of the 2019-nCov outbreak originating in Wuhan, China: A modelling study, *Lancet* **395** (2020) 689–697.
[49] Yang, C. and Wang, J., A mathematical model for the novel coronavirus epidemic in Wuhan, China, *Math. Biosci. Eng.* **17** (2020) 2708–2724.
[50] You, C., Deng, Y., Hu, W., Sun, J., Lin, Q., Zhou, F., Peng, C. H., Zhang, Y., Chen, Z. and Zhou, X. H., Estimation of the time-varying reproduction number of COVID-19 outbreak in China, *Int. J. Hyg. Environ. Health* **228** (2020) 113555.
[51] Zhao, S., Lin, Q., Ran, J., Musa, S. S., Yang, G., Wang, W., Lou, Y., Gao, D., Yang, L., He, D. and Wang, M. H., Preliminary estimation of the basic reproduction number of novel coronavirus (2019-nCoV) in China, from 2019 to 2020: A data-driven analysis in the early phase of the outbreak, *Int. J. Infect. Dis.* **92** (2020) 214–217.
[52] Zhou, T., Liu, Q., Yang, Z., Liao, J., Yang, K., Bai, W., Lu, X. and Zhang, W., Preliminary prediction of the basic reproduction number of Wuhan novel coronavirus 2019-nCoV, *J. Evid. Based Med.* **13** (2020) 3–7.

PART 4
Models of Evolution

Chapter 24

Overview

A broader perspective. From a complex systems perspective, the topic of evolution cannot be reduced to biological systems. The principles and mechanisms of evolution, for instance, mutation, selection, recombination, and adaptation, are not restricted to biological species. Instead, they can be found on various organizational levels, notably in social systems. Cultural evolution, for example, is a relevant topic for *Advances in Complex Systems* with topical issues in 2012 and 2022.

The overarching concept of evolution advocated here has to encompass, on the one end, the evolution of the early universe and the cosmic evolution and, on the other end, the evolution of socio-technical and socio-economic systems. Thus, evolution is one of those fundamental concepts in science that have the potential to connect different disciplines. In turn, these disciplines contributed to a generalized theory of evolution. For instance, the thermodynamics of irreversible processes provides insights into the energetic and entropic conditions for structure formation and self-organization. Optimization procedures are formalized in operations research and computer science. Adaptation and learning processes play a role in machine learning and artificial intelligence. Innovation is a crucial topic in economics and management science.

Advances in Complex Systems has facilitated this overarching view of evolution from its beginning. Many publications address evolutionary concepts in combination with other modeling approaches, for instance, learning in agent-based models, cooperation and coordination in populations, and information processing in networks. Therefore, we will not discuss the best topical assignment for these papers. Instead, in the following, we highlight principal problems of evolution addressed from various scientific perspectives.

Dynamics versus evolution. The question of what distinguishes evolutionary dynamics from other system dynamics is difficult to answer. System dynamics model the trajectory of a system characterized by non-linear feedback and specific control variables. Thus, the system changes over time but does not evolve because the system dynamics itself have not changed. In evolutionary systems, on the other hand, we usually assume that the dynamics also evolve. Feedback from the system level to the interactions of the system elements requires them to change [1]. Hence, we have to model *coevolution* [10, 92, 94, 112]. The fact that agents have to adapt to the structures that emerge from their interactions is sometimes called second-order emergence.

Not all evolutionary models take this bigger perspective. In most cases, they are restricted to investigating particular evolutionary mechanisms, for instance, selection or replication.

Adaptation and learning. Evolution is an open-ended process that never settles down to an equilibrium state. For instance, ongoing environmental changes force the system to respond. This process is generally denoted as adaptation. Ecological systems may adapt to a rising temperature and economic systems to the depletion of a resource. They can undergo a regime shift or a phase transition once a critical value of some control variable is reached, as already discussed for system dynamics. If these changes are not exogenous but endogenous, i.e., caused by the agents, we are back at the problem of coevolution.

But even without this additional feedback, systems adapt in various ways. In biological systems, for instance, the interaction network can be rewired in a self-organized manner [6, 121]. In socio-economic systems, the adaptation process is sometimes guided by a centralized controlling instance, for example, the government or the board [36, 48].

If agents have a specific level of internal complexity, they can also adapt internally [44]. For instance, they can adopt a more beneficial strategy if their market environment has changed [57]. Instead of a sudden change, there can be a *gradual adaptation.* This is related to the concept of *learning* [3, 113]. Different from supervised learning, reinforcement learning allows agents to find the optimal parameters for their adaptation themselves [4, 21, 53, 86]. It also applies to agent-based models where computer simulations need to be calibrated against empirical data [47, 63].

Novelty. Another reason for the open-ended evolutionary dynamics is the spontaneous occurrence of novelties. These can result from *mutations*, i.e., random changes in existing system elements or their interactions [34, 65]. For instance, mutations in the genetic code may lead to new phenotypes. Also, new combinations of already existing system elements would generate novelty. Genetic algorithms, for example, use such mechanisms to generate new solutions that have to compete against established ones. In general, *recombination* requires some kind of *modularity* in the existing elements [87].

Innovation eventually introduces new elements to the system that have not been there before. These can be new rules, e.g., for coordination or division of labor, new ecological nices, new mechanisms for reproduction, new products, and technologies. The term radical innovation is used in management science to distinguish the latter from other types of novelty.

Innovation and adoption. It bears conceptual problems to model something that is not already there, not even known from the outset. While we cannot anticipate the result of an innovation process, we can still model its dynamics. In the simplest case, the spontaneous emergence of new elements can be due to fluctuations [23]. New elements can also be introduced by design, e.g., as a result of research and development activities of firms or based on user requests [108] or as cultural inventions [80]. If these innovations have some evolutionary advantage, such as a higher fitness, they may survive and even replicate [60].

Therefore, the second important problem regards the *adoption*, or the spread or diffusion, of innovations in a population, or a market [12, 110]. Models of innovation often consist of a *stochastic* process to introduce the innovation and subsequent dynamics of their dispersal, which both may occur at different time scales [91].

Replication, reproduction, specialization, and competition. Replication is a fundamental evolutionary process found on all levels of organizations. In microbiology, cells or genomic information replicate. In higher organisms, populations reproduce by having offspring. Also, products or technologies "replicate" if a larger fraction adopts them in a population. The dynamics can be described using autocatalytic or heterocatalytic processes, i.e., amplifications depending on the own, but also on other species [32]. These growth dynamics saturate because of limited resources. Hence, an accelerated growth phase is usually followed by a competition phase, in which only the "fittest" individuals or species survive.

This basic scenario is mitigated by the fact that species can specialize in occupying ecological or technological nices. Thus, instead of an outcome where the winner takes all, we find the coexistence o different species, technologies, or strategies.

Evolutionary optimization. From an evolutionary perspective, optimization can be modeled as a search for better solutions in a search space. It requires solving several problems: (i) Define the search space [23, 64]. Analogous to a phenotype space, the search space is seen as a *fitness* landscape, where every possible realization gets assigned a fitness value [5, 11]. It, of course, depends on how fitness is quantified.

(ii) Propose search algorithms that can efficiently explore high-dimensional and fragmented search spaces. Evolutionary algorithms combine different elements from biological evolution. Genetic algorithms, for example, use *recombination* to generate new solutions from previous ones [70]. They allow for large jumps in the search space. Ant colony optimization builds on a distributed exploration of the search space by many agents [2, 88]. Simulated annealing implements random mutations and accepts slight deteriorations to explore the search space [20].

Evolutionary game theory. Most papers about evolution in *Advances in Complex Systems* discuss models related to game theory. They use reproduction, competition, and selection as evolutionary elements. Agents have a set of strategies to choose from when interacting with other agents, mostly in bilateral encounters. Sometimes the strategies also evolve themselves [46].

Strategies differ from rules in that they are uncertain about the strategy chosen by the counterparty. But the payoff resulting from an interaction depends on the strategies of both partners. Strategies with a higher payoff have a higher replication rate and will establish themselves through competition with other strategies. On the other hand, rules describe how agents interact, for instance, in a classical 2-person game, in one-to-many interactions, or indirectly, via a common resource. One can also use time-series data to construct an interaction network [75]. Rules can evolve as well [68, 71].

Games can be defined differently and are not restricted to the classical game theoretical setup. For instance, chess games [46], card games [68], naming games [8], online games [98] or guessing games [18] are modeled. Still, the Prisoner's Dilemma game is the most common example, but there are also analyses of the ultimatum game [62], the Snowdrift game [102, 118] or the battle-of-the sexes game [116].

In some papers, evolutionary games are used to model risk preferences [69] or the evolution of trust, reciprocity, or altruism [114, 117]. But the dominating topic is the evolution of cooperation.

Application: Evolution of cooperation. In the classical Prisoner's Dilemma game, two strategies exist, defection and cooperation. Given the uncertainty about the counterparty's strategy, rational agents would always choose defection. But empirical studies of biological and social systems report a remarkable degree of cooperation in a population. This raises the question of which mechanisms exist to foster cooperation.

The literature discusses various solutions, and *Advances in Complex Systems* has continuously published about this topic. Here are just a few proposals for model extensions to enhance cooperation: agent heterogeneity [76, 102], social preferences [109], collective learning [39], migration [52, 96], neighborhood and network influence [26, 77], social herding [106], similarity [73], rewards [122], punishment [118], emotions [111], knowledge [100], memory [115]. Some models have also focused on the behavior of primates [74, 84].

Application: Biological evolution. Although evolution is seen as an overarching concept in complex systems research, many papers published in *Advances in Complex Systems* specifically deal with problems of biological evolution. One focus is on genomic information [9, 28, 82, 120], the evolution of regulatory networks [35, 37, 66, 72] and protein-protein interactions [27].

Higher levels of biological organization are also studied, ranging from multicellular [43, 61]. to neuronal structures [38]. Models of early cognition [40] and mammalian evolution [90] are proposed as well. Some models specifically approach fossil records [7] and the Cambrian evolution [11].

A special application are rather abstract models of guided self-organization [105] to explain the emergence of early life [29, 30, 78] and cognition [101, 107].

Application: Evolutionary economics. Based on the insight that the economy is a complex adaptive system, different approaches have been developed to model its evolution. While *evolutionary economics* is an established research field in economics, *artificial economics* mostly addresses agent-based models.

Applications focus on markets, for instance, on strategies in labour markets [14, 33] and on competition [99]. Technological evolution is a core topic of evolutionary economics [15, 23]. The evolution of industrial sectors is often studied from a regional perspective [16, 45] which also considers the consequences for economic integration [58]. Other papers take the firms' perspective instead and model their decisions and their growth [13, 17, 123].

Application: Cultural evolution. When a topic rooted in the Humanities, like culture, becomes subject to formal models, publication opportunities are relatively scarce. *Advances in Complex Systems* has frequently published about such transdisciplinary problems, combining expertise from different areas. These include, on the one hand, spatial models of structured populations [89] and, on the other hand, models of innovations, diffusion of innovations, adaptation, and reproduction. Also, cooperation between populations are considered [24].

Space evidently plays an important role in the evolution of culture [86, 95, 103]. Additionally, such models have to implement the specifics of social organizations in the early history of man [54, 79, 92]. Eventually, environmental dependencies and resource availability have to be taken into account [83, 93]. Such models often combine formal with empirical analyses to provide a broader view of the conditions and constraints in human evolution.

Application: Evolution of languages. The change of languages is also governed by adaptation, adoption, replication, mutation, and selection. Therefore, it could be described similarly to other formal models of evolution. Fortunately, such models can be linked to the many empirical findings about the evolution of languages.

We must distinguish the evolution within a language, i.e., the emergence of new abstractions or the changed meaning of words, from the evolution of a set of languages, e.g., their replication and competition. Studies about the change of language analyze, for instance, how color categorization emerged [97], how the use frequency of words [19, 22, 119] or the ordering or composition of words changed [25, 31, 49, 50, 51]. The dynamics of dialect formation, for example, can be described by simple models of *neutral evolution*, i.e., processes of random copying, similar to genetic drift [67, 81].

Models about the interaction between languages identify, for instance, common structures across languages [56], or pay attention to the adoption of languages in multilingual environments, to estimate the distribution of language usage [42, 55, 59, 124]. Also, the spatial distribution of ethnolinguistic groups is studied [85].

Comments on the reprinted publications

Coevolutionary learning and the design of complex systems [10]. This paper points to a fundamental problem in complex systems research illustrated using a one-dimensional cellular automaton. N agents are characterized by a binary state and influenced by their 3 closest left and right neighbors. Their state in the next time step thus depends on 7 inputs. How many possible outcomes can be generated from this? $2^{2^7} = 2^{128}$, which is the number of possible rules using this input. Now, which of these rules are compatible with the desired outcome? This problem, known as reverse engineering, is computationally exhaustive. Thus, one has to find ways to restrict the search space of possible rules to achieve the desired outcome iteratively.

The procedure proposed in the paper is called coevolutionary learning and uses two populations of agents that compete or cooperate to find better rules. With its cellular automata setup, the Boolean functions for the rules, and the genetic algorithms to modify them, this paper is very typical for the 1990s and represents a whole class of models. At the same time, it does not just propose a possible solution but also discusses which setups won't work. For those who only know the recent advancements in machine learning, e.g., deep neural networks, to discover the best rules to produce a desired macro state, this is a welcome opportunity to look back.

Search processes in complex adaptive landscapes [23]. Evolutionary dynamics is essentially coevolutionary dynamics. Unlike optimization problems where agents explore a rather complex but static search space, in evolutionary models the search space also changes. More precisely, it can adapt depending on the positions of the searchers. This paper proposes to model the search space as a fitness landscape, similar to the phenotype space in biology. Agents try to find local and global fitness maxima in this search space. At the same time, successful agents reproduce. The time-dependent population density feeds back at the shape of this space, which is therefore denoted as an adaptive landscape.

In this publication, different methodological contributions are merged. On the one hand, we have differential equations known from population biology, like the Lotka-Volterra or the Fisher-Eigen equations. On the other hand, evolution is described as a continuous search process in an optimization landscape. The key idea is to combine these two approaches, making the search space dependent on the population density. The application potential for this framework is only sketched in this paper, mentioning technological and economic evolution. Therefore, the paper is worth to be revisited.

Molecular replicator dynamics [32]. The replicator dynamics are key to a proper understanding of evolution. Auto- and heterocatalytic processes form a network of chemical reactions that can, in principle, generate many structures at the molecular level. Similar to the discussion above, the question is about possible kinetic mechanisms to restrict this outcome to functional or meaningful structures.

The paper applies a population dynamics perspective to chemical kinetics, using chemical "species". This allows studying analytically the impact of restrictions on the dynamics, for instance, error or survival thresholds or interactions between different replicating species. A bifurcation analysis of the dynamics shows which kinetic assumptions lead to attractors in the phase space corresponding to, e.g., coexisting species.

As an asset of this publication, its overarching approach helps to structure the plethora of dynamic models for prebiotic evolution in a sensible manner. To not get lost in hyper-cycles, quasi-species, or parabolic growth dynamics, these concepts are related in a systematic way.

Cooperation, collectives formation, and specialization [41]. In this paper, the approach of evolutionary game theory is reviewed and advanced in two main directions: (i) a spatial dimension is considered, which leads to *structured* populations, and (ii) the coexistence of more than two strategies is explored. Loners who can abstain from an interaction are introduced as the third strategy. While the results are illustrated using computer simulations of a two-dimensional cellular automaton, the mathematical analysis allows for identifying critical parameters for various types of dynamics. These include cyclic dominance, i.e., the dominating strategy changes periodically. Further, the conditions for the spatial synchronization of such periodic changes, the emergence of traveling waves, or the formation of clusters to reduce the exploitation by adversaries are studied.

The paper attempts to study setups beyond the classical prisoner's dilemma. A varying degree of cooperation is introduced, which leads to continuous strategies dependent on the "investment level" into cooperation. The discussion is related to empirical examples from biology to illustrate the practical relevance.

Macroevolution as a branching process based on innovations [91]. How can the impact of innovations on the evolution of a set of species be modeled? This paper studies a combination of two evolutionary principles, mutation and recombination. Evolution is seen as a branching process that shapes the phylogenetic tree. Thus, the end nodes of the tree represent the species coexisting at a given time. A rare mutation introduces a new feature to a species, leading to a new tree branch. Also, the loss of existing features is possible. Because of recombination, such events trigger cascades of diversification of species.

On the macro level, these dynamics can be described by a probability distribution over the set of species. Different assumptions about mutations then lead to trees of varying topologies. Simulated trees are compared with those from phylogenetic databases to find out which topologies are relevant empirically.

This paper gives a lucid example of how a complex problem can be mapped to a surprisingly simple evolutionary model, which nevertheless allows for comparison with empirical data. We note that branching models did not only find applications in modeling phylogenetic trees but also in language dynamics. Thus, it is worth studying this class of models.

Self-organizing particle systems [104]. The emergence of organisms from the collective interaction of cells is a central problem of biological evolution. Again, how can we model this process without getting lost in too many biological details? What are the necessary ingredients to obtain self-organization, and how can we measure it? What is the role of space?

The paper proposes a very abstract model of interacting particles to address these questions. It bears similarities to agent-based modeling and molecular dynamics. Specifically, distance-dependent forces between different kinds of particles are proposed. The novelty is in quantifying the resulting structures using an information-theoretic approach. The novel measure, "multi-information" allows characterizing the degree of self-organization and should also apply to other systems. As a general insight, the model demonstrates to which extent the spread of information is a condition for self-organization.

Selected publications

[1] Anderson, C., *Simulation of the Feedbacks and Regulation of Recruitment Dancing in Honey Bees*, Advances in Complex Systems **01** (**1998**) 267–282.

[2] Botee, H. M. and Bonabeau, E., *Evolving Ant Colony Optimization*, Advances in Complex Systems **01** (**1998**) 149–159.

[3] Chang, C. and Gaudiano, P., *Application of Biological Learning Theories To Mobile Robot Avoidance and Approach Behaviors*, Advances in Complex Systems **01** (**1998**) 79–114.

[4] Heusse, M., Snyers, D., Guérin, S., and Kuntz, P., *Adaptive Agent-Driven Routing and Load Balancing in Communication Networks*, Advances in Complex Systems **01** (**1998**) 237–254.

[5] Hordijk, W. and Stadler, P. F., *Amplitude Spectra of Fitness Landscapes*, Advances in Complex Systems **01** (**1998**) 39–66.

[6] Segev, R. and Ben-Jacob, E., *From Neurons To Brain: Adaptive Self-Wiring of Neurons*, Advances in Complex Systems **01** (**1998**) 67–78.

[7] Solé, R. V., Manrubia, S. C., Pérez-Mercader, J., Benton, M., and Bak, P., *Long-Range Correlations in the Fossil Record and the Fractal Nature of Macroevolution*, Advances in Complex Systems **01** (**1998**) 255–266.

[8] Steels, L. and Mcintyre, A., *Spatially Distributed Naming Games*, Advances in Complex Systems **01** (**1998**) 301–323.

[9] Bar-Yam, Y., *Formalizing the Gene Centered View of Evolution*, Advances in Complex Systems **02** (**1999**) 277–281.

[10] Juillé, H. and Pollack, J. B., *Coevolutionary Learning and the Design of Complex Systems*, Advances in Complex Systems **02** (**1999**) 371–393.

[11] Solé, R. V., Salazar-Ciudad, I., and Garcia-Fernández, J., *Landscapes, Gene Networks and Pattern Formation: On the Cambrian Explosion*, Advances in Complex Systems **02** (**1999**) 313–337.

[12] Weisbuch, G. and Boudjema, G., *Dynamical Aspects in the Adoption of Agri-Environmental Measures*, Advances in Complex Systems **02** (**1999**) 11–36.

[13] Ballot, G. and Taymaz, E., *Competition, Training, Heterogeneity Persistence, and Aggregate Growth in A Multi-Agent Evolutionary Model*, Advances in Complex Systems **03** (**2000**) 335–351.

[14] Mueller, G., *Exploring the Dynamics of Social Policy Models: a Computer Simulation of Long-Term Unemployment*, Advances in Complex Systems **03** (**2000**) 433–447.

[15] Arenas, A., Díaz-Guilera, A., Guardiola, X., Llas, M., Oron, G., Pérez, C. J., and Vega-Redondo, F., *New Results in a Self-Organized Model of Technological Evolution*, Advances in Complex Systems **04** (**2001**) 89–100.

[16] Brenner, T. and Weigelt, N., *The Evolution of Industrial Clusters — Simulating Spatial Dynamics*, Advances in Complex Systems **04** (**2001**) 127–147.

[17] Kwasnicki, W., *Firms Decision Making Process in an Evolutionary Model of Industrial Dynamics*, Advances in Complex Systems **04** (**2001**) 101–125.

[18] Pastor-Abia, L., Pérez-Jordá, J. M., San-Fabián, E., Louis, E., and Vega-Redondo, F., *Strategic Behavior and Information Transmission in a Stylized (So-Called Chinos) Guessing Game*, Advances in Complex Systems **04** (**2001**) 177–190.

[19] Zanette, D. H., *Self-Similarity in the Taxonomic Classification of Human Languages*, Advances in Complex Systems **04** (**2001**) 281–286.

[20] Bonzo, D. C. and Hermosilla, A. Y., *Clustering Panel Data Via Perturbed Adaptive Simulated Annealing and Genetic Algorithms*, Advances in Complex Systems **05** (**2002**) 339–360.

[21] Borkar, V. S., *Reinforcement Learning in Markovian Evolutionary Games*, Advances in Complex Systems **05** (**2002**) 55–72.

[22] Cancho, R. F. I. and Solé, R. V., *Zipf'S Law and Random Texts*, Advances in Complex Systems **05** (**2002**) 1–6.

[23] Ebeling, W., Karmeshu, and Scharnhorst, A., *Dynamics of Economic and Technological Search Processes in Complex Adaptive Landscapes*, Modeling Complexity in Economic and Social Systems (**2002**) 79–96.

[24] Kvasnicka, V. and Pospichal, J., *Evolutionary Study of Interethnic Cooperation*, Modeling Complexity in Economic and Social Systems (**2002**) 293–321.

[25] Montemurro, M. A. and Zanette, D. H., *Entropic Analysis of the Role of Words in Literary Texts*, Advances in Complex Systems **05** (**2002**) 7–17.

[26] Schweitzer, F., Behera, L., and Mühlenbein, H., *Evolution of Cooperation in a Spatial Prisoner's Dilemma*, Advances in Complex Systems **05** (**2002**) 269–299.

[27] Solé, R. V., Pastor-Satorras, R., Smith, E., and Kepler, T. B., *A Model of Large-Scale Proteome Evolution*, Advances in Complex Systems **05** (**2002**) 43–54.

[28] Torres, J.-L. and Trainor, L., *Genomic Organization and Hopfield'S Model of Associative Memory*, Advances in Complex Systems **05** (**2002**) 361–369.

[29] Larter, R., Craig, M. G., and Tinsley, R., *Continuing Emergence in Living Systems*, Advances in Complex Systems **06** (**2003**) 93–114.

[30] Shenhav, B., Segrè, D., and Lancet, D., *Mesobiotic Emergence: Molecular and Ensemble Complexity in Early Evolution*, Advances in Complex Systems **06** (**2003**) 15–35.

[31] Smith, K., Brighton, H., and Kirby, S., *Complex Systems in Language Evolution: The Cultural Emergence of Compositional Structure*, Advances in Complex Systems **06** (**2003**) 537–558.

[32] Stadler, B. M. R. and Stadler, P. F., *Molecular Replicator Dynamics*, Advances in Complex Systems **06** (**2003**) 47–77.

[33] Fagiolo, G., Dosi, G., and Gabriele, R., *Matching, Bargaining, and Wage Setting in an Evolutionary Model of Labor Market and Output Dynamics*, Industry and Labor Dynamics (**2004**).

[34] Mortveit, H. S. and Reidys, C. M., *Neutral Evolution and Mutation Rates of Sequential Dynamical Systems*, Advances in Complex Systems **07** (**2004**) 395–418.

[35] Wagner, A. and Wright, J., *Compactness and Cycles in Signal Transduction and Transcriptional Regulation Networks: A Signature of Natural Selection?*, Advances in Complex Systems **07** (**2004**) 419–432.

[36] Brabazon, A., Silva, A., De Sousa, T. F., O'Neill, M., Matthews, R., and Costa, E., *Organizational Strategic Adaptation in the Presence of Inertia*, Advances in Complex Systems **08** (**2005**) 497–519.

[37] Drasdo, D. and Kruspe, M., *Emergence of Regulatory Networks in Simulated Evolutionary Processes*, Advances in Complex Systems **08** (**2005**) 285–318.

[38] Elston, G. N. and Zietsch, B., *Fractal Analysis as a Tool For Studying Specialization in Neuronal Structure: The Study of the Evolution of the Primate Cerebral Cortex and Human Intellect*, Advances in Complex Systems **08** (**2005**) 217–227.

[39] Helbing, D., Schönhof, M., Stark, H.-U., and Hołyst, J. A., *How Individuals Learn To Take Turns: Emergence of Alternating Cooperation in A Congestion Game and the Prisoner's Dilemma*, Advances in Complex Systems **08** (**2005**) 87–116.

[40] Okadome, T., *A Formal Theory of Early Cognition Development*, Advances in Complex Systems **08** (**2005**) 229–260.

[41] Hauert, C., *Cooperation, Collectives Formation and Specialization*, Advances in Complex Systems **09** (**2006**) 315–335.

[42] Schulze, C. and Stauffer, D., *Monte Carlo Simulation of Survival For Minority Languages*, Advances in Complex Systems **09** (**2006**) 183–191.

[43] Vincent, T. L., *Carcinogenesis as an Evolutionary Game*, Advances in Complex Systems **09** (**2006**) 369–382.

[44] Bosse, T., Jonker, C. M., and Treur, J., *Simulation and Analysis of Adaptive Agents: An Integrative Modeling Approach*, Advances in Complex Systems **10** (**2007**) 335–357.

[45] Fleming, L. and Frenken, K., *The Evolution of Inventor Networks in the Silicon Valley and Boston Regions*, Advances in Complex Systems **10** (**2007**) 53–71.

[46] Hauptman, A. and Sipper, M., *Emergence of Complex Strategies in the Evolution of Chess Endgame Players*, Advances in Complex Systems **10** (**2007**) 35–59.

[47] Johansson, A., Helbing, D., and Shukla, P. K., *Specification of the Social Force Pedestrian Model By Evolutionary Adjustment To Video Tracking Data*, Advances in Complex Systems **10** (**2007**) 271–288.

[48] Press, K., *Divide and Conquer? the Role of Governance For the Adaptability of Industrial Districts*, Advances in Complex Systems **10** (**2007**) 73–92.

[49] Cysouw, M., *Linear Order as a Predictor of Word Order Regularities*, Advances in Complex Systems **11** (**2008**) 415–420.

[50] Ferrer-I-Cancho, R., *Some Limits of Standard Linguistic Typology: The Case of Cysouw's Models For The Frequencies of the Six Possible Orderings of S, V and O*, Advances in Complex Systems **11** (**2008**) 421–432.

[51] Ferrer-I-Cancho, R., *Some Word Order Biases From Limited Brain Resources: A Mathematical Approach*, Advances in Complex Systems **11** (**2008**) 393–414.

[52] Helbing, D. and Yu, W., *Migration as a Mechanism To Promote Cooperation*, Advances in Complex Systems **11** (**2008**) 641–652.

[53] Jost, J. and Li, W., *Learning, Evolution and Population Dynamics*, Advances in Complex Systems **11** (**2008**) 901–926.

[54] Kurahashi, S. and Terano, T., *Historical Simulation: a Study of Civil Service Examinations, the Family Line And Cultural Capital in China*, Advances in Complex Systems **11** (**2008**)

187–198.

[55] Malarz, K. and Stauffer, D., *Search For Bottleneck Effects in Penna Ageing and Schulze Language Model*, Advances in Complex Systems **11** (**2008**) 165–169.

[56] Mukherjee, A., Choudhury, M., Basu, A., Ganguly, N., and Chowdhury, S. R., *Rediscovering the Co-Occurrence Principles of Vowel Inventories: A Complex Network Approach*, Advances in Complex Systems **11** (**2008**) 371–392.

[57] Navarro-Barrientos, J.-E., *Adaptive Investment Strategies For Periodic Environments*, Advances in Complex Systems **11** (**2008**) 761–787.

[58] Reyes, J., Schiavo, S., and Fagiolo, G., *Assessing the Evolution of International Economic Integration Using Random Walk Betweenness Centrality: The Cases of East Asia and Latin America*, Advances in Complex Systems **11** (**2008**) 685–702.

[59] Wichmann, S., Stauffer, D., Schulze, C., and Holman, E. W., *Do Language Change Rates Depend on Population Size?*, Advances in Complex Systems **11** (**2008**) 357–369.

[60] Araújo, T. and Mendes, R. V., *Innovation and Self-Organization in a Multi-Agent Model*, Advances in Complex Systems **12** (**2009**) 233–253.

[61] Buck, M. and Nehaniv, C. L., *Looking For Evidence of Differentiation and Cooperation: Natural Measures For The Study of Evolution of Multicellularity*, Advances in Complex Systems **12** (**2009**) 255–271.

[62] Eguíluz, V. M. and Tessone, C. J., *Critical Behavior in an Evolutionary Ultimatum Game with Social Structure*, Advances in Complex Systems **12** (**2009**) 221–232.

[63] Hesse, F., Der, R., and Herrmann, J. M., *Modulated Exploratory Dynamics Can Shape Self-Organized Behavior*, Advances in Complex Systems **12** (**2009**) 273–291.

[64] Joachimczak, M. and Wróbel, B., *Complexity of the Search Space in a Model of Artificial Evolution of Gene Regulatory Networks Controlling 3d Multicellular Morphogenesis*, Advances in Complex Systems **12** (**2009**) 347–369.

[65] Rohlf, T. and Winkler, C. R., *Emergent Network Structure, Evolvable Robustness, and Nonlinear Effects of Point Mutations in an Artificial Genome Model*, Advances in Complex Systems **12** (**2009**) 293–310.

[66] Tang, B., He, L., Jing, Q., and Shen, B., *Model-Based Identification and Adaptive Control of the Core Module in a Typical Cell Cycle Pathway Via Network and System Control Theories*, Advances in Complex Systems **12** (**2009**) 21–43.

[67] Ali, A. and Grosskinsky, S., *Pattern Formation Through Genetic Drift At Expanding Population Fronts*, Advances in Complex Systems **13** (**2010**) 349–366.

[68] Janssen, M. A., *The Evolution of Rules in Shedding-Type of Card Games*, Advances in Complex Systems **13** (**2010**) 741–754.

[69] Roos, P. and Nau, D., *Risk Preference and Sequential Choice in Evolutionary Games*, Advances in Complex Systems **13** (**2010**) 559–578.

[70] Shi, C., Yan, Z., Wang, Y., Cai, Y., and Wu, B., *A Genetic Algorithm For Detecting Communities in Large-Scale Complex Networks*, Advances in Complex Systems **13** (**2010**) 3–17.

[71] Wilkins, J. F. and Thurner, S., *The Jerusalem Game: Cultural Evolution of the Golden Rule*, Advances in Complex Systems **13** (**2010**) 635–641.

[72] Decraene, J. and Mcmullin, B., *The Evolution of Complexity in Self-Maintaining Cellular Information Processing Networks*, Advances in Complex Systems **14** (**2011**) 55–75.

[73] Howley, E. and Duggan, J., *Investing in the Commons: A Study of Openness and the Emergence of Cooperation*, Advances in Complex Systems **14** (**2011**) 229–250.

[74] Kuperman, M. N., *A Model For the Emergence of Social Organization in Primates*, Advances in Complex Systems **14** (**2011**) 403–414.

[75] Murks, A. and Perc, M., *Evolutionary Games on Visibility Graphs*, Advances in Complex Systems **14** (**2011**) 307–315.

[76] Ohdaira, T. and Terano, T., *The Diversity in the Decision Facilitates Cooperation in the Sequential Prisoner's Dilemma Game*, Advances in Complex Systems **14** (**2011**) 377–401.

[77] Antonioni, A. and Tomassini, M., *Cooperation on Social Networks and Its Robustness*, Advances in Complex Systems **15** (**2012**) 1250046.

[78] Arendt, D. and Cao, Y., *Evolutionary Motifs For the Automated Discovery of Self-Organizing Dimer Automata*, Advances in Complex Systems **15** (**2012**) 1250081.

[79] Barton, C. M. and Riel-Salvatore, J., *Agents of Change: Modeling Biocultural Evolution in Upper Pleistocene Western Eurasia*, Advances in Complex Systems **15** (**2012**) 1150003.

[80] Bentley, R. A. and Ormerod, P., *Accelerated Innovation and Increased Spatial Diversity of Us Popular Culture*, Advances in Complex Systems **15** (**2012**) 1150011.

[81] Blythe, R. A., *Neutral Evolution: A Null Model For Language Dynamics*, Advances in Complex Systems **15** (**2012**) 1150015.

[82] Cebrat, S., Waga, W., and Stauffer, D., *The Role of Haplotype Complementation and Purifying Selection in the Genome Evolution*, Advances in Complex Systems **15** (**2012**) 1250041.

[83] Cohen, M. H. and Ackland, G. J., *Boundaries Between Ancient Cultures: Origins and Persistence*, Advances in Complex Systems **15** (**2012**) 1150004.

[84] Cronin, K. A. and Sánchez, A., *Social Dynamics and Cooperation: The Case of Nonhuman Primates and Its Implications For Human Behavior*, Advances in Complex Systems **15** (**2012**) 1250066.

[85] Currie, T. E. and Mace, R., *The Evolution of Ethnolinguistic Diversity*, Advances in Complex Systems **15** (**2012**) 1150006.

[86] Fogarty, L., Rendell, L., and Laland, K. N., *The Importance of Space in Models of Social Learning, Cultural Evolution and Niche Construction*, Advances in Complex Systems **15** (**2012**) 1150001.

[87] Geipel, M. M., *Modularity, Dependence and Change*, Advances in Complex Systems **15** (**2012**) 1250083.

[88] He, D., Liu, J., Yang, B., Huang, Y., Liu, D., and Jin, D., *An Ant-Based Algorithm with Local Optimization For Community Detection In Large-Scale Networks*, Advances in Complex Systems **15** (**2012**) 1250036.

[89] Kandler, A., Perreault, C., and Steele, J., *Cultural Evolution in Spatially Structured Populations: A Review Of Alternative Modeling Frameworks*, Advances in Complex Systems **15** (**2012**) 1203001.

[90] Kandler, A. and Smaers, J. B., *An Agent-Based Approach To Modeling Mammalian Evolution: How Resource Distribution and Predation Affect Body Size*, Advances in Complex Systems **15** (**2012**) 1150014.

[91] Keller-Schmidt, S. and Klemm, K., *A Model of Macroevolution as a Branching Process Based on Innovations*, Advances in Complex Systems **15** (**2012**) 1250043.

[92] Kohler, T. A., Cockburn, D., Hooper, P. L., Bocinsky, R. K., and Kobti, Z., *The Coevolution of Group Size and Leadership: An Agent-Based Public Goods Model For Prehispanic Pueblo Societies*, Advances in Complex Systems **15** (**2012**) 1150007.

[93] Lake, M. W. and Crema, E. R., *The Cultural Evolution of Adaptive-Trait Diversity When Resources Are Uncertain and Finite*, Advances in Complex Systems **15** (**2012**) 1150013.

[94] Lozano, S., Borge-Holthoefer, J., and Arenas, A., *Emerging Cohesion and Individualization in Collective Action: A Co-Evolutive Approach*, Advances in Complex Systems **15** (**2012**) 1250067.

[95] Premo, L. S., *Local Extinctions, Connectedness, and Cultural Evolution in Structured Populations*, Advances in Complex Systems **15** (**2012**) 1150002.

[96] Schweitzer, F. and Behera, L., *Optimal Migration Promotes the Outbreak of Cooperation in Heterogeneous Populations*, Advances in Complex Systems **15** (**2012**) 1250059.

[97] Steingrimsson, R., *Evolutionary Game Theoretical Model of the Evolution of the Concept of Hue, a Hue Structure, and Color Categorization in Novice and Stable Learners*, Advances in Complex Systems **15** (**2012**) 1150018.

[98] Szell, M. and Thurner, S., *Social Dynamics in A Large-Scale Online Game*, Advances in Complex Systems **15** (**2012**) 1250064.

[99] Vasile, A., Costea, C. E., and Viciu, T. G., *An Evolutionary Game Theory Approach To Market Competition and Cooperation*, Advances in Complex Systems **15** (**2012**) 1250044.

[100] Xia, C.-Y., Meloni, S., and Moreno, Y., *Effects of Environment Knowledge on Agglomeration and Cooperation in Spatial Public Goods Games*, Advances in Complex Systems **15** (**2012**) 1250056.

[101] Balduzzi, D., Ortega, P. A., and Besserve, M., *Metabolic Cost as an Organizing Principle For Cooperative Learning*, Advances in Complex Systems **16** (**2013**) 1350012.

[102] Barreira Da Silva Rocha, A. and Laruelle, A., *Evolution of Cooperation in the Snowdrift Game with Heterogeneous Population*, Advances in Complex Systems **16** (**2013**) 1350036.

[103] Callegari, S., Weissmann, J. D., Tkachenko, N., Petersen, W. P., Lake, G., De León, M. P., and Zollikofer, C. P. E., *An Agent-Based Model of Human Dispersals At a Global Scale*, Advances in Complex Systems **16** (**2013**) 1350023.

[104] Harder, M. and Polani, D., *Self-Organizing Particle Systems*, Advances in Complex Systems **16** (**2013**) 1250089.

[105] Polani, D., Prokopenko, M., and Yaeger, L. S., *Information and Self-Organization of Behavior*, Advances in Complex Systems **16** (**2013**) 1303001.

[106] Schweitzer, F., Mavrodiev, P., and Tessone, C. J., *How Can Social Herding Enhance Cooperation?*, Advances in Complex Systems **16** (**2013**) 1350017.

[107] Yaeger, L. S., *Identifying Neural Network Topologies That Foster Dynamical Complexity*, Advances in Complex Systems **16** (**2013**) 1350032.

[108] Geipel, M. M., Press, K., and Schweitzer, F., *Communication in Innovation Communities: An Analysis of 100 Open Source Software Projects*, Advances in Complex Systems **17** (**2014**) 1550006.

[109] Janssen, M. A., Manning, M., and Udiani, O., *The Effect of Social Preferences on the Evolution of Cooperation in Public Good Games*, Advances in Complex Systems **17** (**2014**) 1450015.

[110] Przybyła, P., Sznajd-Weron, K., and Weron, R., *Diffusion of Innovation Within an Agent-Based Model: Spinsons, Independence And Advertising*, Advances in Complex Systems **17** (**2014**) 1450004.

[111] Righi, S. and Takács, K., *Emotional Strategies as Catalysts For Cooperation in Signed Networks*, Advances in Complex Systems **17** (**2014**) 1450011.

[112] Wu, J., Xu, M., and Gao, Z., *Modeling the Coevolution of Road Expansion and Urban Traffic Growth*, Advances in Complex Systems **17** (**2014**) 1450005.

[113] Zgonnikov, A. and Lubashevsky, I., *Unstable Dynamics of Adaptation in Unknown Environment Due To Novelty Seeking*, Advances in Complex Systems **17** (**2014**) 1450013.

[114] Xu, W.-J., Zhong, L.-X., Huang, P., Qiu, T., Shi, Y.-D., and Zhong, C.-Y., *Evolutionary Dynamics in Opinion Formation Model with Coupling of Social Communities*, Advances in Complex Systems **18** (**2015**) 1550003.

[115] Cetin, U. and Bingol, H. O., *The Dose of the Threat Makes the Resistance For Cooperation*, Advances in Complex Systems **19** (**2016**) 1650015.

[116] Chacoma, A., Kuperman, M. N., and Zanette, D. H., *Payoff Nonmonotonic Dynamics in an Evolutionary Game*, Advances in Complex Systems **19** (**2016**) 1650007.

[117] Shi, K. and Ma, H., *Evolution of Trust in a Dual-Channel Supply Chain Considering Reciprocal Altruistic Behavior*, Advances in Complex Systems **19** (**2016**) 1650014.

[118] Da Silva Rocha, A. B., *Cooperation in the Well-Mixed Two-Population Snowdrift Game with Punishment Enforced Through Different Mechanisms*, Advances in Complex Systems

20 (**2017**) 1750010.

[119] Ruck, D., Alexander Bentley, R., Acerbi, A., Garnett, P., and Hruschka, D. J., *Role of Neutral Evolution in Word Turnover During Centuries of English Word Popularity*, Advances in Complex Systems **20** (**2017**) 1750012.

[120] Balaz, I. and Haruna, T., *Evolution of Influenza a Nucleotide Segments Through the Lens of Different Complexity Measures*, Advances in Complex Systems **21** (**2018**) 1850009.

[121] Lou, Y., Sheng, Q., and Chen, Y., *When Relaxation Meets Adaptation in Complex Adaptive Systems: A Computational Study of Tumorigenesis*, Advances in Complex Systems **21** (**2018**) 1750016.

[122] Schweitzer, F., Verginer, L., and Vaccario, G., *Should the Government Reward Cooperation? Insights From an Agent-Based Model Of Wealth Redistribution*, Advances in Complex Systems **23** (**2020**) 2050018.

[123] Yu, Q., Wu, B., and Chen, F., *An Evolution Analysis on the Dynamic Game of Tax Compliance Behavior with Credit Evaluation in Ewa Learning Model*, Advances in Complex Systems **23** (**2020**) 2040001.

[124] De Silva, K., Basheer, A., Antwi-Fordjour, K., Beauregard, M. A., Chand, V., and Parshad, R. D., *The "Higher" Status Language Does Not Always Win: The Fall of English in India And the Rise of Hindi*, Advances in Complex Systems (**2021**) 2050021.

Adv. Complex Systems (1999) **4**, 371–393

Coevolutionary Learning and the Design of Complex Systems

Hugues Juillé†
eurobios
Tour Ernst & Young
92037 La Défense cedex, France

Jordan B. Pollack
Computer Science Department
Volen Center for Complex Systems
Brandeis University
Waltham, MA 02254-9110, USA

ABSTRACT. Complex systems composed of a large number of loosely coupled entities, with no central coordination offer a number of attractive properties like scalability, robustness or massively distributed computation. However, designing such complex systems presents some challenging issues that are difficult to tackle with traditional top-down engineering methodologies.

Coevolutionary learning, which involves the embedding of adaptive learning agents in a fitness environment that dynamically responds to their progress, is proposed as a paradigm to explore a space of complex system designs. It is argued that coevolution offers a flexible framework for the implementation of search heuristics that can efficiently exploit some of the structural properties exhibited by such state spaces. However, several drawbacks have to be overcome in order for coevolutionary learning to achieve continuous progress in the long term. This paper presents some of those problems and introduces a new strategy based on the concept of an "ideal" trainer to address them. This presentation is illustrated with a case study: the discovery of cellular automata rules to implement a classification task. The application of the "ideal" trainer paradigm to that problem resulted in a significant improvement over previously known best rules for this task.

KEYWORDS: coevolutionary learning, cellular automata, evolutionary search, density classification task.

1. Introduction

Methodologies for designing complex systems have been the subject of much research. In particular, the "divide-and-conquer" paradigm has allowed engineering sciences to achieve impressive successes. The space shuttle may be one

† Contact author.

of the most representative example of such achievements. However the design of such complex systems is based on the assumption that a natural decomposition of the initial task into independent sub-problems exists, and that each sub-problem can be addressed recursively in the same manner until all the components (or building blocks) of the system are identified. Such an approach is very successful when the task to address is well understood and its scope well defined. Since this paradigm results in a rigid functional decomposition into components implementing non-overlapping behaviors (meaning that there is limited redundancy across components), such systems are often seen as brittle : the good working of the system relies heavily on each component and a single failure in one component is often enough to cause havoc.
Central coordination is also a typical characteristic of this design strategy. As a result, there is little flexibility to adapt the organization of the system to perform a slightly different function, or to scale the system (in order to improve its performance, for instance).

Living systems in nature operate in a completely different way. Taking inspiration from such living systems, an artificial complex system would be composed of many locally interacting processing units, each embedding limited computing ability, with no central coordination. Such systems are expected to provide multiple advantages like distributed computing (allowing scalability) and robustness to the failure of individual components. Harnessing the collaborative capacity of a population of simple agents and applying this capacity for problem solving is an exciting challenge. Already, social insects like ants and termites are at the origin of new classes of computational procedures that have successfully been applied to problems in artificial intelligence and combinatorial optimization (Bonabeau *et al.*, 1999). Designing such systems is far from trivial. Indeed, the computation performed by a complex system is an emergent behavior resulting from the local interactions of its parts. A complex system is totally specified by the function embedded in its parts and how those parts interact. The name of the game is to determine those interaction rules that will result in the emergence of the desired behavior.

Obviously, new design paradigms are required to make this approach practical. The purpose of this paper is to explore such a paradigm. It is based on an evolutionary approach for exploring the space of designs for a class of complex systems. The motivation for exploiting this approach is that, in many instances, Evolutionary Computation (EC) techniques have demonstrated their ability to tackle ill-structured and poorly understood problems against which traditional artificial intelligence search algorithms fail (Hillis, 1992; Koza *et al.*, 1996; Juillé, 1995). The principle of operation behind EC techniques can be described as a statistical inference process which implements a sampling-based strategy to gather information about the state space, and then exploits this knowledge for controlling search (Peck and Dhawan, 1995).

Searching a state space means that an objective function must be defined in order to assess the quality of any candidate solution. This function determines what properties should be satisfied by a solution. The difficulty in defining an appropriate objective function lies in the fact that complex systems can be exposed to a large range of situations (or test cases). Assessing their quality would require the evaluation of their performance against an extremely large number of test cases. However, to make search tractable, solutions can be evaluated only with respect to a training environment composed of a subset of all the test cases. Usually, this issue is addressed by "engineering" a fixed training environment against which solutions are evaluated. At that stage, a significant amount of knowledge about the problem is explicitly introduced by the designer. Then, the search algorithm follows the gradient implemented in the training environment. However, the performance of solutions relies heavily on this training environment and on the search procedure to explore the state space. If little knowledge is available or if it is is difficult to introduce in the training environment, solutions have poor quality.

The approach presented in this paper to get around that problem considers coevolutionary learning, a paradigm which involves the embedding of adaptive agents (the candidate solutions) in a fitness environment that dynamically responds to their progress. Coevolution is proposed as a framework in which evolving agents would be permanently challenged, eventually resulting in continuous improvement of their performance. For the rest of the paper, we define *coevolutionary learning* as a search procedure involving a population of learners coevolving with a population of problems such that *continuous progress* results from this interaction.

However, in practice, the picture is not that simple. We will discuss the different issues that are involved to achieve coevolutionary learning by considering a particular problem: the discovery of cellular automata rules to implement a classification task. Cellular automata have been studied widely as they represent one of the simplest systems in which complex emergent behaviors can be observed. This model is very attractive as a means to study complex systems in nature. Indeed, the evolution of such systems is ruled by simple, locally-interacting components which result in the emergence of global, coordinated activity. Moreover, this problem presents some interesting properties that provide us with a simple framework to monitor the dynamics of the search resulting from different setups. After identifying impediments to continuous progress, the concept of an "ideal" trainer is introduced as a paradigm which successfully achieves that goal by maintaining a pressure toward adaptability.

This paper is structured as follows. Section 2 describes the majority classification task. In section 3, an experimental analysis illustrates the different impediments to coevolutionary learning. Then, section 4 introduces the "ideal" trainer as a paradigm to address those impediments. Experimental results for the classification problem are presented in section 5. Finally, the paper concludes with a discussion on the relevance of this work to the general problem of designing complex systems.

2. Description of the Problem

2.1. ONE-DIMENSIONAL CELLULAR AUTOMATA

A one-dimensional cellular automaton (CA) is a linear wrap-around array composed of N cells in which each cell can take one out of k possible states. A rule is defined for each cell in order to update its state. This rule determines the next state of a cell given its current state and the state of cells in a predefined

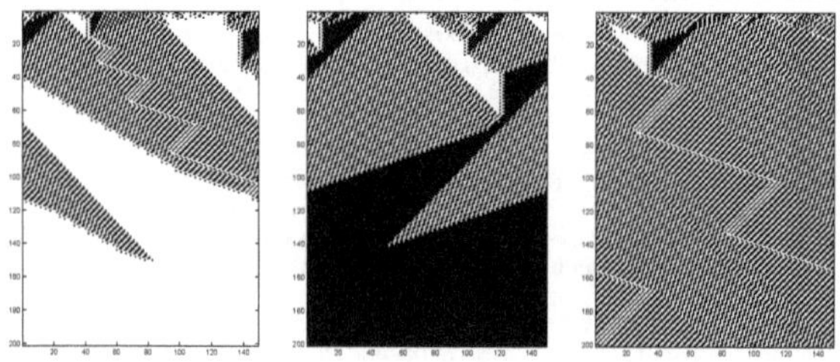

Figure 1: Three space-time diagrams describing the evolution of CA states: in the first two, the CA relaxes to the correct uniform pattern while in the third one it doesn't converge at all to a fixed point.

Table 1: Performance of different published CA rules and a new best rule for the $\rho_c = 1/2$ task.

N	149	599	999
Coevolution	0.863 +/- 0.001	0.822 +/- 0.001	0.804 +/- 0.001
Das rule	0.823 +/- 0.001	0.778 +/- 0.001	0.764 +/- 0.001
ABK rule	0.824 +/- 0.001	0.764 +/- 0.001	0.730 +/- 0.001
GKL rule	0.815 +/- 0.001	0.773 +/- 0.001	0.759 +/- 0.001

neighborhood. For the model discussed in this paper, this neighborhood is composed of cells whose distance is at most r from the central cell. This operation is performed synchronously for all the cells in the CA. From now on, we will consider that the state of cells is binary ($k = 2$), $N = 149$ and $r = 3$. This means that the size of the rule space is $2^{2^{2*r+1}} = 2^{128}$.

2.2. THE MAJORITY FUNCTION

This is a density classification task, for which one wants the state of the cells of the CA to relax to all 0's or all 1's depending on the density of the initial configuration (IC) (whether it has more 0's or more 1's), within a maximum of M time steps. Following (Mitchell *et al.*, 1994), ρ_c denotes the threshold for the classification task (here, $\rho_c = 1/2$), ρ denotes the density of 1's in a configuration and ρ_o denotes the density of 1's in the initial configuration. Figure 1 presents three examples of the space-time evolution of a CA. One with $\rho_0 < \rho_c$ on the left and another with $\rho_0 > \rho_c$ in the middle for which the CA relaxes to the correct configuration. The third one shows an instance for which the CA doesn't relax to any of the two desired convergence patterns. For each diagram, the initial configuration is at the top and the evolution in time of the state of the CA is represented downward.

The $\rho_c = 1/2$ task is known to be difficult. In particular, it has been proven that no rule exists that results in the CA relaxing to the correct state for all possible ICs (Land and Belew, 1995). Indeed, the density is a global property of the initial configuration while individual cells of the CA have access to local information only. Discovering a rule that will display the appropriate computation by the CA with the highest accuracy is a challenge, and the upper limit for this accuracy is still unknown. Table 1 describes the performance for that task for different published rules and different values of N, along with the performance of the new best rule that resulted from the work presented in this paper. The Gacs-Kurdyumov-Levin (GKL) rule was designed in 1978 for a different goal than solving the $\rho_c = 1/2$ task (Mitchell *et al.*, 1994). However, for a while it provided the best known performance. (Mitchell *et al.*, 1994) and (Das *et al.*, 1994) used Genetic Algorithms (GAs) to explore the space of rules. The main purpose of this work was to develop a particle-based methodology for the analysis of the complex behaviors exhibited by CAs. The GKL and Das rules are human-written while the Andre-Bennett-Koza (ABK) rule has been discovered using the Genetic Programming paradigm (Andre *et al.*, 1996). More recently, (Paredis, 1997) describes a coevolutionary approach to search the space of rules and shows the difficulty of coevolving consistently two populations toward continuous improvement. (Capcarrere *et al.*, 1996) also reports that by changing the specification of the convergence pattern, a two-state, $r = 1$ CA exists that can perfectly solve the density problem in $\lceil N/2 \rceil$ time steps.

For the $\rho_c = 1/2$ task, it is believed that the best rules are in the domain of the rule space with density close to 0.5. An intuitive argument to support this hypothesis is presented in (Mitchell *et al.*, 1993). It is also believed that the most difficult ICs are those with density close to 0.5.

3. Models for Coevolutionary Search

In the research literature, Hillis' pioneering work represents an important milestone by showing that coevolution can be used to improve search performance (Hillis, 1992). In his work, a population of sorters (the hosts) coevolve with input vectors (the parasites). The goal of sorters is to construct sequences of comparator-swaps that sort the input vectors that are proposed by the parasites while parasites search for input vectors that are difficult to sort. In a sense, this can be seen as an implementation of a coverage-based heuristic: a construction is sought that sorts correctly every possible input vector, thus resulting in a sorting network. This heuristic adaptively focuses the search for solving problem instances (i.e., input vectors) that are the most difficult for the population of networks. This work has been followed by others using both competitive and cooperative models of coevolution. For instance, Husbands implemented a model similar to Hillis' to address a generalized version of the job-shop scheduling problem (Husbands, 1994). Paredis (Paredis, 1996) used competition between a population of solutions and a population of problems as a search strategy for applications in inductive learning (Paredis, 1994b) and constraint satisfaction problems (Paredis, 1994a). Pursuer/evader games have also been used as a test problem for research in coevolution. In particular, Cliff and Miller (Cliff and Miller, 1995; Cliff and Miller, 1996) developed several tools to track progress and detect loss of traits resulting from the Red Queen effect. Sims' block-creatures (Sims, 1994) and Reynolds' experiments with the game of tag (Reynolds, 1994) are also two successful applications of competitive evolution. Rosin's work on coevolutionary learning (Rosin, 1997) addresses the different issues related to competitive evolution in the context of adversarial problems (e.g., game strategies). The goal of this work is to define a framework for coevolutionary search that results in continuous progress on the long term. In a theoretical analysis (Rosin and Belew, 1996), Rosin and Belew described a coevolutionary environment and proved it allows the discovery of perfect game strategies.

Cooperative models of coevolution have also been implemented. Such models have been used for function optimization (Potter and De Jong, 1994) and for the design of control systems (Potter *et al.*, 1995). Another application is the search of a space of problem decompositions to construct modular solutions (Potter, 1997; Moriarty, 1997). Following a different track, Paredis (Paredis, 1995) designed a model exploiting a symbiotic relationship to coevolve solutions and their representation.

In this paper, a two-population model of coevolution is considered for which the fitness of individuals in each population is defined with respect to the members of the other population. Two cases can be considered in such a framework, depending on whether the two populations benefit from each other or whether they have conflicting interests. Those two modes of interaction are called *cooperative* and *competitive* respectively. In the following sections, those modes of interaction are described experimentally using the $\rho_c = 1/2$ task in order to stress the different issues related to coevolutionary learning.

For the experiments presented in this section, we used an implementation of Genetic Algorithms similar to the one described in (Mitchell *et al.*, 1994). Each rule is coded on a binary string of length $2^{2*r+1} = 128$. One-point crossover is used with a 2% bit mutation probability. The population size is $n_R = 200$ for rules and $n_{IC} = 200$ for ICs. The population of ICs is composed of binary strings of length $N = 149$. The population of rules and ICs are initialized according to a uniform distribution over $[0.0, 1.0]$ for the density. For all the experiments in this paper, the value of M (the maximum number of time steps) is set to 320 and is kept unchanged. At each generation, the top 95% of each population reproduces to the next generation and the remaining 5% is the result of crossover between parents from the top 95% selected using a fitness proportionate rule.

The motivation for implementing this unusually small generation gap is to implement a smooth evolution of the fitness landscape individuals of each population are exposed to. Indeed, a large generation gap would result in a dramatic change in the composition of each population, resulting in important changes in the structure of the fitness landscapes because of the relative definition of the fitness. Evolutionary search can be described as a statistical inference process (Juillé, 1999) and, like any search strategy, the exploration of the search space by evolutionary computation techniques is based on the identification and exploitation of structural regularities. In order to make correct decisions to control search, this inference process must be based on information that is consistent in time. If individuals are evaluated in an ever changing environment, little struc-

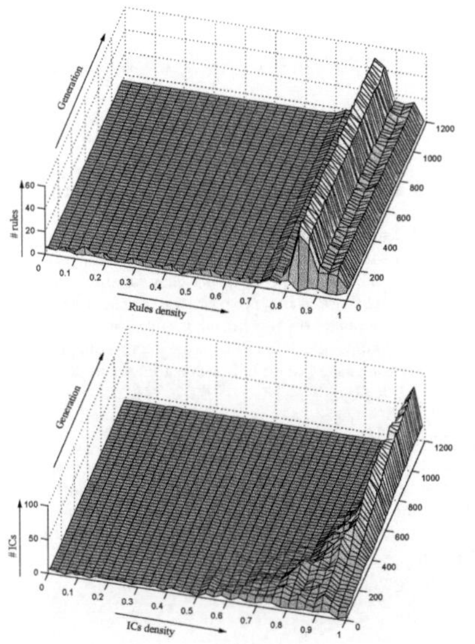

Figure 2: Coevolution of CA rules (top) and ICs (bottom) in a cooperative relationship.

tural regularities are maintained from one generation to the following in general. In that case, the inference process makes decisions based on ever changing information, and cannot usually drive search toward improving solutions.

3.1. COOPERATION BETWEEN POPULATIONS

In this mode of interaction, improvement on one side results in positive feedback on the other side. As a result, there is a reinforcement of the relationship between the two populations. From a search point of view, this can be seen as an *exploitative* strategy. Agents are not encouraged to explore new areas of the search space but only to perform local search in order to further improve the strength of the relationship. In the cooperative model, a natural definition for the fitness of rules (resp. ICs) is the number of ICs (resp. rules) for which the

CA relaxes to the correct state:

$$f(R_i) = \sum_{j=1}^{n_{IC}} covered(R_i, IC_j) \tag{1}$$

$$f(IC_j) = \sum_{i=1}^{n_R} covered(R_i, IC_j) \tag{2}$$

where $covered(R_i, IC_j)$ returns 1 if a CA using rule R_i and starting from initial configuration IC_j relaxes to the correct state. Otherwise, it returns 0.

Figure 2 presents the evolution of the density of rules and ICs for one run using this cooperative model. Without any surprise, the population of rules and ICs quickly converge to a domain of the search space where ICs are easy for rules and rules consistently solve ICs. As a result, there is little exploration of the search space. The convergence configuration depends on the initial populations, some other runs ended up with low density rules and ICs. This experiment confirms that ICs with low or high density are the easiest to classify since a larger number of rules classify them correctly.

3.2. COMPETITION BETWEEN POPULATIONS

In this mode of interaction, the two populations are in conflict. Improvement on one side results in negative feedback for the other population. The fitness of rules and ICs defined in the cooperative case can be modified as follows to implement the competitive model:

$$f(R_i) = \sum_{j=1}^{n_{IC}} covered(R_i, IC_j) \tag{3}$$

$$f(IC_j) = \sum_{i=1}^{n_R} \overline{covered(R_i, IC_j)} \tag{4}$$

where $\overline{covered(R_i, IC_j)}$ returns the inverse of the original function. Here, the goal of rules is to defeat (i.e. cover) ICs, while the goal of ICs is to defeat rules by discovering initial configurations that are difficult to classify. Figure 3 describes a run using this definition of the fitness. Two kind of behaviors can be observed in this experiment. In a first stage, the two populations exhibit a cyclic behavior. It is a consequence of the *Red Queen* effect (Cliff and Miller, 1995): fitness landscapes are changing as a result of agents of each population adapting in response to the evolution of members of the other population. The evaluation of individuals' performance in this changing environment makes continuous progress difficult. A typical consequence is that agents have to learn again what they already knew in the past. In the context of evolutionary search, this means that domains of the state space that have already been explored in the past are searched again. Then, a stable state is reached because both populations have finite size: in this case, the population of rules adapts faster than the population of ICs, resulting in a population focusing only on rules with high density and eliminating all instances of low density rules. Then, low density ICs exploit those

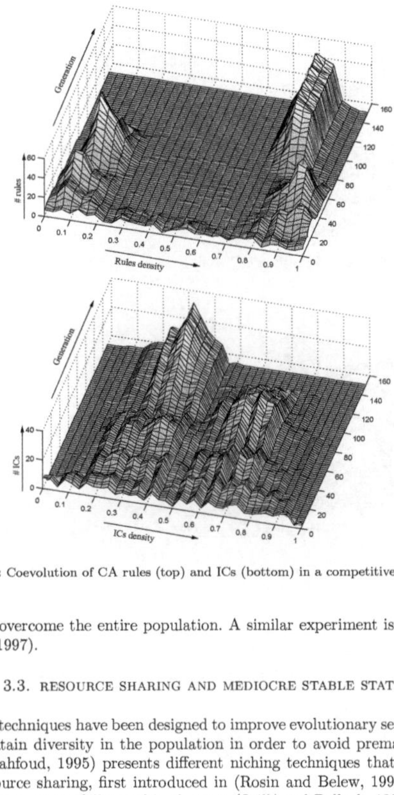

Figure 3: Coevolution of CA rules (top) and ICs (bottom) in a competitive relationship.

rules and overcome the entire population. A similar experiment is described in (Paredis, 1997).

3.3. RESOURCE SHARING AND MEDIOCRE STABLE STATES

Several techniques have been designed to improve evolutionary search. Usually they maintain diversity in the population in order to avoid premature convergence. (Mahfoud, 1995) presents different niching techniques that achieve this goal. Resource sharing, first introduced in (Rosin and Belew, 1995), is a technique that we successfully used in the past (Juillé and Pollack, 1996). Resource sharing implements a coverage-based heuristic by giving a higher payoff to problems that few individuals can solve. Resource sharing can be introduced in the competitive model of coevolution as follows:

$$f(R_i) = \sum_{j=1}^{n_{IC}} weight\_IC_j \times covered(R_i, IC_j) \quad (5)$$

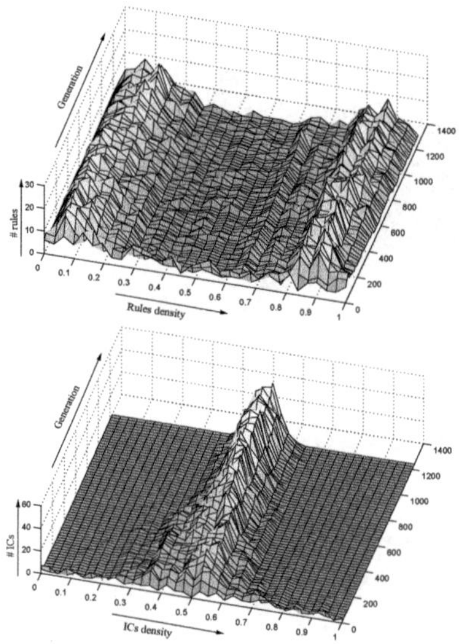

Figure 4: Coevolution of CA rules (top) and ICs (bottom) in a competitive relationship with resource sharing.

where:

$$weight\_IC_j = \frac{1}{\sum_{k=1}^{n_R} covered(R_k, IC_j)} \tag{6}$$

and

$$f(IC_j) = \sum_{i=1}^{n_R} weight\_R_i \times \overline{covered(R_i, IC_j)} \tag{7}$$

where:

$$weight\_R_i = \frac{1}{\sum_{k=1}^{n_{IC}} \overline{covered(R_i, IC_k)}} \tag{8}$$

In this definition, the weight of an IC corresponds to the payoff it returns if a rule covers it. If few rules cover an IC, this weight will be much larger than if a lot of rules cover that same IC. The definition for the weight of rules has the same purpose. This framework allows the presence of multiple niches (or species) in populations. Figure 4 describes one run for this definition of the fitness. The cyclic behavior which was observed in the previous section doesn't occur anymore. Instead, two species coexist in the population of rules: a species for low density rules and another one for high density rules. Those two species drive the evolution of ICs toward the domain of initial configurations that are most difficult to classify (i.e., $\rho_0 = 1/2$). However, the two populations have entered a *mediocre stable state*. This means that multiple average performance niches coexist in both populations in a stable manner. Put in another way, this can be seen as an equilibrium configuration in which a number of suboptimal species have found a way to collude by sharing the total credit between themselves. Usually, this is a consequence of some singularities inherent in the problem definition and/or the search procedure. In our example, ICs are concentrated around the $\rho_0 = 1/2$ threshold and they can be divided into two groups: those with density $\rho_0 < 1/2$ and those with density $\rho_0 > 1/2$. This distribution means that ICs can be exploited consistently by rules with low and high density that both occur in the second population (because a CA implementing a low (resp. high) density rule usually relaxes to all 0's (resp. all 1's) for most ICs). However, this is a mediocre stable state in the sense that evolved rules have poor performance with respect to the $\rho_c = 1/2$ task and there is no pressure toward improvement. The concept of mediocre stable states is also discussed in (Pollack *et al.*, 1996).

3.4. DISCUSSION

We have described different models for the coevolution of two populations. Some of the fundamental impediments to coevolutionary learning have been identified along with some of the reasons why continuous progress is difficult to achieve. It is now clear that none of these approaches can address successfully the problem of coevolutionary learning alone. All the rules discovered in those experiments perform poorly since they never approach the 50% density. The following section proposes a framework to get around those problems.

Each of the canonical models discussed so far implements a single specific strategy. In the literature, there has been some successful applications for both the cooperative and the competitive approaches. However, those works usually introduce some mechanisms to address the problems specific to each model. For instance, a noisy evaluation of the fitness can force exploration in a cooperative model, and an evaluation of individuals with respect to a set of opponents extracted from previous generations can limit the cyclic behavior observed in competitive models (e.g., see the life-time fitness evaluation technique described in (Paredis, 1996) or the "hall of fame" method presented in (Rosin, 1997)). However, those mechanisms usually fail to address entirely the fundamental issues discussed previously.

4. Coevolving the "Ideal" Trainer

4.1. PRESENTATION OF THE APPROACH

From the analysis of the experiments presented in section 3 at least two reasons seem to prevent continuous progress in coevolutionary search.

The first one, identified as the occurrence of mediocre stable states, is that the training environment provided by the population of problems (ICs) returns little information to the population of "learning agents" (evolving rules) because a stable configuration is reached in which the credit is distributed according to a fixed pattern. Therefore, *adaptability*, which represents the ability of evolving agents to exhibit continuous progress, is a central property that must be captured. However, capturing adaptability is possible only if the population of evolving agents is exposed to a gradient. That is, some information is available allowing differentiation between the members of the population. This pinpoints our first requirement for achieving continuous progress : the training environment should facilitate adaptability by always exposing the population of evolving agents to a gradient.
The second reason that prevents continuous progress is that the dynamics of the search performed by the two coevolving populations doesn't drive individuals toward domains of the state space that contain most promising solutions because there is no "high-level" strategy to play this role. The coevolving populations explore the state space with no other goal than performing well in the environment they are exposed to at the current moment. There is no notion of continuous progress with respect to some absolute measure. This is a consequence of the Red Queen effect which usually results in a continuous wandering across the state space. This dynamics is illustrated with the cyclic behavior observed in the competitive model of coevolution. Indeed, continuous progress is a global property: it is defined with respect to some absolute measure of performance. Therefore, to achieve continuous progress (and the emergence of high quality solutions), problems of increasing difficulty should be proposed by the training environment. This constitutes our second requirement which can be achieved only if this absolute notion of difficulty is introduced in the system in the form of a meta-level strategy. The purpose of the meta-level strategy is to prevent cyclic behaviors by providing a direction for the evolution of the training environment.

The idea of the "ideal" trainer is based on the introduction of explicit mechanisms to implement the two requirements identified in the previous paragraph. The purpose of those mechanisms is to control the evolution of the population of problems. In fact, the central strategy underlying the concept of the "ideal" trainer can be described with the following statement:

> *"The best way for adaptive agents to learn is to be exposed to problems that are just a little more difficult than those they already know how to solve."*

This strategy covers exactly the two requirements discussed previously: the need to maintain useful feedback (by exposing agents to problems "a little more difficult" than those they know how to solve) and the need for a meta-level strategy (to propose problems of increasing difficulty). The implementation of this strategy requires the definition of the following terms:

1. a *distance measure* between agents performance and problems difficulty to formalize the concept of "a little more difficult", and

2. a *partial order* over the space of problems in order to control the evolution of the training environment toward problems of increasing difficulty.

It should be noted that the *partial order* defines a fixed structure over the space of problems with respect to an absolute notion of problem difficulty, while the *distance measure* is relative to each agent (a problem that is difficult for one agent may actually be simple for another). As a result of this definition, this search procedure always maintains a gradient to drive the exploration of the space of solutions. That is, the training environment proposes a variety of problems covering a range of difficulty without exposing the evolving agents to problems that are too difficult or too easy. Indeed, if problems are too difficult, none of the agents can solve them. On the contrary, if they are too easy, all the agents can solve them. In both cases, those problems are useless for learning since they provide little information for differentiating members of the population. Then, the "high-level" strategy embedded in the search procedure allows continuous progress by proposing problems of increasing difficulty with respect to an absolute reference, thereby preventing some of the negative effects associated with the Red Queen. By maintaining this constant pressure toward slightly more difficult problems, a race is induced among evolving agents such that agents that adapt better have an evolutionary advantage. The underlying heuristic implemented by this evolutionary race is that *adaptability* for solving increasingly difficult problems is the driving force.

When applying the "ideal" trainer concept to a specific task, multiple difficulties must be overcome in order to implement accurately the different concepts introduced in this section. So far, our methodology has consisted in constructing an explicit topology over the space of problems by defining a partial order with respect to the relative difficulty of problems among each other. In our current work, the concept of "relative difficulty" has been defined by exploiting some *a priori* knowledge about the task. The definition of this topology over the space of problems makes possible the implementation of the two goals required in our coevolutionary learning approach. Indeed, since learners are evaluated against a known range of difficulty for problems, it is possible to monitor their progress and to expose them to problems that are just "a little more difficult". In our work, this last concept has been formalized by defining empirically a distance measure. In this framework, learners are always exposed to a gradient for search and it is possible to control the evolution of the training environment toward problems of increasing difficulty in order to ensure continuous progress.

In the future, our goal is to eliminate some of those explicit components by introducing some heuristics that automatically identify problems that are appropriate for the current set of learners. The work of Rosin (Rosin, 1997) already proposes a few methods to address this issue.

4.2. DISCUSSION

As stated previously, the coevolutionary learning framework introduces a pressure toward adaptability. The central assumption is that individuals that adapt faster than others in order to solve the new challenges they are exposed to are also more likely to solve even more difficult problems. The main difficulty is to

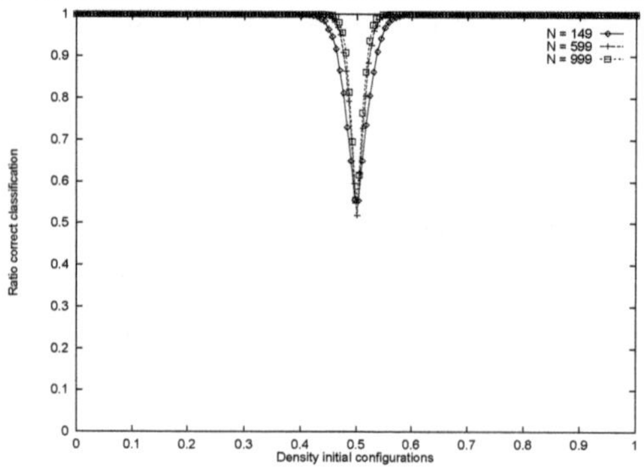

Figure 5: Distribution of performance for the GKL rule for $\rho_0 \in [0.0, 1.0]$.

setup a coevolutionary framework that implements this heuristic accurately and efficiently.

The new contribution of this work is the idea of maintaining a gradient for search as one of the underlying heuristics. In the literature, different approaches have been proposed to address the issues associated with the Red Queen effect (Paredis, 1996; Rosin, 1997). However, to our knowledge, explicit methods to force progress and to prevent mediocre stable states in the context of evolutionary search have never been tried.

The idea of introducing a pressure toward adaptability as the central heuristic for search is not new. Schmidhuber (Schmidhuber, 1995) proposed the Incremental Self-Improvement system in which adaptability is the measure that is optimized. The contribution of our paper lies in the formulation of this heuristic in the framework of evolutionary search. The advantages of this approach are twofolds: robustness of search, and the potential for an efficient parallel implementation.

The concept of an ideal trainer is also discussed in (Epstein, 1994) in the context of game learning. However, this work addresses the issue of designing the "ideal" training procedure which would result in high quality players rather than coevolving the training environment in response to the progress of learners.

5. Application to the Discovery of CA Rules

5.1. Experimental Setup

The approach described in the previous section is applied to the $\rho_c = 1/2$ task. It is believed that ICs become more and more difficult to classify correctly as their density gets closer to the ρ_c threshold. This hypothesis is supported by the distribution of the performance for the GKL rule for $\rho_0 \in [0.0, 1.0]$ presented in figure 5. Therefore, our idea is to construct a framework that adapts the distribution of the density for the population of ICs as CA-rules are getting better to solve the task. The following definition for the fitness of rules and ICs has been used to achieve this goal.

$$f(R_i) = \sum_{j=1}^{n_{IC}} weight\_IC_j \times covered(R_i, IC_j) \tag{9}$$

where:

$$weight\_IC_j = \frac{1}{\sum_{k=1}^{n_R} covered(R_k, IC_j)} \tag{10}$$

and

$$f(IC_j) = \sum_{i=1}^{n_R} weight\_R_i' \times E(R_i, \rho(IC_j)) \times \overline{covered(R_i, IC_j)} \tag{11}$$

where:

$$weight\_R_i' = \frac{1}{\sum_{k=1}^{n_{IC}} E(R_i, \rho(IC_k)) \times \overline{covered(R_i, IC_k)}} \tag{12}$$

This definition implements the competitive relationship with resource sharing. However, a new component, namely $E(R_i, \rho(IC_j))$, has been added in the definition of the ICs' fitness. The purpose of this new component is to penalize ICs with density $\rho(IC_j)$ if little information is collected with respect to the rule R_i. Indeed, we consider that if a rule R_i has a 50% classification accuracy over ICs with density $\rho(IC_j)$ then this is equivalent to random guessing and no payoff should be returned to IC_j. On the contrary, if the performance of R_i is significantly better or worse than the 50% threshold for a given density of ICs this means that R_i captured some relevant properties to deal with those ICs. Once again, the idea is that the training environment should be composed of ICs that provide useful information to identify good rules from poor ones. In order to allow continuous progress, our implementation exploits an intrinsic property of the $\rho_c = 1/2$ task. Indeed, it seems that CA-rules that cover ICs with density $\rho_0 < 1/2$ (resp. $\rho_0 > 1/2$) with high performance will also be very successful over ICs with density $\rho_0' < \rho_0$ (resp. $\rho_0' > \rho_0$). Therefore, as ICs become more difficult, their density is approaching $\rho_0 = 1/2$ but rules don't have to be tested against easier ICs. Following this idea, we defined $E()$ as the complement of the entropy of the outcome between a rule and ICs with a given density:

$$E(R_i, \rho(IC_j)) = \log(2) + p\log(p) + (1-p)\log(1-p) \tag{13}$$

where: p is the probability that an IC with density $\rho(IC_j)$ defeats the rule R_i. $E()$ implements the distance measure discussed in section 4.1. Its purpose is to maintain the balance between the search for more difficult ICs and ICs that can be solved by rules. In practice, the entropy is evaluated by performing some simple statistics over the population of ICs : p is the number of ICs in the population with density $\rho(IC_j)$ that rule R_i classifies correctly divided by the total number of ICs in the population with density $\rho(IC_j)$.

Table 2: Description of the current best rule and published rules for the $\rho_c = 1/2$ task.

Coevolution	00010100	01011111	01000000	00000000
	00010111	11111100	00000010	00010111
	00010100	01011111	00000011	00001111
	00010111	11111111	11111111	11010111
Das rule	00000111	00000000	00000111	11111111
	00001111	00000000	00001111	11111111
	00001111	00000000	00000111	11111111
	00001111	00110001	00001111	11111111
ABK rule	00000101	00000000	01010101	00000101
	00000101	00000000	01010101	00000101
	01010101	11111111	01010101	11111111
	01010101	11111111	01010101	11111111
GKL rule	00000000	01011111	00000000	01011111
	00000000	01011111	00000000	01011111
	00000000	01011111	11111111	01011111
	00000000	01011111	11111111	01011111

5.2. EXPERIMENTAL RESULTS

Experiments were performed with different sizes for the population of rules and ICs. The best rule whose performance is reported in table 1 resulted from the experiments that used the largest population size. In those experiments, 6 runs were performed for 5,000 generations, using a size of 1,000 for the two populations. Each rule is coded on a binary string of length $2^{2*r+1} = 128$. One-point crossover is used with a 2% bit mutation probability. The population of rules is initialized according to a uniform distribution over [0.0, 1.0] for the density. Each individual in the population of ICs represents a density $\rho_0 \in [0.0, 1.0]$. This population is also initialized according to a uniform distribution over $\rho_0 \in [0.0, 1.0]$. At each generation, each member generates a new instance for an initial configuration with respect to the density it represents. All rules are evaluated against this new set of ICs. The motivation for having evolving individuals represent a density instead of an actual instance of an IC is to favor the emergence of robust rules, that is rules that perform consistently against a variety of ICs. The generation gap is 5% for the population of ICs (i.e., the top 95% ICs reproduce to the next generation). There is no crossover nor mutation. The new 5% ICs are the result of a random sampling over $\rho_0 \in [0.0, 1.0]$ according to a uniform probability distribution. The generation gap is 80% for the population of rules. New rules are created by crossover and mutation. Parents are randomly selected from the top 20%. All runs consistently evolved some rules that score above 82%. Table 2 describes lookup tables for the current best CA rule and other rules discussed in the literature. The leftmost bit corresponds to the result of the rule on input 0000000, the second bit corresponds to input 0000001, ... and the rightmost bit corresponds to input 1111111.

Figure 6 describes the evolution of the density of rules and ICs for one run. As rules improve, their density gets closer to 1/2 and the density of ICs is distributed on two peaks on each side of $\rho_c = 1/2$. In that particular run, it is only after

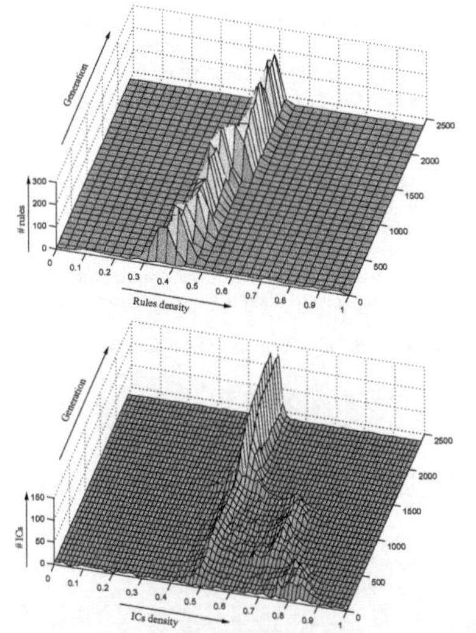

Figure 6: Coevolutionary learning between CA rules (top) and ICs (bottom).

1, 300 generations that a significant improvement is observed for rules and that, in response, the population of ICs adapts dramatically in order to propose more challenging initial configurations. This shows that our strategy to coevolve the training environment and the learners has been successfully implemented in the definition of the fitness functions.

5.3. PERFORMANCE COMPARISON: FIXED VS. ADAPTING SEARCH ENVIRONMENT

The purpose of this section is to illustrate experimentally that the search performed in a framework where the environment responds appropriately to the progress of individuals does result in an improved performance. This analysis is composed of a set of three experiments. In all experiments, a population size of 200 is used for rules. For each set of experiments, the population of ICs is generated as follows:

Experiment 1: rules are evaluated in an environment composed of 200 ICs drawn at each generation from an *unbiased* distribution. This means, that the distribution of the density of ICs is binomial centered on 0.5.

Experiment 2: rules are evaluated in an environment composed of 200 ICs drawn at each generation from a *biased* distribution, such that the distribution for the density of ICs is uniform.

Experiment 3: rules are coevolving with a population of ICs composed of 200 individuals. The "ideal" trainer framework is implemented to control the evolution of the population of ICs.

Because the following experiments are computer intensive, the population size has been reduced to 200 individuals (compared to 1000 in the experiments discussed in section 5.2).
For all those experiments, the definition for the fitness of rules is identical:

$$f(R_i) = \sum_{j=1}^{n_{IC}} weight\_IC_j \times covered(R_i, IC_j) \tag{14}$$

where:

$$weight\_IC_j = \frac{1}{\sum_{k=1}^{n_R} covered(R_k, IC_j)} \tag{15}$$

This definition implements resource sharing in order to allow multiple species in the population of rules. In the coevolutionary experiments, the definition for the fitness of ICs is the same as the one introduced in section 5.1:

$$f(IC_j) = \sum_{i=1}^{n_R} weight\_R'_i \times E(R_i, \rho(IC_j)) \times \overline{covered(R_i, IC_j)} \tag{16}$$

where:

$$weight\_R'_i = \frac{1}{\sum_{k=1}^{n_{IC}} E(R_i, \rho(IC_k)) \times \overline{covered(R_i, IC_k)}} \tag{17}$$

and:

$$E(R_i, \rho(IC_j)) = \log(2) + p\log(p) + (1-p)\log(1-p) \tag{18}$$

The representation for rules and ICs is also identical to the one in previous experiments. In the coevolutionary experiments, each individual in the population of ICs represents a density $\rho_0 \in [0.0, 1.0]$. At each generation, each member generates a new instance for an IC with respect to the density it represents. For each set of experiments, 100 runs were performed. For the three sets of experiments, the generation gap is 80% for the population of rules (i.e., the top 20% reproduces in the next generation). For the third set of experiments, the generation gap is 3% for the population of ICs.

At each generation, the top 10 rules are evaluated against 5,000 ICs drawn randomly according to the unbiased distribution. After 500 and 1,000 generations respectively, the average of the performance for those top 10 rules is computed. An average over a window of size 10 is considered in order to smooth the noisy evaluation for rules performance. Figures 7 and 8 describe the evolution of the

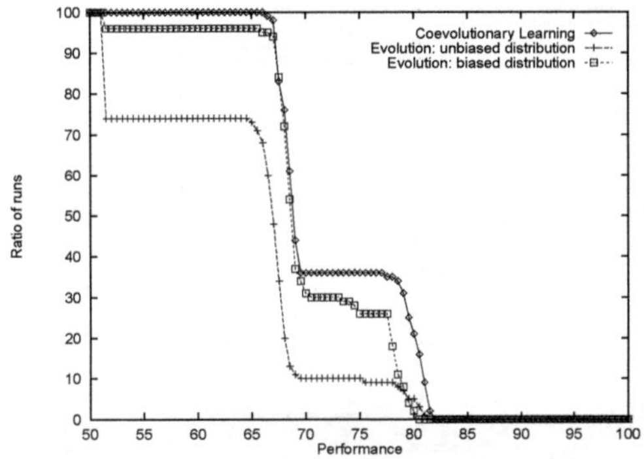

Figure 7: Distribution of the ratio of runs achieving a specific performance after 500 generations.

ratio of runs for which this average is above a given performance (represented on the horizontal axis).

The following observations can be made from those figures. First, the "ideal" trainer framework always results in a higher ratio of success. In particular, there is a 100% success rate for generating a rule of performance above 67% (about 96% of success with the biased distribution and about 74% of success with the unbiased distribution). Moreover, coevolution also results in the discovery of rules of higher performance (a few rules with performance above 82% were discovered).

Second, the experiments with the biased distribution resulted in a higher success rate than the experiments with the unbiased distribution. Indeed, in the case of an unbiased distribution, rules are exposed to ICs with density close to 0.5, which are the most difficult. As a result, there is little information about the gradient and a large number of runs don't even discover rules that do significantly better than random guessing. However, for those few runs for which good rules have been discovered, the unbiased setup proposes a better environment to continue improvement than the biased setup. The reason is that the biased environment doesn't offer a challenging environment anymore to the population of rules because most of the ICs are covered by those rules. As a result, there is almost no gradient to drive the search. On the contrary, the unbiased environment still offers a gradient for search and results in the discovery of rules that have better performance than the best rules discovered in the experiments with the biased distribution, both after 500 and $1,000$ generations.

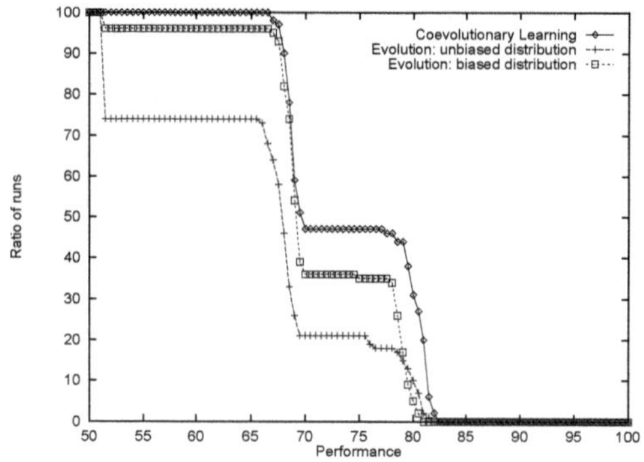

Figure 8: Distribution of the ratio of runs achieving a specific performance after 1000 generations.

6. Conclusion

This paper explores the use of coevolution as a paradigm for the design of complex systems that can implement non-trivial tasks requiring the coordinated action of all their parts. The concept of *coevolutionary learning* has been defined as a scheme from which continuous progress for the performance of evolving agents can be observed. After proposing different setups for controlling coevolution between a population of agents and a population of problems, two fundamental requirements have been identified in order to achieve the objective of continuous progress:

1. The training environment should facilitate adaptability by always exposing the population of evolving agents to a gradient, and
2. Problems of increasing difficulty with respect to some absolute measure should be proposed by the training environment.

The concept of an "ideal" trainer described in this paper implements those two requirements by exploiting the definition of a topology over the space of problems. We applied this new paradigm to the problem of evolving CA rules for a classification task. Our experiments resulted in new rules whose performance improves very significantly over previously known rules for that particular task.

Coevolution is a paradigm that is particularly attractive because it proposes a solution to the problem of designing a training environment for searching a space of complex system designs. Indeed, the inductive bias that results from the evaluation of candidate solutions in a static training environment relies on the introduction of explicit knowledge about the structural properties of the

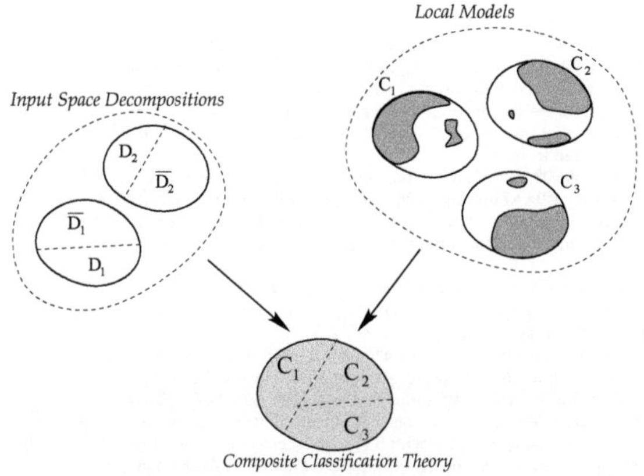

Figure 9: The space of decompositions and the space of local models are explored simultaneously. If some decompositions characterizing the domain of specialization of some local models (represented in dark) are discovered, then a composite classification theory may be constructed. In that example, the composite solution is defined as follows: **if** $(x, y) \in D_2$ **then** C_1 **else if** $(x, y) \in D_1$ **then** C_3 **else** C_2

search space. However, in the current state of the art, the relationship between interaction rules in a complex system and the emergent behavior resulting from those interactions are still poorly understood. Therefore, the structural properties of the phenotype space associated with a space of complex system designs are extremely difficult to capture in a formal scheme. As a result, the design of any static training environment is based only on subjective choices that can rarely be substantiated with solid arguments. The goal of this paper is to provide some insights on the use of coevolutionary approaches to address such issues. By providing a methodology in which candidate solutions are evaluated in a changing environment, more elaborate heuristics can be implemented for search. In the case of the "ideal" trainer paradigm, the central underlying heuristic introduces a pressure toward adaptability, which is a fundamental condition for achieving continuous progress. The experimental results presented in this paper does support the idea that this paradigm can be effective in tackling the task of designing complex systems.

The authors have also applied the concept of an "ideal" trainer to propose a modular approach to inductive learning (Juillé, 1999). To put it in a nutshell, Modular Inductive Learning (MIL) is a bottom-up approach to inductive learning which explores at the same time a space of local models and a space of decompositions of the input space. Two goals underlie the strategy implemented in MIL for the exploration of those two spaces. The first one is to discover some local models that propose an accurate approximation of the training data over some domain of the input space, along with decompositions of the input space that capture the domains of "specialization" of those local models. The central idea motivating that strategy is that a composite classification theory with good accuracy over the entire input space may then be constructed. This strategy is illustrated with the diagram in Figure 9.

The second goal consists in discovering a composite solutions using as few components as possible while still exhibiting a high accuracy with respect to the training data. The motivation for that goal is to induce classification theories that also generalize to input data outside the training set. This is a direct application of Occam's razor.

The challenge of Modular Inductive Learning is to implement the right balance between accuracy and the complexity of the final classification theory (that is, the number of components in its description).

MIL is a natural extension of the "ideal" trainer paradigm in which learning agents correspond to local models (or concepts) while problems correspond to domains of the input space. Indeed, by maintaining a pressure toward adaptability, the "ideal" trainer paradigm favors the emergence of agents that capture some intrinsic properties of the changing environment. This side effect of adaptability is fundamental to the notion of generalization because it allows the construction of compact descriptions. The central idea of the MIL system is to exploit that feature as the fundamental strategy to control search in the space of decompositions and the space of local models. A detailed description of the MIL system is presented in (Juillé, 1999).

In conclusion, the authors believe in the capacity of the "ideal" trainer paradigm to address a large range of problems that are characterized by a ill-structured search space and for which limited knowledge is available. In particular, our goal is to confirm that this paradigm can contribute to solve some of the many challenges relative to the design of complex systems by introducing general-purpose heuristics to replace the current problem-specific strategies.

References

Andre, D., Bennett III, F. H. and Koza, J. R. Evolution of intricate long-distance communication signals in cellular automata using genetic programming. In: *Proceedings of the Fifth Artificial Life Conference*, pp. 16–18 (1996).

Bonabeau, E., Dorigo, M. and Theraulaz, G. *Swarm Intelligence : From Natural to Artificial Systems*. Oxford University Press (1999).

Capcarrere, M. S., Sipper, M. and Tomassini, M. Two-state, r=1 cellular automaton that classifies density. *Physical Review Letters* **77** (24), 4969–4971 (1996).

Cliff, D. and Miller, G. F. Tracking the red queen: Measurements of adaptive progress in co-evolutionary simulations. In: *The Third European Conference on Artificial Life*, pp. 200–218. Springer-Verlag (1995). LNCS 929.

Cliff, D. and Miller, G. F. Co-evolution of pursuit and evasion II: Simulation methods and results. In: *Proceedings of the Fourth International Conference on Simulation of Adaptive Behavior* (Maes, P., Mataric, M. J., Meyer, J.-A., Pollack, J. and Wilson, S. W., eds.), pp. 506–515. Cambridge, Massachusetts: MIT Press (1996).

Das, R., Mitchell, M. and Crutchfield, J. P. A genetic algorithm discovers particle-based computation in cellular automata. In: *Parallel Problem Solving from Nature III, LNCS 866*, pp. 344–353. Springer-Verlag (1994).

Epstein, S. L. Toward an ideal trainer. *Machine Learning* **15**, 251–277 (1994).

Hillis, W. D. Co-evolving parasites improve simulated evolution as an optimization procedure. In: *Artificial Life II* (Langton, C. *et al.*, eds.), pp. 313–324. Addison Wesley (1992).

Husbands, P. Distributed coevolutionary genetic algorithms for multi-criteria and multi-constraint optimisation. In: *Proceedings of Evolutionary computing, AISB Workshop Selected Papers* (Fogarty, T., ed.), pp. 150–165. Springer-Verlag (1994). LNCS 865.

Juillé, H. Evolution of non-deterministic incremental algorithms as a new approach for search in state spaces. In: *Proceedings of the Sixth International Conference on Genetic Algorithms* (Eshelman, L. J., ed.). San Mateo, California: Morgan Kaufmann (1995).

Juillé, H. *Methods for Statistical Inference: Extending the Evolutionary Computation Paradigm*. Ph.D. thesis, Brandeis University, Waltham, Massachusetts, USA (1999).

Juillé, H. and Pollack, J. B. Co-evolving intertwined spirals. In: *Proceedings of the Fifth Annual Conference on Evolutionary Programming*, pp. 461–468. MIT Press (1996).

Koza, J. R., Bennett, F. H., Andre, D. and Keane, M. A. Four problems for which a computer program evolved by genetic programming is competitive with human performance. In: *Proceedings of the Third IEEE International Conference on Evolutionary Computation* (1996).

Land, M. and Belew, R. K. No perfect two-state cellular automata for density classification exists. *Physical Review Letters* **74** (25), 1548–1550 (1995).

Mahfoud, S. W. *Niching Methods for Genetic Algorithms*. Ph.D. thesis, University of Illinois at Urbana-Champaign (1995). IlliGAL Report No. 95001.

Mitchell, M., Crutchfield, J. P. and Hraber, P. T. Evolving cellular automata to perform computations: Mechanisms and impediments. *Physica D* **75**, 361–391 (1994).

Mitchell, M., Hraber, P. T. and Crutchfield, J. P. Revisiting the edge of chaos: Evolving cellular automata to perform computations. *Complex Systems* **7**, 89–130 (1993).

Moriarty, D. E. *Symbiotic Evolution of Neural Networks in Sequential Decision Tasks*. Ph.D. thesis, University of Texas at Austin, USA (1997).

Paredis, J. Co-evolutionary constraint satisfaction. In: *Parallel Problem Solving from Nature - PPSN III, LNCS 866* (Davidor, Y., Schwefel, H.-P. and Manner, R., eds.), pp. 46–55. Springer-Verlag (1994a).

Paredis, J. Steps towards co-evolutionary classification neural networks. In: *Artificial Life IV* (Brooks and Maes, eds.), pp. 102–108. MIT Press (1994b).

Paredis, J. The symbiotic evolution of solutions and their representations. In: *Proceedings of the Sixth International Conference on Genetic Algorithms* (Eshelman, L. J., ed.), pp. 359–365. Morgan Kaufmann (1995).

Paredis, J. Coevolutionary computation. *Artificial Life* **2** (4) (1996).

Paredis, J. Coevolving cellular automata: Be aware of the red queen! In: *Proceedings of the Seventh International Conference on Genetic Algorithms* (Bäck, T., ed.), pp. 393–400. Morgan Kaufmann (1997).

Peck, C. C. and Dhawan, A. P. Genetic algorithms as global random search methods: An alternative perspective. *Evolutionary Computation* **3** (1), 39–80 (1995).

Pollack, J. B., Blair, A. and Land, M. Coevolution of a backgammon player. In: *Proceedings of Artificial Life V* (Langton, C., ed.). MIT Press (1996).

Potter, M. A. *The Design and Analysis of a Computational Model of Cooperative Coevolution*. Ph.D. thesis, George Mason University, Fairfax, Virginia (1997).

Potter, M. A. and De Jong, K. A. A cooperative coevolutionary approach to function optimization. In: *Parallel Problem Solving from Nature - PPSN III, LNCS 866*, pp. 249–257. Springer-Verlag (1994).

Potter, M. A., De Jong, K. A. and Grefenstette, J. J. A coevolutionary approach to learning sequential decision rules. In: *Proceedings of the Sixth International Conference on Genetic Algorithms* (Eshelman, L. J., ed.), pp. 366–372. San Mateo, California: Morgan Kauffmann (1995).

Reynolds, C. W. Competition, coevolution, and the game of tag. In: *Artificial Life IV* (Brooks and Maes, eds.). MIT Press (1994).

Rosin, C. D. *Coevolutionary Search Among Adversaries*. Ph.D. thesis, University of California, San Diego (1997).

Rosin, C. D. and Belew, R. K. Methods for competitive co-evolution: Finding opponents worth beating. In: *Proceedings of the Sixth International Conference on Genetic Algorithms* (Eshelman, L. J., ed.). San Mateo, California: Morgan Kauffmann (1995).

Rosin, C. D. and Belew, R. K. A competitive approach to game learning. In: *Proceedings of the Ninth Annual ACM Conference on Computational Learning Theory* (1996).

Schmidhuber, J. Discovering solutions with low kolmogorov complexity and high generalization capability. In: *Machine Learning: Proceedings of the twelfth International Conference* (Prieditis, A. and Russell, S., eds.), pp. 188–196. Morgan Kaufmann (1995).

Sims, K. Evolving 3d morphology and behavior by competition. In: *Artificial Life IV* (Brooks and Maes, eds.), pp. 28–39. MIT Press (1994).

Advances in Complex Systems, Vol. 4, No. 1 (2001) 71–88

DYNAMICS OF ECONOMIC AND TECHNOLOGICAL SEARCH PROCESSES IN COMPLEX ADAPTIVE LANDSCAPES

W. EBELING

Humboldt University Berlin, Institute of Physics,
Invalidenstrasse 110, D-10115 Berlin, Federal Republic of Germany
E-mail: ebeling@physik.hu-berlin.de

KARMESHU

Jawaharlal Nehru University, School of Computer and Systems Sciences,
New Delhi-110067, India
E-mail: karmeshu@jnuniv.ernet.in

A. SCHARNHORST*

Wissenschaftszentrum Berlin für Sozialforschung,
Reichpietschufer 50, D-10785 Berlin, Federal Republic of Germany
E-mail: as@medea.wz-berlin.de

Received 10 June 2000
Accepted 5 July 2000

We investigate the dynamics of economic evolution and technological change as hill-climbing in an adaptive landscape over a continuous characteristics space. A technology/firm is described by a large number of attributes or characteristics representing technology-inherent aspects, financial, organizational and economic features. These parameters span a characteristics space, which is a real Euclidean vector space, in analogy to the phenotype space in biology. Further we define a real-valued multimodal fitness function/functional and a population density over the characteristics space. The evolutionary dynamics including competition and mutations/innovations is modeled by reaction-diffusion equations of Fisher–Eigen or Lotka–Volterra type. We demonstrate the potential of such models, which in certain aspects go beyond the widespread applications of discrete replicator dynamics. Concerning technological change the emergence of technological populations as the result of a search process in an adaptive landscape will be investigated. In particular, the relation between incremental and radical innovations will be considered, especially the apparent paradox of a discrete continuum of technological change. Further, an application of the developed framework to the assessment of firms in the stock market is discussed.

Keywords: Evolutionary economics, technological populations, characteristics space, adaptive fitness landscape, continuous evolutionary model.

*Corresponding author.

1. Introduction

The analogies of economic and biological evolution are extensively discussed in the literature. Approaches that describe technological or economic change from a biophysical perspective include concepts like competition and selection and corresponding mathematical models [51, 49, 38, 53, 23, 37, 11, 73, 62, 70, 7, 39, 71, 21, 74, 44, 66, 17, 42, 59, 40, 65]. Self-organization theories from the physical research tradition including irreversibility, non-linearity and fluctuations have mainly influenced this direction. Recently, from statistical physics concepts like fractals, self-organized criticality and scaling have been applied successfully to economic problems, e.g., firm growth and financial market dynamics [68, 57, 46, 67, 45]. In this paper we describe the evolution of technological or firm populations as hill-climbing process in an adaptive fitness landscape over a continuous characteristics space. The relevance of the concept of an adaptive landscape for economic and technological evolution has been discussed notably by Allen [3–5]. Meanwhile, the idea of evolutionary search processes in an abstract fitness landscape can be found in different socio-economic models [13, 31, 36, 41, 64, 72]. Following this line we develop in this paper a framework for continuous models of technological and economic evolution. We demonstrate the potential of such models, which in certain respects, goes beyond the widespread application of discrete replicator dynamics.

1.1. *Conceptual background*

A technology can be described by a large number of attributes, features or characteristics representing inherent technology aspects (performance, size, chemical composition) and economic parameters (input coefficients or certain product attributes) [53]. Often technological change is visualized by means of the temporal evolution of a single characteristic or some corresponding indicator. Some approaches extend such a one-dimensional description to a multidimensional mapping of technological evolution. Sahal [54, 55] used a classification of design variables (e.g., the stroke length of an engine) and performance variables (e.g., the fuel consumption) to build a "topography of technological evolution". Based on this concept technological models occupy different loci in a parameter space. The probability distribution of certain parameters forms a surface over the space, and the movement of the occupied regions indicates technological change. Saviotti and Metcalfe [60] elaborated a methodological framework for a characteristics space of product technologies. They use two sets of output indicators, viz. technical characteristics and service characteristics, and describe processes like substitution, specialization and innovation in terms of changes in these parameter sets (see also [61, 58, 63]).

A similar description may be developed for the evolution of firms and, in particular, their evaluation in the finance market. Indeed, the price of shares of a firm in the stock market depends on a multitude of characteristic features like the capital stock, the investment rate, the management, the working skills, the technological profile and R&D strategies. The assessment of a firm through the market environment depends not only on inherent properties of firm but also on the interaction with other firms (e.g., competitors, suppliers) and on general properties of economic and political networks like information flows.

In this paper we develop a framework of continuous evolutionary models combining characteristics representation of a firm or a technology with an additional dimension corresponding to valuation or fitness function in an evolutionary dynamics. This dynamics incorporates various interactions between the participating competitors and feedback into the valuation function. In the following we propose the model framework mainly in the context of technological evolution. We want to emphasize, however, that the model framework itself is more universal and has a broad spectrum of applications. Therefore, an application of the modeling framework to market dynamics is also proposed. Some possible implications of such a type of models for the relation of firm behavior and financial markets are brought out in the conclusions.

Following the above mentioned concepts for mapping technological evolution we assume that the characteristics space is formed by three sets of indicators: technical characteristics X_1, X_2, $X_3, \dots, X_l$, service characteristics Y_1, Y_2, $Y_3, \dots, Y_m$ and financial characteristics Z_1, Z_2, $Z_3, \dots, Z_k$. Then, each product model is represented by a point in this space described by a certain vector $q = (q_1, q_2, \dots, q_d) = (X_1, X_2, \dots, X_l, Y_1, Y_2, \dots, Y_m, Z_1, Z_2, \dots, Z_k)$ with $d = l + m + k$. All real vectors span the characteristics space Q which is a real Euclidean vector space. This space is seen in analogy to the phenotype space in biology. Obviously, only a subset of all possible combinations of real numbers q_i is realized in the real evolutionary process. Most of the regions of Q are empty. Some combinations of parameter values are self-contradictory (see e.g., [34]). Therefore, the set of possible technological populations Q is restricted to certain compact areas of Q.

Next we define a population density $x(q, t) = x(q_1, q_2, \dots, q_d, t)$ as a function over the characteristics space. Morphologically oriented investigations of technologies show that the dominant or the finally established technology is typically selected among different possibilities [34, 26]. In general, the value of the function stands for the number or frequency a certain product model is realized (more precisely, $x(q)dq$ stands for the number of product models in the interval dq while $x(q, t)$ is the corresponding population density function). A more precise definition will affect the focus of the competition and selection considered. Here, we refer to the number of commodities produced and sold as a quantitative expression of the population density. In this case, following Dosi [22], competition is clearly realized through the market. Further, in order to understand the movement of technological populations in the characteristics space and to follow their trajectory, especially between locations which are able to compete, it seems to be useful to include also models that have been proposed but not yet market-proven. In light of this phenotypic approach the realized technological populations result from a competition and selection process through which the shape of the population density in terms of its modality changes. The occurrence and movement of such relatively stable populations corresponds clearly to concepts of *technological regimes* and *natural trajectories* [52], *dominant design* [69, 1], *technological guideposts* [54, 56]) and *technological paradigms/technological trajectories* [22].

These approaches are relevant in more than one respect for the modeling framework developed in this paper.

Further utilizing the analogy to biophysical evolution, we assume that the dynamics of competing economic or technological populations corresponds to a fitness function that spans a landscape over a multidimensional characteristics space. The success of a firm or technology is defined in economic terms through market environments [53]. To define characteristics of the external environment in a quantitative way would seem to be extremely complex. Moreover, it appears theoretically not possible to know the structure of this fitness landscape; however, from the existing economic/technological populations we can draw some conclusions about its structure. Looking at the multiplicity of technologies, the landscape, globally considered, will be clearly multimodal. Even in a certain restricted region of similar characteristics, specialization or the existence of niches can also be understood as a sign of multimodality. Further, we can assume that changes in the landscape will be smooth. Besides multimodality there are reasons to believe that the landscape has a very complicated structure. One reason is that economic structures and technologies normally represent compromises between several needs. Finding optimal solution under conflicting conditions is a so-called "frustrated problem" with a chaotically shaped landscape [9]. However, using recently developed mathematical techniques, it is possible to draw qualitative conclusions about characteristics of evolutionary processes under such conditions.

In this paper we use the conceptual setting developed to implement a continuous evolutionary model framework as a new approach to economic/technological evolution. If we see fitness as an unknown landscape over the characteristics space, economic and technological change is mainly a search and learning-by-doing process. This is quite consistent with most of the evolutionary theories of technological change and economic development [25]. Analogously to phenotypical evolution in biology, we describe the search process for economic and technological improvements as hill-climbing process in an adaptive evolutionary fitness landscape. In biophysics this type of model, originally proposed by Wright [75] and Conrad [19], was mathematically developed by Feistel and Ebeling [33] (see also [28]).

The central point of the theory of the adaptive evolutionary landscape is to link the growth and movement of populations to changes in the fitness. This feedback can lead to an increase in the survival possibilities for some parts of the population but to a decline for others. In technological evolution the exclusive character of technological paradigms (or lock-in) [10] seems to indicate such behavior.

2. Models of a Search Process in Complex Adaptive Landscapes

A formal model of an evolutionary search in a phenotype space reads in a continuous formulation [27, 33, 28]:

$$\partial_t x(q,t) = x(q,t)w(q;\{x\}) + Mx(q,t)\,, \tag{1}$$

where w is a function of q, and possibly also a functional of $x(q)$. M is the mutation operator. The function w describes the rate of growth or decline of the population density $x(q,t)$ (e.g., output of product models). In biophysical models w stands for the process of self-reproduction (replication). The "reproduction rate" $w(q)$ defines an evolutionary landscape superimposed on a high-dimensional space Q.

2.1. *The Fisher Eigen model of technological evolution*

In the rather simple Fisher–Eigen model $w(q)$ is given as the difference between a self-reproduction function $E(q)$ and its (only time-dependent) ensemble average [32]:

$$w(q,t) = E(q) - \langle E \rangle\,. \tag{2}$$

Then, the linear functional $w(q;x)$ is given as

$$w = E(q) - N^{-1}\int dq'\, E(q')x(q')\,, \tag{3}$$

where

$$N = \int x(q',t)dq'$$

is the overall population size. The population average $\langle E \rangle$ is defined by

$$\langle E \rangle = \frac{\int dq'\, E(q')x(q',t)}{\int dq'\, x(q',t)}\,. \tag{4}$$

The integral is extended over the whole phenotype space. Since the shape of the fitness functional will be unknown in general we assume that $E(q)$ (and consequently $w(q)$) has the form of a correlated random landscape. Then, in analogy with solid state physics we may consider $E(q)$ as a "correlated random potential". Several specific features of the search in high-dimensional landscapes ($d \gg 1$) have been analyzed by Conrad and Ebeling [20]. In particular, it has been shown that in high-dimensional correlated landscapes the probability that a given stationary point is a saddle point increases with the dimension. In other words, the searcher (or searching population) is more often confronted with the problem of leaving a saddle point than of escaping from a proper relative maximum.

The second term in (1) denotes a mutation operator. In the simplest case it may be modeled as a diffusion term:

$$Mx(q,t) = D\Delta x(q,t)\,, \tag{5}$$

where Δ is the Laplace operator and D is the diffusion coefficient. In a more general form we introduce the diffusion matrix D_{ij}

$$Mx(q,t) = \sum_i \partial_i \sum_{j=1}^{d} D_{ij}\partial_j x(q,t)\,. \tag{6}$$

In a still more general setting M denotes a linear operator

$$Mx = \int [A(q,q')x(q',t) - A(q',q)x(q,t)]dq'\,. \tag{7}$$

Here the transition matrix may take into account also long range transitions (as e.g. Cauchy or Levy-type transitions).

So far, a full analysis is available only for the simplest case given by Eqs. (1), (3) and (5) [33, 28]. The dynamic properties of equations of this type were investigated by Zeldovich *et al.* [78] and others [29, 12]. Mathematically speaking, they are closely related to the Bloch equations of statistical physics (Schrödinger equations with imaginary time). The problem reduces more or less to the solution of a Schrödinger eigenvalue problem for the potential $U(q) = -E(q)$. It may be considered as a great advantage for this kind of models, that many results of the quantum mechanics for more or less complicated potentials are available.

As shown in Fig. 1, in the Fisher–Eigen approach the self-reproduction function $E(q)$ can be considered as a landscape over the characteristics space. The region of technological populations follows the maxima of this function which lie above the ensemble average. The function $E(q)$ can be considered as valuation of different locations. This function is static, i.e. time-independent. Moreover, it does not depend on the interaction of the members of the populations. The fitness function in the narrower sense $w(q)$ undergoes temporal changes only through the ensemble average $\langle E(q)\rangle$. This means, the fitness landscape will be shifted only and does'nt change its shape. Models of this kind can be used to understand, e.g., concentration processes and the relation between continuity and discontinuity in technological evolution.

For the problem of economic and technological evolution it seems to be very interesting to consider the case when changes of the fitness landscape are endogenously determined and are related to the movement of the populations itself. "In

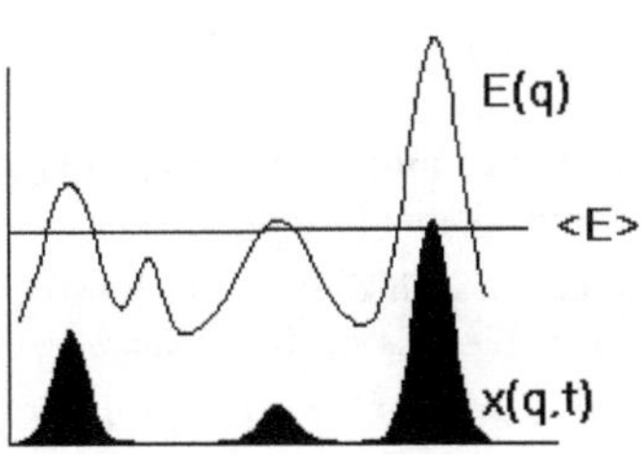

Fig. 1. One-dimensional representation of a fitness landscape and locations of technological populations.

changing the relative significance of competing technologies, selection also results in changes in the price structures that evaluate performance characteristics, so reshaping the selection environment. Indeed one of the central themes of the evolutionary approach to competition is that technologies and their selection environments co-evolve." ([48], p. 157). The dependence of the fitness landscape on growth and movement of populations is central to the theory of the adaptive evolutionary landscape. We will overcome the disadvantage of the models considered so far for the modeling of such a kind of feedback in more advanced models exhibiting co-evolution of the competitors.

2.2. *Lotka–Volterra dynamics of technological evolution*

Such a more flexible dynamics results if $w(q)$ is considered as a linear functional in the following way

$$w\big(q; x(q,t)\big) = a(q) + \int b(q,q')x(q',t)dq' \,. \tag{8}$$

The special case

$$a(q) = E(q)\,, \qquad b(q,q') = -E(q') \tag{9}$$

brings us back to the Fisher–Eigen case (3). In the general setting we find nonlinear equations of the type

$$\partial_t x(q,t) = \left[a(q) + \int b(q,q')x(q',t)\right] x(q,t) + Mx(q,t)\,. \tag{10}$$

So far, the dynamics of equations of this type which resemble Hartree eqs. with imaginary time is not well investigated; only the initial steps have been carried out in [33, 28]. First important result for the case $D = 0$ is due to Musher *et al.* [50] in connection with the theory of weak Langmuir turbulence. Zakharov proposed recently to use equations of the type of Eq. (10) for the modeling of complicated nonlinear evolution processes including economic and social processes [76]. In particular, it can be shown that nonlinear growth rates as modeled by Eq. (10) lead to solutions with self-accelerating character [28, 16]. This entails an effect which sometimes is called the Matthew effect: *The rich become richer and the poor become poorer* or the principle of cumulative advantage [47, 14, 15].

3. Technological Trajectories and Continuity versus Discontinuity in the Process of Technological Change

To demonstrate the capability of such kind of models we consider in more detail the case of the Fisher–Eigen dynamics. As starting point we use the following equation:

$$\partial_t x(q,t) = x(q,t)[E(q) - F(t)] + D\Delta x(q,t)\,. \tag{11}$$

This is a slightly generalized form of the Fisher–Eigen dynamics (see Eqs. (1) and (2)) because $F(t)$ is not be identical to the population average of $E(q)$. Further, the mutation operator is modeled as diffusion term (cf. Eq. (5)). We now compare Eq. (11) with the famous replicator model due to Fisher–Eigen:

$$\frac{d}{dt}x_i(t) = (A_i - D_i)x_i + \sum_j (A_{ij}x_j - A_{ji}x_i) - F(t)x_i \quad i = 1, \ldots, n\,. \tag{12}$$

Here, A_i is the self-reproduction rate, D_i is that for decline processes and $F(t)$ includes boundary conditions like a constant overall population. The summation term includes error reproduction or mutations which in the case of social systems can be understood as transition or exchange processes [16, 17].

In the continuous model framework the typological classification of populations used in discrete replicator models is replaced by a characteristics representation. Subsequently, the population density x_i is replaced by function $x(q,t)$ in a continuous phenotype-like space. If the population density is concentrated in certain regions of Q ("islands") then these "islands" can be related to the original classified populations. The "selective value" $E(q)$ is linked to the net reproduction rate $(A_i - D_i)$. The choice of a diffusion-like mutation operator (cf. Eqs. (5) and (11)) corresponds to the assumption, that the mutation rates A_{ij} are symmetric, homogeneous and of short range. For technological evolution — and for social processes in general — it seems to be of particular interest to consider inhomogeneous and "directed" mutations. The emergence of a new technology in the system is related to a stepwise process ranging from research at the "pure science"-level, to applied R&D level to production relevance. The final introduction of a new technology, understood as a mutation in the system of established technologies, is the result of a multi-level process. Different institutions are the carriers of these processes [35]. At each level decisions about selection between variants are taking place [22]. These are influenced by feedback mechanisms. This kind of "contextual pre-selection" can be understood as "selection of the 'mutation generating' mechanisms" [22] and can be modeled by means of more elaborate mutation operators. Here, for the sake of mathematical simplicity we restrict ourselves to the diffusion approach.

In contrast to discrete models, the vanishing, merging, division and emergence of technologies are expressed by changes in the shape of the function $x(q)$, without having to consider changes in the taxonomy of the model. This results in a greater mathematical complexity of the model.

As mentioned above, the population density follows the shape of the fitness landscape. If we assume $E(q)$ to be a random function, then the shape of $x(q,t)$ is sensitive to statistical properties of this function given by the probability density functional $P[E(q)]$.

In Eq. (11) the term $x(q,t)[E(q) - F(t)]$ describes the selection process. This becomes evident if we consider the temporal evolution of populations without mutations:

$$\partial_t x(q,t) = x(q,t)[E(q) - F(t)]\,. \tag{13}$$

For the Gaussian distribution $P[E(q)]$, this was investigated in [77] using percolation theory. The result shows that with increasing time the population is concentrated in islands which correspond to particularly high values of the random function $E(q)$. These islands of high density are surrounded by regions of low density. This means, that the selection process leads to a concentration of the distribution around the maxima. The diffusion process, on the other hand, leads to a widening of the distribution and extending of its tails.

The mathematical solution of Eq. (11) in the presence of mutations is more complicated. Using the analogy with the Schrödinger equation for an electron in a random field some approximate expression for the time dependent solution $x(q,t)$ was given by Ebeling *et al.* [27] (see also [33, 28]). It can be shown that the existence of technological trajectories and technological populations correspond to the problem of the existence of localized states (according to the localization problem in random potentials). A localized state can be understood as a distribution of product models around a dominant design belonging to a single technological population. Technological trajectories can be understood as being formed by the movement of the localization centers. It is well known that for high-dimensional spaces ($d \geq 4$) and δ-correlated potentials $E(q)$, there are no localized states at all. Therefore, the existence of a correlation length greater than zero seems to be a necessary condition for the emergence of distinguishable parts (or populations) in the population density function. This is in accordance with the smoothness postulate of an adaptive landscape formulated by Conrad [18, 19]. An evolutionary system can only develop a strategy for search processes in a landscape with correlations. Furthermore, this entails that we have to distinguish between lack of information about the environment (which can be modeled by means of a random fitness function with certain statistical properties) and complete irregularity (stochasticity) where in principle no extrapolation from the local knowledge is possible. From the existence of localized states in a random unrestricted extension of the function $E(q)$ (res. $w(q)$) the following statement for the time evolution of $x(q,t)$ can be made: with increasing time the density becomes concentrated in localized states with decreasing localization radius (for mathematical details see [33, 28]). This process can continue endlessly.

The concentration of populations in regions of the fitness landscape with high valuation is one effect. For the continuation of evolution it seems to be very interesting to ask for possibilities to abandon such regions for new ones. Interesting characteristics of this transition process are the mean distance between the starting point q' and the nearest localization center as also the mean transition time between successive steps. The most important result of the mathematical analysis [28] is, that there exists a characteristic finite jump — the so-called "evolutionary quantum".

If we consider the transition from one localized state around a fitness maximum w_0 to the next nearest state with another maximum w_n, the transition time t as a

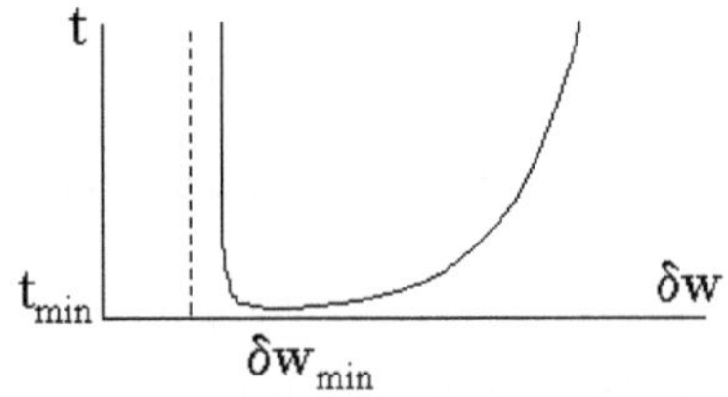

Fig. 2. Schematic representation of the transition time as a function of the target fitness value.

function of the value of δw, has a specific form (see Fig. 2). The existence of a minimum $t_{\min}$ can be understood from the dynamics of the system. Transitions to much higher maxima of w require several successful mutations. Such big jumps are relatively rare. This is expressed by high values of the transition time for $\delta w \gg \delta w_{\min}$. On the other hand, shifts to maxima with similar fitness can be achieved much quicker by mutations but the following selection process needs more time because of the minor improvement of the new area (population). This means that the transition time increases also for $\delta w \ll \delta w_{\min}$. Therefore, $\delta w_{\min}$ can be understood as an optimal step of improvement — also called the "quantum of evolution". In the characteristics space this discontinuity corresponds to a step-like behavior.

Evolution in correlated valuation landscapes proceeds jump-like. This means that longer periods of smooth evolution of technologies are sometimes interrupted by jumps. Therefore, "incremental" and "radical" innovations are both part of the system dynamics driven by mutations. From the perspective of the system the "big jumps" occur as the radical innovations changing the composition of the system. The drift stands for continuous change (incremental innovations). It was shown by Zhang, Engel and one of the present authors [79, 30] that there exists a definite scaling between the distance $|\delta q|$ of the jump in the characteristics space and the characteristic time τ for the jump

$$|\delta q| = \frac{\tau}{[\ln \tau]^{1/2}}\,. \tag{14}$$

4. Applications to Market Dynamics

The dynamics of the market is central to the competition and selection processes in the real economy. In spite of the fact that real dynamics is extraordinarily complicated let us try to model typical features with the tools developed above. So far, in the context of technological evolution we have considered the market as a selective environment. Now we consider the market itself in more detail. First we consider the market to continue as an environment of a technological or economic system and assign the fluctuating properties to the market itself. We start with the Fisher–Eigen equation. As mentioned above, a difficult question is the determination of the valuation function. So far, we have modeled principal uncertainty in terms of spatial fluctuations of the valuation function over the characteristics space. Now, we introduce time-related fluctuations. In some earlier work [33, 28] the hypothesis has been developed that the replication rate of technological populations is proportional to the profit which a particular technology can generate in the market

$$E(q,t) = \text{const. } Pf(q,t)\,. \tag{15}$$

Here, the profit $Pf(q)$ is determined as difference of the price $P_c(q)$ prevailing in the market and the production cost $C(q)$. This leads to:

$$E(q,t) = \text{const. } \big(P_c(q,t) - C(q,t)\big) \tag{16}$$

and in the simplest case $F(t)$ is determined by the mean

$$F(t) = \text{const. } \frac{\int dq\, x(q,t)[P_c(q,t) - C(q,t)]}{\int dq\, x(q,t)}\,, \tag{17}$$

If we consider firms as carriers of the evolutionary process (cf. [17]) then $E(q,t)$ can be understood as net growth rate of firms with certain characteristics q (including the technology in use). In this case the characteristic space has to be re-constructed conceptually in terms of the characteristics of firms. According to recent findings [8] we may assume that the company growth rate consists of a systematic part and a fluctuating part

$$E(q,t) = E(q) + \delta E(q,t)\,, \tag{18}$$

where the fluctuations obey certain scaling rules [8].

In another extension of the model framework being discussed we will consider the market itself as an evolving system. The interplay of sellers and buyers in the stock market serves as an example. So far, our analysis was entirely concentrated on the Fisher–Eigen dynamics. Much less work has been devoted to economic applications of the continuous Lotka–Volterra equation, in spite of numerous applications of its discrete counterpart (cf. [43]). As far as we see, continuous Lotka–Volterra equations are quite appropriate to model the market dynamics expressed by the interplay between buyers and sellers. Recently, Takayasu *et al.* [68, 57] have succeeded in describing the stock market dynamics by discrete stochastic models. In the Sato–Takayasu model the dealer i in a stock market is characterized by a minimum price B_i and by an interval $L_i = \delta B_i$, giving the selling price $S_i = B_i + L_i$. If he succeeds to find another dealer in the market with the buying price B_j such that

$$B_i < B_j \leq B_i + \delta B_i \tag{19}$$

he can sell his product and make a profit. In case

$$B_j < B_i \leq B_j + \delta B_j \tag{20}$$

the dealer i may buy a product from the dealer j. We are not going to describe the details of the Sato–Takayasu model which includes stochastic elements and describes correctly the fluctuations observed in the stock markets.

A plausible continuous variant of the Sato–Takayasu model may be formulated as follows. The interplay of buyers and sellers changes the amount and the distribution of money in the system. The amount of money forms a population over a space of characteristics of financial products. For each point of this space we define buying prices $B(q)$ and price intervals $L(q)$. Then, the market dynamics is described by the Lotka–Volterra equations ((5) and (10)) with a correlated random growth function $a(q,t) = E(q,t)$. The act of selling and buying is consequently expressed in the exchange term $b(q,q')$ which has to be antisymmetric. This kernel should be a functional of $B(q)$ and $L(q)$. A possible choice for the kernel is the derivative of a Gaussian

$$b(q,q') = \frac{1}{\tau}\frac{B(q)-B(q')}{\sqrt{L(q)}}\exp\left[-\frac{(B(q)-B(q'))^2}{2L(q)^2}\right]. \tag{21}$$

Then the dynamics works out as follows. If a dealer $B(q)$ succeeds in finding a partner with a higher buying price $B(q')$ such that $B(q) \leq B(q') \leq B(q) + L(q)$ he may sell his product and has a chance to grow. Otherwise, he still has a chance to buy from a dealer in the corridor below his own buying price and the possibility to sell the asset in the next turn.

5. Conclusions

In this paper a modeling framework for the co-evolution of economic or technological competitors in a continuous phenotype-like space is developed. The introduction of a continous characteristics space allows us, in principle, to distinguish rather easily between drifts and jumps in the technological progress. Further, the emergence, vanishing, differentiation or merging of different economic structures or technologies can be described. One purpose of the present paper is to demonstrate the versatility of continuous models for such processes. In this paper, the Fisher–Eigen dynamics and the Lotka–Volterra dynamics are introduced. In both cases there are different kinds of feedback mechanisms between the population density and the shape of the fitness landscape.

Concerning technological evolution, the adaptive landscape concept seems to be important for the understanding of an "innovative environment" [24] and the relation between "blind" and "directed" search strategies or the nature of change [6]. "Irreversibility is not only the result of imperfect information and sequentiality of decisions but is due to the fact that the world genuinely changes and it changes as a consequence of the very actions of the agents." ([24], p. 146). The adaptive landscape concept is related to a Lotka–Volterra dynamics which expresses the mutual dependence of changes in the population density and changes in the fitness landscape. Inasmuch as the fitness functional w depends both on the characteristics values q *and* the population density $x(q,t)$ (cf. Eq. (8)), the following circle can be observed: self-reproduction according to the fitness value w leads to a change in the population density (cf. Eqs. (1) and (10)). As the fitness functional depends on the population density, the shape of the fitness landscape changes simultaneously and, this way, the conditions for self-reproduction change. In this manner, self-referentiality is introduced in the model.

A second objective of the present paper is to link the concept of an adaptive landscape to complex search processes. Therefore, the structure of the fitness landscape and, in particular, its different statistical properties must be considered in greater detail. The fitness function is assumed to be a correlated random potential. Thus links to mathematical techniques used in statistical physics can be established. In this paper some implications from recent research trends for technological evolution are discussed. In particular, for the Fisher–Eigen dynamics the occurrence of localized states and the jump-like character of evolution have been considered. On a short time scale the hill-climbing character of evolution due to the interaction of selection and diffusion processes leads to the development of an island structure in the characteristics space. This corresponds to empirically observable technological populations. In the continuous approach the existence of such populations is not an initial assumption (like in discrete descriptions) but the result of search processes in a random landscape. On a long time scale hill-climbing proceeds by small but discrete steps. Transitions from an initial region with a certain given fitness w_0 to regions whose fitness is better just by a "quantum" are preferred. Surprisingly, this discontinuous character of evolution is the result of a continuous approach. This indicates that the stepwise character of technological evolution as observed is not a direct consequence of the discreteness of mutations but rather a general feature of the selection-mutation processes. Concerning the relation between incremental and radical innovations in the model framework developed here, shifts can be explained in terms of diffusion as well as jumps in terms of relocation of populations. The occurrence of an optimal step-width of improvement is probably related to the phenomenon of discrete steps of technological change described in the literature as the apparent paradox of a " 'discrete continuum' of technological change" [26]. Further, a scaling behavior between distance and characteristic time of jumps can be observed.

Thus, the modeling framework developed here seems to be useful for the description of economic and technological systems which behave in a complicated unknown environment. The units of evolution can be described by a set of varying characteristics. In this paper we mainly consider an approach which describes technological evolution in terms of the movement of technological populations. In this case, product models or technological processes are taken as carriers of the evolution process. Another widespread approach to technological evolution considers the firms themselves as carriers of the search process for technological improvement and market success [53, 6, 17]. In such approaches and models, characteristics of firms like size, age, belonging to a sector or technology, capital etc. appear as variables, parameters and classification features. A possible extension of the model framework consists in the inclusion of temporal fluctuations in the valuation function (fitness landscape). Therefore, links to observed fluctuations in firms growth rates can be established. According to the framework developed above, we can imagine that a choice of such economic, financial, organizational and technological characteristics can be used to construct a characteristics space, in which firms occupy different regions, and populations are built by groups of firms with similar characteristics. Evaluations of these populations are performed through market environments. The stock market can be seen as a special case. The price of shares of a firm on the stock market depends on a multitude of characteristic features, but also on the interactions between the firms and on general properties of economic and political networks like information flows. The concept of the evolution in an adaptive landscape can probably be used as an instrument to describe interactions between the firm strategies and the reactions in the stock market. In this paper we consider as an example the relevance of Lotka–Volterra models for the stock market dynamics. In particular, a continuous formulation of the Sato–Takayasu model is introduced.

In this paper we hope to have given some starting-points for further research on learning, problem solving and adaptation in economic and technological evolution, leading to a mathematical formulation of the market dynamics.

References

[1] Abernathy W. J., J. M. Utterback, Patterns of industrial innovation, *Technological Review* **50**, 41–47 (1978).

[2] Allen P. M., Evolution, innovation and economics, in *Technological Change and Economic Theory*, eds. G. Dosi, C. Freeman, R. R. Nelson, G. Silverberg, L. Soete (Pinter Publishers, London, 1988), pp. 95–119.

[3] Allen P. M., Evolutionary complex systems: models of technological change, in *Evolutionary Economics and Chaos Theory*, eds. L. Leydesdorff, P. van den Besselaar (Pinter Publishers, London, 1994), pp. 1–17.

[4] Allen P. M., Intelligent, self-organizing models in economics and in finance, in *Intelligent Systems for Finance and Business*, eds. S. Goonatilake, P. Treleaven (John Wiley, Chichester, 1995), pp. 289–311.

[5] Allen P. M., M. Lesser, Evolving complexity in social science, in *Systems — New Paradigms for the Human Sciences*, eds. G. Altman, W. A. Koch (Walter de Gruyter, Berlin, London, 1998).

[6] Allen P. M., M. Lesser, Evolutionary human systems: learning, ignorance and subjectivity, in *Evolutionary Theories of Economic and Technological Change*, eds. P. P. Saviotti, J. S. Metcalfe (Harwood Academic Publishers, Chur, 1991), pp. 160–171.

[7] Allen P. M., N. Clark, F. Perez-Trejo, Strategic planning of complex economic systems, *Review of Political Economy* **4** (3), 275–290 (1992).

[8] Amaral L. A. N., S. V. Buldyrev, S. Havlin, H. Leschhorn, P. Maass, M. A. Salinger, H. E. Stanley, M. H. R. Stanley, Scaling behavior in economics: I. empirical results for company growth, *Journal de Physique I* (*France*) **7** (4), 621–633 (1997); S. V. Buldyrev, L. A. N. Amaral, S. Havlin, H. Leschhorn, P. Maass, M. A. Salinger, H. E. Stanley, M. H. R. Stanley, Scaling behavior in economics: II. modeling the company growth, *Journal de Physique I* (*France*) **7** (4), 635–650 (1997).

[9] Anderson P. W., Suggested model for prebiotic evolution — the use of chaos, *Proceedings of the National Academy of Sciences of the United States of America — Biological Sciences* **80** (11), 3386–3390 (1983).

[10] Arthur B. W., Competing technologies: An overview, in *Technological Change and Economic Theory*, eds. G. Dosi, C. Freeman, R. R. Nelson, G. Silverberg, L. Soete (Pinter Publishers, London, 1988), pp. 590–607.

[11] Arthur B. W., Competing technologies, increasing returns, and look-in by historical events, *The Economic Journal* **99**, 116–131 (1989).

[12] Asselmeyer T., W. Ebeling, H. Rose, Smoothing representation of fitness landscapes — the genotype-phenotype map of evolution, *BioSystems* **39** (1), 63–76 (1996).

[13] Axelrod R. M., *The Complexity of Cooperation* (Princeton University Press, Princeton, 1997).

[14] Bonitz M., The scientific talents of nations, *Libri* **47** (4), 206–213 (1997).

[15] Bonitz M., E. Bruckner, A. Scharnhorst, Characteristics and impact of the Matthew effect for countries, *Scientometrics* **40** (3), 407–422 (1997).

[16] Bruckner E., W. Ebeling, A. Scharnhorst, Stochastic dynamics of instabilities in evolutionary systems, *System Dynamics Review* **5** (2), 176–191 (1989).

[17] Bruckner E., W. Ebeling, M. A. Jiménez-Montaño, A. Scharnhorst, Nonlinear stochastic effects of substitution — an evolutionary approach, *Journal of Evolutionary Economics* **6**, 1–30 (1996).

[18] Conrad M., Evolution of the adaptive landscape, in *Theoretical Approaches to Complex Systems*, eds. R. Heim, G. Palm (Springer, Heidelberg, 1978), pp. 147–169.

[19] Conrad M., *Adaptability. The Significance of Variability from Molecule to Ecosystems* (Plenum Press, New York, 1983).

[20] Conrad M., W. Ebeling, *M. V. Volkenstein* evolutionary thinking and the structure of fitness landscapes, *BioSystems* **27**, 125–128 (1992).

[21] Day R., C. Ping Chen (eds.), *Nonlinear Dynamics and Evolutionary Economics* (Oxford University Press, Oxford, 1993).

[22] Dosi G., Technological paradigms and technological trajectories, *Research Policy* **11**, 147–162 (1982).

[23] Dosi G., C. Freeman, R. R. Nelson, G. Silverberg, L. Soete (eds.), *Technological Change and Economic Theory* (Pinter Publishers, London, 1988).

[24] Dosi G., J. S. Metcalfe, On some notions of irreversibility in economics, in *Evolutionary Theories of Economic and Technological Change* (Harwood Academic Publishers, Chur, 1991), pp. 133–159.

[25] Dosi G., O. Marsili, L. Orsenigo, R. Salvatore, Learning, market selection and the evolution of industrial structures, *Small Business Economics* **7**, 411–436 (1995).

[26] Durand T., Dual technological trees: assessing the intensity and strategic significance of technological change, *Research Policy* **21**, 361–380 (1992).

[27] Ebeling W., A. Engel, B. Esser, R. Feistel, Diffusion and reaction in random media and models of evolution processes, *Journal of Statistical Physics* **37** (3-4), 369–384 (1984).

[28] Ebeling W., A. Engel, R. Feistel, *Physik der Evolutionsprozesse* (Akademie, Berlin, 1990).

[29] Engel A., W. Ebeling, Interaction of moving interfaces with obstacles, *Phys. Lett.* **A122**, 20 (1987a).

[30] Engel A., W. Ebeling, Comment on Diffusion in a random potential: hopping as a dynamical consequence of localization, *Phys. Rev. Lett.* **59** (17), 1979 (1987b).

[31] Ebeling W., A. Scharnhorst, M. A. Jiménez-Montaño, Karmeshu, Evolutions und Innovationsdynamik als Suchprozess in komplexen adaptiven Landschaften, in *Komplexe Systeme und Nichtlineare Dynamik in Natur und Gesellschaft*, ed. K. Mainzer (Springer, Berlin, 1999), pp. 446–473.

[32] Feistel R., W. Ebeling, Models of Darwin processes and evolution principles, *BioSystems* **15**, 291–299 (1982).
[33] Feistel R., W. Ebeling, *Evolution of Complex Systems* (Kluwer, Dordrecht, 1989).
[34] Foray D., A. Grübler, Morphological analysis, diffusion and lock-out of technologies: ferrons casting in France and the FRG, *Research Policy* **19**, 535–550 (1990).
[35] Freeman C., *The Economics of Industrial Innovation* (Penguin Books, Harmondsworth, 1974).
[36] Frenken K., L. Marengo, M. Valente, Interdependecies, nearly-decomposability, and adaptation, in *Computational Techniques to Model Learning in Economics*, ed. T. Brenner (Penguin Books, Harmondsworth, Kluwer Publisher, Boston, 1998).
[37] Hanusch H. (ed.), *Evolutionary Economics. Applications of Schumpeter's Ideas* (Cambridge University Press, Cambridge, Massachusetts, 1988).
[38] Jiménez-Montaño M. A., W. Ebeling, A stochastic evolutionary model of technological change, *Collective Phenomena* **3**, 107–114 (1980).
[39] Karmeshu (ed.), Mathematical-modelling in science and technology studies, *Journal for Scientific and Industrial Research*, Special volume 51 (1992).
[40] Karmeshu, V. P. Jain, Modelling innovation diffusion, in *Innovative Behaviour in Space and Time*, eds. C. S. Bertuglia, S. Lombardo, P. Nijkamp (Springer, Berlin, 1997), pp. 64–74.
[41] Krugman P., *The Self-Organizing Economy* (Blackwell Publishers, Cambridge, Boston, 1996).
[42] Kwasnicki W., Innovation regimes, entry and market structure, *Journal of Evolutionary Economics* **6**, 375–409 (1996).
[43] Levy M., S. Solomon, Power laws are logarithmic Boltzmann laws, *International Journal of Modern Physics* **C7** (4), 595–601 (1996); S. Solomon, M. Levy, Spontaneous scaling emergence in generic stochastic systems *International Journal of Modern Physics* **C7** (5), 745–751 (1996).
[44] Leydesdorff L., P. van den Besselaar (eds.), *Evolutionary Economics and Chaos Theory* (Pinter Publishers, London, 1994).
[45] Lubashevsky I. A., V. V. Gafiychuk, Yu. L. Klimontovich, Model for a hierarchically organized market functioning perfectly, *Matematicheskoe Modelirovanie* **9** (5), 3–16 (1997) (in Russian).
[46] Mantegna R. N., H. E. Stanley, Scaling behaviour in the dynamics of an economic index, *Nature* **376** (6 July), 46–49 (1995).
[47] Merton R. K., The matthew effect in science, *Science* **159** (3810), 56–63 (1968).
[48] Metcalfe J. S., M. Gibbons, Technology, variety and organization: a systematic perspective on the competitive process, in *Research on Technological Innovation, Management and Policy*, Vol. 4, eds. R. S. Rosenbloom, R. .A. Burgelman (JAI Press Inc., Greenwich, London, 1989), pp. 153–193.
[49] Montroll E., K. E. Shuler, Dynamics of technological evolution: random walk model for the research enterprise, *Proceedings of the National Academy of Sciences of the United States of America — Applied Mathematical Sciences* **76** (12), 6030–6034 (1979).
[50] Musher S. L., A. M. Rubenchik, V. E. Zakharov, Weak Langmuir turbulence, *Physics Reports* **252** (4), 178–274 (1995).
[51] Nelson R. R., S. G. Winter, H. Schuette, Technical change in an evolutionary model, *Quarterly Journal of Economics* **90**, 90–118 (1976).
[52] Nelson R. R., S. G. Winter, In search of useful theory of innovation, *Research Policy* **5**, 36–76 (1977).

[53] Nelson R. R., S. G. Winter, *An Evolutionary Theory of Economic Change* (The Belknap Press of Harvard University Press, Cambridge, Massachusetts, 1982).
[54] Sahal D., *Patterns of Technological Innovation* (Addison-Wesley, Reading, Massachusetts, 1981).
[55] Sahal D., Foundation of technometrics, *Technological Forecasting and Social Change* **27**, 1–37 (1985a).
[56] Sahal D., Technological guideposts and innovation avenues, *Research Policy* **14**, 61–82 (1985b).
[57] Sato A., H. Takayasu, Dynamic numerical models of stock market price: from microscopic determinism to macroscopic randomness, *Physica* **A250**(1–4), 231–252 (1998).
[58] Saviotti P. P., An approach to the measurement of technology based on the hedonic price method and related methods, *Technological Forecasting and Social Change* **27**, 309–334 (1985).
[59] Saviotti P. P., *Technological Evolution, Variety and the Economy* (Edward Elgar, Cheltenham, 1996).
[60] Saviotti P. P., J. S. Metcalfe, A theoretical approach to the construction of technological output indicators, *Research Policy* **13**, 141–151 (1984).
[61] Saviotti P. P., A. Bowman, Indicators of output of technology, in *Science and Technology Policy in the 1980s and Beyond*, eds. Gibbons, M., P. Gummett, B. Udgaonkar (Longman, London, 1984), pp. 117–147.
[62] Saviotti P. P., J. S. Metcalfe (eds.), *Evolutionary Theories of Economic and Technological Change* (Haarwood Academic Publishers, Chur., 1991).
[63] Saviotti P. P., G. S. Mani, Competition, variety and technological evolution: a replicator dynamics model, *Journal of Evolutionary Economics* **5**, 369–392 (1995).
[64] Scharnhorst A., Citations — networks, science landscapes and evolutionary strategies, *Scientometrics* **43**, 95–106 (1998).
[65] Schweitzer F. (ed.), *Self-Organization of Complex Structures — From Individual to Collective Dynamics* (Gordon and Breach Publishers, London, 1997).
[66] Silverberg G., B. Verspagen, Collective learning, innovation and growth in a boundedly rational, evolutionary world, *Journal of Evolutionary Economics* **4**, 207–226 (1994).
[67] Stanley M. H. R., L. A. N. Amaral, S. V. Buldyrev, S. Havlin, H. Leschhorn, P. Maass, M. A. Salinger, H. E. Stanley, Scaling behaviour in the growth of companies *Nature* **379** (29 February), 804–806 (1996).
[68] Takayasu H., H. Miura, T. Hirabayashi, K. Hamada, Statistical properties of deterministic threshold elements — the case of market price, *Physica* **A184** (1–2), 127–134 (1992).
[69] Utterback J. M., W. J. Abernathy, A dynamic model of process and product innovation, *Omega* **3** (6), 639–656 (1975).
[70] Weidlich W., Physics and social science — the approach of synergetics, *Physics Reports* **204** (1), 1–163 (1991).
[71] Weidlich W., M. Braun, The master equation approach to nonlinear economics, *Journal of Evolutionary Economics* **2**, 233–265 (1992).
[72] Westhoff F. H, B. V. Yarbrough, R. M. Yarbrough, Complexity, organization, and Stuart Kauffman's the origin of order, *Journal of Economic Behaviour and Organization* **29**, 1–25 (1996).
[73] Witt U. (ed.), *Studien zur Evolutorischen Ökonomik I* (Duncker & Humblodt, Berlin, 1990).
[74] Witt U., Emergence and dissemination of innovations — some problems and principles of evolutionary economics, in *Nonlinear Dynamics and Evolutionary*

Economics, eds. Day, R., C. Ping (Oxford University Press, Oxford, 1993), pp. 91–100.

[75] Wright S., The roles of mutation, inbreeding, crossbreeding and selection in evolution, *Proceedings of the Sixth International Congress on Genetics*, **1** (6), 356–366 (1932).

[76] Zakharov V. E., Models of competition, *Invited Lecture at the Conference on Self-organization* (Souzdal, June (mimeo), 1995).

[77] Zeldovich Ya. B., Disturbances leading to the disconnected domains, *Soviet Physics Doklady* **280**, 490 (1983); *Doklady Akademii Nauk SSSR* **270**, 1369–1372 (1983) (in Russian).

[78] Zeldovich Ya. B., S. A. Molchanov, A. A. Ruzmaikin, D. D. Sokolov, Intermittency in random media, *Uspekhi Fizicheskikh Nauk* **152** (1), 3–32 (1985).

[79] Zhang Y. C., Diffusion in a random potential: hopping as a dynamical consequence of localization, *Physical Review Letters* **56** (20), 2113–2116 (1986).

Advances in Complex Systems, Vol. 6, No. 1 (2003) 47–77

World Scientific
www.worldscientific.com

MOLECULAR REPLICATOR DYNAMICS

BÄRBEL M. R. STADLER
Max Planck Institute for Mathematics in the Sciences,
*Inselstraße 22-26, D-04103 Leipzig, Germany**
and
Institut für Theoretische Chemie und Strukturbiologie, Universität Wien,
Währingerstraße 17, A-1090 Wien, Austria
stadler@mis.mpg.de

PETER F. STADLER
Lehrstuhl für Bioinformatik, Institut für Informatik, Universität Leipzig,
Kreuzstraße 7a, D-04103 Leipzig, Germany
Institut für Theoretische Chemie und Strukturbiologie, Universität Wien,
Währingerstraße 17, A-1090 Wien, Austria
and
The Santa Fe Institute, 1399 Hyde Park Rd., Santa Fe NM 87501, USA
peter.stadler@bioinf.uni-leipzig.de

Received 28 October 2002
Accepted 7 November 2002

Template-dependent replication at the molecular level is the basis of reproduction in nature. A detailed understanding of the peculiarities of the chemical reaction kinetics associated with replication processes is therefore an indispensible prerequisite for any understanding of evolution at the molecular level. Networks of interacting self-replicating species can give rise to a wealth of different dynamical phenomena, from competitive exclusion to permanent coexistence, from global stability to multi-stability and chaotic dynamics. Nevertheless, there are some general principles that govern their overall behavior. We focus on the question to what extent the dynamics of replication can explain the accumulation of genetic information that eventually leads to the emergence of the first cell and hence the origin of life as we know it. A large class of ligation-based replication systems, which includes the experimentally available model systems for template directed self-replication, is of particular interest because its dynamics bridges the gap between the survival of a single fittest species to the global coexistence of everthing. In this intermediate regime the selection is weak enough to allow the coexistence of genetically unrelated replicators and strong enough to limit the accumulation of disfunctional mutants.

Keywords: Self-replication; ligation; autocatalytic network; quasispecies; hypercyle; replicator equation; chemical kinetics; emergence.

*Address for correspondence.

1. Molecular Replicators

The notion of a *replicator* — originally invented by Richard Dawkins [19, pp. 13–21] — is now used in biology for "*an entity that passes on its structure largely intact in successive replications*" [138]. The origin of life is characterized by the emergence of heritable information that, through the interplay of selection and variation, leads to Darwinian evolution [58]. The appearance of the first replicator thus marks the transition between the worlds of prebiotic chemistry and primitive biology. The uniformity of the biochemistry in all known living organisms suggests that all modern organisms descend from a *last common ancestor*, which according to a detailed analysis of protein sequences had a complexity comparable to that of a simple modern bacterium and lived some 3.2–3.8 Gyr ago.

The formation of the Solar System began about 4.6 Gyr ago in the aftermath of a local supernova explosion. The inner planets were probably formed by collision of some 500 Moon-sized planetesimals [140]. It is unlikely that Earth could have sustained life before about 4.2 Gyr or even 4.0 Gyr because of meteorite bombardment [108]. Depending on the details of the prebiotic environment a large variety of organic compounds may have been available about 4.0 Gyr ago, including amino acids, hydroxy acids, sugars, purines, pyrimidines and fatty acids [12, 68, 147]. Geological evidence [78], on the other hand, shows that it is certain that life has been present on Earth for at least 3.5 Gyr, and it is probable that life began before 3.8 Gyr [89, 93].

There are strong reasons to conclude that the Last Common Ancestor was preceded by simpler life forms that were based primarily on RNA. This era, during which the genetic information resided in the sequence of RNA molecules and the phenotype derived from the catalytic properties of RNA, has been termed the *RNA world* [36, 37]. In this scenario, the translation of RNA into proteins and, finally, the usage of DNA [35] as an information storage device are later inventions. A rather detailed model of the steps leading from the RNA world to modern cellular architectures is discussed in Ref. 84. The RNA world scenario is supported both by the wide range of catalytic activities that can be realized by relatively small ribozymes [52, 57, 58, 64, 131], Fig. 1, and by the usage of RNA catalysis at crucial points in modern cells [22, 56, 76]. Plausible ribozyme catalyzed pathways for a late-stage ribo-organism are discussed in Ref. 58, the role and evolution of co-enzymes in the RNA world is explored in Ref. 53.

The ribozymes in Fig. 1, in particular the replicase (a), are probably still too large to arise spontaneously; it seems much more plausible that molecules such as these arose from smaller, simpler molecules with less sophisticated functionalities.

Julius Rebek and others [29, 79, 129] have demonstrated that autocatalytic chemical reaction systems can be assembled in which organic molecules undergo self-replication. Autocatalysis in such systems of small molecules is certainly interesting because it reveals some mechanistic details of molecular recognition. However, these systems are hardly at the basis of biological replication since they cannot store

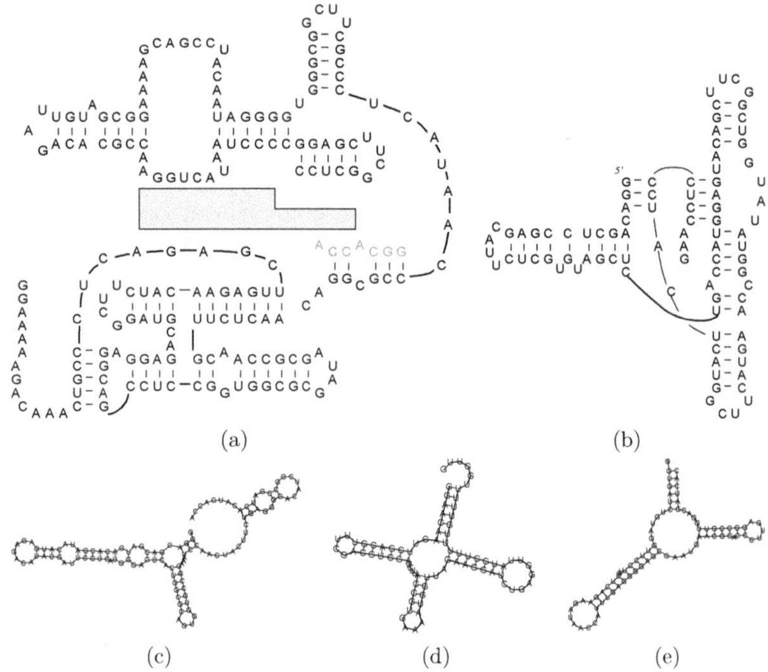

Fig. 1. Examples of artificial ribozymes with catalytic functions that are interesting in the context of the RNA world. (a) RNA replicase 18 [57], (b) 16.min ligates RNA to protein [7], (c) a ribozyme that activates amino acids [61], (d) ATrib [64] can self-aminoacylate, and (e) a ribozyme that catalyzes tRNA aminoacylation [64].

a sizable amount of information. Consequently, there is no room for molecular variants, "mutations," and hence these systems cannot sustain Darwinian evolution.

The simplest way of overcoming this limitation is to envision some kind of polymer as the carrier of genetic information, just as DNA takes on this role in present day cells. RNA has the same capacity, which is actually made use of by a large class of viruses. Polypeptides are the obvious alternative candidates.

Template-induced synthesis of longer RNA molecules from monomers in the absence of an enzyme, however, has not been successfully achieved so far [82]. Autocatalytic template-induced synthesis of oligonucleotides from smaller oligonucleotide precursors, however, has been successful: a hexanucleotide through ligation of two trideoxynucleotide precursors was carried out by Günter von Kiedrowski [135] (see also Ref. 106 and 137). More recently, it has been shown that a peptide comprised of 32 amino acids can undergo exactly the same type of autocatalytic

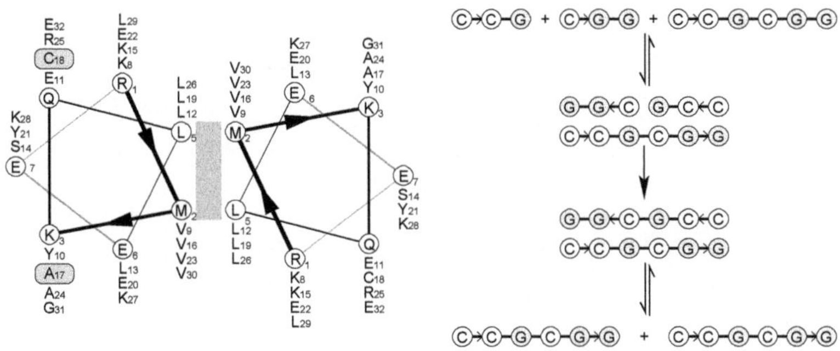

Fig. 2. Reza Ghadiri's [63] self-replicating peptide $\mathfrak{T}$ acts as a template for ligation of the two fragments $\mathfrak{E}$ and $\mathfrak{N}$, where

$$\mathfrak{E} = \mathrm{Ar} \cdot \mathsf{RMKQLEEKVYELLSKVA} \cdot \mathrm{CO} \cdot \mathrm{S} \cdot \mathrm{Benzyl}\,,$$
$$\mathfrak{N} = \mathrm{H_2N} \cdot \mathsf{CLEYEVARKJJKGE} \cdot \mathrm{CO} \cdot \mathrm{NH_2}\,,$$
$$\mathfrak{T} = \mathsf{RMKQLEEKVYELLSKVA : CLEYEVARKJJKGE}$$

and Ar = 4 − acetamidobenzoyl. The ligation site is located between A_{17} and C_{18}. Günter von Kiedrowski's [135] self-replicating palindromic RNA reproduces by means of ligation of trinucleotides to the template hexanucleotide.

synthesis from two roughly equal fragments [62], Fig. 2. A recent experimental study using nucleic acids [54] emphasizes the importance of ligation-based replication mechanisms for the origin of life.

An interesting feature of oligopeptide self-replication concerns easy formation of higher replication complexes: coiled-coil formation is not restricted to two interacting helices, triple helices and higher complexes are known to be very stable too. Autocatalytic oligopeptide formation may thus involve not only a template and two substrates but, for example, a template and a catalyst that form a triple helix together with the substrates [63, 145]. We will return to such higher order replication systems in Sec. 6.

A major shortcoming of peptide replication, however, is the fact that only a very small fraction of all possible sequences fold into three-dimensional structures that are suitable for leucine zipper formation and hence a given autocatalytic oligopeptide is very unlikely to retain the capability of template action on mutation. The (rather large) leucine zipper sequences of the building block must themselves be somehow instructed somewhere in the system. Peptides are therefore only *occasional templates*; their evolutionary adaptability is crippled by the fact that mutants are unlikely to be templates themselves. In contrast to the volume filling principle of protein packing, specificity of catalytic RNAs is provided by base pairing and to a lesser extent by tertiary interactions. Both are the results of hydrogen bond specificity. Template action of nucleic acid molecules, being the basis for replication, results directly from the structure of the double helix. It

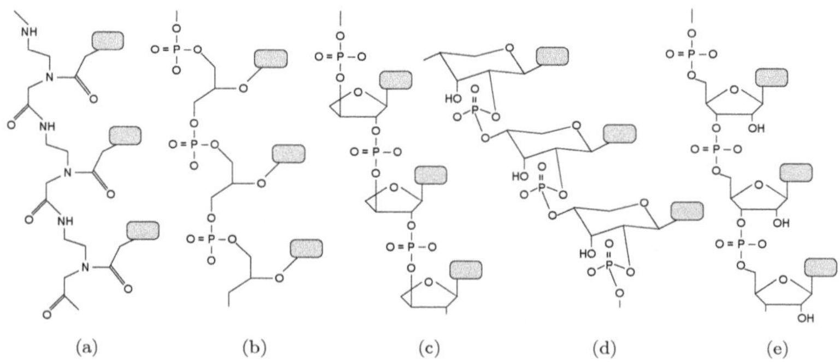

Fig. 3. Alternatives to RNA. A pre-RNA world may have existed where genetic information was stored in alternative pre-bio-polymers. (a) PNA (peptide nucleic acid) [77], (b) glycerol-derived nucleic-acid analogue [17], (c) TNA (threose nucleic acid) [92], (d) pyranosyl-RNA [83], (e) RNA for comparison.

requires an appropriate backbone provided by the anti-parallel ribose-phosphate or 2′-deoxyribose-phosphate chains and a suitable geometry of the complementary purine-pyrimidine pairs. All RNA (and DNA) molecules, however, share these features which, accordingly, are independent of sequence. Every RNA molecule has a uniquely defined complement. Nucleic acid molecules, in contrast to proteins, are therefore *obligatory templates*. This implies that mutations are conserved and readily propagated into future generations. A somewhat different classification of replicators in terms of their combinatorial limitations can be found in Ref. 124.

While some classes of non-instructed polymers, such as random oligopeptides of the proteinoid type [32], are plausible products of prebiotic chemistry, this is not the case for RNA [60, 100, 105]. Heritable molecular information therefore most likely has its origins in a different chemical system. Two widely different scenarios have been proposed. One possibility is that nucleic acid like polymers with different backbones and maybe different side chains predated the RNA world. There is no shortage of plausible candidates, including tetrose based nucleic acids and peptide nucleic acids, Fig. 3.

Another alternative is the *container first* hypothesis that assumes the existence of autocatalytically replicating lipid micelles or vesicles, a possibility that was demonstrated experimentally by Luisi and co-workers [6, 69, 133, 139]. Doron Lancet and Daniel Segré suggest that such "proto-cells" might have a complex lipid composition that is faithfully transferred when the vesicle duplicates, see e.g. [102, 103, 104]. Of course, a sizable amount of information can be accumulated only if a very large set of chemically distinct lipids can be brought together in a single vesicle. In this *Lipid World* scenario, the catalytic activities of the complex lipid phase would eventually instruct the synthesis of peptides and nucleic acids.

For the purpose of this article, fortunately, the precise chemical instantiation of the earliest replicons is largely irrelevant, because their basic kinetic properties depend more on the logics of the replication mechanism than on the molecular details. In this contribution we shall focus on the consequences of the peculiar chemical reaction kinetics that is implied by replication. We will consider two times two classes of models in some detail: template induced replication with and without an additional self-replicating catalyst, both with simple mass action kinetics and with a chemically more realistic ligation-like mechanism. The common theme of these models is the emergence and persistence of a sufficient amount of genetic information for biological evolution to take off.

2. The Molecular Quasispecies and "Survival of the Fittest"

Let us begin with the simplest possible replication scheme:

$$\mathsf{A} + \mathsf{C} \xrightarrow{\mathsf{E}} 2\mathsf{C} + \mathsf{A}' \tag{1}$$

in which some energy rich building material A is used to produce copies of the template C. This describes the logic, but not necessarily the details of the reaction kinetics, of the examples in Fig. 2.

The first successful attempts to study RNA evolution *in vitro* were carried out in the late sixties by Sol Spiegelman and his group [75, 109]. They created a "protein assisted RNA replication medium" by adding an RNA replicase isolated from *E. coli* cells infected by the RNA bacteriophage Qβ to a medium for replication that also contained the four ribonucleoside triphosphates in a suitable buffer solution. Qβ RNA and some of its smaller variants start instantaneously to replicate when transferred into this medium.

Extensive studies on the reaction kinetics of RNA replication in the Qβ replication assay were performed by Christof Biebricher in Göttingen [8]. These studies revealed consistency of the kinetic data with a multi-step reaction mechanism. Depending on concentration the growth of template molecules allows one to distinguish three phases of the replication process: (i) at low concentration all free template molecules are instantaneously bound by the replicase which is present in excess and therefore the template concentration grows exponentially, (ii) excess of template molecules leads to saturation of enzyme molecules, then the rate of RNA synthesis becomes constant and the concentration of the template grows linearly, and (iii) very high template concentrations impede dissociation of the complexes between template and replicase, and the template concentration approaches a constant in the sense of product inhibition. We neglect plus-minus complementarity in replication by assuming stationarity in relative concentrations of plus and minus strand [25] and consider the plus-minus ensemble as a single species. Then, RNA replication may be described by the over-all mechanism:

$$\mathsf{A} + \mathsf{C_i} + \mathsf{E} \underset{\bar{k}_i}{\overset{k_i}{\rightleftharpoons}} \mathsf{A} + \mathsf{C_i} \cdot \mathsf{E} \xrightarrow{a_i} \mathsf{C_i} \cdot \mathsf{E} \cdot \mathsf{C_i} \underset{\bar{k}'_i}{\overset{k'_i}{\rightleftharpoons}} \mathsf{C_i} \cdot \mathsf{E} + \mathsf{C_i}\,. \tag{2}$$

This simplified reaction scheme reproduces all three characteristic phases of the detailed mechanism and can be readily extended to replication and mutation.

Evolution of molecules based on replication and mutation exposed to selection at constant population size has been formulated and analyzed in terms of chemical reaction kinetics [25, 26, 27, 73]. Error-free replication and mutation are parallel chemical reactions:

$$\mathsf{A} + \mathsf{C_i} \xrightarrow{Q_{ji}a_i} \mathsf{C_j} + \mathsf{C_i} \tag{3}$$

and form a network which, in principle, allows every RNA genotype to form as a mutant of any other genotype. The material A required for or consumed by RNA synthesis is continuously replenished. The quantities of interest are the relative concentrations $x_k = [\mathsf{C_k}]/c_0$ where $c_0 = \sum_k [\mathsf{C_k}]$ is the total concentration of replicating material. The reaction scheme 3 translates into the kinetic equations:

$$\dot{x}_k = x_k\big(a_k Q_{kk} - \Phi(t)\big) + \sum_{j \neq k} Q_{kj} a_j x_j \,. \tag{4}$$

The diagonal elements of Q are the replication accuracies, i.e. the fractions of correct replicas produced on the corresponding templates. The time dependent excess productivity which is compensated by the flow in the reactor is the mean value $\Phi(t) = \sum a_j x_j$. The *selective value* of $\mathsf{C_k}$ is the diagonal element w_{kk} of matrix W with the entries $w_{kj} = a_k Q_{kj}$.

The selective value of a genotype is tantamount to its fitness in the case of vanishing mutational back-flow and hence the genotype $\mathsf{C_m}$ with maximal selective value, $w_{mm} = \max_j w_{jj}$, dominates a population after it has reached the selection equilibrium. It is called the *master* sequence. The notion of a *quasispecies* was introduced for the stationary genotype distribution in order to point at its role as the genetic reservoir of the population. In the simplest case one considers only point mutations. This leads to the explicit expression:

$$Q_{kj} = (1-p)^n \left(\frac{p}{1-p}\right)^{d_{kj}}, \tag{5}$$

where d_{kj} is the Hamming distance of the two sequences $\mathsf{C_k}$ and $\mathsf{C_j}$, and p is the per-digit error rate.

The stationary frequencies x_k can be computed explicitly in many cases, e.g. when the master sequence is derived from the single peak model landscape that assigns a higher replication rate to the master and identical values to all others [1, 121, 125]. The master sequence vanishes at some finite replication accuracy that depends on the superiority and the length n of the replicating sequences. The critical value $p_{\max}$ of the mutation rate is known as the *error threshold.* Above the threshold no stationary distribution of sequences is formed. Instead, the population drifts randomly through sequence space. This implies that all genotypes have only finite life times, inheritance breaks down, and evolution becomes impossible. The effects of finite population sizes on the error threshold are considered, e.g. in Refs. 2, 15 and 80.

Variations in the accuracy of *in vitro* replication can indeed be easily achieved because error rates can be tuned over many orders of magnitude [65, 67]. The range of replication accuracies that are suitable for evolution is limited, however, by the maximal accuracy that can be achieved by the replication machinery and the minimum accuracy determined by the error threshold. Populations in constant environments have an advantage when they operate near the maximal accuracy because then they lose as few copies through mutation as possible. In highly variable environments the opposite is true: it pays to produce as many mutants as possible because then the ability to cope successfully with chance is largest.

In order to check the relevance of the error threshold for the replication of RNA viruses the minimum accuracy of replication can be transformed into a maximum chain length $n_{\max}$ for a given error rate p. The condition for stationarity of the quasispecies reads

$$n < n_{\max} = -\frac{\ln \sigma}{\ln(1-p)} \approx \frac{\ln \sigma}{p}, \tag{6}$$

where σ is the superiority of the master species, see e.g. Ref. 27. The populations of most RNA viruses were indeed shown to live near the critical value of replication accuracy [20, 21]. Moreover, the genome size n of any organism is roughly the inverse mutation rate per site and replication [23, 24], see Fig. 4.

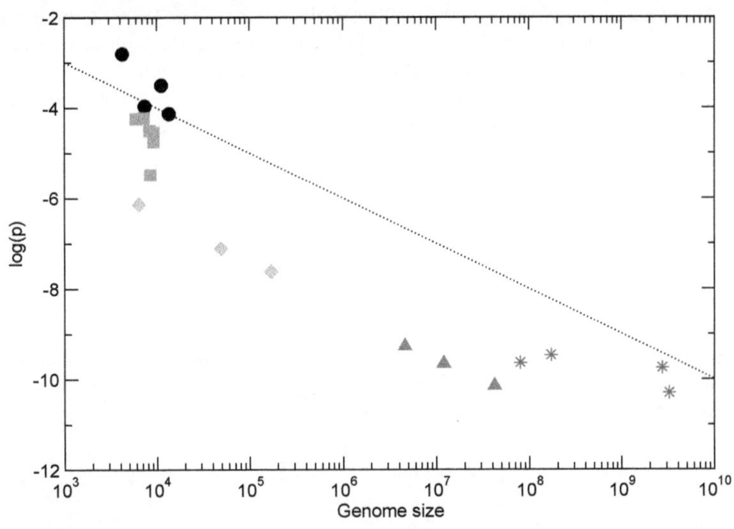

Fig. 4. Relationship between genome length and mutation rate: • single stranded RNA viruses, ■ retro-transcribing viruses and transposons, ♦ DNA viruses, ▲ Bacteria, ∗ Eukarya. Data are taken from Ref. 24. The dotted line is $p = 1/n$.

Equation (6) imposes a fundamental limitation on the size of the self-replicating biopolymers. In the following sections we shall encounter mechanisms by which this limitation can be circumvented.

3. Replicator Equations

So far we have neglected any direct interactions between the replicating species. In the most general case, we have

$$\mathsf{C_k} \xrightarrow{\mathsf{C_1,C_2,\ldots,C_n}} 2\mathsf{C_k}\,, \tag{7}$$

where the replication rate f_k of $\mathsf{C_k}$ is influenced by the concentrations of all the species $\mathsf{C_1}$ through $\mathsf{C_n}$. Most papers on autocatalytic reaction networks (including much of the work of the group in Vienna) use the *replicator equation* [47, 48, 96]:

$$\dot{x}_k = x_k \left[f_k(x) - \sum_{j=1}^{n} x_j f_j(x) \right] \tag{8}$$

despite the fact that a *continuously stirred tank reactor* (CSTR) would in many cases be a more appropriate description of the biological or chemical situation. This simplification is motivated by the equivalence of the two models for the special case of homogeneous interaction functions [97]. Furthermore, it can be shown that the CSTR dynamics converges to the replicator dynamics in the limit of small flux rates [40]. We remark that the quasispecies model yields, in the limit of vanishing mutation rates, a replicator equation with constant growth functions $f_k(x) = a_k$.

Let us now return to the case of interacting replicators. The simplest example assumes reactions that are catalyzed by a single replicator, i.e.

$$\mathsf{C_k} + \mathsf{C_j} \to 2\mathsf{C_k} + \mathsf{C_j}\,. \tag{9}$$

Assuming, furthermore, that these reaction follow mass action kinetics, the growth functions f_k in Eq. (8) are linear,

$$f_k(x) = \sum_{j=1}^{n} a_{kj} x_j \tag{10}$$

with $a_{kj} \geq 0$. These *2nd-order replicator equations* also describe the dynamics of strategies in evolutionary games [126]. Hofbauer [42] showed that they are topologically equivalent to the Lotka–Volterra equations:

$$\dot{y}_k = y_k \left(r_k + \sum_{j=1}^{n-1} b_{kj} y_j \right), \tag{11}$$

which are a standard model in mathematical ecology. Indeed, recent experimental research in the group of John McCaskill [74, 142] is dealing with *molecular ecologies* of strongly interacting molecular replicators.

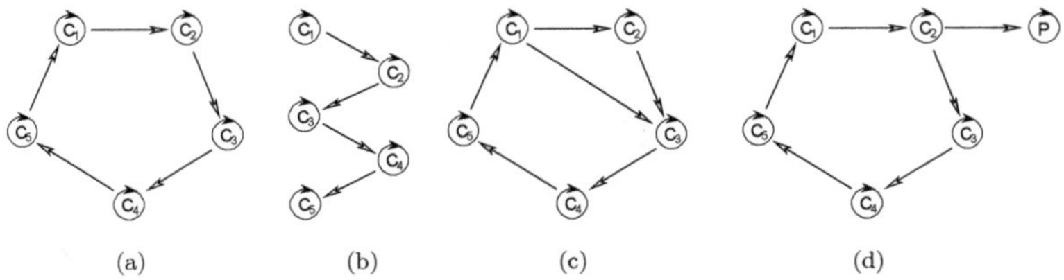

Fig. 5. Architectures of catalytic networks: (a) hypercycle, (b) catalytic chain, (c) hypercycle with "short-cut", and (d) hypercycle with a parasite.

A particularly important class of second-order replicator equations are the *catalytic networks* [99] characterized by $a_{kj} \geq a_{jj}$, i.e. the networks in which the catalytic assistance for all other replicators is at least as strong as self-catalysis. The interaction structure of catalytic networks is conveniently represented as a directed graph $\Gamma(\mathbf{A})$ in which the vertices are the replicating species and there is an edge $[\mathsf{C_j}] \to [\mathsf{C_k}]$ if $a_{kj} - a_{jj} > 0$, i.e. if $[\mathsf{C_j}]$ advances the replication of $[\mathsf{C_k}]$, Fig. 5.

The dynamics of a catalytic network is at least in part determined by its graph structure. For example, coexistence is impossible in linear chains, Fig. 5(b). The *hypercycle*, on the other hand, was introduced as a possibility to overcome the limitations of the error threshold [28]. The positive feedback cycle, in which each template promotes the replication of the next one guarantees the permanent coexistence of all its members. While hypercycles are stable against dynamical perturbations, they are unstable against structural perturbations of the networks. Both "short cuts," Fig. 5(c), and parasites, Fig. 5(d), may lead to dynamically unstable networks.

An important concept in the theory of replicator equations is that of a *saturated fixed point* [44]. Let F_K be the face of the concentration simplex S_n whose (relative) interior is defined by $x_k = 0$ for all species $k \in K$. An eigenvector of (the Jacobian matrix of) the vector field at a point $x^K \in F_K$ that points out of this plane is called a transversal eigenvalue. We have $\lambda_k(x^K) = f_k(x^K) - \Phi$ for $k \in K$ and $x^K \in F_K$. A fixed point $\hat{x} \in F_K$ is saturated if none of its transveral eigenvalues is positive.

So far, we have excluded mutation from our considerations. Mutation can be treated as a perturbation at least in the limit of small mutation rates [118]. Since mutation by construction points inward on the boundary of the simplex, one obtains the *rest point migration theorem*: If a fixed point is saturated, it "moves" into the interior of the simplex when mutation is switched on; all other equilibria are pushed into the non-physical exterior. Mutation thus may simplify the phase portraits considerably. Larger mutation rates, however, may lead to complicated bifurcation patterns, an increase in the number of competing equilibria, and even the appearance of limit cycles [43, 116].

Molecular Replicator Dynamics 57

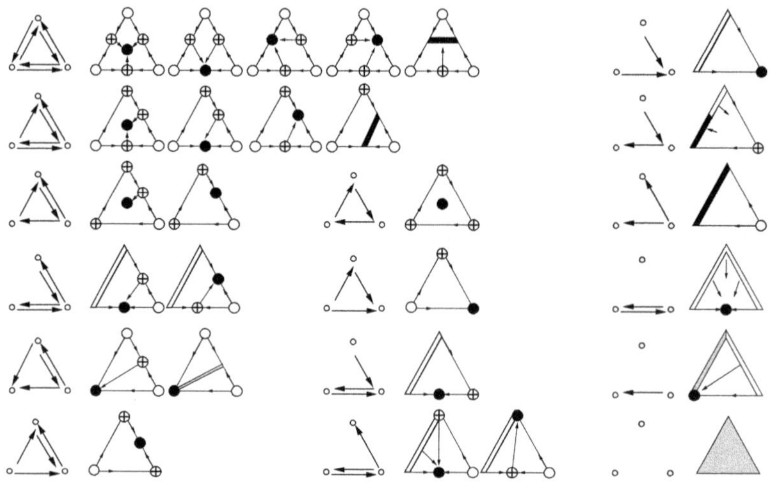

Fig. 6. Classification of autocatalytic networks with $n = 3$ species. The interactions are represented as graphs with a directed edge iff $a_{kj} - a_{jj} > 0$.

The dynamics of second order replicator equations can be extremely complicated despite the rather simple form of the differential equation. In the case of two independent variables ($n = 3$, the state space is an equilateral triangle) there are 35 different generic phase portraits [10, 117]. It is interesting to note that this rather involved classification remains essentially unchanged under monotonous transformations of the linear growth functions, i.e. for $f_k(x) = \vartheta([\mathbf{A}x]_k)$, see Ref. 111. The classification of three-species catalytic networks is shown in Fig. 6. In the case of three independent variables, i.e. $n = 4$ species, there are heteroclinic orbits, multiple limit cycles, and strange attractors [5, 13, 38, 49, 91, 132], Fig. 3.

Often one is not interested in all details of a dynamical system or in the structure of its ω-limit sets; less detailed knowledge may well be sufficient. Probably the most important question is: *Can all species coexist in the system for arbitrarily long time? Or will some species die out in the long run?* Schuster *et al.* [98] introduced the notion of *permanence* (permanent coexistence) to formalize this question. A variety of different notions of cooperation, the first of which is now called *weak persistence* [34], have been proposed by various authors, e.g. Refs. 33 and 51.

A replicator equation (8) is said to be permanent if there is a compact subset C of the interior of the state space S_n, such that any trajectory starting in the interior of S_n will eventually end up in C. In other words, there is a (possibly very thin but finite) repulsive "skin" on the boundary of the state space.

A sufficient (but for $n \geq 5$ not necessary) criterion for permanence of second order replicator equations was derived by Jansen [55]:

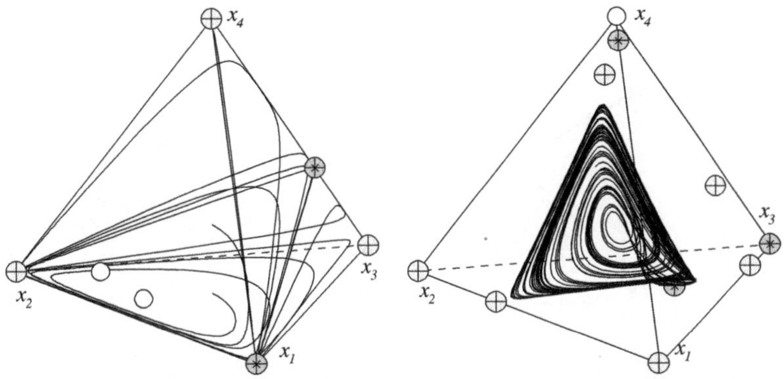

Complex dynamics in second-order replicator equations. Left: an attracting heteroclinic orbit in the central plane [110]. Right: A chaotic attractor of class described in Ref. 5. The interaction matrices are

$$\mathbf{A} = \begin{pmatrix} 0 & -2 & 7 & 8 \\ 20 & 0 & -7 & -6 \\ -20 & 1 & 0 & 2 \\ -20 & 1 & 1 & 0 \end{pmatrix}, \qquad \mathbf{A} = \begin{pmatrix} 0 & 0.5 & -0.1 & 0.1 \\ 1 & 0 & -0.5 & 0 \\ -0.5 & 1.0 & 0 & 0 \\ 1.5 & -0.8 & -0.2 & 0 \end{pmatrix}.$$

Fixed points are distinguished by the number of stable directions: sources without stable direction $\bigcirc$, one stable direction $\oplus$, and $\circledast$ two stable directions.

Theorem 3.1. *If there is a vector* $p \in \mathrm{int}S_n$ *such that for every isolated rest point* $x^K \in F_K$ *on the boundary of the simplex holds* $\sum_{k\in K} p_k \cdot \lambda_k(x^K) > 0$ *then the second order replicator equation is permanent.*

This condition is equivalent to a linear program and hence computationally accessible [55].

In some cases the graph structure of the catalytic network allows conclusions about permanence. For example: If the catalytic network is permanent, then its graph $\Gamma(\mathbf{A})$ is strongly connected [107]. In particular, two hypercycles cannot coexist [45]. For $n \leq 5$ the graph of a permanent catalytic network must contain a Hamiltonian circuit, i.e. a closed path visiting every vertex exactly once [3, 47]. Conversely, if Γ contains a Hamiltonian circuit then there is a non-negative matrix $\mathbf{A}$ with $\Gamma(\mathbf{A}) = \Gamma$ such that the second-order replicator equation is permanent [119].

4. The Evolution of Coexistence

A fundamental necessary condition for any kind of coexistence is posed by the *Exclusion Principle* [50]:

Theorem 4.1. *If a second order replicator network does not have an interior fixed point, then all orbits converge to the boundary of the state space.*

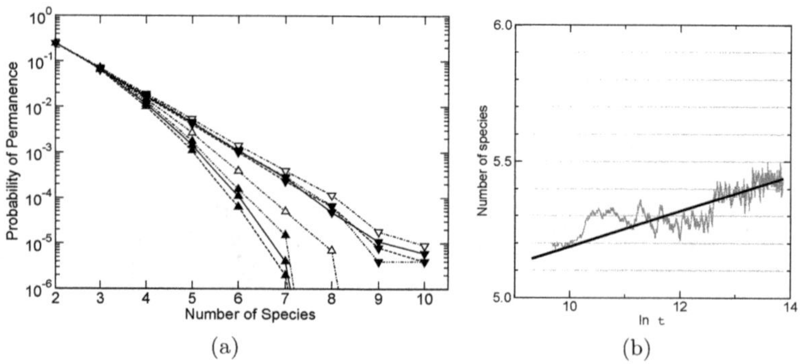

Fig. 8. (a) Probability of generating a permanent replicator network by chance. Upper bounds are from algebraic conditions on the interaction matrix **A**, the lower bound is obtained by checking Jansen's criterion (Theorem 3.1). Data are for Gaussian, uniform, Laplace, and Cauchy distributions for the coefficients $a'_{ij} = a_{ij} - a_{jj}$. Long tails as in the Cauchy distribution seem to favor permanence [114]. (b) The average number of species increases very slowly by incorporation of mutants. The resulting networks are permanent, highly connected, and (at least approximately) catalytic networks [39].

Since the existence of a positive solution of $\mathbf{A}x = \Phi\vec{1}$ is a rather restrictive condition, we have to interpret the Exclusion Principle as a rather pessimistic statement about the prospects of coexistence. Permanence requires an even more restrictive condition: If the system is permanent it has no saturated fixed point on the boundary [44, 46].

A numerical survey [114] for small and moderate size replicator networks therefore shows that permanence — and cooperativity in general — is a very rare phenomenon, except for very small networks with $n = 2$, 3 or 4 members. The probability of finding cooperative behavior in random autocatalytic networks of replicator or Lotka–Volterra type decreases at least as 4^{-n} with the number of species involved, maybe even faster than exponential, Fig. 8(a). Permanence is more likely in catalytic networks and decreases dramatically with the number of inhibitory interactions $a_{kj} - a_{jj} < 0$. Even in catalytic networks, however, the probability decreases exponentially.

In the limit of small mutation rates one can model the long time evolution of replicator networks. To this end the differential equation is numerically integrated until the ω-limit is reached. In some cases the appropriate fixed points can also be obtained analytically. Then the interaction matrix **A** is modified by (i) removing all rows and columns that correspond to extinct species $x_k = 0$ and (ii) by adding an additional row and column. In the second step either randomly chosen entries are used to produce a new species that is unrelated to the existing network (an *invader*) or a row and corresponding column is copied and a small amount of noise is added.

May and Nowak [70] analyzed the evolution of a simple Lotka–Volterra type model of super-infection in host-parasite associations. Interestingly, the same type of equations were obtained in a study on *competition and biodiversity in spatially structured habitats* [128]. The total number of species in their system slowly increases, $n(\tau) \propto \ln \tau$, where τ is the number of mutation events. More recently, a square root law $n(\tau) \sim \sqrt{\tau}$ was found in a co-infection model that neglects competition of different parasites within the same host [71]. For general replicator networks, we found that mutation leads to a slow increase of diversity consistent with a logarithmic increase of the number of species with the number of mutation events [39], Fig. 8(b), while unrelated invaders repeatedly lead to a complete collapse of the network and hence preclude any long-term growth. The evolved networks have a specific structure in the mutation case: all strong interactions are positive, the interaction matrix is nearly symmetric, and the connectivity is very high.

The mutation mechanism employed in these simulations might not be entirely realistic, however. It seems entirely plausible that some mutants completely lose their catalytic activities while retaining their template properties. Such mutants are *parasites* [14, 72].

Despite their importance for the theory of prebiotic evolution, there does not seem to be any systematic treatment of parasitic interactions in replicator networks. A classification of types of parasites in terms of the structure of the networks is also missing; a distinction between "short-cuts" and "true parasites" appears to be useful, see Fig. 5. True parasites might be defined as dead ends (sinks) in the network graph $\Gamma(\mathbf{A})$, i.e. they do not contribute anything to the self-maintenance of the network. In the most general case one might want to consider a species P that dominates another network species $\mathsf{C_k}$ in the following sense:

$$\mathsf{P} \succ \mathsf{C_k} \Leftrightarrow f_{\mathsf{P}}(x) \geq f_k(x) \text{ and } f_{\mathsf{P}}(x) > f_k(x) \text{ whenever } f_k(x) > 0\,. \tag{12}$$

Equation (8) satisfies the "quotient rule":

$$\frac{\mathrm{d}}{\mathrm{d}t} \ln \frac{x_k}{x_j} = f_k(x) - f_j(x)\,, \tag{13}$$

from which one can deduce that a dominated species will always die out. The argument proceeds by integrating the quotient rule over time and observing that $\ln \frac{x_k}{x_j}$ must converge to either $+\infty$, or $-\infty$ unless both $f_k(x)$ and $f_j(x)$ converge to zero. A parasitic mutant that dominates its ancestor will therefore cause the extinction of the ancestor. In the example of the hypercycle with a parasite, Fig. 5(d), this changes the network graph to a catalytic chain, Fig. 5(b); hence all other hypercycle members will also die out. Since the parasite at the end of the chain cannot catalyze its own replication, it will finally disappear as well.

So far we have not considered the detrimental effects of mutation itself on catalyzed replication. Simplified model equations use a single dynamical variable to describe the "error tail" that consist of all "lethal" mutants [4, 16, 81, 115]. Beyond a critical mutation rate this error tail dominates. Christian Forst [31] shows that there is an error threshold also for permanent catalytic networks also without this simplification. Hence, genome size is limited by replication accuracy also in the case of cooperating replicators. Once the mutation rate is small enough to sustain an RNA molecule with replicase activity below the error threshold, however, selection can act to improve the copying fidelity. This allows for an increased sequence length and thus opens up the possibility for a further decrease of the per-base mutation rate [87].

5. Coexistences by Means of Product Inhibition

Eörs Szathmáry [123] pointed out that a simple modification of the replicator equation to a "parabolic" growth law:

$$\dot{x}_k = b_k x_k^a - x_k \Phi \tag{14}$$

with $0 < a < 1$ leads to unconditional coexistence. We remark that a variant of Eq. (14) with an explicit decay term $-d_k x_k$ yields the same qualitative behavior [88]. The case $a = 1/2$ is obtained as a limiting case of von Kiedrowski's "Minimal Replicator Theory" [136]. Equation (14) is, however, not physically meaningful on the boundary of the simplex where the Jacobian of the vector field diverges. In Refs. 113, 136 and 141 chemically realistic mechanisms are considered that produce Eq. (14) as a limiting case. The full ligation mechanism of both von Kiedrowski's and Ghadiri's examples, Fig. 2 in the introduction, consists of the following steps:

$$\mathsf{A}+\mathsf{C} \underset{\bar{a}}{\overset{a}{\rightleftharpoons}} \mathsf{AC}\,, \quad \mathsf{AC}+\mathsf{B} \underset{\bar{h}}{\overset{h}{\rightleftharpoons}} \mathsf{ABC}\,, \quad \mathsf{ABC} \overset{r}{\rightarrow} \mathsf{C}_2\,,$$

$$\mathsf{B}+\mathsf{C} \underset{\bar{b}}{\overset{b}{\rightleftharpoons}} \mathsf{BC}\,, \quad \mathsf{BC}+\mathsf{A} \underset{\bar{g}}{\overset{g}{\rightleftharpoons}} \mathsf{ABC}\,, \quad \mathsf{C}_2 \underset{\bar{d}}{\overset{d}{\rightleftharpoons}} 2\mathsf{C}\,.$$

Using mass action kinetics this translates into the following system of differential equations:

$$\begin{aligned}
\frac{\mathrm{d}[\mathsf{C}]}{\mathrm{d}t} &= \bar{a}[\mathsf{AC}] - a[\mathsf{A}][\mathsf{C}] + \bar{b}[\mathsf{BC}] - b[\mathsf{B}][\mathsf{C}] + 2d[\mathsf{C}_2] - 2\bar{d}[\mathsf{C}]^2\,,\\
\frac{\mathrm{d}[\mathsf{AC}]}{\mathrm{d}t} &= a[\mathsf{A}][\mathsf{C}] - \bar{a}[\mathsf{AC}] + \bar{h}[\mathsf{ABC}] - h[\mathsf{B}][\mathsf{AC}]\,,\\
\frac{\mathrm{d}[\mathsf{BC}]}{\mathrm{d}t} &= b[\mathsf{B}][\mathsf{C}] - \bar{b}[\mathsf{BC}] + \bar{g}[\mathsf{ABC}] - g[\mathsf{A}][\mathsf{BC}]\,,\\
\frac{\mathrm{d}[\mathsf{ABC}]}{\mathrm{d}t} &= g[\mathsf{A}][\mathsf{BC}] + h[\mathsf{B}][\mathsf{AC}] - (\bar{g} + \bar{h} + r)[\mathsf{ABC}]\,,\\
\frac{\mathrm{d}[\mathsf{C}_2]}{\mathrm{d}t} &= \bar{d}[\mathsf{C}]^2 - d[\mathsf{C}_2] + r[\mathsf{ABC}]\,.
\end{aligned} \tag{15}$$

For the total template concentration:

$$c = [\mathsf{C}] + [\mathsf{AC}] + [\mathsf{BC}] + [\mathsf{ABC}] + 2[\mathsf{C}_2]\,, \tag{16}$$

we obtain the simple growth law:

$$\frac{\mathrm{d}c}{\mathrm{d}t} = r[\mathsf{ABC}]\,. \tag{17}$$

The quasi-steady state approximation (QSSA), see e.g. Refs. 11 and 101, can now be used to express the concentrations of intermediate products in terms of the free template concentration. The total template concentration can also be represented in this way:

$$\begin{aligned} c &= [\mathsf{C}]\left[\left(1+\frac{a[\mathsf{A}]}{\bar{a}+h[\mathsf{B}]}+\frac{b[\mathsf{B}]}{\bar{b}+g[\mathsf{A}]}+Z\right)\left(\frac{\bar{h}}{\bar{a}+h[\mathsf{B}]}+\frac{\bar{g}}{\bar{b}+g[\mathsf{A}]}+1+\frac{2r}{d}\right)\right]+[\mathsf{C}]^2\frac{2\bar{d}}{d}\,,\\ Z &= \frac{[A][B](gb(\bar{a}+h[\mathsf{B}])+ha(\bar{b}+g[\mathsf{A}]))}{[\mathsf{A}]g\bar{h}(\bar{b}+g[\mathsf{A}])+[\mathsf{B}]h\bar{g}(\bar{a}+h[\mathsf{B}])-(\bar{g}+\bar{h}+r)(\bar{a}+h[\mathsf{B}])(\bar{b}+g[\mathsf{A}])}\,. \end{aligned} \tag{18}$$

Thus the total template concentration is related to the free template concentration by the quadratic equation:

$$c = X[\mathsf{C}]+Y[\mathsf{C}]^2\,, \qquad [\mathsf{C}] = \frac{\sqrt{1+4cY/X^2}-1}{2Y/X}\,, \tag{19}$$

where X and Y are the complicated functions of the elementary rate constants defined in Eq. (18). With the QSSA Eq. (17) becomes

$$\frac{\mathrm{d}c}{\mathrm{d}t} = rZ[\mathsf{C}]\,. \tag{20}$$

This can be rewritten in the form:

$$\frac{\mathrm{d}c}{\mathrm{d}t} = \alpha c\psi(\beta c) \quad \text{with} \quad \psi(u) = \frac{2}{u}\left(\sqrt{1+u}-1\right), \tag{21}$$

where $\alpha = rZ/X$ and $\beta = 4Y/X^2$. The function ψ has the Taylor series expansion $\psi(u) = 1 - u/4 + u^2/8 + \mathcal{O}(u^3)$ for small arguments u and satisfies $\psi(u) = 2/\sqrt{u} + \mathcal{O}(u^{-1})$ for large u. The interpretation of the parameters α and β is straightforward: The *Darwinian fitness* α describes the growth rate in dilute solution, while β describes the *strength of product inhibition.*

Equation (21) is surprisingly general. If we consider three or more fragments that have to be ligated together, we eventually find the same functional form of the growth law, of course with even more complicated expressions for α and β.

A quite different mechanism of replication proceeds via DNA triple helices [66], Fig. 9. Here, a DNA duplex $\mathsf{C}\cdot\mathsf{C}$ is replicated by first forming an adduct $\mathsf{C}\cdot\mathsf{C'DE}$ with triple helix geometry, where the template strand forms standard Watson–Crick pairs, while the fragments D and E are attached via Hoogsteen pairs. The fragments are ligated and then the resulting $\mathsf{C}\cdot\mathsf{C'C}$ complex dissociates along the weaker Hoogsteen pairs. Finally, the single stranded template sequence is ligated

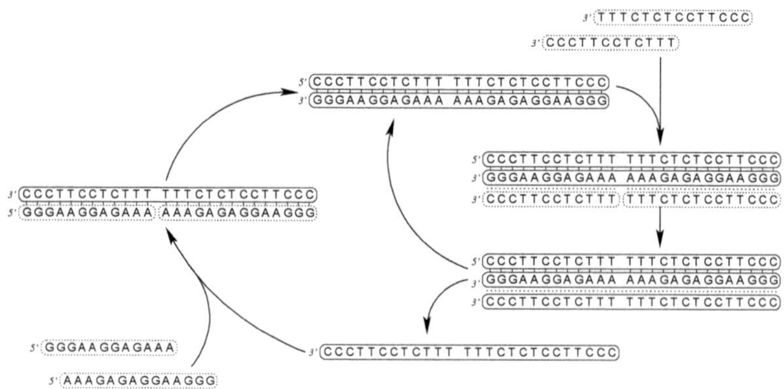

Fig. 9. Self-replication of duplex DNA via a triple-helix stage. In the experiment by Li and Nicolaou [66] a palindromic DNA template was used. In the general case, a suitable second cycle would be necessary. Watson–Crick base pairs are indicated by short lines, Hoogsteen pairs are shown as dots.

with fragments of its complements and forms a copy of the original duplex DNA. The reaction mechanism can be summarized as follows:

$$\mathsf{C}\cdot\mathsf{C}+\mathsf{D}+\mathsf{E}\underset{\bar{b}}{\overset{b}{\rightleftharpoons}}\mathsf{C}\cdot\mathsf{C'DE}\,,\quad \mathsf{C}\cdot\mathsf{C'DE}\overset{r}{\longrightarrow}\mathsf{C}\cdot\mathsf{C'C}\,,\quad \mathsf{C}\cdot\mathsf{C'C}\underset{\bar{d}}{\overset{d}{\rightleftharpoons}}\mathsf{C}\cdot\mathsf{C}+\mathsf{C}\,,$$

$$\mathsf{C}+\mathsf{A}+\mathsf{B}\underset{\bar{a}}{\overset{a}{\rightleftharpoons}}\mathsf{C}\cdot\mathsf{AB}\,,\qquad \mathsf{C}\cdot\mathsf{AB}\overset{s}{\longrightarrow}\mathsf{C}\cdot\mathsf{C}\,.$$

In this simple form the mechanism works only with palindromic DNA since the original and the copied $\mathsf{C}\cdot\mathsf{C}$ complex are reversed. One can easily envision another round of triple helix formation and dissociation, however. For the purpose of the kinetic analysis we stick to the palindromic case. The kinetic differential equations read:

$$\begin{aligned}
\frac{\mathrm{d}[\mathsf{C}\cdot\mathsf{C}]}{\mathrm{d}t} &= -b[\mathsf{C}\cdot\mathsf{C}][\mathsf{D}][\mathsf{E}]+\bar{b}[\mathsf{C}\cdot\mathsf{C'DE}] \\
&\quad + s[\mathsf{C}\cdot\mathsf{AB}]+d[\mathsf{C}\cdot\mathsf{C'C}]-\bar{d}[\mathsf{C}\cdot\mathsf{C}][\mathsf{C}]\,, \\
\frac{\mathrm{d}[\mathsf{C}\cdot\mathsf{C'DE}]}{\mathrm{d}t} &= b[\mathsf{C}\cdot\mathsf{C}][\mathsf{D}][\mathsf{E}]-\bar{b}[\mathsf{C}\cdot\mathsf{C'DE}]-r[\mathsf{C}\cdot\mathsf{C'DE}]\,, \\
\frac{\mathrm{d}[\mathsf{C}\cdot\mathsf{C'C}]}{\mathrm{d}t} &= r[\mathsf{C}\cdot\mathsf{C'DE}]-d[\mathsf{C}\cdot\mathsf{C'C}]+\bar{d}[\mathsf{C}][\mathsf{C}\cdot\mathsf{C}]\,, \\
\frac{\mathrm{d}[\mathsf{C}]}{\mathrm{d}t} &= d[\mathsf{C}\cdot\mathsf{C'C}]-\bar{d}[\mathsf{C}][\mathsf{C}\cdot\mathsf{C}]-a[\mathsf{C}][\mathsf{A}][\mathsf{B}]+\bar{a}[\mathsf{C}\cdot\mathsf{AB}]\,, \\
\frac{\mathrm{d}[\mathsf{C}\cdot\mathsf{AB}]}{\mathrm{d}t} &= a[\mathsf{C}][\mathsf{A}][\mathsf{B}]-\bar{a}[\mathsf{C}\cdot\mathsf{AB}]-s[\mathsf{C}\cdot\mathsf{AB}]\,,
\end{aligned}\qquad (22)$$

and we have the following expression for the total template concentration:

$$c = [\mathsf{C}] + [\mathsf{C}\cdot\mathsf{AB}] + 2[\mathsf{C}\cdot\mathsf{C}] + 2[\mathsf{C}\cdot\mathsf{C'DE}] + 3[\mathsf{C}\cdot\mathsf{C'C}]\,. \tag{23}$$

The net growth law is therefore:

$$\dot{c} = s[\mathsf{C}\cdot\mathsf{AB}] + r[\mathsf{C}\cdot\mathsf{C'DE}]\,. \tag{24}$$

Using the quasi-steady state approximation, we obtain

$$\dot{c} = 2\frac{as}{a+s}[\mathsf{A}][\mathsf{B}][\mathsf{C}] \tag{25}$$

and $c = X[\mathsf{C}] + Y[\mathsf{C}]^2$ with

$$\begin{aligned} X &= 1 + \frac{a}{\bar{a}+s}\left(1 + 2\frac{s}{r} + 3\frac{s}{d}\right)[\mathsf{A}][\mathsf{B}] + 2\frac{(\bar{b}+r)as}{(\bar{a}+s)br}\frac{[A][B]}{[D][E]}\,, \\ Y &= 3\frac{\bar{d}}{d}\frac{(\bar{b}+r)as}{(\bar{a}+s)br}\frac{[\mathsf{A}][\mathsf{B}]}{[\mathsf{D}][\mathsf{E}]}\,. \end{aligned} \tag{26}$$

The over-all kinetics of Eq. (22) is therefore again described by Eq. (21), this time with the parameters $\alpha = 2as[\mathsf{A}][\mathsf{B}]/[X(\bar{a}+s)]$ and $\beta = 4Y/X^2$.

Let us now turn to the competition of different strains of templates $[C_k]$. In Ref. 141 we show that in the absence of direct interactions between the templates and with "correct instruction", i.e. in the absence of mutation, the evolution of a mixture of templates is described by the replicator equation:

$$\dot{x}_k = x_k\left(\alpha_k\psi(c_0\beta_k x_k) - \Phi\right). \tag{27}$$

This follows from Eq. (21) and a transformation to relative concentrations. In addition one obtains an equation for the total concentration c_0, which we assume here to be regulated such that it varies only slowly. In this case one can regard c_0 as an additional tunable parameter.

The analysis of Eq. (27) is in essence based on Theorem A from Ref. 42, which can be restated in the following form:

Theorem 5.1. (i) *There is a unique fixed point $\hat{x}$ which is the ω-limit of all orbits in the interior of the simplex S_n. (ii) If $\hat{x}$ lies in the interior of a face, then it is also the ω-limit of all orbits in the interior of its face.* (iii) *If the species are labeled such that $\alpha_1 \geq \alpha_2 \geq \cdots \alpha_n$, then there is an index $m \geq 1$ such that $\hat{x}$ is of the form $\hat{x}_i > 0$ if $i \leq m$ and $\hat{x}_i = 0$ for $i > m$.* (iv) *If $\min\{\beta_k\}$ is large enough, then $m = n$ and $\hat{x}$ is a uniquely determined interior fixed point.*

It is shown in Ref. 47 that Eq. (27) is a Shashahani gradient system. Furthermore, $V(x) = \prod_k x_k^{z_k}$ is a Ljapunov function, when z_k are the coordinates of the globally stable fixed point.

It is not hard to verify that the condition for survival of species k, item (iii) in Theorem 5.1, is explicitly

$$\alpha_k > \Phi(\hat{x})\,. \tag{28}$$

If we sort the replicating species according to decreasing values of the Darwinian fitness parameters, $\alpha_1 \geq \alpha_2 \geq \cdots \geq \alpha_n$, then there is an index m such that $\hat{x}$ is of the form $\hat{x}_k > 0$ if $k \leq m$ and $\hat{x}_k = 0$ for $k > m$. In other words, m species survive while the $n - m$ least efficient replicators die out. This behavior is completely analogous to the reversible exponential competition model discussed in Ref. 97, where the rate constants a_k play the role of our Darwinian fitness parameters α_k.

The threshold value $\Phi(\hat{x})$ can be computed explicitly in the form:

$$\Phi(\hat{x}) = \frac{1}{c_0} \sum_k^m \frac{\alpha_k}{\beta_k} \left[\sqrt{1 + c_0 \left(\sum_k^m \frac{\alpha_k^2}{\beta_k} \right) \Big/ \left(\sum_k^m \frac{\alpha_k}{\beta_k} \right)^2} - 1 \right], \tag{29}$$

where the sum runs over the surviving species only. It is interesting to note that the Darwinian fitness parameters α_k determine the order in which species reach extinction, whereas the concentration-dependent values $\beta_k c_0$ collectively influence the flux term and hence set the "extinction threshold." In contrast to Szathmáry's model equation [123] the extended replicator kinetics leads to both competitive selection and coexistence of replicators depending on total concentration and kinetic constants. For small c_0 it behaves like a quasispecies model, while product inhibition and therefore unconditional coexistence dominates in a "thick soup," Fig. 10.

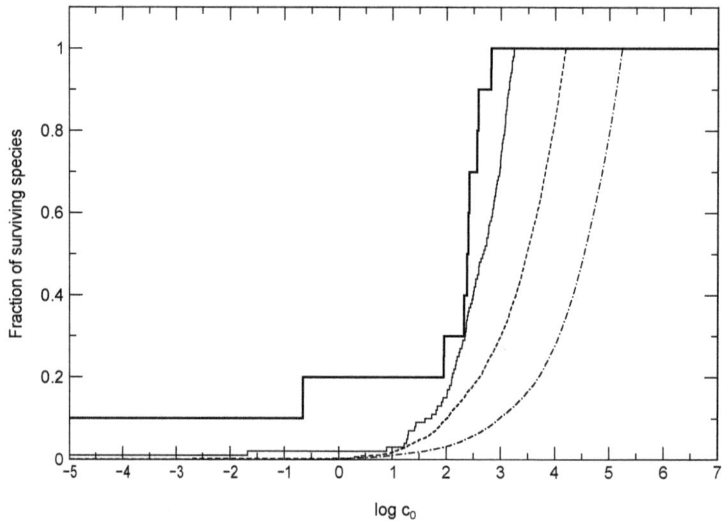

Fig. 10. Fraction of surviving species as a function of the total concentration c for $n = 10$, 100, 1000, 10000 species (from left to right) with $\alpha_k \sim \exp(-k/n)$ and β_k chosen from Uniform $[0, 1]$. Data are taken from Ref. 113.

Systems of competing species replicating via template-directed ligation therefore can meet two of the criteria required for effective evolution: (i) strong selection leading to the extinction of some or even almost all species, and (ii) susceptibility to invasion by new, advantageous species. The origin of life requires a mechanism of chemical replication in which strong selection enables some species to outgrow others, the "losers" which die out. On the other hand, the coexistence of more than one "master sequence" is required for functional specialization and cooperation to emerge because of the limitations on heritable information that are imposed by the error threshold. Replicators which reproduce through duplex formation utilizing a Michaelis–Menten type of mechanism fulfil these conditions much more easily than replicators needing the strong and specific catalytic interactions characteristic of hypercyclic cooperation [28, 119]. Therefore, they are good candidates for the first molecules which may have been selectively amplified in the prebiotic environment.

6. Higher Order Systems

In this section, we briefly consider plausible molecular mechanisms for ligation-based replication involving both a template and a second "catalyst" that acts as a "replicase ribozyme." Experimentally such a system is at least reasonable since multiple helices are well-known structures in both nucleic acids and coiled-coil peptides. Alternatively, one may think of the catalyst as a version of Johnston's [57] RNA replicase shown in Fig. 1(a).

The generic features of the resulting reaction networks are largely unexplored despite some efforts towards understanding the dynamics of chemically realistic models of "prebiotic" replication in Ref. 41. On the other hand, the limiting case of strong selection is well understood. When complex formation and product inhibition play no role the dynamics is described by the second order replicator equations which we have considered in Sec. 3. In the spirit of the previous section we consider a template species C that can be obtained by the concatenation (ligation) of two building blocks A and B provided in the environment. Again, a generalization to multiple building blocks is straightforward even though explicit expressions for the aggregate parameters in terms of the elementary rate constants become too complicated to be useful. We restrict ourselves to a rather qualitative discussion here, more technical details can be found in Ref. 112.

The crucial step of the replication mechanism is the formation of an intermediate species ABCC, which we may visualize as a triple helix consisting of the template C, the building material A and B properly aligned along C so that the ligation of the parts A and B can readily proceed, and an additional copy of C that might be thought either as a specific catalyst that is involved in the ligation reaction, or its role might be seen in facilitating the release of the product C. In either case, we consider the reaction scheme of Fig. 11.

It is instructive to compare the reaction scheme in Fig. 11 with the triple-helix ligation model of Li and Nicolaou in Fig. 9. The main difference is that in the present

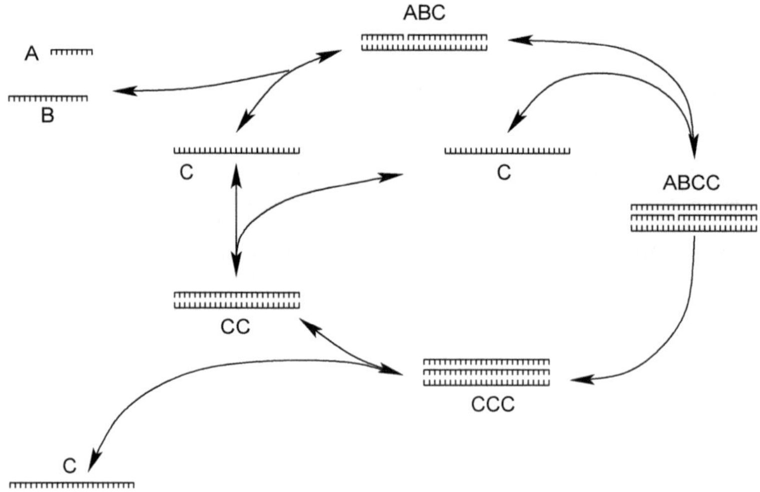

Fig. 11. Scheme of a ligation-replication mechanism with higher order catalysis.

model the catalyst [C] is attached after the ligation complex [ABC] is formed. The consequences of this difference on the dynamical behavior are rather dramatic. A tedious computation [112] yields

$$c = \frac{p_4[\mathsf{C}]^4 + p_3[\mathsf{C}]^3 + p_2[\mathsf{C}]^2 + p_1[\mathsf{C}]}{q_2[\mathsf{C}]^2 + q_1[\mathsf{C}] + q_0}, \tag{30}$$

$$\frac{\mathrm{d}c}{\mathrm{d}t} = \frac{\mathrm{d}[\mathsf{C}]}{\mathrm{d}t} = a[\mathsf{ABCC}] = [\mathsf{A}][\mathsf{B}][\mathsf{C}]^2\frac{\alpha + \gamma[\mathsf{C}]}{\beta + \delta[\mathsf{C}]}, \tag{31}$$

where the QSSA is used to obtain the last equality. All coefficients here are complicated combinations of the elementary rate constants, all of which are positive. For small values of $c \ll 1$ we have $c \approx p_1[\mathsf{C}]/q_0$ and hence

$$\dot{c} \sim \frac{q_0^2\alpha}{p_1^2\beta}[\mathsf{A}][\mathsf{B}]c^2, \tag{32}$$

i.e. we have second order autocatalytic or hyperbolic growth.

For large values of c we have $c \approx (p_4/q_1)[\mathsf{C}]^3$ and hence

$$\dot{c} \sim \sqrt[3]{\frac{q_1^2}{p_4^2}\frac{\gamma}{\delta}}[\mathsf{A}][\mathsf{B}]c^{2/3}. \tag{33}$$

The concentration grows polynomially in this case.

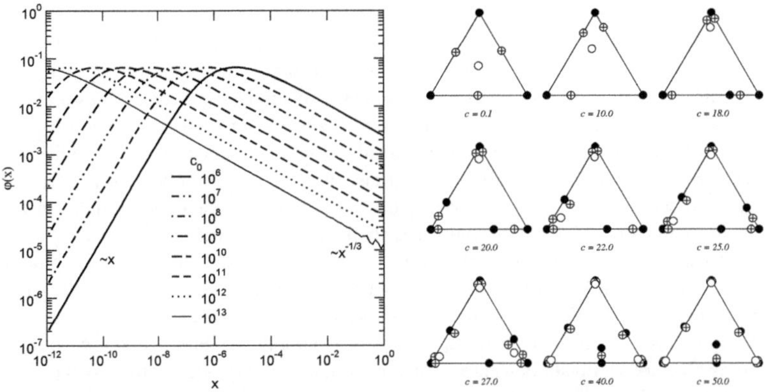

Fig. 12. L.h.s.: Growth function φ, R.h.s.: Bifurcation diagram as a function of c_0.

If $q_2[\mathsf{C}]^2$ cannot be neglected, we have $c \approx (p_4/q_2)[\mathsf{C}]^2$ for small values of c, and hence

$$\dot{c} \propto \frac{q_2\gamma}{p_4\delta}c, \tag{34}$$

i.e. we observe exponential growth.

Let us now consider the case of multiple templates and assume that each of them catalyzes only its own replication. It can be shown [112] that the relative concentrations again follow a replicator equation, which now is of the form shown in the l.h.s. of Fig. 12:

$$\dot{x}_k = x_k\left[\varphi_k(x_k; c_0) - \Phi\right] \tag{35}$$

with the growth functions φ_k of the form:

$$\varphi_k(x_k) \sim \begin{cases} c_0 \dfrac{q_{k;0}^2\alpha_k}{p_{k;1}^2\beta_k}[\mathsf{A_k}][\mathsf{B_k}]x_k & c_0 \to 0\,, \\ \sqrt[3]{\dfrac{q_{k;1}^2}{c_0 p_{k;4}^2}}\dfrac{\gamma_k}{\delta_k}[\mathsf{A_k}][\mathsf{B_k}]x_k^{-1/3} & c_0 \to \infty\,. \end{cases} \tag{36}$$

As a consequence, we recover two well-studied models as limiting cases: In dilute solution, $c_0 \to 0$, we obtain Schlögl's model of competing exponentially growing autocatalysts [90], in saturated solution, $c_0 \to \infty$, we recover Szathmáry's model of parabolic growth, Eq. (14) [123], with the exponent $a = 2/3$. The transition between the two limiting cases proceeds via the series of saddle-node bifurcations shown on the r.h.s. of Fig. 12. Similarly, one obtains second order replicator equations in the case of cross-catalysis in dilute solution. More details can be found in Ref. 112.

7. Discussion

Template induced replication is the basis of multiplication and reproduction in nature and *in vitro*, hence a detailed understanding of its mechanisms in terms of chemical reaction kinetics is an important facet of understanding evolution at the molecular level. Simple single step mechanisms of independent replication lead to exponential growth while the explicit consideration of dissociation of the replication complex reduces the net rate of synthesis and yields parabolic growth. Only a detailed mechanistic analysis reveals that both types of behavior result from the same chemically plausible mechanisms, albeit under different conditions.

Evolution is limited in both cases. An error threshold limits the amount of heritable information in the selection dominated exponential growth regime, while parabolic growth essentially switches off selection because all replicators coexist. Cooperatively coupled replicators, such as the hypercycle [28], are one way to overcome the error threshold. These networks, however, are susceptible to parasitic mutants and invaders. Furthermore, it is not clear at this point that uncatalyzed replication can sustain RNAs that are large enough to efficiently assist in RNA replication.

Models with significant product inhibition bridge this gap. A "survival threshold" allows the coexistence of not only the fittest molecule but of an ensemble of possibly unrelated fit replicators. These might collectively accumulate sufficient information for active catalysts and hence lead to the onset of a positive feedback loop that leads to the evolution of efficient ribozymes. In this scenario selection is still an important force that removes unfit variants from the pool. On the other hand, it appears to at least alleviate the parasite problem that plagues the second order replicator models.

Of course, a homogeneous "soup" environment is not the only possibility. The idea to consider a spatial organization as a means of overcoming the parasite problem was introduced by Boerlijst and Hogeweg [9], who investigated hypercycles with parasites using a two-dimensional cellular automaton. The authors observed spirals that stabilized the hypercycles in competition with the parasite. A similar model based on partial differential equations produced comparable results [120]. Later work showed, however, that the stabilizing effect caused by the spiral patterns is highly model-dependent [18]. Coexistence of quasispecies as a consequence of spatial structure and limited diffusion is described in Ref. 127.

A different scenario is replication in open chaotic flows such as those near hydrothermal vents. The associated fractal dynamics leads to singularly enhanced concentrations, resulting in a growth law for the replicating material of the form:

$$\frac{\mathrm{d[C]}}{\mathrm{d}t} = -\kappa[\mathrm{C}] + \nu[\mathrm{C}]^{-\beta}, \qquad (37)$$

which is completely different from the behavior in homogeneous solution or on a surface [59, 130]. Similarly to models with parabolic growth, it leads to coexistence.

Another interesting class of models, which is however beyond the scope of this paper, assumes that the genetic material is already encapsulated in a protocell. This protocell duplicates with a rate that depends on the type and concentration of the enclosed material. Both the *stochastic corrector model* [122] and hypercycles in compartments can explain the accumulation of information and show significant resistance against parasites of various kinds, as shown in recent computer simulations [146]. It is interesting to note that the genes that are enclosed in the compartment of the stochastic corrector model form a catalytic network that would be unstable in homogeneous solution. Information accumulation through compartmentalization would therefore seem to require that a very short "genome" can already significantly influence the duplication of the compartment. This possibility is supported to a certain extent by the fact that RNA molecules can specifically interact with phospholipid bilayers and regulate their permeability [134]. The particular RNA molecules described in Ref. 134, however, are by far too large to be replicated without enzymes or ribozymes. Once there is a replicase-ribozyme that can replicate itself (which is pre-supposed in both the hypercycle and the stochastic corrector model) there is a selection pressure even in a homogeneous medium to improve the replication accuracy, which in turn allows for larger genomes and hence opens up additional possibilities for increasing the accuracy [87]. Compartmentalization *per se* is therefore neither necessary nor sufficient to overcome the information bottle-neck imposed by the error-threshold. Since it is by no means clear that compartments could facilitate the evolution of the first replicase-ribozyme, we cannot use replication kinetics to deduce whether compartmentalization should have occurred before or after the invention of catalyzed replication. The kinetics of uncatalyzed ligation-based replication linked to a compartment will be explored elsewhere [85].

In this contribution we have restricted ourselves to chemical kinetics and its consequences, i.e. to models of "chemical population dynamics." A more complete model of evolution would have to include a description of the mutants that are accessible from a given population [30] and a way to construct the kinetic constants from a mechanistic model. In the simplest cases this leads to the notion of a fitness landscape [143, 144] on which the evolving population moves. The properties of this landscape then determine the long-term outcome of evolution, see e.g. Refs. 94 and 95. While the theory of fitness landscapes [86] is quite well developed, the generalization to networks of interacting replicators [31] has not yet led to a concise theory of the long-term evolution of strongly interacting replicators.

Acknowledgments

This work was supported in part by the Austrian *Fonds zur Förderung der Wissenschaftlichen Forschung*, Project No. P-13887-MOB.

References

[1] Alves, D. and Fontanari, J. F., Population genetics approach to the quasispecies model, *Phys. Rev.* **E54**, 4048–4053 (1996).

[2] Alves, D. and Fontanari, J. F., Error threshold in finite populations, *Phys. Rev.* **E57**, 7008–7013 (1998).
[3] Amann, E., *Permanence in Catalytic Networks*, Ph.D. thesis, Universität Wien, Institut für Mathematik (1989).
[4] Andrade, M. A., Nuño, J. C., Morán, F., Montero, F. and Mptisos, G., Complex dynamics of a catalytic network having faulty replication into error-species, *Physica* **D63**, 21–40 (1993).
[5] Arneodo, A., Coullet, P. and Tresser, C., Occurence of strange attractors in three-dimensional Volterra equations, *Phys. Lett.* **A79**, 259–263 (1980).
[6] Bachmann, P. A., Luisi, P. L. and Lang, J., Autocatalytic self-replicating micelles as models for prebiotic structures, *Nature* **357**, 57–59 (1992).
[7] Baskerville, S. and Bartel, D. P., A ribozyme that ligates RNA to protein, *Proc. Natl. Acad. Sci. USA* **99**, 9154–9159 (2002).
[8] Biebricher, C. K. and Eigen, M., Kinetics of RNA replication by $Q\beta$ replicase, in *RNA Genetics. Vol. I: RNA Directed Virus Replication*, Domingo, E., Holland, J. J. and Ahlquist, P. eds., pp. 1–21 (CRC Press, Boca Raton, FL, 1988).
[9] Boerlijst, M. C. and Hogeweg, P., Spiral wave structure in pre-biotic evolution: Hypercycles stable against parasites, *Physica* **D48**, 17–28 (1991).
[10] Bomze, I., Lotka-Volterra equations and replicator dynamics: A two-dimensional classification, *Biol. Cybern.* **48**, 201–211 (1983).
[11] Borghans, J. A. M., de Boer, R. J. and Segel, L. A., Extending the quasi-steady state approximation by changing variables, *Bull. Math. Biol.* **58**, 43–63 (1996).
[12] Brack, A., ed., *The Molecular Origins of Life: Assembling Pieces of the Puzzle* (Cambridge University Press, Cambridge, UK, 1999).
[13] Brannath, W., Heteroclinic networks on the tetrahedron, *Nonlinearity* **7**, 1367–1384 (1994).
[14] Bresch, C., Niesert, U. and Harnasch, D., Hypercycles, parasites, and packages, *J. Theor. Biol.* **85**, 399–405 (1980).
[15] Campos, P. R. A. and Fontanari, J. F., Finite-size scaling of the error threshold transition in finite population, *J. Phys. A: Math. Gen.* **32**, L1–L7 (1999).
[16] Campos, P. R. A., Fontanari, J. F. and Stadler P. F., Error propagation in the hypercycle, *Phys. Rev.* **E61**, 2996–3002 (2000).
[17] Chaput, J. C. and Switzer, C., Nonenzymatic oligomerization on templates containing phosphoester-linked acyclic glycerol nucleic acid analogues, *J. Mol. Evol.* **51**, 464–470 (2000).
[18] Cronhjort, M. B. and Blomberg, C., Hypercycles versus parasites in a two dimensional partial differential equations model, *J. Theor. Biol.* **169**, 31-49 (1994).
[19] Dawkins, R., *The Selfish Gene* (Oxford University Press, Oxford, 1976).
[20] Domingo, E., Biological significance of viral quasispecies, *Viral Hepatitis Rev.* **2**, 247–261 (1996).
[21] Domingo, E. and Holland, J. J., RNA virus mutations and fitness for survival, *Ann. Rev. Microbiol.* **51**, 151–178 (1997).
[22] Doudna, J. A. and Cech, T. R., The chemical repertoire of natural ribozymes, *Nature* **418**, 222–228 (2002).
[23] Drake, J. W., Rates of spontaneous mutation among RNA viruses, *Proc. Natl. Acad. Sci. USA* **90**, 4171–4175 (1993).
[24] Drake, J. W., Charlesworth, B., Charlesworth, D. and Crow, J. F., Rates of spontaneous mutations, *Genetics* **148**, 1667–1686 (1998).
[25] Eigen, M., Selforganization of matter and the evolution of biological macromolecules, *Naturwissenschaften* **58**, 465–523 (1971).

[26] Eigen, M., McCaskill, J. and Schuster, P., Molecular quasi-species, *J. Phys. Chem.* **92**, 6881–6891 (1988).
[27] Eigen, M., McCaskill, J. S. and Schuster, P., The molecular quasi-species, *Adv. Chem. Phys.* **75**, 149–263 (1989).
[28] Eigen, M. and Schuster, P., *The Hypercycle* (Springer-Verlag, New York, Berlin, 1979).
[29] Famulok, M., Nowick, J. and Rebek Jr., J., Self-replicating systems, *Act. Chim. Scand.* **46**, 315–324 (1992).
[30] Fontana, W. and Schuster, P., Continuity in evolution: On the nature of transitions, *Science* **280**, 1451–1455 (1998).
[31] Forst, C. V., Molecular evolution of catalysis, *J. Theor. Biol.* **205**, 409–431 (2000).
[32] Fox, S. W. and Dose, H., *Molecular Evolution and the Origin of Life* (Academic Press, New York, 1977).
[33] Freedman, H. and Moson, P., Persistence definitions and their connections, *Proc. Amer. Math. Soc.* **109**, 1025–1033 (1990).
[34] Freedman, H. and Waltman, P., Mathematical analysis of some three-species food-chain models, *Math. Biosc.* **33**, 257–276 (1977).
[35] Freeland, S. J., Knight, R. D. and Landweber, L. F., Do proteins predate DNA? *Science* **286**, 690–692 (1999).
[36] Gesteland, R. F. and Atkins J. F., eds. *The RNA World* (Cold Spring Harbor Laboratory Press, Plainview, NY, 1993).
[37] Gilbert, W., The RNA world, *Nature* **319**, 618 (1986).
[38] Gilpin, M. E., Spiral chaos in a predator prey system, *Amer. Nat.* **133**, 306–308 (1978).
[39] Happel, R. and Stadler, P. F., The evolution of diversity in replicator networks, *J. Theor. Biol.* **195**, 329–338 (1998).
[40] Happel, R. and Stadler, P. F., Autocatalytic replication in a CSTR and constant organization, *J. Math. Biol.* **38**, 422–434 (1999).
[41] Hecht, R., Happel, R., Schuster, P. and Stadler, P.F., Autocatalytic networks with intermediates I: irreversible reactions. *Math. Biosc.* **140**, 33–74 (1997).
[42] Hofbauer, J., On the occurrence of limit cycles in Volterra-Lotka equations, *Nonlin. Anal.* **5**, 1003–1007 (1981).
[43] Hofbauer, J., The selection mutation equation, *J. Math. Biol.*, **23** (1985).
[44] Hofbauer, J., Saturated equilibria, permanence, and stability for ecological systems, in *Mathematical Ecology: Proceedings on the Second Autumn Course on Mathematical Ecology*, Trieste, Italy, Hallham, T. G., Gross, L. J., Lewin, S. A. eds. (World Scientific, Singapore, 1986), pp. 625–642.
[45] Hofbauer, J., Competitive exclusion of disjoint hypercycles, *Z. Phys. Chem.* **216**, 35–39 (2002).
[46] Hofbauer, J. and Sigmund, K., Permanence for replicator equations, *Lecture Notes Econ. Math.* (Springer-Verlag Berlin, Heidelberg, New York, 1987), Vol. 287, pp. 131–146.
[47] Hofbauer, J. and Sigmund, K., *Dynamical Systems and the Theory of Evolution* (Cambridge University Press, Cambridge, U.K., 1988).
[48] Hofbauer, J. and Sigmund, K., *Evolutionary Games and Population Dynamics* (Cambridge University Press, Cambridge, U.K., 1998).
[49] Hofbauer, J. and So, J., Multiple limit cycles for three-dimensional Lotka-Volterra equations, *Appl. Math. Lett.* **7**, 65–70 (1994).
[50] Hutson, V. and Moran, W., Persistence of species obeying difference equations, *Bull. Math. Biol.* **15**, 203–212 (1982).

[51] Hutson, V. and Schmitt, K., Permanence and the dynamics of biological systems, *Math. Biosc.* **111**, 1–71 (1992).
[52] Illangasekare, M. and Yarus, M., A tiny RNA that catalyzes both aminoacyl-RNA and peptidyl-RNA synthesis, *RNA* **5**, 1482–1489 (1999).
[53] Jadhav, V. R. and Yarus, M., Coenzymes as coribozymes, *Biochimie* **84**, 877–888 (2002).
[54] James, K. D. and Ellington, A. D., The fidelity of template-directed oligonucleotide ligation and the inevitability of polymerase function, *Orig. Life Evol. Biosph.* **29**, 375–390 (1999).
[55] Jansen, W., A permanence theorem for replicator and Lotka-Volterra systems, *J. Math. Biol.* **25**, 411–422 (1986).
[56] Jeffares Daniel, C., Poole, A. M. and Penny, D., Relics from the RNA world, *J. Mol. Evol.* **46**, 18–36 (1998).
[57] Johnston, W. K., Unrau, P. J., Lawrence, M. J., Glasner, M. E. and Bartel, D. P., RNA-catalyzed RNA polymerization: Accurate and general RNA-templated primer extension, *Science* **292**, 1319–1325 (2001).
[58] Joyce, G. F., The antiquity of RNA-based evolution, *Nature* **418**, 214–221 (2002).
[59] Károlyi, G., Péntek, Á., Scheuring, I. and Toroczkai, Z., Chaotic flow: the physics of species coexistence, *Proc. Natl. Acad. Sci.* **97**, 13661–13665 (2000).
[60] Keefe, A. D. and Miller, S. L., Are polyphosphates or phosphate esters prebiotic reagents? *J. Mol. Evol.* **41**, 693–702 (1995).
[61] Kumar, R. K. and Yarus, M., RNA-catalyzed amino acid activation, *Biochemistry* **40**, 6998–7004 (2001).
[62] Lee, D. H., Granja, J. R., Martinez, J. A., Severin, K. and Ghadiri, M. R., A self-replicating peptide, *Nature* **382**, 525–528 (1996).
[63] Lee, D. H., Severin, K., Yokobayashi, Y. and Ghadiri, M. R., Emergence of symbiosis in peptide self-replication through a hypercyclic network, *Nature* **390**, 591–594 (1997).
[64] Lee, N., Bessho, Y., Wei, K., Szostak, J. W. and Suga, H., Ribozyme-catalyzed tRNA aminoacylation, *Nat. Struct. Biol.* **7**, 28–33 (2000).
[65] Leung, D. W., Chen, E. and Goeddel, D. V., A method for random mutagenesis of a defined DNA segment using a modified polymerase chain reaction, *Technique* **1**, 11–15 (1989).
[66] Li, T. and Nicolaou, K. C., Chemical self-replication of palindromic duplex DNA, *Nature* **369**, 218–221 (1994).
[67] Martinez, M. A., Vartanian, J. P. and Wain-Hobson, S., Hypermutagenesis of RNA using human immunodeficiency virus type 1 reverse transcriptase and biased dNTP concentrations, *Proc. Natl. Acad. Sci. USA* **91**, 11787–11791 (1994).
[68] Mason, S. F., *Chemical Evolution. Origin of the Elements, Molecules, and Living Systems* (Clarendon Press, Oxford, UK, 1991).
[69] Mavelli, F. and Luisi, P. L., Autopoietic self-reproducing vesicles: A simplified kinetic model, *J. Phys. Chem.* **100**, 16600–16607 (1996).
[70] May, R.M. and Nowak, M. A., Superinfection, metapopulation dynamics, and the evolution of diversity, *J. Theor. Biol.* **170**, 95–114 (1994).
[71] May, R. M. and Nowak, M. A., Coinfection and the evolution of parasite virulence, *Proc. R. Soc. London* **B261**, 209–215 (1995).
[72] Maynard Smith, J., Hypercycles and the origin of life, *Nature* **280**, 445–446 (1979).
[73] McCaskill, J. S., A localisation threshhold for macromolecular quasispecies from continuously distributed replication rates, *J. Chem. Phys.* **80**, 5194 (1984).

[74] McCaskill, J. S., Spatially resolved *in vitro* molecular ecology, *Biophys. Chem.* **66**, 145–158 (1997).

[75] Mills, D. R., Peterson, R. L. and Spiegelman, S., An extracellular Darwinian experiment with a self-duplicating nucleic acid molecule, *Proc. Natl. Acad. Sci. USA* **58**, 217–224 (1967).

[76] Moore, P. B. and Steitz, T. A., The involvement of RNA in ribosome function, *Nature* **418**, 229–235 (2002).

[77] Nelson, K. E., Levy, M. and Miller, S. L., Peptide nucleic acids rather than RNA may have been the first genetic molecule, *Proc. Natl. Acad. Sci. USA* **97**, 3868–3871 (2000).

[78] Nisbet, E. G. and Sleep, N. H., The habitat and nature of early life, *Nature* **409**, 1083–1091 (2001).

[79] Novick, J. S., Feng, Q. and Tjivikua, T., Kinetic studies and modeling of a self-replicating system, *J. Amer. Chem. Soc.* **113**, 8831–8838 (1991).

[80] Nowak, M. and Schuster, P., Error thresholds of replication in finite populations. Mutation frequencies and the onset of Muller's ratchet, *J. Theor. Biol.* **137**, 375–395 (1989).

[81] Nuño, J. C., Andrade, M. A., Morán, F. and Montero, F., A model of an autocatalytic network formed by error-prone self-replicative species, *Bull. Math. Biol.* **55**, 385–415 (1993).

[82] Orgel, L. E., The origin of life — a review of facts and speculations, *Trends Biochem. Sci.* **23**, 491–495 (1999).

[83] Pitsch, S., Krishnamurthy, R., Bolli, M., Wendeborn, S., Holzner, A., Minton, M., Lesueur, C., Schlonvogt, I., Jaun, B. and Eschenmoser, A., Pyranosyl-RNA (p-RNA): base-pairing selectivity and potential to replicate, *Helv. Chim. Acta* **78**, 1621–1635 (1995).

[84] Poole, A., Penny D. and Sjäberg B. M., Methyl-RNA: an evolutionary bridge between RNA and DNA? *Chem. & Biol.* **7**, R207–R216 (2000).

[85] Rasmussen, S., Chen, L., Stadler, B. M. R. and Stadler, P. F., Proto-organism kinetics: Evolutionary dynamics of lipid aggregates with genes and metabolism, *Orig. Life Evol. Biosph.* (2003), in press.

[86] Reidys, C. M. and Stadler, P. F., Combinatorial landscapes, *SIAM Review* **44**, 3–54 (2002).

[87] Scheuring, I., Avoiding catch-22 of early evolution by stepwise increase in copying fidelity, *Selection* **1**, 135–145 (2000).

[88] Scheuring, I. and Szathmáry, E., Survival of replicators with parabolic growth tendency and exponential decay, *J. Theor. Biol.* **212**, 99–105 (2001).

[89] Schidlowski, M., A 3,800 million-year old record of life from carbon in sedimentary rocks, *Nature* **333**, 313–318 (1988).

[90] Schlögl, F., Chemical reaction models for non-equilibrium phase transitions, *Z. Physik* **253**, 147–161 (1972).

[91] Schnabl, W., Stadler, P. F., Forst, C. and Schuster, P., Full characterization of a strange Attractor, *Physica* **D48**, 65–90 (1991).

[92] Schöning, K. U., Scholz, P., Guntha, S., Wu, X., Krishnamurthy, R. and Eschenmoser, A., Chemical etiology of nucleic acid structure: The α-threofuranosyl-(3'$\to$2') oligonucleotide system, *Science* **290**, 1347–1351 (2000).

[93] Schopf, J. W. and Packer, B. M., Early archean (3.3 billion to 3.5 billion year old) microfossils from Warrawoona Group, Australia, *Science* **237**, 70–73 (1987).

[94] Schuster, P., Landscapes and molecular evolution, *Physica* **D107**, 351–365 (1997).

[95] Schuster, P., Fontana, W., Stadler, P. F. and Hofacker, I. L., From sequences to shapes and back: A case study in RNA secondary structures, *Proc. Roy. Soc. London* **B255**, 279–284 (1994).

[96] Schuster, P. and Sigmund, K., Replicator dynamics, *J. Theor. Biol.* **100**, 533–538 (1983).

[97] Schuster, P. and Sigmund, K., Dynamics of evolutionary optimization, *Ber. Bunsen-Gesellsch. Phys. Chem.* **89**, 668–682 (1985).

[98] Schuster, P., Sigmund, K. and Wolff, R., Dynamical systems under constant organization III: Cooperative and competitive behaviour of hypercycles, *J. Diff. Eqns.* **32**, 357–368 (1979).

[99] Schuster, P., Sigmund, K. and Wolff, R., Mass action kinetics of selfreplication in flow reactors, *J. Math. Anal. Appl.* **78**, 88–112 (1980).

[100] Schwartz, A. W., Speculation on the RNA precursor problem, *J. Theor. Biol.* **187**, 523–527 (1997).

[101] Segel, L. A. and Slemrod, M., The quasi-steady state assumption: A case study in perturbation, *SIAM Rev.* **31**, 446–477 (1989).

[102] Segré, D., Ben-Eli, D., Deamer, D. W. and Lancet, D., The lipid world, *Orig. Life. Evol. Biosph.* **31**, 119–145 (2001).

[103] Segré, D., Ben-Eli, D. and Lancet, D., Compositional genomes: Prebiotic information transfer in mutually catalytic noncovalent assemblies, *Proc. Natl. Acad. Sci. USA* **97**, 4112–4117 (2000).

[104] Segré, D., Shenav, B., Kafri, R. and Doron, L., The molecular roots of compositional inheritance, *J. Theor. Biol.* **213**, 481–491 (2001).

[105] Shapiro, R., Prebiotic cytosine synthesis: A critical analysis and implications for the origin of life, *Proc. Natl. Acad. Sci. USA* **96**, 4396–4401 (1999).

[106] Sievers, D. and Von Kiedrowski, G., Self-replication of complementary nucleotide-based oligomers, *Nature* **369**, 221–224 (1994).

[107] Sigmund, K. and Schuster, P., Permanence and uninvadability for deterministic population models, in *Stochastic Phenomena and Chaotic Behaviour in Complex Systems*, Peter Schuster, ed., Vol. 21 of *Synergetics* (Springer-Verlag, New York, 1984), pp. 186–205.

[108] Sleep, N. H., Zahnle, K. J., Kasting, J. F. and Morowitz, H. J., Annihilation of ecosystems by large asteroid impacts on the early Earth, *Nature* **342**, 139–142 (1989).

[109] Spiegelman, S., An approach to the experimental analysis of precellular evolution, *Quart. Rev. Biophys.* **4**, 213–253 (1971).

[110] Stadler, B. M. R., Segregation distortion and heteroclinic cycles, *J. Theor. Biol.* **183**, 363–379 (1996).

[111] Stadler, B. M. R. and Stadler, P. F., Dynamics of small autocatalytic reaction networks III: Monotonous growth functions, *Bull. Math. Biol.* **53**, 469–485 (1991).

[112] Stadler, B. M. R., Stadler, P. F. and Schuster, P., Dynamics of autocatalytic replicator networks based on higher order ligation reactions, *Bull. Math. Biol.* **62**, 1061–1086 (2000).

[113] Stadler, B. M. R., Stadler, P. F. and Wills, P. R., Evolution in systems of ligation-based replicators, *Z. Phys. Chem.* **216**, 21–33 (2001).

[114] Stadler, P. F. and Happel, R., The probability for permanence, *Math. Biosc.* **113**, 25–50 (1993).

[115] Stadler, P. F. and Nuño, J. C., The influence of mutation on autocatalytic reaction networks, *Math. Biosci.* **122**, 127–160 (1994).

[116] Stadler, P. F., Schnabl, W., Forst, C. V. and Schuster, P., Dynamics of small autocatalytic reaction networks II: Replication, mutation, and catalysis, *Bull. Math. Biol.* **57**, 21–61 (1995).
[117] Stadler, P. F. and Schuster, P., Dynamics of small autocatalytic reaction networks I: Bifurcations, permanence and exclusion, *Bull. Math. Biol.* **52**, 485–508 (1990).
[118] Stadler, P. F. and Schuster, P., Mutation in autocatalytic networks — an analysis based on perturbation theory, *J. Math. Biol.* **30**, 597–631 (1992).
[119] Stadler, P. F. and Schuster, P., Permanence of sparse autocatalytic networks, *Math. Biosc.* **131**, 111–134 (1996).
[120] Streissler, C., *Autocatalytic networks under diffusion*, Ph.D. thesis, Universität Wien, Austria, 1992.
[121] Swetina, J. and Schuster, P., Self-replication with errors — A model for polynucleotide replication, *Biophys. Chem.* **16**, 329–345 (1982).
[122] Szathmáry, E. and Demeter, L., Group selection of early replicators and the orgin of life, *J. Theor. Biol.* **128**, 463–486 (1987).
[123] Szathmáry, E. and Gladkih, I., Sub-exponential growth and coexistence of non-enzymatically replicating templates, *J. Theor. Biol.* **138**, 55–58 (1989).
[124] Szathmáry, E. and Maynard-Smith, J., From replicators to reproducers: The first major transitions leading to life, *J. Theor. Biol.* **187**, 555–571 (1997).
[125] Tarazona, P., Error-tresholds for molecular quasi-species as phase transitions: From simple landscapes to spinglass models, *Phys. Rev.* **A45**, 6038–6050 (1992).
[126] Taylor, P. D. and Jonker, L. B., Evolutionary stable strategies and game dynamics, *Math. Biosc.* **40**, 145–156 (1978).
[127] Tereshko, V., Selection and coexistence by reaction-diffusion dynamics in fitness landscapes, *Phys. Lett.* **A260**, 522–527 (1999).
[128] Tilman, D., Competition and biodiversity in spatially structured habitats, *Ecology* **75**, 2–16 (1994).
[129] Tjivikua, T., Ballester, P. and Rebek Jr., J., A self-replicating system, *J. Am. Chem. Soc.* **112**, 1249–1250 (1990).
[130] Toroczkai, Z., Károlyi, G., Péntek, A., Tél, T. and Grebogi, C., Advection of active particles in open chaotic flows, *Phys. Rev. Lett.* **80**, 500–503 (1998).
[131] Unrau, P. J. and Bartel, D. P., RNA-catalysed nucleotide synthesis, *Nature* **395**, 260–263 (1998).
[132] Vance, R. R., Predation and resource partitioning in one predator – two prey model communities, *Amer. Nat.* **112**, 797-813 (1978).
[133] Veronese, A. and Luisi P. L., An autocatalytic reaction leading to spontaneously assembled phosphatidyl nucleoside giant vesicles, *J. Amer. Chem. Soc.* **120**, 2662–2663 (1998).
[134] Vlassov, A., Khvorova, A. and Yarus, M., Binding and disruption of phospholipid bilayers by supramolecular RNA complexes, *Proc. Natl. Acad. Sci. USA* **98**, 7706–7711 (2001).
[135] von Kiedrowski, G., A self-replicating hexadeoxynucleotide, *Angew. Chem. Int. Ed. Engl.* **25**, 932–935 (1986).
[136] von Kiedrowski, G., Minimal replicator theory I: Parabolic versus exponential growth, in *Bioorganic Chemistry Frontiers*, Vol. 3 (Springer-Verlag, Berlin, Heidelberg, 1993), pp. 115–146.
[137] von Kiedrowski, G., Wlotzka, B. and Helbing, J., Sequence dependence of template directed syntheses of hexadeoxynucleotide derivatives with 3'–5' pyrophosphate linkage, *J. Angew. Chem. Int. Edn. Engl.* **28**, 1235–1237 (1989).

[138] Vrba, E. S., Levels of selection and sorting with special reference to the species level, in *Oxford Surveys in Evolutionary Biology*, P. H. Harvey and L. Partridge eds., Vol. 6 (Oxford University Press, Oxford, 1989), pp. 114–115.

[139] Walde, P., Wick, R., Fresta, M., Mangone, A. and Luisi, P. L., Autopoietic self-reproduction of fatty acid vesicles, *J. Amer. Chem. Soc.* **116**, 11649–11654 (1994).

[140] Wetherill, G. W., Formation of the Earth, *Annu. Rev. Earth Planet. Sci.* **18**, 205–256 (1990).

[141] Wills, P. R., Kauffman, S. A., Stadler, B. M. and Stadler, P. F., Selection dynamics in autocatalytic systems: Templates replicating through binary ligation, *Bull. Math. Biol.* **60**, 1073–1098 (1998).

[142] Wlotzka, B. and McCaskill, J. S., A molecular predator and its prey: Coupled isothermal amplification of nucleic acids, *Chemistry & Biology* **4**, 25–33 (1997).

[143] Wright, S., The roles of mutation, inbreeding, crossbreeeding, in *Proceedings of the Sixth International Congress on Genetics.* Jones, D. F. ed. (Brooklyn Botanic Gardens, New York, 1932), Vol. 1, pp. 356–366.

[144] Wright, S., "Surfaces" of selective value, *Proc. Nat. Acad. Sci. USA* **58**, 165–172 (1967).

[145] Yao, S., Ghosh, I., Zutshi, R. and Chmielewski, J., Selective amplification by auto- and cross-catalysis in a replicating peptide system *Nature* **396**, 447–450 (1998).

[146] Zintzaras, E., Santos, M., Szathmáry, E., "Living" under the challenge of information decay: The stochastic corrector model vs. hypercycles, *J. Theor. Biol.* **217**, 167–181 (2002).

[147] Zubay, G. and Mui, T., Prebiotic synthesis of nucleotides, *Origin Life Evol. Biosp.* **31**, 87–102 (2001).

Advances in Complex Systems, Vol. 9, No. 4 (2006) 315–335

World Scientific
www.worldscientific.com

COOPERATION, COLLECTIVES FORMATION AND SPECIALIZATION

CHRISTOPH HAUERT

Program for Evolutionary Dynamics, Harvard University,
One Brattle Square, Cambridge MA 02138, USA
christoph hauert@harvard.edu

Received 11 September 2005
Revised 30 April 2006

Cooperation in spatial evolutionary game theory has revealed various interesting insights into the problem of the evolution and maintenance of cooperative behavior. In social dilemmas, cooperators create and maintain a common resource at some cost to themselves while defectors attempt to exploit the resource without contributing. This leads to classical conflicts of interest between the individual and the community with the prisoner's dilemma as the most prominent mathematical metaphor to describe such situations. The evolutionary fate of cooperators and defectors sensitively depends on the interaction structure of the population. In spatially extended populations, the ability to form clusters or collectives often supports cooperation by limiting exploitation to the cluster boundaries but often collectives formation may also inhibit or even eliminate cooperation by hindering the dispersal of cooperators. Another attempt at resolving the conflict of interest allows individuals to drop out of unpromising public enterprises and hence changes compulsory interactions into voluntary participation. This leads to a cyclic dominance of cooperators, defectors and loners that do not participate and gives rise to oscillatory dynamics which again subtly depends on the population structure. Here we review recent advances in the dynamics of cooperation in structured populations as well as in situations where cooperative investments vary continuously. In such continuous games, the evolutionary dynamics driven by mutation and selection can lead to spontaneous diversification and specialization into high and low investing individuals which provides a natural explanation for the origin of cooperators and defectors.

Keywords: Evolutionary game theory; spatial structure; social dilemmas; synchronization; oscillations; evolutionary branching; tragedy of the commons; tragedy of the commune; public goods games; prisoner's dilemma; voluntary participation; continuous games.

1. Introduction

The emergence of cooperative behavior has challenged evolutionary biologists for decades. According to Darwinian selection, any behavior that benefits others but not the individual itself should be doomed and vanish. This is in obvious contrast with the abundance of cooperation in nature ranging from bacterial colonies to animal and human societies [5,11]. Recently, the problem of cooperation even made it onto the prominent list of the 25 most important unresolved scientific questions published in *Science* [50]. Following the seminal works of Trivers [59] as well as Axelrod and Hamilton [1] much theoretical effort has been expended on the understanding of the evolution of cooperative behavior based on the game theoretical model of the prisoner's dilemma. In the prisoner's dilemma, two interacting individuals can either cooperate, which provides a benefit β to the co-player at a cost γ to the cooperator (with $\beta > \gamma$), or defect, which neither provides benefits nor incurs costs. Thus, irrespective of what the co-player does, it is always advantageous to choose defection. Consequentially the two players end up with nothing instead of the favorable reward for mutual cooperation — and hence the dilemma.

The prisoner's dilemma can be easily extended to interactions among groups of N individuals. Such interactions are usually referred to as public goods games [31] — a term that originated in economics — but often they are also termed N-player prisoner's dilemmas [3, 27], the *Tragedy of the Commons* [17], or free-rider problems [42]. In a typical public goods experiment a group of six players is endowed with one dollar each. Every player is then offered the opportunity to invest the dollar into a common pool knowing that the total amount in the common pool will be tripled by the experimenter and equally shared among all members of the group irrespective of their contributions. Realizing that each invested dollar returns only 50 cents to the investor, rational players will withhold their contributions and attempt to free ride on the other players' contributions. Consequentially, the group foregoes the public good and fails to increase the initial capital even though they could have tripled it if everybody had cooperated. This is in stark contrast to experimental findings. For example, humans display a surprisingly high readiness to cooperate in prisoner's dilemma [40, 64] or public goods interactions [12]. This calls for a better theoretical understanding of the problem in order to identify mechanisms that can help to overcome the dilemma.

In formal terms, the payoffs for defectors and cooperators in a group with k cooperators are given by $P_D(k) = rkc/N$ and $P_C(k) = P_D(k) - c$, respectively, where c denote the costs of cooperation and r the multiplication factor of the common pool, i.e. of the total contributions. A true public goods interaction is characterized by the facts that (i) groups of cooperators must outperform groups of defectors but (ii) in mixed groups defectors must outperform cooperators. This requires $1 < r < N$. Otherwise, for $r < 1$ groups of cooperators would be worse off than groups of defectors, whereas for $r > N$ the dilemma would be resolved because every invested dollar returns more than one dollar to the investor and thus cooperation would become the dominant solution. However, also note that defectors still outperform cooperators in mixed groups but now they can further improve their payoff by also switching to cooperation. For pairwise interactions ($N = 2$) the prisoner's dilemma and public goods approach become identical with the following transformation: $\beta = rc/2$ and $\gamma = (r/2 - 1)c$, i.e. γ specifies the net or effectively arising costs from the act of cooperation [28].

In order translate these game theoretical considerations into an evolutionary setting, Maynard Smith and Price [38] simply and ingeniously related payoffs from interactions with other members of the population to fitness, i.e. to the reproductive success of individuals. Consider a large population with a fraction x cooperators and $1-x$ defectors. In well-mixed populations, i.e. where individuals interact in randomly formed groups of size N, the average payoff of cooperators f_C and defectors f_D is given by

$$f_C = \sum_{k=0}^{k=N-1} \binom{N-1}{k} x^k (1-x)^{N-1-k} P_C(k+1), \tag{1a}$$

$$f_D = \sum_{k=0}^{k=N-1} \binom{N-1}{k} x^k (1-x)^{N-1-k} P_D(k). \tag{1b}$$

The evolutionary fate of cooperators, i.e. the time evolution of the frequency of cooperators x is then determined by the replicator dynamics [30, 58]:

$$\dot{x} = x(f_C - \bar{f}), \tag{2}$$

where $\bar{f} = xf_C + (1-x)f_D$ denotes the average population payoff. In prisoner's dilemma and public goods interactions $f_C < \bar{f}$ holds for all x. For pairwise interactions this is readily verified for $f_C = c(xr-1)$ and $\bar{f} = cx(r/2(x+1)-1)$ with $r < 2$. Thus, in the absence of supporting mechanisms selection works against cooperators and drives them to extinction. This outcome changes upon considering more sophisticated strategies which are able to condition their behavior on past interactions with the same individuals (direct reciprocity [59]) or on the reputation of other individuals (indirect reciprocity [47]; reward and punishment [52]) but these are other lines of research that will not be considered here.

The failure of cooperation is also based on the assumption that populations are unstructured such that individuals interact in randomly formed groups. This conclusion no longer holds if interactions are restricted to a limited local neighborhood such as in spatially structured populations [44]. Spatial extension enables cooperators to thrive by forming compact clusters or collectives and thereby reducing exploitation by defectors [18, 62]. This can lead to the emergence of intriguing spatio-temporal patterns and critical phase transitions [56] as reviewed in Sec. 2. Further reasons why cooperation fails are the compulsory interactions and lack of alternatives. In nature, individuals often have the capacity to refuse to participate in common enterprises. Such voluntary prisoner's dilemma or public goods games can be modeled by introducing a third strategic type, the loners [23]. Loners are risk averse and refuse to participate in the public endeavor and instead prefer to rely on some autarkic resource. This results in a rock-paper-scissors type cyclic dominance of the three strategic types of cooperators, defectors and loners. In spatial settings, the cyclic dominance drives fascinating spatio-temporal patterns, which includes traveling waves, but, moreover, raises interesting issues related to synchronization of oscillations across spatial dimensions [55]. This is reviewed in Sec. 3.

The prisoner's dilemma and public goods games represent the most stringent form of social dilemmas [7] because cooperation is dominated by defection. Social dilemmas are characterized by a conflict of interest between the individuals and the community but the actual severity of the dilemma can vary. In general, cooperation is not necessarily dominated by defection such that cooperators and defectors can co-exist in snowdrift type interactions [54] even in unstructured populations. Introducing spatial extension again enables cooperators to form clusters but the cluster shape and dynamics under these relaxed conditions are surprisingly different from spatial prisoner's dilemma or public goods games. In fact, spatial structure often turns out to be detrimental to cooperation by reducing the equilibrium fraction of cooperators as compared to well-mixed populations [24]. This is reviewed in Sec. 4 together with a generalized framework to study cooperation in social dilemmas, which is based on discounted and synergistically enhanced values of accumulated cooperative benefits [26].

In biological systems interactions tend not to be as clear-cut as black and white or cooperate and defect and thus in many situations it might be more appropriate to allow for a continuous range of degrees of cooperation. Evolutionary changes in the degree of cooperation are driven by mutation and selection. In prisoner's dilemma and public goods games this change has little effect on the evolutionary outcome in unstructured populations: cooperative investments decrease over time and eventually disappear. However, as before, this changes when introducing spatial extension such that higher investors can survive by forming compact clusters and thereby minimizing exploitation by lower investors. Eventually, this process converges to an equilibrium level of cooperative investments [35, 36]. In contrast, considering snowdrift type interactions in a setting with continuous degrees of cooperative investments leads to much richer dynamics. Based on the fact that cooperators and defectors co-exist in the traditional setting, one might naively expect that in unstructured populations intermediate investment levels would evolve. Although this is one possible outcome, a far more intriguing scenario can occur where two distinct phenotypic clusters of high and low investors evolve and co-exist. This suggests a natural explanation for the evolutionary origin of cooperators and defectors [10]. Thus, in some situations, natural selection may actually favor clear-cut behavioral patterns. Interestingly, this outcome with highly distinct degrees of cooperative investments runs against the accepted notion of fairness in human interactions and may give rise to social tension — a scenario that was termed the *Tragedy of the Commune* [10]. This is reviewed in Sec. 5.

Section 6 concludes this review by putting the theoretical results into perspective with experimental findings and by suggesting potential applications and verifications of the theory in biologically relevant settings with an outlook on promising directions for future theoretical and empirical investigations.

2. Collectives Formation and Phase Transitions

In unstructured populations cooperators readily disappear in prisoner's dilemma ($N = 2$) or public goods games. Recall that for true public goods games, the multiplication factor r must be smaller than the interaction group size N (for $r > N$ cooperation becomes dominant). In contrast, in structured populations cooperators can thrive already for $r < N$ in both prisoner's dilemmas [29, 44, 45,] and public goods games [28]. The limited local interactions enables cooperators to form collectives or clusters and thereby reducing exploitation by defectors (see Fig. 1).

The evolutionary dynamics in structured populations can be modeled as follows: first, a focal individual is randomly drawn and its payoff or fitness is determined by a single interaction within its neighborhood. If the interaction group size N is smaller than the neighborhood size (including the focal individual), then $N - 1$ random neighbors are selected for the interaction (plus the focal individual). Second, randomly pick one of the focal individual's neighbors and similarly determine its fitness from a single interaction within its respective neighborhood. Finally, a probabilistic comparison of the two payoffs determines which individual's offspring replaces the focal individual. In the following we use a particularly simple comparison where the neighbor's offspring replaces the focal individual with a probability proportional to the payoff difference provided that the neighbor performed better than the focal individual and with probability zero otherwise (for other updating mechanisms see, for example, Refs. 19, 21 and 56). This represents an individual

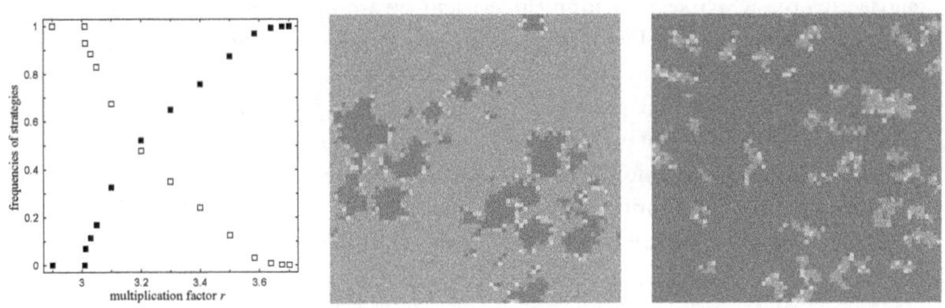

(a) (b) (c)

Fig. 1. (Color online) Spatial structure with limited local interactions enables cooperators to thrive in public goods interactions by forming compact clusters and thereby reducing exploitation by defectors. (a) Frequency of cooperators (■) and defectors (□) as a function of the multiplication factor r of the public good. Individuals interact on a square lattice (100 × 100) within the Moore neighborhood (the eight neighbors reachable by a chess-kings-move) in groups of $N = 5$. For $r < r_C \approx 3.01$ cooperators vanish, for $r_C < r < r_D \approx 3.68$ cooperators and defectors co-exist in dynamical equilibrium and for $r > r_D$ cooperation dominates and reaches fixation. (b, c) Typical snapshots of lattice configurations where cooperators (blue/dark grey) and defectors (red/light grey) co-exist. (b) For small r, i.e. close to the extinction threshold of cooperators ($r = 3.1$) and (c) for large r close to the extinction threshold of defectors ($r = 3.65$).

based analogy of the replicator dynamics [30] and actually recovers it in the limit of infinite population size and increasing neighborhood sizes. All simulations presented in this article are based on this update procedure. Unstructured populations correspond to networks where every individual is linked to every other member of the population (fully connected graph).

In spatial public goods games, defection reigns for low r just as in unstructured populations. For r above a critical threshold, r_C, cooperators quickly increase in frequency and co-exist with defectors until above another critical threshold, $r_D < N$, defectors are driven to extinction (Fig. 1(a)). In unstructured populations, the transition from dominant defection to dominant cooperation occurs only for $r > N$. Near the extinction threshold of cooperators and defectors, the rare strategy forms isolated clusters (Fig. 1(b, c)). The clusters move randomly across the lattice and can coalesce or divide. This suggests that the system becomes equivalent to a branching and annihilating random walk [4] which exhibits a critical phase transition belonging to the universality class of directed percolation [37]. This has been confirmed for both spatial prisoner's dilemmas [55, 57] and spatial public goods games [56].

While critical phase transitions are exiting for physicists, their presence alone may not be exceedingly important in biologically relevant scenarios. However, they do have substantial implications with far reaching consequences. For example, this demonstrates that small changes in one parameter can have tremendous effects on the equilibrium state of a system. But more importantly still, this indicates that in vulnerable systems it might be difficult, if not intrinsically impossible, to define characteristic scales in time and space that allow to fully understand the systems dynamics in empirical settings because both spatial and temporal correlation lengths diverge when approaching the critical threshold. This might be particularly relevant in conservation biology dealing with species interactions at the edge of extinction.

The results for spatial prisoner's dilemmas and public goods games promoted and supported the general conclusion that spatial structure is capable of promoting and maintaining cooperation. The next section, however, demonstrates that spatial structure may not be as universally beneficial because in many biologically relevant situations spatial structure can actually be detrimental to cooperation.

3. Synergy and Discounting of Cooperation

The abundance of the puzzle of cooperative behavior in human and animal societies is reflected in numerous game theoretical models to address this problem. Apart from the aforementioned prisoner's dilemma, public goods games, free-rider problems [42], or the *Tragedy of the Commons* [17], closely related scenarios are described by Snowdrift games [54] or by-product mutualism [6]. Despite the variety, all models actually share a common pattern: all represent conflicts of interest between the individuals and the group — only the severity of the conflict varies. Such situations are generally referred to as social dilemmas [7]: cooperators produce a valuable common good at some cost to themselves while defectors attempt to exploit the resource without contributing. Thus, groups of cooperators are better off than groups of defectors. However, in any mixed group, defectors outperform cooperators and hence the dilemma.

An encompassing framework to model cooperation in social dilemmas has recently been proposed based on synergistic enhancement or discounting of the value of accumulated benefits [26]. In groups of N interacting individuals, cooperators produce a beneficial common resource b at a cost c to themselves. The common resource is evenly shared among all group members regardless of whether they contributed or not. If there are several cooperators in a group, the actual value of additional benefits may be synergistically enhanced or discounted by a factor w. The first cooperator produces a benefit b/N for every member of the group, the second increases everyones benefit by b/Nw, the third by b/Nw^2, etc. Thus, the total payoffs for defectors and cooperators in a group with k cooperators are

$$P_D(k) = \frac{b}{N}(1 + w + w^2 + \cdots + w^{k-1}) = \frac{b}{N}\frac{1 - w^k}{1 - w}, \tag{3a}$$

$$P_C(k) = P_D(k) - c. \tag{3b}$$

Note that neither discounting nor synergy involve temporal components referring to potential future benefits (temporal discounting is often considered in economics [13]). Discounting of benefits occurs, for example, in foraging yeast cells. They produce and secret an enzyme to lyse their environment and thereby produce a valuable common food resource, which is prone to exploitation by other cells. A single cell may be better off producing the enzyme (prevent starvation) if no one else does. However, if an increasing number of cells secrete the enzyme the value of the additional benefits decreases until further increases become useless because of the cells limitations of food intake [15]. Similarly, accumulated benefits can be synergistically enhanced in situations where cooperators produce substances for chemical reactions [16].

In unstructured populations, this system can be fully analyzed by inserting Eq. (3) into Eqs. (1) and (2). Apart from the two trivial fixed points $x_0^* = 0$ and $x_1^* = 1$ there may exist a third interior fixed point $x_2^* = [1 - (cN/b)^{1/(N-1)}]/[1 - w]$, depending on the parameter values. This results in a natural classification of social dilemmas which recovers the four basic scenarios of evolutionary dynamics [48]. (i) Defection is dominant if $cN/b > 1$ and $cN/b > w^{N-1}$ holds. This corresponds to interactions of the type of prisoner's dilemmas or public goods games. (ii) Cooperators and defectors co-exist at an equilibrium level x_2^* if $1 > cN/b > w^{N-1}$. The fact that cooperators persist, reflects the relaxed conditions of the social dilemma as represented by snowdrift type interactions where the rare type is always favored. (iii) Cooperation dominates if $cN/b < 1$ and $cN/b < w^{N-1}$ holds. Even though the social dilemma is completely relaxed, it still holds that defectors outperform cooperators in mixed groups. However, defectors would be even better off by switching

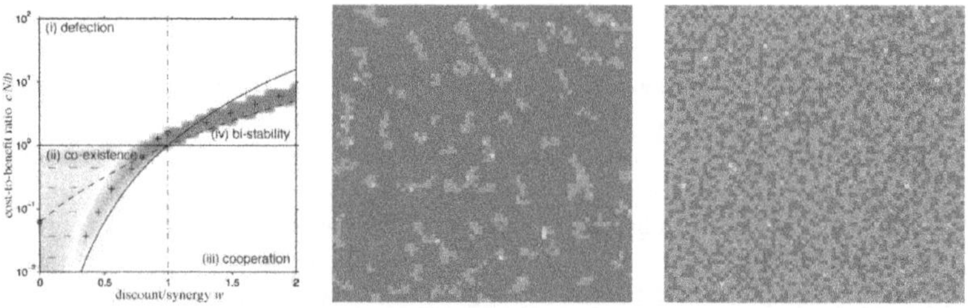

(a) (b) (c)

Fig. 2. (Color online) Effects of spatial structure in social dilemmas. (a) Phase plane diagram depicting parameter regions where spatial structure (square lattice, Moore neighborhood) promotes (blue, +) or inhibits (red, −) cooperation or has no effect (white) for interactions in groups of size $N = 5$ as compared to unstructured populations. The saturation of the colors indicates the strength of the effect. The phase plane is divided into four regions corresponding to social dilemmas of different severity. The dash-dotted line separates regions where benefits are discounted (left) and synergistically enhanced (right). Region (i) corresponds to public goods type interactions where defection dominates. For a small region, spatial clustering enables cooperators to thrive (○ marks the settings in Fig. 1(b)). In the region of snowdrift type interactions (ii) cooperators and defectors co-exist in unstructured populations. The dashed line indicates equilibrium states with equal proportions of cooperators and defectors in well-mixed populations (○ marks the settings in panels (b, c)). (b) typical lattice configuration of cooperators (blue/dark grey) and defectors (red/light grey) where spatial structure can again be beneficial to cooperation and is boosted to 90% instead of 50% in unstructured populations ($b = 7.62, c = 1, w = 0.8$). (c) spatial structure often turns out to be detrimental to cooperation with 40% cooperators as compared to 50% in unstructured populations ($b = 80, c = 1, w = 0$). In region (iii) cooperation is dominant in the form of by-product mutualism. Spatial structure has no effect on the evolutionary success of cooperation. In the region of bi-stability (iii) the basin of attraction of the cooperative state is considerably enhanced. Variations of the group size N do not alter the qualitative results and essentially only change the boundary between (ii) and (iii) or (i) and (iv), respectively. For $N = 2$ the boundary is simply a vertical line such that all regimes have equal size, whereas for $N \to \infty$ the boundary converges to the horizontal line with $cN/b = 1$ such that the co-existence and bi-stability regions disappear.

to cooperation (just as everybody else). This corresponds to by-product mutualism. (iv) If $w^{N-1} > cN/b > 1$, the system is bi-stable and the social dilemma presents itself as a coordination problem such that cooperation is favored only if it is already common ($x > x_2^*$). These four dynamical domains are shown in Fig. 2(a).

This encompassing framework for modeling social dilemmas provides an ideal vantage point to investigate effects of population structure on the evolutionary fate of cooperators from a broader and more general perspective (Fig. 2) [21]. If defection dominates, i.e. in public goods type interactions, spatial structure enables cooperators to survive by forming clusters in order to reduce exploitation by defectors, which is in perfect agreement with the previous section (c.f. Fig. 1). However, it also becomes clear that the clustering advantage of cooperators is rather limited such that cooperators survive only within a restricted parameter range where the cooperative benefits significantly exceed the incurring costs (e.g. $b \gtrsim cN/2$ for $w = 1$). Conversely, if cooperation dominates, spatial structure has no effect on the evolutionary outcome and cooperation invariably reaches fixation. In the region of bi-stability, spatial structure significantly extends the basin of attraction of the cooperative end state. The intuitive reason for this increase is the fact that in spatial settings cooperators need to exceed the threshold frequency only locally. If this condition is satisfied anywhere, a cooperative cluster forms that grows and eventually takes over the entire population. Most interesting effects of spatial structure, however, are observed whenever cooperators and defectors co-exist in unstructured populations. Not only does it lead to complex dynamics [33, 34] but more generally, affects the equilibrium levels of cooperators and defectors. Under these relaxed conditions of the social dilemma, spatial structure can be advantageous for cooperators, too, but quite intriguingly it often turns out to be detrimental to cooperation by lowering the fraction of cooperators as compared to unstructured populations or by even eliminating cooperation altogether [24]. Ironically, the ultimate reason for the inhibition of cooperation actually lies in the maintenance of cooperation in unstructured populations, i.e. because in snowdrift type interactions it is best to adopt a different strategy than the co-players. This mechanism prevents the formation of compact clusters of cooperators and, instead, cooperators congregate in small filament like patterns.

Near the extinction threshold of cooperators, the relevant mechanisms driving the pattern formation process can be intuitively summarized as follows: in public goods type interactions cooperators lower exploitation by forming compact clusters that "minimize" interactions with defectors. Conversely, in snowdrift type interactions, cooperators form filament-like clusters that "maximize" interactions with defectors.

4. Cyclic Dominance and Synchronization

In nature, individuals often have the capability to abstain from or refuse to participate in joint endeavors by relying on a modest autarkic resource. Such risk averse individuals can be modeled by introducing a third strategic type, the loners [23]. Loners prefer a small but fixed income

$$P_L = \sigma c \tag{4}$$

with $0 < \sigma < r - 1$, such that the loners payoff is better than the payoff for mutual defection but worse than the payoff for mutual cooperation, $P_C(N) = (r - 1)c$. Thus, loners provide an escape hatch out of states of mutual defection and economic stalemate, which operates under full anonymity and without requiring preferential assortment or sophisticated strategic responses involving anticipation, conditioning or reward and punishment mechanisms [52]. In the traditional formulation of public goods games (see Sec. 1 and 2) the loner option leads to a variable number

$S \leq N$ of actual participants in the public goods interaction [22]. A single individual willing to participate in the public goods game has no choice but to act as a loner.

In evolving populations, the replicator dynamics describes the evolutionary change of the frequencies of the strategies cooperate, defect and loner with relative abundances x, y and z ($x + y + z = 1$):

$$\dot{x} = x(f_C - \bar{f}), \tag{5a}$$
$$\dot{y} = y(f_D - \bar{f}), \tag{5b}$$
$$\dot{z} = z(f_L - \bar{f}), \tag{5c}$$

where the average payoffs for cooperators, defectors and loners are given by

$$f_D = \sigma c z^{N-1} + r\frac{x}{1-z}\left(1 - \frac{1-z^N}{N(1-z)}\right), \tag{6a}$$
$$f_C = f_D - c\left(1 + (r-1)z^{N-1} - \frac{r}{N}\frac{1-z^N}{1-z}\right), \tag{6b}$$
$$f_L = \sigma c, \tag{6c}$$

and an average population payoff of

$$\bar{f} = \sigma c - c\left((1-z)\sigma - (r-1)x\right)(1 - z^{N-1}). \tag{7}$$

For a detailed derivation, see Ref. 22. A close inspection of f_C and f_D reveals that for $r > 2$ cooperation can be favorable ($f_C > f_D$) for sufficiently large z. This results in a rock-scissors-paper type dominance of the three strategies: if cooperators abound, defection is favored; whereas if defectors prevail, it pays to abstain and choose the loner option; but if loners dominate, small interaction groups can form and re-establish cooperation. Experiments with students have confirmed these results [51] and similar findings were reported for closely related scenarios focussing on social welfare [49] and conflict resolution [41].

Interestingly, for prisoner's dilemma interactions ($N = 2$) the loner option is unable to maintain cooperation in infinite and unstructured populations (Fig. 3). Random shocks may lead to brief intermittent burst of cooperation but the dynamics drives the system back to a state with all loners. In contrast, in public goods games ($N > 2$) a fixed point Q appears in the interior of the simplex S_3 spanned by the three strategies. Q is a center (neutrally stable) and surrounded by closed orbits which lead to periodic oscillations of cooperators, defectors and loners. The period and amplitude of the oscillations is determined by the initial configuration. Along any trajectory, the average payoffs for cooperators, defectors and loners are all equal and reduce to the loner's payoff σ [22]. Thus, in the long run everybody fares equally well than if the public goods game had not existed. However, also note that if the loner's option had not existed, then everybody would have been worse off and ended up with nothing at all.

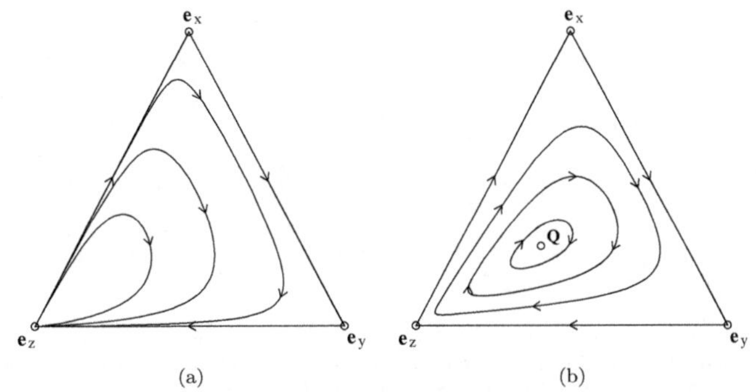

Fig. 3. The rock-scissors-paper type dominance of the three strategies cooperation, defection and loner is reflected in the heteroclinic cycle along the boundary of the simplex S_3. (a) For voluntary prisoner's dilemma interactions ($N = 2$) in unstructured populations, the interior of S_3 consists of homoclinic cycles starting and ending in the homogenous state with all loners ($r = 1.8, c = 1, \sigma = 1$). (b) In public goods games ($N > 2$) a neutrally stable interior fixed point Q appears, which is surrounded by closed periodic orbits ($N = 5, r = 3, c = 1, \sigma = 1$). Thus, in unstructured, infinite populations cooperators, defectors and loners co-exist and their frequencies oscillate periodically.

All oscillatory trajectories are structurally unstable and any kind of stochastic disturbances (e.g. arising from finite population sizes) are sufficient to eventually drive the system to the boundary of S_3 and end in one of the three homogenous states along the heteroclinic cycle. Thus, the long-term maintenance of cooperation requires additional stabilization. This can be achieved in different ways [23], e.g. through modifications of the dynamics, changes in the updating procedure, or by considering structured populations (Fig. 4). The spatial structure of lattice populations suppresses global oscillations and replaces them with uncorrelated local oscillations. This prevents the build up of fluctuations that can lead to extinctions. In addition, the loner option largely extends the range where cooperators persist (c.f. Figs. 1(a) and 4(a)). For small multiplication factors $r < \sigma + 1$ loners are clearly the best option. For $\sigma + 1 < r < r_L$ all three strategies co-exist in dynamical equilibrium. Note that in compulsory public goods games, i.e. in absence of loners, such values of r would almost invariably end up with everybody defecting. The cyclic dominance of cooperators, defectors and loners gives rise to traveling waves propagating across the lattice. For clusters of cooperators the loners provide protection against exploitation restricting the detrimental impact of defectors to one side while cooperators can expand into the loners' territory. Above the threshold r_L loners vanish and cooperators are able to thrive through cluster formation alone. Thus, the dynamics reverts the voluntary public goods game back to a compulsory interaction. The extinction of loners exhibits another critical phase transition in the universality class of directed percolation [56]. The robustness of directed percolation

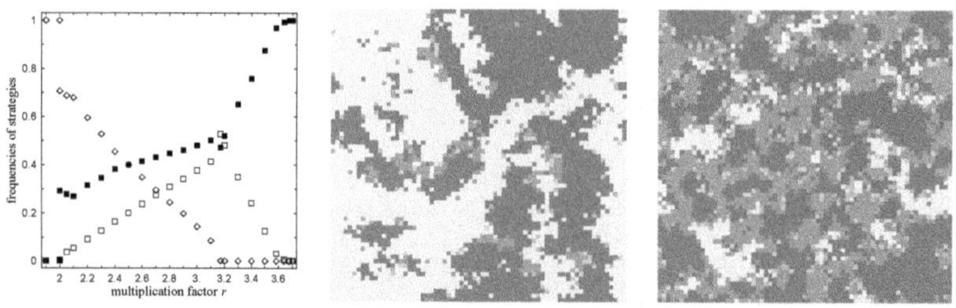

(a) (b) (c)

Fig. 4. (Color online) Spatial structure is capable of stabilizing the co-existence of cooperators, defectors and loners in prisoner's dilemma as well as public goods interactions. (a) Frequency of cooperators (■), defectors (□) and loners (◇) as a function of the multiplication factor r of the public good. Individuals interact on a square lattice (100×100) within the Moore neighborhood in groups of $N = 5$ with the loners payoff $\sigma = 1$. For $r < \sigma + 1 = 2$ loners dominate because they outcompete even groups of cooperators. All three strategies co-exist in dynamical equilibrium for $\sigma + 1 < r < r_L \approx 3.18$. Note that in absence of loners, cooperation would be doomed and vanish (cf. Fig. 1). For $r > r_L$ the persistence of cooperation no longer hinges on the presence and protection of loners and cooperators thrive through cluster formation. Finally, for $r > r_D \approx 3.68$ cooperators dominate and drive defectors to extinction. (b, c) Typical snapshots of lattice configurations where cooperators (blue/dark grey), defectors (red/light grey) and loners (yellow/white) co-exist. (b) Near the extinction threshold of defectors ($r = 2.1$) and (c) close to the threshold where cooperators can survive on their own (cf. Fig. 1) but still crucially depend on the presence of loners ($r = 2.9$).

transitions is nicely demonstrated by noting that the extinction of cooperators and defectors in the compulsory public goods game (see Sec. 1) left a homogenous state behind, whereas the extinction of loners occurs on an inhomogeneous and fluctuating background of cooperators and defectors.

Interestingly, spatial structure is even able to stabilize cooperation in the voluntary prisoner's dilemma [29, 55]. It is important to note, however, that the stabilizing effects are intrinsically linked to the actual structure of the population. In particular, the above lattice configuration prevents synchronized oscillations because the spatial separation allows for uncorrelated fluctuations in different areas of the lattice. This can be compromised in spatial structures that include long range interactions such as small-world networks [63] where far reaching connections can induce global synchronization [29, 55]. To exemplify this, consider the extreme case of random regular graphs [2]. Such graphs are generated by randomly assigning neighbors to each site under the constraint that every site ends up with the same number of neighbors while excluding self and double connections. For small multiplication factors r the results for random regular graphs (Fig. 5(a)) are in close agreement with lattice populations (cf. Fig. 3(a)). However, in contrast to the lattice results, the co-existence of the three strategies persists only for a small range of r ($2 < r < 2.2$ for $N = 5$) and above defectors reign. Only for significantly larger r, cooperators reappear and co-exist with defectors. The frequency of cooperators quickly increases

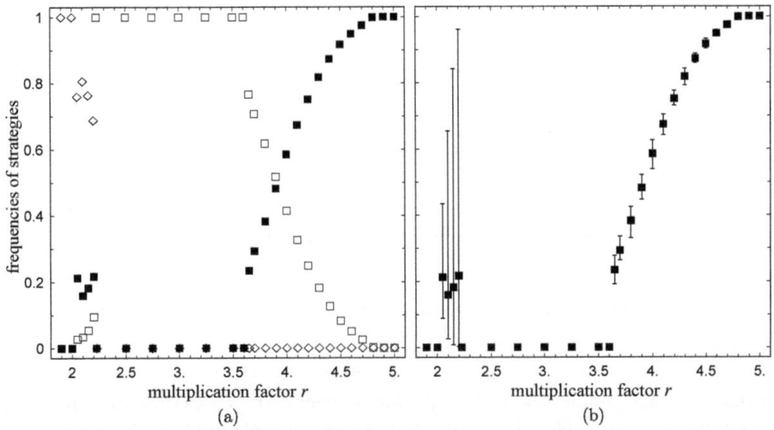

Fig. 5. Long-range interactions in structured populations can promote global synchronization of periodic fluctuations induced by the cyclic dominance of the three strategies cooperation, defection and loner in voluntary public goods games. (a) Frequency of cooperators (■), defectors (□) and loners (◇) as a function of the multiplication factor r of the public good. Individuals are arranged on a random regular graph (10^4 sites) where each site has eight neighbors and interacts in groups of $N = 5$ with the loners payoff $\sigma = 1$. As on square lattices (cf. Fig. 4), loners reign for $r < \sigma + 1 = 2$ and for $r > \sigma + 1$ the three strategies co-exist. However, already for $r > r_L \approx 2.2$ the co-existence breaks down and leads to a homogenous state with all defectors. Only for $r > r_C \approx 3.6$ cooperators can persist and co-exist with defectors but, interestingly, loners are absent. Finally, for $r > r_D \approx 4.6$ cooperation again becomes dominant and reaches fixation. (b) Average, minimal and maximal frequency of cooperators for the same settings as in panel (a). For $\sigma + 1 < r < r_L$ the fluctuations in the frequency of cooperators rapidly increases with r, indicating global synchronization of the oscillations in the frequencies of the three strategies. For $r > r_L$ the fluctuations eventually lead to the extinction of one strategy with another strategy following swift and resulting in a homogenous state. In most cases defectors reach fixation but occasionally the system could also end up in states with all defectors or all loners, depending on which strategy vanished first. All simulations are randomly initialized with strategy frequencies close to the interior equilibrium point Q of the unstructured model (cf. Fig. 3), if it exists, or equal proportions otherwise.

with r until they eventually displace the defectors. These surprising results are better understood by considering the fluctuations of the strategies (Fig. 5(b)). In the region where all three strategies co-exist, the minima and maxima of the strategy frequencies rapidly increase with r. This clearly indicates global synchronization of the local fluctuations that are driven by the cyclic dominance of cooperators, defectors and loners. The building up of these global fluctuations eventually leads to the extinction of one strategy and the system approaches a homogeneous absorbing state. Which strategy disappears first is essentially random but the basin of attraction of the three absorbing states depends on the parameters, the initial configuration and the population size. For example, for the configuration in Fig. 5 in all cases loners disappeared first, which seals the fate of cooperators and leads to an end state with all defectors. Other outcomes where cooperators or loners prevailed were rarely observed in other simulation runs. Also note that this setting is quite different from a system of coupled oscillators because individuals will alter their strategy only in response to changes in their neighborhood.

Apart from altering the stabilization versus synchronization capabilities, the details of the population structure also affect the characteristics of strategy extinctions, i.e. the type of phase transitions. For random regular graphs and small world networks, the lack of spatial correlations results in a linear decrease of cooperators in the prisoner's dilemma indicating a mean-field transition [29].

5. Specialization and the Origin of Cooperators and Defectors

Instead of focussing on pure cooperators and pure defectors, it often seems more natural to assume continuously varying degrees of cooperative investments. This could be, for example, the time and effort expended by cooperators in producing the common good, such as for the enzyme production and secretion in yeast cells (see Sec. 3). The strategy of an individual is then given by a real number x between zero and an upper limit $x_{\max}$ which specifies the individual's investment in the common enterprise. The theory of social dilemmas is easily translated to settings with continuous strategies. For example, in the continuous prisoner's dilemma, an x-strategist facing a y-strategist obtains the payoff $Q(x, y) = B(y) - C(x)$, where $B(y)$ determines the benefits that accrue to the x-strategist as a function of the co-player's investment level y and $C(x)$ denotes the costs incurring to the x-strategist based on its own investment level x. $B(x)$ and $C(x)$ are both assumed to be monotonously increasing functions with $B(0) = C(0) = 0$, i.e. zero investments (pure defection) incur no costs but produce no benefit either, and $B(x) > C(x)$ at least for small x. The latter condition creates the classical prisoner's dilemma for players with different investment levels. Obviously, the investment levels decrease over time and converge to zero, because an individual can only improve its payoff by reducing the incurring costs and hence by decreasing the investment level x. In spatially structured populations, however, this conclusion no longer holds and instead intermediate investment levels can evolve [35]. This approach is not further pursued in the following but, in lieu, a slightly different model is considered that proves to be more general and displays much richer dynamics.

Returning once more to the foraging yeast cells, it is evident that the continuous prisoner's dilemma fails to appropriately capture these interactions because every yeast cell also profits from their own investments into enzyme production and secretion. In this situation, the cooperative benefits depend on the investments of all participants. In order to keep the model simple and transparent, the following exposition focusses on pairwise interactions but it is important to note that the theory is easily generalized to interacting groups of N individuals [10]. For pairwise interactions, the payoff of an x-strategist against a y-strategist is then given by $P(x, y) = B(x + y) - C(x)$. In that case, the naïve expectation would be that intermediate investment levels evolve (because of $B(x) > C(x)$ for small x) but the dynamics turns out to be much more interesting.

The evolution of the investment levels can be analyzed by using adaptive dynamics [8, 14, 39, 46]. The two strategies x and y represent the cooperative investment levels of residents and mutants, respectively. In the framework of adaptive dynamics, i.e. in the limit of small mutations, the mutant is either unable to invade or it invades and replaces the resident. For rare mutations, mutants always face a homogeneous resident population. The replicator dynamics [30] then states that the growth rate or invasion fitness of a rare mutant y in a resident population x is given by $f_x(y) = P(y,x) - P(x,x)$. Note that because y is rare, mutant-mutant and resident-mutant interactions, i.e. $P(y,y)$ and $P(x,y)$, can be neglected. The evolutionary change of trait x is then governed by the selection gradient $D(x) = \partial f_x/\partial y|_{y=x} = B'(2x) - C'(x)$ which leads to the canonical equation of adaptive dynamics:

$$\dot{x} = mD(x), \tag{8}$$

where m depends on the population size and accounts for the mutational process driving changes in x. For constant population sizes m reduces to a constant and setting $m = 1$ merely rescales time. Trait values where the selection gradient vanishes ($D(x^*) = 0$) are termed singular strategies x^*. If no singular strategy exists, then the investment level x either always decreases ($D(x) < 0$) or always increases ($D(x) > 0$) until the boundaries of the investment interval $[0, x_{\max}]$ are reached. The former reflects prisoner's dilemma type interactions whereas the latter corresponds to a continuous variant of by-product mutualism.

If a singular strategy x^* exists, it is convergent stable and hence an attractor of the evolutionary dynamics if $dD(x)/dx|_{x=x^*} = 2B''(2x^*) - C''(x^*) < 0$. Conversely, if $dD(x)/dx|_{x=x^*} > 0$ holds, x^* is a repellor and the investment levels evolve away from x^* (Fig. 6(c)). Depending on the initial configuration x_0, the investment level increases ($x_0 > x^*$) or decreases ($x_0 < x^*$). Assuming that an attractor x^* exists, then all initial investment levels in the vicinity of x^* converge to x^*. Subsequent evolutionary changes, however, depend on whether x^* is also evolutionarily stable, i.e. whether it represents a maximum or a minimum of the invasion fitness $f_x(y)$. If it is a maximum ($\partial^2 f_x(y)/\partial y^2|_{y=x^*} = B''(2x^*) - C''(x^*) < 0$), then x^* is evolutionarily stable (Fig. 6(b)). This outcome corresponds to the naïve expectation where stable intermediate investment levels evolve. Yet, if x^* turns out to be a fitness minimum ($\partial^2 f_x(y)/\partial y^2|_{y=x^*} > 0$) then x^* is called an evolutionary branching point because a population of x^*-strategists can be invaded by both higher and lower investing mutants (Fig. 6(a)). Eventually, the population splits into two types of individuals characterized by high investments (cooperators) and low investments (defectors). It is important to note that this diversification does not require any external tuning because the dynamics keeps the population at the fitness minimum until suitable mutants occur to finally escape that unfortunate state. For quadratic cost and benefit functions, this system can be fully analyzed [10]. In the case of evolutionary branching, the two investment levels continue to diverge until both reach the boundaries of the range of investments. This spontaneous social diversification and specialization provides a natural explanation for the evolutionary origin of cooperators and defectors.

A celebrated avatar of social dilemmas is the *Tragedy of the Commons* [17] which basically states that any common resource is bound to be overexploited — at least unless rigorous control mechanisms are introduced. In the present context, whenever evolution favors co-existence of cooperators and defectors over homogenous populations with uniform intermediate investments, another kind of dilemma emerges, which has been termed the *Tragedy of the Commune* [10]. In communal enterprises, unequal contributions to the common good are against the accepted notion of fairness. Low investors are perceived as social parasites which creates tensions and has a high potential for conflicts.

According to preliminary simulations, spatial structure has similar effects on the dynamics in continuous games [25] as in games with a discrete strategy set (Sec. 3) [26]. In agreement with simulations of the continuous prisoner's dilemma [35], spatial structure promotes and supports cooperation for parameter ranges where defection dominates, i.e. investment levels converge to zero, in unstructured

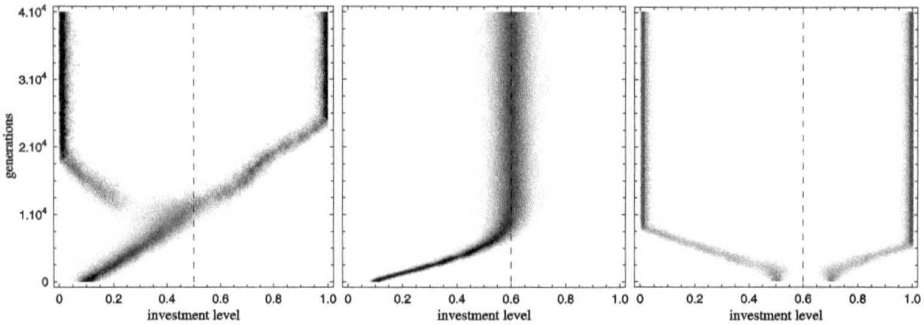

Fig. 6. Evolutionary dynamics of investment levels for continuously varying degrees of cooperation in unstructured populations (darker shades indicate higher frequencies of certain investment levels). Individuals interact in pairs and the benefits depend on the joint investment levels but the costs are determined solely by the individuals' investment. (a) Evolutionary branching results in two distinct phenotypic clusters of high investors (cooperators) and low investors (defectors). First the trait evolves towards the convergent stable singular investment level x^* (dashed line) but x^* is not evolutionarily stable and thus adjacent strategies can invade and induce phenotypic diversification of the strategies. (b) Stable intermediate investment levels evolve if x^* is both convergent and evolutionarily stable. (c) If x^* is a repellor, the evolutionary end state depends on the initial conditions and the population either evolves to full cooperation or full defection (two simulations shown). Parameters: quadratic cost and benefit functions $B(z) = b_2z^2 + b_1z, C(x) = c_2x^2 + b_1x$ where $z = x + y$ denotes the joint investment of both players. (a) $b_2 = -1, b_1 = 7.5, c_2 = -1.5, c_1 = 7$ resulting in $x^* = 0.5$; (b) $b_2 = -1.5, b_1 = 7, c_2 = -1, c_1 = 4.6$ leading to $x^* = 0.6$; (c) $b_2 = -0.5, b_1 = 3.4, c_2 = -1.5, c_1 = 4$ again yielding $x^* = 0.6$.

(a) (b) (c)

Fig. 7. Evolutionary dynamics of the distribution of investment levels for continuously varying degrees of cooperation in spatially structured lattice populations with Moore neighborhood (darker shades indicate higher frequencies of certain investment levels). Apart from the spatial structure, the setting is identical to Fig. 6(a). (a) Just as in unstructured populations, the investment level first evolves towards the singular level x^*, which is predicted to be shifted to $x^* = 0.729$ due to the spatial structure. Once the investment levels approach x^*, selection pressure decreases and the distribution widens and eventually again splits into two strategic types of high investors (cooperators) and low investors (defectors). (b) Snapshot of a typical lattice configuration after $t = 3 \times 10^4$ generations (lighter shades indicate higher investors). The average investment level is close to x^* with 0.7005 (range: 0.42–0.84) and the population is about to undergo evolutionary branching. (c) Lattice configuration after branching at $t = 10^5$ generations. In this evolutionary end state, the high and low investing types essentially engage in traditional snowdrift type interactions.

populations. The stabilizing effects of spatial structure reduces the parameter range where evolutionary branching occurs, but quite intriguingly, the range now includes areas where defection dominates otherwise. This indicates that evolutionary branching also occurs in the spatial continuous prisoner's dilemma. In the case of evolutionary branching, the average investment level first converges to the singular strategy x^*. Once the population is near x^*, the variance in investment levels increases and eventually the population again splits into distinct high and low investors (Fig. 7). At this point, the population essentially consists of two strategic types and similar pattern formation processes are observed as discussed for the discrete strategy set of cooperators and defectors (Secs. 2 and 3). The cluster shape and distribution is again determined by the corresponding interactions in unstructured populations.

6. Summary and Conclusions

Evolutionary game theory applied to the problem of cooperation in populations with different kinds of interaction structures represents a fascinating field for studying emergent phenomena including pattern formation and specialization in a behavioral context. Population structures with limited numbers of interaction partners have profound effects on the evolutionary fate of cooperators and defectors. For the most stringent form of social dilemmas, spatial structure promotes cooperation by enabling cooperators to form clusters and thereby reducing exploitation (prisoner's dilemma or public goods type interactions). However, under relaxed conditions of the social dilemma, spatial structure often inhibits cooperation by actually preventing the formation of compact clusters (snowdrift type interactions) and, instead, favors highly dispersed and filament-like patterns. Microorganisms appear to be ideal model systems to test these hypotheses. Promising candidates include production of replication enzymes in RNA phages [60, 61], foraging in yeast [15] or antibiotic resistance in bacteria [43].

In voluntary public goods interactions, the cyclic dominance of the three behavioral strategies (cooperate, defect and loner) induces oscillatory dynamics. The efficiency of the loner option has been demonstrated in human experiments [51]. Population structure again supports cooperation but the details of the structure have intricate consequences on the dynamics. Long-range connections (such as on small world networks and random regular graphs) can induce and promote global synchronization, which sometimes leads to disastrous amplification of local oscillations. This often prevents co-existence of all three strategies and usually eliminates cooperation. Cooperators thrive only under more favorable conditions that allow co-existence with defectors in absence of loners, i.e. returning to compulsory interactions. Conversely, lattice structures stabilize the system and support co-existence of all three strategies. The spatial separation of the individuals prevents global synchronization and, instead, gives rise to traveling waves sweeping across the lattice. Promising experimental systems, which naturally exhibit this rock-scissors-paper type dominance, include mating strategies in the side-blotched lizard (*Uta stansburiana*) [53] and competition in *E. coli* strains [32].

Game dynamics in structured populations suggests interesting links to condensed matter physics. Specifically, this includes cluster motion and dynamics that relate to random walks, critical phase transitions in the universality class of directed percolation as well as diverging fluctuations that are characteristic to the Ising model.

In nature, cooperation may not always be an all or nothing decision but rather a continuous range of cooperative investment levels. The surprisingly rich evolutionary dynamics of such continuous games is nicely captured by the framework of adaptive dynamics. Most importantly, however, evolution may lead to specialization and spontaneous splitting of the population into high and low investing individuals, i.e. emerging cooperators and defectors, instead of homogeneous populations with uniform investment levels. This represents the *Tragedy of the Commune* because, at least in humans, unequal contributions in a communal enterprise are considered unfair and result in a superimposed dilemma with quite some potential for escalating conflicts. This contrasts with the *Tragedy of the Commons*, which states that public resources are doomed to be overexploited. Again, microbial systems appear to be promising candidates to study cooperative investment levels including cooperative polymorphisms [9].

For systems with such complex dynamics, it is often challenging to get an intuitive understanding of the relevant processes. Visualization of the spatio-temporal patterns or the dynamic changes of continuous strategies helps to understand the collectives formation and evolutionary dynamics as well as to provide insights and inspirations for further explorations. A suite of interactive tutorials allows to reproduce and verify essentially all results reported in this review and moreover to check the robustness of the conclusions by altering numerous parameters [20].

Acknowledgments

Helpful comments on the manuscript by Martin A. Nowak are gratefully acknowledged.

References

[1] Axelrod, R. and Hamilton, W. D., The evolution of cooperation, *Science* **211**, 1390–1396 (1981).
[2] Bollobás, B., *Random Graph* (Academic Press, New York, 1995).
[3] Boyd, R. and Richerson, P. J., The evolution of reciprocity in sizeable groups, *J. Theor. Biol.* **132**, 337–356 (1988).
[4] Cardy, J. L. and Täuber, U. C., Field theory of branching and annihilating random walks, *J. Stat. Phys.* **90**, 1–56 (1998).
[5] Colman, A. M., *Game Theory and Its Applications in the Social and Biological Sciences* (Butterworth-Heinemann, Oxford, 1995).
[6] Connor, R. C., Altruism among non-relatives: Alternatives to the 'Prisoner's Dilemma', *TREE* **10**(2), 84–86 (1995).
[7] Dawes, R. M., Social dilemmas, *Ann. Rev. Psychol.* **31**, 169–193 (1980).
[8] Dieckmann, U. and Law, R., The dynamical theory of coevolution: A derivation from stochastic ecological processes, *J. Math. Biol.* **34**, 579–612 (1996).
[9] Doebeli, M. and Hauert, C., Models of cooperation based on the prisoner's dilemma and the snowdrift game, *Ecol. Lett.* **8**, 748–766 (2005).
[10] Doebeli, M., Hauert, C. and Killingback, T., The evolutionary origin of cooperators and defectors, *Science* **306**, 859–862 (2004).
[11] Dugatkin, L. A., *Cooperation Among Animals: An Evolutionary Perspective* (Oxford University Press, Oxford, 1997).
[12] Fehr, E. and Gächter, S., Altruistic punishment in humans, *Nature* **415**, 137–140 (2002).
[13] Fudenberg, D. and Maskin, E., The folk theorem in repeated games with discounting or with incomplete information, *Econometrica* **54**(3), 533–554 (1986).
[14] Geritz, S. A. H., Kisdi, E., Meszéna, G. and Metz, J. A. J., Evolutionarily singular strategies and the adaptive growth and branching of the evolutionary tree, *Evol. Ecol.* **12**, 35–57 (1998).
[15] Greig, D. and Travisano, M., The Prisoner's Dilemma and polymorphism in yeast SUC genes, *Biol. Lett.* **271**, S25 – S26 (2004).
[16] Hammes, G. G., *Enzyme Catalysis and Regulation*, (Academic Press, New York, 1982).
[17] Hardin, G., The tragedy of the commons, *Science* **162**, 1243–1248 (1968).
[18] Hauert, C., Fundamental clusters in spatial 2×2 games, *Proc. Roy. Soc. London B* **268**, 761–769 (2001).

[19] Hauert, C., Effects of space in 2×2 games, *Int. J. Bifurc. Chaos* **12**, 1531–1548 (2002).

[20] Hauert, C., Virtuallabs: Interactive tutorials on evolutionary game theory, http://www.univie.ac.at/virtuallabs (2005).

[21] Hauert, C., Spatial effects in social dilemmas, *J. Theor Biol.* (in print) (2006).

[22] Hauert, C., De Monte, S., Hofbauer, J. and Sigmund, K., Replicator dynamics in optional public goods games, *J. Theor. Biol* **218**, 187–194 (2002).

[23] Hauert, C., De Monte, S., Hofbauer, J. and Sigmund, K., Volunteering as red queen mechanism for cooperation in public goods games, *Science* **296**, 1129–1132 (2002).

[24] Hauert, C. and Doebeli, M., Spatial structure often inhibits the evolution of cooperation in the snowdrift game, *Nature* **428**, 643–646 (2004).

[25] Hauert, C. and Doebeli, M., Adaptive dynamics and game theory in structured populations, (*in preparation*) (2005).

[26] Hauert, C., Michor, F., Nowak, M. and Doebeli, M., Synergy and discounting of cooperation in social dilemmas, *J. Theor. Biol.* **239**, 195–202 (2006).

[27] Hauert, C. and Schuster, H. G., Effects of increasing the number of players and memory size in the iterated prisoner's dilemma: A numerical approach, *Proc. Roy. Soc. London B* **264**, 513–519 (1997).

[28] Hauert, C. and Szabó, G., Prisoner's dilemma and public goods games in different geometries: Compulsory versus voluntary interactions, *Complexity* **8**(4), 31–38 (2003).

[29] Hauert, C. and Szabó, G., Game theory and physics, *Am. J. Phys.* **73**(5), 405–414 (2005).

[30] Hofbauer, J. and Sigmund, K., *Evolutionary Games and Population Dynamics* (Cambridge University Press, Cambridge, 1998).

[31] Kagel, J. H. and Roth, A. E., (eds.), *The Handbook of Experimental Economics* (Princeton University Press, Princeton, 1995).

[32] Kerr, B., Riley, M. A., Feldman M. W. and Bohannan, B. J. M., Local dispersal promotes biodiversity in a real-life game of rock-paper-scissors, *Nature* **418**, 171–171 (2002).

[33] Killingback, T. and Doebeli, M., Spatial evolutionary game theory: Hawks and doves revisited, *Proc. Roy. Soc. London B*, **263**, 1135–1144 (1996).

[34] Killingback, T. and Doebeli, M., Self-organized criticality in spatial evolutionary game theory, *J. Theor. Biol.* **191**, 335–340 (1998).

[35] Killingback, T., Doebeli, M. and Knowlton, N., Variable investment, the continuous prisoner's dilemma, and the origin of cooperation, *Proc. Roy. Soc. London B*, **266**, 1723–1728 (1999).

[36] Le Gaillard, J.-F., Ferrière, R. and Dieckmann, U., The adaptive dynamics of altruism in spatially heterogenous populations, *Evolution* **57**, 1–17 (2003).

[37] Marro, J. and Dickman, R., *Nonequilibrium Phase Transitions in Lattice Models* (Cambridge University Press, Cambridge, 1999).

[38] Maynard Smith, J. and Price, G., The logic of animal conflict, *Nature* **246**, 15–18 (1973).

[39] Metz, J. A. J., Geritz, S. A. H., Meszena, G., Jacobs, F. J. A. and van Heerwaarden, J. S., Adaptive dynamics: A geometrical study of the consequences of nearly faithful replication, in *Stochastic and Spatial Structures of Dynamical Systems*, eds. van Strien, S. J., and Verduyn Lunel, S. M. (North Holland, Amsterdam, 1996), pp. 183–231.

[40] Milinski, M. and Wedekind, C., Working memory constrains human cooperation in the prisoner's dilemma, *Proc. Natl. Acad. Sci.* **95**, 13755–13758 (1998).

[41] Miller, R. R., No play: A means of conflict resolution, *J. Pers. Soc. Psych.* **6**(2), 150–156 (1967).

[42] Molander, P., The prevalence of free riding, *J. Conflict Resolution* **36**, 756–771 (1992).

[43] Neu, H. C., The crisis in antibiotic resistance, *Science* **257**, 1064–1073 (1992).

[44] Nowak, M. A. and May, R. M., Evolutionary games and spatial chaos, *Nature* **359**, 826–829 (1992).

[45] Nowak, M. A. and May, R. M., The spatial dilemmas of evolution, *Int. J. Bifurc. Chaos* **3**(1), 35–78 (1993).

[46] Nowak, M. A. and Sigmund, K., The evolution of stochastic strategies in the prisoner's dilemma, *Acta Appl. Math.* **20**, 247–265 (1990).

[47] Nowak M. A. and Sigmund, K., Evolution of indirect reciprocity by image scoring, *Nature* **393**, 573–577 (1998).

[48] Nowak, M. A. and Sigmund, K., Evolutionary dynamics of biological games, *Science* **303**, 793–799 (2004).

[49] Orbell, J. H. and Dawes, R. M., Social welfare, cooperators' advantage, and the option of not playing the game, *Ame. Soc. Rev.* **58**, 787–800 (1993).

[50] Pennisi, E., How did cooperative behavior evolve?, *Science* **309**, 93 (2005).

[51] Semmann, D., Krambeck, H.-J. and Milinski, M., Volunteering leads to rock-paper-scissors dynamics in a public goods game, *Nature* **425**, 390–393 (2003).

[52] Sigmund, K., Hauert, C. and Nowak, M. A., Reward and punishment, *Proc. Natl. Acad. Sci.* **98**, 10757–10762 (2001).

[53] Sinervo, B. and Lively, C. M., The rock-paper-scissors game and the evolution of alternative male strategies, *Nature* **380**, 240–243 (1996).

[54] Sugden, R., *The Economics of Rights, Co-operation and Welfare* (Blackwell, Oxford and New York, 1986).

[55] Szabó, R. and Hauert, C., Evolutionary prisoner's dilemma with optional participation, *Phys. Rev. E* **66**, 062903 (2002).

[56] Szabó, G. and Hauert, C., Phase transitions and volunteering in spatial public goods games, *Phys. Rev. Lett.* **89**, 118101 (2002).

[57] Szabó, G. and Tőke, C., Evolutionary prisoner's dilemma game on a square lattice, *Phys. Rev. E* **58**, 69–73 (1998).

[58] Taylor, P. and Jonker, L., Game dynamics and evolutionarily stable strategies, *Math. Bio. Sci.* **40**, 145–156 (1978).

[59] Trivers, R. L., The evolution of reciprocal altruism, *Q. Rev. Biol.* **46**, 35–57 (1971).

[60] Turner, P. E. and Chao, L., Prisoner's dilemma in an RNA virus, *Nature* **398**, 441–443 (1999).

[61] Turner, P. E. and Chao, L., Escape from prisoner's dilemma in RNA phage $\phi 6$, *Am. Nat.* **161**, 497–505 (2003).

[62] van Baalen, M. and Rand, D. A., The unit of selection in viscous populations and the evolution of altruism, *J. Theor. Biol.* **193**, 631–648 (1998).

[63] Watts, D. J. and Strogatz, S. H., Collective dynamics of 'small world' networks, *Nature* **393**, 440–442 (1998).

[64] Wedekind, C. and Milinski, M., Human cooperation in the simultaneous and the alternating prisoner's dilemma: Pavlov versus generous tit-for-tat, *Proc. Natl. Acad. Sci.* **93**, 2686–2689 (1996).

Advances in Complex Systems
Vol. 15, No. 7 (2012) 1250043 (16 pages)

DOI: 10.1142/S0219525912500439

World Scientific
www.worldscientific.com

A MODEL OF MACROEVOLUTION AS A BRANCHING PROCESS BASED ON INNOVATIONS

STEPHANIE KELLER-SCHMIDT* and KONSTANTIN KLEMM†
Bioinformatics, Institute of Computer Science,
University Leipzig, Härtelstr. 16-18, 04107 Leipzig, Germany
**keller-schmidt@informatik.uni-leipzig.de*
†klemm@bioinf.uni-leipzig.de

Received 22 November 2011
Revised 16 February 2012
Published 24 May 2012

We introduce a model for the evolution of species triggered by generation of novel features and exhaustive combination with other available traits. Under the assumption that innovations are rare, we obtain a bursty branching process of speciations. Analysis of the trees representing the branching history reveals structures qualitatively different from those of random processes. For a tree with n leaves generated by the introduced model, the average distance of leaves from root scales as $(\log n)^2$ to be compared to $\log n$ for random branching. The mean values and standard deviations for the tree shape indices depth (Sackin index) and imbalance (Colless index) of the model are compatible with those of real phylogenetic trees from databases. Earlier models, such as the Aldous' branching (AB) model, show a larger deviation from data with respect to the shape indices.

Keywords: Macroevolution; phylogenetic tree; branching model; imbalance; depth scaling; Colless index.

1. Introduction

Since the seminal work by Darwin [11], the evolution of biological species has been recognized as a complex dynamics involving broad distributions of temporal and spatial scales as well as stochastic effects, giving rise to so-called frozen accidents. There is vast exchange and overlap of concepts and methods between the theory of evolution and the foundations of complex systems such as fitness landscapes [35, 12, 20] and neutral networks [19], the evolution of cooperation [3] and self-organized criticality [4] to name but a few.

A striking feature of biological macroevolution is its burstiness. The temporal distribution of speciation and extinction events is highly inhomogeneous in time [28]. As described by the theory of punctuated equilibrium [13], a connection between punctuated equilibrium in evolution and the theory of self-organized criticality [4] is established through the model by Bak and Sneppen [5, 29]. Ecology, i.e., the system of trophic interactions and other dependencies between species' fitnesses, is driven to a critical state. Then, minimal perturbations cause relaxation cascades of broadly distributed sizes.

Rather than through ecological interaction across possibly all species, bursty diversification may also be due to *adaptive radiation* as a rapid multiplication of species in one lineage after a triggering event. About 200 million years ago, a novel chewing system with dedicated molar teeth evolved in the lineage of mammals, allowing it to rapidly diversify into species using vastly distinct types of nutrition [33]. There are many more examples where a single *innovation* triggers adaptive radiation such as the tetrapod limb morphology caused by a binary shift in bone arrangement [32] and the homeothermy as a key innovation by the group of mammals [14, 21]. Adaptive radiation is observed also when a species is confronted with a change in environmental conditions e.g., when entering a new geographical area. The diversity of finch species on Galapagos islands is the famous example first studied by Darwin. Spontaneous phenotypic or genetic innovations and those caused by the pressure to adapt to a change in environment are treated on the same footing for the modeling purposes in this contribution. Though being a central concept in the theory of evolution, the term innovation has not been ascribed a unique definition so far [23].

Here we study a branching process to mimic the evolution of species driven by innovations. The process involves a separation of time scales. Rare innovation events trigger rapid cascades of diversification where a feature combines with previously existing features. We call this newly defined branching process *innovation model.*

How can the validity of models of this kind be assessed? The evolutionary history of species is captured by phylogenetic trees. These are binary trees where leaves represent extant species, alive today, and inner nodes stand for ancestral species from which the extant species have descended. By comparing the shapes of these trees [26, 17, 8, 31], in particular their degree of imbalance [9, 22], with trees generated by different evolutionary mechanisms [2, 6, 15], a selection of realistic models is possible.

2. Stochastic Models of Macroevolution

We consider models of macroevolution within the following formal framework. At each point in time t, there is a set of species $S(t)$. Evolution proceeds as follows. A species $s \in S(t)$ is chosen according to a probability distribution $\pi(s,t)$ on $S(t)$. Speciation of s means replacing s by two new species s' and s'' such that

$$S(t+1) = S(t) \backslash \{s\} \cup \{s', s''\} \tag{1}$$

is the set of species at time $t+1$. The initial condition (at $t = 1$) is a single species. Therefore, discrete time t and number of species n are identical, $n = |S(t)| = t$.

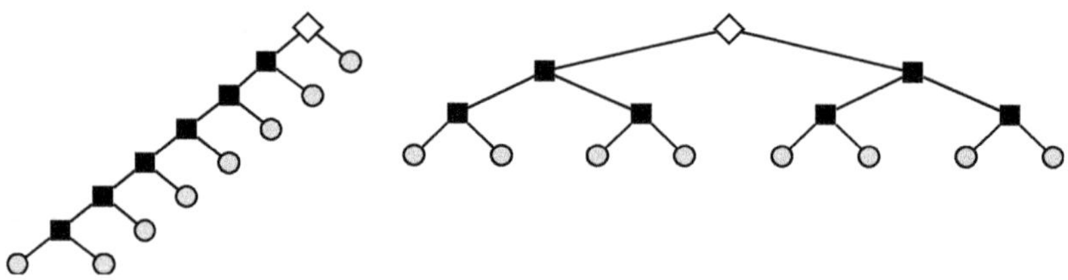

Fig. 1. (Color online) Comparison of tree shapes. Each tree of size eight consists of a root (white diamond), a set of inner nodes (black squares) and a set of leaves (gray circles). The left tree is totally imbalanced, also called comb tree, with depth $d = 35/8 = 4.375$ and Colless index $c = 21/21 = 1$. The right tree is a complete binary tree with depth $d = 24/8 = 3$ and Colless index $c = 0/21 = 0$.

2.1. *Trees*

The evolutionary history of organisms is represented by a phylogenetic tree. For the purpose of this contribution, a phylogenetic tree is a rooted strict binary tree T: a tree with exactly one node (the root) with degree two or zero, all other nodes having degree three (inner node) or one (leaf node), see e.g., illustrations in Fig. 1. For such a tree T with root w, a subtree T' is obtained as the component not containing w after cutting an edge $\{i, j\}$ of T. T' is again a rooted strict binary tree. Since this contribution focuses on tree shape, all edges have unit length. The distance between nodes i and j on a tree T is the number of edges contained in the unique path between i and j.

From the evolutionary dynamics, an evolving phylogenetic tree $T(t)$ is obtained as follows. At each time step t, the leaves of $T(t)$ are the species $S(t)$. When s undergoes speciation, two new leaves s' and s'' attach to a leaf s. After this event, s is an inner node and no longer a leaf of the tree. In this way, each model of speciation dynamics also defines a model for the growth of a binary tree by iterative splitting of leaves.

2.2. *Yule model*

In the simplest case, the probability of choosing a species is uniform at each time step, $\pi(s, t) = 1/t$. This is the Yule model or ERM model. It serves as a null model of evolution.

The model corresponds to a particularly simple probability distribution on the set of generated trees. For a tree with $n \geq 2$ leaves generated by the Yule model and $i \in \{1, 2, \ldots, n-1\}$, let $p_{\mathrm{ERM}}(i \,|\, n)$ be the probability that exactly i leaves are in the left subtree of the root. Then $p_{\mathrm{ERM}}(i \,|\, n) = 1/(n-1)$. This is shown inductively as follows. Obtaining exactly i leaves at step n, either they were already present at the previous step and the speciation took place in the right subtree, or the number increased from $i-1$ to i by speciation in the left subtree. Addition of these products of probabilities for the two cases yields

$$p_{\rm ERM}(i\,|\,n) = \frac{n-1-i}{n-1} p_{\rm ERM}(i\,|\,n-1) + \frac{i-1}{n-1} p_{\rm ERM}(i-1\,|\,n-1). \tag{2}$$

With the induction hypothesis $p_{\rm ERM}(j\,|\,n-1) = 1/(n-2)$ for all j, we obtain

$$p_{\rm ERM}(i\,|\,n) = \frac{(n-1-i)+(i-1)}{(n-1)(n-2)} = \frac{1}{n-1}. \tag{3}$$

The induction starts with $p_{\rm ERM}(1\,|\,2) = 1$ which holds because a tree with two leaves has one leaf each in the left and in the right subtree. Thus, the uniform selection of species turns into a uniform distribution on the number of nodes in the left or right subtree. Note that the same distribution applies to each subtree of an ERM tree. Therefore, $p_{\rm ERM}$ fully describes the statistical ensemble of ERM trees. The probability of obtaining a particular tree is the product of $p_{\rm ERM}$ terms taken over all subtrees. This becomes particularly relevant for modifications of the model taking p nonuniform, see the following section.

2.3. *Aldous' branching (AB) model*

The class of beta-splitting models defines a distribution of trees by the probability

$$p_\beta(i\,|\,n) = \frac{1}{\alpha_\beta(n)} \frac{\Gamma(\beta+l+1)\Gamma(\beta+n-l+1)}{\Gamma(l+1)\Gamma(n-l+1)}, \tag{4}$$

with appropriate normalization factor $\alpha_\beta(n)$. Analogous to $p_{\rm ERM}$ of the previous section, $p_\beta(i\,|\,n)$ is the probability that a tree has i out of its n leaves in the left subtree. In order to build a tree with n leaves, one first decides according to $p_\beta(i\,|\,n)$ to have i leaves in the left and $n-i$ leaves in the right subtree. Then the same rule is applied to both subtrees with the determined number of leaves. The recursion into deeper subtrees naturally stops when a subtree is decided to have one leaf.

The parameter $\beta \in [-2; +\infty]$ in Eq. (4) tunes the expected imbalance. By increasing β, equitable splits with $i \approx n/2$ become more probable. The probability distribution of trees from the Yule model is recovered by taking $\beta = 0$. The case $\beta = -1.5$ is called Proportional to Distinguishable Arrangements (PDA). It produces a uniform distribution of all ordered (left-right labeled) trees of a given size n [25, 24, 30, 10].

Another interesting case is Aldous' branching (AB) model [1, 2] obtained for $\beta = -1$, where Eq. (4) reads

$$p_{-1}(i\,|\,n) \propto \frac{1}{i(n-i)}. \tag{5}$$

Blum and François have found that $\beta = -1.0$ is the maximum-likelihood choice of β over a large set of phylogenetic trees [6]. Therefore we use it as a standard of comparison. The AB model does not have an interpretation in terms of macroevolution, as noted by Blum and François [6]. In particular, it is unknown if its probability distribution of trees can be obtained by stochastic processes of iterated speciation as introduced at the beginning of this section.

2.4. *Activity model*

In the activity model [15], the set of species $S(t)$ is partitioned into a set of active species $S_A(t)$ and a set of inactive species $S_I(t)$. At each time step, a species $s \in S_A(t)$ is drawn uniformly if $S_A(t)$ is nonempty. Otherwise $s \in S_I(t)$ is drawn uniformly. The two new species s' and s'' independently enter the active set $S_A(t+1)$ with probability p. The activation probability p is a parameter of the model. For $p = 0.5$ a critical branching process is obtained. Otherwise the model is similar to the Yule model. A variation of the activity model has been introduced by Herrada *et al.* [16] in the context of protein family trees.

2.5. *Age-dependent speciation*

In the *age model* [18], the probability of speciation is inversely proportional to the age of a species. At each time, a species $s \in S(t)$ is drawn with probability

$$\pi_s(t) \propto \tau_s(t)^{-1} \tag{6}$$

normalized properly. The age τ_s is the number of time steps passed since creation of species s.

2.6. *Innovation model*

In the *innovation model*, each species s is defined as a finite set of features $s \subseteq \mathbb{N}$. Features are taken as integer numbers in order to have an infinite supply of symbols. We denote by $F(t)$ the set of all features existing at time t, that is $F(t) = \bigcup_{s \in S(t)} s$. Each speciation occurs as one of two possible events.

An *innovation* is the addition of a new feature $\phi \in \mathbb{N} \backslash F(t)$ not yet contained in any species at the given time t. One of the resulting species carries the new feature, $s' = s \cup \{\phi\}$. The other species has the same features as the ancestral one, $s'' = s$.

A *loss* event generates a new species by the disappearance of a feature. A feature ϕ is drawn from $F(t)$ uniformly. The loss event is performed only if $s \backslash \{\phi\} \notin S(t)$ such that elimination of ϕ from s actually generates a new species. In this case, the resulting species are the one having suffered the loss, $s' = s \backslash \{\phi\}$ and the species $s'' = s$ remaining unaltered. Otherwise, ϕ is not present in s or its loss would lead to another already existing species, so nothing happens.

We assume that creation of novel features is significantly less abundant than speciation by losses. This separation of time scales is implemented by the rule that an innovation event is only possible when no more losses can be performed. In order to facilitate further studies with the model, we provide a pseudocode description in Algorithm 1. Figure 2 shows an example of the dynamics.

Forbidding duplication of species by loss events is a crucial ingredient of the model. If arbitrary loss events were allowed, the cascade of speciations after a single innovation would not come to a stop. The statistics of tree shapes would be similar to that of the ERM model, where all extant species undergo speciation with uniform probability.

Algorithm 1: Pseudocode for the innovation model

```
set t = 1, F(0) = ∅, S(0) = {∅};
while |S(t)| < N do                    // N as final size of simulated tree
    if S(t)\{s\{φ} : s ∈ S(t), φ ∈ F(t)} ≠ ∅ then
        // loss event
        draw φ ∈ F(t) uniformly;
        draw s ∈ S(t) uniformly;
        if s\{φ} ∉ S(t) then
            S(t + 1) = S(t) ∪ {s\{φ}};
            F(t + 1) = F(t);
            increment t;
    else
        // innovation event
        draw s ∈ S(t) uniformly;
        set φ = 1 + max(F(t) ∪ {0});
        set S(t + 1) = S(t) ∪ {s ∪ {φ}};
        set F(t + 1) = F(t) ∪ {φ};
        increment t;
```

3. Comparison of Simulated and Empirical Data Sets

Now let us compare the tree shapes obtained by the models with those of evolutionary trees in databases. The TreeBASE [27] database contains phylogenetic information about the evolution of species whereas the database PANDIT [34] contains phylogenetic trees representing protein domains. Analyzing the properties with reference to the tree shape of both data sets and applying a comparative study with statistical data sets of different models, one can conclude how well a growth model constructs "real" trees.

Comparison by simple inspection of trees from real data and models may already reveal substantial shape differences. Figure 3 shows an example. The trees in panels (a) and (b) are less compact than that of panel (c) of Fig. 3.

For an objective and quantitative comparison of trees, we use the following two measures of tree shape. The *depth* (or Sackin index) [26] is the average distance of leaves from root,

$$d = \frac{\sum_{i=1}^{n} d_i}{n}, \tag{7}$$

where d_i is the number of edges on the unique path between a leaf i and the root node.

The *Colless index* measures the average imbalance of a tree [9]. The imbalance at an *inner node* j of the tree is the absolute difference $c_j = |l_j - r_j|$ of leaves in

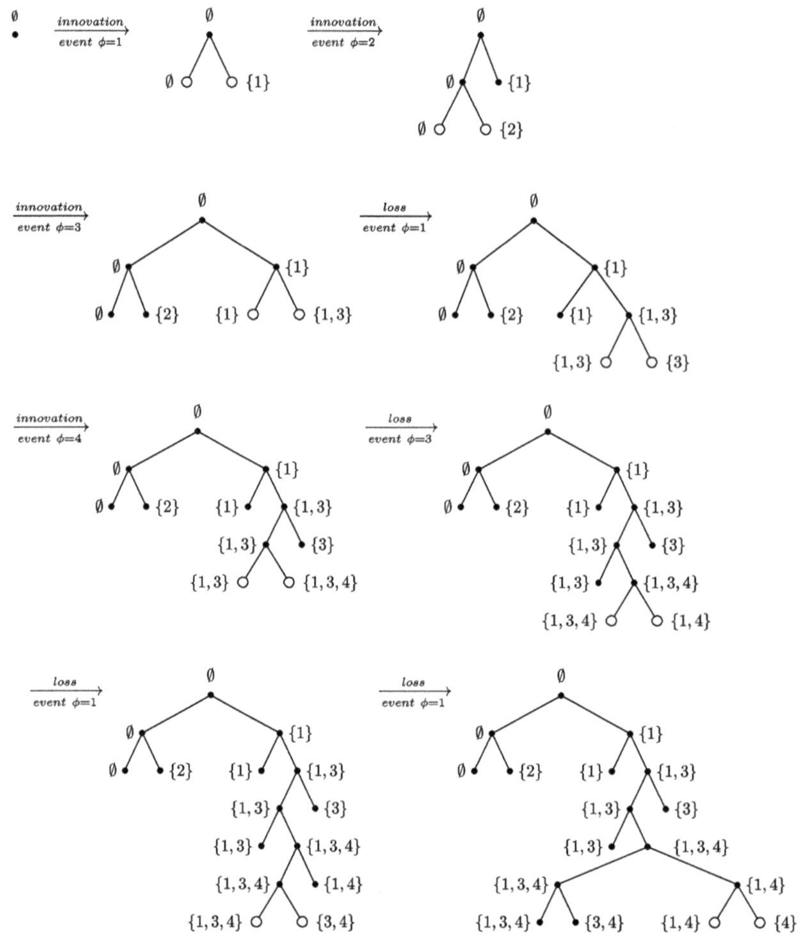

Fig. 2. A tree growing by the innovation model to a size of $n = 9$ leaves. The root node labeled with the empty feature set $\emptyset$ speciates by an innovation event adding the feature 1 to the feature set. This results in the species $\emptyset$ and $\{1\}$. Innovation events are performed, generating features until a loss event is possible. The first loss event generates the species $\{3\}$ by removing the feature 1 from $\{1,3\}$. Now no further loss events are possible, because removal of a feature from any of the extant species (leaves) $\emptyset, \{2\}, \{1,3\}, \{2\}$ does not create an additional species. The following innovation event creates the species $\{1,3,4\}$ with the new feature $\phi = 4$, now allowing for three subsequent loss events.

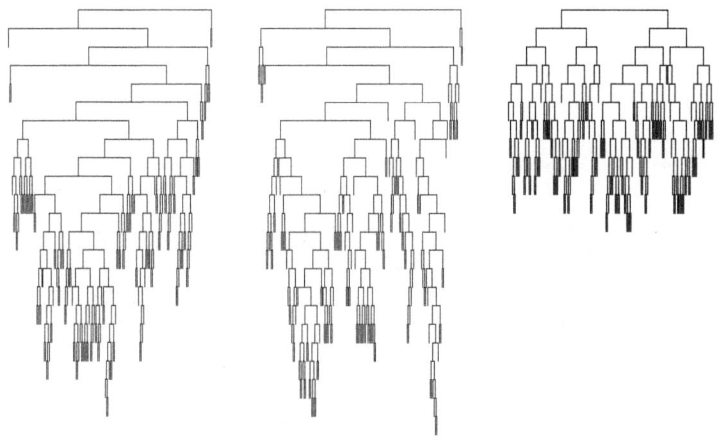

(a) (b) (c)

Fig. 3. Empirical and simulated trees. The depicted phylogenetic tree in (a) is from the database TreeBASE (Matrix ID M2957, relationships in Rosids based on mitochondrial matR sequences), (b) is a tree created as a realization of the innovation model and (c) a tree from the ERM (Yule) model. Each of the trees has 161 leaves.

the left and right subtree rooted at j. Then the average of imbalances

$$c = \frac{2}{(n-1)(n-2)} \sum_{j=1}^{n-1} c_j \tag{8}$$

with appropriate normalization is the Colless index c of the tree. The index j runs over all $n-1$ inner nodes including the root itself. We find $c = 0$ for a totally balanced tree and $c = 1$ for a comb tree, see also Fig. 1.

Ensemble mean values and standard deviations of these indices are shown in Fig. 4. Comparing the results of three models (ERM, AB and innovation) to those of trees from two databases, the least discrepancy is obtained between the innovation model and the trees from TreeBASE, representing macroevolution. In Fig. 5, the averages of the two indices are shown after rescaling to facilitate the comparison. Of all models, the values of the innovation model are also best matching those of PANDIT.

4. Depth Scaling in the Innovation Model

4.1. *Subtree generated by an innovation*

Suppose the ith innovation, generating feature i, affects a species s with f features. Then s is removed from the set S of extant species, turning into an inner node in the tree. Two new species s' and s'' are attached, having feature sets $s' = s$ and

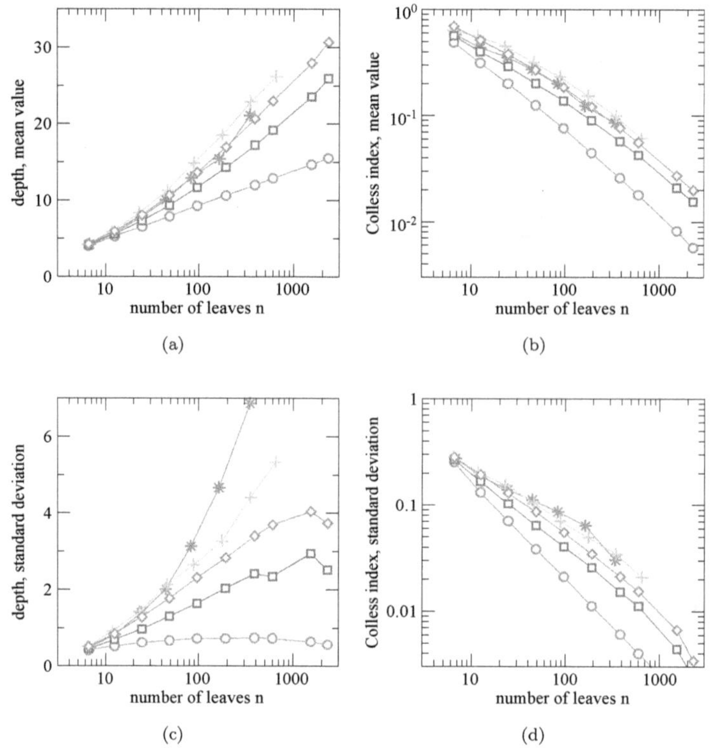

Fig. 4. (Color online) Comparison of size-dependent summary statistics for models and real trees. Symbols distinguish the ERM model (○), the AB model (□) and the innovation model (◇) and the data sets TreeBASE ($*$) and PANDIT ($+$). The data sets were preprocessed by solving monotomies and polytomies randomly as well as removing the outgroups as proposed by [6]. The mean values of depth, and Colless index, panels (a) and (b) are binned logarithmically as a function of tree size n. The same procedure is applied to the standard deviations, panels (c) and (d). The analyzed TreeBASE data set has been downloaded from http://www.treebase.org on June, 2007 containing 5,087 trees of size 5 to 535 after preprocessing. The PANDIT data set has been downloaded from http://www.ebi.ac.uk/goldman-srv/pandit on May 2008 and includes 36,136 preprocessed trees of size 5 to 2,562. The simulated data set comprises for each model (AB model, ERM model and innovation model) 1,000 trees for each tree size from 5 to 535 and 10 trees for each tree size from 536 to 2,562.

$s'' = \{i\} \cup s$. In subsequent loss events, a subtree T_i is built up with 2^f leaves, each of which is a species $\sigma \subseteq s \cup \{i\}$. Call $D(T_i)$ sum of the distances of all the leaves in T_i from the root of T_i.

Let us now estimate the expectation value $\langle D(T_i) \rangle$, which only depends on the number of features of f. Trivially, $D(T_i)$ is lower bounded by $f2^f$ since the most

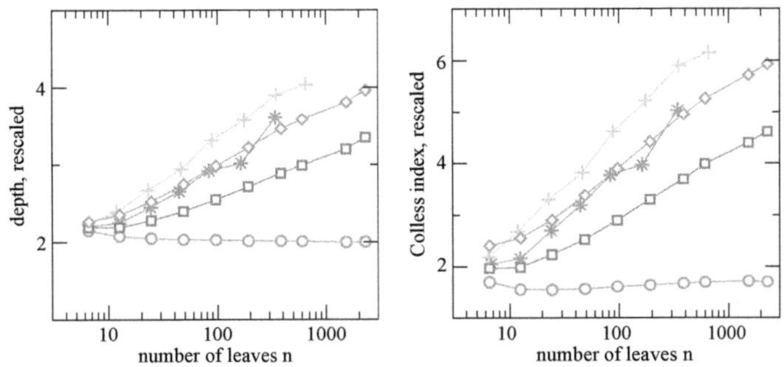

(a) (b)

Fig. 5. (Color online) The same values of depth and Colless index as in Figs. 4(a) and 4(b) with an n-dependent rescaling. (a) Average depth divided by $\ln n$. (b) Average Colless index divided by $n^{-1} \ln n$. These factors are chosen such that the rescaled values for the ERM model asymptotically approach a constant. See Ref. [7] for the scaling of the indices of the ERM model.

compact tree is the fully balanced one with all nodes at distance f from root. In particular, we conjecture

$$f2^f < \langle D(T_i) \rangle < D_{\mathrm{ERM}}(2^f). \tag{9}$$

The second inequality is corroborated by the plots in Fig. 6. We make it plausible as follows. Similar to the ERM model, a leaf is chosen in each time step when executing loss events. Here, however, the loss event is performed only if the chosen leaf carries the chosen feature and the reduced feature set is not yet present in the tree. Thus, the probability of accepting a proposed loss event at a leaf s is anticorrelated with the number of features $|s|$ at s. The expected number of features carried by a leaf decreases with its distance from root. Therefore we argue that the present model adds new nodes preferentially to leaves closer to root than average, resulting in trees with an expected depth increasing more slowly than in the ERM model.

4.2. *Approximation of depth scaling*

We study a tree growth that is derived from the innovation model by two simplifying assumptions. (i) Each innovation is introduced at the leaf with the largest number of features in the tree. (ii) Introducing an innovation at a leaf with f features triggers the growth of a subtree that is a perfect (complete) binary tree with 2^f leaves at distance f from the root of this subtree.

This leads us to consider the following *deterministic growth* starting with a single node and $i = 0$. Choose a leaf s at maximum distance from root; split s obtaining new leaves s' and s''; take s'' as the root of a newly added subtree that is a perfect tree with 2^i leaves; increase i by one and iterate. Figure 7 illustrates the first few steps of the growth.

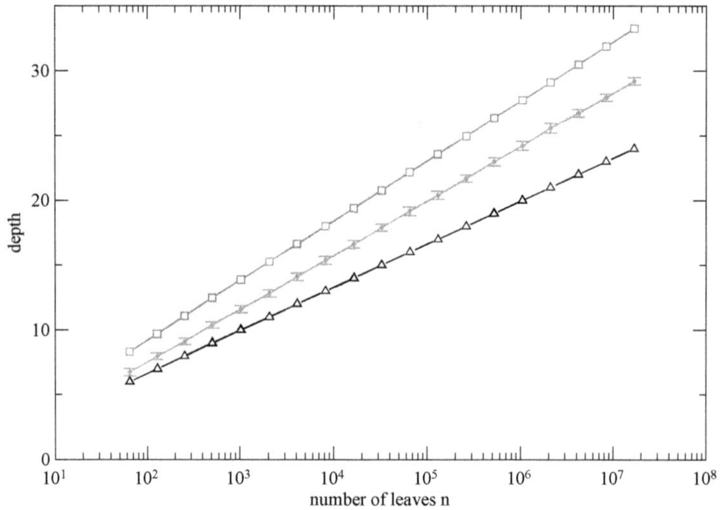

Fig. 6. (Color online) Average depth in dependence of the number of leaves n in trees generated with stochastic loss events (dots with error bars). Each data point is an average over 100 realizations with error bars indicating standard deviations. For comparison, the expected depth for the ERM model (□) and for complete binary trees (△) are shown.

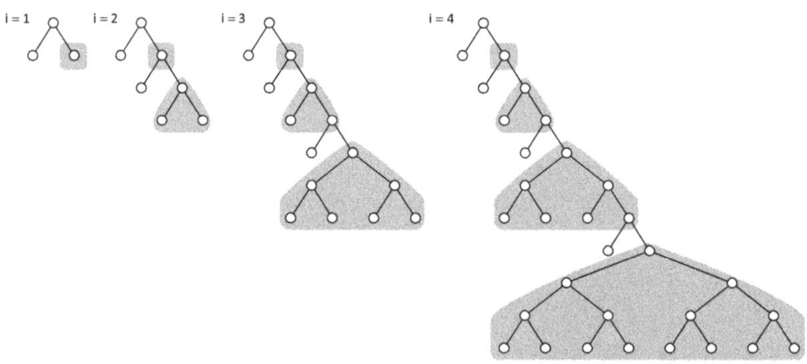

Fig. 7. The deterministic growth of a tree considered as an approximation of the innovation model. Each subtree generated by an innovation is indicated as a shaded area.

After i steps, the number of leaves added to the tree most recently is 2^{i-1}. Therefore, the total number of leaves after step i is

$$n(i) = 1 + \sum_{j=1}^{i} 2^{j-1} = 2^i, \tag{10}$$

because the procedure starts with a single leaf at $i = 0$.

The leaves of the subtree added by the jth innovation have distance

$$\sum_{k=1}^{j} k = \frac{j(j+1)}{2}, \tag{11}$$

from root because these leaves are j levels deeper than those generated by the previous innovation. Therefore the sum of all leaves' distances from root is

$$D(i) = i + \sum_{j=1}^{n} 2^{j-1}[j(j+1)/2], \tag{12}$$

after the ith innovation has been performed. The first term i arises because the innovation itself renders one previously existing leaf at a distance increased by one, see e.g., the leaves outside the shaded areas in Fig. 7. In performing the sum of Eq. (12), we use the equality

$$\sum_{j=0}^{i} x^{j-1}[j(j+1)] = 2^i[i^2 - i + 2] - 2, \tag{13}$$

to arrive at

$$D(i) = i + 2^{i-1}[i^2 - i + 2] - 1. \tag{14}$$

We substitute $n(i) = 2^i$, i.e., $i = \log_2 n$, and divide D by n to arrive at the depth

$$d(n) = \frac{1}{2}[(\log_2 n)^2 - (\log_2 n) + 2] + \frac{(\log_2 n) - 1}{n}, \tag{15}$$

of the tree with n leaves generated by deterministic growth. For large n, the depth scaling is

$$d(n) \sim (\log n)^2. \tag{16}$$

4.3. *Comparison between innovation model and deterministic growth*

We compare the expected n-dependence of depth from deterministic growth with simulation results from the innovation model as defined in Sec. 2.6. In Fig. 8, straight lines in the logarithmic-linear plot of $\sqrt{d}$ versus n indicate a scaling $d \sim (\log n)^2$. For the data points (circles) of the innovation model, a least-squares fit of the form $\sqrt{d}(n) = a + b \ln n$ with free parameters a and b results in a correlation coefficient of 0.99988 and a slope $b = 0.603 \pm 0.003$. For the deterministic growth,

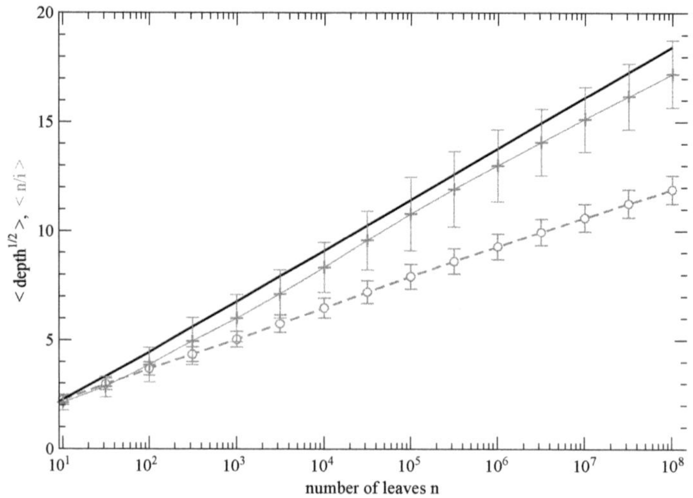

Fig. 8. (Color online) Depth as a function of tree size n for the innovation model ($\circ$) and for the deterministic growth (thick solid curve) according to Eq. (15). Note that square root of depth is plotted such that a straight line in the plot indicates a depth scaling $d(n) \sim (\log n)^2$. Small symbols (+) connected by thin lines give $\langle n/i \rangle$, the average number of leaves per innovation. For each size n, the plotted points ($\circ$, +) are averages over $\sqrt{d(n)}$ and i/n for 100 independently generated trees. Error bars give the standard deviation.

the asymptotic slope is $\sqrt{1/2}/\ln 2 \approx 1.020$ according to Eq. (15). Thus the increase of depth with the number of leaves is slower in the innovation model than in the deterministic growth process. In the innovation model, most innovations hit a leaf with a nonmaximal number of features and therefore trigger the growth of a lower subtree than assumed by deterministic growth.

Let us consider the dependence of the number of leaves n on the number of innovations i. Simulation of the model yields $n/i \sim \ln n$, plotted as +-symbols in Fig. 8. Thus $i \sim n/\ln n$. The number of innovations required to build up a tree with n leaves is weakly sublinear in n, i.e., linear with a logarithmic correction. This dependence is different for deterministic growth. Here the number of leaves doubles by each innovation, since $i = \log_2 n$ according to Eq. (10).

Nevertheless, depth scales as $(\log n)^2$ both in the innovation model and the deterministic growth process and, as shown above, the ratio of the coefficients of this leading order is approximately 0.6. We hypothesize that the deterministic growth captures the essential mechanism leading to the depth scaling of the innovation model: Tree shape in the innovation model is mostly determined by large bursts of speciations following an innovation. Table 1 provides an overview of the scaling of average depth with the number of leaves for various models.

Table 1. Depth scaling of models.

Innovation model	$(\log n)^2$
β-Splitting [1]	$\begin{cases} \log n & \text{if } \beta > -1, \text{ includes } \mathbf{ERM}\ (\beta = 0) \\ (\log n)^2 & \text{if } \beta = -1, \mathbf{AB} \text{ model} \\ n^{-\beta-1} & \text{if } \beta < -1, \text{ includes } \mathbf{PDA}\ (\beta = -1.5) \end{cases}$
Age model [18]	$(\log n)^2$
Activity model [15]	$\begin{cases} n^{0.5} & \text{if } p = 0.5, \\ \log n & \text{otherwise.} \end{cases}$
Complete tree	$\log n$
Comb tree	n

5. Discussion

The innovation model establishes a connection between the burstiness of macroevolution and the observed imbalance of phylogenetic trees. Bursts of diversification are triggered by generation of new features and combination with the repertoire of existing traits. In order to keep the model simple, the diversification after an innovation is implemented as a sequence of random losses of features. More realistic versions of the model could be studied where combinations of traits are enriched by re-activation of previously silenced traits or horizontal transfer between species. Furthermore, the model as presented here neglects the extinction of species and their influence on the shapes of phylogenetic trees.

Regarding the robustness of the model, the depth scaling would have to be tested under modifications. In particular, the infinite time scale separation between rare innovations and frequent loss events could be given up by allowing innovations to occur at a finite rate set as a parameter.

In summary, we have defined a well-working, biologically motivated model which nevertheless is sufficiently simple to allow for further enhancement regarding biological concepts such as sequence evolution and genotype–phenotype relations.

Acknowledgments

The authors thank Kathrin Lembcke, Maribel Hernández Rosales and Nicolas Wieseke for a critical reading of the draft. This work was supported by Volkswagen Stiftung through the initiative for Complex Networks as a Phenomenon across Disciplines.

References

[1] Aldous, D., Probability distributions on cladograms, in *Random Discrete Structures*, eds. Aldous, D. and Pemantle, R. (Springer, 1996), pp. 1–18.

[2] Aldous, D., Stochastic models and descriptive statistics for phylogenetic trees, from yule to today, *Stat. Sci.* **16** (2001) 23–34.

[3] Axelrod, R., *The Evolution of Cooperation* (Basic Books, 1984).

[4] Bak, P., *How Nature Works: The Science of Self-Organised Criticality* (Copernicus Press, New York, 1996).
[5] Bak, P. and Sneppen, K., Punctuated equilibrium and criticality in a simple model of evolution, *Phys. Rev. Lett.* **71** (1993) 4083–4086.
[6] Blum, M. G. B. and François, O., Which random processes describe the tree of life? A large-scale study of phylogenetic tree imbalance, *Syst. Biol.* **55** (2006) 685–691.
[7] Blum, M. G. B., François, O. and Janson, S., The mean, variance and limiting distribution of two statistics sensitive to phylogenetic tree balance, *Ann. Appl. Probab.* **16** (2007) 2195–2214.
[8] Campos, P., de Oliveira, V. and Maia, L., Emergence of allometric scaling in genealogical trees, *Adv. Complex Syst.* **7** (2004) 39–46.
[9] Colless, D. H., Phylogenetics: The theory and practice of phylogenetic systematics, *Syst. Zool.* **31** (1982) 100–104.
[10] Cotton, J. A. and Page, R. D., The shape of human gene family phylogenies, *BMC Evol. Biol.* **6** (2006) 66.
[11] Darwin, C., *On the Origin of Species* (John Murray, 1859).
[12] Gavrilets, S., *Fitness Landscapes and the Origin of Species* (Princeton University Press, 2004).
[13] Gould, S. J. and Eldredge, N., Punctuated equilibrium comes of age, *Nature* **366** (1993) 223–227.
[14] Heard, S. B. and Hauser, D. L., Key evolutionary innovations and their ecological mechanisms, *Historical Biol.* **10** (1995) 151–173.
[15] Hernández-García, E., Tugrul, M., Herrada, A. E., Eguíluz, V. M. and Klemm, K., Simple models for scaling in phylogenetic trees, *Int. J. Bif. Chaos* **20** (2010) 805–811.
[16] Herrada, A., Eguíluz, V. M., Hernández-García, E. and Duarte, C., Scaling properties of protein family phylogenies, *BMC Evol. Biol.* **11** (2011) 155.
[17] Herrada, A. E., Tessone, C. J., Klemm, K., Eguíluz, V. M., Hernández-García, E. and Duarte, C. M., Universal scaling in the branching of the tree of life, *PLoS ONE* **3** (2008) e2757.
[18] Keller-Schmidt, S., Tugrul, M., Eguíluz, V. M., Hernández-García, E. and Klemm, K., An age dependent branching model for macroevolution (2010), http://arXiv.org/abs/1012.3298.
[19] Kimura, M., *The Neutral Theory of Molecular Evolution* (Cambridge University Press, Cambridge, 1983).
[20] Klemm, K. and Stadler, P. F., Rugged and elementary landscapes, in *Theory and Principled Methods for Designing Metaheustics*, ed. Borenstein, Y. (2012), accepted.
[21] Liem, K. F. and Nitecki, M. H., *Key Evolutionary Innovations, Differential Diversity, and Symecomorphosis* (University of Chicago Press, 1990), pp. 147–170.
[22] McKenzie, A. and Steel, M., Distributions of cherries for two models of trees, *Math. Biosci.* **164** (2000) 81–92.
[23] Pigliucci, M., What, if anything, is an evolutionary novelty? *Phil. Sci.* **75** (2008) 887–898.
[24] Pinelis, I., Evolutionary models of phylogenetic trees, *Proc. R. Soc. Lond. B* **270** (2003) 1425–1431.
[25] Rosen, D. E., Vicariant patterns and historical explanation in biogeography, *Syst. Zool.* **27** (1978) 159–188.
[26] Sackin, M., Good and bad phenograms, *Syst. Zool.* **21** (1972) 225–226.
[27] Sanderson, M. J., Donoghue, M. J., Piel, W. and Eriksson, T., TreeBASE: A prototype database of phylogenetic analyses and an interactive tool for browsing the phylogeny of life, *Am. J. Botany* **81** (1994) 183.

[28] Sepkoski, J. J., Ten years in the library; new data confirm paleontological patterns, *Paleobiology* **19** (1993) 43–51.
[29] Sneppen, K., Bak, P., Flyvbjerg, H. and Jensen, M. H., Evolution as a self-organized critical phenomenon, *Proc. Natl. Acad. Sci.* **92** (1995) 5209–5213.
[30] Steel, M. and McKenzie, A., Properties of phylogenetic trees generated by Yule-type speciation models, *Math. Biosci.* **170** (2001) 91–112.
[31] Stich, M. and Manrubia, S. C., Topological properties of phylogenetic trees in evolutionary models, *Eur. Phys. J. B* **70** (2009) 583–592.
[32] Thomson, K. S., Macroevolution: The morphological problem, *Society* **32** (1992) 106–112.
[33] Ungar, P. S., *Mammal Teeth: Origin, Evolution, and Diversity* (The Johns Hopkins University Press, 2010).
[34] Whelan, S., de Bakker, P., Quevillon, E., Rodriguez, N. and Goldman, N., Pandit: An evolution-centric database of protein and associated nucleotide domains with inferred trees, *Nucleic Acids Res.* **34** (2006) D327–D331.
[35] Wright, S., The roles of mutation, inbreeding, crossbreeding and selection in evolution, in *Proc. Sixth Int. Congress on Genetics*, ed. Jones, D. F., Vol. 1 (Brooklyn Botanic Gardens, New York, 1932), pp. 356–366.

Advances in Complex Systems
Vol. 16, Nos. 2 & 3 (2013) 1250089 (24 pages)

DOI: 10.1142/S0219525912500890

SELF-ORGANIZING PARTICLE SYSTEMS

MALTE HARDER* and DANIEL POLANI†

Adaptive Systems Research Group,
University of Hertfortshire,
College Lane, Hatfield, AL10 9AB, United Kingdom
**m.harder@herts.ac.uk*
†d.polani@herts.ac.uk

Received 21 February 2012
Revised 28 July 2012
Accepted 6 September 2012
Published 22 October 2012

The self-organization of cells into a living organism is a very intricate process. Under the surface of orchestrating regulatory networks there are physical processes which make the information processing possible, that is required to organize such a multitude of individual entities. We use a quantitative information theoretic approach to assess self-organization of a collective system. In particular, we consider an interacting particle system, that roughly mimics biological cells by exhibiting differential adhesion behavior. Employing techniques related to shape analysis, we show that these systems in most cases exhibit self-organization. Moreover, we consider spatial constraints of interactions, and additionaly show that particle systems can self-organize without the emergence of pattern-like structures. However, we will see that regular pattern-like structures help to overcome limitations of self-organization that are imposed by the spatial structure of interactions.

Keywords: Self-organization; information theory; morphogenesis.

1. Introduction

The development of organisms is one of the most prominent examples of self-organization and the emergence of shapes. The process of forming shapes is usually an interplay between environmental dynamics (e.g., global physical rules), and agent actuations (e.g., a change of local properties) regulated through complex networks.

In the early stage of laying out body plans, morphological changes are induced mainly due to control of cell adhesion, cell motility and oriented cell division. In particular, differential cell adhesion prevents areas consisting of different tissues to mix and starts an automatic sorting process. This happens, if for example cells have been forced to mix in a solution [47]. Gastrulation, the process of rearranging a ball of cells in the early stage of embryonic development into a more complex body structure, can be simulated by contractions in cell shape that then lead to an automatic rearrangement of cells forming an inner structure [26].

One important aspect of all these processes is that, in many cases, the information processing capabilities of the individual cells (i.e., agents) are severely limited, especially in scenarios that consider large collectives. In these cases the environmental dynamics dominate the process of organization while individual agents actively guide the process. Cells for example can change adhesion properties or partially contract. *Morphogenesis*, the formation of shapes, as we will see, can be achieved purely by environmental dynamics up to certain limits. The process of shape formation can be seen as a selection of a configuration which fulfills certain properties. Thus, the course of a given process typically leads to a reduction of entropy. In the context of this paper, we would like to reinterpret this as saying that there are information processing capabilities in the environment. This is justified by the view of the controlled dynamics of a system as an entropy reduction mechanism [43], more below.

Note that these capabilities are often rooted in the structure of the space and the physical laws that govern it. In Ref. 31 it has been shown that consistency in the embodiment of agents reduces cognitive load, lack of such consistency increases it. This implies that, consistency or homogeneity of the space also can increase the information processing capabilities of the environment, as a reduction of cognitive load means that the information needs to be processed elsewhere. Information processing/entropy reduction capabilities that a system provides can also be used by nonreactive systems (for example, we consider particles here instead of autonomous agents). In particular they can be a driving force of self-organization.

1.1. *Self-organizing particle systems*

In order to investigate the information processing capabilities of a morphogenetic process, we will use a model of particle collectives similar to the models in [10, 11, 35] that mimics features of cell adhesion and cell motility to a certain grade. A human observer easily detects organizational patterns in simulation runs of this model. In many cases the resulting particle configurations even resemble the morphology of biological structures, showing features that look like membranes or nuclei, see Fig. 1 for an example. However, a human observer is a quite subjective measure and not transferable. The question is: Is there an objective way to characterize whether formation of particles is a result of mere chance, some driving force or is it due to the underlying dynamics that did let the system self-organize? Using quantitative methods it is possible to investigate such a formation process in greater detail. In particular, information theory will serve us as a useful tool as it offers a universal and model agnostic way to investigate self-organization in arbitrary dynamical systems.

In self-organizing processes, individual parts of the whole system usually interact with each other, this is the case in particular in the particle model considered here. Interactions have been the basis for information theoretic investigations before [19], and can be closely linked to information storage and transfer [44, 24, 5].

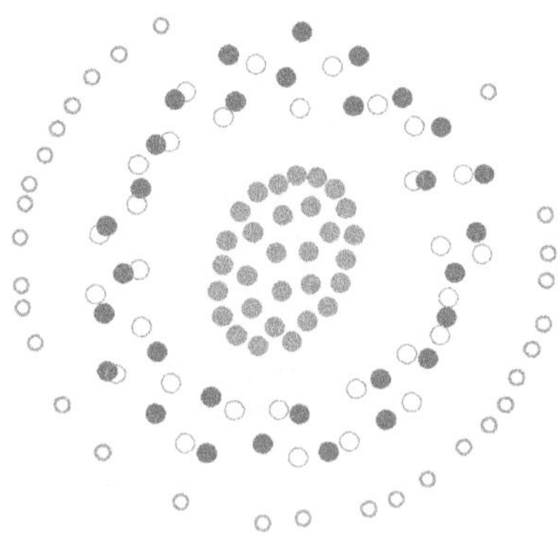

Fig. 1. Example of a particle configuration.

A requirement that organization can occur is the spread of information through the system, which in turn requires interaction between individual parts of the system [40].

2. Information Theory

Our studies will be based on information theory [38]. For self-containedness, we introduce here the basic notions of information theory, for a detailed account see Thomas and Cover [6]. A fundamental measure in information theory is the *entropy* of a probability distribution, measured in bits. It is defined by

$$H(X) := -\sum_{x\in\mathcal{X}} p(x)\log_2 p(x), \tag{1}$$

where X denotes a finite-valued random variable with values in $\mathcal{X}$ and $p(x)$ the probability that X takes on the value $x \in \mathcal{X}$. Entropy measures the uncertainty of the outcome of a random variable. Now *mutual information* is defined by $I(X;Y) := H(X)+H(Y)-H(X,Y)$. It is symmetric in X and Y and can be interpreted as the information the random variables mutually encode about each other. For continuous random variables, differential entropy is defined as

$$h(X) := \int_{\mathcal{X}} p(x)\log_2 p(x)dx. \tag{2}$$

Differential entropy differs in important aspects from its discrete counterpart and is, strictly spoken, not a true, but rather a "renormalized" entropy [16]. The analogous concept of differential mutual information, however as difference of entropies $I(X;Y) := h(X) + h(Y) - h(X,Y)$, still retains its character as a mutual information and discrete variables can be seamlessly replaced by continuous ones in the case of the latter.

2.1. *Multi-information*

Multi-information, also called total correlation [17] or integration [42], is one generalization of mutual information to the multivariate case. It is defined as

$$I(X_1, \ldots, X_n) := \sum_{i=1}^{n} H(X_i) - H(X_1, \ldots, X_n), \tag{3}$$

where the sum goes over the entropies of the marginal variables. The continuous case is defined analogously. Multi-information measures the statistical dependence between the individual parts of the whole system. Multi-information has been used as a complexity measure [1] and incorporates the idea that changes in one part of the system are reflected in the overall state.

There are several other multivariate generalizations of mutual information which retain certain properties of the bivariate case. For a thorough discussion of most of these measures, especially in the context of analyzing time series, see Ref. 17. Here, we will use multi-information to measure the amount of self-organization that is present in the system.

3. Quantifying Self-Organization

Although it is quite easy for humans, from a visual standpoint, to point out whether we consider a system as self-organizing ("I know it when I see it"), there are surprisingly few quantitative characterizations of self-organization. Let us assume we observe some pattern that seems to self-organize over time, but we only see a single instance of this time series. Our intuition about self-organization implicitly assumes that there is a causal relation between the observed time-steps. However, this could be the improbable but possible observation of a time series of i.i.d random variables. In this case we do not want to quantify the system as self-organizing. By choosing a statistical or information theoretic formulation of self-organization we are able to account for this causal relationship.

Let us first consider what self-organization is not. Most importantly, there should not be an external or central force that controls the process in a "top-down" fashion. Also, if there is a system that is causally dependent [28, 2] on another self-organizing system, we do not want to consider the former as self-organizing. This means that the system should be autonomous, which is not a trivial task to determine [3], and external influences need to be separable, which can be investigated using the measure of causal information flow [2]. However, we will not consider these problems here, they are more apparent if the underlying model is unknown. Here, the model is designed to fulfill these criteria.

This still leaves the question of how to quantify organization. Organization means an increase of structure over time, whereby structure in this context usually is considered to be a spatial correlation between parts of the whole system. One account for a definition of self-organization can be found in Ref. 37, where self-organization is defined as the increase of statistical complexity of a system over time via ϵ-machines. Measuring self-organization via statistical complexity has the limitation and advantage that it assumes no structure on the space underlying the time-series. It is an advantage because the measure is very versatile and does not need to be changed for different spatial structures. On the other hand it is a limitation, because the spatial structure will be implicitly encoded in the states of the ϵ-machine, which makes it less accessible for further analysis [37]. There is an extension of ϵ-machines called spatial ϵ-machines which have a structural assumption about the space they are defined on, but this has only been done for discrete spaces so far. Thus, we will describe a method which has been suggested as alternative to the measure of self-organization via statistical complexity (see Ref. 30 for a detailed survey on the foundation of self-organization).

3.1. *Self-organization via observers*

We define self-organization as an increase of multi-information between observer variables $I(X_1, \ldots, X_n)$ during the run of the process [29]. A collection of random variables $X_1, \ldots, X_n$ are called *observers* of a system X if the system is fully determined by the collection of random variables, that means $H(X \mid X_1, \ldots, X_n)$ vanishes, and all variables only depend on X, meaning $H(X_1, \ldots, X_n \mid X)$ vanishes as well. The observers impose a "coordinate system" on the observed random variable and can be chosen freely. The additional specification is not very problematic in practice, as there are often natural choices for observer variables, as many systems are just a collection of individual random variables.

For a completely random process this measure never detects any self-organization, as there is no increase in correlation between observer variables. Note that there could be correlation between observers because they observe the same part of X, but because they exclusively depend on X, there is no hidden common cause between observers that could increase the correlation. The other extreme case is that the entropy of the whole system vanishes, in which case there also cannot be an increase of multi-information between the observers. So, to achieve self-organization, the system requires some remaining degree of freedom.

Interestingly, this definition also gives the opportunity to build hierarchies by considering coarse to fine grained observers, which then leads to a decomposition of self-organization. If k-observers are grouped to one coarse-grained observer variable $\tilde{X}_i$ we get that the following multi-information term

$$I(\underbrace{X_1, \ldots, X_{i_1}}_{\tilde{X}_1}, \underbrace{X_{i_1+1}, \ldots, X_{i_2}}_{\tilde{X}_2}, \ldots, \underbrace{X_{i_{k-1}+1}, \ldots, X_{i_k}}_{\tilde{X}_k}) \tag{4}$$

can be decomposed into $k+1$ multi-information terms (see Ref. 30 for a derivation)

$$= I(\tilde{X}_1, \ldots, \tilde{X}_k) + \sum_{j=1}^{k} I(X_{i_{j-1}}, \ldots, X_{i_j}). \tag{5}$$

The decomposition now allows the separation of organization that is apparent within individual parts of the system, where each part is a coarse-grained observer, and organization, that can only be explained as an interaction between coarse-grained observers.

This allows us to see whether there are parts of the system that dominate the process of self-organization. For example by grouping individual observers by common properties of the observed variables, it is possible to see whether a specific property has a higher contribution to the self-organization or whether most of the self-organization is a result of interaction between different coarse-grained observers. We will now use this formulation of self-organization to investigate spatial self-organization in particle collectives.

4. Particle Collectives and Self-Organisation

4.1. *The particle model*

There are numerous models of morphogenesis and pattern formation including reaction-diffusion models [25], cellular automata [46], diffussion-limited aggregation [45], L-systems [33] and agent based models [4]. The particle model we will describe is based on the model by [10, 11], and shares some similarities with the Swarm Chemistry model [35]. It mimics the way biological cells stick together by cell adhesion, allowing different types to recognize each other.

In our model, each particle interacts with all particles within a certain cut-off radius r_c. For reasons of simplicity, as well as to be able to have long range interactions, we ignore a cell-like tessellation (as opposed to [10]), where interactions can only take place between direct neighbors of the tessellation. The equation of motion for each particle is given by

$$\dot{z}_i = \sum_{j \in N_{r_c}(i)} -F_{\alpha\beta}(\|\Delta z_{ij}\|_2)\Delta z_{ij} + w, \tag{6}$$

where $\Delta z_{ij} = z_i - z_j$, $N_{r_c}(i)$ denotes the set of indices of particles within radius r_c of particle i and $F_{\alpha\beta}$ is a force-scaling function, α is the type of particle i, β is the type of particle j and w an additive white Gaussian noise term, where $w \sim \mathcal{N}(0, 0.05)$ throughout all experiments. Note that the velocity is proportional to the force applied and thus the dynamics are studied in the strong limit of friction. This assumption holds for example, for the motion of insects and cellular motility, in contrast to the movement of larger animals and humans which can build up momentum.

Now, the equation of motion can be solved using Euler–Maruyama integration [20, 32]. We used the following two force-scaling functions, see Fig. 2 for a plot of both functions.

$$F^1_{\alpha\beta}(x) = k_{\alpha\beta}\left(1 - \frac{r_{\alpha\beta}}{x}\right) \tag{7}$$

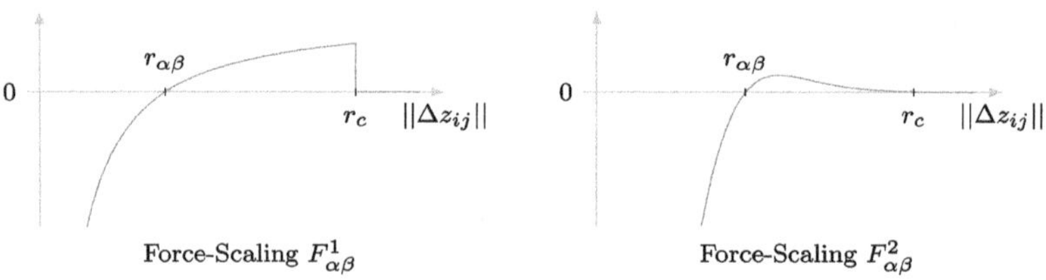

Fig. 2. Comparison of both force-scaling functions, $r_{\alpha\beta}$ denotes the prefered distance between particles of type α and β. For $F^{(1)}$ this radius can be directly specified. The long range attraction of $F^{(1)}$ is only cut-off by the radius r_c.

and

$$F^2_{\alpha\beta}(x) = k_{\alpha\beta}\left(\frac{1}{\sigma^2_{\alpha\beta}}e^{-\frac{x^2}{2\sigma_{\alpha\beta}}} - e^{-\frac{x^2}{2\tau_{\alpha\beta}}}\right). \tag{8}$$

The matrices $k_{\alpha\beta}, r_{\alpha\beta}$ as well as $\sigma_{\alpha\beta}$ and $\tau_{\alpha\beta}$ define the interactions between the particles and have a strong impact on the dynamics of the experiment. Values for the parameters were chosen from the following ranges: $k_{\alpha\beta} \in [1.0, 10.0]$, $r_{\alpha\beta} \in [0.0, 1.0]$ and $\tau_{\alpha\beta} \in [1.0, 10.0]$ with $\sigma_{\alpha\beta} = 1$ throughout all experiments. Note that choosing a nonsymmetric matrix often leads to unstable dynamics or cycling patterns as the preferred distance is mutually different, we therefore only consider symmetric matrices in what follows. The force-scaling function defines how much attraction or repulsion the particles show among each other depending on the type and distance between particles. The first force-scaling function F^1 shows stronger attraction compared to repulsion than F^2. For each type, the force-scaling functions result in a preferred distance of particles of other types, denoted $r_{\alpha\beta}$. By using smaller diagonal values than the off-diagonal elements in $k_{\alpha\beta}$ or $r_{\alpha\beta}$ it is possible to force clustering of particles of the same type.

In Fig. 3 are three examples of equilibrium states of particle collectives. For the particle collective with only one type, a simple disc shaped pattern can be seen.

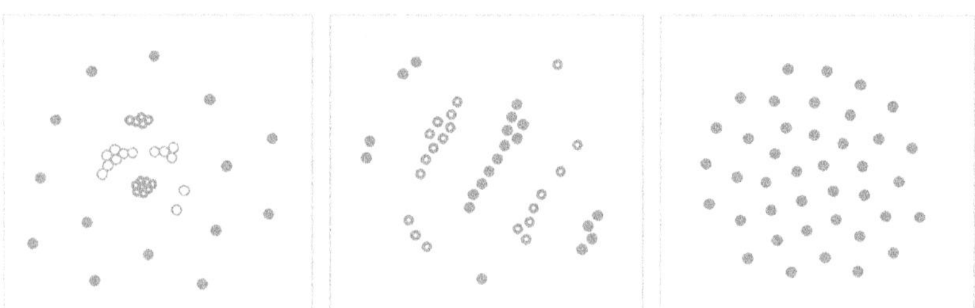

Fig. 3. Examples of equilibrium states of particle collectives with different number of types.

The collective is considered to be in equilibrium, if for several time-steps the sum of the L^2 norm of the sum of all forces acting on each particle is below a specific threshold.

4.2. *Measuring organization in particle collectives*

To measure self-organization within a particle collective using multi-information as introduced in Sec. 3.1 we need to define observer variables. A natural choice would simply be the collection of variables denoting the positions of each individual particle. However, we have to consider that certain transformations of the configuration leave the shape of the particle collective invariant. So, if we do not consider these invariants, the measured multi-information can be different from what we do want to consider as organization toward a shape. But even if, in the stochastic limit, rotations and translations are equidistributed, by factoring them out we reduce the sparsity of samples in the space of possible configuration of particles.

There are several accounts on spatial statistics and stochastics [36, 23], however in these references interacting particle systems are defined as (continuous time) Markov processes on discrete domains while our experiments are in the continuous domain. In the area of geo-information systems and medical image processing, there is a large interest in statistical models of shapes, and there is a large body of literature on shape models [9, 39, 12]. One particular problem, the alignment of overlapping images or shapes, is similar to the problem of reducing our experiment samples (i.e., the simulations) to an invariant representation.

Rotation, translation as well as permutation of particles of the same type leave the observable shape, as well as the dynamics involved, invariant. Let $ISO^+(2)$ denote the group of direct isometries (rotation, translation and identity) of the euclidean plane. This group now acts on the space of particle configurations $\mathcal{Z}$ by rigid body motions:

$$\mathcal{Z} \times ISO^+(2) \to \mathcal{Z}. \qquad (9)$$

To account for permutations, let S_n denote the permutation group of n-elements, which also naturally acts on the space of samples by permuting the particle vectors for all time-steps. Now we consider the subgroup $S_n^* \subset S_n$ that permutes only particles of the same type. The direct product $F = ISO^+(2) \times S_n^*$ now classifies all shape invariant transformations.

It is important to note that these transformation also have the property that they leave the dynamics of the system invariant. Let $\mathbf{z}^{(t)} \in \mathcal{Z}$ denote the configuration of the particle collective at time t, then

$$p(\mathbf{z}^{(t)} \mid \mathbf{z}^{(t-1)}) = p(f\mathbf{z}^{(t)} \mid f\mathbf{z}^{(t-1)}) \quad \text{for all } f \in F \text{ and all } \mathbf{z}^{(t)}, \mathbf{z}^{(t-1)}. \qquad (10)$$

This means that a configuration that is transformed will lead to a distribution of configurations in the future, that is equivalent to the distribution of the transformed future states of the original configuration.

In the case that additionally the initial state is invariant under the action of this transformation group, we also have $p(\mathbf{z}^{(t)}) = p(f\mathbf{z}^{(t)})$ for all t. Thus, we can easily factor out the transformation group, and get random variables over the space of shapes (transformation invariant particle configurations). Factoring out all symmetries F from $\mathcal{Z}$ then leads to a reduced space of particle configurations $\mathcal{W}$ which then allows us to define a random variable $\mathbf{W}^{(t)}$ (the whole collective at time t) and corresponding observer variables $\mathbf{W}_1^{(t)}, \ldots, \mathbf{W}_n^{(t)}$ for a collective with n particles. The indistinguishableness of particles of the same type means that no observer $\mathbf{W}_i^{(t)}$ can be used to make predictions about the future of a specific particle. Measuring multi-information on these derived random variables $\mathbf{W}_1^{(t)}, \ldots, \mathbf{W}_n^{(t)}$ ignores certain degrees of freedom, i.e., rotation, permutations of particles of the same type and translation. Now we can express every configuration of particles $\mathbf{z}$ as a permutation, translation and rotation of invariant coordinates $\mathbf{w}$, i.e., for all $\mathbf{z}$ there exists $\mathbf{w}$ and $f \in F$ such that $\mathbf{z} = f\mathbf{w}$. Due to the group structure of F and the invariance of the states (at all times) under transformations of F we have

$$I(\mathbf{Z}_1, \ldots, \mathbf{Z}_n) = \int_{\mathcal{Z}} p(\mathbf{z}_1, \ldots, \mathbf{z}_n) \log \frac{p(\mathbf{z}_1, \ldots, \mathbf{z}_n)}{p(\mathbf{z}_1) \cdots p(\mathbf{z}_n)} d\mathbf{z} \tag{11}$$

$$= \int_F \int_{\mathcal{W}} p(f(\mathbf{w}_1, \ldots, \mathbf{w}_n)) \log \frac{p(f(\mathbf{w}_1, \ldots, \mathbf{w}_n))}{p(f\mathbf{w}_1) \cdots p(f\mathbf{w}_n)} d\mathbf{w} df \tag{12}$$

$$= \int_F \int_{\mathcal{W}} p(\mathbf{w}_1, \ldots, \mathbf{w}_n) \log \frac{p(\mathbf{w}_1, \ldots, \mathbf{w}_n)}{p(\mathbf{w}_1) \cdots p(\mathbf{w}_n)} d\mathbf{w} df \tag{13}$$

$$= I(\mathbf{W}_1, \ldots, \mathbf{W}_n). \tag{14}$$

Therefore, factoring out the transformation group F does not change the multi-information of the observers, in the case of an invariant system. However, that means that in practice we would have to sample the initial state with full support on $\mathbb{R}^2$ which would lead to a very sparse sampling and because of the finite cut-off radius to almost no interactions between particles. To avoid this, we use an initial distribution of particles, which is uniform within a certain radius around the origin, so that particles are initially placed uniformly on a centered disc. This initial distribution is still invariant with respect to rotation and permutation, but not translation invariant. Even though the equality from above thus does not hold anymore, we argue, that due to the symmetry of the forces ($r_{\alpha\beta}$, $k_{\alpha\beta}$, $\sigma_{\alpha\beta}$, $\tau_{\alpha\beta}$ are symmetric in all cases) and the independence of the noise, there is no information coming from translation in the original coordinates $\mathbf{z}^{(t)}$ that is relevant for the organization of the collective, so that we can simply consider the factored coordinates $\mathbf{w}^{(t)}$ to measure multi-information and successively self-organization.

4.2.1. *Indistinguishable particles*

If we make predictive statements about a particle we need to be able to identify particles through time, otherwise the statistics about the future of a particular particle are skewed. That the interchangeability of variables has an impact on information processing and measurements has been considered before in terms of recoding equivalence [7]. By reordering the particles, we lose the information to identify the same particle over time and they become indistinguishable. To measure self-organization of shapes we actually want indistinguishable particles (if they have the same type) and therefore we introduced the permutation group S_n^* as one set of shape invariant transformations. Distinguishing them would mean that there can be an event that increases the measurement of self-organization, but is not reflected in the shape and structure of the particle configuration. For example there could be a permutation of two particles of the same type that is always reflected by a permutation of two particles of same type elsewhere in the system. This would then be taken into account by the multi-information, but has no impact on the shape that is formed.

On the other hand we also do not want to equate particles which have a different type, and show different interactions. Particles of different types should be distinguishable as permutations of particles of different type actually change the shape of the configuration. Additionally, if they would be indistinguishable this would contradict the assumption made in (10) that the dynamics are invariant to permutations of indistinguishable particles.

The problem of indistinguishable particles and the related change of entropy is also a problem in thermodynamics where it is known as Gibbs phenomenon and Mixing paradox [13, 18]. By only making a distinction between particles that show observably different behavior we are in line with the solution to this problem in physics [18].

5. Methods

This section describes how we derive estimates of multi-information from simulation samples of particle dynamics. The reader not interested in the computational details may skip this section to Sec. 6 without loss of understanding.

5.1. *Particle and sample space*

Each simulation runs with a fixed number of n particles, l different types and each particle gets a fixed type assigned at the start of the simulation. The particles are located in the infinite two-dimensional plane $\mathbb{R}^2$ and are initialized with a uniform distribution on a disc of fixed radius. Each particle is of a specific type. The types can vary between different experiments, but the properties ($r_{\alpha\beta}$, etc.) of each type are fixed for all simulation samples of one experiment. The assignment of a type to a particle is fixed over the time of the simulation run. Each simulation run is a sample and is denoted by

$$\bar{\mathbf{z}} = (\mathbf{z}^{(1)}, \ldots, \mathbf{z}^{(t_{\max})}), \tag{15}$$

where each time-step is a vector of particle coordinates

$$\mathbf{z}^{(t)} = (z_1^{(t)}, \ldots, z_n^{(t)}). \tag{16}$$

To gather statistics for an experiment, we need to run the simulation multiple times. The collection of all m samples is denoted

$$\mathfrak{z} = (\bar{\mathbf{z}}_1, \ldots, \bar{\mathbf{z}}_m)^\top = (\mathfrak{z}^{(1)}, \ldots, \mathfrak{z}^{(t_{\max})}). \tag{17}$$

Now, let the space of all particle vectors $\mathbf{z} = (z_1, \ldots, z_n)$ be denoted $\mathcal{Z}$ and $\mathbf{Z}^{(t)}$ the random variable over $\mathcal{Z}$ at time-step t, so all $\mathbf{z}^{(t)} \in \mathfrak{z}^{(t)}$ are samples of $\mathbf{Z}^{(t)}$.

5.2. *Factoring out symmetries*

Now we factor out the symmetries for each time-step as introduced in Sec. 4.2. The samples $\mathbf{z}^{(t)} \in \mathfrak{z}^{(t)}$ for each time-step t, the raw output of the simulations, are still with respect to a common coordinate system. We will now proceed to factor out translations, rotations and permutations resulting in processed samples $\mathbf{w}^{(t)} \in \mathfrak{w}^{(t)}$ for each time-step t. In practice this is done by expressing all particle configuration samples $\mathbf{z}^{(t)} \in \mathfrak{z}^{(t)}$ with respect to its centroid. This is followed by aligning all configuration samples $\mathbf{z}^{(t)}$ for each time-step using an ICP (Iterative Closest Point) algorithm [48, 34]. For the application of the alignment the particle configuration is transferred to a three-dimensional representation where the third coordinate of each particle is represented by its type, where the type coordinates are scaled by a factor a magnitude larger than the diameter of the collective. Thus, the alignment respects the type of the particles. After the alignment the coordinates of all particle are reordered by types and correspondences. Correspondences between particles of different samples, but of the same type, are found using a nearest neighbor search within the ICP algorithm (implementation from the point cloud library [34]). This means that particles close to each other in different samples at the same time are considered to represent the same particle. Note that the notion of same particle establishes a correspondence between different samples at a specific time-step. The correspondence between particles of the same sample, but different time-steps is, however, lost in this process.

Equipped with this preprocessing, we reach an isometry- and permutation-reduced representation of the particle collective in terms of processed samples $\mathbf{w}^{(t)} \in \mathfrak{w}^{(t)}$. We can now use the statistics of these samples to calculate the multi-information $I(\mathbf{W}_1^{(t)}, \ldots, \mathbf{W}_n^{(t)})$.

The invariant representation also has the advantage that the samples are much denser in the space of possible configurations which improves the quality of the estimates. It is important to note, that for statistics that need to track particles over time, we cannot use the permutation reduced representation because we would then lose any correspondence of particles over time, e.g., Ref. 21.

5.3. *Estimation of multi-information*

To estimate the multi-information we used the Kraskov–Stögbauer–Grassberger Estimator [22]. The estimate is based on a k-nearest-neighbor search. We chose $k = 5$ for all experiments. The estimate is not very sensitive for changes of k and we get similar results with $k = 2$ or $k = 10$. If k is chosen too large the resulting estimate vanishes for almost all samples. The estimator for m samples and n variables is given by

$$I(\mathbf{W}_1^{(t)}, \ldots, \mathbf{W}_n^{(t)}) \triangleq \psi(k) + (n-1)\psi(m) - \langle \psi(c_1) + \psi(c_2) + \cdots + \psi(c_n) \rangle_{\mathbf{w}^{(t)}}, \tag{18}$$

where ψ is the digamma function and the brackets denote the average taken over all samples. The c_i are dependent of $\mathbf{w}^{(t)}$ and defined as follows: let $N_k(\mathbf{w}^{(t)})$ denote the kth neighbor of the sample $\mathbf{w}^{(t)}$ using the following metric:

$$\|\mathbf{w}' - \mathbf{w}\| := \max_{i \in \{1,\ldots,n\}} \|\mathbf{w}'_i - \mathbf{w}_i\|_2, \tag{19}$$

then c_i is defined as

$$c_i = |\{\mathbf{w}'^{(t)} \in \mathfrak{w}^{(t)} : \|\mathbf{w}'^{(t)}_i - \mathbf{w}^{(t)}_i\|_2 < \|N_k(\mathbf{w}^{(t)})_i - \mathbf{w}^{(t)}_i\|_2\}| - 1. \tag{20}$$

The idea is that a high correlation between the variables leads on average to a low count of samples per variable that, for each sample, are closer to the sample itself than the kth neighbor over all variables, thus maximizing the estimator.

In our tests this approximation shows less variance and is much faster to calculate than other methods. This holds especially in the higher-dimensional setting with more than ten particles (20 dimensional) and only few samples (500 to 1000). We compared the method to a kernel based approach which was multiple orders of magnitudes slower and showed a larger variance in higher dimensions [41]. We also compared it to a shrinkage type binning estimator [15], which overestimated the multi-information in higher dimension due to the sparse sampling, so much that almost no change in information could be seen.

5.3.1. *A further approximation*

For large collectives, the alignment of samples and the estimation of the multi-information can still be a computationally expensive task. Therefore, we can reduce the dimensionality of the problem by introducing mean random variables. We perform a k-means clustering on the particles of each type and thus recover $l \cdot k$ mean variables $\hat{\mathbf{W}}_1^{(t)}, \ldots, \hat{\mathbf{W}}_{lk}^{(t)}$, where l is the number of types. Now we use $I(\hat{\mathbf{W}}_1^{(t)}, \ldots, \hat{\mathbf{W}}_{lk}^{(t)})$ as an approximation measure for the multi-information $I(\mathbf{W}_1^{(t)}, \ldots, \mathbf{W}_n^{(t)})$.

This must be done carefully, because the clustering process itself can introduce structure into the collective of particles, and thus can lead to a higher measurement of multi-information than actually is present. On the other hand, the clustering ignores all small scale self-organization processes, and hence the measured multi-information is less than the exact value.

6. Results

In the following sections all experiments were run with a sample size of $m = 500$ to $m = 1000$, $t_{\max} = 100$ to $t_{\max} = 250$ and $k = 4$ for the knn-selection of the multi-information estimator. For systems with more than 60 particles, the k-means clustering approximation was used to reduce the number of dimensions. The first observation is that this method can be used in practice to detect self-organization and there is a visual correlation between the formation process and the increase of the multi-information estimate as depicted in Fig. 4. In the beginning the sum of the marginal entropies $H(\mathbf{W}_i^{(t)})$ is as large as the overall entropy of the system because there is no correlation between particles at all. Over time, the marginal entropies decrease, however the overall entropy decreases even faster as the variations of individual particles are correlated. This then leads to an increase of multi-information over time. In Fig. 6 we see snapshots of different samples of one specific experiment run. The final shapes show a certain variety, and there are two visually distinguishable categories of shapes. One has a triangular cluster of dark particles in the center whereas the other one has a cluster of light and outlined particles, where the light particles are sandwiched between the outlined particles. So even though there are several final states, they can be grouped into several shapes with specific distinctive visual features.

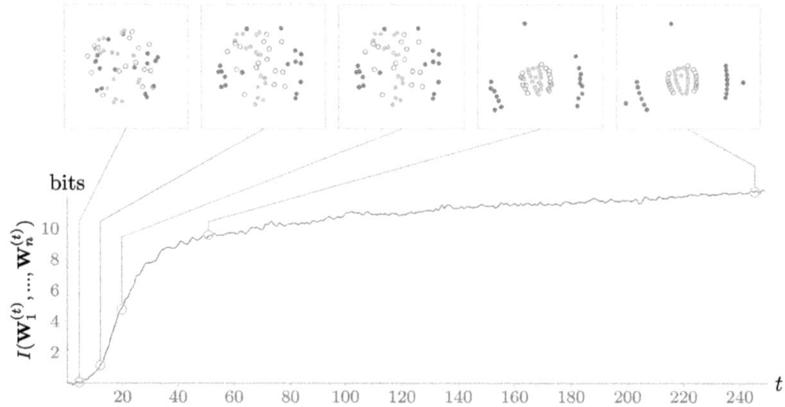

Fig. 4. Multi-information between particles plotted against time with $n = 50$, $l = 3$, $r_c = 5.0$ and $r_{\alpha\beta} = \{\{2.5, 5.0, 4.0\}, \{5.0, 2.5, 2.0\}, \{4.0, 2.0, 3.5\}\}$. The increase of multi-information correlates with the visual organization shown by snapshots of one particular sample at different times $t = \{0, 10, 20, 50, 249\}$.

If the types are restricted to a single type, which means each particle interacts with each other in the same way, the resulting equilibrium configuration is always a regular grid for force-scaling function F^2 (see rightmost panel in Fig. 3), and the self-organization is very low. This is due to two effects: First, the regular grid is also always in the form of a disc, there is no variety in shapes, so the entropy for each particle is already very low in general (after alignment), and second, small perturbations in the grid structure are local and do not spread through the grid.

Interestingly, this is not always the case when we used the force-scaling function $F^{(1)}$ which show a different behavior. For example, if the cut-off radius r_c is larger than $2r_{\alpha\alpha}$ (we consider only one type) and there are 20 particles, then the particles configure into two concentric regular polygons where the rotation of the inner polygon with regard to the outer polygon shows one degree of freedom (see Fig. 7). This already leads to a relatively high amount (compared to all other experiments we analyzed) of self-organization with just one type, as shown in Fig. 5. This already foreshadows the insight from Sec. 6.1 that the amount of self-organization a system can exhibit depends strongly on how individual particles interact with each other. In the next section we will explore this relationship in terms of interaction radius, number of types and self-organization in greater detail.

It can be seen in Fig. 5 that the multi-information is still increasing at time-step 250, even though a visual inspection of the simulation samples shows that there is hardly any movement except a slight expansion of the configuration. We constrained our simulations in general to 250 time-steps because of the limitation of computational capacity. In most of our experiments the equilibrium was attained well before the maximum of 250 time-steps was reached. When this was not the case the system was either still slowly expanding, but already formed most of its final shape, or it reached a limit cycle with a periodic dynamic. In the case of the periodic limit cycle, the multi-information entered a plateau before the maximum

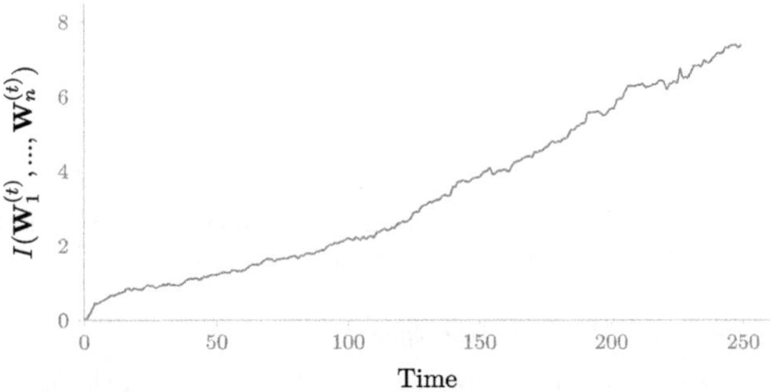

Fig. 5. Multi-information plotted against time using force-scaling function $F^{(1)}$, 20 particles of one type, 500 samples.

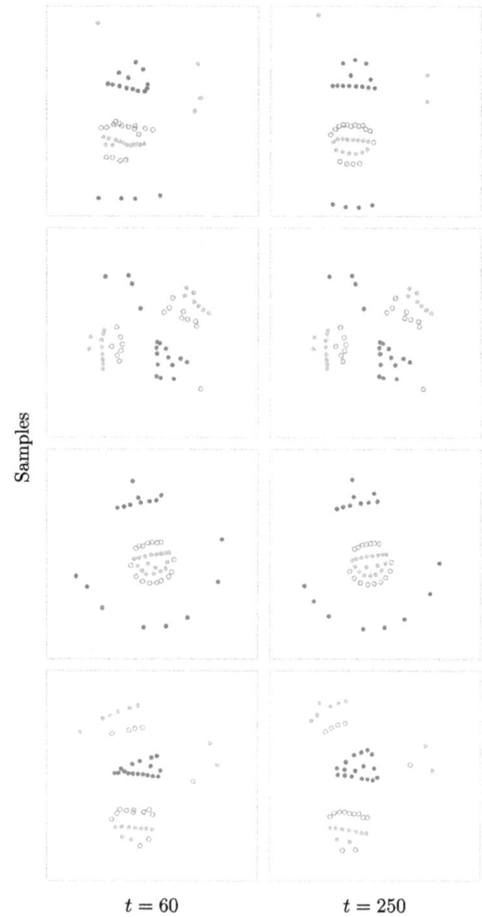

Fig. 6. Snapshots of different samples of the experiment shown in Fig. 4 at $t = 60$ (left) and $t = 250$ (right).

of 250 steps was reached, however, the equilibrium stopping criterion which requires nearly vanishing forces for several time-steps was not fulfilled at any time.

6.1. *Comparison of interaction types*

Simulations with $l = 3$ to 5 types and $n = 20$ to 120 particles almost always show quantifiable self-organization reflected in multi-information (see Fig. 4 for a typical example). Observing an increase in multi-information is independent of the

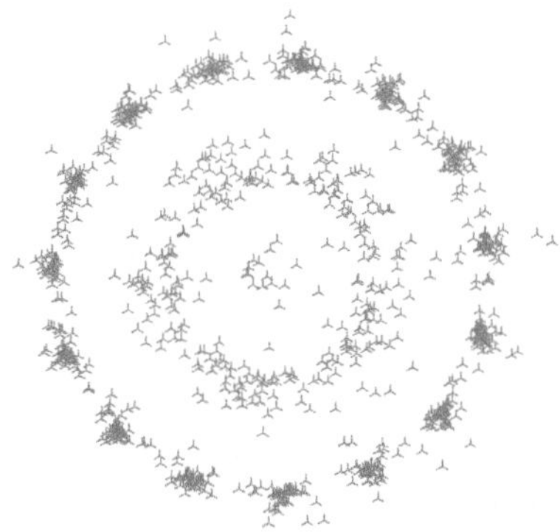

Fig. 7. Plot of all particles of all samples at time $t = 250$ using force-scaling function $F^{(1)}$, 20 particles of one type per sample. It can be seen that the outer ring has been much better aligned so that for each particle samples match more closely (denser clusters), while this is not possible for the inner ring of particles as their alignment related to the outer ring is a degree of freedom.

employed force-scaling function, and only vanishes in cases where the interactions between the different types are very similar (however as mentioned earlier, this is not a sufficient condition). Initially we hypothesized that increasing the number of types above a certain ratio of distinct types to particles will generally lead to a decrease of self-organization. It can be seen in Fig. 8 that there is a decrease in self-organization with increasing number of types (for a fixed number of particles), when $F^{(2)}$-scaling is used with randomly generated type matrices. However, the assumption does not hold for $F^{(1)}$ scaling.

In Fig. 9 it can be seen that with increasing cut-off radius r_c, the self-organization increases even in the case where there are the same number of types

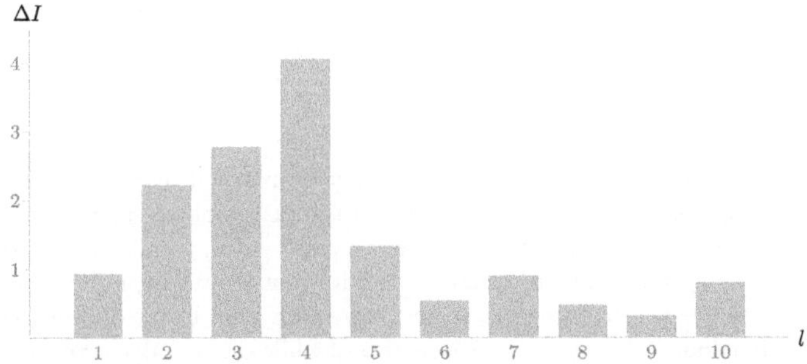

Fig. 8. Increase of multi-information between $t = 0$ and $t = 250$ using force-scaling function $F^{(2)}$, for different numbers of types. Average over 10 randomly generated types with mutual preferred distance radii $r_{\alpha\beta}$ between 1.0 and 5.0.

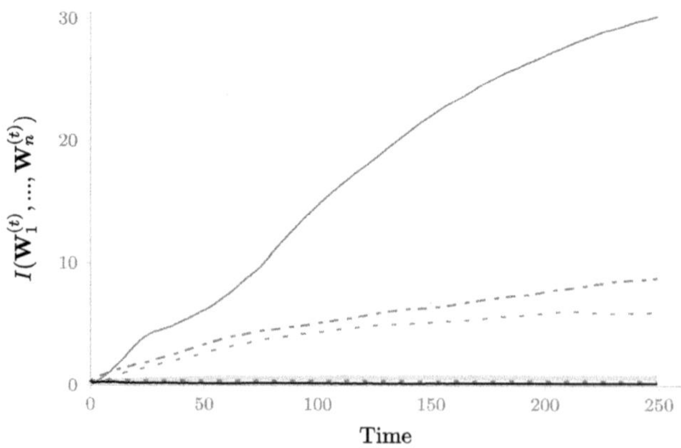

Fig. 9. Multi-information plotted against time using force-scaling function $F^{(1)}$, 20 particles, 20 types, averaged over 10 samples of random types where $r_{\alpha\beta} \in [2.0, 8.0]$, $k_{\alpha\beta} = 1$ for different cut-off radii r_c: ▪ ▪ ▪ 2.5, —— 5.0, —— 7.5, - - - 10.0, - — - 15.0, —— ∞.

as there are particles $l = m$. This result seems counter intuitive at first: the resulting particle configurations look arbitrary and unstructured, though the multi-information is increasing over time. But note: we have to consider that spatial regularities are not a necessary condition for self-organization, but that the mutual interactions define possible attractors to which the particles then organize. Because of the large number of different types compared to particles the structure is not (and cannot expected to be) regular. On the other hand, if the interactions are locally limited, either because of a small cut-off radius or because of a decreasing force-scaling function like $F^{(2)}$, the self-organization is limited as well (Fig. 9, $r_c \leq 7.5$). If we compare this to the self-organization exhibited by systems with the same amount of particles, same local limitations on interactions, but considerably fewer different types, we can make the following observation: The increase of multi-information over time in these systems is much higher than in those being local and having as many types as particles (see Fig. 10).

An increase of multi-information can either be reached by the decrease of the overall entropy of the system, the increase of the marginal observer entropies or a mixture of both. In the case of a decrease of overall entropy and constant observer entropies, the system looks the same for each observer and it is quite clear that the correlation between the particles must have increased. In the opposite case, the degree of freedom each observer sees is increased while the overall entropy of the system stays constant, therefore there must exist some correlation between the particles that removes the individual degrees of freedom when looking at the system as a whole.

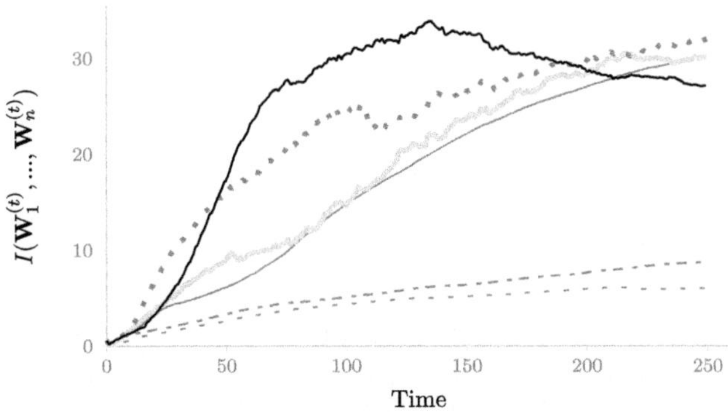

Fig. 10. Multi-information of 20 particles plotted against time using force-scaling function $F^{(1)}$, averaged over 10 samples of random types where $r_{\alpha\beta} \in [2.0, 8.0]$, $k_{\alpha\beta} = 1$ for different cut-off radii r_c and different number of types: - - - $l = 20$, $r_c = 10$, - — - $l = 20$, $r_c = 15$, —— $l = 20$, $r_c = \infty$, —— $l = 5$, $r_c = 10$, ▪ ▪ ▪ $l = 5$, $r_c = 15$, —— $l = 5$, $r_c = \infty$.

To reach an increase in correlation (i.e., multi-information) among the particles information needs to spread through the collective [40]. And hence, it is not surprising that long-range interactions lead to a lot of self-organization. What is, however, quite interesting is, that this is also possible if the interactions are local but limited in variety. In these cases, where interactions are local, there are almost always smaller clusters interacting with each other. Each cluster shows a very regular structure and consists of particles of one type. We will further discuss this observation in Sec. 7.2.

6.1.1. *Localization of organization*

If we have a cluster structure with spatially confined subsystems, there is a natural question: Is it possible to locate where the largest contribution to the organization is made? In Sec. 3.1, we showed that it is possible to decompose the multi-information of the observer variables into the several multi-information terms, that each measure the multi-information of a subset of the observer variables, and one term that measures the multi-information between these coarse-grained joint observer variables. We now consider the joint random variable of all observers of each type of particles as coarse-grained observers $\tilde{\mathbf{W}}_1, \ldots, \tilde{\mathbf{W}}_l$ (see Sec. 3.1), and calculate the multi-information individually. The results using this decomposition do not draw a clear picture: A general observation is that in every experiment we were able see organization on all levels. If we normalize the decomposition with respect to the total multi-information for each time-step, it can be seen that in the beginning of the experiment that the relative contribution of each decomposition term still varies, while after a few steps the relative contribution values will settle, even though the whole multi-information is still increasing and the system is still organizing (see

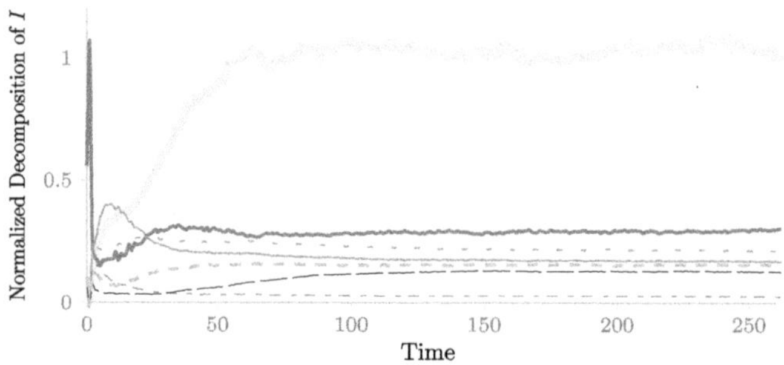

Fig. 11. Contribution of the different terms of the decomposition normalized with the multi-information in each time-step (is a normalized plot of the multi-information between all particles). —— $I(\tilde{\mathbf{W}}_1, \ldots, \tilde{\mathbf{W}}_l)$, multi-information between all particles of —— type 1, - - - type 2, - - type 3, — – type 4, - - - type 5. This is a decomposition of one of the sampled types with $l = 5, r_c = 15$ from Fig. 10.

Fig. 11). However, there is no common pattern in the decomposition that we were able to relate to the dynamics of the system. Almost all experiments show this pattern of an early phase where the decomposition varies by large amounts, and then a phase where the decomposition settles down. But, at this stage we were not able to link the dominance of a specific type or the inter-type multi-information, to specific observable behaviors.

7. Discussion

We used multi-information as a measure for self-organization and applied it to experiments of interacting particle systems. Estimations of multi-information were obtained using the Kraskov–Stögbauer–Grassberger estimator [22]. The approach seems to be in general transferable to other discrete-time dynamical systems that share the same invariants (or the same method can be adapted to other invariant transformations).

As mentioned earlier, defining a measure for self-organization is not a straightforward task. With our definition one has to be careful in the choice of the observer variables. However, the results show that particle/type-based observers are a practicable approach to measure self-organization in spatial systems. This measure also has the advantage that we do not need to assume stationarity of the system we investigate.

Regarding the specific model we chose, a core observation is that the choice of the force-scaling function, and therefore the form of interaction between particles, has a strong influence on the resulting dynamics. We have studied particular aspects of collective organization; the lessons drawn from may have implications for the understanding of biological collective self-organization in space.

7.1. *Uniform collectives*

Our first observation was, that a uniform collective (only one type) when forming regular grids only shows a small amount of measurable self-organization. Recalling that the particles are initially uniformly distributed in a disc of certain radius, and then form an almost unique equilibrium configuration, we see that the measured self-organization is in agreement with the statistical complexity notion of self-organization. The time-dependent statistical complexity [8] of the initial state vanishes as it is completely random, but also once the system reaches the equilibrium state there was no spatial or temporal variance, and so the statistical complexity vanishes as well. On the other hand, the process from randomly distributed states to a regular grid structure is similar to crystals or paracrystallines which often are considered as self-organized [14]. So, it could be that there is a small increase in multi-information in the beginning that then should vanish again, when the system settled on the equilibrium state. However, we were not able to show an early increase of multi-information with the estimator. It is possible that the amount of early self-organization is smaller than the bias of the estimator.

These discussions lead to an interesting related question. Suppose we observe two processes X_t, Y_t, both starting with a uniform distribution over all states, going into a phase of self-organization where the multi-information of some observer variables increases. But then, the first process X_t goes back to a uniform distribution, whereas Y_t cools down to a deterministic periodic process. In both cases there is no difference between the multi-information at the initial-state as well as the multi-information after the organization phase. Though there are qualitative differences in the process and we would consider X_t as temporarily organized while Y_t should be considered as self-organized even after the cooling down phase. Therefore, it is helpful to look at the evolution of entropy over time as well, which helps here to distinguish both cases, even though it does not serve as a measure for self-organization in itself.

7.2. *Long range interactions*

Our main observation of the self-organization of particle systems concerned the variation of the cut-off radius r_c and the number of types in the particles. Here, we were able to see that given unconstrained interactions ($r_c = \infty$) the self-organization can be very high even if the particles all have distinct types with different mutual interactions. This was surprising insofar that the particle configuration in these settings do not show much spatial structure, and there is generally no emergent description in terms of clusters interacting with each other. However, the configurations show a lot of statistical structure, i.e., correlations, that the multi-information is able to detect. This can be related to the retrieval of spatial configuration of sensors using information-distance [27]. The distances are in this case represented by the $r_{\alpha\beta}$ radii and the experiment with $r_c = \infty$ is equivalent to the relaxation procedure that was used in Ref. 27 for the reconstruction of spatial structure. Another interesting point

Fig. 12. Examples of emergent structures in particle collectives.

here is, that self-organization can occur without exhibiting a visually emergent spatial structure, this could support the idea put forward in [37] that self-organization and emergence are separate concepts.

Now, decreasing the cut-off radius r_c also decreases the observable self-organization if the number of types is held constant. This supports another assumption about self-organization: Information spread through the system is a crucial property of self-organizing systems. By limiting the cut-off radius, we are constraining the particles ability to transfer information through the system and therefore its ability to organize.

Surprisingly however, if the number of types is decreased (with fixed small value of r_c), the self-organization increases and we can observe emergent structures like balls enclosed in circles, layers of different types (see Fig. 12). It seems that the common co-occurence of self-organization and emergence of clustered structures is a result of the way a system can achieve higher overall complexity when interactions are locally constrained. Even with limited r_c, the homogeneity of the space as well as the homogeneity of local structures allow long-range structural interactions between groups of particles, which in turn allows to produce to a higher amount of self-organization of the whole system.

7.3. *Future work*

The methods developed in Ref. 24 promise to furnish tools to investigate the information dynamics between individual particles over time. We tried to measure the information transfer between particles, but so far the results are still inconclusive and this is part of ongoing research.

Our investigations showed that the multi-information is decomposed without clear preference in the initial phase, but settle to a fixed amount of contribution to the total amount of multi-information. So it is already possible to distinguish qualitatively different phases of the self-organization process. Having a measure like the interaction measure devised by Ref. 19, it might be possible to use a continuous version of this measure to investigate the interaction dynamics and autonomy of sub-processes in the particle collectives, relate it to the amount of self-organization that is exhibited, and such get even more insight to process of self-organization.

So far we only considered systems where particles have fixed dynamics. The system simply self-organizes according to fixed "physics". Already with this, we can reach quite a variation of patterns and configurations by merely changing the parameters of the dynamics. However these changes, if seen in the context of biological systems, will be limited by evolutionary speed and quite inflexible. If we want to get systems exhibiting higher complexity there needs to be some "informed" local guidance, either through external top-down intervention in the organization process, or by considering reactive agents instead of particles that just follow rules. As part of the rich toolbox of information-theoretic concepts, the present methods plug in immediately into the existing arsenal of information-theoretic tools to model agents in their environment and this scenario opens many doors for further investigations of guided self organization.

Acknowledgments

The authors thank the anonymous reviewers for very helpful comments.

References

[1] Ay, N., Olbrich, E. and Bertschinger, N. and Jost, J., A unifying framework for complexity measures of finite systems, in *Proceedings of ECCS06* (European Complex Systems Society, 2006).

[2] Ay, N. and Polani, D., Information flows in causal networks, *Adv. Complex Syst.* **11** (2008) 17–41.

[3] Bertschinger, N., Olbrich, E., Ay, N. and Jost, J., Autonomy: An information theoretic perspective, *Biosystems* **91** (2008) 331–345.

[4] Bonabeau, E., From classical models of morphogenesis to agent-based models of pattern formation, *Artif. Life* **3** (1997) 191–211.

[5] Ceguerra, R., Lizier, J. and Zomaya, A., Information storage and transfer in the synchronization process in locally-connected networks, in *2011 IEEE Symp. Artificial Life (ALIFE)* (IEEE, 2011), pp. 54–61.

[6] Cover, T. M. and Thomas, J. A., *Elements of Information Theory*, 2nd edn., Wiley Series in Telecommunications and Signal Processing (Wiley-Interscience, 2006).

[7] Crutchfield, J., Information and its metric, *Nonlinear Structures in Physical Systems–Pattern Formation, Chaos and Waves* (Springer, 1990), pp. 119–130.

[8] Crutchfield, J., Semantics and thermodynamics, in *Santa Fe Institue Studies in the Sciences of Complexity*, Vol. 12 (Addison-Wesley Publishing Co, 1992), pp. 317–317.

[9] Davies, R., Twining, C. and Taylor, C., *Statistical Models of Shape: Optimisation and Evaluation* (Springer-Verlag, New York Inc, 2008).

[10] Doursat, R., Organically grown architectures: Creating decentralized, autonomous systems by embryomorphic engineering, *Organic Computing* (Springer, 2008), pp. 167–199.

[11] Doursat, R., Programmable architectures that are complex and self-organized: From morphogenesis to engineering, *Artif. Life* (2008) 181–188.

[12] Dryden, I. and Mardia, K., *Statistical Shape Analysis*, Vol. 4 (John Wiley & Sons, New York, 1998).

[13] Gibbs, J., *On the Equilibrium of Heterogeneous Substances* (Transactions of the Connecticut Academy, 1874).

[14] Grom, G., Lockwood, D., McCaffrey, J., Labbe, H., Fauchet, P., White Jr., B., Diener, J., Kovalev, D., Koch, F., Tsybeskov, L. *et al.*, Ordering and self-organization in nanocrystalline silicon, *Nature* **407** (2000) 358–361.

[15] Hausser, J. and Strimmer, K., Entropy inference and the James-Stein estimator, with application to nonlinear gene association networks, *The Journal of Machine Learning Research* **10** (2009) 1469–1484.

[16] Haykin, S., *Communication Systems* (Wiley-India, 2008).

[17] James, R., Ellison, C. and Crutchfield, J., Anatomy of a bit: Information in a time series observation, arXiv:1105.2988 (2011).

[18] Jaynes, E., *The Gibbs Paradox* (Kluwer Academic, Dordrecht, 1992).

[19] Kahle, T., Olbrich, E., Jost, J. and Ay, N., Complexity measures from interaction structures, *Phys. Rev. E* **79** (2009) 11.

[20] Kloeden, P., Platen, E. and Schurz, H., Stochastic differential equations, *Numerical Solution of SDE Through Computer Experiments* (Springer, 1994), pp. 63–90.

[21] Kondor, I., *Group Theoretical Methods in Machine Learning* (Columbia University, 2008).

[22] Kraskov, A., Stögbauer, H. and Grassberger, P., Estimating mutual information, *Phys. Rev. E* **69** (2004) 066138.

[23] Liggett, T., *Interacting Particle Systems*, Vol. 276 (Springer-Verlag, 1985).

[24] Lizier, J., Prokopenko, M. and Zomaya, A., Local information transfer as a spatiotemporal filter for complex systems, *Phys. Rev. E* **77** (2008) 026110.

[25] Meinhardt, H., *Models of Biological Pattern Formation* (Academic Press London, 1982).

[26] Odell, G., Oster, G., Burnside, B. and Alberch, P., A mechanical model for epithelial morphogenesis, *J. Math. Biol.* **9** (1980) 291–295.

[27] Olsson, L., Nehaniv, C. and Polani, D., Sensory channel grouping and structure from uninterpreted sensor data, in *Proc. 2004 NASA/DoD Conf. Evolvable Hardware* (IEEE, 2004), pp. 153–160.

[28] Pearl, J., *Causality: Models, Reasoning and Inference* (Cambridge University Press, 2000).

[29] Polani, D., Measures for the organization of self-organizing maps, *Stud. Fuzziness Soft Comput.* **78** (2002) 13–44.

[30] Polani, D., Foundations and formalizations of self-organization, *Advances in Applied Self-Organizing Systems* (Springer, 2008), pp. 19–37.

[31] Polani, D., An informational perspective on how the embodiment can relieve cognitive burden, in *2011 IEEE Symp. Artificial Life (ALIFE)* (IEEE, 2011), pp. 78–85.

[32] Press, W., Flannery, B., Teukolsky, S., Vetterling, W. *et al.*, *Numerical Recipes*, Vol. 547 (Cambridge University Press, 1986).

[33] Prusinkiewicz, P., Visual models of morphogenesis, *Artif. Life* **1** (1993) 61–74.

[34] Rusu, R. and Cousins, S., 3d is here: Point cloud library (PCL), in *2011 IEEE Int. Conf. Robotics and Automation (ICRA)* (IEEE, 2011), pp. 1–4.

[35] Sayama, H., Swarm chemistry, *Artif. Life* **15** (2009) 105–14.

[36] Schinazi, R., *Classical and Spatial Stochastic Processes* (Springer, 1999).

[37] Shalizi, C. R., *Causal Architecture, Complexity and Self-Organization in Time Series and Cellular Automata*, Ph.D. thesis (2001).

[38] Shannon, C. E., A mathematical theory of communication, *Bell Syst. Tech. J.* **27** (1948) 379–423.

[39] Small, C., *The Statistical Theory of Shape* (Springer-Verlag, 1996).

[40] Steudel, B. and Ay, N., Information-theoretic inference of common ancestors, arXiv:1010.5720 (2010).
[41] Suzuki, T., Sugiyama, M., Sese, J. and Kanamori, T., Approximating mutual information by maximum likelihood density ratio estimation, in *JMLR Workshop and Conference Proceedings*, Vol. 4 (Citeseer, 2008), pp. 5–20.
[42] Tononi, G., Sporns, O. and Edelman, G., A measure for brain complexity: Relating functional segregation and integration in the nervous system, *Proc. Natl. Acad. Sci.* **91** (1994) 5033.
[43] Touchette, H. and Lloyd, S., Information-theoretic approach to the study of control systems, *Physica A, Stat. Mech. Appl.* **331** (2004) 140–172.
[44] Wang, X. R., Miller, J. M., Lizier, J. T., Prokopenko, M. and Rossi, L. F., Measuring information storage and transfer in swarms, in *Proceedings of the Eleventh European Conference on the Synthesis and Simulation of Living Systems* (2011), pp. 838–845.
[45] Witten Jr., T. and Sander, L., Diffusion-limited aggregation, a kinetic critical phenomenon, *Phys. Rev. Lett.* **47** (1981) 1400–1403.
[46] Wolfram, S., *Theory and Applications of Cellular Automata* (World Scientific, 1986).
[47] Wolpert, L., Beddington, R., Jessell, T., Lawrence, P., Meyerowitz, E. and Smith, J., *Principles of Development*, Vol. 3 (Oxford University Press, New York, 2002).
[48] Zhang, Z., Iterative point matching for registration of free-form curves and surfaces, *Int. J. Comput. Vis.* **13** (1992) 119–152.

Lightning Source UK Ltd.
Milton Keynes UK
UKHW030618151022
410523UK00004B/276